3rd International Conference on Engineering Technology for Sustainable Development (ICET4SD 2019)

IOP Conference Series: Materials Science and Engineering
Volume 722

Yogyakarta, Indonesia
23 - 24 October 2019

ISBN: 978-1-7138-2581-4
ISSN: 1757-8981

Printed from e-media with permission by:

Curran Associates, Inc.
57 Morehouse Lane
Red Hook, NY 12571

CURRAN ASSOCIATES INC. proceedings.com

Some format issues inherent in the e-media version may also appear in this print version.

Institute of Physics
Dirac House, Temple Back
Bristol BS1 6BE UK

Phone: 44 1 17 929 7481
Fax: 44 1 17 920 0979

techtracking@iop.org

Additional copies of this publication are available from:

Curran Associates, Inc.
57 Morehouse Lane
Red Hook, NY 12571 USA
Phone: 845-758-0400
Fax: 845-758-2633
Email: curran@proceedings.com
Web: www.proceedings.com

TABLE OF CONTENTS

IOP Conf. Series: Materials Science and Engineering **722** (2020) 011001 doi:10.1088/1757-899X/722/1/011001

Message from Conference's Chair

Distinguished guests, respected colleagues, and ladies and gentlemen. On behalf of the 3rd International Conference on Engineering Technology for Sustainable Development (ICET4SD) 2019 Committee, I would like to welcome you to this biannual conference held by the Faculty of Industrial Technology, Universitas Islam Indonesia Yogyakarta. Th is conference also a joint event with the 1st International Biomedical Instrumentation and Technology Conference (IBITeC) 2019, a part IEEE conference on biomedical engineering and related fi eld in collaboration with UII IEEE Student Branch and IEEE Indonesia Section.

Both ICET4SD and IBITeC aims to facilitate the research sharing and discussions of engineering disciplines that cover and accommodate the growth of the Internet of Th ings (IoT) to help and solve the healthcare and welfare problems around the world. Th is year, ICET4SD and IBITeC 2019 conferences are co-organized with our partners; Kumamoto University, BKSTI, CBIOM3S Diponegoro University, and Universiti Teknologi Malaysia (UTM).

The committee is delighted with the positive response of researchers to this conference. We received more than 200 submissions from Japan, Portugal, Sri Lanka, China, Iraq, Malaysia, Th ailand, and our own Indonesia as well. Th e papers were blindly reviewed by our reviewers that came from several countries to maintain the quality of this conference. Th e Acceptance rate of the 3rd ICET4Sd and 1st IBITeC 2019 are around 70% and 60%, respectively.

The accepted and presented papers of ICET4SD will be published in Proceedings series of IOP, and selected papers will be considered to be published in selected journals (JMES (Q2) and JAMT (Q3)). Whereas the accepted and presented paper of IBITeC 2019 will be forwarded to the IEEE Explore Digital Library for the publication. All of the publication schemes will be submitted for major international indexing databases to ensure the visibility internationally of the articles presented in ICET4SD and IBITeC 2019.

We are grateful for the contributions of our invited speakers. We will have six keynotes total from the two conferences that we believe could spread the new insight for engineering disciplines to follow the industry 4.0 needs.

The organization of a conference is very much a team eff ort. I want to thanks to all committee, editorial team, event-organizer, reviewers, and other parties who have carried a vast and complicated workload.

ICET4SD and IBITeC strive to offer plenty of opportunities, especially networking. Authors and participants have the chance to meet and interact with each other to share and transfer their knowledge in similar fields. We hope that you can take the benefit of this conference during discussions and, most importantly, networking among our peers. We hope that this conference will be unforgettable moments and experiences.

Thank you.

ICET4SD

Yogyakarta October 2019 Alvin Sahroni, Ph.D.

Conference chair of the 3rd ICET4SD and 1st IBITeC 2019

IOP Conf. Series: Materials Science and Engineering **722** (2020) 011001 doi:10.1088/1757-899X/722/1/011001

Conference or Sponsor Logo

3rd ICET4SD 2019

CO-ORGANIZED WITH

熊本大学
Kumamoto University

BADAN KERJASAMA
BKSTI
PENYELENGGARA PENDIDIKAN TINGGI TEKNIK INDUSTRI INDONESIA

IOP Conf. Series: Materials Science and Engineering **722** (2020) 011001 doi:10.1088/1757-899X/722/1/011001

List of Committees

Advisors

Hari Purnomo

Dwi Ana Ratna Wati

Sisdarmanto Adinandra

Conference Chairman

Alvin Sahroni

Deputy Chairman

Harwati

Secretary

Elvira Sukma Wahyuni

Treasurer

Almira Budiyanto

Technical Committee

Annisa Uswatun Khasanah

Muflih Arisa Adnan

Donny Suryawan

Supporting Staff

Slamet Puji Astuti

Risky Falahiya

Ana Yuliani

Scientific Committee:
Tomohiko Igasaki, Kumamoto University, Japan
Norlaili Binti Mat Safri, Universiti Teknologi Malaysia,Malaysia
Nasser Najibi, City University of New York, United States
Radu Godina, Faculty of Science and Technology (FCT), New University of Lisbon, Portugal
Sanjeevikumar Padmanaban, Aalborg University, Denmark
Vladimir Strezov, Macquarie University, Australia
Emmanuel Abdul, Portland State University, United States
Prathamesh Churi, NMIMS University, School of Engineering and Technology Management, India
Mu-Song Chen, Da-Yeh University, Taiwan
Addy Wahyudie, United Arab Emirates University, United Arab Emirates

IOP Conf. Series: Materials Science and Engineering **722** (2020) 011001 doi:10.1088/1757-899X/722/1/011001

Moslem Yousefi, Islamic Azad University, Roudehen Branch, Iran
Roman Voliansky, Dniprovsk State Technical University, Ukraine
Giedrė Streckienė, Vilnius Gediminas Technical University, Lithuania
Hasan AL Dabbas, Philadelphia University, Jordan
Leonel E Hernandez, ITSA University Institution, Colombia
Makbul Anwari, King Abdulaziz University, Saudi Arabia
Nelson Martins, Universidade de Aveiro, Portugal
Mohamed Serrhini, University Mohammed Premier Oujda, Morocco
Raoudha Chaabane, Uuniversity of Monastir/ENIM/LESTE, Tunisia
Laura Piedra-MUÑOZ, University of Almeria, España
José António Beleza Carvalho, Instituto Superior de Engenharia do Porto, Portugal
Raman Singh, Monash University, Australia
Darunee Wattanasiriwech, Mae Fah Luang University, Thailand
Bonnin Michele, Politecnico di Torino, Italy
Amelia Kurniawati, Telkom University, Indonesia
Suyitno, Nanobioenergy Laboratory, Universitas Sebelas Maret, Indonesia
Nazaruddin SInaga, Diponegoro University, Indonesia
Boy Arief Fachri, Universitas Jember, Indonesia
Andreas Handojo, Petra Christian University, Indonesia
Nasser Najibi, City University of New York, United States
Safanah M. Raafat, University of technology, Iraq
Sudhanshu Suhas Gonge, Marathwada Mitra Mandal's College of Engineering, Karvenagar , Pune affiliated to Savitribai Phule Pune University, Pune., India
Ipseeta Nanda, NIIT University, India
Yutika Amelia Effendi, Universitas Airlangga, Indonesia
Omar Ibrahim Alsaif, Northern Technical University, Iraq
Qasim Yahya Rahawi, University of Mosul, Iraq
Emmanuel Abdul, Portland State University, United States
Khalil Azha Mohd Annuar, Universiti Teknikal Malaysia Melaka, Malaysia
Natrayan, VIT University, India
Yasser Gaber, Beni-Suef university, Egypt
Nazaruddin Sinaga, Diponegoro University, Indonesia
Erwin B. Daculan, University of San Carlos, Philippines
Yahya, Universidade de Aveiro, Portugal
Adriano Peres, Federal University oficial Santa Catarina, Brasil
Ashok Kaushal, Concordia University, Canada
Bachir ACHOUR, Research Laboratory in Subterranean and Surface Hydraulics (LARHYSS)-University of Biskra, Algeria
Ferry Doringin, Bina Nusantara University, Indonesia
Marcelo Martín Marciszack, National Technological University, Argentina
Abdul Hadi, Universiti Teknologi MARA, Malaysia
Nico F. Declercq, Georgia Institute of Technology, USA-France-Belgium
Pattaraporn Posoknistakul, Mahidol University, Thailand

IOP Conf. Series: Materials Science and Engineering **722** (2020) 011001 doi:10.1088/1757-899X/722/1/011001

Subhasis Roy, University of Calcutta, India
Tutun Nugraha, IULI (International University Liaison Indonesia), Indonesia
Suthee Wattanasiriwech, Mae Fah Luang University, Thailand
Kishor Sarkar, University of Calcutta, India
Siti Shawalliah Idris, Universiti Teknologi MARA, Malaysia
Ren-Jieh Kuo, National Taiwan University of Science and Technology, Taiwan
Yalun Arifin, Universitas Prasetiya Mulya, Indonesia
Zahari Taha, Universiti Malaysia Pahang, Malaysia
Mohd Rozi Ahmad, Universiti Teknologi MARA, Malaysia
Shi-Woei Lin, National Taiwan University of Science and Technology, Taiwan
Jamari, Diponegoro University, Indonesia
Soorathep Kheawhom, Chulalongkorn University, Thailand
Agung Sutrisno, Sam Ratulangi University, Indonesia
Siti Shawalliah Idris, Universiti Teknologi MARA, Malaysia
Wahyu Pamungkas, Institut Teknologi Telkom Purwokerto, Indonesia
Seyed Amidedin Mousavi, Islamic Azad University, Zanjan Branch, Iran
Emmanuel Abdul, Portland State University, United States
Qasim Yahya Rahawi, University of Mosul, Iraq
Tan Xiao Jian, University Malaysia Perlis, Malaysia
Arpitha G R, Presidency University Bangalore, India
Shashikant Patil, SVKMs NMIMS Mumbai, India
Yun Ii, Go, Heriot-Watt University Malaysia, Malaysia
Alias Mohd Saman, Universiti Teknologi MARA (UiTM), Malaysia
S.Velliangiri, CMR Institute of Technology, Hyderabad, India
Nazaruddin Sinaga, Diponegoro University, Indonesia
Joshua A. Abolarinwa, Federal University of Technology Minna, Nigeria
Nikhil Khanna, Motilal Nehru College, University of Delhi, India
Akeel Abbas Mohammed, Al-Furat Al-Awsat Technical University, Iraq
Adid Adep Dwiatmoko, Indonesian Institute of Sciences (LIPI), Indonesia
Tri Handhika, Gunadarma University, Indonesia
Adid Adep Dwiatmoko, LIPI, Indonesia
Nuno Domingues, ISEL, Portugal
Christiono Utomo , ITS , Indonesia
Nyoman Sutapa , Petra Christian University , Indonesia
Hussien Al-Hmood , University of Thi-Qar , Iraq
dr nor azlan othman , universiti teknologi mara , malaysia
amitava das , NSHM Knowledge Campus Durgapur India , INDIA
F. Punnya Priya , Anna University , India
Agung Ari Wibowo , Politeknik Negeri Malang , Indonesia
Chunhui Guo , San Diego State University , United States
Kittipong Tripetch , Rajamangala University of Technology Suvarnabhumi , Thailand
Siti Akhtar Mahayuddin , Universiti Teknologi MARA, Perak Branch , Malaysia
Qasim Yahya Rahawi , University of Mosul , Iraq

IOP Conf. Series: Materials Science and Engineering **722** (2020) 011001 doi:10.1088/1757-899X/722/1/011001

Birinderjit Singh Kalyan , Chandigarh University , India
Muhammad Syafrudin , Dongguk University , Korea
Niloy Sarkar , Maulana Abul Kalam Azad university of technology , India
Nuha , University of Basrah , Iraq
Triwiyanto , Politeknik Kesehatan Kementerian Kesehatan Surabaya , Indonesia

IOP Conf. Series: Materials Science and Engineering **722** (2020) 011002 doi:10.1088/1757-899X/722/1/011002

Photographs

IOP Conf. Series: Materials Science and Engineering **722** (2020) 011002 doi:10.1088/1757-899X/722/1/011002

Photographs

IOP Conf. Series: Materials Science and Engineering **722** (2020) 011002 doi:10.1088/1757-899X/722/1/011002

Photographs

IOP Conf. Series: Materials Science and Engineering **722** (2020) 011002 doi:10.1088/1757-899X/722/1/011002

Photographs

IOP Conf. Series: Materials Science and Engineering **722** (2020) 011002 doi:10.1088/1757-899X/722/1/011002

Photographs

IOP Conf. Series: Materials Science and Engineering **722** (2020) 011003 doi:10.1088/1757-899X/722/1/011003

Peer review statement

All papers published in this volume of *IOP Conference Series: Materials Science and Engineering* have been peer reviewed through processes administered by the proceedings Editors. Reviews were conducted by expert referees to the professional and scientific standards expected of a proceedings journal published by IOP Publishing.

IOP Conf. Series: Materials Science and Engineering **722** (2020) 012001 doi:10.1088/1757-899X/722/1/012001

Studies on Nickel-based Bimetallic Catalysts for the Hydrodeoxygenation of Stearic Acid

A Rafiani[1,2], F Aulia[1], A A Dwiatmoko[1]*, N Rinaldi[1], Nurhasni[2], R R Widjaya[1]

[1]*Department of Chemistry, UIN Syarif Hidayatullah Jakarta*
[2]*Research Center for Chemistry, Indonesian Institute of Sciences (LIPI)*

*adid.adep.dwiatmoko@lipi.go.id

Abstract. Fatty acids, which are contained in vegetable oils, can be converted into alkanes through the hydrodeoxygenation (HDO) reaction, as an alternative fuel. In this study, the HDO of stearic acid was carried out in the presence of nickel-based bimetallic catalysts supported on Silica-Alumina (SiAl) in an autoclave batch reactor using decane as a solvent. Various metals, including Fe, Cu, and Co, were added into Ni catalysts using conventional wet impregnation method. Among others, additional Cu on Ni/SiAl increased the performance of the catalyst for the reaction.

1. Introduction

Fatty acids, one of renewable raw materials for biomass-derived fuels, are attractive to be upgraded into transport fuels [1, 2]. Commonly, biomass-derived fuels needs to be upgraded due to their high oxygen content[3]. Oxygen has a significant effect on physical and chemical properties of the derived fuel, such as miscibility with fossil fuels or stability during storage[4, 5]. Compared to bio-oils produced from lignocellulose[6], these fatty acids have low oxygen contents. Fatty acids can be converted into high-yield hydrocarbon fuels using the hydrodeoxygenation (HDO) process[7-9]. HDO is a series of reactions, which convert molecular structure of carboxylic acids into straight chain hydrocarbons, by removing oxygen as water in the presence of catalysts and hydrogen[10, 11].

Several studies reported HDO of various types of fatty acids, such as dodecanoic acid[12], linoleic acid[13, 14], stearic acid[15-17], oleic acid[18-20]. Yoosuk et al. reported that $NiMoS_2$ catalyst showed high catalytic HDO acivities for oleic acid (100% conversion with 78.8% n-C_{18} selectivity) and palmitic acid (95.2% conversion with 78.5% n-C_{16} selectivity). The results also found that the high pressure favored HDO pathway, while high temperature strongly affected the decarboxylation and decarbonylation pathways[21]. Recently, nitrogen-doped carbon–alumina hybrid supported iron (Fe-N-C@Al_2O_3) was used as a catalyst for HDO of stearic acid producing *n*-octadecane and *n*-heptadecane with yields of 91.9% and 6.0%, respectively[11].

Nickel-based catalysts has been employed for many types of reactions, including HDO[22-24]. Kumar et al. reported that complete conversion of stearic acid during HDO was accomplished with more than 80% selectivity to *n*-heptadecane at reasonable reaction temperature of 290 °C after 4 h of reaction using 15 wt.% Ni/γ-Al_2O_3 catalyst[25]. On the other hand, nickel-based bimetallic catalysts have attracted interest many researchers for a long time. For example, non-noble and non-expensive metals, such as iron (Fe)[26, 27], copper (Cu)[28, 29], and cobalt (Co)[30] are extensively used as second metal in nickel (Ni) catalysts due to their enhanced performances and stabilities. These second metals decrease reduction temperatures of Ni catalysts, and subsequently increase the performance of the catalysts for HDO reaction[31]. Guo et al. reported high catalytic performance of Ni-Cu/ZrO_2 on the HDO of algae bio-oils with HDO efficiency were equal to 82–90%[32].

In this work, we studied the effects of non-noble metals (including Fe, Cu, and Co) addition as second metals on on the catalytic activity of Silica-Alumina (SiAl)-supported Ni-based catalyst in the HDO of fatty acids. SiAl is generally applied as a catalyst support since it has good mechanical stability and thermal stability. Stearic acid were employed for the catalytic test reaction.

2. Experimental

2.1. Materials

All chemicals were used without further purification unless indicated otherwise. Silica-Alumina catalyst support, Grade 135, decane (99%), eugenol (99%) were purchased from Sigma-Aldrich (St. Louis, Missouri, USA). Nickel (II) nitrate hexahydrate (≤100%), iron (II) nitrate nonahydrate (99%), copper (II) nitrate trihydrate (99.5%), cobalt (II) nitrate hexahydrate (99%), stearic acid (97%), N,O-Bis(trimethylsilyl)trifluoroacetamide (BSTFA, 98%), and dichloromethane (99.9%) were obtained from Merck (Kenilworth, New Jersey, USA).

2.2. Catalyst Preparations

Supported Ni on Silica-Alumina (Ni/SiAl) and Cu on Silica-Alumina (Cu/SiAl) catalysts were prepared by wet impregnation. SiAl powder was impregnated with $Ni(NO_3)_2.6H_2O$ as a precursor for Ni (Ni loading = 20 wt%), calcined in a flow of air at 500 °C for 2 h, and then reduced in a flow of H_2 at 500 °C for 4 h. Finally, the prepared catalysts were stored under ambient conditions[33]. Similarly, supported Ni-based bimetallic catalysts (Ni-M/SiAl) catalysts were prepared by the same method. SiAl powder was simultaneously impregnated with $Ni(NO_3)_2.6H_2O$ as a precursor for Ni (Ni loading = 10 wt%) and $Cu(NO_3)_2.3H_2O$ as a precursor for Cu (Cu loading = 10 wt%) or $Co(NO_3)_2.6H_2O$ as a precursor for Co (Co loading = 10 wt%) or $Fe(NO_3)_2.9H_2O$ as a precursor for Fe (Fe loading = 10 wt%), then calcined in a flow of air at 500 °C for 2 h, and then reduced in a flow of H_2 at 500 °C for 4 h. Finally, the prepared catalysts were stored under ambient conditions[33].

2.3. Catalyst Characterizations

Transmission electron microscopy (TEM; Tecnai G2 20S-Twin, FEI, Hillsboro, Oregon, USA) images of as-prepared catalysts were obtained at an acceleration voltage of 200 kV at Research Center for Physics, Indonesian Institute of Sciences (LIPI). Powder X-ray diffraction (XRD; X-ray diffractometer, Rigaku, Tokyo, Japan) results were recorded using a CuKα (λ = 0.15406 nm) radiation source operated at 40 kV and 30 mA at Research Center for Physics, LIPI. Pore structures were calculated from N_2-physisorption isotherms recorded using a Micromeritics Tristar II instrument. Prior to these measurements, samples were evacuated at 200 °C for 3 h. Thermogravimetry analysis (TGA) of pre-calcined catalysts was done on a Linseis instrument (STA PT 1600, Germany) from 25 to 600 °C at a heating rate of 10 °C/min in air atmosphere by taking ca. 15 mg of sample. Catalyst acidiy was studied by ammonia temperature-programmed desorption (NH_3 -TPD) using Micromeritics instrument (Chemisorb 2750, USA) equipped with TCD. Prior to adsorption, synthesized catalysts were pretreated in a flow of He (15 mL/min) at 400 °C for 30 min. NH_3 adsorption, a sample was subjected to a flow of diluted NH_3 (15 mL/min) at 100 °C for 30 min. The physisorbed NH_3 was removed in a flow of He at the same temperature for 30 min. Finally, TPD experiments were done in the temperature range of 100–600 °C with heating rate of 10 °C/min. Hydrogen temperature-programmed desorption (H_2-TPR) experiments were carried out in a flow of 5% H_2 in Ar mixtures (40 mL/min) in the temperature range 100–600 °C. Prior to analysis, the catalysts were pretreated in a flow of He (15 mL/min) at 200 °C for 1 h.

2.4. Catalytic Conversion of Fatty Acids
Catalytic hydrodeoxygenation of fatty acids over Ni-based catalysts was carried out using a 160 mL stainless steel autoclave batch reactor equipped with a magnetic stirrer. Typically, a mixture of fatty acid (0.4 g), n-decane (10 mL) and pre-reduced catalyst (0.05 g) was put into the reactor. After sealing, the reactor was flushed three times with He from a high pressure gas cylinder and charged with H_2 (40 bar) at room temperature. Then, the reactor was heated under stirring (200 rpm) at 300 °C for 3 h. After the reaction, the catalyst was separated from the mixture, and the liquid products were analyzed by gas chromatography–mass spectrometry (GC/MS; Agilent 7890A instrument featuring a 5975C inert MS XLD and equipped with an HP-5MS capillary column (60 m × 0.25 mm × 250 μm))[34]. Prior to analysis, the obtained product (100 μL) was derivafftized using 50 μL of BSTFA which was heated at 70 °C for 30 min. After cooling, the mixture was then diluted with 350 μL of dichloromethane.

3. Results and Discussion
TGA measures weight loss of materials at a certain temperature to know their thermal stability. The weight loss was due to decomposition of the material at high temperatures. Based on the observed TGA results of pre-calcined catalysts in Fig. 1, there was no further decomposition of materials at the calcination temperature of 500 °C, indicating that all nitrate precursors has been converted into their corresponding oxides.

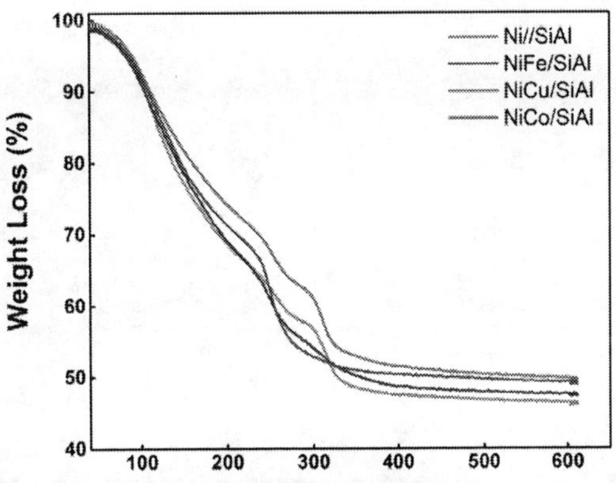

Fig. 1. TGA results of pre-calcined SiAl-supported Ni-based catalysts

N_2-physisorption analysis was carried out to determine the pore structure of the catalysts, and the results were shown in Table 1. According to the results, all the synthesized catalysts showed mesoporous property, which the pore size was in the range of 2-50 nm[35]. SiAl support showed high surface area (415.6 m^2/g). After impregnation, all the catalysts obtained a lower surface area, which was in the range of 74–268 m^2/g, depending on the type of metals. The decrease of surface area after impregnation indicates that there was partially blocking of the pore of the support by transition metals. The results of the measurement of the area of the bimetal catalyst have increased compared to the area of the monometal catalyst Ni / SiAl. NiFe / SiAl catalysts and NiCo / SiAl catalysts are bimetal catalysts which have the highest increase in surface area compared to other bimetal catalysts. NiFe/SiAl and NiCo/SiAl catalysts also have the highest micropore area compared to other bimetal catalysts, this indicates that NiFe/SiAl catalysts and NiCo/SiAl catalysts have more micro (<2 nm) pores.

IOP Conf. Series: Materials Science and Engineering **722** (2020) 012001 doi:10.1088/1757-899X/722/1/012001

Table 1. Pore properties of SiAl-supported Ni-based catalysts

Catalyst	N_2 physisorption			
	BET Surface Area (m2/g)	*Micropore Surface Area (m2/g)*	*Pore Volume (cm3/g)*	*Pore size (nm)*
SiAl	415.6	4.4	0.70	2-22
Ni/SiAl	74.3	4.9	0.10	2-12
NiFe/SiAl	268.7	20.0	0.32	2-12
NiCo/SiAl	244.4	12.3	0.32	2-13
NiCu/SiAl	231.8	10.0	0.36	2-13

The surface morphologies of the supported Ni-based catalysts were determined by transmission electron microscopy (TEM) and shown in Fig. 2. Metal nanoparticles, imaged as small black spots, were observed on the surface of SiO_2-Al_2O_3. Average size of metal particles was calculated by measuring diameter of black spots (at least 50 spots) on the TEM images as nickel particles or a combination of nickel and second metals (Fe, Cu, Co) dispersed on the SiAl. All catalysts showed wide distribution of metal particle size, in which most of the size was in the diameter range of 11–20 nm. Average metal particle size were 31.2, 23.7, 46.4, and 36.4 for Ni/SiAl, NiFe/SiAl, NiCu/SiAl, NiCo/SiAl, respectively. Average metal particle size was less affected when Co was introduced on Ni catalyst. In the presence of Cu, the average size increased significantly, indicating that the metal exhibited lower dispersion. On the other hand, Fe improved the dispersion of metal nanoparticles, indicated from the smaller size of average metal particles.

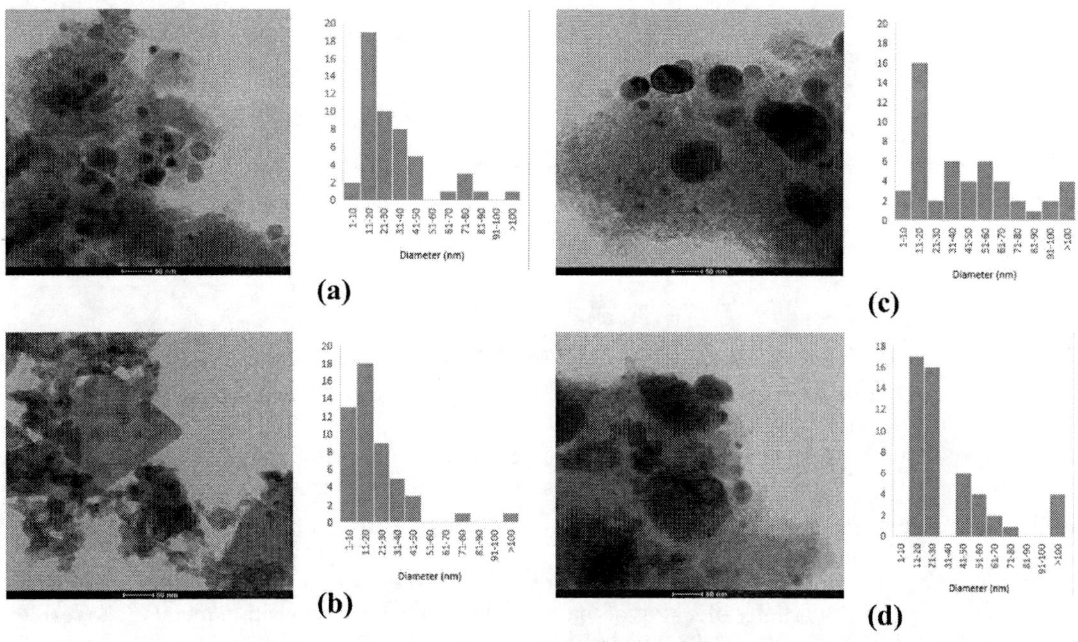

Fig. 2. TEM images of (a) Ni/SiAl, (b) NiFe/SiAl, (c) NiCu, and (d) NiCo/SiAl catalysts

Analysis using NH_3-TPD aims to determine the effect of acidity of a catalyst if added to the metal, based on NH_3 gas adsorption. Acid site will adsorb NH_3, and then during desorption, this desorbed NH_3 gas was then detected by the detector. The acidity results can be seen in the following Table 2. The results show that the addition of metals has an influence on the acidity of the Ni/SiAl catalyst. Highest catalyst acidity was exhibited by Ni/SiAl with a total acidity of 0.23 mmol/g. The catalysts impregnated with Fe, Cu, and Co metals showed decreased acidity, indicating that impregnated metals cover the acidic side of the Ni / SiAl catalyst.

Table 2. Acidity of SiAl-supported Ni-based catalysts

Catalyst	Acidity (mmol/g)		
	Weak Acid	Strong Acid	Total
Ni/SiAl	0.06	0.17	0.23
NiFe/SiAl	0.04	0.04	0.08
NiCo/SiAl	0.03	0.007	0.037
NiCu/SiAl	0.05	0.02	0.07

HDO reaction was carried out to determine the effect of the addition of Fe, Cu, and Co metals on nickel metal activity. HDO stearic acid reaction using catalysts Ni/SiAl, NiFe/SiAl, NiCu/SiAl, and NiCo/SiAl with a decane solvent. GC-MS analysis of liquid products showed that stearic acid was still detected after the reaction (Fig. 3), indicating that the full conversion of the acid was not achieved under our reaction conditions. Note that, the conversion value was difficult to calculate, since stearic acid was not fully soluble in *n*-decane at room temperature. In the presence of Fe and Ni, the conversion of stearic acid increased, with the minimum peak area of stearic acid after the reaction was obtained from the NiCu/SiAl catalyst. This indicating that the highest conversion of stearic acid was achieved.

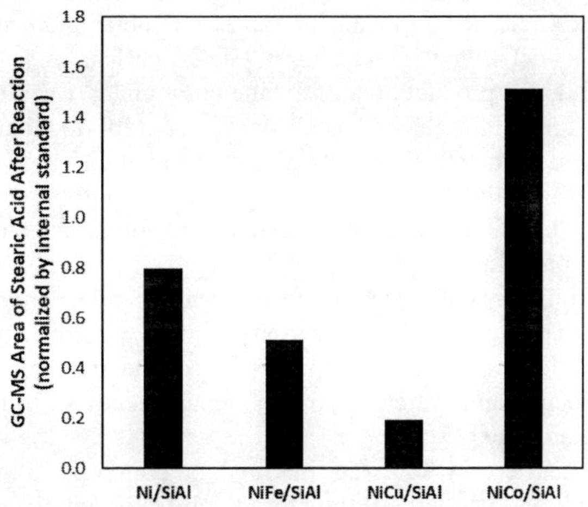

Fig. 3. Normalized peak area of stearic acid in the liquid product after reaction. The area was normalized with internal standard peak.

Ni/SiAl catalyst is a catalyst that can form the largest heptadecane product compared to bimetal catalysts. This was proven by the research of Foraita et al., (2016) regarding HDO stearic acid using a Ni / ZrO2 catalyst, which produced a heptadecane product that was dominant compared to octadecane products. The addition of Fe, Cu, and Co metals to the Ni / SiAl catalyst can encourage the formation of octadecane products, as evidenced by the increase in the percentage of the octadecane product area on the bimetal catalyst. Ni / SiAl catalysts and Ni-M / SiAl bimetal catalysts (M: Fe, Cu, Co) used in the HDO reaction of stearic acids produce heptadecane and octadecane products with different area values. The product areas of heptadecane and octadecane resulting from HDO reaction of stearic acid can be seen in the following Fig. 4.

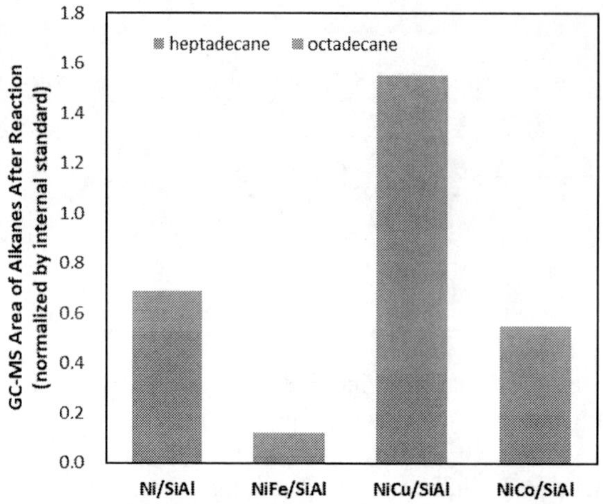

Fig. 4. Normalized peak area of alkanes in the liquid product after reaction. The area was normalized with internal standard peak

Fig. 4 shows that the four catalysts have different product areas. NiFe/SiAl catalyst is the catalyst with the smallest product area compared to other catalysts. The small value of the product area on the NiFe/SiAl catalyst indicates that the Fe metal is not an active metal as a second metal for Ni metals, so the NiFe/SiAl catalyst is not active for the reaction of HDO stearic acid.

The catalyst with the highest product area of alkane compounds was the NiCu/SiAl catalyst. Cu is an active metal as a second metal for Ni/SiAl catalysts. Li et al reported that the addition of Cu metal to Ni-based catalysts produced high selectivity, compared to the addition of Co and Sn metals[27]. The product area from GC-MS analysis is directly proportional to the product concentration, so the NiCu/SiAl catalyst that has the highest product area of alkane compounds indicates that the catalyst has a large product concentration value.

Heptadecane formed from the HDO reaction of stearic acid can be converted directly by stearic acid through the decarboxylation pathway or decarbonation pathway. Stearic acid undergoes decarboxylation by eliminating one CO_2 molecule, so heptadecane is formed. Stearic acid can also be decarbonated by eliminating one CO molecule and water, so that 1-heptadecene is formed, then 1-heptadecene is hydrogenated to form heptadecane[25].

The HDO reaction of stearic acid forms an intermediate compound in the form of octadecanol, in which the octadecanol compound undergoes transformation from two different pathways. Octadecanol is formed from stearic acid which is reduced by releasing H_2O groups, where in the first path the octadecanol undergoes dehydrogenation to form an aldehyde group, namely 1-octadecanal. The 1-octadecanal group undergoes decarbonilation by eliminating one CO molecule to form heptadecane. The transformation of octadecanol in the second pathway is dehydrated to form 1-octadecene alkene groups, which then 1-octadecene undergoes hydrogenation to form octadecane. The results of data from the GC-MS analysis resulted that the NiCu/SiAl catalyst was the most active catalyst for the HDO reaction of stearic acid.

4. Conclusions

Metals, Fe, Cu, and Co influence Ni / SiAl catalysts. Fe, Cu, and Co metals are able to increase the area of the Ni / SiAl catalyst, and are able to influence the particle size of the Ni / SiAl catalyst. Fe, Cu, and Co metals are also able to influence the acidity of the Ni/SiAl catalyst. Based on the results of the reaction, the NiCu/SiAl catalyst gives the best effect on the performance of the reaction. NiCu / SiAl bimetal catalyst is the best catalyst for the HDO reaction of stearic acid.

Acknowledgment
This work was done by the financial support of Research Center for Chemistry, Indonesian Institute of Sciences (LIPI). Authors would like to thank Research Center for Physics, Indonesian Institute of Sciences (LIPI) for providing instruments for catalyst characterizations.

References
[1] Hossain, M.Z., et al., *Continuous low pressure decarboxylation of fatty acids to fuel-range hydrocarbons with in situ hydrogen production.* 2018. **212**: p. 470-478.
[2] Liu, M., et al., *Upgrading of palmitic acid and hexadecanamide over Co-based catalysts: Effect of support (SiO2, γ-Al2O3 and H-ZSM-22).* 2019: p. 105726.
[3] Iliopoulou, E.F., et al., *Overview of catalytic upgrading of biomass pyrolysis vapors toward the production of fuels and high-value chemicals.* 2019. **8**(1): p. e322.
[4] Løhre, C., et al., *The effect of solvent and input material pretreatment on product yield and composition of bio-oils from lignin solvolysis.* 2016. **119**: p. 208-216.
[5] Mullen, C.A. and A.A.J.F. Boateng, *Mild hydrotreating of bio-oils with varying oxygen content produced via catalytic fast pyrolysis.* 2019. **245**: p. 360-367.
[6] Koike, N., et al., *Upgrading of pyrolysis bio-oil using nickel phosphide catalysts.* 2016. **333**: p. 115-126.
[7] Jeništová, K., et al., *Hydrodeoxygenation of stearic acid and tall oil fatty acids over Ni-alumina catalysts: Influence of reaction parameters and kinetic modelling.* 2017. **316**: p. 401-409.
[8] Ma, B., et al., *Interconnected hierarchical HUSY zeolite-loaded Ni nano-particles probed for hydrodeoxygenation of fatty acids, fatty esters, and palm oil.* 2016. **4**(29): p. 11330-11341.
[9] Kon, K., et al., *Hydrodeoxygenation of Fatty Acids, Triglycerides, and Ketones to Liquid Alkanes by a Pt–MoOx/TiO2 Catalyst.* ChemCatChem, 2017. **9**(14): p. 2822-2827.
[10] Sotelo-Boyás, R., F. Trejo-Zárraga, and F.J.J.H. Hernández-Loyo, *Hydroconversion of triglycerides into green liquid fuels.* 2012. **338**: p. 338.
[11] Li, J., et al., *Chemoselective Hydrodeoxygenation of Carboxylic Acids to Hydrocarbons over Nitrogen-Doped Carbon–Alumina Hybrid Supported Iron Catalysts.* 2019. **9**(2): p. 1564-1577.
[12] Brillouet, S., et al., *Deoxygenation of decanoic acid and its main intermediates over unpromoted and promoted sulfided catalysts.* 2014. **148**: p. 201-211.
[13] Santikunaporn, M., S.J.S. Danphitak, and T. Asia, *Hydrodeoxygenation of Linoleic Acid on Ni-Mo Catalyst.* 2010: p. 1-6.
[14] Singh, H.K.G., et al., *Production of gasoline range hydrocarbons from catalytic cracking of linoleic acid over various acidic zeolite catalysts.* Environmental Science and Pollution Research, 2018: p. 1-8.
[15] Denk, C., et al., *Rate enhancement by Cu in Ni x Cu 1− x/ZrO 2 bimetallic catalysts for hydrodeoxygenation of stearic acid.* 2019.
[16] Wu, G., et al., *Construction of Bifunctional Co/H-ZSM-5 Catalysts for the Hydrodeoxygenation of Stearic Acid to Diesel-Range Alkanes.* ChemSusChem, 2018.
[17] Arora, P., et al., *Kinetic study of hydrodeoxygenation of stearic acid as model compound for renewable oils.* 2019. **364**: p. 376-389.
[18] Coumans, A.E. and E.J.M. Hensen, *A model compound (methyl oleate, oleic acid, triolein) study of triglycerides hydrodeoxygenation over alumina-supported NiMo sulfide.* Applied Catalysis B: Environmental, 2017. **201**: p. 290-301.
[19] Bhattacharjee, S. and C.-S. Tan, *Hydrodeoxygenation of oleic acid in hexane containing pressurized CO2 using Fe/SBA-15 as catalyst.* Journal of Cleaner Production, 2017. **156**: p. 203-213.
[20] Nikulshin, P.A., et al., *The use of CoMoS catalysts supported on carbon-coated alumina for hydrodeoxygenation of guaiacol and oleic acid.* Catalysis Today, 2016. **271**: p. 45-55.

[21] Yoosuk, B., et al., *Hydrodeoxygenation of oleic acid and palmitic acid to hydrocarbon-like biofuel over unsupported Ni-Mo and Co-Mo sulfide catalysts.* Renewable Energy, 2019. **139**: p. 1391-1399.

[22] Jang, M.S., et al., *Metallic nickel supported on mesoporous silica as catalyst for hydrodeoxygenation: effect of pore size and structure.* 2018. **44**(6): p. 3723-3735.

[23] Gamliel, D.P., et al., *Nickel impregnated mesoporous USY zeolites for hydrodeoxygenation of anisole.* 2018. **261**: p. 18-28.

[24] Jahromi, H. and F.A.J.A.C.B.E. Agblevor, *Hydrodeoxygenation of pinyon-juniper catalytic pyrolysis oil using red mud-supported nickel catalysts.* 2018. **236**: p. 1-12.

[25] Kumar, P., et al., *Kinetics of hydrodeoxygenation of stearic acid using supported nickel catalysts: Effects of supports.* Applied Catalysis A: General, 2014. **471**: p. 28-38.

[26] Tezel, E., H.E. Figen, and S.Z.J.I.J.o.H.E. Baykara, *Hydrogen production by methane decomposition using bimetallic Ni–Fe catalysts.* 2019. **44**(20): p. 9930-9940.

[27] Li, C., et al., *Hydrogenation of biomass-derived ethyl levulinate into γ-valerolactone by activated carbon supported bimetallic Ni and Fe catalysts.* 2017. **203**: p. 23-31.

[28] Basnet, P., et al., *Synergistic bimetallic Ni/Ag and Ni/Cu catalysis for regioselective γ, δ-diarylation of alkenyl ketimines: Addressing β-H elimination by in situ generation of cationic Ni (II) Catalysts.* 2018. **140**(46): p. 15586-15590.

[29] Kannapu, H.P.R., et al., *Catalytic transfer hydrogenation for stabilization of bio-oil oxygenates: Reduction of p-cresol and furfural over bimetallic Ni–Cu catalysts using isopropanol.* 2015. **137**: p. 220-228.

[30] Ay, H. and D.J.A.C.B.E. Üner, *Dry reforming of methane over CeO2 supported Ni, Co and Ni–Co catalysts.* 2015. **179**: p. 128-138.

[31] Khromova, S.A., et al., *Anisole hydrodeoxygenation over Ni–Cu bimetallic catalysts: The effect of Ni/Cu ratio on selectivity.* 2014. **470**: p. 261-270.

[32] Guo, Q., et al., *Catalytic hydrodeoxygenation of algae bio-oil over bimetallic Ni–Cu/ZrO2 catalysts.* 2015. **54**(3): p. 890-899.

[33] Foraita, S., et al., *Controlling Hydrodeoxygenation of Stearic Acid to n-Heptadecane and n-Octadecane by Adjusting the Chemical Properties of Ni/SiO2–ZrO2 Catalyst.* ChemCatChem, 2017. **9**(1): p. 195-203.

[34] de Souza, P.M., et al., *Hydrodeoxygenation of phenol over Pd catalysts. Effect of support on reaction mechanism and catalyst deactivation.* ACS Catalysis, 2017. **7**(3): p. 2058-2073.

[35] Thommes, M., *Physical Adsorption Characterization of Nanoporous Materials.* Chemie Ingenieur Technik, 2010. **82**(7): p. 1059-1073.

IOP Conf. Series: Materials Science and Engineering **722** (2020) 012002 doi:10.1088/1757-899X/722/1/012002

The aluminum powder size' effect on rice plant fiber reinforced composite to hardness, wear and coefficient of friction of brake lining

A S Darmawan[1,*], P I Purboputro[1], B W Febriantoko[1]

[1]Department of Mechanical Engineering, Faculty of Engineering, Universitas Muhammadiyah Surakarta, Jl. Ahmad Yani, Tromol Pos 1 Pabelan, Surakarta 57162, Indonesia

*Corresponding author: Agung.Darmawan@ums.ac.id

Abstract. Composite particles are composites that contain particle or powder-shaped reinforcing materials. Particles as reinforcing materials greatly determine the mechanical properties of composites because they carry the load distributed by the matrix. The particle size, shape, and material are factors that influence the mechanical properties of particle composites. The friction component material that is intended for brake lining applications is strongly influenced by the size of the particles and the direction of loading. On the other hand it is necessary to step up the particle size optimization which will affect braking performance, which includes: friction numbers, wear, friction time and friction temperature produced. Therefore this study is aimed at optimizing the particle size against hardness, wear resistance and friction of aluminum powder in rice plant fiber reinforced composite. In this study, the materials used in the experimental process were rice plant fiber, fiberglass, aluminum powder and polyester resin. After the material is processed into the composite, hardness is tested using the durometer shore hardness method, after that, wear and friction testing were conducted. The results show the smaller the size of the aluminum powder, the harder the composite. The smaller the size of aluminum powder, the less wear of the composites. The failure of the adhesive was seen in the specimen tested for wear.

1. Introduction

The development of increasingly modern automotive makes the market demand for spare parts products even bigger. Brake lining is one component of a motorized vehicle that functions to stop motor vehicles. Brakes function to slow or stop the movement of the wheels so that the wheels move slowly. The kinetic energy lost from a moving object is converted to heat due to friction. If the brakes are not functioning properly, an accident is possible. The safety factor of the driver is very important. So that every motor vehicle manufacturer designs the system and uses brake components (brake lining) that are in accordance with the capabilities of the vehicle. Asbestos type brake lining is widely circulating at a cheap price and guarantees the durability of brake shoes, but the result of friction powder in the form of small particles is very dangerous for human health.

In the braking process, two materials will rub against each other and convert the motion energy into heat energy. At the time of friction, wear will occur in both materials. The constituent material for brakes that is widely used is asbestos. Asbestos powder as a result of wear and tear can enter the body of a person so that it can cause disease. For this reason, research is needed on materials that can be used as a substitute for asbestos. Composite material has the potential to be used as a substitute material.

The availability and accessibility of abundant plant is the main reason for the emergence of new interests in sustainable technology, including interest in composite materials. When focusing on composite materials, the main points to consider are environmentally friendly and lightweight, non-toxic, with high mechanical properties [1-4].

There is an urgency to overcome concerns in the production of new materials for the environment and economy and hence in this case new materials based on plant and bio-resins must be able to produce green materials that are environmentally friendly [5]. The Indonesian government is currently promoting the achievement of the results of the nation's technology that can be applied or downstream. This becomes a policy which benefits: can minimize the import component, dependence on foreign products, develop the capabilities of its researchers, reduce imports, the existence and independence of the nation's technology. Along with the times and the emergence of government regulations, the public is made aware of the preservation of the environment.

The strength of the brake particle composite material is greatly influenced by the particle size, material of the matrix and the manufacturing process. The strength of particle composites was obtained at a maximum size of 0.01 to 0.1 mm and strength of surface bonding, pressing, and sintering [6]. The process is very rarely explained by friction lining producers, so there is a need for research on: materials and processes that are scientifically standard so that friction can be easily made, so it is necessary to develop this research continuously. When developing suitable brakes and friction pairs, engineers must consider various parameters such as braking interface temperature, braking applied, and speed that affect the process of surface friction in rough contact [7].

A composite is a combination of two materials in which one material, called a reinforcement phase, is in the form of fibers, sheets, or particles, and is embedded in another material called the matrix phase. The reinforcing material and matrix material can be metal, ceramic or polymer. Composites usually have a fiber or particle phase that is more rigid and stronger than the continuous matrix phase and serves as the main burden that carries members. The matrix acts as a medium to transfer loads between fibers or particles, and in less ideal cases where the load is complex. The fiber or particle is more brittle than the matrix, so the matrix acts as a source of composite toughness. This matrix also functions to protect fibers or particle from environmental damage before, during and after composite processing. When properly designed, new combined materials show better strength than each individual material. Composites are used not only for their structural properties, but also for electricity, thermal, tribology and environmental applications. Composite is a multifunctional material system that provides characteristics that cannot be obtained from discrete material. They are cohesive structures made by physically combining two or more compatible materials, differing in composition and characteristics and sometimes in form. Composites should not be considered simple as a combination of two ingredients. In a broader sense; the combination has its own characteristics. In terms of strength for heat resistance or some other desired quality, it is better than one of its own components or very different from both [8, 9].

Composite is a compound material that is different from alloy by the fact that individual components retain their characteristics but once incorporated into composites get better properties. Composite material as a heterogeneous material consisting of two or more solid phases, which are in intimate contact with each other on a microscopic scale. They can also be regarded as homogeneous material on a microscopic scale in the sense that each part of it will have the same physical properties [10].

Interest in natural composite materials is growing rapidly both in terms of industrial applications and fundamental research. They can be updated, cheap, fully or partially recyclable, and can be decomposed. Their availability, renewable power, low density and price and satisfying mechanical properties make them an attractive ecological alternative.

Composite materials are actually abundant in nature, because composite materials can consist of organic and inorganic materials such as wood, bamboo, and leaves. Various types of composites have been unconsciously made. Clay strengthened with straw, is a composite that has long been known.

Manufacturing natural fiber composites is one of the real industry challenges today. This type of material offers economic and ecological advantages to promote sustainable development by producing recyclable products [11, 12].

Composite material composition affects mechanical properties. Zhao and friends [13] examined the effect of adding carbon nano tube to the coefficient of friction of carbon nano tube - hydroxyapatit composites. Addition of carbon nano tube 20% to hydroxyapatit will reduce the friction coefficient from 0.05 to 0.28. Xin and friends [14] found that the composite friction and wear resistance would be optimal if the ratio of sisal receipt and fiber was 3: 4.

Shalwan and Yousif [15] found that surface characteristics, volume fractions, physical properties and fiber directions had a significant effect on the mechanical properties of composites. In other words, the selection and modification of fibers is needed to improve the mechanical properties of composites.

From the considerations above, the researchers investigated the effect of the size of aluminum powder on the hardness, wear and coefficient of friction of aluminum in the rice plant fiber reinforced composite.

2. Materials and Methods

In this study, the materials used were carbon particle from rice plant, fiberglass, Polyester Resin and Aluminum powder. The materials were mixed to produce aluminum and carbon particle reinforced composite. The composition of composite can be seen in table 1.

Table 1. Composition of Composite

No	Al powder size	Al powder (%)	Fiberglass (%)	Polyester Resin (%)	Rice plant fiber (%)
1	Mesh 40	10	10	40	40
2	Mesh 50	10	10	40	40
3	Mesh 60	10	10	40	40

2.1. Methods

This research was conducted in several stages. The stages that were passed:

1. Prepare tools and materials
2. Calculate the composition of the brake lining materials
3. The materials were weighed using digital scales according to the composition and then manually mixed until the material is evenly mixed.
4. Place a mixture of material in the mold.
5. The next step is to press the mold using a press machine. Pressing was conducted by giving heat at a temperature of 90 °C. Then it is sintered at 180 °C for 15 minutes.
6. After that, micro photo is taken.
7. Hardness is measured using a durometer shore D hardness tester. Like other hardness tests, that durometer shore D hardness measures the depth of indentation in the material made by a force applied to a standard presser foot. This depth depends on the hardness of the material namely the viscoelastic properties, the shape of the presser foot, and the time of testing. These basic tests require the application of a stable force, indentation depth and absence of jerking. For example, the hardness tested has time to apply the force and then read the hardness from the object.
8. Perform friction testing to find the wear value and coefficient of friction. Wear is generally defined as progressive loss of material or the transfer of a number of materials from a surface as a result of

the relative movement between the surface and other surfaces. The explanation of wear mechanisms in the material is closely related to friction and lubrication against external systems (surface contact). Any material can experience wear due to various mechanisms.

9. Analyze data and find conclusions

3. Results and Discussion

Brake linings made from the composites are shown in figure 1. These aluminum powders are filtered with mesh 40, 50 and 60. The larger the mesh values the smaller the size of aluminum powder.

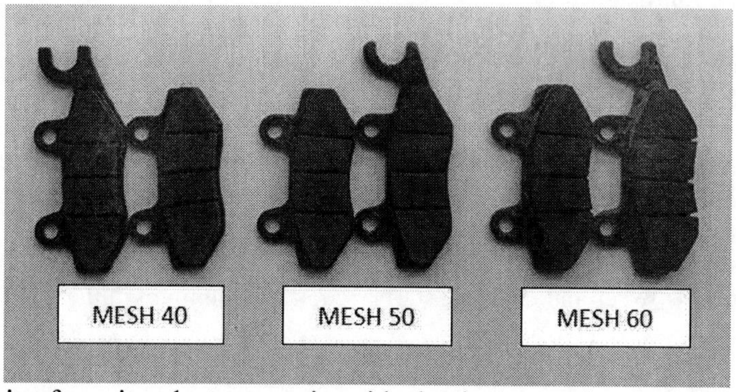

Figure 1. Brake lining form rice plant composite with aluminum powder produced from filter of mesh 40, 50 and 60.

Figure 2 shows aluminum powder mesh 40 on a rice plant fiber reinforced composite, while figure 3 shows aluminum powder mesh 50 and figure 4 shows aluminum powder mesh 60. Seen in the picture, the size of aluminum gets smaller as the mesh gets bigger.

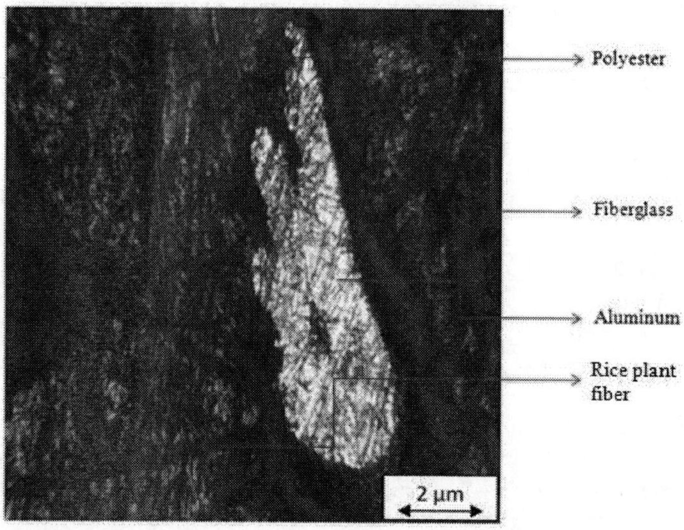

Figure 2. Micro Photograph of aluminum powder mesh 40.

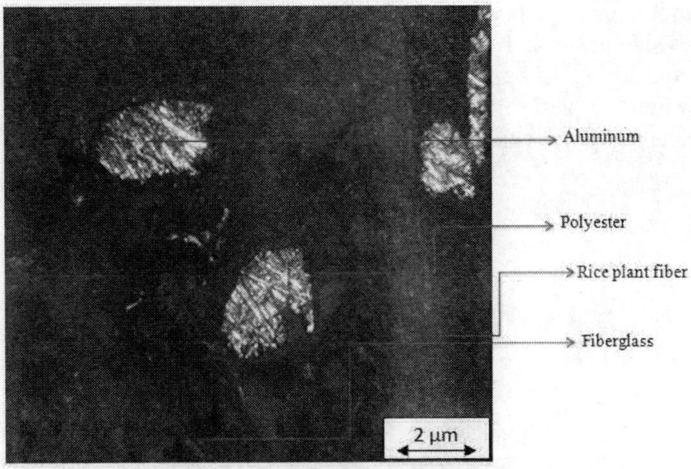

Figure 3. Micro Photograph of aluminum powder mesh 50.

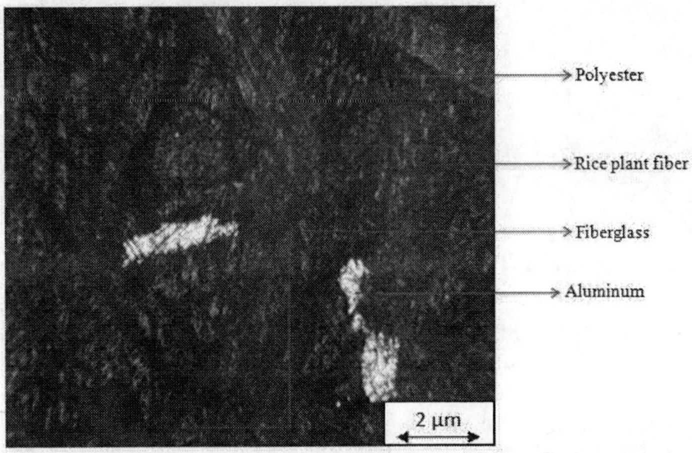

Figure 4. Micro Photograph of aluminum powder mesh 60.

The test results show a tendency to increase hardness when the size of aluminum powder gets smaller as illustrated in figure 5. Aluminum powder with mesh of 40 produces a hardness of 79.5 HD, mesh of 50 produces hardness of 82.7 HD and mesh of 60 produces hardness of 84.6 HD.

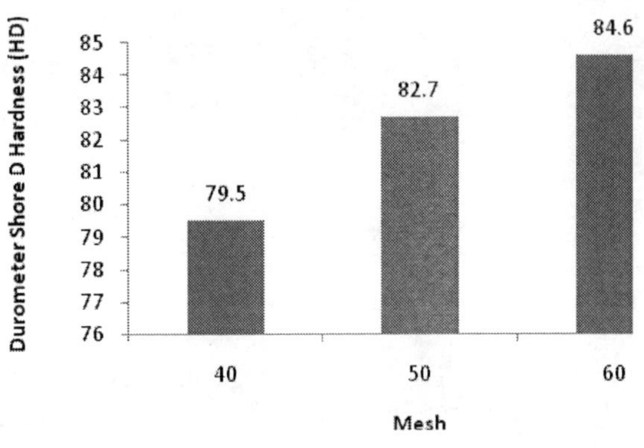

Figure 5. Effect of aluminum powder size to material's hardness.

Wear experienced by brake lining shows a declining trend, when the size of aluminum powder is getting smaller, both for dry conditions and conditions with brake oil as shown in figure 6. From aluminum powder size of mesh 40 to 60, the wear decrease from 112.5 mm^3/hour to 81.25 mm^3/hour in dry condition and from 78.12 mm^3/hour to 68.34 mm^3/hour in brake oil environment. The figure 6 also shows that when the brake lining works in the brake oil environment, the wear that occurs will decrease compared to when working in dry conditions.

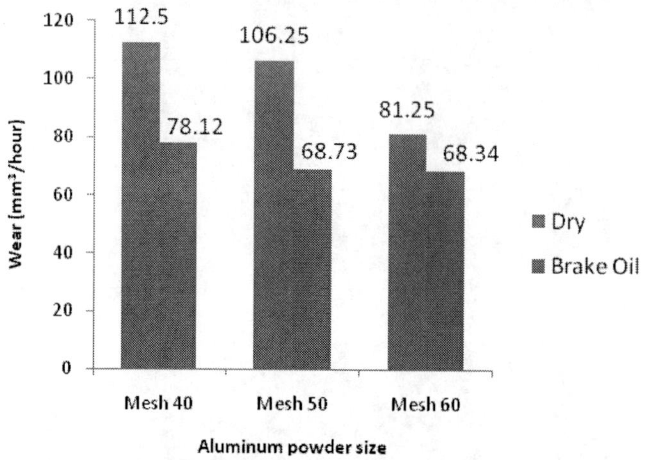

Figure 6. Relationship between aluminum powder size and wear.

The relationship between the coefficient of friction and the size of aluminum powder is shown in figure 7. The smaller the size of the aluminum powder the greater the coefficient of friction will be. This trend is shown both in dry conditions and conditions with brake oil.

However conditions with brake oil show a smaller coefficient of friction than dry conditions. Dry conditions produce a friction coefficient of 0.5799, 0.6017 and 0.6302 for mesh of 40, 50 and 60, respectively. While conditions with brake oil produce a friction coefficient of 0.5568, 0.5738 and 0.5862 for mesh of 40, 50 and 60, respectively.

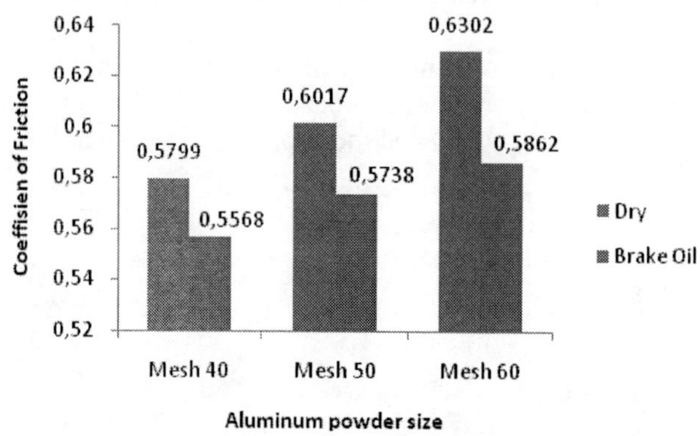

Figure 7. Relationship between aluminum powder size and wear.

Adhesive failure is an interfacial bond failure between the adhesive and the adherent. Figure 8 until figure 10 show the bond failure that occurs in the specimen. Adhesive failure was seen in all three specimens that had been tested for wear.

IOP Conf. Series: Materials Science and Engineering **722** (2020) 012002 doi:10.1088/1757-899X/722/1/012002

Figure 8. Failure of adhesive bonding in rice plant fiber reinforced composite with aluminum powder mesh of 40.

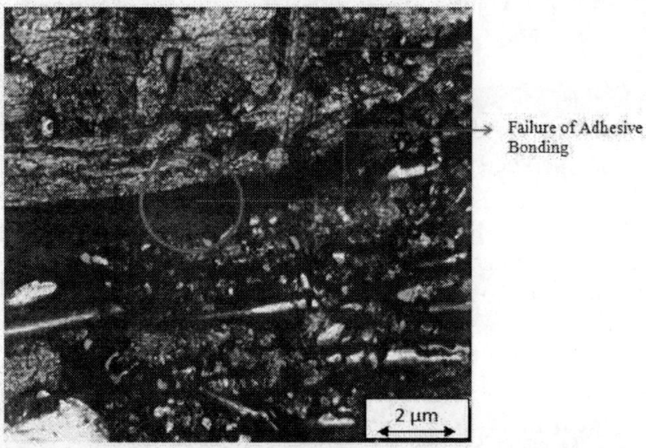

Figure 9. Failure of adhesive bonding in rice plant fiber reinforced composite with aluminum powder mesh of 50.

Figure 10. Failure of adhesive bonding in rice plant fiber reinforced composite with aluminum powder mesh of 60

4. Conclusion

From the test results, conclusions can be taken as follows:

1. The smaller the size of the aluminum powder, the harder the rice plant fiber reinforced composite. From aluminum powder size of mesh 40 to 60, the hardness increase from 79.5 HD to 84.6 HD
2. The smaller the size of aluminum powder, the less wear of rice plant fiber reinforced composites. From aluminum powder size of mesh 40 to 60, the wear decrease from 112.5 mm^3/hour to 81.25 mm^3/hour in dry condition and from 78.12 mm^3/hour to 68.34 mm^3/hour in brake oil environment.
3. The smaller the size of aluminum powder, the higher coefficient of friction of rice plant fiber reinforced composites. From aluminum powder size of mesh 40 to 60, the coefficient of friction increase from 0.5799 to 0.6302 in dry condition and from 0.5568 to 0.5862 in brake oil environment.
4. The failure of the adhesive is seen in the specimen after being tested for wear.

Acknowledgment

This research was funded by The Directorate of Research and Community Service, The Ministry of Research, Technology and Higher Education of the Republic of Indonesia [Project Number = 199.24/A.3-III/LPPM/V/2019]. The author would like to thank the Mechanical Engineering Laboratory of Universitas Muhammadiyah Surakarta and Universitas Gajah Mada for mechanical testing carried out in this research

References

[1] Zhang C W, Li F Y, Li J F, Wang L M, Xie Q, Xu J and Chen S 2017 A new biodegradable composite with open cell by combining modified starch and plant fibers *Materials & Design* **120** 222-229.

[2] Ramesh M, Palanikumar K and Reddy K H 2017 Plant fibre based bio-composites: Sustainable and renewable green materials *Renewable and Sustainable Energy Reviews* **79** 558-584.

[3] Zhou Y, Fan M and Li C 2016 Interface and bonding mechanisms of plant fibre composites: An overview *Composites Part B: Engineering* **101** 31-45.

[4] Fernandes E M, Mano J F and Reis R L 2013 Hybrid cork–polymer composites containing sisal fibre: Morphology, effect of the fibre treatment on the mechanical properties and tensile failure prediction *Composite Structures* **105** 153–162.

[5] Ramesh M, Palanikumar K and Reddy K H 2013 Mechanical property evaluation of sisal–jute–glass fiber reinforced polyester composites *Composites: Part B* **48** 1–9.

[6] Callister W D 2014 *Material Science and Engineering ed. 9E* Mc. Graw Hill London United Kingdom 637.

[7] Rehman A, Das S and Dixit G 2012 Analysis of stir die cast Al–SiC composite brake drums based on coefficient of friction *Tribology International* **51** 36–41.

[8] Elanchezhian C, Ramnath B V, Ramakrishnan G, Rajendrakumar M, Naveenkumar V and Saravanakumar M K 2018 Review on mechanical properties of natural fiber composites *Materials Today: Proceedings Part 1* **5(1)** 1785–1790.

[9] dos Reis J M L 2012 Effect of temperature on the mechanical properties of polymer mortars *Materials Research* **15(4)** 645–649.

[10] Hassan M L, Rowell R M, Fadl N A, Yacoub S F and Christainsen A W 2000 Thermoplasticization of Bagasse. II. Dimensional Stability and Mechanical Properties of Esterified Bagasse Composite *Journal of Applied Polymer Science* **76** 575–586.

[11] Chegdani F and Mansori M E 2018 Friction scale effect in drilling natural fiber composites *Tribology International* **119** 622–630.

[12] Dittenber D B and Rao H V S G Critical review of recent publications on use of natural composites in infrastructure *Composites: Part A* **43(11)** 1419–1429.

IOP Conf. Series: Materials Science and Engineering **722** (2020) 012002 doi:10.1088/1757-899X/722/1/012002

[13] Zhao Q, Shen Y, Ji M, Zhang L, Jiang T and Li C 2014 Effect of carbon nanotube addition on friction coefficient of nanotubes/hydroxyapatite composites *Journal of Industrial and Engineering Chemistry* **20(2)** 544-548.
[14] Xin X, Xu C G and Qing L F 2007 Friction properties of sisal fibre reinforced resin brake composites *Wear* **262(5–6)** 736-741.
[15] Shalwan A and Yousif B F 2013 In State of Art: Mechanical and tribological behaviour of polymeric composites based on natural fibres *Materials & Design* 48 14–24.

Blood Collection Strategy: A Lesson Learned from Indonesian Blood Supply Chain

A Mansur[1,2], I Vanany[2] and NI Arvitrida[2]

[1] Industrial Engineering Department, Universitas Islam Indonesia, Yogyakarta Indonesia
[2] Industrial Engineering Department, Institut Teknologi Sepuluh Nopember, Surabaya, Indonesia

E-mail: agusmansur@uii.ac.id, vanany@its.ac.id, niniet@its.ac.id

Abstract. Blood collection management consists of complex processes that impact the blood outdated rate. This research intends to build an effective collection strategy for the blood bank so expiration rates could be reduced. According to the observed system complexity, a suitable method in this research is discrete event simulation. Three scenarios are used in this research is the strategy of PMI in collecting the blood supply, which is a Mobile Unit Model (MUM) strategy. With the 95% minimum target of order fulfilment rate, this research is succeeded in decreasing expiration rates by 4%, by changing the supply rate of MUM, from 40 hours of time between arrivals to be 44 hours.

1. Introduction

Blood is body fluid that has specific function to distribute oxygen, nutrition, and the remains of metabolism. For specific reasons, such as accident, postpartum haemorrhage, and specific disease will cause a person in the urgency of accessing others' assistance to fulfill his requirement for blood. Generally, the process of blood's intake from the donor could be performed with the conventional model (whole blood) and apheresis model (specific component).

Blood that is given by donors in Indonesia is mostly in the form of whole-blood (blood that has complete components). Following donation, whole-blood may be processed to be several components like Packed Red Cell (PRC), Platelet (TC), Fresh Frozen Plasma, and others. There is also blood donation process that only aims to get a specific blood component, like the use of apheresis method for Platelet (TC) composition transfusion.

A blood bank is an independent organization that has the authority to manage whole-blood donation and its distribution for humanitarian needs. Whole-blood supply can be grouped by the blood types and rhesus. According to the type, whole-blood can be grouped to A, B, AB, and O. When grouped by rhesus, whole-blood can be grouped into R(+) and R(-). The organization in Indonesia that is legally authorized to collect blood from donors is Indonesian Red Cross (PMI). One important indicator of blood bank performances is the accuracy in determining the blood supply. The exact determination of supplies numbers will affect the expected service level achievement and blood's outdated rate, so collecting management is a critical point that requires careful consideration.

Blood Collecting Management is a complex process. The complexity is caused by several factors, those are: 1. The blood supply source of UDD PMI is an independent supplier with its own uncertainty and stochastic pattern. 2. The demands for blood to the blood bank also has its own uncertainty and stochastic pattern. 3. Blood is a perishable product because it has a short shelf life. The decision uncertainty from blood bank in managing blood collection will lead to the excessive or lack of supplies.

The problem that is often faced by a blood bank is the high shortage rate and the high expiration rate of certain types of blood component. This anomaly is experienced by a blood bank when determining the inaccuracy target and supply strategy that also involves donors. This research aims to build a blood collection strategy for a blood bank by determining the blood supply composition so that shortage rate and expiration rate can be suppressed.

2. Literature Review

Blood collection management is one of the main echelons in blood supply chain that has a great impact on the blood supply chain management performance [1]. Blood collection problem is prevalent and must be solved in order to sustain blood availability [2] because such a problem may cause shortage and high expiration rates.

Managing the rate of expired blood products in a blood bank is one of the challenges in blood management in Indonesia [3]. This is caused by the high rate of supply fluctuation from donors and the uncertain blood demands. Good coordination between parties is needed to create an effective method of blood collection. The parties include those responsible for forecasting the amount of supply, those who manage blood, those distribute blood, and those who need the blood [4]. The effectiveness of blood collection procedure can be determined from the decision accuracy that can fulfil blood demands in the right time with right amount [5]. Factors that affect the performance of blood collection are the resources and the collection strategy effectiveness [6] and [7]

Early research on blood collecting strategy is done by [8], which built a forecasting model for the blood supply planning by using Markov model. The rationale of using the method is from the stochastics condition of the involved entities in the blood collection system. [9] created the blood supply planning procedure in a natural disaster condition by using a simulation. The simulation approach was also used by [10] to adjust the rate of blood supply to the apheresis donor system without considering the variation of blood age needed by hospitals. This research will present the management of blood collecting by considering supply and demand fluctuate. The current research's novelty is that the blood supply from donors is still in the form of whole-blood and the demands are allowed blood age preference. By considering the system complexity that involved in this research, hence simulation model approach will be employed as tools in accomplishing the existing problems.

3. Problem Review

The problem that is going to be solved in this research is policy planning in a blood bank to determine the blood supply rate from donors. Blood supply model in a blood bank consists of two main schemes, which are Mobile Unit Model (MUM) and Fixed Location Model (FLM). The application of MUM is donor activity by sending blood bank's employees to a place where the donors are gathered. Meanwhile, in FLM, donors are the ones who visit the blood bank collection. The model is illustrated in Figure 1.

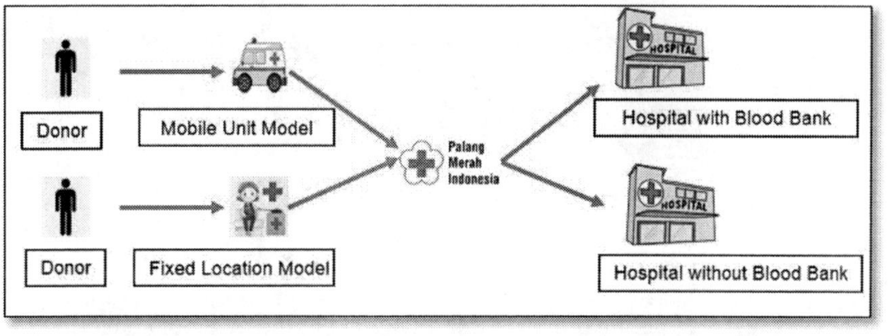

Figure 1. The illustration of Indonesia blood supply chain system

According to Fig. 1, the demands of blood come from two customer groups, which are Hospital Blood Bank (BDRS) and Non-Hospital Blood Bank (NBDRS). The differences in the demands from each

IOP Conf. Series: Materials Science and Engineering **722** (2020) 012003 doi:10.1088/1757-899X/722/1/012003

group is the blood age. BDRS needs much shorter blood type than the NBDRS. The demand from BDRS has the maximum of 7 days WB and PRC age, while the maximum age of the demands from NBDRS is 35 days.

4. The Modelling Approach

The method in this research is the discrete simulation method to determine the amount of whole-blood that is targeted from MUM donors. The consideration of using simulation in BSCM is because of its capability in modelling a complex system (Edgar Alfonso et al., 2012). The model considers the blood products queue in the inventory system before distributing them to the customers, so following [11] the appropriate model simulation is the discrete event simulation.

4.1. System Components

System components identification aims to understand the changes in the system, from the input arrival until the distribution of output. [12] and [13] stated that the elements or important components that should be taken into account to build a simulation model are entities, resources, activities, controls, and queues in the system.

An entity is the object observed in the system. An entity has dynamic characteristics because it can experience displacement, transformation, status change, and can affect and be affected by other entities in the system. The entity that is involved in this research is Whole Blood (WB), Packed Red Cell (PRC), and Platelet (TC) to all blood types (A, B, AB, O) and Rhesus (+,-).

Control Variable or decision variable is a sequence of procedures or acts to arrange how the simulation activities run. In this research, the control variable is the procedure in scheduling of blood collection from the donors in the MUM model.

4.2. Simulation Flowchart

Flow chart diagram was used as a tool to simplify the real system. This will help make it easier to build models in computer simulation. The flow chart diagram shown in figure 2.

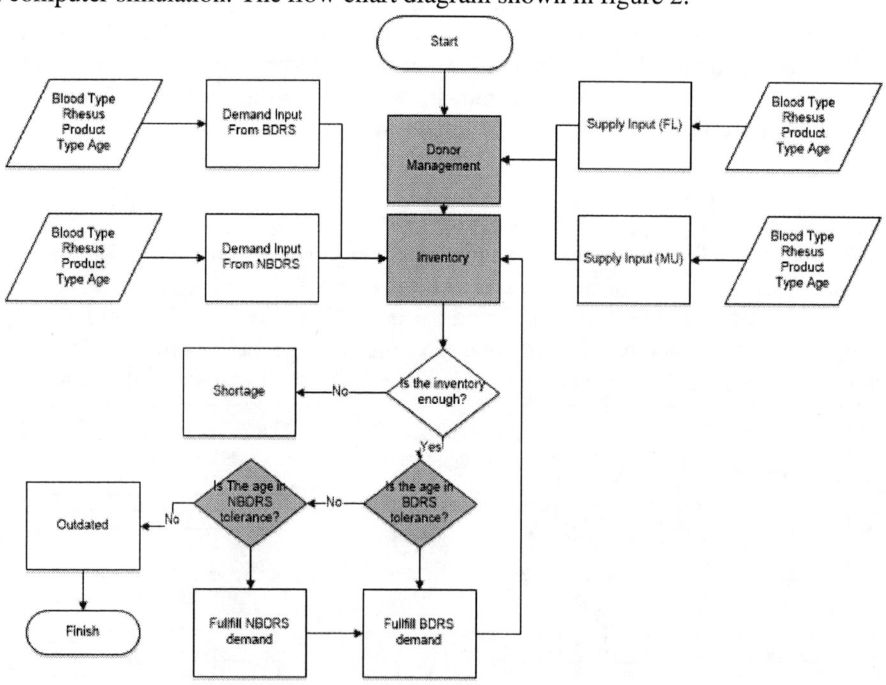

Figure 2. Simulation flow chart diagram in blood supply chain

IOP Conf. Series: Materials Science and Engineering **722** (2020) 012003 doi:10.1088/1757-899X/722/1/012003

Research stages as illustrated in Figure. 2, could be explained as follows:

1. Donor Analysis
   The activities that covered at this stage is examining the donor behaviour (Mobile Unit Model & Fixed Location Model), so it could be identified the statistic distribution started from donor's time between arrival and donor's entity per arrival. The portion of donor's blood type will be identified also at this stage.
2. Demand Analysis.
   The activities that covered at this stage is observing the behaviour of blood's demand from hospital, therefore, it could be identified statistic distribution time between arrival of demand and entity per arrival of demand. It is also could be notified the proportion of blood's component type requested and the age of blood that tolerated by the consumers.
3. The Measurement of Shortage Rate
   The activities that covered at this stage is performing blood's supply analysis that will be reduced with the upcoming demand. Shortage will be experienced if the demand from consumers cannot be accomplished. In this research, assumption on back order towards unfulfilled demand will be excluded.
4. The measurement of Expired Rate
   The activities that performed at this stage is identifying the undistributed numbers of blood's component that reach its tolerated limitation of age.
5. The arrangement of collecting strategy scenario
   At this stage, alternative of actions will be carried out in managing the supplies, hence, blood supply will fulfil its minimal value of shortage and outdated.

5. Result and Discussions

5.1. Current State Model

The current state model is built to test if the proposed model represents the current system effectively. In the current state, the researchers input some stochastic values from the system based on observation data collected in a year period in one of the blood banks in Indonesia. The uncertainty variable in the current state model is presented in Table 1.

Table 1. Uncertainty Variables on The Current System

No	Variable Name	Expression
1	Time Between Arrival MUM	0.5 + EXPO(1.21)
2	Entities Per Arrival MUM	ANINT(NORM(37, 17.7)
3	Time Between Arrival FLM	Constant, Value: 1
4	Entities Per Arrival FLM	ANINT(NORM(7.88, 3.95))
5	Time Between Arrival BDRS's Demand	0.5 + EXPO(0.84)
6	Entities Per Arrival BDRS's Demand	ANINT(0.5 + ERLA(8.75, 2))
7	Time Between Arrival NBDRS's Demand	0.5 + EXPO(0.51)
8	Entities Per Arrival NBDRS's Demand	ANINT(NORM(6.06, 3.02))

EXPO in the column of expression at Table 1 refers to exponential distribution. While the word of NORM refers to normal distribution, while ERLA refers to erlang distribution. ANINT means the value of entities that is assumed in the integer number. For example, time between arrival MUM has expression 0.5 + EXPO (1.21) it means that Time between arrival at MUM has exponential distribution with mean 1.21 days.

To prove that the built model has represented the current system, a statistical test by using T-student test was conducted. Hypothesis at this test is H_0: There is no difference between simulation model with current system. H_1: There is a difference between simulation model with current system. On the 95% level of confidence with error rate tolerance of 5%, the result of t Stat is 0.97476 and the value of t Critical two-tail is 2.178813 so the H_0 was not to reject. This means that the built model has represented the current system.

5.2. Scenario Design

The focus of this research is to build a scenario of the most precise number blood supply that to be targeted by a blood bank per month. The supply source uses one of the two models, which are mobile unit and fixed location. In the scenario planning, an intervention is made in the mobile unit supply. This is because the MUM behaviour is more controllable by the blood bank and it is assumed that the FL behaviour will be stable during the simulation period.

The scenario of the supply design is based on the blood bank strategy in controlling the supply from MU. In Scenario 1, the blood bank used an offensive strategy by a marketing campaign to increase the supply from MU. In Scenario 2 and 3, the blood bank party negotiated to delay the supply rate from MU. The difference in the scenarios is shown from the level of TBA. In the existing condition, the average time between arrival is 41.04 hours or 1.71 days. The scenario detail is shown in Table 2.

Table 2. Supply Strategy Planning Scenario of Mobile Unit Model

Supply Source	Scenario 1	Scenario 2	Scenario 3
MUM	0,8 x TBA MU Existing	1,1 x TBA MU Existing	1,5 x TBA MU Existing

The simulation was done in 90 days. The average and variance of the expired items in accumulation is presented in Table 3.

Table 3. The Average and Variance of Outdate Value Comparison

Groups	Sum	Average	Variance
EXISTING	1810	181	28276.89
SCENARIO 1	4137	413.7	36082.23
SCENARIO 2	1132	113.2	10056.18
SCENARIO 3	148	14.8	187.2889

From the ANOVA test within the scenarios, the obtained F value is F (15.436) > F crit (2.866), which indicates the difference of outdated rate caused by the current condition, scenario 1, scenario 2, and scenario 3.

The blood bank requires that the tolerated number of order fulfilment is minimum of 95% from total demands. Table 4 describes the levels of shortages and order fulfilment rate from the current condition and proposed scenarios.

Table 4. Shortages Number and Order Fulfilment Rate

Groups	*Sum*	*Average*	*Variance*	*Order fulfilment*	
EXISTING	593	59.3	347.5667	0.96	Acceptable
SCENARIO 1	547	54.7	225.1222	0.96	Acceptable
SCENARIO 2	728	72.8	318.8444	0.95	Acceptable
SCENARIO 3	1353	135.3	10803.12	0.90	Unacceptable

Referring to the low expiration rate, the third scenario is the best scenario. However, if it is based on the number of fulfilment rate that is tolerated by the blood bank, scenario 3 cannot meet the specified standards.

6. Conclusion and Recommendation

The blood collection strategy gives impacts on the number of expired blood bags in the blood bank inventory system. The effort to decrease an outdated rate should consider the shortage caused by inventory reduction. In the order fulfilment target of 95%, the best scenario to reduce expiration is scenario 2 that lessens the expiration rate by 4% (the average of outdated rate on current system is 181 blood bags, while on the result of proposed scenario 2, the average of outdated rate decreases to 113 blood bags). To implement the scenario 2, the blood bank should regulate the campaign intensity in executing the donor activity for mobile unit model. So, activity of MUM is recommended as every 45,1 hours or 1,88 days.

This research's limitation is that it only involves supply arrangement from MUM. Recommendation for further research is to decrease the expiration rate by is considering the production composition from the blood bank, the inventory target policies, and the inventory cost calculation in ordering and storing products.

7. References

[1] Taylor, P., Osorio, A. F., Brailsford, S. C., & Smith, H. K.. A structured review of quantitative models in the blood supply chain : a taxonomic framework for decision-making, (February), 2015, pp. 37–41
[2] Pierskalla, W. P. Supply chain management of blood banks William, (v), 2005, pp 1–32.
[3] Mansur, A., Vanany, I., & Indah Arvitrida, N. (2018). Challenge and opportunity research in blood supply chain management: A literature review. In MATEC Web of Conferences , 2018,Vol. 154.
[4] Güre, S. B., Carello, G., Lanzarone, E., & Yalçındağ, S. (2017). The Management of Operations Unaddressed problems and research perspectives in scheduling blood collection from donors. Production Planning & Control, 7287(August), 1–7.
[5] Zahiri, B., Torabi, S. A., Mousazadeh, M., & Mansouri, S. A. . Blood collection management : Methodology and application. In Applied Mathematical Modelling, 2015, (Vol. 39, pp. 7680–7696). Elsevier Inc.
[6] Alfonso, E., Xie, X., Augusto, V., & Garraud, O. Modeling and simulation of blood collection systems. Health Care Management Science, 15(1), 2012, 63–78.
[7] Pitocco, C., & Sexton, T. R.. Alleviating blood shortages in a resource-constrained environment. Transfusion, 45(7), 2005, 1118–1126.
[8] Cumming, P.D., Kendall, K.E., Pegels, C.C., Seagle, J.P. an Shubsda, J.F., "A collections planning model for regional blood suppliers: description and validation", Management Science, Vol. 22, 1976, pp. 962-71
[9] Fahimnia, B., Jabbarzadeh, A., Ghavamifar, A., & Bell, M. Supply chain design for efficient and effective blood supply in disasters. In International Journal of Production Economics, 2015, (pp. 1–10). Elsevier
[10] Alfonso, E., Xie, X., Augusto, V., & Garraud, O.. Modelling and simulation of blood collection systems: Improvement of human resources allocation for better cost-effectiveness and reduction of candidate donor abandonment. Vox Sanguinis, 104(3), (2013), pp. 225–233.
[11] Robinson, S.. Simulation - The Practice of Model Development and Use. Chicester: Jhon Wiley & Sons Ltd., 2004
[12] Banks, J., Nelson, B. L., Carson, J. S., & Nicol, D. M. (2010). Discrete-Event System Simulation. PrenticeHall International Series in Industrial and Systems Engineering, 640.
[13] Law, A. M.. Simulation Modeling and Analysis (Fifth). New York: Mc-Graw-Hill Education, 2015

Acknowledgments

Foremost, I would like to express my very profound gratitude to Kementerian Riset Teknologi dan Pendidikan Tinggi Republik Indonesia (Ministry of Research, Technology and Higher Education of the Republic of Indonesia), Kementerian Keuangan Republik Indonesia (Ministry of Finance, Republic of

IOP Conf. Series: Materials Science and Engineering **722** (2020) 012003 doi:10.1088/1757-899X/722/1/012003

Indonesia), and Palang Merah Indonesia for generous support in this research and I am honoured to be one of the recipients of BUDI-DN_ LPDP scholarship.

IOP Conf. Series: Materials Science and Engineering **722** (2020) 012004 doi:10.1088/1757-899X/722/1/012004

Platelet Inventory Management System Using Monte Carlo Simulation

A Mansur[1], FI Mar'ah[2], P Amalia[3]

[1,2,3]Department of Industrial Engineering, Islamic University of Indonesia, Yogyakarta, Indonesia

E-mail: [1]agusmansur@uii.ac.id, [2]14522437@students.uii.ac.id, [3]17522035@students.uii.ac.id

Abstract. Improper of blood management will lead to wastage due to the nature of blood that is not durable and easily damaged, so if the blood is stored too much and not used until the expired date, the blood will be damaged. Platelet components have a limited lifetime for 5 days. With a short blood age, every month there will be two possibilities, one is shortage of blood inventory (stockout) or two, over blood inventory (overstock) in the storage area. From the data obtained, the expired rate in 2017 reached 743 blood bags. Until now Indonesian Red Cross has not been able to predict the level of blood demand because of the uncertain level of demand. In this study, simulation will use Monte Carlo to develop a number of policies that can optimize inventory levels and minimize inventory costs.

1. Background

The uncertainty in blood availability caused by unpredictable blood demand and supply. In other word supply and demand in blood supply chain have stochastics or uncertainty characteristics [1]. According to the World Health Organization (WHO) recommendation that the minimum number of donor in one community is around 2% from the total population. However, the facts of the blood supply have decreased in the current decade, whereas the blood demand has increasing tendency [2]. In other word, currently donors are rare assets in society [3].

Improper blood management will create wastage because of the blood characteristics that are not durable and can be damaged easily, so if the blood inventory is in a big number of amount and not be used until the expiration date, then the blood will be broken. As specified by the annual report of a blood bank in Kabupaten Sleman in 2017, there are 743 blood bags that should be annihilated due to its expiration date. The expired rate in 2017 is explained in Figure 1.

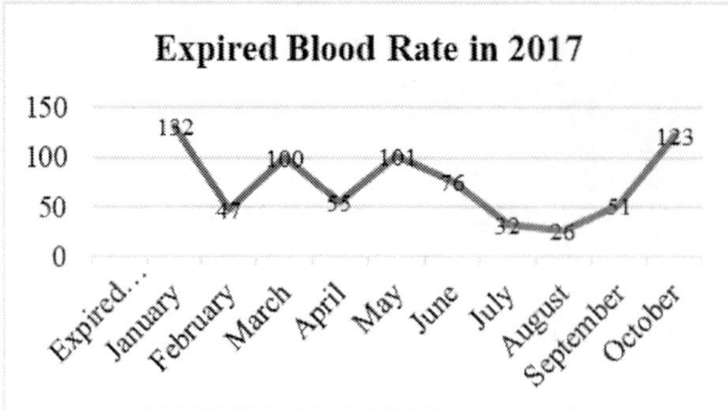

Figure 1. Expired Blood Rate in 2017

In accordance with the limited blood age that is 5 days for the TC component (Platelet), it needs a complex blood inventory management [5]. With a short age of blood, then there will be 2 possibilities that are stockout or overstock in each month in the inventory. If stockout, then the worst possibility is death. While if overstock happens, it will make the blood wasted if the saving time is exceeding its expiration date. Blood demand has uncertainty characteristics and the supply of blood also not constant. The demand pattern from the blood bag is different from another general product. In compliance with [4], blood product demand will be equal with the population growth rate, but the number of donors will be reduced. Blood demand across Indonesia could be classified into two; normal condition demand and special condition demand. The normal condition is a condition when there is no extraordinary situation that could cause a sudden spike in blood demand. A special condition is a certain condition that affects blood demand and directly spiked the blood demand, in example endemic disease outbreak such as dengue fever [6] or disaster [7].

Blood supply quantity is based on the number of donors that come to blood bank. The blood bank usually held blood donor events or sponsored many events to get supplies from donors [8] . Besides that, blood bank also reminds the permanent donors and non-permanent donors to do blood donors by broadcast messages or through social media. However, with all of the efforts, blood bank cannot predict the number of blood products that obtained will be on target. Therefore, with the lack of blood supply that can damage the patients' life when they need the blood product.

There are a lot of researches that develops many kinds of inventory model that taking into account various factors like product age, discount, increasing-price even up to inflation rate. One of the researches that develops inventory optimization model is developed stochastics model to determine the optimum quantity from blood inventory and [9] considered an inventory system with discrete event and safety stock in platelet ordering case. In this research, six policies will be built and simulate with Monte Carlo in 30 days. The result of the simulation will decide the appropriate policy to minimize the total cost of inventory in the blood bank Kabupaten Sleman case study. Monte Carlo simulation will be done by Microsoft Excel.

2. Literature Study

The previous research studied blood inventory optimization or minimizing the operational cost of blood inventory with a heuristic approach solution has been done by previous researchers until now. Researches about blood inventory that easily broken have developed many methodologies to solve that case. There was a research developed in the challenge of dealing with demand uncertainty in problems from inventory management. Previous research in blood inventory usually assumed that the uncertainty demand can be modelled with Poisson distribution or Normal distribution. Because of the proposed approach to overcome the model limitation is not rely on the assumption about how the demand uncertainty can be modelled.

The characteristic of the blood product that is easy to be broken is difficult to be modelled in blood supply management. Blood inventory that is easy to be broken will cause the heap of stocks because of uncertainty demand will make the long-time blood releasement and related costs incurred. Moreover, blood that received from the donor also limited. Hence, efficient blood inventory in blood bank is very important. Whereas [9] deliberated an inventory system with discrete event and safety stock for platelet ordering case.

In the last 10 years, the blood supply chain has gained significant attention, there are several methods that can be implemented to do observation, like the queuing theory and Markov chain, statistics analysis, simulation, and optimization [6]. Previous research has made inventory mathematical model that considering the product age and discount factors. After that a research was made a model with considering all unit discount factor and repair the algorithm functions from model that previously made. Moreover, there is an inventory model that made for multiple items that considering product age factor and all unit discount where the demand for the product is deterministic.

3. Research Methods

This research is based on the case study in one of regional blood bank in Yogyakarta. The purpose of this research is to minimize the inventory cost that can optimize the blood inventory platelet component using Monte Carlo Simulation.

This research uses Microsoft Excel as the research instrument to run the Monte Carlo simulation. Therefore, a mathematical function is needed to determine the total inventory cost to solve the blood inventory problem so the optimum inventory rate can be known. The total inventory cost function that is developed can be explained as:

3.1. Simulation Parameter and Index

Parameters that used in the total inventory cost are as follow:
Index
t = simulation period index
p = incoming blood index
h = inventory index
s = shortage index
E = expired index
Parameters
Cp= ordering cost
Ch= holding cost
Cs= shortage cost
CE= expired cost
Q_t^s= rate of blood shortage in t period
Q_t^E= rate of expired blood in t period
It= rate of blood inventory in t period
Pt= rate of blood supply in t period

3.2. Total Inventory Cost Function

The function of total inventory cost is as follows:

$$\sum_{t=1}^{30} Q_t^s . C_s + \sum_{t=1}^{30} Q_t^E . C_E + \sum_{t=1}^{30} I_t . C_h + \sum_{t=1}^{30} P_t . C_p \qquad (1)$$

Equation (1) is a function to minimize the total inventory cost. This total cost is influenced by the rate of blood supply variable, the blood inventory rate, shortage rate and also expired rate. In the above

IOP Conf. Series: Materials Science and Engineering **722** (2020) 012004 doi:10.1088/1757-899X/722/1/012004

function, the inventory total cost consists of the ordering cost, holding cost, expired cost, and shortage cost components.

4. Result and Discussion

4.1. Validation Test
The validation test that used in this research is the mean comparison test and variance comparison test.

4.1.1. Mean Comparison Test. The mean comparison test is used to compare the performance between the real system with the simulation model from the value of output mean on both population data. Because the data is from two populations, so the test used is double-sided test. The t-value from the mean comparison test in the demand of O blood is 0, and the t-value in the supply of O blood is -0.7376. Since the range of acceptance is between -2.048 and 2.048, so the result is H_0 is accepted, which means that the simulation data is in line with the real system data.

4.1.2. Variance Comparison Test. To make sure that the assumption of both populations has the same variance, the variance comparison test is done. According to the calculation, the result of the F-value is 1. Since the range of acceptance is between 0.476 and 2.101, then the H_0 is accepted or the simulation result is in line with the real system.

4.2. Scenarios Development
From Monte Carlo simulation that has been done, then several scenarios were made to search the optimum total inventory cost. According to the observation result and interviews, there are several scenarios as follow:

4.2.1. Scenario 1. The first scenario is presenting the current condition in a blood bank in Kabupaten Sleman. It can be known that the current condition in blood bank in Kabupaten Sleman has not determined the safety stock policy.

4.2.2. Scenario 2. The next scenario is the policy to have safety stock for TC blood products. Since the product TC age is only for 5 days, so the inventory management will be more complex. In this scenario, the safety stock is 1 blood bag each day. This applied when the total inventory of blood is less than 1 blood bag, so on the next day, the supply will increase 1 blood bag. The policy of safety stock is based on the average of the shortage replication data that happened each day. This is caused because the blood bank has no shortage data for the TC component.

4.2.3. Scenario 3. Unlike the second scenario that regulates the safety stock, the third scenario makes policy to decrease the supply of the TC component. This is based on the demand data for the TC component that is rare, but there is daily supply each day. The number of blood demand cannot be predicted, and the blood bank is striving to get donors every day. Therefore, the expired product rate will be high. In the third scenario, a policy to decrease the supply rate will be built to make the supply rate near to the demand rate. This scenario is based on the average comparison of supply and demand on each day for a number of replications to determine the percentage of supply declination rate.

4.2.4. Scenario 4. In line with the second scenario, this scenario 4 applies the safety stock to decrease the shortage faced by blood bank. The number of safety stock in scenario 4 policy is 2 blood bags in every day. This policy is implemented if the number of bloods in inventory is less than 2 blood bags, so 2 blood bags will be added in the next day.

4.2.5. Scenario 5. Scenario 5 contains a policy that has a purpose to decrease the supply rate of the TC component. This is based on the low level of demand for TC component, while the TC supply is increasing every day. The reduction of the TC supply is to make the rate of supply is close to the rate of demand, so the number of expired blood products can be decreased. The decrease of supply in this scenario is 90% of the previous supply.

4.2.6. Scenario 6. Correlated with the previous scenario, the researcher in scenario 6 simulate if the policy of supply rate decrease is implemented with a 95% rate of reduction. It is intended to know where is the equilibrium point to make no shortage and a low number of expired products, so the inventory cost can be minimized.

Table 1. Total Cost Comparison

	Stock	Holding Cost	Don ors	Ordering Cost	Shortage	Shortage Cost	Expired	Expired Cost	Total Inventory Cost
Sc1	33692	185306000	251	3765000	2	720000	207	41400000	231191000
Sc2	32874	180807000	255	3825000	1	360000	212	42400000	227392000
Sc3	5307	29188500	60	900000	9	3240000	42	8400000	41728500
Sc4	33569	184629500	254	3810000	0	0	212	42400000	230839500
Sc5	4886	2687300	60	900000	11	3960000	37	7400000	39133000
Sc6	3975	21862500	60	900000	13	4680000	30	6000000	33442500

4.3. Total Inventory Cost Comparison

This research is expected to know the most optimum scenario to decrease the total inventory cost. Total inventory cost consists of ordering cost, holding cost, expired cost, and shortage cost. Based on the 30 replications that have been done, the result of the total inventory cost is as in Figure 4.1.

From the simulation result with 30 times replications, different inventory costs are shown depends on the scenarios. From the first scenario which represents the current system, the total inventory cost spent on a month is Rp231.191.000,00. The mean of the inventory cost is Rp7.706.366,67 with Rp1.315.407 variance. The result of scenario 2 which adding 1 product as a safety stock has the total inventory cost as Rp227.392.000,00 with the mean of inventory cost is Rp7.579.733,33 and Rp1.581.354 of variance. The scenario 3 policy is to decrease the supply rate as 88% to make the supply number close with the demand, and the result of the simulation is Rp41.728.500,00 as the total inventory cost. The mean and variance of the total inventory cost are Rp1.390.950,00 and Rp192.160.

In scenario 4, 2 blood bags are used as safety stock. The total inventory cost in scenario 4 is Rp230.839.500,00 with the mean of Rp7.694.650,00 and Rp1.116.856 as the variance. The result of inventory cost in scenario 5 is Rp39.133.000,00, applying declination of supply rate by 95%. The mean of inventory cost in scenario 5 is Rp1.304.433,33 and Rp227.069 as the variance. Scenario 6 was built to know the impact if the declination rate changed to 90%. The result of the simulation showed that the total inventory cost is Rp33.442.500,00 with Rp1.114.750,00 and Rp205.661 as mean and variance respectively.

5. Conclusion

To get the appropriate policy in order to optimize the platelet component inventory and minimize the total inventory cost, Anova and Bonferroni test has been done to test the simulation result. In the Anova test, the result of the F-value is 385,59 with F-critical value is 2,266. Since the F-value is higher than F-critical, the scenarios result from scenario 2,3,4,5, and 6 has different mean with the current system or scenario 1. From the Bonferroni test, the output from the first scenario with the second scenario and fourth scenario has no difference. The outputs of scenario 3 with scenario 5 and scenario 2 with scenario 4 are also have no difference. However, the others are having different outputs. From the total inventory cost result, the minimum cost is from the scenario 6 with a declining supply rate on 95% and resulting in the total inventory cost Rp33.442.500. The rate of cost decreased is 14.45%. Thus, it can be concluded that the best scenario to minimize the total inventory cost and optimize the blood inventory system is the scenario 6. For the next research, it is better if considering the blood production not only platelet, considering the seasonal demand and considering if the demand from customers are different age request.

References

[1] Cohen, M. d. Perishable Inventory Theory and Its Application to Blood Bank Management. National Center for Health Service Researce and Development, 1974.
[2] Heinrich, J. Blood Supply: Availability of Blood. United States: Health Education and Human Services Division Report, 2012.
[3] Belien J., F. H. Supply Chain Management of Blood Product: A Literature Review. European Journal of Operational Research, Vol. 217, pp. 1-16, 2012.
[4] Greinacher, A. F. Impact of demographic changes on the blood supply: Mecklenburg-West Pomerania as a model region for Europe. Transfusion 47(3), pp. 395-401, 2007.
[5] Alfonso, E., Xie, X., Augusto, V., & Garraud, O.. Modelling and simulation of blood collection systems: Improvement of human resources allocation for better cost-effectiveness and reduction of candidate donor abandonment. Vox Sanguinis, 104(3), (2013), pp. 225–233.
[6] Mansur, Agus & Vanany, Iwan & Indah Arvitrida, Niniet. Challenge and Research Opportunity in Blood Supply Chain Management: A Literature Review, 2018.
[7] Fahimnia, B., Jabbarzadeh, A., Ghavamifar, A., & Bell, M. Supply chain design for efficient and effective blood supply in disasters. In International Journal of Production Economics, 2015, (pp. 1–10). Elsevier
[8] Allamsyah, Z., & Mansur, A. System design of blood supply chain management based on Supplier Customer Relationship Management (SCRM) approach. In *MATEC Web of Conferences* (Vol. 154), 2018.
[9] Civelek, I. K.-W. Blood Platelet Inventory Management with Protection Levels. Eur. J. Oper. Res. 243 (3), 826-238, 2015.

Concentration of Exopolysaccharides (EPS) Produced from Fermentation of Hydrolysate Biomass of Tuni Sagoo (*Metroxylon sp.*) Starch by *Lactobacillus plantarum* FU-0811 in Semi Pilot Scale using Nanofiltration

A Susilowati*, Aspiyanto, Y Maryati and A Dinoto

[1]Research Center for Chemistry
Indonesian Institute of Sciences (LIPI), South Tangerang, Banten
[2] Research Center for Biology,
Indonesian Institute of Sciences (LIPI), Cibinong, Bogor

*Coresponding author : agustine_1408@yahoo.co.id

Abstract. Nanofiltration (NF) system is one of the concentration process step to recover exopolysaccharides (EPS) concentrate (retentate) as food aditive (thickener, stabilizer and emultion). Feed was biomass containing EPS as a result of growth of *Lactobacillus plantarum* FU-0811 culture under 37 °C for 72 hours in semi pilot scale (20 L). Biomass was a growth media containing *sagoo sugar* as a result of step enzymatic hydrolysis on tuni *sagoo (Metroxylon sp.) starch* using commercial α–Amylase and Amyloglucosidase (AMG) enzymes and other active compounds. Concentration process was conducted by pump motor frequency of 20 Hz, room temperature, and pressure of 10, 15 and 20 bar for 0, 30, 60, 90 and 120 minutes. Result of experimental work showed that NF system has potential use to concentrate biomass containing higher composition than that permeate. Based on optimum EPS, NF system that results a successfully separation of EPS with rejection of 95.24 % was reached at pressure of 20 bar and 60 minutes to retain EPS in concentrate (retentate) of 2.454 % (dry weight) and freely pass in permeate of 0.0097 % (dry weight) with permeate flux value 37.5 L/m^2.hour. In this optimum process condition, NF system was able to increase of EPS in retentate of 56.46 % from prior to process (feed) and after process concentration.

1. Introduction

Fermentation process of *Lactobacillus plantarum* FU-0811 using sagoo (*Metroxylon sp.*) sugar media as glucose substitution and other active agents in semi pilot (20 L) scale is an initiation effort to get exopolysaccharides (EPS) in production scale. EPS is produced as metabolite of *Lactobacillus plantarum* FU-0811 in fermentation process at 37 ºC for 72 hours [1] and as carrier agent of active compound of oral insulin in pharmaceutical industry [2] and as food aditive (thickener, stabilizer and emultion). EPS is a compound produced by microbes through metabolism process excreted to outer cell [3]. Selection on hydrolysate of tuni sagoo starch obtained through subsequent enzymatic hydrolysis using α-Amylase and Amyloglucosidase (AMG) enzymes caused by hydrolysate recovery containing more specific glucose so that it produces more optimum EPS when compared by hydrolysis using acid (HCl) [4]. Sagoo sugar is use as diversification of carbon source materials, beside glucose, which is utlized to growth of *Lactobacillus plantarum* FU-0811 or other components, such as tryptone, pepton, yeast extract, etc.).

In semi pilot scale, fermentation is efforted to recover their metabolite and composition with indifferent composition on laboratory scale under optimum condition of *Lactobacillus plantarum* FU-0811 growth. Nevertheless, larger process scale produces generally different composition with laboratory scale (200 – 300 mL). Factors affecting on product recovery in larger scale are stability of environmental condition (temperature, humidity), growth of *Lactobacillus plantarum* FU-0811 relating with metabolism and nutritive value, but main factor is ability of microbe adaptation on media in larger volume.

To recover higher EPS concentration from fermentation process in 20 L scale, it is efforted to concentrate biomass through NF membrane in order to separate EPS to retentate with higher purity. NF ability in this process is caused by utilizing NF membrane having pore size range of 1 – 100 nanometer (nm) supported by operation pressure of 7 – 30 bars, so that it is able to pass particles of biomass. NF membrane is able to separate compounds with molecular weight (MW) of 150 Dalton (Da.), such as monovalent anion and di- & multivalent anions, organic compounds and retain particle of sugar solute and multivalent salt, but passes freely a mount of monovalent salt, mineral and water [5]. Membrane separation process has an ability to transport selected components from solution other better components so that component is separated from initial solution. Membrane ability to separate certain components is caused by its presence of driving force, such as pressure, concentration and temperature, and electric potential. In NF-based membrane was used pressure as driving force [6,7].

The goal of this experimental activity was to find out process optimization of EPS concentration as a result of fermentation of hydrolysate of tuni sagoo starch by *Lactobacillus plantarum* FU-0811 in 20 L scale through NF membrane on composition of concentration result under fixed condition (pump motor frequency of 20 Hz and room temperature) and different time and operation pressure..

2. Materials and methods

2.1. Materials and Equipments

Materials used in these activities were tuni sago starch (Seram, Maluku), *Lactobacillus sp.* FU-0811 culture (Pusat Penelitian Biologi – LIPI, Cibinong), α-Amylase and Amyloglucosidase (AMG) (NOVO), chemicals for analysis and preparation of induction medium, and 2 sheets of NF Thin Layer Composite membranes (NF-PE-99, Danish Separation Systems, Denmark) and 2 sheets of MF 0.15 μm [Anonymous, 2000]. Equipments utilized in this work were High Frequency Separation, membrane modul of LabUnit M20 (DSS, Denmark) Fermentor in 20 L scale, equipments used in microbiology process, Instrument of Spectrofotometer UV-1201 and glass ware for chemistry analysis..

2.2. Research Design

This experiment was performed with variation : (1) concentration process time of 0, 30, 60, 90 and 120 minutes and (2) operation pressure of 10, 15 and 20 bar through NF membrane under pump motor frequency of 20 Hz (flow rate 7.5 L/minute) and room temperature. Analysis was conducted on biomass composition, such as total solids (Gravimetric method), EPS (Phenol-Sulphate method), reducing sugar (Somogyi-Nelson method) and Total Acids (Titration method) [8] and NF membrane performance, such as permeate flux value and rejection.

2.2.1. Enzymatic hydrolysis on tuni sagoo (Metroxylon sp.) starch. A mount of tuni sagoo starch was subsequent diluted in aquadest, adjusted to pH of 6, heated at 90 – 95 °C, agitated and added α-Amylase (Thermamyl 60 L) enzyme of 0.08 % (v/w, dry starch) for 60 minutes to produce suspension. Suspension produced (dextrin) was then autoclved at 121 °C for 30 minutes and inactivated at 70 °C and pH 4 for 15 minutes. Suspension was subsequent allowed cool, adjusted pH to 4.5, added with *Amyloglucosidase* (AMG) enzyme of 0.08 % (v/v), and re-heated at 60 °C for 48 hours to get hydrolysate of tuni sagoo starch. Hydrolysate was filtered by 200 mesh sieve and introduced to MF system at pump motor frequency of 20 Hz, room temperature, and pressure of 10 bar for 120 minutes to result retentate and permeate. Permeate (filtrate or extract) as sagoo sugar is raw material of biomass to get EPS through fermentation process by *Lactobacillus plantarum* FU-0811 culture.

2.2.2. Inoculum preparation and fermentation of Lactobacillus plantarum FU-0811. Preparation of *Lactobacillus plantarum* FU-0811 inoculum was conducted by inoculating *Lactobacillus plantarum FU-0811 isolate in glyserol stock using MRS medium.* Grown culture was then re-inoculated in MRS A via spread plate method and incubated aerobic at 37 °C for 48 hours. After single colony is grown, this culture is then inoculated in induction medium of MRS Broth by modifying formation components consisting of sagoo hydrolysate, peptone, beef extract, Sodium Acetate, yeast extract, Ammonium Citrate, Di-Sodium Hydrogen Phosphate, Magnesium Sulphate, Mangaan Sulphate, Tween-80, L-cystiene-HCl and Silicon anti foaming [2]. From this process is reached *Lactobacillus plantarum* FU-0811 inoculum. Fermentation process is performed by inoculating 2 % (v/v) of inoculum in induction medium (20 L) in fermentor aseptically at 37 ºC for 72 hours with control of biomass of 12 hours.

2.2.3. Concentration process of biomass through Nanofiltration (NF) membrane. Feed fluid of biomass of sagoo hydrolysate by *Lactobacillus sp.* FU-0811 culture passing via 200 mesh sieve placed in feed tank of 9 L of capacity were subsequently pumped through 200 µm of sieve, heat exchanger, membrane modul and retentate outlet. Retentate was re-circulated to feed tank. During circulation process, cooling water at 23 – 24 °C was flown to heat exchanger to stable fluid temperature in tank at room temperature (± 25 °C). After stable process condition, pump motor frequency was arranged to 20 Hz (flow rate of ~ 7.5 L/minute) and operation pressure was fixed at 10 bar. Fluid passing freely per membrane area unit (permeate) via permeate housing was collected, measured its volume and recorded its time to determine permeate flux value. Investigation and sampling were conducted on permeate and retentate for 0, 30, 45, 60, 90 and 120 minutes. Similar process procedure was employed under pump motor frequency of 20 Hz (flow rate of 7.5 L/minute), room temperature and operation pressures of 15 and 20 Bar.

3. Result and Disccusion

3.1. Characteristic of sago starch hydrolysate, biomass and biomass extract as feed

Hydrolysate of tuni sagoo starch was produced through hydrolysis process by α-Amylase and Amyloglucosidase enzymes, which is brown turbid suspension with less thick and sweet taste, as demonstrated in **Figures 1**.

(a) (b) (c)

Figure 1. Sagoo starch hydrolysate (a), biomass of fermentation result of *Lactobacillus plantarum* FU-0811 in 20 L scale (b) and biomass as a result of 200 mesh sieve as feed (c).

Feed composition in concentration process through NF membrane indicated that contents of reducing sugar of 1.4871 mg/mL, total solids of 5.1389 %, total acids of 0.033 %, and EPS of 1.0685 % (dry weight). This composition is different with prior to filtration of 200 mesh sieve equipped by High Frequency Separator or after fermentation process (37 °C for 72 hours). It had been shown, the composition decreases to 10 – 15 % from initial biomass, but this step is carried out to anticipate its occurence of plugging in NF membrane system and get more optimal process efficiency [Anonymous, 2000]. The process is initial separation to sieve anorganic components contained in biomass of fresh hydrolysate, such as dextrin residues (non-hydrolyzed sagoo component by amyloglucosidase enzyme).

3.2. Effect of concentration process condition on NF membrane performance and composition

Permeate flux value. Permeate flux value (J) is a mount of permeate produced per membrane area unit (A) per time unit (t) and expressed as J = 1/ (A x t). Permeate flux value is one of the most important parameter of membrane performances and becomes a reference of membrane process acting good [9]. Biomass concentration process of *Lactobacillus plantarum* FU-0811 culture produced from fermentasi in semi pilot (20 L) scale through NF membrane under pump motor frequency of 20 Hz (flow rate of 7.5 L/minute), room temperature and operation pressure of 10, 15 and 20 bar for long time process generated low permeate flux value, as shown in Figure 2.

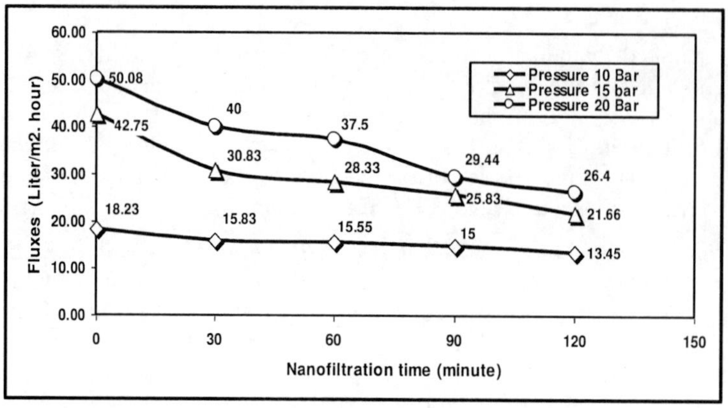

Figure 2. Relationship of time and permeate flux value in concentration biomass as a result of fermentation of sagoo sugar by *Lactobacillus plantarum* FU-0811 culture.

Concentration process at operation process of 20 Bar gave higher value of permeate flux than that 10 and 15 bar. Decrease of permeate flux value is caused by tangential flow of feed fluid on the top membrane surface, in which tangential flow do not give a chance on solute particles to accumulate on the top membrane surface [6]. It had been shown that pump motor frequency affects generally on flow rate of fluid, while operation pressure affects on permeate flow rate penetrating through membrane, so that high flow rate of feed fluid and permeate flux value becomes more and more high [7]. It had been showed that sharply decrease of permeate flux value is occurred under operation pressures of 15 bar and 20 bar between 0 – 30 minutes, followed by gradually decline between 30 – 60 minutes, and strongly drop to end concentration process. It is predicted that process time of 0 – 30 minutes is not occurred 'fouling' on membrane yet.

Fouling is a phenomenon of its occurrence of particle 'cake' layer on the top membrane surface, which prevent solutes to pass freely through membrane so that hinders and declines permeate flow rate and permeate flux value [6,7]. After 30 to 120 minutes (end process), permeate flux value drops more gradual because it starts to be happened fouling phenomena causing its prevent of permeate to pass through membrane, but this gradual decrease of permeate flux value is running from 30 to 60 minutes only, and sharp decline to final concentration process because part of components pass already through membrane.

Exopolysaccharides (EPS) (mg/mL). EPS is a polysaccharide produced and secreted out from microbe cells, and get from biomass of microorganism, such as bacteria, micro algae, and fungi [10]. It had been known that *Lactobacillus* plantarum FU-0811 is able to produce EPS through its growth in biomass of sagoo gel get through gelatinization process at 37 °C [2]. In fermentation of sagoo sugar by *Lactobacillus plantarum*-FU 0811 in 20 L scale is generated through enzimatic hydrolysis using α-Amylase and re-hydrolyzed using *Amyloglucosidase (*AMG) to glucose alternative. Fermentation process in higher scale produces lower EPS recovery (1.0685 %, dry weight) when compared with fermentation process in 300 – 500 mL of laboratory scale (5.5922 %, dry weight) [4]. Environmental factors affect possibility on EPS recovery, such as process conditions (temperature, time, internal factors of *Lactobacillus plantarum* FU- 0811, such as the growth level of *Lactobacillus plantarum*-FU 0811).

Concentration process by means of NF system showed that NF system demonstrates an EPS increase in retentate (concentrate) relating with long time of concentration under operation pressure of 10, 15 and 20 bar, while in permeate for similar condition tends to be constant. Nevertheless, NF system at pressure of 20 bar increases EPS concentration relating with long time of concentration, as shown in Figure 3.

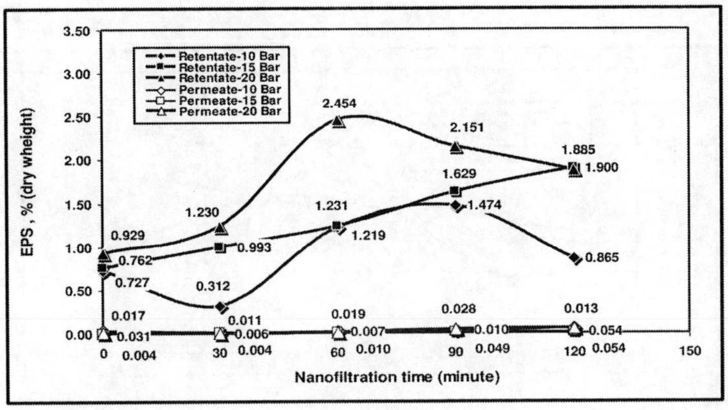

Figure 3. Relationship between time and EPS contents in retentate and permeate in concentration biomass as a result of fermentation of sago sugar by *Lactobacillus plantarum* FU-0811 culture.

Difference in products are possibility caused by difference in driving force interacting with time. To result optimal EPS concentration, in higher operation pressure will need faster process time. High operation pressure will become more and more high driving force so that it enables to compress and bind EPS particles by adhesi force, such as material components and its deficite of water mass during concentration process.

On membrane performance of NF-PE-99, concentration process of EPS at different pressure for 0 - 120 minutes generated different rejection coefficient of EPS. It had been seen that high operation pressure would become more and more high rejection coefficient of EPS. In other words, selectivity level in NF system at operation pressure of 20 bar was better condition both 15 bar and 10 bar, as shown in Figure 4.

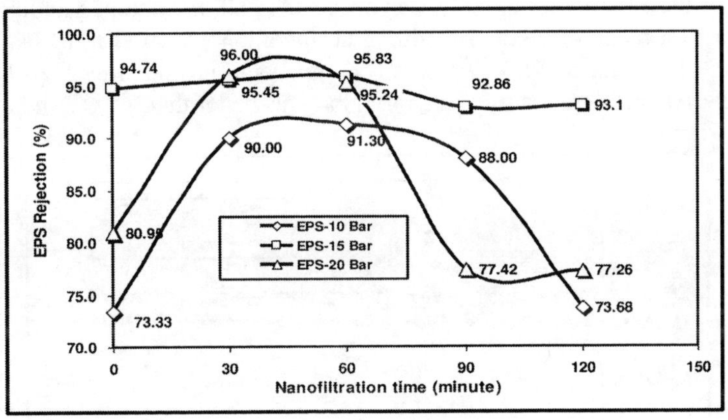

Figure 4. Relationship between time and EPS rejection in concentration biomass as a result of fermentation of sago sugar by *Lactobacillus plantarum* FU-0811 culture.

Rejection of the solute, expressed as rejection of the membrane. Based on membrane performance relating with optimalization of concentration process, the best process condition was reached at operation pressure 20 bar for 30 minutes with the observed EPS rejection of 96 % (EPS concentration of 1.23 %, dry weight). There is a difference in process optimalization based on the highest concentration of EPS. This case is possibility caused by heteropolymer property of EPS, variouus MW and its present of other components with similar MW, such as organic acids, causing a shearing selectivity or optimum performance of membrane, besides sensivity of NF membrane.

Total Solids (%). Total solids are all components in biomass consisted of soluble and insoluble components, such as fermentation products, especially lactic acid, mineral, carbohydrate, organic compounds as a hydrolysis product and microbes. Total solids becoming parameter in concentration process using NF system affects on permeate flow rate. Concentration of EPS produced as shown in Figure 5.

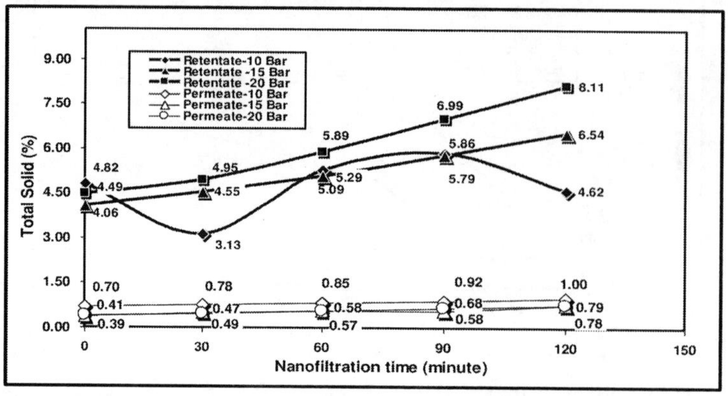

Figure 5. Relationship between time and total solids concentration in retentate and permeate in concentration of EPS produced from fermentation of hydrolysate biomass of Tuni sago (Metroxylon sp.) starch by *Lactobacillus plantarum* FU-0811.

Concentration processes, optimum concentration of total solids in retentate at operation pressure of 20 bar for 120 minutes was reached in 8.11 % and 0.79 % in permeate. There is an increase from prior to process as feed 5.1389 % to 8.11 % (36.63 %). This matter indicated that high operation pressure caused its occurence of high driving force, so that it compacted particles on the top membrane surface, whereas water mass would pass through membrane pores. Using NF-99-PE membrane enabled to raise concentration of solutes due to its specification (material kind, selectivity), in spite of membrane has mainly maximum separation tolerance (30 % of total material) [7].

This optimization level was supported by performance of NF membrane yielding observed rejection value of total solids in range of 78.26 - 85.48 % at operation pressure of 10 bar, 88.15 - 90.5 % at operation pressure of 15 bar, and 90.19 - 99.91 % at operation pressure of 20 bar. In other words, selectivity in NF system at operation pressure of 20 bar was better than 15 bar and 10 bar (20 > 15 > 10 bar), as shown in Figure 6.

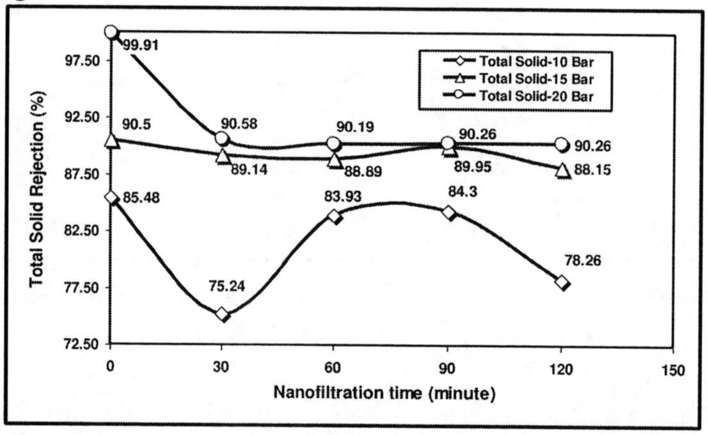

Figure 6. Relationship between time and total solids rejection in concentration of EPS produced from fermentation of hydrolysate biomass of Tuni sago (Metroxylon sp.) starch by *Lactobacillus plantarum* FU-0811.

It had been shown that this selectivity was tend to lower total solids relating with long concentration time, in which under operation pressure of 20 bar gave the best performance of NF membrane or almost completely observed rejection value (> 90 %) when compared with operation pressures of 15 and 10 bar. Long concentration time in NF system would pass freely solutes in permeate side, so that total

IOP Conf. Series: Materials Science and Engineering **722** (2020) 012005 doi:10.1088/1757-899X/722/1/012005

solids in permeate side and retentate side ratio becomes more and more small. This matter would decrease selectivity in NF system, especially for concentration treatment at operation pressures of 10 and 15 bar. This case was possibility caused by internal factors (type of material or feed, membrane material) and external factors (pressure, flow rate, temperature and time).

Reducing Sugar (mg/mL). Reducing sugar in biomass of sago sugar by using *Lactobacillus plantarum* FU-0811 culture is all sugars, which has an ability to reduce sugar to carbon source on the growth of *Lactobacillus plantarum* FU-0811 like Lactic Acid Bacteria (LAB). NF system is able to concentrate reducing sugar in retentate under optimum operation pressure and time, in which separation is shown almost completely, so that reducing sugar concentration in retentate is higher compared with in permeate, as shown in Figure 7. High operation pressure and long process time in NF system are able to retain reducing sugar on the top membrane surface, while passed reducing sugar in permeate tends to increase, as well. This case indicated that high driving force of sugar particles will be accumulated on the top membrane surface to form 'cake' layer, so that it indicates its occurence of 'fouling'.

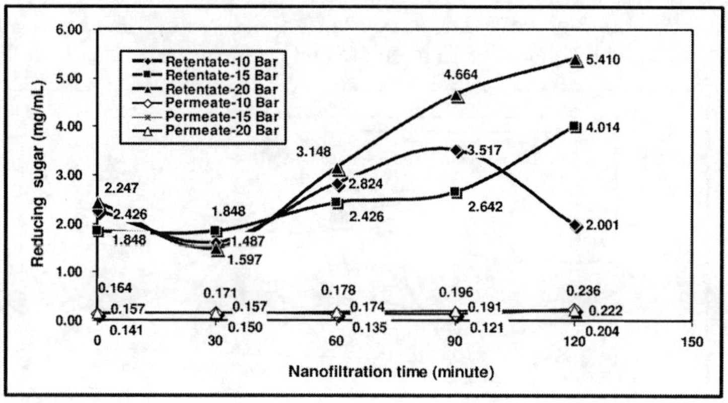

Figure 7. Relationship between time and reducing sugar concentration in retentate and permeate in concentration of EPS produced from fermentation of hydrolysate biomass of Tuni sago (Metroxylon sp.) starch by *Lactobacillus plantarum* FU-0811.

This optimum condition was supported by performance of NF membrane generating observer rejection value of reducing sugar in range of 88.23 - 96.57 % at operation pressure of 10 bar, 90.74 - 94.91 % at operation pressure of 15 bar and 89.45 - 95.89 % at operation pressure of 20 bar, respectively. In other words, selectivity in NF system at operation pressure of 20 bar was better than 15 bar, and operation pressure of 15 bar was better than dan 10 bar (20 > 15 > 10 bar), as shown in Figure 8.

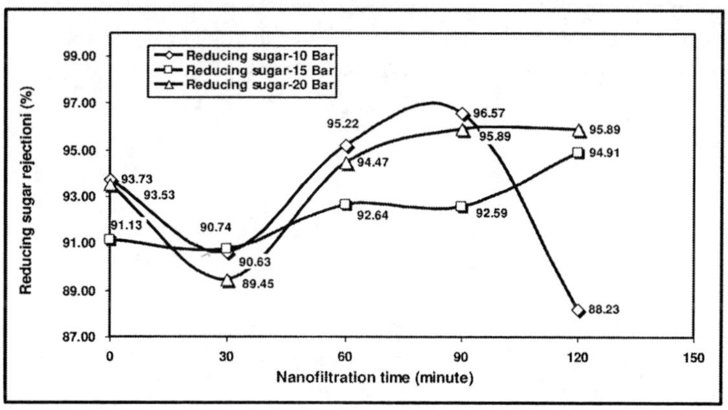

Figure 8. Relationship between time and observed rejection value of reducing sugar in concentration of EPS produced from fermentation of hydrolysate biomass of Tuni sago (Metroxylon sp.) starch by *Lactobacillus plantarum* FU-0811.

Observer rejection of reducing sugar fluctuates at operation pressure of 10 bar, in which observer rejection of reducing sugar drops for 30 minutes of process, increases and is optimum for 90 minutes followed by a decrease of observer rejection to end process (120 minutes). While, operation pressures of 15 and 20 bar increase to end process. This condition is possibility caused at initial process (0 - 30 minutes), NF system do not indicate fouling, so that ratio of reducing sugar concentrations in retentate and permeate is not showing in difference yet. Long time of reducing sugar concentration process will become more and more much that is retained and pass. Whereas, thick suspension become more and more deficite due to lack of water. High operation pressure and short time tends to result observer rejection of reducing sugar, so that the whole NF system will become more and more selective to separate sugar particles.

Total Acids (%). Total acids are organic compounds with low MW, such as lactic acid, malic acid as Lactobacillus plantarum FU-0811 metabolite. This fermentation process uses carbohydrate (glucose, maltose, fructose) from sagoo sugar media as a source of carbon and is expressed as titratable total acids. The high concentration of total acids is a parameter, in which its occurrence of homofermentative metabolism process[12]. Long process time at operating pressures of 10, 15 and 20 bar increase lactic acid concentration as titratable total acids in retentate and permeate, as indicated in Figure 9.

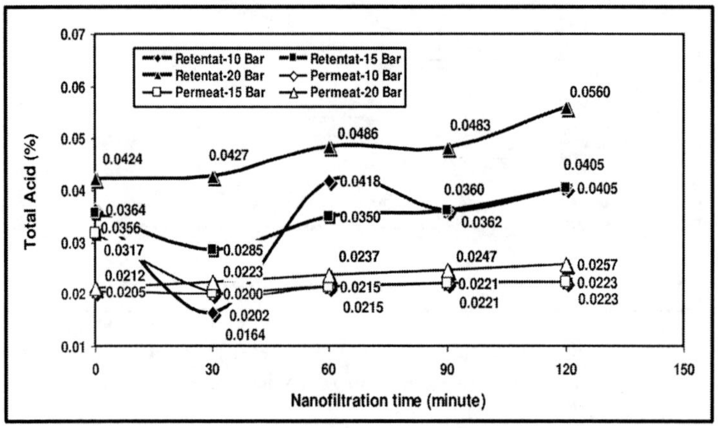

Figure 9. Relationship between time and total acids concentration in retentate and permeate in concentration of EPS produced from fermentation of hydrolysate biomass of Tuni sago (Metroxylon sp.) starch by Lactobacillus plantarum FU-0811 .

This optimum condition is not supported by performance of NF system showed by less completely observer rejection or selectivity of NF membrane (smaller than 90 %). (Figure 10). At operation pressures of 10, 15 and 20 bar for 0 - 60 minutes, NF system gave a decline of observer rejection of total acids. In other words, membrane system tends to reject or retain solute of total acids in order to pass membrane, so that ratio of total acids concentration in retentate and total acids concentration in permeate is similar or imperfect observer rejection of total acids.

IOP Conf. Series: Materials Science and Engineering **722** (2020) 012005 doi:10.1088/1757-899X/722/1/012005

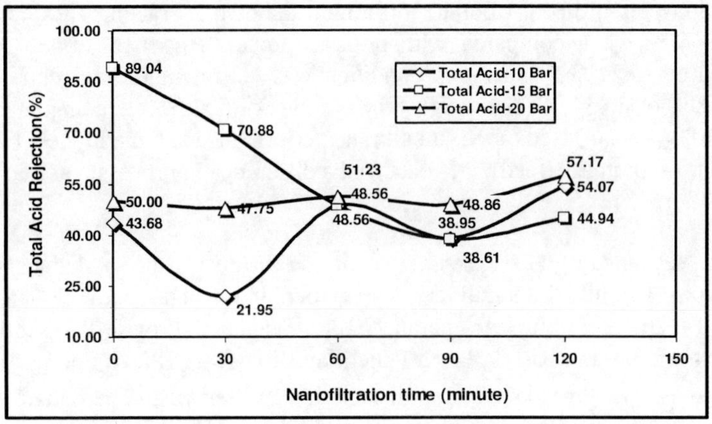

Figure 10. Relationship between time and observed rejection value of total acids in concentration of EPS produced from fermentation of hydrolysate biomass of Tuni sago (Metroxylon sp.) starch by *Lactobacillus plantarum* FU-0811.

From the whole concentration process of biomass as a result was reached the best process condition at operation pressure of 20 bar for 60 minutes and gave retentate as thick brown suspension, while permeate was clear liquid like water, as showed in Figure 11. Based on optimization of EPS concentration, the best process condition from the whole concentration process of biomass as a result of fermentation of sago sugar by *Lactobacillus plantarum* FU-0811 culture in semi pilot scale (20 L) are reached at operation pressure of 20 bar for 60 minutes by resulting retentate as thick brown suspension, whereas permeate is clear liquid like water, as shown in Figure 13a and 13b, respectively.

(a) (b)

Figure 11. Retentate (a), permeate (b) as a result of concentration of EPS produced from fermentation of hydrolysate biomass of Tuni sago (Metroxylon sp.) starch by *Lactobacillus plantarum* FU-0811.

In this condition is reached increasing EPS in retentate of 56.46 % from prior to process (feed) of 1.0685 % to 2.454 % for 60 minutes of concentration process with observer rejection of 95.24 %. Based on investigation on selectivity optimization of NF membrane performance, the best process condition was reached at operation pressure of 20 bar for 30 minutes with observer rejection value of optimum EPS (96 %) by increasing EPS of 13.17 % from prior to process (feed) (1.0685 %) to 1.23 % (dry weight) for 0 - 30 minutes. Based on membrane performance, selection of optimum condition is an important effort to get selected process based on process safety and life time on NF system.

4. Conclusion

Concentration process of biomass as a result of fermentation of sago sugar by Lactobacillus plantarum FU-0811 culture in semi pilot scale (20 L) through NF system gave optimization in process condition on each component based on component property, interaction between operation pressure and time, selectivity aspect or rejection ability and membrane material. Long concentration time would decline permeate flux values for the whole processes and raise EPS concentration, total solids, total acids, and reducing sugar. Operation pressure of 20 bar gave better composition and component rejection than that operation pressure of 10 and 15 bar. Based on the highest EPS recovery, optimum condition was reached at operation pressure of 20 Bar for 60 minutes shown by raising EPS content in concentrate (retentate) of 56.46 % when compared with prior to process (feed) of 1.0685 % to 2.454 % with observer rejection of 95.24 %. Under this condition was get retentate (concentrate) of biomass with compositions of EPS of 2.454 % (dry weight), total solids of 5.89 %, reducing sugar of 3.148 mg/mL, and total acids of 0.0486 %. Based on NF membrane performance and the highest EPS concentration by various components in biomass (feed), it is occurred a difference in process condition optimalization.

5. References

[1]. Dinoto, A., et al. Produksi Eksopolisacharida bakteri usus berbahan baku tepung sagu Metroxylon sp.) untuk drug delivery sistem berbentuk nano partikel dan hidrogel, Laporan Kegiatan Tahap II Tahun Anggaran 2011, Kegiatan Program Insentif, Lembaga Ilmu Pengetahuan Indonesia - Ristek, Pusat Penelitian Biologi – LIPI, Cibinong, Bogor.

[2]. Dinoto, A., *et al.* Produksi Eksopolisacharida bakteri usus berbahan baku tepung sagu Metroxylon sp.) untuk drug delivery sistem berbentuk nano partikel dan hidrogel, Laporan Kegiatan Tahap I Tahun Anggaran 2010, Kegiatan Program Insentif 2010., Lembaga Ilmu Pengetahuan Indonesia - Ristek, Pusat Penelitian Biologi – LIPI, Cibinong, Bogor.

[3]. Tallon R, Bressollier P, Urdaci MC. Isolation and characterization of two exopolysaccharides produced by Lactobacillus plantarum EP56. *Res Microbiol.* 2003;154(10):705–12.

[4]. Susilowati, A., Aspiyanto dan Dinoto, A. Proses Pemekatan Eksopolisakarida dari Fermentasi Lactobacillus plantarum FU-0811 pada gula sagu hasil hidrolisis asam pati sagu tuni (Metroxylon sp) melalui sistem nanofiltrasi. Prosiding Seminar Nasional Kimia dan Workshop Instrumentasi Kimia oleh Jurusan Kimia, Fakultas MIPA, Universitas Sebelas Maret, Surakarta, 2011.

[5]. Van der Bruggen B, Schaep J, Wilms D, Vandecasteele C. Influence of molecular size, polarity and charge on the retention of organic molecules by nanofiltration. J Memb Sci. 1999;156(1):29–41.

[6]. Mulder J. Basic principles of membrane technology. Springer Science & Business Media; 2012.

[7]. Michael, A. S. Handbook of Industrial Membrane Technology, Noyes Publications, Park Ridge, U.S.A. 1989.

[8]. A.O.A.C. 1995. Official Methods of Analysis, Association of Official Analytical Chemists, Washington, D.C.

[9]. Cheryan M. Ultrafiltration and microfiltration handbook. CRC press; 1998.

[10]. Poli A, Anzelmo G, Nicolaus B. Bacterial exopolysaccharides from extreme marine habitats: production, characterization and biological activities. Mar Drugs. 2010;8(6):1779–802.

IOP Conf. Series: Materials Science and Engineering **722** (2020) 012006 doi:10.1088/1757-899X/722/1/012006

Evaluation of the Centralized HVAC Chilled Water Pumps of Suranaree University of Technology Hospital Main Building

A Prasetyadi[1,2] and A Koonsrisuk[1]

[1] School of Mechanical Engineering, Suranaree University of Technology, Nakhon Ratchasima 30000, Thailand

[2] Department of Mechanical Engineering, Universitas Sanata Dharma, Yogyakarta 55281, Indonesia

E-mail: pras@usd.ac.id

Abstract. Pumps are the main issue of the chilled water distribution in a centralized system. In purpose of evaluating the chiller pump system of the centralized HVAC for Suranaree University of Technology Hospital (SUTH) main building, a model comprised evaporator, pump and pipe network was proposed. An analysis of energy rate losses due to pressure losses was applied in a model composed of vertical and horizontal distribution pipes. Pressure loss in pipes was assumed to follow Darcy-Weisbach equation. The resistance of the systems were also calculated. The results show that distribution of losses in vertical shaft and stories were unequal. The pressure loss and energy rate loss were function of flow-rate. But the resistance was an inverse function of the flow-rate. The difference distribution emphasized the need of reconfiguring chilled water pump for accommodating flexibility usage.

1. Introduction

Centralized HVAC is a common method for large building as it has high cooling load. The centralized HVAC works more efficiently than split system when the system needs high heat rejection at the same time. A centralized system works with less moving parts, higher heat working fluid capacity and higher COP [1]. Less moving parts mean less maintenance. Higher heat working fluid capacity can handle more heat rejection load. The COP tells energy consumption over heat rejection. Therefore, the centralized HVAC system is preferable for large system. However, the high capacity system has vulnerability of load variation range.

Variation of load in a HVAC system can degrade the efficiency of the system. Out of designed working point, the efficiency of the system mostly decreases. Because the centralized HVAC depends on pumps for transferring heat among the components, increasing load above the designed load forces the pumps to increase the flow rate of working fluid. It implies higher flow rate and less efficient motor. When the load is less than it designed, the pump can work better, but the motor might work at less efficient condition.

Variable speed motor is proposed for variable load system. Variable speed motor can work over long range speed with high efficiency [2]. Therefore, load variation does not affect the performance of motor. When the load increases, the motor can adjust the speed to provide higher pressure and vice versa. To set at high performance, control system of the motor is important. Unfortunately, the variable speed motor is still has problem of the high range variety and the price could be very high for a high capacity motor and its controller.

Constructal law emphasizes a design to follow the nature design in order to get high efficiency. Combination of few larges and many smalls is considered the best design as the nature shows in many nature flows [3]. It has been studied in many engineering area such as cooling system [4]–[6], heat storage [7], and energy harnessing [8]. These studies focused on design base and not for retrofitting the established system.

The objective of the study is to evaluate the chilled water pumps of HVAC system in a multi-story building, the main building of Suranaree University of Technology Hospital (SUTH). Evaluation of centralized HVAC system pump can be considered as the compromise between high variety of load and large system [1]. Moreover, it is also a new field of constructal law area to be more practical in established system. In order to evaluate the HVAC system in large building with high load variety in term of constructal study, a case of SUTH main building is proposed.

The structure of the article is mainly composed of introduction, methodology, result and discussion, and conclusion. At the introduction, the state of the problem is presented to show the research gap of the proposed topic. The methodology part consists of description of the HVAC system of SUTH main building and its performance records. In this part, the model of the system will be presented for evaluating the performance. Result will have some data that show analysis of the performance at the current condition. Conclusion will close the paper with some main trend found.

2. Methodology

2.1. Centralized HVAC of SUTH main building

Rattanajapevat is the operational main building of Suranaree University of Technology Hospital. It has 12 floors which has been functioned as the administration building, OPD clinics, wards, operation rooms, emergency, pharmacy, laboratories and some support systems. The clinics areas are distributed in 1st, 2nd, and 3rd floors. The administration office is centered at 4th floor. The operation rooms are located in 5th floor. The 6th floor is dedicated for delivery, while the 7th floor is the place for intensive care units. Hemodialysis machines and its recovery rooms are located in 8th floors. The rest floors; they are 9th – 11th, are wards for women, men, and VIP, respectively. The ground floor is the place for some support systems, they are engineering, CCDS, pharmacy stocks, securities, and morgue.

Total HVAC system capacity of the building is 750 TR which consists of 3 chillers with 250 TR each. The system is also supported by 3 x 30 kW CDPs, 3 x 37 kW CHPs and 3 x 3.8 million BTU/hr. The coil units capacity of the building is 7.2 million BTU/hr. The chilled water flow-rate capacity of the building is 83.9 l/s. The chilled water is designed to work at 7-13.5°C. The chilled water output pressure is set to be 65 psi. The condenser water pumps have to manage 40 psi pressure head loss compensation.

The chilled water is distributed to the building by 3 x 37 kW pumps operated alternately. Every time, only two of the three pumps work. These pumps are connected to vertical main shaft of 10" pipe. The shaft is connected to distribution pipes in every floor. Different sizes of pipes are applied for the distribution pipes. Some floors have equal size of the supply and return pipes. Some of them have different sizes for the reason of price. The size of the pipes for the HVAC and the chilled water flow-rate of the system are provided in table 1. The designation of the stories follows the building condition. The 0 and 12 are used for the basement and rooftop, respectively. This chilled water flow-rates also show the cooling capacity in respective floors.

The chilled water is assumed to be distributed in proportional velocity at steady state condition. The valves are set to provide proportional water flowing at every floor according the capacity of the cooling system. The relationship of the chilled water flow-rate and pressure difference at the evaporator is determined by equation (1) as provided by the provider company [9]. The pressure difference is in psi and the flow-rate is in gallon per minute. The model of the chiller system is

IOP Conf. Series: Materials Science and Engineering **722** (2020) 012006 doi:10.1088/1757-899X/722/1/012006

provided at figure 1. As the input of the pump is set to be 65 psi, the pressure loss of the pipes network follows equation (2).

$$Q_{chw} = -0.0226\Delta p_{ev}^2 + 8.9826\Delta p_{ev} + 136.65 \tag{1}$$

$$\Delta p_n = -3\times10^{-10}Q^4 + 4\times10^{-7}Q^3 + 1\times10^{-4}Q^2 + 0.0309Q + 0.6721 \tag{2}$$

Table 1. Story, chilled water flow-rate, and pipe sizes.

Story		Water	Inhabitant	pipe size	pipe size
Name	**Designation**	**(l/min)**	**(day equivalent)**	**S (inch)**	**R (inch)**
Ground	0	77.84	18	3	3
1	1	236.10	55	4	4
2	2	203.43	23	4	4
3	3	137.43	23	2	3
4	4	110.60	7	2	3
5	5	120.34	13	2	3
6	6	123.26	5	2	3
7	7	100.65	14	2	3
8	8	86.66	17	2	3
9	9	22.10	25	2	3
10	10	22.10	29	2	3
11	11	74.36	17	2	3
R	12	15.00	1	2	3

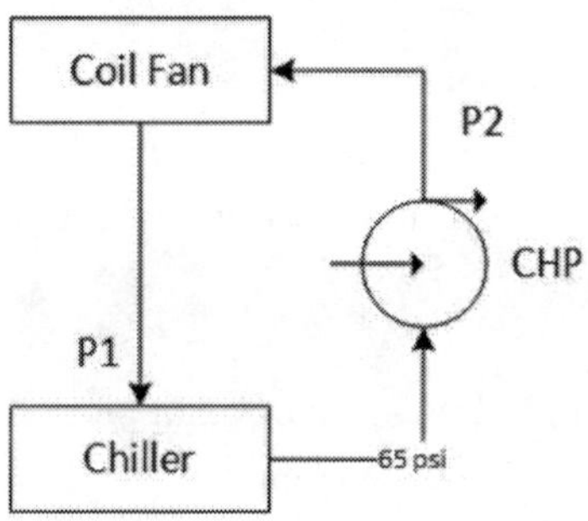

Figure 1. Model of the chilled water loop of the HVAC.

2.2. The evaluation of the chilled water pump configuration

There are two questions of the configuration in chilled water distribution problem. The first question is distributing cooling water as load needed. The second question is minimum energy to flow the chilled water to the load which has horizontal and vertical part. The first question can be considered as sufficiency problem of the chilled water. Assuming load as proportional to number of people and its density is proportional to the area, it can inferred that the scale of the load is L^2. L is length scale. At the same time the chilled water flow-rate expresses the cooling capacity of the network in a story. Higher water flow-rate means higher cooling capacity and vice versa.

To evaluate the chilled water pump, distribution of chilled water is mapped. To simplify the system, a model as shown in figure 2 is proposed. The model consists of evaporator, pump, and pipe networks which their flow rates and pressure difference. The pipes network is composed of shafts (main pipes), and story distribution pipes. The story pipe line is described as a single equal diameter line connecting inlet main shaft and outlet main shaft. It is assumed to be a single segment which has pressure difference between its inlet and outlet. Every main shaft segment connecting two adjacent stories is assumed to be a single segment. In the model, the segment has its own pressure difference and flow rate.

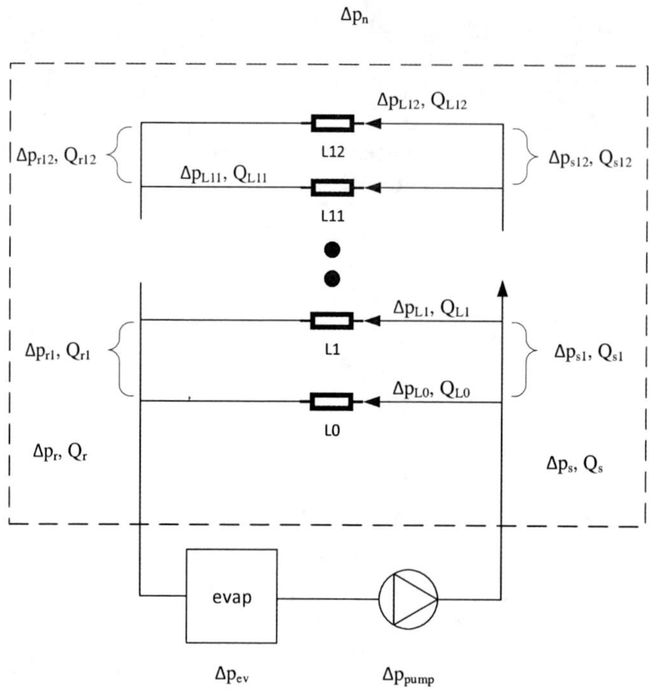

Figure 2. The model of chilled water loop and its pressure and flow-rate.

The pressure of the loop can be shown in equation (3) with η as the total efficiency of pump and motor. As the figure 2 shows, the Δp_n is the pressure of the network and the Δp_{ev} is the pressure difference of evaporator inlet-outlet.

$$\Delta p_n + \Delta p_{ev} = \eta \Delta p_{pump} \tag{3}$$

The pressure of network is composed of pressure of main pipes (Δp_s and Δp_r) and pressure of a story load (Δp_L), that generally can be formed as equation (4).

$$\Delta p_n = \Delta p_s + \Delta p_r + \Delta p_L \tag{4}$$

The total load pressure is equal to load pressure at ground Δp_{L0} and accumulation of other load pressure added by pressure loss at shaft between the ground and the 1st floor. Equations (5) and (6) show the relationship of the pressures.

$$\Delta p_L = \Delta p_{s1} + \Delta p_{r1} + \Delta p_{n1} \tag{5}$$

$$\Delta p_L = \Delta p_{L0} \tag{6}$$

At the branches, equations (7) and (8) take place.

$$Q_{Li} + Q_{S(i+1)} = Q_{Si} \tag{7}$$

$$\Delta p_{Li} = \Delta p_{si} + \Delta p_{ri} + \Delta p_{ni} \tag{8}$$

It can be inferred, that equation (4) can be formed as equation (9).

$$\Delta p_n = \sum_i \Delta p_{si} + \sum_i \Delta p_{ri} + \Delta p_{L12} \tag{9}$$

The flow-rate can be also shown as equation (10).

$$Q_S = \sum_i Q_{Li} \tag{10}$$

According the Darcy Weisbach equation for the pipe, $\Delta p = f_D \rho 8LQ^2 / \pi^2 D^5$, therefore it can be formed equations (11) and (12) that resembles the electricity equation with the current as volume flow-rate rather than mass flow-rate as Bejan did [10] for the reason of incompressibility of the fluid.

$$\dot{E} = \frac{f_D \rho 8LQ^3}{\pi^2 D^5} \tag{11}$$

$$R = \frac{f_D \rho 8LQ}{\pi^2 D^5} \tag{12}$$

The pressure head difference of the adjacent stories can be inferred as equation (13). The first term is the height effect and the second is length effect.

$$\Delta p_{si} = \rho g \Delta h_i + \frac{f_D \rho 8 \Delta h_i Q_i^2}{\pi^2 D^5} \tag{13}$$

The equation (13) emphasizes that the pressure difference of the shaft is proportional to the acceleration of the fluid flow along the vertical shaft. It also happens in horizontal pipes. While at the vertical shaft the potential part is in opposite direction between supply and return, therefore this part collapses in calculation.

Derivation of the pumping power against the pressure different can be shown as in equation (14).

$$\frac{\mathrm{d}\dot{E}}{\mathrm{d}\Delta p} = 3C_1 \Delta p^{\frac{1}{2}} \tag{14}$$

Equation (14) implies that power of pump is a function of pressure difference by power of 3/2. The derivation of power against flow-rate gives equation (15). The derivation of flow-rate against pressure difference is shown in equation (16).

$$\frac{\mathrm{d}\dot{E}}{\mathrm{d}Q} = C_2 Q^2 = 3\Delta p \tag{15}$$

$$\frac{\mathrm{d}Q}{\mathrm{d}p} = C_1 \Delta p^{-\frac{1}{2}} \tag{16}$$

For vertical pipe in multistory building, the flow-rate of the vertical main pipe between two adjacent stories can be assumed as iterative function as equation (17).

$$Q_{si} = \sum_{j=i}^{n} Q_j \tag{17}$$

Equation (13) implies that the pressure difference of vertical pipe is proportional to the height difference and the acceleration. Therefore, it can be formed into equation (18).

$$\frac{\mathrm{d}\dot{E}}{\mathrm{d}Q} = C_2 Q^2 + C_4 \tag{18}$$

The first term comes from the pressure different of the kinetic and the second from the potential.

3. Result and Discussion

At current operational condition the power of pumping is the function of the total flow-rate. The total flow-rate is determined from equation (1). The total flow-rate is quadratic function of the pressure difference at evaporator. Therefore a pressure difference at evaporator determines power of the pump for distributing the chilled water. Applying single primary pump only for the system, the power of the pump for distributing the chilled water according the flow-rate is shown in figure 3. The function of the pump can determined from equations (1), (2) and (3).

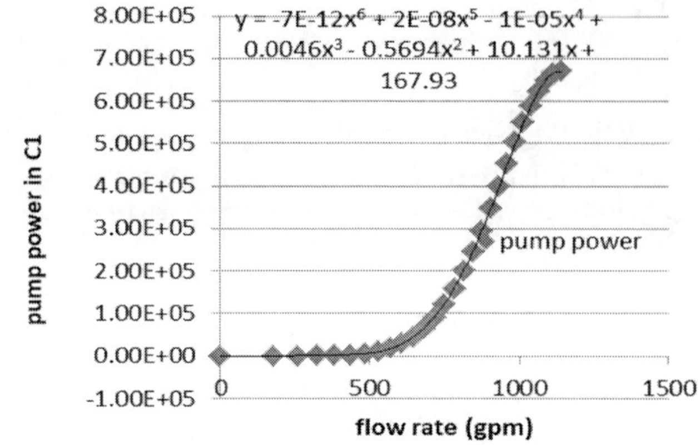

Figure 3. Pump power as function of chilled water flow-rate. The power is scaled into C1.

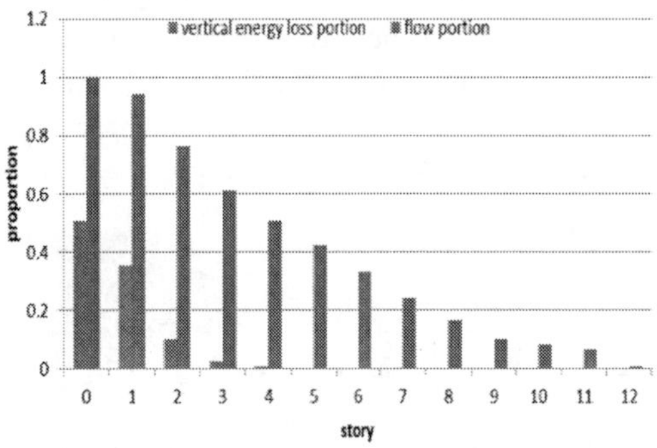

Figure 4. The distribution of energy loss in vertical shaft.

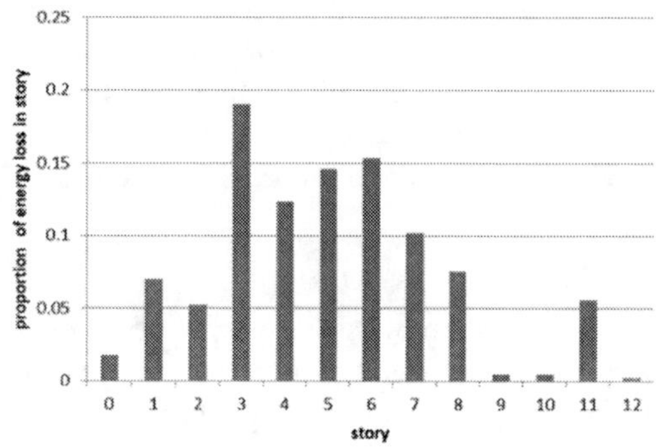

Figure 5. The distribution of energy loss in story.

The energy loss at vertical shaft and at the stories can be traced from equations (13) and (15). Assuming the proportions of the flow rate are equal to the maximum capacity, the energy lost at vertical shaft can be shown in figure 4. Transferring chilled water from the evaporator to the ground story chilled water network is the place where most of the vertical energy loss happened, while transferring chilled water to the roof-top is the least energy consumer.

The energy loss at vertical shaft depends on flow-rate and the height. The distribution of the energy loss according the story shows that vertical effect is far less than energy loss at the story. The pressure loss at vertical shaft for the ground story is ranging from 0.0% - 0.25% of the total pressure loss. Flow rate is more dominant in determining energy loss in distribution of the chilled water than the height.

The maximum delta pressure of the story took place at the ground floor, but the minimum delta pressure happened at the rooftop. Range of the delta pressure between the inlet and outlet at the story is 565.1 – 568.1 kPa at 50 liter/second flow rate of the chilled water. The resistance of the flow at the equal chilled water flow-rate shows different trend. The range of the resistance is 64.39 kPa s/m^3 at the 1st story until 1012.68 kPa s/m^3 at the rooftop. It is clear that the chilled water flow-rate at the 1st story is the highest and the lowest took place at roof-top. The complete pressure difference among inlet and outlet of the stories, flow-rates of the chilled water at the stories, and the respective resistance are shown in table 2.

The pressure loss at stories is function of the chilled water flow-rate. Figure 6 shows the relation of the pressure loss and the flow-rate as a linear function. Increasing the chilled water flow-rate increases

IOP Conf. Series: Materials Science and Engineering **722** (2020) 012006 doi:10.1088/1757-899X/722/1/012006

the pressure loss at the stories linearly. The normalization of the system shows that the increasing takes place proportionally according to the flow-rate.

Horizontal energy rate loss does not depend on the story position. But it depends on the flow-rates and diameter of the pipes as shown in table 1 and table 2. The relation of the energy rate loss and the flow-rate of chilled water is also presented in figure 7. The resistance of the story is also a function of flow-rate as shown in figure 8. Considering relation of equation (13), it shows that the coefficient is a function of flow-rate which is confirming figure 6. This condition emphasizes necessity of considering the constructal configuration at the story as mentioned by Xia [11].

Table 2. The pressure, flow-rate, and resistance of the stories at 50 liter/second flow-rate.

Story	Δp_{Li} (Pa sec/m^3)	Q_{Li} (liter/sec)	R_{Li} (Pa sec/m^3)
0	5.66 10^5	2.90 10^0	1.95 10^5
1	5.66 10^5	8.78 10^0	6.44 10^4
2	5.65 10^5	7.57 10^0	7.47 10^4
3	5.65 10^5	5.11 10^0	1.11 10^5
4	5.65 10^5	4.11 10^0	1.37 10^5
5	5.65 10^5	4.48 10^0	1.26 10^5
6	5.65 10^5	4.59 10^0	1.23 10^5
7	5.65 10^5	3.74 10^0	1.51 10^5
8	5.65 10^5	3.22 10^0	1.75 10^5
9	5.65 10^5	8.22 10^{-0}	6.87 10^5
10	5.65 10^5	8.22 10^{-0}	6.87 10^5
11	5.65 10^5	2.77 10^0	2.04 10^5
12	5.65 10^5	5.58 10^{-0}	1.01 10^6

Unequal distribution of the vertical and story energy loss shows the importance of reconfiguring the pumps of the chilled water in order to minimize the effect of loss. Controlling single primary pump as the current operational condition assumed equal distribution of the vertical and story losses. Increasing power of the pump will increase the water flow-rate proportionally according to the valves setting. At this condition, losses at stories are assumed in accordance with the position. Adjustment of valves should be conducted to allow the change of flow rates. Optimum condition is hardly achieved as the reason of the distribution. In this case, flexibility is important due to interrelation as shown in figure 6, figure 7, and figure 8. Secondary pumps in stories are suggested [2].

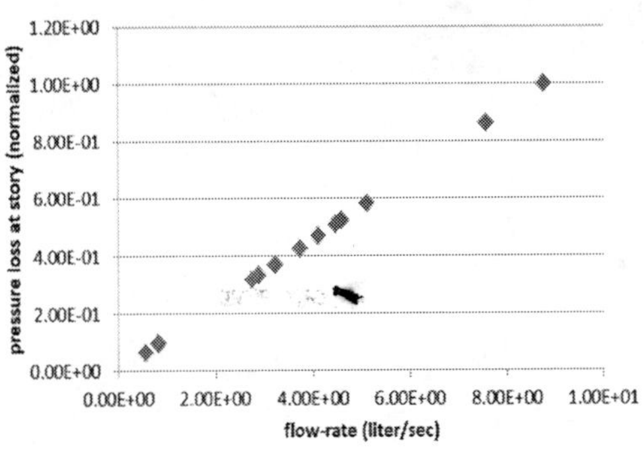

Figure 6. Pressure loss at stories as function of flow-rate.

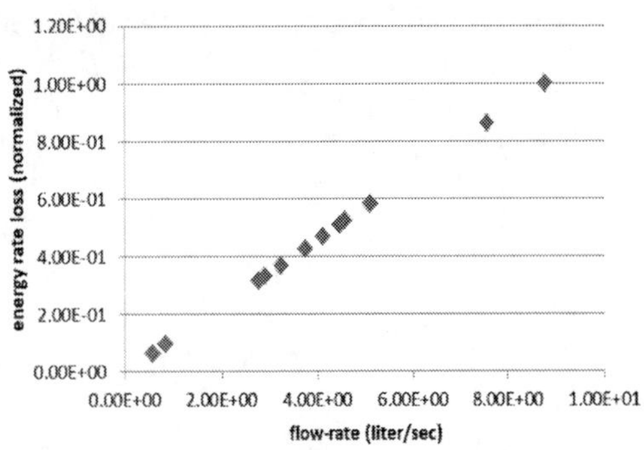

Figure 7. The energy rate loss as function of flow-rate.

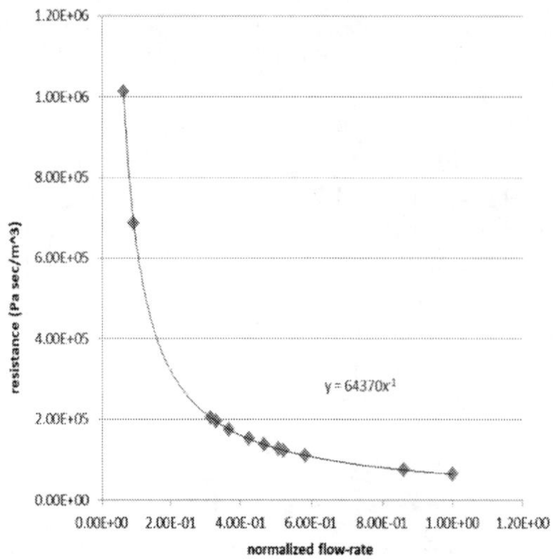

Figure 8. Resistance of the stories as function of normalized flow-rate.

4. Conclusion

The evaluation of the centralized HVAC system at SUTH main building was conducted in order to consider retrofitting the pump configuration. The work shows that distribution of losses at vertical shaft and stories are unequal. The energy losses and pressure losses at the stories are linear function of flow-rate in normalized way. The resistances of stories are inverse function of normalized flow-rate. Simple configuration using single primary pump for the system has drawback of losses at some stories which do not need it. It is suggested applying secondary pumps in every story.

Acknowledgement

The authors would like to express gratitude to SUT for providing the Postgraduate Scholarships for Foreign Students to the first author and Yayasan Sanata Dharma for the support attending the conference. The authors also express their sincere thanks to the SUTH director, Dr. Niwatchai Namvichaisirikul, and the staff of the Engineering and Maintenance Department at SUTH, particularly Mr. Watcharin Singtana, Mr. Songkrit Jitpakdeebordin, Mr Chaleomkiat Kaewkurn, Mr. Prashya Suranachit, and Ms. Naphatthira Decheerawas.

References

[1] Perez-Lombard L, Ortiz J and Maestre I R 2011 *Appl. Energy* **88** 5020–31
[2] Tirmizi S A, Gandhidasan P and Zubair S M 2012 *Appl. Energy* **100** 238–48
[3] Bejan A and Lorente S 2011 *Phys. Life Rev* **8** 209–40
[4] Ghodoossi L and Eğrican N 2004 *Energy Convers. Manag.* **45** 811–828
[5] Adewumi O O, Bello-Ochende T and Meyer J P 2013 *Int. J. Heat Mass Transf* **66** 315–23
[6] Kalbasi R and Salimpour M R 2015 *Appl. Therm. Eng.* **84** 339–49
[7] Miguel A F 2008 *Energy Build.* **40** 1020–30
[8] Lorente S, Koonsrisuk A and Bejan A 2010 *Int. J. Green Energy* **7** 577–92
[9] TRANE 2012 *Installation, Operation and Maintenance Series R ® Rotary Liquid Chillers Water-Cooled and Compressor-Chillers* (Ingersoll Rand) p 215
[10] Bejan A and Lorente S 2008 *Design with Constructal Theory* (New Jersey: John Wiley and Son) p 517
[11] Xia L, Lorente S and Bejan A 2011 *Int. J. Energy Res.* **35** 806–12

IOP Conf. Series: Materials Science and Engineering **722** (2020) 012007 doi:10.1088/1757-899X/722/1/012007

Categorization of Reliability Performance Indicators of Humanitarian Response Supply Chain

A Sutrisno[1], D Handayani[2], W Caesarendra[3] and I Gunawan[4]

[1]Department of Mechanical Engineering, Sam Ratulangi University, Indonesia
[2] Department of Industrial Engineering, Universitas Islam Indonesia, Indonesia
[3]Faculty of Integrated Technologies, Universiti Brunei Darussalam,Brunei Darussalam
[4]Center of Entrepreneurship and Innovation, The University of Adelaide, Australia

[1,2,3,4]Email: agungsutrisno@unsrat.ac.id, dwihandayani@uii.ac.id, wahyu.caesarendra@ubd.edu.bd, indra.gunawan@adelaide.edu.au

Abstract. To minimize a number of disaster victims due to certain disaster, identifying performance indicators representing reliable disaster response operation are very necessary to drive success in humanitarian action. However, literature in disaster response managements that focused on reliability performance measures are very few. To fill in this gap, in this paper presents categorization of humanitarian supply chain reliability performance indicators. This method consists of both hard and soft indicators. Our classification enables humanitarian practitioners to possibly develop unified reliability estimation indices containing of both qualitative and quantitative reliability performance indicators. At the end, conclusions and new research directions from the preliminary study are presented.

1. Introduction

When disastrous event occurred, an immediate response operation is required to help the victims with necessary aids. From this point of view, the existence of reliable disaster response operation is necessary for ensuring all activities in response operation are working as intended. Nevertheless, as it is undertaken in an immediate, dynamic and uncertain situation, likelihood of successful response operation is heavily affected by its reliability. In this regard, reliability of humanitarian response operation is very important in endeavor to save disaster victims. Moreover, unreliable humanitarian response operation will cause increasing disaster victims, crisis escalation and increasing recovery cost. Therefore, improving understanding on what metric representing reliability dimension of disaster response operation is very important for decision makers to take any necessary action to improve it. In spite of its importance, references discussing on reliability performance indicator issues in the context of humanitarian supply chain are very limited in literature [1] as most of reliability performance metric is dedicated to profit-oriented operation [2]. In filling in the research gap, this paper presents a classification method of reliability performance indicators of humanitarian response operation from theoretical lens. The objective of this paper is to study classification of humanitarian supply chain reliability parameters and their performance indicators using hard and soft measures. The structure of the paper is as follows. In section 2, an overview and characteristics of humanitarian response supply chain and characteristics of humanitarian supply chain failure is presented. A brief description on structural model of disaster response supply chain reliability and followed by reliability parameters in supply chain is discussed in section 3. In section 4, a categorical model of reliability parameters of disaster response supply chain is presented and discussed. In section 5 conclusions and potential research direction from this initial study are presented.

2. Failure Characteristics in Humanitarian Response Supply Chain

2.1. Humanitarian Response Supply Chain

In dealing with adverse impact of disaster, collaborative effort to tackle the occurrence of disaster is accomplished among humanitarian response organizations. The humanitarian supply-chain management (HSCM) involves managing the different factors in the system such as goods and materials, information, manpower, political authorities, available infrastructure, etc. to reduce the impact of a disaster to the people. According to [3], humanitarian supply chain management can be defined as activities related to planning, executing, controlling and monitoring the flow of goods, service and information from the suppliers to the disaster victims in a cost effective and efficiently manner. Operated to fulfil the need of beneficiaries, humanitarian response supply chain is having similarities and differences with commercial supply chains [4]. Similarities of both supply chains are related on goals in delivery supplies to the beneficiaries, complexity of the operations and demand uncertainty. Meanwhile, difference between those supply chains are depends on involvement of parties, network configurations and demand predictability. Detailed comparative attributes between commercial and humanitarian response supply chain can referred to [5]. Different from commercial supply chain which is operated on basis to fulfill satisfaction of the customers, humanitarian supply chain operates on the basis of humanity and saving human life is becoming its priority. In similar with commercial supply chain, the applicability of the Supply Chain Operational Reference model (SCOR) as means to analyze the supply chain process is applicable to both types of supply chain.

2.2. Failure Characteristics in Humanitarian Response Supply Chain

Failure is defined as the inability of an entity to perform operating function as intended. In relation to reliability terminology. It is likelihood to perform operation as intended. Failure is the cause of unreliable operational event occurrence. In this regard, it is inevitable to consider failure modes of supply chain in reliability evaluation of humanitarian response operation. Considering its locus of occurrences, failures in disaster response supply chain could occur at any point of operation. And also, it can be in the form of hard failures such the unavailability of response equipment and lack of supplies. The second mode of failure, the soft failure is in the form of dysfunctionalities such as bad accountability, poor planning and lack sense of crisis among humanitarian officers. Within humanitarian supply chain context, supply chain failure is related with incapability of a typical humanitarian response supply chain to deliver all needed supplies of goods, services and information to the beneficiaries [6]. In accordance with disaster relief cycle, failures are starting from failure in disaster hazard analysis, failure in emergency response and failure in disaster management itself. In connection with the role of disaster response supply chain which delivers supplies to the disaster victims, failure characteristics in disaster response supply chain can be manifested by the 7W terminology (the wrong product, the wrong quantity, wrong condition, wrong places, wrong time, wrong customers and at wrong cost). Those terminology can be used to propose method to calculate quantitative measure of delivery reliability indices.

3. Humanitarian Supply Chain Reliability

Supply chain reliability according to [7] is defined as the probability of a typical supply chain to deliver the amount of goods and services to the critical transfer points in the systems. In similar, [8] stated that supply chain reliability is defined as the likelihood of incoming materials and product flow will arrive at the transfer points under specified conditions. And within supply chain context, [9] defined reliability as probability that a supply chain enables to deliver the amount of ordered goods or service to the end customer within due date and under specified condition.

Despite varying definitions on what supply chain reliability are existing in literature, the terminology of supply chain reliability having universal contents related to capability to fulfill logistical needs of the beneficiaries under specified conditions and at a certain time interval. Supply chain reliability as an aspect in undertaking humanitarian action must be taken into consideration since it affects to recovery costs and resource will be spent in emergency operation. With its goal to deliver supplies at transfer points, the structural model and characteristics of the supply chain affect to reliability of supply chain. Typically, there are five structural models describing reliability of supply chain [10]; Series, Parallel, Parallel-Series, Series-Parallel, and G-out of K. Different structure of reliability configuration is having different reliability formulation indices.

3.1. Factors Affecting Humanitarian Supply Chain Reliability

In performing intended function to deliver good and service to the disaster victims, many internal and external factors affect to the operational capability of disaster response supply chain. Those factors must be considered to create effective and efficient humanitarian response operation [11]. The internal factors are related to the disaster characteristics, demand uncertainty, availability of local suppliers, donation types, attitude of local government, law and access to the beneficiaries. The external factors affecting to successfulness in undertaking response operations are related to the mission of humanitarian response organizations and coordination among them.

4. Results and Discussion

In order to identify humanitarian supply chain reliability parameters, one dimensional approach is taken by using literature survey to bibliographical databases. The reference database consulted are Emerald, ScienceDirect and IEEExplore using keywords, "supply chain reliability", "humanitarian supply chain reliability", and "supply chain reliability performance indicators". Then the content of papers found to the above subjects are analyzed and tabulated as in Table 1.

4.1. Reliability Performance Indicator of Humanitarian Response Supply Chain – A Categorical Model

Altogether with other performance indicators in humanitarian response operation such as delivery, flexibility, responsiveness, reliability performance indicator representing possibility a typical humanitarian supply chain will perform its function in delivering goods and services to the victims. Regarding that humanitarian supply chain is representation of a kind of socio-technical system, its performance indicators should consider both of technical and non-technical measures. The first category, the technical parameters represent quantitative measures of response reliability. It is called hard measures since its occurrence will totally terminating reliability and causing failure in the eyes of beneficiaries. The latter, non-technical parameters represent qualitative indicator affecting partially failure event occurrence in performing response operation. The soft measure shows qualitative attribute indicating how well a humanitarian response operation would be. The second parameters which is called soft factor represent some non-technical parameters in which their existential affecting to reliability of disaster response operation. Classification of both hard and soft supply chain reliability performance indicators from various references are presented in Table 1.

Table 1. Categorization of Humanitarian Supply Chain Performance Indicators

Reliability Performance Indicator		
Category	Performance Indicator	Reference
Hard Measure		
Delivery Reliability	Percentage of order delivered in full	[10] and [12]
	Delivery performance commitment date	
	Order fulfillment Rate	
	Accuracy on purchasing document	
	Percentage of perfect goods condition	
	Existential on risk mitigation plan	
Disaster Response Transportation Reliability	Fleet availability	[13]
	Travel Time Duration	
	Transportation Capacity	
	Reliability of Transportation Connectivity	
Disaster Response Infrastructures reliability	Communication and Information infrastructure Reliability	[14]
	Quality and adequacy of goods storage (warehouse)	[11]
	Availability of skilled humanitarian personnel	[15]
Soft Measure	Performance Indicator	
Governmental /political factors	Regulation related to humanitarian organization	[14] and [11]
	Law	[11]
	Political Constraint	[14]
Socio-Economic Situational Factor	Availability of local suppliers	[15]
	Religion and local culture	[11]
	Security Issues	[11]
	Demand Uncertainty	[11]
Interorganizational factor	The existence of contingency planning of response operation	[11]
	Type of Donation availability	[11]
	Quality of coordination among agencies involved in humanitarian operation.	[11] and [16]
	Information Transparency and trust across humanitarian supply chain	[15]

4.2. Discussions

In the first category, hard reliability performance indicators of humanitarian supply chain are consisting of three reliability indicators; delivery, transportation and disaster response infrastructure reliability. Those three categories are having quantitative measures as depicted in Table 1. Delivery reliability relates to probability that the aids can be delivered in the correct amount, time and to the right beneficiaries. Furthermore, delivery reliability also relates to the existential of contingency planning considering the occurrence of delivery risks. The second reliability parameters, transportation reliability defined as the possibility that the owned fleet to deliver the aids can perform its function in the assigned time. The last parameter, infrastructure reliability represent possibility that the available infrastructure will be functioning as intended. The second reliability performance measures are consisting of some qualitative indicators ranging from the existence of contingency planning of response operation until the interest of stakeholders. Different from the first category, the second reliability performance indicator, soft reliability measures are counted based on qualitative indicators consisting of governmental, socio-economic situational factors and the interorganizational factors. Governmental factors are factors related to the governmental regulations or politics adopted by a certain government which can accelerate or hinder the execution of humanitarian operations. Socio-economic situational factors relate to typical factors related to the culture of the people in affected disasters, the existential of suppliers' supply logistics to the victims and also the demand uncertainty and security in the affected disaster area. The interorganizational factors closely connected to the contingency planning in humanitarian response operation, typological of donation, coordination and transparency and trust among response operations. Inclusion on using soft reliability measures enable decision makers consider both of tangible (hard measure) and intangible (soft measure) attribute of reliable response operation commonly encountered by humanitarian practitioners. The new classification of reliability performance indicators as above enable decision makers to possibly build new holistic logistics reliability indices from earlier model as presented by [17].

5. Conclusion

Identifying performance indicators of reliable disaster response supply chain is very important factor for reducing the number of casualties when disastrous event occurred. However, it was rarely deeply discussed in disaster management discipline. In narrowing down this research gap, in this paper, factors affecting reliability of humanitarian supply chain and a categorical model of qualitative and quantitative performance indicators of humanitarian response supply chain reliability is presented. Inclusion of soft reliability performance indicators enables decision makers consider qualitative measures commonly encountered in practical situation. Stating as an initial effort toward a more comprehensive reliability calculation indices, future studies are opened in the following paths. At first, refining our conceptual reliability classification model by adding example metrics for empirical calculation testing. The second research focus relates to the development of unified humanitarian supply chain reliability indices consisting of both of hard and soft performance indicators which is still missing in literature.

Acknowledgement
This research is financially supported by Indonesian Ministry of Research, Technology and Higher Education under Basic Research Scheme 2019 under contract number 134 / UN12.13/LT/2019.

References

[1] J Zhang Z Wang and F Ren 2019, *Ann. Op. Res.***1** 1-22
[2] S S Bhamare O P Yadav and A Rathore 2007, *Int. J. Reliab. and Saf.***1**377-410.
[3] L John A. Ramesh and R. Sridharan 2012, *Int. J. Serv. and Op. Man.*, **13**236-272.
[4] L John and A Ramesh 2012 *J. of Adv. in Man. Mod.***9** 217-235.
[5] I N Pujawan N Kurniati N A Wessiani 2013 Int. J. of Log. Sys. and Man.**10**,51-60.
[6] A Maghsoudi 2016 Int. J. of Str. Ch. Man.**7**43-68.
[7] I Antai C Mutshida R. Owusu 2015 *J. of Hum. Log. and Supp. Ch. Man.***5**234-252.
[8] M U Thomas 2012 *Proc. of Ann. Rel. and Maint. Conf.* **52** 62-67, 2012.
[9] B A Diaz C Mena S G Carbajal M Liechty 2012 *Supp. Ch. Man.: An Int. J.***17**263-276.
[10] C Ha H-B Jun C Ok 2014 *Comp. and Ind. Eng.***27**346-366
[11] H.Taghizadeh and E.Hafezi 2012 *Ind. Eng. Int. J.***8**2-10.
[12] I Dolinskaya M Besiau S G Garcia 2018 *J. of Hum. Log. and Supp. Ch. Man.***8**199-226.
[13] Q Lu M Goh R D Souza 2016 *J. of Hum. Log. and Supp. Ch. Man.***6** 222-239.

IOP Conf. Series: Materials Science and Engineering **722** (2020) 012007 doi:10.1088/1757-899X/722/1/012007

[14] A Edrisi M Nourinedjad M J Rooda 2015*Trans. Res. Part E* **56**56-73.
[15] J K Kinyua 2012, "*Supply Chain Performance in Humanitarian Organization in Kenya*," MSc Thesis, University of Nairobi, Kenya.
[16] C W Kamau 2013, "*Humanitarian Supply Chain in Kenya*," MSc Thesis, University of Nairobi, Kenya.
[17] T Nowakowski 2008 *Tot. Log. Man.***1**128-136.

The mental workload analysis of gojek drivers

A D Prabaswari[1], A N Hamid[2], and H Purnomo[3]

[123]Department of Industrial Engineering, Universitas Islam Indonesia, Indonesia

[1]Atyanti.dyah@uii.ac.id

Abstract. Gojek is a service that is very helpful for people's lives by opening jobs for the community and also making it easier for the community itself to live their lives. So many users of the GOJEK application, it also does not rule out the possibility of excessive passengers during rush hour. The needs of passengers who want to be fast and the needs of the company cause gojek drivers to arise workloads on the driver itself. Therefore a study was conducted by analyzing the relationship between mental workload and work stress experienced by gojek drivers using the NASA-Task Load Index (TLX) method. 30 gojek drivers were respondents in this study. By observing the indicators on NASA TLX the results obtained are as many as 3 respondents categorized as having a very high mental workload, 26 respondents having a high category and 1 respondent having a rather high category. it is necessary to improve immediately to make the work become not hampered. the workload is a significant stressor that raises a type of destructive psychological reaction.

1. Introduction

Gojek is one of the breakthroughs of online-based transportation that also combines transportation services with communication technology in Indonesian transportation [1]. Gojek is very helpful in people's lives that require movement and movement in every variety of activities. The phenomenon that currently occurs is that people prefer to use younger services to be sought and used. The nature of urban society which is generally dynamic always moves and adapts to changes or circumstances that exist. Gojek is a company engaged in technology as a social service provider and aims to improve the welfare of workers in various sectors [2].

The presence of Gojek in helping improve the welfare of workers is one of them in opening jobs as a Gojek driver. By becoming a motorcycle taxi driver, it will help the Gojek driver's family economy a little more. Gojek is the most widely used online transportation. This is based on the global research institute Growth for Knowledge (GfK) proving that Gojek online application users reached 21.6% [1]. With so many users of the Gojek application, it also does not rule out the possibility of excessive passengers during rush hour which sometimes makes customers disappointed. Like the problems faced by mass transportation without having a fixed schedule, roads that are crowded with vehicles, how to drive carelessly vehicles and customers who act arbitrarily like canceling orders can create a very high mental workload for GOJEK drivers themselves [2].

According to [3], the workload is the proportion of capacity possessed by humans to meet demands, thus producing a reaction in the form of certain performance. In general, the workload can be defined as the human capacity that is released, both physically and mentally, to fulfill a number of task demands according to predetermined targets. The amount of mental burden received by workers makes workers work not optimal so that measurement and evaluation be required to improve the work process to be more optimum. Improving work process to be more optimum can make increase productivity. One of the subjective measurement methods commonly used to analyze the mental workload of workers is the

NASA TLX method. This method has six factors, that is mental needs, physical needs, time requirements, performance, level effort, and the level of frustration [4].

2. Literature Review

2.1. Definition of mental workload

The mental workload concept has been defined, in the contexts of interest for this review, as "the portion of an individual's limited mental capacity that is actually required by task demands" [5]. Such an assessment could avoid the occurrence of sustained periods of mental overload during driving, which eventually may result in a drastic decrease of performance with possibly dangerous consequences [6].

Mental workload is a psychological construct, difficult to define a difficult to assess [7]. The workload is an effort that must be issued by someone to fulfill the "request" of the job. The workload in question is the size (portion) of the limited operator capacity needed to do certain work. According to [8] the number of activities that must be completed by a person or group of people during a certain period of time under normal circumstances namely workload. The other definition [9] workload is the difference between the workload demands of a task and the maximum capacity of a person's mental burden in a motivated condition. Excessive mental workload will result in work stress. According to Lazarus [10] said that work stress is events around work which are hazards or threats such as fear, anxiety, guilt, anger, sadness, despair, boredom, and the emergence of work stress due to the workload received exceeds the limit.

2.2. NASA TLX

NASA TLX is a method for measuring mental workloads. NASA TLX method has six factors there are, mental needs, physical needs, time needs, frustration, performance, and the last is the level of effort [11]. There is a scale that respondents must later fill in the factors of mental needs, physical needs, time requirements, and frustration levels. The scale used is low to high whereas for performance measurement used well to bad scale.

The NASA Task Load Index (TLX) has been shown to meet these criteria [12]. It is a subjective workload assessment technique that relies on a multidimensional construct to derive an overall workload score [13]. The NASA task-load index (TLX) was developed more than 20 years ago to measure workload in aviation [14]. The NASA TLX is one of the most widely used instruments to assess overall subjective workload [4]. NASA TLX method was developed by Sandra G. Hart of NASA-Ames Research Center and Lowell E. Staveland of San Jose State University in 1981 based on the emergence of the need for subjective measurements consisting of nine factors (task difficulties, time pressure, type of activity, physical effort , effort mental, performance, frustration, stress and fatigue). These nine factors are simplified into 6 factors, there is Mental Demand (MD), Physical Demand (PD), Temporal Demand (TD), Performance (P), Effort (E), Frustration Level (FR) [14].

After taking data from 30 respondents, then the calculation is carried out through the following formulation:

$$\text{Product Value} = \text{Rating} \times \text{Weight Factor or rating comparison} \tag{1}$$

$$\text{Total Value} = \frac{\sum Product}{15} \tag{2}$$

Then the results obtained are interpreted in the following categories:

Table 1. Interpretation score

Score WWL	Category
0 – 9	Low
10 – 29	Rather High

IOP Conf. Series: Materials Science and Engineering **722** (2020) 012008 doi:10.1088/1757-899X/722/1/012008

30 – 79	High
80 – 100	Very High

2.3. Regression and correlation test

Regression analysis is a method of analysis in statistics that is used to determine the causal relationship between one variable with another variable [15]. Regression analysis is also used to understand the pattern of relationships between independent variables and dependent variables and is used to predict or estimate the value of a variable against other variables.

Correlation analysis or association test is used in knowing the closeness of the relationship between two variables and knowing the direction of the relationship that occurs [16]. The closeness of the relationship is expressed by the name of the correlation coefficient (or can be called correlation only). Correlation and regression correlation analysis (both simple and multiple) is often the analytical tool used in association tests. The purpose of the correlation analysis is to find out whether there is a relationship between two or more variables, while the regression analysis predicts how far they affect. this test will be carried out on the SPSS software.

3. Research Method

This research was conducted in Yogyakarta. This activity includes an interview phase for respondents with professions as GOJEK drivers totaling 30 people to fill the NASA TLX weighting and rating table.

The tools used in this study are a form of NASA TLX, SPSS software, and Microsoft Excel Software. The steps taken in this study are as follows:

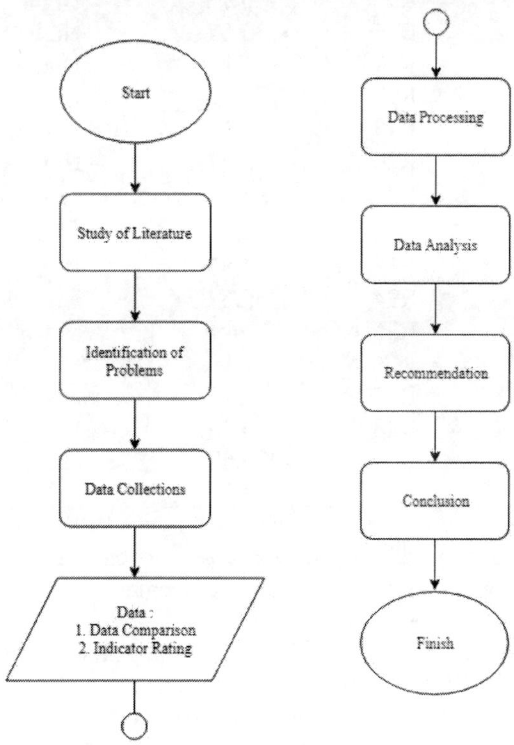

Figure 1. The research stage

IOP Conf. Series: Materials Science and Engineering **722** (2020) 012008 doi:10.1088/1757-899X/722/1/012008

4. Result and Discussion

4.1. NASA TLX Score

NASA TLX is one of the information for researchers or owners to find out the total value of the WWL. The following is data from NASA TLX with the category:

Table 2. Interpretation score

Name	Score	Category
R1	36	High
R2	65.33	High
R3	52.67	High
R4	52	High
R5	68.67	High
R6	68	High
R7	84.67	Very High
R8	78	High
R9	62	High
R10	64	High
R11	58.67	High
R12	54.67	High
R13	47.33	High
R14	62.67	High
R15	70	High
R16	30.67	High
R17	54	High
R18	75.33	High
R19	28	Rather High
R20	50.67	High
R21	82.67	Very High
R22	53.33	High
R23	74.67	High
R24	56.67	High
R25	94	Very High
R26	64	High
R27	58	High
R28	66.33	High
R29	76.67	High
R30	52	High

The results of the classification of mental workloads indicate that all operators fall into different categories. However, all operators need further improvement because they are included in the high (86%) and very high (11%) category.

Based on the Mental Workload final score or the WWL average, the score is classified into which category the score is. Based on data taken from 30 data, only 1 person has a BKM score below 30 which is still tolerable, but 29 other respondents have scores above 30 which according to the table the interpretation of scores is included in the category of high to very high.

Based on the graph of the results of calculations using the NASA TLX method, the workload that has the highest score is the respondent named Rawdah which is equal to 94. One of the activities that make the respondent burdened is in terms of Mental Demand. Where the respondent named Rawdah is a woman who is required to work as a Gojek driver who is always faced with a busy and hot Yogyakarta highway to deliver the interests of her passengers.

4.2. *Classic assumption test*

The following are the results of classical assumption test calculations using SPSS software:

One-Sample Kolmogorov-Smirnov Test

		Unstandardized Predicted Value
N		30
Normal Parameters[a,b]	Mean	61.3896667
	Std. Deviation	15.13023456
Most Extreme Differences	Absolute	.106
	Positive	.053
	Negative	-.106
Test Statistic		.106
Asymp. Sig. (2-tailed)		.200[c,d]

a. Test distribution is Normal.

Figure 2. Residual normality test results

From the results of the SPSS calculation in Figure 2, the significance of > 0.05 then the residual data has a Normal distribution. The data used already represents the intended population. The tolerance > 0.1 and VIF <10, multicollinearity does not occur. Figure number 4 that all aspects of tolerance > 0.1 and VIF < 10 which indicate that there is no multicollinearity in the data used. Significance values of all variables> 0.05, it can be concluded that the regression model does not occur heteroscedasticity.

4.3. *Regression and correlation test*

The following are the results of regression tests and correlations using SPSS software:

Coefficients[a]

Model		Unstandardized Coefficients		Standardized Coefficients	t	Sig.	Collinearity Statistics	
		B	Std. Error	Beta			Tolerance	VIF
1	(Constant)	80.494	102.625		.784	.441		
	PD	.642	.991	.130	.648	.523	.694	1.441
	TD	-1.496	1.135	-.272	-1.318	.200	.660	1.516
	OP	-1.715	.982	-.395	-1.747	.093	.548	1.826
	EF	1.070	.829	.248	1.292	.209	.757	1.321
	FR	2.318	1.162	.394	1.995	.058	.716	1.396

a. Dependent Variable: Rata_Rata_WWL

Figure 3. Regression test

The table above illustrates the regression equation that appears for the results of the WWL mean value with all factors at NASA TLX.

$$Y = 80{,}494 + 0{,}642X1 - 1.496X2 - 1{,}715X3 + 1070X4 + 0{,}67X5 + 2318X6 \quad (3)$$

The results of the SPSS calculation, a significance value of 0.441 is obtained. The result of the correlation test states the magnitude of the relationship between the WWL mean value variables and the dimensions that are at NASA TLX. Theoretically, the greater the correlation value, the stronger the linkages. The level of significance above explains whether or not the correlation is real. The Summary model shows the value of R which is an explanation of how much a variable affects other variables. The number of R square in the table above is 0.811. This means that 81.1% of the variation in the WWL mean value can be explained by the MD variable, PD TD, OP, EF, and FR. While the rest (100% - 81.1% = 18.9%) are explained by other causes. R square ranges from numbers 0 to 1, with a note that

IOP Conf. Series: Materials Science and Engineering **722** (2020) 012008 doi:10.1088/1757-899X/722/1/012008

the smaller the R square number, the weaker the relationship between the two variables. So that in this table it is found that the relationship between the two variables is tight.

5. Conclusion

To get the value of the mental workload of 30 respondents, the calculation was done using the NASA TLX method. Scores from NASA TLX are classified according to the NASA TLX classification table. From the results of the classification, it was found that 30 respondents had a very high level of the workload of 3 people, who had a high mental workload of 26 people, who had a rather high mental workload of 1 people.

With a level of mental workload that is high on average or even in a very high classification, it is necessary to improve immediately to make the work become not hampered. the workload is a significant stressor that raises a type of destructive psychological reaction [17].

To find out the correlation or relationship between variables or indicators at NASA TLX regression tests and correlations were performed using SPSS software. So that the results show that there is a significant relationship between variables as much as 81.1%.

Acknowledgment

Researchers thank the support of the Industrial Engineering Department, Faculty of Industrial Technology, Universitas Islam Indonesia.

References

[1] Amajida F D 2016 *Digital Creativity in Urban Risk Communities: A Study of "Gojek" Online in Jakarta, Informasi Kajian Ilmu Komunikasi* pg. 46
[2] Damayanti S A S 2017 *Online Application Based Transportation: Gojek as Transportation Public for the City of Surabaya* Fisip Unair
[3] D De Waard 1996 *The Measurement of Driver's Mental Workload*
[4] P Hoonakker et al 2011 *Measuring workload of ICU nurses with a questionnaire survey: the NASA Task Load Index (TLX) IIE Trans. Healthc. Syst. Eng.* vol. 1 no. 2 pp. 131–143
[5] R D O'Donnell 1986 *Workload assessment methodology* Cogn. Process. Perform
[6] A Holm 2010 *Developing neurophysiological metrics for the assessment of mental workload and the functional state of the brain*
[7] P Minister 2009 *IET Road Pricing 2008 Conference, London Transp. Res.* no. April, pp. 2008–2010
[8] R H 2010 *Kesehatan Kerja*
[9] H R Jex 1998 *Human Mental Workload.* New York, USA: Elsevier Science Publisher B V
[10] Fraser 1992 *Stres dan Kepuasan Kerja* Jakarta Pustaka Binawan Pressindo
[11] L Young G and Zavelina 2008 *Assessment of Workload Using NASA Task Load Index in Perianesthesia Nursing*
[12] S Rubio, E Díaz, J Martín, and J M Puente 2004 *Evaluation of subjective mental workload: A comparison of SWAT, NASA-TLX, and workload profile methods Appl. Psychol.* vol. 53 no. 1 pp. 61–86
[13] A Cao, K K Chintamani, A K Pandya, and R D Ellis 2009 *NASA TLX: Software for assessing subjective mental workload Behav. Res. Methods* vol. 41 no. 1 pp. 113–117
[14] S G Hart 2006 *NASA-Task Load Index (NASA-TLX), 20 years later Hum. Factors Ergon. Soc.* pp. 904–908
[15] R L Charles and J Nixon 2019 *Measuring mental workload using physiological measures: A systematic review Appl. Ergon.* vol. 74 no. September 2016 pp. 221–232
[16] A Şeker 2014 *Using outputs of NASA-TLX for building a mental workload expert system* vol. 27
[17] M A Kusnadi 2014 *The Relationship between Workload and Self Efficacy with Work Stress at University X Lecturer at Surabaya University Ilm. Mhs. Univ. Surabaya* vol. 3

Lighting Quality Improving Work Thoroughness of Sorting Operators

Anizar[1*], I Siregar[1], E Putri[1], I Maududi[2], and W Habibi[1]

[1]Department of Industrial Engineering, Faculty of Engineering, Universitas Sumatera Utara, Padang Bulan, Medan 20155, Indonesia
[2] Department of Mechanical Engineering, Faculty of Mechanical and Aerospace Engineering, Institut Teknologi Bandung, Bandung, Indonesia

[*]Email: anizar_usu@usu.ac.id

Abstract. This study observes work thoroughness of sorting operators in quality control station at drinking bottle industry. Operators sort the bottles manually using naked eyes to decide whether the bottles have some defects or not. Dirty bottles, cracked, chipped or have water volume that is different from other bottles will be taken by operators. The low light intensity in quality control station causes high numbers of defect bottles to pass the inspection. The standard of Government of Indonesia Health Ministerial Decree No 1405 The Year 2002 that work performed continuously needs 200 lux of lighting. Illuminance level measurement is conducted for five days at 09.00, 11.00, 13.00, and 15.00 o'clock using lux meter on four measurement points using a grid of 1m x 1m. In this study, the average light intensity in quality control station 1 and 2 is 59.30 lux and 52.15 lux. The number of defective products that pass the inspection in a quality control station for 5 days of observation is 2,676 bottles or 55.47% from the total defect products, while the number allowed by the company is only 5%. The correlation between illumination factor in quality control station 1 and defect products that pass the inspection is calculated moment Pearson correlation product and the results are $\hat{y}$ = 292,885-1,317x and r = -0,3147, while in quality control station 2 the results are $\hat{y}$ = 524,077-4,919x and r = -0,6329. The higher illumination level, the lower the number of defective products that pass the inspection, which means operators' work thoroughness improve.

1. Introduction

This study is conducted at a beverage factory, which main products are sarsaparilla drink and soda water, to identify the relation between defect products inspected and lighting quality. Indonesian beverages industries have already been using automation for the production process, yet workers still performed inspection manually. A quality control station, workers inspect the products using naked eyes and taking the defect products if the products do not meet the criteria set by the company. Workers at quality control station 1 are in charge of observing bottles already washed in sterilization machine whether bottles are clean, have no crack and dent. A quality control station 2, workers are in charge of observing the result from workers at quality control station 1 whether the bottles are clean, have no crack and dent, and the water inside is clean and the volume is different. Types of defects at a quality control station in the beverage industry the dirty bottle, cracked bottle, dented bottle, unclear water, different volume water. The quality control station 1 and 2 work results are the focus of this study, as work detail at two stations are higher than other work stations, so appropriate illuminance is needed as recommended. However, the number of defect products that pass the inspection is still higher 5%, which is the number allowed by the company. There is an indication that low illuminance leads to a decrease in thoroughness, affecting the number of defect products passing the inspection. The measured lighting in quality control station is 72 lux, and this is lower than the standard number set by the government, i.e. 200 lux.

Lighting is one of the most important work environment factors and is of a risk factor, meaning that if light intensity does not adequate, productivity will decrease. Dim lighting at work area causes workers to maximize their vision, leading to eye strain, muscle strain, and nerve strain, which result in eye fatigue, mental fatigue, headache, decrease in concentration, thinking speed, and intellectual ability. Work environment factors that affect workers productivity are noise, lighting, color, air quality. Air quality comprised of temperature, humidity, ventilation, and cleanliness. Inadequate lighting significantly affects workers physically and psychologically, that can be noticed from a health problem that occurs such as a headache, etc [1].

Low illuminance causes activity can not be performed correctly, therefore more windows should be added to let more natural light enters the room. Adding more lamps must be done accordingly with work characteristics performed in a certain room [2]. Work environment affects comfortability, health, morale, efficiency, and work productivity [3]. Lighting at the work environment affects work performance, productivity, and quality. Bad lighting causes low productivity, eye strain, headache, and increase accident number. Lighting improvement in the work environment can be done by designing artificial lighting [4,5]. Light color, light intensity, and colored text significantly affect the answer regarding reading understanding, which means affect visual work [6]. An energy-efficient light bulb will consume the lowest wattage, yet producing a high intensity of visible light. Fluorescent tube and Compact Fluorescent Lamp (CFL) are the energy-saving light bulb of phosphorescent type [7].

Illuminance designed without glare and using efficient lighting will improve workers comfort level because danger can be detected immediately [8]. Lighting has an effect on workers mental load, where the lighting level below standard will affect work performance and disrupt visual ability. Even though this will not cause eyes problem permanently, it will increase workload, speed up tiredness, make workers take more rest, reduce work time, reduce work satisfaction, decrease production quality, increase error frequency, disrupt concentration, and reduce worker productivity.

Humans need lighting to identify an object visually and this comprises of 90% work activity, especially for works requiring sharp vision. Eyes ability to see an object with clear, quickly, without an error is greatly affected by lighting in the work environment [9]. Workers complain of visual fatigue and eye strain is a hidden danger [10]. Visual fatigue and eye strain will affect thoroughness and work quality of workers [11]. A sustainable act to improve productivity, efficiency, and workers skill in an industry should also consider workers comfort. There is a relation between illuminance and productivity. Lighting greatly affects workers' visual perception when handling the process on a production floor. White light increases eyes strain more than other color lights [12]. High color temperature inhibits melatonin secretion and helps increase workers response speed and the opposite is true for low color temperature [13]. Blue light usage is better than white light to alleviate visual fatigue, eye strain, inability to focus, concentration disruption, and blurred vision [14]. Substitution of lights is not only concerned with reducing the burden, but there must be policies and regulations related to the negative impacts caused [15].

2. Method

Illuminance at the production floor is measured using lux meter for 5 days at 09.00-16.00, divided into several intervals during the 1st shift. Each measurement represents a certain time interval with the condition that the 1st measurement conducted at 09.00 represents the time interval of 08.00-10.00, the 2nd measurement conducted at 11.00 represents the time interval of 10.00-12.00, the 3rdmeasurement conducted at 13.00 represents the time interval of 12.00-14.00, and the 4th measurement conducted at 15.00 represents time interval of 14.00-16.00. Defect products that pass the inspection are calculated twice, the 1st time window represents defect products that pass the inspection at 08.00-12.00 and the 2nd time window defect products that pass the inspection 13.00-16.00. Measured points are chosen using a reference of *SNI 16-7062-2004* regarding light intensity measurement in workspace using certain size grids according to an area that will be measured its illuminance [16]. Production floor has 2 quality stations, quality control station 1 and quality control station 2, both with an area of 9.92 m^2. Since both areas are less than 10 m^2, a grid with a size of 1 m x 1 m is used so that there are 4 points of measurement. There are 4 points measured in both quality control station 1 and 2 as shown in figure 1.

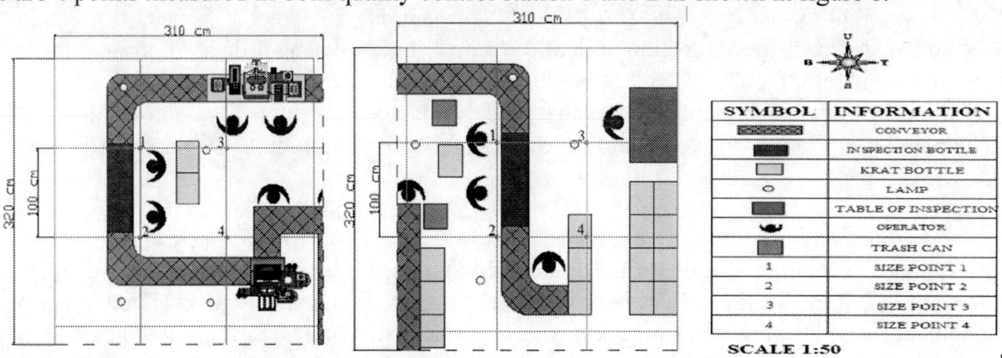

Figure 1. The layout of the Illumination point of measurement in quality control station 1 and 2

Illuminance data and the number of defective products that pass the inspection will be tested its normality using Kolmogorov-Smirnov and its sufficiency. Regression equation and correlation coefficient test are used to predict defect products that pass the inspection based on illumination level and relation degree between illumination level with a quality control station work result. The correlation coefficient is calculated to know whether there is a correlation between illuminance variable with quality control work result.

3. Results and Discussions

In these parts, the results of average illuminance, defect products that pass the inspection, regression equation, and correlation coefficient will be discussed. The result of the Kolmogorov Smirnov test on illuminance and defect products that pass the inspection at quality control station shows that data is distributed normally, since $D_{max} \geq D_{\alpha}$. The result of the sufficiency test on illuminance and defect products that pass the inspection at quality control station shows that N'<N, which means the data is sufficient.

3.1. The average illuminance in a quality control station

Illuminance measured at quality control station 1 and 2 is performed at 4 points. Measurement conducted at 09.00, 11.00, 13.00, and, 15.00 o'clock for five days shows that the average illuminance at quality control station 1 is 59.30 lux and at quality control station 2 is 52.15 lux. Recapitulation of average illuminance at quality control station 1 is shown in table 1.

Table 1. The average illuminance in quality control station

Quality Control Station	Time	Iluminance (lux)					Average
		Day-1	Day-2	Day-3	Day-4	Day-5	
1	09.00	54.90	57.59	58.05	56.82	60.11	
	11.00	55.45	57.03	60.52	62.23	57.78	
	13.00	67.08	57.27	60.22	62.40	57.65	
	15.00	67.89	58.13	59.60	58.56	56.47	
	Average	61.33	57.51	59.60	60.01	58.01	59.30
2	09.00	50.36	55.55	52.33	49.12	50.12	
	11.00	52.61	55.93	52.81	49.18	50.38	
	13.00	52.92	56.68	53.16	49.25	50.27	
	15.00	53.01	56.53	53.12	49.00	50.41	
	Average	52.23	56.18	52.86	49.14	50.30	52.15

The comparison between illuminance at quality control station 1 and 2 on the same day implies that illuminance at quality control station 1 is higher, yet the value is far below the standard illuminance for the corresponding work, 200 lux, as set by Indonesian Government Health Ministerial Decree No 1405 The Year 2002 [17]. Illumination on the production floor is affected by lamp wattage, lamp distance toward the work field, and room area.

3.2. Calculation of Defect Products That Pass The Inspection
Defect bottles that pass the inspection at quality control station during 5 days of observation are 2,676 bottles, while the total number of defect bottles at both quality control station 1 and 2 is 4,824 bottles. Therefore, the percentage of defect bottles that pass the inspection at a quality control station is 55.47%. This number is very large compared to the allowed number of 5% and this, of course, inflicts a financial loss to the company.

3.3. Relation Between Illuminance and Defect Products That Pass The Inspection
Regression is a measuring tool that can also be used to measure whether there is a correlation between variables or not. Regression analysis is performed to predict the average value of dependent variables based on independent variables value known. This correlation aims to measure the relation degree between two variables or more. Work results data used from both station is defect products that pass the inspection. Data distribution of dependent and independent variables at quality control station 1 and 2 and the corresponding regression line are shown in figure 2.

Regression line equation of illuminance and defect products that pass the inspection at quality control station 1 is $\hat{y}$=292,885-1,317x and Pearson correlation test result gives correlation coefficient value (r) of -0.3174 (figure 2a). This value means there is a low inverse relation between illuminance and defect products that pass the inspection at quality control station 1. In other words the higher illuminance at quality control station 1, the lesser defect products that pass the inspection.

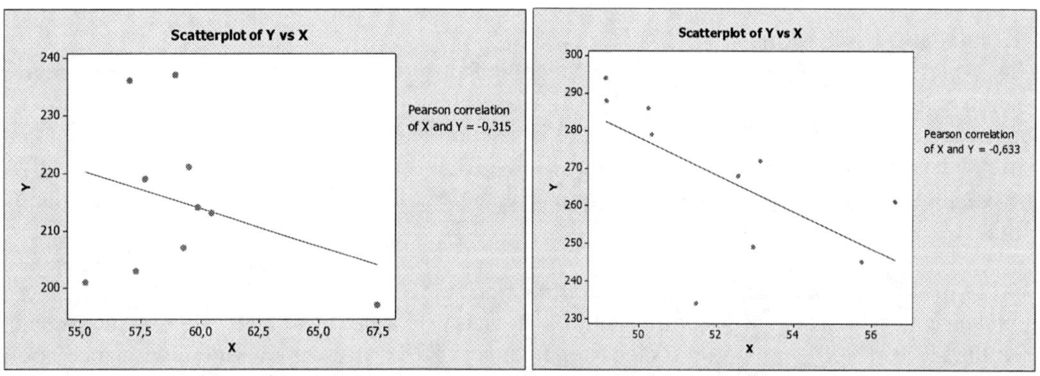

(a) (b)

Figure 2. Scatter diagram dan regression line at quality control station 1 and 2

Regression line equation of illuminance and defect products that pass the inspection at quality control station 2 as shown in figure 2b is $\hat{y}$=524,077-4,919x and Pearson correlation test result gives correlation coefficient value (r) of -0.6329. This value means there is a strong inverse relation between illuminance and defect products that pass the inspection at quality control station 1. In other words the higher illuminance at quality control station 1, the lesser defect products that pass the inspection. In order to obtain optimum illuminance (x), assume $\hat{y}$=0 in a regression equation, then we obtain x = 222.39 lux at quality control station 1 and x = 106.54 lux at quality control station 2.

Illuminance needed at quality control station 1 differs from Illuminance needed quality control station 2, as workers at quality control station 1 are older than at quality control station 2, affecting eyes ability to observe drink bottles. Correlation coefficient values between illuminance at quality control with defect products that pass the inspection are calculated and the obtained values are r_{qc1} = -0,3174 and r_{qc2}= -0,6329. Both values show an inverse relation between illuminance at a quality control station and the number of defective products that pass the inspection. The higher the illumination level becomes, the lesser defect products that pass the inspection. Lighting quality improvement at quality control station can minimize defect products that pass the inspection.

3.4. The Reflectance of Object Materials

The reflectance of object materials measured is the wall, ceiling, and machine (washer machine and conveyor). The reflectance of object materials at quality control station is varied, where the wall, ceiling, washer machine, and conveyor have larger reflectance number than the recommended. It is shown in figure 3.

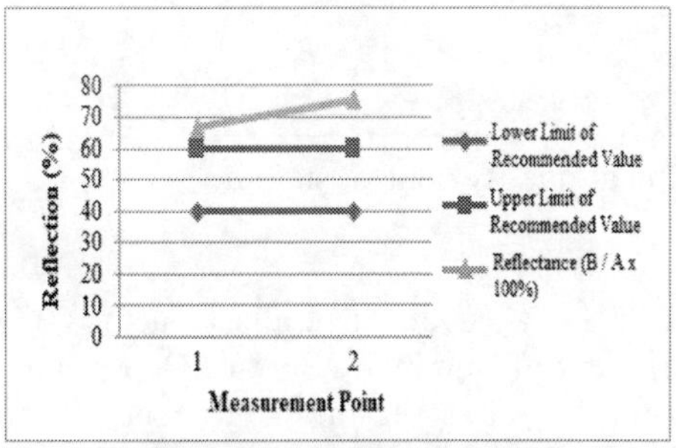

Figure 3a. Wall reflectance

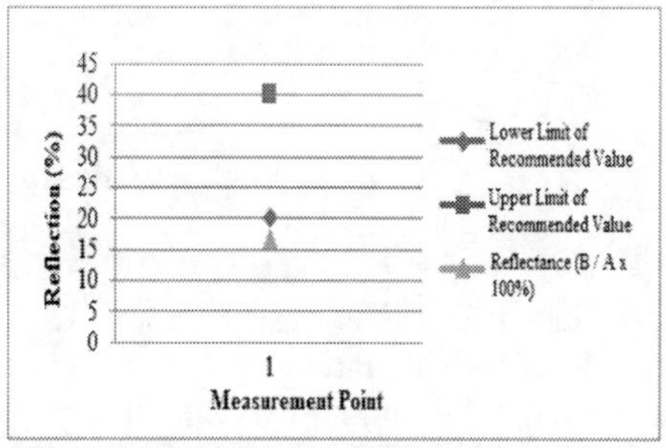

Figure 3b. Floor reflectance

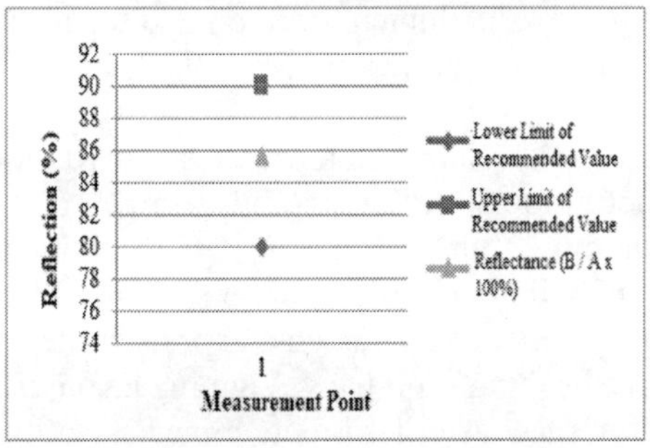

Figure 3c. Ceiling reflectance

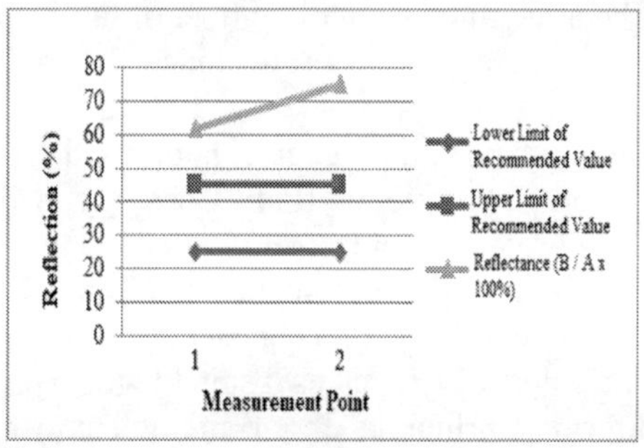

Figure 3d. Engine reflectance

Based on the graph in Figure 3a to 3d there are two material objects whose material reflectance values exceed the recommended values, namely the reflectance figures of the wall and the washer and conveyor machine. Wall material is made of bright ceramic, increasing its reflectance number. There is a lamp installed on the wall at a sorting area, causing reflectance number of washer machine and conveyor to increase. However, ceiling reflectance number still meets the recommended range, i.e. 85.6%, with material made of the white ceiling. Floor reflectance number is still far below the recommended range, i.e. 16.7%. The reason is the floor is dirty and enclosed by puddles so that light can not be well-reflected. Cleaning the floor regularly and building good drainage water will make floor surface clean. Therefore water flowing from the washer machine will not make floor filled by puddles and slippery, allowing workers to work comfortably.

3.5. Type and Number of Lamps Mapping

The actual number of lamps on production floor does not the needed number and the type of lamp used is also not suitable for indoor usage, as it has closed luminaire that is more suitable for outdoor usage. Data used to calculate the number of light bulb on production floor are room length (L) 6.2 m; room width (W) 3.2 m; room height (H) 10 m; coefficient of reflectance for ceiling (ρ_c) 0.86; coefficient of reflectance for wall (ρ_w) 0.72; coefficient of reflectance for floor (ρ_f) 0.17; distance from luminaire plane to ceiling (h_c) 0,25 m; distance from work plane to floor (h_f) 0.75 m; distance from luminaire plane to work plane (h_r) 7 m. The division of the cavity is shown in figure 4.

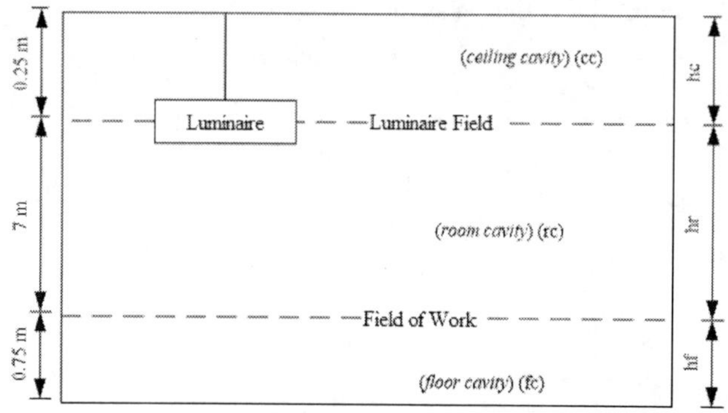

Figure 4. Cavity distribution at quality control station

The coefficient of Light Loss Factor (LLF) is 0.71, based on the Ballast Factor (BF) of 1; the coefficient of Voltage Variation (VV) is 1; the coefficient of Luminaire Surface Depreciation (LSD) is 1; the coefficient of Luminaire Ambient Temperature (LAT) is 1; the coefficient of Lamp Lumen Depreciation (LLD) 0.85; the coefficient of Room Surface Dirt Depreciation (RSDD) is 0.92; the coefficient of Luminaire Dirt Depreciation (LDD) is 0.95; the coefficient of Lamp Burnout (LBO) is 0.95. The minimum standard number of lumen required is 14,708 lumen based on E 200; area (A) 19.84 m^2; Effective Floor Cavity reflectance (ρ_{fc}) 0.84; coefficient of utilization (CU) 0.38; light loss factor (LLF) 0.71.

Changing type and number of lamp used on the production floor will meet the standard requirement of lumen on all production floor. Currently, 8 Compact Fluorescent Lamp (CFL) lamps of 18 Watt are used, each producing only 1100 lumen, so that in total only give 8800 lumen. The recommended type of lamp to be used is Fluorescent Tube, each producing 3350 lumens, therefore only 5 bulbs are needed. This Fluorescent Tube can consistently produce a high number of the lumen and prevent the lamp from fading its color, which makes it suitable to be used in the beverage company. Lighting layout is set using the uniformity principle, by arranging lamp based on spacing criteria determined. Quality Control Station has an area of 19.84 m^2 and if 5 light bulb is to be used, the maximum distance between each light bulb is 1.99 m with the layout as shown in figure 5.

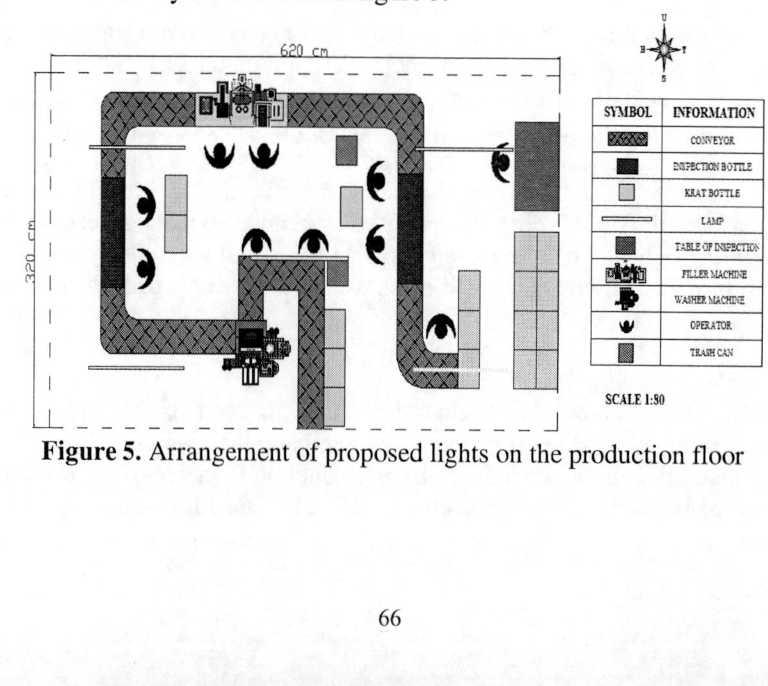

Figure 5. Arrangement of proposed lights on the production floor

4. Conclusions

This article aims to understand the effect of lighting on workers thoroughness during a manual inspection on defect products at a quality control station. The main parameters of this study are average illuminance, defect products that pass the inspection, regression coefficient, and correlation coefficient. The average illuminance at quality control station is far below the standard value set by The Indonesian Government of Health Ministerial Decree Number 1405 The Year 2005, i.e. 200 lux. The number of defective products that pass the inspection at quality control station during 5 days of observation reaches 55.47% of the total defect products, far higher than the company allowed number. The illuminance at quality control station 1 has a low and medium inverse relation with defect products that pass the inspection, meaning that the higher illuminance becomes the lesser defect products will pass.

References

[1] Sarode AP, Shirsath M 2012 *International Journal of Science and Research* (IJSR)**3** (11) 2735-2737

[2] Bella L, Spada G, Pearce A, Fragliasso F 2015 *Energy Procedia***78** 3138-3143

[3] Ajala EM 2012 *The African Symposium***12** (1) 141-149

[4] L Espinoza LA, Najera JM 2010 *Research Journal of the Costa Rican Distance Education University (Online Edition)***2** (1) 63-68

[5] Anizar, Erwin 2017 IOP Conf. Ser.: Mater. Sci. Eng. 180 012122

[6] Lin CC, Huang KC 2014 *International Journal of Applied Science and Engineering***12** (3) 193-202

[7] Chumaidy A 2017 Sinusoida**19** (1) 1-8

[8] Pal N, Krishnan SV, Sadhu PK 2013 *International Journal of Engineering Science & Research Technology***2** (2) 240-247

[9] Dey AK, Mann DD 2011 *Journal of Agricultural of Safety and Health***7** (2) 91-110

[10] Osuke GO, Okolie STA, Anyaeche CO 2015 *International Journal of Innovative Science & Technology***2** (7) 34-37

[11] Seegehalli PJ 2016 *International Journal of Computer Sciences and Engineering (IJCSE)***8** (3) 94-100

[12] Kumar R, Pandey S, Jana P, Panghal D 2016 *International Journal of Electrical, Electronics, and Data Communication***4** (2) 9-12

[13] Jiandong Z, Shuo M 2016 *International Journal of Signal Processing, Image Processing and Pattern Recognition***9** (6) 437-446

[14] Viola AU, James LM, Schlangen LJM, Dijk DJ 2008 *Scand Journal Work Environ Health***34** (4) 297-306

[15] J Ogunyemi, I.A. Adejumobi 2018 *Lautech Journal of Engineering and Technology***12** (2) 33-40

[16] Indonesian National Standard 2004 Lighting Intensity Measurement at Work Place

[17] The Government of Indonesia Health Ministerial Decree No 1405 The Year 2002, 2002, Health Requirements of Work Environment at Office Complex and Industry

Modelling the continuance intention towards the use of mobile hospital appointment system

A R Taufiq[1], A Widyanti[1], K Muslim[1], T Wijayanto[2], F Trapsilawati[2], H M Arini[2] and A D Wibawa[3]

[1] Department of Industrial Engineering, Bandung Institute of Technology, Bandung, Indonesia
[2] Department of Mechanical and Industrial Engineering, Gadjah Mada University, Yogyakarta, Indonesia
[3] Department of Computer Engineering, Institut Teknologi Sepuluh November, Surabaya, Indonesia

E-mail: widyanti@mail.ti.itb.ac.id

Abstract. Effective management of patient flow in the outpatient unit of a hospital is the key to achieving operational excellence. Hospital X, as one of the major hospitals in Indonesia, launched a mobile hospital appointment system (X-Go) to increase the effectiveness of patient flow management. Although the application has a good rating, only less than 28% of the patients are still using it. Aim of this study is to observe the factors affecting the continuance intention towards X-Go, using the Technology Acceptance Model. A questionnaire consisted of 15 questions, using 7 point Likert Scale was used to develop the model. Ninety eight respondents were involved voluntarily in this study by filling out the questionnaire. Results showed that relationship quality has the most significant direct influence on the continuance intention to use the system. Meanwhile, perceived ease of use has the most significant indirect influence on the continuance intention to use the system.

1. Introduction

Effective management of patient flow in the outpatient unit of a hospital is the key to achieving operational excellence [1, 2]. In real conditions, most hospitals in Indonesia have been administering patient registration procedure that is quite time-consuming. Such procedure requires patients to be present at the hospital, taking a queue number and wait to be called to the registration desk. Often times, patients have to wait several hours, even before the opening hours, just to pick a queue number. This condition is often worse in outpatients units (i.e., defined as the medical services provided for patients which are not in the form of hospitalization [3]) in major hospitals in Indonesia, due to very large volume of patients visit with mixed cases of disease [2]. One of the major hospitals that experience this problem is Hospital X, a national referral hospital in Bandung city.

To overcome this problem and improve patient registration, Hospital X has launched a mobile application for outpatient appointment called X-Go. This app utilized internet, accessible from Android smartphones. In addition to being able to speed up service and reduce patient waiting time, this app can help hospital staffs to reduce file preparation time. The app only involves one hospital—Hospital X. The app has been downloaded more than 11,000 times and received a high rating by the users. Although the app has the potential to improve the quality of existing health services, it has not been widely used by outpatient patients. Only less than 28% of the total outpatient patients use the app daily. Most importantly, the app has a very low number of repeat uses, showing the fact that most app users have been reluctant to use the app for the second time. In other words, the continuance intention to use the app has been low.

Many factors influence the adoption of technology, including mobile application, by a society [4, 5, 6, 7, 8]. One reliable approach that can be used to observe factors influencing the continuance intention to use such application is technology acceptance model (TAM). TAM was first developed by Davis in 1989 as an information system theory that modeled how users understood, tried, used, and accepted technology [9, 10]. This theory was first developed from the theory of reasoned action (TRA) and was used as a basis for determining the relationships that occur between six key beliefs, namely the external factors, perspective ease of use (PEOU), perceived usefulness (PU), attitude towards application, behavioral intention to use, and actual system use [11]. Considering the importance of outpatient registration application to improve hospital X service performance, this research aimed to evaluate the level of continuance intention to use such application. The discussion utilized the TAM developed by Chen, et.al. [5], as can be seen in Table 1.

2. Method

A questionnaire was developed to observe factors that influence the continuance intention to use the X-Go app., consisted of 15 items with 7 Likert scales of which value 1 states "highly disagree" and value 7 states "highly agree". Back-translation procedure was used to translate the questionnaire to Bahasa Indonesia. A pilot testing was done before carrying out the main data [12]. The TAM framework of this research can be seen in Figure 1. The definition of each construct can be seen in Table 1.

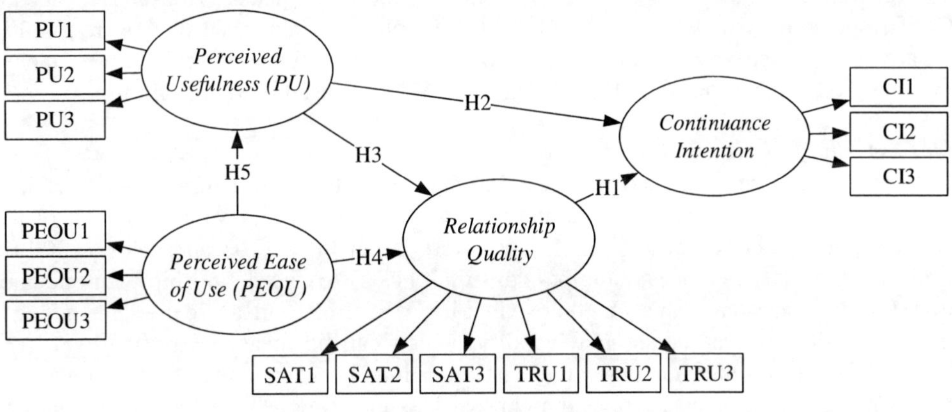

Figure 1. TAM framework by Chen et al. [5].

Table 1. Construct of TAM framework by Chen et al. [5].

Construct	Definition
Perceived Usefulness (PU)	A belief that the hospital e-appointment system useful in improving life performance
Perceived Ease Of Use (PEOU)	A believe that the hospital e-appointment system is easy to use.
Relationship Quality	The final outcome from interactions between two parties, and consists of two aspects, that are satisfaction and trust
Continuance Intention (CI)	An intention to continue to use the hospital's e-appointment system

Based on the research framework, this research proposed several hypotheses:

- H1 = Relationship quality (consist of satisfaction and trust) has a positive influence on continuance intention of X-Go users.
- H2 = PU influence continuance intention of X-Go users positively
- H3 = PU influence relationship quality of X-Go users positively
- H4 = PEOU influence relationship quality of X-Go users positively
- H5 = PEOU influence PU of X-Go users positively

To analyze the result of TAM, this study used Structural Equation Modelling (SEM), a multivariate method to incorporate unobservable variables through indicator variables [13, 14, 15, 16, 17]. Two types of SEM include covariance-based SEM (CB-SEM, that is used to determine how well a proposed theoretical model can estimate the covariance matrix for a sample data set [13]) and PLS-SEM (i.e., used to develop theories in exploratory research [13]). The goal of the latter is predicting key target constructs or identifying key "driver" constructs [13]. Because this study aims to evaluate the level of continuance intention based on the influencing factors, this study used the PLS-SEM method.

3. Data Analysis and Results

3.1. Descriptive Statistics Analysis

This study collected 98 copies of questionnaires in total. This number exceeds the minimum total sample needed to conduct a PLS-SEM analysis [18, 19]. The following is the respondent's demographic data. In terms of gender, 40 (40.82%) were male and 58 (59.18%) were female. In terms of age, 9.18% of the respondents were aged <=20, 38.78% were aged 21-30, 37.76% were aged 31-40, and 13.27% were aged 41-50. In terms of education, 2.04% have master's degree, 15.31% have bachelor's degree, 6.12% have diploma degree, 50% were high school graduates, 23.47% were junior high school graduates, and 3.06% were elementary school graduates. Housewives accounted for 36.73% of respondents, students accounted for 6.12% of respondents, 51.02% were employed, and 6.12% were unemployed. In regard to the level of experience of the app users, 31.63% have used the app 2-5 times, 10.20% have used 6-10 times, and 58.16% have used more than 10 times.

3.2. Measurement Model Analysis

The first criterion to be evaluated was the internal consistency reliability (IRC), which can be measured by the Cronbach's alpha value and the composite reliability value [13]. All of the survey questions were acceptable because all the values were larger than 0.708 and below 0.900 (Table 2). This implies that each indicator variable per construct, has a strong relationship with one another and has values that are consisten with each other so that it can be used to measure the same latent variable.

The second criterion was convergent validity, which can be measured by the average variance extracted (AVE) and/or outer loading value. The outer loading value above 0.708 and AVE value of 0.5 or higher indicates that on average the construct explains more than half of the variance of its indicators [13]. Based on results in Table 2, all the requirements have been met. Therefore, the measurement model is acceptable in terms of convergent validity. The result of these two methods imply that each indicator variable has a high correlation with other indicator variables, in the same construct.

Table 2. Internal consistency reliability and convergent validity analysis (a).

Construct	Composite Reliability	Cronbach's α	AVE	Items	Outer Loading			
					CI	PEOU	PU	RQ
CI	0,890	0,886	0.736	PU1	**0,854**	0,499	0,435	0,668
				PU2	**0,908**	0,550	0,557	0,571
				PU3	**0,807**	0,521	0,428	0,521

Table 3. Internal consistency reliability and convergent validity analysis (b).

Construct	Composite	Cronbach's α	AVE	Items	Outer Loading

	Reliability				CI	PEOU	PU	RQ
PEOU	0,889	0,856	0.774	PEOU1	0,348	**0,802**	0,191	0,418
				PEOU2	0,594	**0,917**	0,392	0,571
				PEOU3	0,612	**0,916**	0,453	0,571
PU	0,828	0,820	0.634	SAT1	0,367	0,270	**0,782**	0,443
				SAT2	0,490	0,372	**0,873**	0,665
				SAT3	0,451	0,336	**0,727**	0,380
				TRU1	0,523	0,584	0,360	**0,744**
				TRU2	0,475	0,556	0,453	**0,736**
				TRU3	0,574	0,474	0,614	**0,823**
RQ	0,742	0,711	0.639	CI1	0,541	0,393	0,484	**0,779**
				CI2	0,633	0,416	0,624	**0,847**
				CI3	0,543	0,473	0,485	**0,842**

There are three methods to asses discriminant validity. The first one is by examining the cross loading of the indicators [13]. All the indicator's outer loading on the associated construct were greater than all of its loadings on the other constructs, thus exhibit discriminant validity (Table 2). The second method is using the Fornell-Larcker criterion in which it compares the square root of the AVE values with the latent variable correlations [13]. Based on the result, the square root AVE of each construct exceeded its highest correlation with other construct, thus it supported discriminant validity (Table 4). The third method is Heterotrait-Monotrait (HTMT) correlation ratio. An HTMT ratio close to 1 indicates the model doesn't have a sufficient discriminant validity. A model is considered to have a sufficient discriminant validity if each of the relationships between the latent variables has a HTMT ratio with a value equal to or less than 0.85 and has a confidence interval that does not exceed 1. Based on results in Table 4, all the requirements have been met. Therefore, the measurement model is acceptable in terms of convergent validity.

Table 4. Fornell-Larcker and HTMT ratio analysis.

	Correlation				$\sqrt{AVE}$	HTMT Ratio (2,5%; 97,5%)		
	CI	PEOU	PU	RQ		CI	PEOU	PU
CI	1				0,858			
PEOU	0,609	1			0,880	0,704 (0,410; 0,903)		
PU	0,552	0,413	1		0,796	0,717 (0,420; 0,894)	0,498 (0,239; 0,733)	
RQ	0,689	0,601	0,636	1	0,800	0,801 (0,623; 0,908)	0,682 (0,476; 0,833)	0,772 (0,528; 0,923)

3.3. Structural Model Analysis

The first step for structural model analysis was to measure the collinearity of the constructs. The research result showed that each construct's tolerance, variance inflation factors (VIF) value were all lower than 5 (Table 5) therefore the stuctural model does not have any collinearity issues. The path coefficient supported all hypothesize relationship except H2 (Table 5). In addition to direct impacts caused by the structural relations in the model, an analysis of mediation variables was also carried out by looking at the total effects and indirect effects (Table 6) found in the model. Evaluation on the coefficient of determination (R^2), predictive relevance (Q^2), f^2 effect size, and q^2 effect size of the model was also conducted. The R^2 was used to determine the predictive power of the model. On the other hand, Q^2 is an indicator of out-of-sample predictive power or predictive relevance (Table 7). The Q^2 test was carried out using a blindfolding procedure. The f^2 effect size is a measure used to evaluate whether an exogenous construct has a substantial impact on its endogenous construct, while the q^2 effect size was used to measure the relative strength or impact of Q^2 (Table 7). A value of 0.02, 0.15, and 0.35 for the effect sizes each represents a low, moderate, and high value [13].

Table 5. VIF analysis and hypothesis test result.

Relation		VIF	Path coeff.	t-val	p-val	2.5%	97.5%	Hypothesis Result
H5	PEOU → PU	1,000	0,413	3,928	0,000	0,176	0,595	Accepted
H4	PEOU → RQ	1,206	0,408	4,876	0,000	0,252	0,588	Accepted
H2	PU → CI	1,678	0,192	1,911	0,056	-0,008	0,382	Not accepted
H3	PU → RQ	1,206	0,467	4,752	0,000	0,234	0,628	Accepted
H1	RQ → CI	1,678	0,567	6,632	0,000	0,386	0,718	Accepted

Table 6. Total effects and indirect effects analysis.

Relation	Total effect	t-val	p-val	2.5%	97.5%	Indirect effect	t-val	p-val	2.5%	97.5%
PEOU → CI	0.420	4.828	0.000	0.236	0.577	0.420	4.762	0.000	0.230	0.572
PEOU → PU	0.413	3.928	0.000	0.176	0.595					
PEOU → RQ	0.601	6.850	0.000	0.395	0.774	0.193	3.407	0.001	0.084	0.306
PU → RQ	0.467	4.752	0.000	0.234	0.628					
RQ → CI	0.567	6.632	0.000	0.386	0.718					
PU → CI						0.265	4.073	0.000	0.147	0.401

Table 7. R^2, Q^2, f^2 and q^2 analysis.

Construct	R^2 value	Indication	Q^2 value	Indication	Relation	f^2 effect size	q^2 effect size
CI	0.496	Moderate	0.314	Accepted	PEOU → PU	0.206	0.008
PU	0.171	Low	0.092	Accepted	PEOU → RQ	0.301	0.04
RQ	0.542	High	0.298	Accepted	PU → CI	0.043	0.003
					PU → RQ	0.395	0.05
					RQ → CI	0.380	0.066

4. Conclusion

Factors of "perceived usefulness" and "perceived ease of use" determined the "continuance intention" of the X-Go app through relationship quality. It is interesting that "perceived usefulness" was not significantly related directly to "continuance intention". This indicates that the X-Go's ability to improve patient's medical services cannot increase the patient's intention to continue to use the app to make appointments in the future, if the patients do not have a satisfactory experience of using the app, trust the quality and reliability of the app, nor trust the hospital that provided the app. This could be the reason why the percentage of daily users were relatively low compared to the number of daily outpatient patients, even though the app had received a high rating and number of downloads. Based on the interview, there were also a number of current patients who stopped using the app after several times of usage because of trust and satisfactory issues. Based on Kettinger et al. (1994) and Woo et al.

(1999) in Chen et al. (2013), a stable internet connection will increase user satisfaction and trust. Therefore, Chen et al. (2013) proposed a relationship between PU latent variables and PEOU so that the model can focus more on the quality of the relationship felt by customers. On the other hand, X-Go application users will use the application to book a schedule for 2 days to 5 months ahead. Hence, X-Go users still have plenty of time available that can be used to re-use the application if the internet connection is not satisfactory. Therefore, the assessment of respondents on satisfaction and trust for latent RQ variables is no longer influenced by these stable internet connection factors. In addition, the object used in the research of Chen et al. (2013) is an application that involves more than one hospital. This causes if the application user was dissatisfied and did not trust a hospital, there is still an opportunity for the user to book a schedule with other hospitals available in the application. Hence, there are still opportunities where users still have the intention to continue using the application. This is different from the X-Go application which only involves one hospital, namely Hospital X. If the users doesn't feel satisfied or trust in X Bandung then the user has no other hospital choice.

With the exception of the result above, the result of this research basically showed similar findings with the prior study by Chen et al. [5]. "Perceived ease of use" have a positive impact on "Perceived usefulness". The findings indicated that the patient who felt that the app was easy to use might have better understanding that the app can improve their medical service quality. In contrast, the patient who felt that the app was difficult to use would not have the chance to acquire the usefulness of the app. Thus, it is an important indicator that the app must be easy to use. This finding can also be explained by the education demographic data. Since 76.53% of the respondents were high school graduates or lower education levels, the learning difficulty may exist among the respondents. The second finding was that "perceived usefulness" and "perceived ease of use" were positively related to "relationship quality". This indicated that the more the patient felt the app was easy and worth it to use, the more satisfied and trusting they were with the app. The last finding was that "relationship quality" positively impacted on "continuance intention". This indicated that in order to use the app, patients must first had a good relationship with the app and the hospital. These findings imply that X-Go needs to provide a user interfaces that are easier to understand and use. A number of patients mentioned they need a bigger font size and notification for error handlings. X-Go also needs to provide useful contents and functions for patients. In addition the main function as an appointment application, hospital X can add various features that can support this main function such as completing the outpatient service schedule to the operating hours of each outpatient polyclinic. Finally, this study may have some insufficient aspects that it lacks a high value of coefficient of determination and the use of PLS-SEM, instead of CB-SEM for confirmatory study. Thus, highly recommended that future studies could overcome these limitations.

References

[1] Cayirli T and Veral E 2009 Outpatient scheduling in health care: a review of literature *Int. J. Oper. Prod. Manage.* **12** 519–49
[2] Mardiah F P and Basri M H 2013 The analysis of appointment system to reduce outpatient waiting time at Indonesia's public hospital *Human Resource Management Research* **3** 27–33
[3] Mariyanti S and Citrawati A 2011 Burnout pada perawat yang bertugas di ruang rawat inap dan rawat jalan RSAB harapan kita *Jurnal Psikologi Volume* **9** 48
[4] Mouakket S and Al-hawari M A 2012 Examining the antecedents of e-loyalty intention in an online reservation environment *The Journal of High Technology Management Research* **23** 46–57
[5] Chen S-C, Liu S-C, Li S-H and Yen D C 2013 Understanding the mediating effects of relationship quality on technology acceptance: an empirical study of e-appointment system *J. Med. Syst.* **37** 9981
[6] Jaradat M I R M, Mohammed Issa Riad, Moh' Z and Smadi d. A 2013 Applying the technology acceptance model to the introduction of mobile healthcare information systems *International Journal of Behavioural and Healthcare Research* **4** 123
[7] Lai Y-H, Huang F-F and Yang H-H 2015 A study on the attitude of use the mobile clinic registration system in Taiwan *Technol. Health Care* **24 Suppl 1** S205–11

IOP Conf. Series: Materials Science and Engineering **722** (2020) 012010 doi:10.1088/1757-899X/722/1/012010

[8] Dillon A and Morris M G 1996 User acceptance of new information technology: theories and models (Medford, NJ: Information Today)
[9] Davis F D 1989 Perceived Usefulness, Perceived Ease of Use, and User Acceptance of Information Technology *Miss. Q.* **13** 319–40
[10] Davis F D 1993 User acceptance of information technology: system characteristics, user perceptions and behavioral impacts *Int. J. Man. Mach. Stud.* **38** 475–87
[11] Legris P, Ingham J and Collerette P 2003 Why do people use information technology? A critical review of the technology acceptance model *Information & Management* **40** 191–204
[12] Nielsen J 1994 *Usability Engineering* (California: Elsevier)
[13] Hair J F Jr, Hult G T M, Ringle C and Sarstedt M 2016 *A Primer on Partial Least Squares Structural Equation Modeling (PLS-SEM)* (SAGE Publications)
[14] Suhr D 2006 The basics of structural equation modeling *Presented: Irvine, CA, SAS User Group of the Western Region of the United States (WUSS)*
[15] Lomax R G and Schumacker R E 2004 *A beginner's guide to structural equation modeling* (New Jersey: Lawrence Erlbaum Associates, Inc)
[16] Kline R B 2015 *Principles and Practice of Structural Equation Modeling, Fourth Edition* (New York: Guilford Publications)
[17] Chin W W and Others 1998 The partial least squares approach to structural equation modeling *Modern methods for business research* **295** 295–336
[18] Sekaran U and Bougie R 2016 *Research Methods For Business: A Skill Building Approach* (Chichester: John Wiley & Sons)
[19] Cohen J 1992 A power primer *Psychol. Bull.* **112** 155–9

Acknowledgments
Authors thank to World Class University (WCU) research scheme PPKI ITB, UGM, and ITS for funding this research.

IOP Conf. Series: Materials Science and Engineering **722** (2020) 012011 doi:10.1088/1757-899X/722/1/012011

The Smart Village Model for Rural Area (Case Study: Banyuwangi Regency)

A A Aziiza [1,2] and T D Susanto [1,2]

[1] Information System Department
[2] Institut Teknologi Sepuluh Nopember, Surabaya, Indonesia, 60111

E-mail: icha.rizia@gmail.com

Abstract Smart Village refers to a concept developed in rural area that provides solutions to problems occurred and improves the quality of life. The main problems faced by rural areas are cover poverty, low level of education, and limited access to technology. Smart village concept emerged due to some different characteristics between rural and urban areas. Banyuwangi Regency is one of regions that created smart concept starting from rural area, called smart *kampung*. So far, smart *kampung* only focused on public services, which included only a small part of smart city concept. Hence, this research was intended to propose the model of smart village examined through initial interview in village sample of Banyuwangi, literature reviews related to smart city, smart village, and smart rural. Then, the results were confirmed and adjusted to support local regulations. This research created a smart village model that was capable to be a guide for each village to develop towards better future. The proposed smart village model was categorized into 6 dimensions including 1) Governance, (2) Technology, (3) Resources, (4) Village Service, (5) Living, and (6) Tourism. This research is expected to be applied to villages in other Regencies by adjusting the characteristics of each region.

1. Introduction

Smart City-based development has become a trend of city development around the world. The development of Smart City is not just to improve the efficiency of the bureaucracy by utilizing information and communication technology (ICT), but also how to develop the community by making ICT infrastructure and facility as supporting factors or enablers [1]. Smart city is defined as an innovative city on the use of Information and Communication Technology (ICT) and other means to improve its quality of life, efficiency of urban services, and competitiveness, as well as sustainability [2].

As a form of government support on Smart City development in Indonesia, Ministry of Communications and Informatics of Republic of Indonesia in collaboration with Ministry of Home Affairs, Ministry of Administrative and Bureaucratic Reform, Indonesia Ministry of National Development Planning/*BAPENNAS*, Ministry of Public Works and Human Settlement, Presidential Staff Office initiated the idea of 100 Smart City Movement of "Gerakan menuju 100 smart city". In 2017, 25 cities/regencies were chosen to be accompanied and mentored by the experts in implementing Smart City concept. In 2019, this program targets as many as 100 cities/regencies in Indonesia that have implemented Smart City development and become models for other regions in Indonesia.

Indonesia is a country with large area, consisting of cities and regencies, urban and rural areas. The need for smart city in urban and rural areas was different, but there were some standards that must be met [3]. The development of the village area was realized through smart village concept [4]. Smart city and smart village are two different cases where the level of the broad are different regional size and problems, so that not all elements in Smart City were able to be applied to all regions in Indonesia [5]. Based on law number 6 of 2014 about the village, the total of the villages is around 73.000 villages in regencies and around 8.000 villages in cities [6]. Villagers were empowered to realize their unique ambitions by picking and choosing the aspects of modernity they wished to incorporate to their communities [7].

Villages were more than cities, so they must take smart way for the improvement and development of the country as a whole [8]. Smart Village was a new concept developed by the researchers from India, N. Viswanadham, and Sowmya Vedula. This concept was developed in 2010, by describing the ecosystem for a village and mapping the procedures of integrated design to build a smart village [9]. The concept of smart village became an alternative for the development of rural areas in the current global era [10]. "Smart village" was a solution to improve people's quality of life in rural areas [11]. In developing a smart village, the village government must set a goal, what must be developed first and what will become the supporter. The motivation behind "smart village" was technology must act as the development, bring through the education, local business opportunities, improve health and welfarea, increase democratic involvement, and improve the villager's quality of life [12]. This concept can be used as a basis for rural development in Indonesia.

The regency in Indonesia that has begun to innovate in the development of rural area is Banyuwangi Regency. In 2016, the District Government of Banyuwangi launched "Smart *Kampung*" program which was established through Regent Regulation Number 18 of 2016 concerning the Integration of Village/Urban-Based Work Programs Through Smart *Kampung* of Banyuwangi. This program designed a village to have an integrated program framework combining the use of optical fiber-based ICT, productive economic activities, creative economic activities, health-education improvement, and poverty alleviation efforts. The governance scope of "smart *kampung*" program was the implementation of public services for the community.

In 2017, the enactment of Regent Regulation Number 60 of 2017 concerning the Implementation of Smart City Master Plan through Banyuwangi Smart *Kampung* was done. This regulation was attached to Banyuwangi Smart City Master Plan discussing the reference to short-term, medium-term and long-term development programs. This Master Plan adopted the smart city model developed by Ministry of Communications and Informatics covering 6 dimensions. However, Master Plan had not provided detailed explanation related to the application of "smart *kampung*" in each village in Banyuwangi Regency.

Some legal products related to the development of "smart *kampung*" on http://jdih.banyuwangikab.go.id/ have not shown any models related to how "smart *kampung*" concept can be implemented. One legal product found was only Public Service Standard (SPM) from several villages, precisely there were 7 out of 189 villages whose SPM were on the page, including Ketapang, Cluring, Wringinrejo, Sumbersari, Plampangrejo, and Sempu Villages. Ketapang Village was the fastest growing village compared to other villages, with 35 services in 5 minutes as average standard of service. There were some differences in the number of services provided by each village, and the speed of service also varied. It showed that there were no standards used by all villages in the provision of services. In 2018, "smart *kampung*" program brought Banyuwangi Regency to the third place among 25 cities/regencies in the evaluation of phase 1 on "Gerakan menuju 100 smart city". It became the reason behind the selection of Banyuwangi Regency as the object of research. The concept of "smart *kampung*" have same subject with smart village concept. It has an innovation focusing on rural areas.

This research was intended to propose the model of smart village for villages in Banyuwangi Regency. The arrangement of this model was adjusted with the problems and characteristics of the research object, Banyuwangi Regency. The model developed was certainly supported by policies and

IOP Conf. Series: Materials Science and Engineering **722** (2020) 012011 doi:10.1088/1757-899X/722/1/012011

regulations existed in order to be implemented optimally and brought positive impacts for the development of villages in Banyuwangi Regency.

2. Methodology

The methods used to arrange a model in this research were done through preliminary interview, literature review of previous studies, books and analysis of supporting regulations. Literature review was taken by selecting journals according to the definition of the model used and the examples of its applications. Then it was continued by an analysis of local regulations dealing with the research object, Banyuwangi Regency. The research methodology can be seen on the figure 1

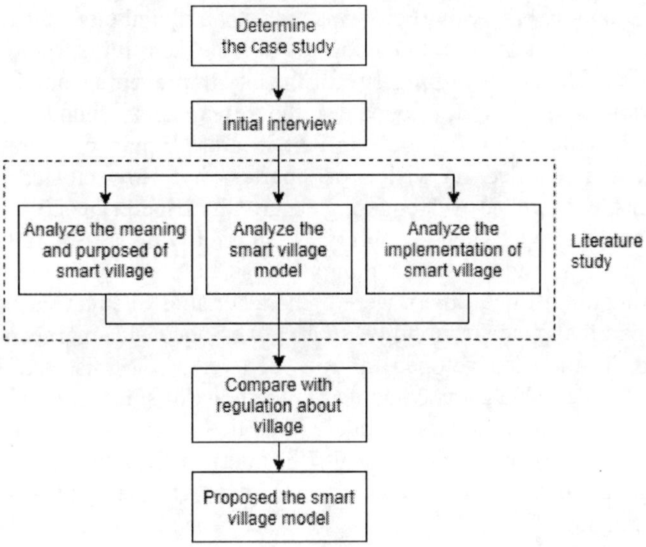

Figure 1. Research Methodology

This research carried out the preliminary interview to obtain information related to the general overview of smart *kampung* implementation in Banyuwangi Regency. The criteria of the informants were Head of Village, TI Staff or service staff and head of department or staff at related department in Banyuwangi Regency. Literature review was done afterwards to find out the definition, model and implementation of smart *kampung*.

Literature review of this research was taken from the database of electronic journals such as Scopus, IEEE, Emerald Insight, Springer, Science Direct and Google Scholar. The keywords used to find the literature were 'smart village' and 'smart rural'. The keywords used were not specified by the year. The "smart village" keyword shows that there were 78 research documents of Science direct database, 87 research documents of IEEE, 1680 research documents of Google Scholar, 84 research documents of Springer database and 36 research documents of Emerald insight database. Another keyword, "smart rural"; showed that there were 23 research documents of Science direct database, 10 research documents of IEEE, 578 research documents of Google Scholar, 127 research documents of Springer database and 8 research documents of Emerald insight database. In a total, there were 2711 research document search on "smart village" and "smart rural" around the world.

Based on these results, not every document was analyzed. The researcher verified the redundant documents and filtered based on its titles. The selection results collected 138 journal documents referring "smart village" and "smart rural". These documents then were read and analyzed based on definition, model and examples of application. There were 25 journal documents selected in the final analysis. These documents were then analyzed by using mapping method concerning the objective,

IOP Conf. Series: Materials Science and Engineering **722** (2020) 012011 doi:10.1088/1757-899X/722/1/012011

model and examples of its application. The results were confirmed by related regulations to produce smart village model of this research

3. Literature Review

This literature review was taken from books, journals, previous studies related to the smart city and smart village or smart rural.

3.1. Smart City

Smart city was an adaptive city, possessing high capacity to react; the key was on the adaptation and learning capacity, in which the citizens as the main roles in reacting, listening and receiving learning itself, this learning must be done within groups [13]. Smart city concept was operated in complex urban area, combined several complex infrastructures, human behavior, technology, social structure and politic as well as economy [14]. Smart city was more than digital city as it was able to connect the capital city physically with its social and develop the services and infrastructures of a better city by combining IT and politic vision to clear program for the city improvement and its services [15].

Smart city had various models developed by the researchers around the world. The models developed were diverse, adjusted to the needs in certain areas. Smart city firstly was introduced by IBM in 2009 on a journal entitled "A vision of smarter city: How city can lead the way into a prosperous and sustainable future" [16]. Hongkong Smart City Blue Print (2017) states that there were 6 areas of smart cities covering "Smart Mobility", "Smart Living", "Smart Environment", "Smart People", "Smart Government" and "Smart Economy". Besides, there were other models mentioned by the researcher, according to Giffinger, there were 6 characteristics of smart city covered smart people, smart environment, smart living, smart mobility, smart governance and smart economy [17].

In addition, smart city model developed in cities of overseas, Ministry of Communications and Informatics launched a model which was contained in the book of smart city guidance. This smart city model had a slight difference from the other model which lied on smart branding dimension to make a city able to display the face of each city through the tourism they had. There were 6 dimensions including: 1) smart government 2) smart economy 3) smart environment 4) smart living 5) smart society and 6) smart branding [1].

3.2. Smart Village

Smart Village was one of concepts for the developed villages in India. This concept was developed by Viswanadham and Vedula in their book entitled "Design of Smart Village" [9]. A smart village model followed a model from smart city as an effect of integrated technology changes to be implemented in the remote areas [18]. The aim of smart village was to help it solve all problems through the implementation of ICT (Information and Communications Technology) and GIS (Geographic Information System) [11].

Nowadays, Indonesia has implemented the concept of "*nawacita*" for regional development, this program prioritized the development in rural area [10]. Smart village concept focused on the role of technology in building governance and public services. Technology used by Muke and Nilesh in their research was able to be used by people lived in rural area in order to improve their quality of life [12]. The use of technology utilized by villagers was able to make them become more responsive [11]. Smart village model based on the concept of "Access to Information for Everybody" in which ICT (Information and Communications Technology) service was reached easily by villagers through IIIC program [19].

Smart village model developed by N. Viswanadham and S. Vedula was called as smart village ecosystem covering 4 aspects; 1) Institution, 2) Resources, 2) Service Chain and 4) Service delivery technologies & mechanism [9]. Besides, there were 7 focus areas in smart village including economy, ICT, people, governance, environment, living and energy [20]. Smart village existed because of ICT awareness that was able to be utilized as the instrument as the efforts of local economic development [21]. The use of technology became a main factor in creating smart village [22]. Based on Guzal-Dec, technology in smart village had important roles such as investing in the infrastructures, business development, human resources, potential and community building [23].

4. Result and Discussion

This section discussed about the preliminary interview, supporting regulation, and the proposed smart village model. Below is the description about the results.

4.1. Preliminary Innterview

The preliminary data collection was carried out by the unstructured interviews with 3 respondents consisting of Chief of village, village operator, and government staff in Banyuwangi Regency. Chief of village and IT staff were selected from village that had implemented "smart *kampung*". According to the Chief of village, "Smart *kampung*" related to online services, for example, the management of birth and death certificates. The form of smart *kampung* was a system. An indicator of smart *kampung* was the existence of a Wi-Fi network at the village office, which could be accessed freely by the villagers. Meanwhile, IT staff explained that "Smart *Kampung*" program was a program that helped to accelerate and facilitate services to the community like services that was accessed both online and offline. So that people did not need to go to Regency office which made the community easier in assessing the online and offline public services. The next respondent was the government staff, who explained that "Smart *kampung*" was a regency program that directed villages to be IT literate. Each village had its own autonomy, so that the regency government only provided infrastructure and direction for improving IT, as if the village did not have IT staff, it was permitted to carry out special recruitment of IT staff.

Based on the results of the preliminary interview, it was found that Smart *kampung* was a Banyuwangi government program for villages that already had supporting regulations. Smart *kampung* program was under the authority of Community Empowerment Service and Banyuwangi Regency Village. Smart *kampung* program was closely related to online services to the community. This program accelerated and facilitated services to the community. Banyuwangi Regency Government acted as a provider in supporting infrastructure, while smart *kampung* program was developed adjusted with the creativity and ability of each village.

4.2. The Regulation about Smart Village

The development of the smart village model was not only compiled from the literature review in previous studies, but there was a need for regulations supporting the implementation of smart *kampung*. Banyuwangi regency government issued the Regent Regulation No.18 of 2016 concerning the Integration of Village-Based Work Program Through Smart *Kampung*. Smart *kampung* was the concept of community development within the community itself to do something smart/wise in overcoming various problems through the availability of resources efficiently in the area inhabited by people who formed their own community associated with local custom and the norm being applied. This regulation states that the scope of work program in Banyuwangi Regency had seven criteria covering (1) public services, (2) economic empowerment, (3) health, (4) education, arts, and culture, (5) improvement human resource capacity, (6) poverty, and (7) legal information [24]. These aspects were taken as guidelines in developing the proposed smart village model.

4.3. Proposed Smart Village Model

The proposed smart village model was analyzed based on 30 selected journals and village regulations. The smart village model consisted of six dimensions: 1) Governance, 2) Technology, 3) Resources, 4) Village Service, 5) Living, and 6) Tourism. The proposed model can be seen on Figure 2.

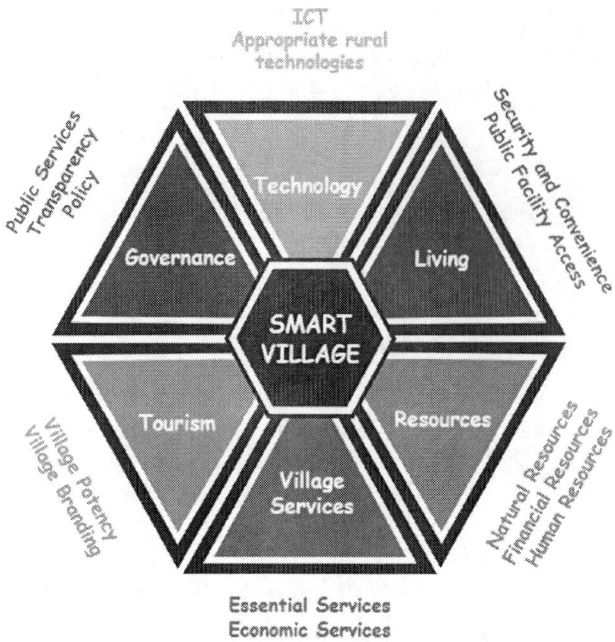

Figure 2. Proposed Smart Village Model

The model proposed on Figure 2 was analyzed based on six Smart City missions of Banyuwangi Regency as stipulated in the Regent Regulation Number 60 of 2017 concerning the Implementation of Smart City Master Plan through Banyuwangi Smart *Kampung*. The first dimension was governance. This dimension referred to Mission 1, which was to realize the effective, efficient, communicative, and innovative governance of Banyuwangi Regency government [25]. This dimension focused on developing good governance in internal and inter-village governance. Government played an important role in this dimension because he had a duty to provide good public services [20]. Governance was related to electronic services and social media in order to increase the empowerment and involvement of citizens in public management and transparency of the decision-making process [21]. The implementation of smart village for village development in Indonesia was depend on governance models that were able to implement planned strategies [26]. The governance dimension had three (3) aspects, covering public service, transparency, and policy.

Public services involved the use of ICT to provide services to the public [21]. One of the utilizations of information technology in government was e-government, which aimed to improve the quality and quantity of public services provided to the public, so that the services provided were faster [27]. In addition to public services, the governance dimension must be transparent. Public transparency refers to the openness of information accessed by the public. All information related to the village, resources, potential, budget, agenda, production results, tourism, etc [21]. Public transparency included openness of information and financial transparency. In village governance, an important role of the village head was needed as a decision maker. Leadership, chief of village played a role in encouraging community interaction and coordinating with various institutions. Community interaction was manifested in public participation in policy decision making at the village level [21].

In addition to the government, technology also played an important role in the smart village [22]. The Regent Regulation No. 60 of 2017 considers that smart cities are a concept of city management by utilizing information and communication technology effectively and efficiently to maximize services to the community [25]. Therefore, technology in smart village had important roles such as investing in infrastructure, business development, human resources, potential and building communities [23]. The existence of smart villages was due to the awareness of ICT that could be used as instruments for local economic development efforts [21]. In determining technology for smart villages, technology must have efficient, durable, inexpensive, easy to use and easy to maintain. Therefore, the use of technology was not only for optimal utilization of resources, but for carrying out sustainable village development [26]. The technologi dimension was divided into two aspects, namely ICT (Information and Communications Technology) and technology that was suitable for rural areas. The technology implemented in a smart village must be in accordance with the needs and abilities of the village itself. ICT was closely related to the concept of "smart". ICT was the core of various smart city and smart village concepts. While the technology used must be suitable to the needs of each village [26].

Besides the quantity of the implementation with technology as the vehicle, status or quality of the resource was the important factor that should be considered in building smart village [28]. According to Regulations no.6 of 2014 article 8 about Village, Village formation should have potential that covers natural resources, human resources, and economic supporting resources [6]. Resources were the village's resources consisting natural, water and energy, human, economic, as well as infrastructure resources [9] [26]. The quality of human resources in the city influenced the quality of the city [21]. In the proposed model, resource dimension was divided into three aspects that were natural, human and economic resources. The important thing that became the main focus of the village resources was the quality of sand, water, energy, and human resources [28]. Human was the essential part of smart village since they planned the social view in the village through personality, skill, creativity, and social relationship with others [29]. According to research carriead out by Rachmawati, there should be a community empowerment activity in the village development [30]. In the aspect of human resource, there were some important things that should be given attention like the education which was expected to be the encouragement of smart village [26]. In addition, the openness of information helped the village become more open to the outside world [29].

The potential of the village resources could be converted into economy involving capital, economic organization, added value, and prosperity in term of economy [31]. Whereas, the focus area in economical services should be developed under the objective to earn village income [20]. In the context of the implementation of village regulation for example, *BUMDes* (village-owned business entity) would be strong if it was build and managed by village community who had been tested in term of value and moral, and had strong social capital, was able to develop creativity and power to reach capital, network and information [31]. Besides economic services or local economic potential-based services, a series of services were provided in smart village, that were, the basic service aspect [26]. The village basic service was the main service provided to the village community such as education and health services [26] [32].

Smart living was related to the quality of life such as health condition, housing quality, education facilities, social cohesion. Quality Smart living supported civil community and social inclusion in the village [17] [21]. The dimension of *living* explained about the aspect of life that could be developed in smart village. Living was related to each individual life, this dimension focused on how someone lived their daily-life [20]. In accordance with mission 4 that was actualizing a decent, comfortable, and efficient neighborhood [25], living dimension was divided into two aspects attempting to create safety and comfort, as well as easiness to access public facility.

Smart city of Banyuwangi regency has the 3rd mission of improving regional competitiveness by structuring the face of the city and regional tourism marketing in national and international scopes [25]. Based on that mission, tourism is proposed as one of the dimensions discussed in the proposed smart village model. Tourism was the development of tourism and village promotion. The development of tourism brought a new chance of job, infrastructure, and social connection that more alive into the area [20]. If a village had potential in tourism, smart village should provide service in tourism object, accommodation, and supporting facilities to develop the tourist village [26]. The dimension of tourism was divided into two aspects that were village potential and village branding. The village identity was able to be actualized through the development of the resources starting from food production or even craft [21]. Ministry of Communications and Informatics suggested developing platform and promoting conducive and comfortable trading ecosystem as well as building and promoting the product and service of regional creative industry in marketing the face of the city [1].

IOP Conf. Series: Materials Science and Engineering **722** (2020) 012011 doi:10.1088/1757-899X/722/1/012011

Based on the explanation above, there were some indicators that should be fulfilled in developing smart village model. The dimensions, aspects, and indicators, from the proposed smart village model for Banyuwangi regency can be seen on Table 1

Table 1. The Dimension, Aspect, and Indicators of Smart Village Model

No	Dimensions	Aspects	Indicators
1	Governance	Public Services	Adminitration services
			Utilization of ICTs to provide services to the community
			Complaint Service
		Transparency	Government Information transparency
			Financial transparency
		Policy	Leadership
			Public participation
2	Technology	ICT	Internet availability
			IT infrastructure
		Appropriate rural technologies	Sensor
			Cloud Computing
3	Resources	Natural resources	Land condition
			Water availability
			Energy uses
		Financial resources	Agriculture
			Fishing
			Livestock
		Human resources	Rural community
			Level of education
			Openness
4	Village Services	Essential Services	Health services
			Education services
		Economic Services	Enterpreneurship
			Job availability
			Economic Institutions
			Distribution/Logistic facilities
5	Living	Security and Convenience	Waste management
			Environmental protection
			Public Safety
			Disaster management
		Public Facility Access	Green space facilities
			Sport area facilities
			Banking facilities
			Road and bridge facilities
6	Tourism	Village Potency	Village identity
			Tourist destinations
		Village Branding	Village branding platform
			Culture and traditions

5. Conclusion and Future Work

Based on the analyzed literature above, it can be concluded that Smart Village was the concept of village development that provided solution of village problems by utilizing technology to provide services, ease information access and provide energy access effectively and efficiently which was based on the regulation and village regulation to make the community life easier and improve its economy. In conducting the smart village model, there was a need for supporting the regulation to create solid foundation and to be implemented for the entire villages in Banyuwangi. The analysis was also conducted by doing related study to Masterplan Smart City of Banyuwangi Regency. The development of smart village should have specific models that were classified into dimensions, aspects, and indicators to reach the success of the implementation. The detail of the proposed model is on Table 1. The proposed smart village model consisted of 6 dimensions; 1) Governance, 2) Technology, 3) Resources, 4) Village Service, 5) Living, and 6) Tourism. The Banyuwangi District Government as a decision-maker can use this proposed model to develop the entire village so that it will have the same progress. Every village had its own characteristics, so its development model could not be linked to each other. Deep analysis is required toward the needs and problems within the area. Therefore, the model that was developed solved the existing problem and improved the quality of life. This proposed model is expected to be a reference for smart city developers specifically to develop a smart village model in a village that is certainly adapted to the needs of the rural area.

References

[1] Ministry of Communications and Informatics 2017 *Buku Panduan Penyusunan Masterplan Smart City 2017* (Jakarta: Organisasi)
[2] S. P. Mohanty 2017 *Everything You Wanted to Know About Smart Cities* (USA)
[3] Fajrillah, Z. Mohamad, and W. Novarika 2018 *Smart city vs smart village* (*Jurnal Mantik Penusa* vol 22) no. 1 pp. 1–6
[4] R. Novi and S. Ella 2019 *Pengembangan Model Smart Rural Untuk Pembangunan Kawasan Perdesaan di Indonesia* (*Jurnal Borneo Adm.* vol. 15) no. 1, pp. 41–58
[5] R. Rachmawati 2018 *Pengembangan Smart Village Untuk Penguatan Smart City Dan Smart Regency* (*Jurnal Sistem Cerdas* vol 01) no. 02 pp. 12–18
[6] Republic of Indonesia 2014 *Law Number 6 of 2014 about Village* pp. 1–103.
[7] J. Holmes and M. Thomas 2015 *Introducing the Smart Villages Concept* (*International Journal on Green Growth and Development* vol. 1) no. 2 pp. 151–154
[8] G. Natarajan and D. L. A. Kumar 2017 *Implementation of IoT based Smart Village for the Rural Development* (*International Journal of Mechanical Engineering and Technology* vol 8) no. 8, pp. 1212–1222
[9] N. Viswanadham and S. Vedula 2010 *Design of Smart Villages* (ISB: The Centre for Global Logistics and Manufacturing Strategies)
[10] E. Syaodih 2018 *Smart Village Development* (*The 9th International Conference of Rural Research and Planning Group*) pp. 22–33.
[11] J. Ahlawat 2017 *Smart Villages , Information Communication Technology and Geographical Information System* (*International Journal of Current Trends in Science and Technology* vol. 7) no. 8, pp. 20232–20238
[12] A. M. Muke and U. Nilesh S 2017 *Use of Advanced Technology in Developing Smart Villages* (*International Journal of Research in Engineering Science and Technology* vol. 03) no. 04, pp. 1–6
[13] A. Bozzon, J. Houtkamp, F. Kresin, N. De Sena, and M. de Weerdt 2015 *From Needs to Knowledge. A reference framework for smart citizens initiatives* (Amsterdam Institute for Advanced Metropolitan Solutions)
[14] A. Gaur, B. Scotney, G. Parr, and S. McClean 2015 *Smart city architecture and its applications based on IoT* (*Procedia Computer Science* vol. 52) no. 1, pp. 1089–1094
[15] K. A.-L. Correia, Luis M., Wünstel 2011 *Smart Cities Applications and Requirements* (Net!Works European Technology Platform)
[16] S. Dirks and M. Keeling 2009 *A vision of smarter cities: How cities can lead the way into a prosperous and sustainable future* (New York)

[17] R. Giffinger 2007 *Smart cities Ranking of European medium-sized cities* (*Center of Regional Science,* Vienna University of Technology vol. 16) pp. 13–18

[18] P. Abinash and J. Josephine 2018 *Internet of Things (IoT) for Smart Village* (*International Conference on Advancements in Engineering, Technology and Sciences (ICAETS)*) pp. 813–819.

[19] T. Kamal, F. J. Tuli, M. Hassan, T. H. Rupam, and B. W. Habib 2018 *Information , Innovation and Implementation Center (IIIC): Concept towards Smart Village* (Researchgate)

[20] M. Mishbah, B. Purwandri, and D. I. Sensuse 2018 *Systematic Review and Meta-Analysis of Proposed Smart Village Conceptual Model : Objectives, Strategies, Dimensions, and Foundation* (*2018 International Conference on Information Technology Systems and Innovation (ICITSI)*) pp. 127–133.

[21] A. D. Santoso *et al.*, 2019 *Desa Cerdas: Transformasi Kebijakan dan Pembangunan Desa Merespon Era Revolusi Industri 4.0.* (Yogyakarta: Center for Digital Society UGM)

[22] A. Singh and M. Patel 2018 *Achieving Inclusive Development Through Smart Village* (*PDPU Journal of Energy and Management* vol. 3) no. 1, pp. 37–43

[23] D. Guzal-Dec 2018 *Intelligent Development Of The Countryside – The Concept Of Smart Villages: Assumptions , Possibilities And Implementation Limitations* (*Economic And Regional Studies (Sciendo)* vol. 11) no. 3, pp. 32–49, 2018.

[24] Banyuwangi District Government 2016 *Local Government Regulation Number 18 of 2016 about the Integration of Village/Urban-Based Work Programs Through Smart Kampung,* pp. 1–13.

[25] Banyuwangi District Government 2017 *Local Government Regulation Number 60 of 2017 about the Implementation of Smart City Master Plan through Banyuwangi Smart Kampung*, pp. 1–5.

[26] S. Ella and R. N. Andari 2018 *Developing a Smart Village Model for Village Development in Indonesia (International Conference on ICT for Smart Society (ICISS))*, pp. 1–6.

[27] A. B. Marlintha 2017 *Design And Implementation Of Smart Village Mapping Geographic Information System Based Web In The Cinunuk Village* (*The 2017 IEEE Asia Pacific Conference on Wireless and Mobile (APWiMob))*, pp. 66–71

[28] Ramachandra, G. Hegde, S. C. M. D, T. A. Kumar, and V. Swamiji 2015 *SMART Ragihalli: Effort towards Self-reliant & Self-sufficient system empowering Man power (rural youth) with Appropriate Rural Technologies* (Bangalore: ETR 90, Energy & Wetlands Research Group, CES)

[29] R. Santhiyakumari, N., Shenbagapriya, M., Hemalatha 2016 *A Novel Approach in Information and Communication Villages* (Humanitarian Technology Conference (R10-HTC), 2016 IEEE Region 10)

[30] D. Rahmawati, H. Sulistyarso, P. G. Ariastita, M. Yusuf, and D. A. Paramasatya 2018 *Smart kampung for Surabaya smart city : Criteria redefined* (IOP Conf. Series: Earth and Environmental Science 202) pp. 1–7.

[31] Ministry of Village, Develpoment of Disadvantaged Regions, and Transmigration, Republic of Indonesia 2015 *Indeks Desa Membangun*

[32] N. Viswanadham and S. Kameshwaran 2013 *Smart Villages and Cities* (*Ecosystem-Aware Global Supply Chain Management*) (Bangalore, India: World Scientific Publishing) pp. 175–192.

Comparison Clustering Performance Based on Moodle Log Mining

C Pradana, S S Kusumawardani, and A E Permanasari

Department of Electrical Engineering and Information Technology, Universitas Gadjah Mada, Yogyakarta 55281, Indonesia

E-mail: christian.pradana.1@mail.ugm.ac.id, suning@ugm.ac.id, adhistya@ugm.ac.id

Abstract. The use of digital data now saves a lot of information, but still raw and not in the form of knowledge that can be directly seen, therefore we need a data processing to get useful particular understanding from raw data. Data Mining has an important role to process and find useful information from data. This process is also called Knowledge Discovery. Educational Data Mining is a Knowledge Discovery process in the world of education using data mining techniques. One method used in data mining is clustering. By using clustering analysis techniques, data can be grouped into groups without the need for prior knowledge (previous knowledge), so that the data is grouped based on similarity of patterns. This paper compares the K-means, Hierarchical and Louvain clustering methods to see the most appropriate clustering technique in analyzing log activity data in Moodle Learning Management System (LMS). The results of clustering are measured using the Silhouette Coefficient, and then we compare the values and distribution between clusters. In conclusion, Hierarchical clustering produces the highest Silhouette Coefficient value and also this algorithm can detect outlier data as new cluster. Louvain clustering perform very well to find cluster groups in new dataset as the algorithm does not required the number of clusters to be specified before. Louvain clustering can divide more evenly and precision compare to K-means and hierarchical cluster techniques.

1. Introduction

Data is now increasingly important and widely used. Almost all aspects of digital store data in database storage. Data stores a variety of valuable information, but the data still consists of various kinds of diverse information. To find the important information, a data collection technique known as data mining is needed. One area that can use data mining is education. User visits on a page and how long users spend on a page is closely related to user interest in using the website. Web Usage Mining (WUM) is a field that utilizes statistics, machine learning, and data mining algorithms in various types of educational data (1). Various types of variables can be analyzed, for example how long students can complete an assignment and how students interact with e-learning, choosing an analytical method must be done carefully (2). The availability of abundant data raises, so it needs to be able to utilize the information and knowledge. A way to obtain the information from data is using data mining techniques. Therefore, the process of data mining is also called the process of finding knowledge or we know as Knowledge Discovery (KD) (3). The world of education cannot be separated from the use

IOP Conf. Series: Materials Science and Engineering **722** (2020) 012012 doi:10.1088/1757-899X/722/1/012012

of technology, from the Learning Management System (LMS), students study report information, to the administration of student data. Modular Object-Oriented Development Learning Environment (Moodle) is a web-based LMS that is commonly used to help learning. All of used of this kind of technology produces very large data, and can be utilized to seek knowledge from these data, therefore Educational Data Mining is needed. The use of data mining techniques on student data also can help the university to do data analysis and determine the promotion strategy plan for marketing to recruit new student (4). By using clustering techniques, the results of grouping students can be obtained, and the results can be used to determine the accurate promotion strategy.

This paper will discuss clustering algorithms for analyzing logs in the Moodle Learning Management System (LMS). Literature review from various aspects from the use of data mining in the field of education, discusses methods, tools, as well as processes commonly used in Web Usage Mining, and also process of finding information through data mining from website data logs in Moodle LMS. The final result of analyzing log data we can search for patterns and explore user groups using LMS.

This paper consists of six parts, in the initial part discussing the role of using and explaining data mining in the field of education. The second part discusses the explanation and description of clustering algorithms and evaluating clustering methods using the Silhouette Coefficient. Henceforth the third section discusses surveys from the literature paper relating to clustering in Educational Data Mining. And in the fourth section discusses methodology and testing of clustering algorithms on the datasets that have been selected in this study. In the fifth section is the result of the research paper, and finally, discussion in section six discusses the conclusion of the results and identification for future work and those related to research development.

2. Background

Data is not always in an ideal condition for analysis, therefore special techniques for data analysis are needed, so it becomes information and knowledge. One of the techniques commonly used is clustering, this technique divides by partitioning a group of data into groups. Data grouped into clusters based on similarity features, and distinct characteristics from other clusters. Before analyzed process, data must first be normalized or original value depend on the case, after the process we can continue analyzed the dataset. There are various types of tools used for Web Usage Mining (WUM), these tools can be used for data normalization, statistical data analysis, after that the results of the data that has been processed can be visualized. In general, there are three processes carried out for WUM, the first is data manipulation, algorithmic analysis, data visualization.

1. Data manipulation, in this section raw data must first be prepared, data must be cleaned (data cleaning), organized (organizing data), and create data formats (creating data). Tools are used to manipulate data are Microsoft excel/google sheets, EDM workbench, Phyton and Jupyter notebook, SQL, logstash.
2. Algorithm Analysis, after data preparation process has been completed, data ready for analyzed process, some tools are used at this stage are Rapid Miner, WEKA, SPSS, KNIME, Orange3, KEEL, Spark MLLib, elasticsearch.
3. Data visualization, after clean data already completely processed, we applied the algorithm and get visual form of data so it is more easily analyzed. Tools used for data visualization include Tableau, D3.js, Orange3 Kibana.

2.1. Clustering Method

Clustering method is a data mining technique that used to analyze data in an unsupervised manner, this technique will divide data into sub data that have been grouped based on similarity relationships. Clustering can make groups of data based on the similarity of data types into one group and separate them from data that has differences from another group of data. In this paper we focus on three types of techniques, Partition, Hierarchical proposed by Kim and Han(5) and Han *et al*(6) which have been used frequently in various studies, and Modularity Method proposed by Blondel *et al*(7) which is used to detect groups in large networks in this study will be compared with Partition and Hierarchical clustering algorithms.

1. Partition: This algorithm will create a group of objects o into a partition c (c <= o) where each partition represents the same object, and which is not the same as the object in another cluster. Each cluster is represented by a centroid.
2. Hierarchical: This hierarchical method will create groups of data, into a cluster tree or hierarchy.
3. Modularity Method is using community detection algorithm, this method assign node as a communities by evaluating how much more densely connected the nodes within the community to compare with other random network, and then maximize modularity score for each community in large network (7).

2.2. Clustering Algorithm

1. K-Means is a clustering algorithm using a partitioning method to analyze data that involves grouping data by partitioning. This algorithm will make the object group o into the c position (c <= o).
2. Hierarchical Clustering is a clustering algorithm that uses an agglomerative hierarchical approach and has good performance on large datasets with multiple large dimensions.
3. Louvain Clustering is clustering algorithm using modular communities for detecting large network, it maximized modularity score and partition size for each community(7).

2.3. Clustering Validation

Cluster validation technique evaluate the performance of a clustering algorithm, consisting of internal and external evaluations. External evaluation is done by comparing the results and ground truth, for example comparing the results of clustering with public class labels to determine how well the cluster technique performs with actual data groups on the available dataset(8). Internal techniques evaluate the similarity of one object in a cluster and measuring the difference between one cluster with another cluster. Silhouette Coefficient (SC) is an internal validity measure that evaluates clustering performance base on the pairwise difference between and within cluster distance(9). The results are between -1 and 1, the closer to 1, the closer to the similarities between entities, conversely, if -1, the less the same between entities.

3. Previous Work

In section three provides an overview various kind of research using data mining methods that are applied to Moodle logs using Web Usage Mining techniques. Research collected from paper journals and conferences. From the literature review it can be seen, certain methods can be used in Moodle logs and the results are used to gain knowledge. Generally, the clustering method is widely used to classify student data based on certain attributes(4)(10)(11)(12)(13). And the results of this grouping can be used for promotion strategies(14) or predictions of student grades. In addition, the Vector Space Model algorithm method can be used to view learning behavior of Learning Management System users(11). Apriori Association algorithm can be used to see the connectedness of data, an example of application in the world of education can be seen in the influence of the relationship between student data and student grades(15). To make predictions of the value of user behavior can be done using the Linear Temporal Logic algorithm, so that the relationship between the two attributes can be seen(16).

Analysis of student access time can be done using the Time Series Cross Section clustering method. By using this method can be seen how many users at each unit of time. So, in the end it can be seen user behavior based on access time(17)(18). For the initial stage of analysis, the data can be analyzed using statistics(19), to see the characteristics of the data, so that the advanced method can be chosen based on the initial analysis of the data. Linear Regression and Multiple Regression methods can be used to get predictive results of the final grade and the number of students taking a course(20)(21). Taking data values from LMS, then the value is used to determine the classification of student groups based on the scores obtained. Values that do not meet the limit, will get a module that is different from the value reaching the limit. Less value groups will be given a module to help to increase grades. After that, a cluster is created based on the value group. After that the results from the cluster students with the lowest grades are given additional tutoring. From the table below, the literature review grouped by method using the Moodle log dataset.

Table 1. Survey of Clustering Analysis for EDM

SURVEY PAPER LIST				
	Paper	Year	Method	Dataset
1	Bovo et al(22)	2013	10-Fold Cross Validation combine with Expectation maximization hierarchical clustering, Simple K-means, X-means	e-learning training (administrative, on-site training, contact, communication history) and logs from Moodle database
2	Maheswari et al(23)	2014	Greedy clustering using belief function	Collage website log
3	Pradas et al(15)	2015	Multiple regression Analysis	Moodle interaction data log in master degree program at Universidad a Distancia de Madrid
4	Gokhan(24)	2015	K-means clustering algorithm	Student Moodle logs
5	Mlynarska et al (25)	2016	Time series clustering with dynamic time warping	A year of Moodle activity log for 16 computer science courses University College Dublin
6	Alvares et al (8)	2016	Linear Temporal Logic	Moodle log data from course at The University of Zaragoza
7	Dobashi(9)	2017	Time series cross section analysis	Moodle course log
8	Estacio et al (10)	2017	Vector Space Model algorithm	Moodle log data from various courses at Jose Rizal University
9	Kurniawati(14)	2017	K-means clustering	Students data from Institut Sains dan Teknologi Al-Kamal
10	Hervianti (4)	2018	K-means clustering	Students data Universitas Gunadarma
11	Kadoic et al (11)	2018	Statistical analysis included ANOVA, Chi-square test, linear regression	Moodle course log from "Business Decision Making" at University of Zagreb
12	Prastyo et al (12)	2018	K-means clustering	Student data from mahasiswa Politeknik Negeri Ujung Pandang
13	Ikraith-Informatika(13)	2018	Association Apriori Algorithm	Kalam Kudus Student data

4. Methodology

The data taken contained 96,127 logs, generated from 863 users in 139 courses, which were available between August 1, 2017 and December 31, 2017. To find information from these log data, clustering analysis was carried out on several courses. Data consists of columns, index 1, index 2, course, time, user, action, info as shown in Figure 1.

INDEX 1	INDEX 2	COURSE	TIME	USER	ACTION	INFO
0	-	C001	2017-08-01 19:48:40	name1	assignment upload	info1
2	2	C001	2017-08-26 05:27:49	name2	resource view	info2
3	0	C002	2017-09-01 23:36:00	name3	assignment view	info3
..	...	...	..	...	..	...
33587	21546	C003	2017-09-30 00:00:28	name4	course view	info4
..	...	...	...	....	...	...
96126	19970	C005	2017-12-31 08:43:22	name5	resource view	info5

Figure 1. Moodle Log

Not all available courses have a view assignment and an upload assignment. The courses are chosen based on the number of activities, interactions between lecturers who give assignments and students who complete the task to see the performance of each clustering algorithm. Four courses with the highest activity were selected in this study, C001 with 137 users and 3873 actions, C002 with 97 users and 5928 actions, C003 with 94 users and 3970 actions, C004 with 79 users and 3067 actions.

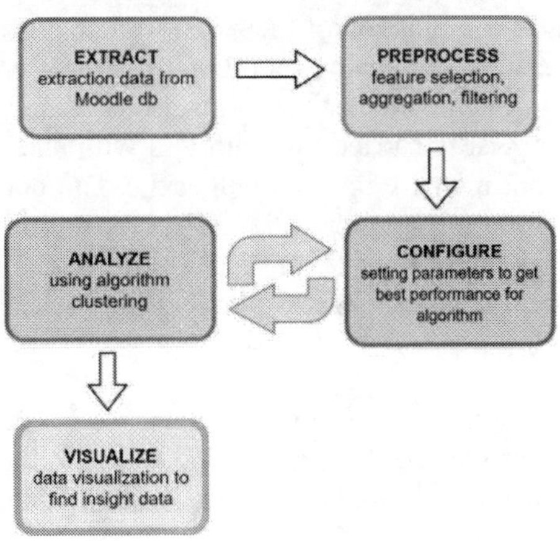

Figure 2. Step for Performing Clustering Evaluation

First, we extract data from log activity from Moodle LMS as shows in Figure 2. At the pre-processing stage, data collection, feature selection, and data filtering are selected. At the collection stage, data from each user activity on one course is collected to produce a feature vector, which consists of resource view, course view, assignment view, upload assignment. The activity from this dimension was chosen to see every interaction between lecturers in giving assignments and students in completing assignments. Each of these activities becomes the dimension of every user in new data table.

Table 2. Feature vector generated from data summarization of Moodle log data

Dimension	Description
assignment upload	Aggregate all uploading file action for each user within the course.
assignment view	Aggregate all assignment view action for each user within the course.
course view	Aggregate all course view action for each user within the course.
resource view	Aggregate all resource view action for each user within the course.

Table 3. Four-dimension data after pre-processing

index	user	assignment upload	assignment view	course view	resource view
1	user1	2	9	16	21
2	user2	3	13	9	13
3	user3	6	10	5	9
...	...	...	...	...	...
50	user50	2	6	10	23
...	...	...	...	...	...
94	user94	6	9	52	10

The study was conducted by using a combination of these four dimensions in Table 2, then the feature selection was made into the dimensions that can be found in Table 3. It became a new dimension that could be included in the calculation of the cluster algorithm for the analysis phase of the test. In the next step, an algorithm comparison test is carried out with and without data normalization to see the comparison result for each algorithm method, to see whether the data normalization has a significant effect on the test results. The function of data normalization is to minimize the value that shows big differences. It will help clustering algorithm to produce more accurate values. The results are documented and displayed in the next chapter in the results section, in the form of tables and graphs.

5. Result

The results table in this chapter shows the results of K-means, Hierarchical, Louvain clustering algorithm performance tests on courses C001, C002, C003, C004. Each algorithm is tested with normalization and without normalization on the dataset. The results table is displayed for comparing each other with normalized and non-normalized algorithm.

Table 4. Result of comparing clustering performance

		K-Means	K-Means Normalization	Hierarchical	Hierarchical Normalization	Louvain	Louvain Normalization
C001	C1	0.480	0.497	0	0	0.484	0.664
	C2	0.673	0.687	0	0	0.235	0.241
	C3	0.397	0.498	0.812	0.668	0.729	0.545
	C4	0	0	0	0	0.365	0.590
C002	C1	0.585	0.522	0	0	0.545	0.339
	C2	0.438	0.401	0.593	0.408	0.344	0.398
	C3	0.465	0.375	0.674	0.650	0.483	0.439
C003	C1	0.582	0.556	0	0	0.203	0.209
	C2	0	0	0	0	0.547	0.400
	C3	0.588	0.371	0.594	0.560	0.474	0.332
	C4	0.466	0.480	0.479	0.627	0.487	0.445
C004	C1	0.502	0.475	0.593	0.408	0.485	0.465
	C2	0.530	0.414	0	0	0.363	0.420
	C3	0.386	0.344	0.674	0.650	0.550	0.463

Table 4 shows on the C001 course, K-Means does not produce a significant difference in results between non-normalized and normalized data. In hierarchical there is a decrease in the value of the silhouette if the data is normalized. The effect of normalization on the Louvain algorithm can be seen in cluster 1 (C1) and cluster 4 (C4), but there is a decrease in value in cluster 3 (C3).

On the C002 and C003 courses as shown in Table 4 there is no significant difference between normalized and non-normalized data. Slightly differences in value of C002 is seen in Louvain clustering where there is a decrease in normalized value in cluster 1 (C1). In course C003, decrease value occurs in K-Means algorithm with normalized in cluster 3 (C3), Louvain with normalized in cluster 2 (C2), cluster 3 (C3). But increase value found in hierarchical with normalized in cluster 4 (C4). The silhouette coefficient course C004, decrease value found in all normalized algorithm, while in Louvain increase value found only in cluster 2 (C2) with normalized data.

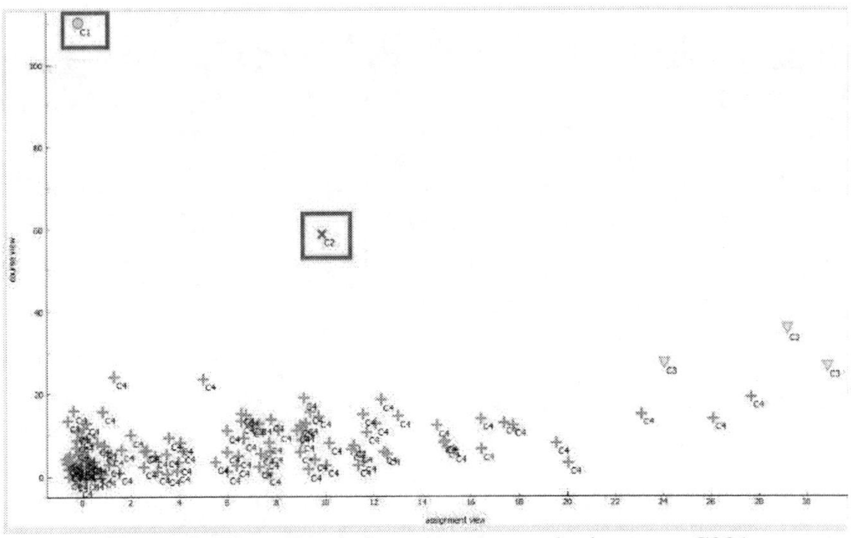

Figure 3. Hierarchical clustering scatter plot in course C001

In the Figure 3 shows the results of the hierarchy clustering in course C001 display with scatter plot, detect the outlier data into separate clusters C1 and C2 with less activity in assignment view and upload but high activity in resource view, so that it can be used for detecting lecturers or students who have different or anomaly activities.

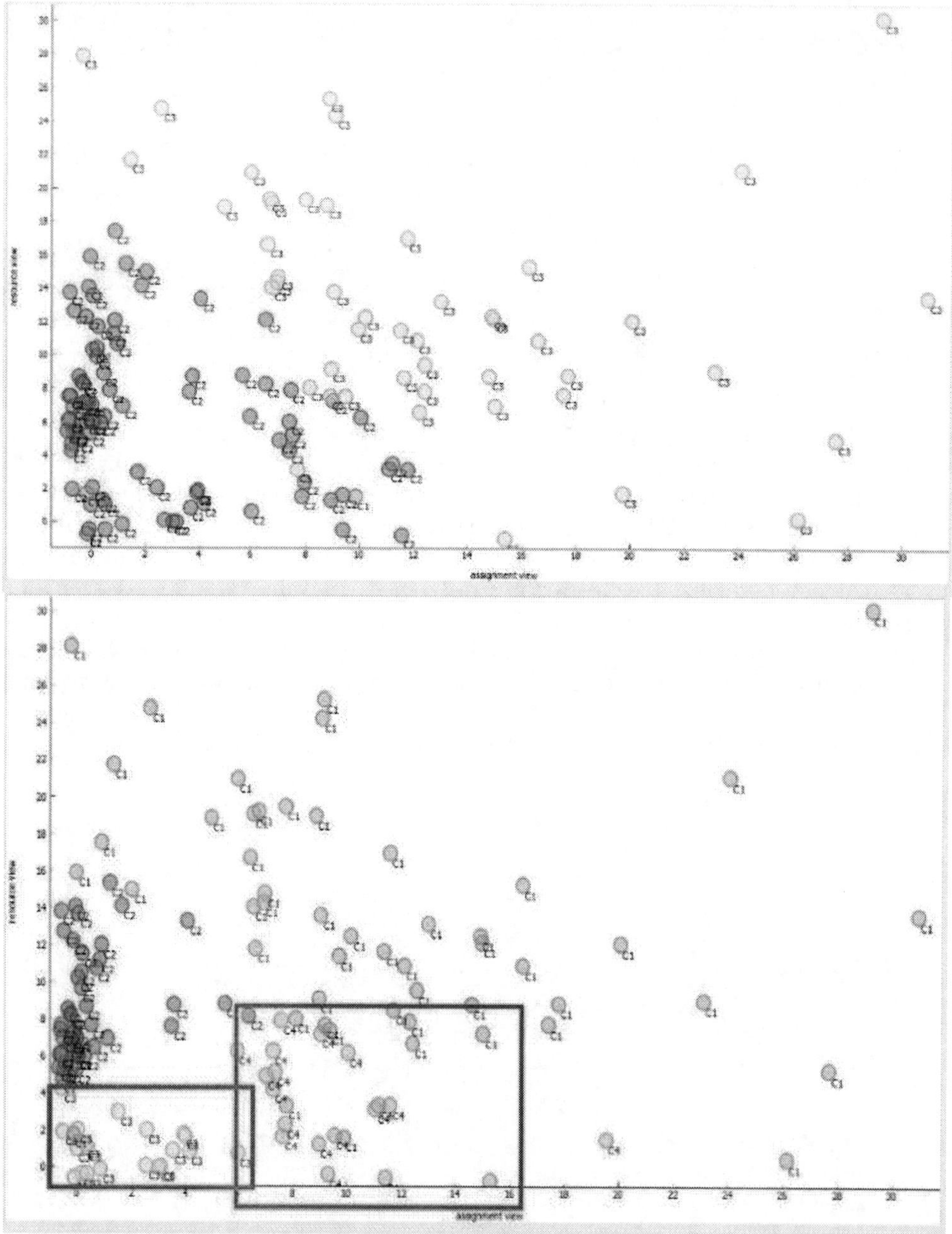

Figure 4. Comparison K-Means and Louvain clustering in scatter plot in course C001

Louvain algorithm can make the cluster more evenly distributed compared to K-Means algorithm in every course as display in Figure 4. Louvain's algorithm can create new clusters in groups of data that are not detected by the K-means algorithm.

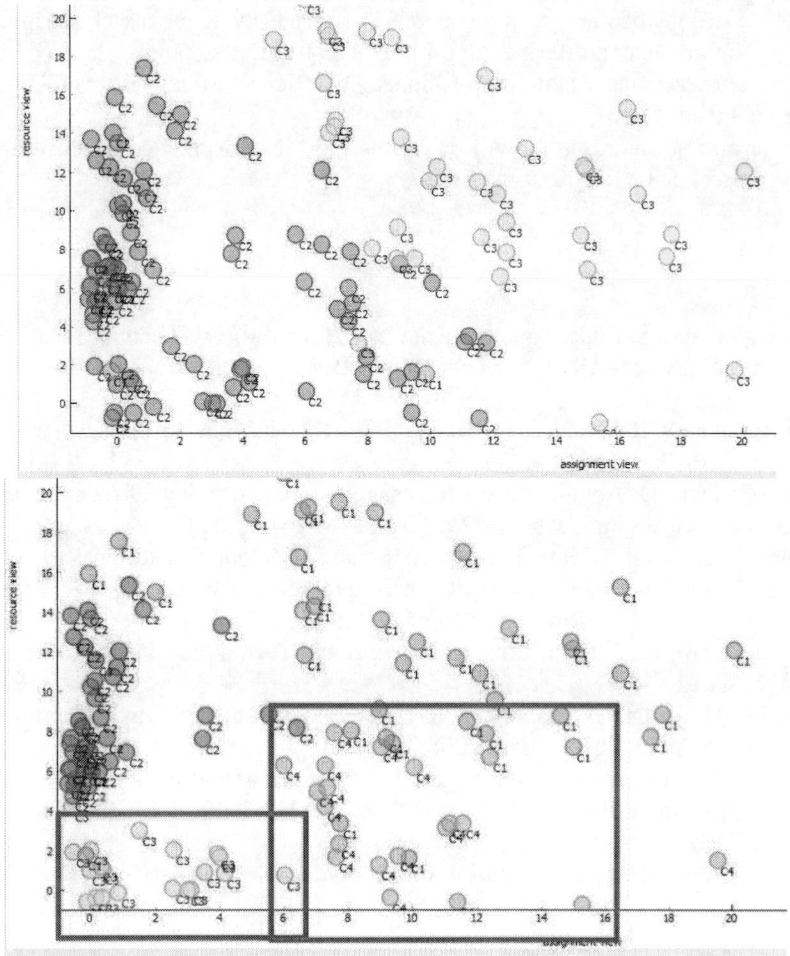

Figure 5. Comparison K-Means and Louvain clustering in scatter plot detail in course C001

Figure 5 above displays, for example user n has ten assignment view features, six resource views, two upload assignments, two course views. By using K-Means algorithm user n is a part of cluster 2 (C2), while using Louvain algorithm user n is a part of cluster 3 (C3). Louvain algorithm detected clusters more evenly, as seen in cluster 3 (C3) and cluster 4 (C4). Louvain algorithm created cluster 3 (C3) with less assignment and resource view activity and cluster 4 (C4) with medium assignment view and less until medium activity resource view, but K-Means algorithm created only two cluster. Normalized and non-normalized data does not become significant to the results of clustering.

6. Conclusion

To find information from a set of data, analysis is needed so that we need analysis technique, data mining can help find information from abundant data set. Cluster analysis can be used to analyze data without the need for previous knowledge. However, choosing the right clustering algorithm determined by knowledge of the data structure, the type of analysis desired, and the size of the dataset being evaluated.

The purpose of this study is to look at the performance of clustering algorithms by comparing the three types of clustering algorithms. The experiment result confirm that Louvain clustering algorithm performed better than Hierarchical and K-means clustering algorithm. Louvain clustering has an accuracy to detect groups of data, which cannot be detected by K-means clustering. The interesting performance of Hierarchical clustering algorithm, it can point anomaly data or outliers to create cluster. K-means clustering can obtain high silhouette coefficient values on several clusters that have a number of adjacent data.

A future extension of this research will be comparing the performance of clustering on a database with a number of users and a longer usage time.

References

[1] Romero C, Ventura S. Educational data mining: A review of the state of the art. Vol. 40, IEEE Transactions on Systems, Man and Cybernetics Part C: Applications and Reviews. 2010. p. 601–18.

[2] Dutt A, Ismail MA, Herawan T. A Systematic Review on Educational Data Mining. Vol. 5, IEEE Access. 2017. p. 15991–6005.

[3] Omar R, Md Tap AO, Abdullah ZS. Web usage mining: A review of recent works. 2014 5th Int Conf Inf Commun Technol Muslim World, ICT4M 2014. 2014;

[4] Hervianti SP. Universitas Gunadarma Menggunakan Metode Clustering. 2010;

[5] Kim M-S, Han J. A particle-and-density based evolutionary clustering method for dynamic networks. Proc VLDB Endow. 2014;2(1):622–33.

[6] Han J, Kamber M, Pei J. Data Mining: Concepts and Techniques. Data Mining: Concepts and Techniques. 2012.

[7] Blondel VD, Guillaume JL, Lambiotte R, Lefebvre E. Fast unfolding of communities in large networks. J Stat Mech Theory Exp. 2008;2008(10).

[8] Alvarez P, Fabra J, Hernandez S, Ezpeleta J. Alignment of teacher's plan and students' use of LMS resources. Analysis of Moodle logs. 2016 15th Int Conf Inf Technol Based High Educ Training, ITHET 2016. 2016;

[9] Dobashi K. Automatic data integration from Moodle course logs to pivot tables for time series cross section analysis. Procedia Comput Sci [Internet]. 2017;112:1835–44. Available from: http://dx.doi.org/10.1016/j.procs.2017.08.222

[10] Estacio RR, Raga Jr RC. Analyzing students online learning behavior in blended courses using Moodle. Asian Assoc Open Univ J [Internet]. 2017;12(1):52–68. Available from: http://www.emeraldinsight.com/doi/10.1108/AAOUJ-01-2017-0016

[11] Kadoic N, Oreski D. Analysis of student behavior and success based on logs in Moodle. 2018 41st Int Conv Inf Commun Technol Electron Microelectron MIPRO 2018 - Proc. 2018;654–9.

[12] Prastyo PH, Saharuna Z, Abduh I. Analisis Data Atribut Mahasiswa untuk Menentukan Strategi Promosi Kampus Analisis Data Atribut Mahasiswa untuk Menentukan Strategi Promosi Kampus Menggunakan Metode Data Mining. 2018;(October).

[13] Sekolah N, Apriori A. Penerapan algoritma apriori untuk menemukan hubungan data murid dengan nilai sekolah. 2018;2(18):7–12.

[14] Kurniawati I, Indrajit RE, Fauzi M. Peran Bussines Intelligence Dalam Menentukan Strategi Promosi Penerimaan Mahasiswa Baru. 2017;1(2):70–9.

[15] Iglesias-Pradas S, Ruiz-De-Azcárate C, Agudo-Peregrina ÁF. Assessing the suitability of student interactions from Moodle data logs as predictors of cross-curricular competencies. Comput Human Behav [Internet]. 2015;47:81–9. Available from: http://dx.doi.org/10.1016/j.chb.2014.09.065

[16] Lerche T, Kiel E. Predicting student achievement in learning management systems by log data analysis. Comput Human Behav [Internet]. 2018;89(June):367–72. Available from: https://doi.org/10.1016/j.chb.2018.06.015

[17] Riofrio-Luzcando D, Ramirez J, Berrocal-Lobo M. Predicting Student Actions in a Procedural

IOP Conf. Series: Materials Science and Engineering **722** (2020) 012012 doi:10.1088/1757-899X/722/1/012012

Training Environment. IEEE Trans Learn Technol [Internet]. 2017;10(4):1–1. Available from: http://ieeexplore.ieee.org/document/7833102/

[18] Appalla P, Kuthadi VM, Marwala T. An efficient educational data mining approach to support e-learning. Wirel Networks. 2017;23(4):1011–24.

[19] Psaromiligkos Y, Orfanidou M, Kytagias C, Zafiri E. Mining log data for the analysis of learners' behaviour in web-based learning management systems. Oper Res. 2011;11(2):187–200.

[20] Zaffar M, Hashmani MA, Savita KS. Performance analysis of feature selection algorithm for educational data mining. 2017 IEEE Conf Big Data Anal [Internet]. 2017;7–12. Available from: http://ieeexplore.ieee.org/document/8284099/

[21] Kavitha G, Raj L. Educational Data Mining and Learning Analytics Educational Assistance for Teaching and Learning. Int J Comput Organ Trends [Internet]. 2017;41(1):21–5. Available from: http://www.ijcotjournal.org/archive/ijcot-v41p304

[22] Bovo A, Sanchez S, Heguy O, Duthen Y. Clustering moodle data as a tool for profiling students. 2013 2nd Int Conf E-Learning E-Technologies Educ ICEEE 2013. 2013;4(3):121–6.

[23] Maheswari BU, Sumathi P. A new clustering and preprocessing for web log mining. Proc - 2014 World Congr Comput Commun Technol WCCCT 2014. 2014;25–9.

[24] Gökhan A. Profiling {Students}' {Approaches} to {Learning} through {Moodle} {Logs}. Proc {Multidisciplinary} {Academic} {Conference} {Education}, {Teaching} {E}-learning {Prague} 2015, {Czech} {Republic} ({MAC}-{ETeL} 2015) [Internet]. :7. Available from: https://books.google.co.ma/books?id=SkAmCwAAQBAJ

[25] Młynarska E, Greene D, Cunningham P. Time series clustering of Moodle activity data. CEUR Workshop Proc. 2016;1751:104–15.

IOP Conf. Series: Materials Science and Engineering **722** (2020) 012013 doi:10.1088/1757-899X/722/1/012013

Effect of listening to Quran recitation on workload and driving performance: A car simulator study

H Purnomo [1], and D Setiawan [1]
[1] Department of Industrial Engineering, Faculty of Industrial Technology, Universitas Islam Indonesia

E-mail: Danang.setiawan@uii.ac.id

Abstract. Both mental overload and mental underload can affect driver performance and lead to traffic accidents. This study examined the effect of Quran recitation, one of the alternative voices often heard during driving, against the mental workload. The frontal lobe EEG, i.e., AF3 and AF4, was used as an indicator of the mental workload. Experiments were carried out in four scenarios, i.e., simple road with Quran recitation, simple road without Quran recitation, complex road with Quran recitation and complex road without Quran recitation. EEGLAB toolbox was used to process raw EEG and to perform statistical parametric test. Paired t-test statistics (p-value 0.05) were applied to compare the scenarios. The statistical test showed that there was statistically significant difference in the alpha and beta wave when the driver was given the Quran recitation both in simple and complex road. An increase in alpha wave was an indication of the increased in focus and relaxation, which can lead to improved driver performance. This finding can be used as an alternative of secondary task that is save for drivers. quality of dried Bilih fish can be determined on the basis of its water and protein content.

1. Introduction

Research showed that 40% of accidents were caused by inappropriate secondary tasks carried out during driving [1]. Inappropriate secondary tasks have an impact on increasing mental workload and lead to mental overload [2]. Humans have limited cognitive abilities, so that the occurrence of mental overload will result in a decrease in the performance of the main activities [3].

On the other hand, the absence of secondary tasks, can lead to mental fatigue, drowsiness and saturation which causes 30% of accidents [4]. Sleepiness and saturation have an impact on decreasing alertness [5], reaction time [6] and overall driver performance [7]. Research showed that the main cause of mental fatigue, drowsiness and saturation is mental underload [8].

As described above, mental overload and mental underload, both can lead to traffic accidents. If the driver is in a situation that leads to mental underload, the driver's mental workload can be increased by active or passive intervention. However, inappropriate intervention performed by the driver, e.g., using mobile phone, can lead to mental overload and decreased driver's performance [2].

Research in the field of traffic safety using driving simulator has been carried out extensively, where one of the objectives is to look for interventions that can lead to improved driver performance. Some studies tried to evaluate active interventions, e.g., drinking coffee, smoking, chewing candy, exercise, listening to audiobooks, listening to music and singing. Some studies also evaluate the impact of passive interventions, e.g., cabin lighting, alarms, music, air and aroma.

[1] To whom any correspondence should be addressed.

This study aims to evaluate the effect of voice intervention on mental workload and driver performance. Voice intervention was chosen because most of the drivers listen to music or radio while driving [9]. Previous research has evaluated some types of voice intervention during driving, e.g., music [10], [11], [12], radio [4] and audiobooks [8].

This study evaluates the effect of the Quran recitation on workload and driving performance. The Quran intervention was chosen because the Quran recitation is an alternative voice that often heard during driving, especially in Muslim-majority countries. Based on the study literature performed, there are no studies that evaluate the effect of Qur'an recitation on driving. Research found that listening to Quran recitation can increase alpha wave and lead to relax condition compared with listening to slow and hard music [13]. This research evaluated the Quran recitation on two road conditions, i.e., simple and complex roads. A simple or monotonous road is a representation of a mental underload condition that leads to sleepiness and saturation. While the complex road is a representation of road conditions that require high concentration and focus.

2. Research method

2.1. Research participant

Five students, between the ages of 18-21 years, be involved as research participants. Participants were required to have a driving license for at least 1 year and driving average an hour per day. Participants were also required not to have hearing loss. From these requirements, it was assumed that all respondents had the same driving ability.

Before participants were involved in the experiments, they had to fill out a pre-experiment questionnaire. The questionnaire consisted of general respondents' information, driving experience, and simulator sickness questionnaires (SSQ). The SSQ questionnaire was intended to obtain the SSQ symptoms experienced by respondents, before and after the experiment.

2.2. Experimental Design

This study used a combination of two treatments, i.e., without Quran and with Quran, each performed on two roads condition, i.e., simple and complex roads. The combination of the voice intervention and the road condition produced four experimental scenarios. For each scenario, participants were asked to drive for 35 minutes, referring to the previous research [8]. To avoid the possibility of a learning effect, the participants were randomly assigned to the certain route. The Quran recitation was run at a moderate sound level - 70 dBA. This number referred to the previous study which states that there is no significant difference between loud (85 dBA) and moderate (70 dBA) sound to driver performance [11].

Experiment was carried out at the Laboratory of Work Systems and Ergonomics, Industrial Engineering, UII. The experiment used a car simulator facility, which had the following specifications: Logitech G29, 49" monitor super ultra-wide 32: 9 curved, and simulation software City Car Driving 1.5.7. Emotive Insight, a wireless and 5 channel EEG tools, was used as a measuring devices of Electroencephalograph (EEG).

EEG waveforms are generally classified according to their frequency, altitude, shape and position of the electrodes on the scalp [14]. The familiar classifications of waveforms are based on signal frequency, such as delta (0.1-4Hz), theta (4-8 Hz), alpha (8-13 Hz), beta (13-30 Hz) and gamma (30-100 Hz). Brain wave frequency differs corresponds to different behaviour and mental states of the brain. Delta wave is the slowest brain wave that represents stages of sleep. Theta wave is related to subconscious activity that founds in deep relaxation and meditation. Alpha wave indicates awake condition but relaxed. Beta wave occurred in conscious state such as talking, problem solving and decision making [15].

Center of awareness and alertness in the human brain is the most easily detected in the frontal lobe [17]. The frontal lobe controls human cognitive skills and their performance is influenced by the presence of fatigue [18], [19]. On this basis, recording and analyzing the EEG data in this study is focused on the frontal lobe, i.e., AF3 and AF4 channels. Data collection process using car simulator and Emotive Insight channels location are shown in Figure 1.

IOP Conf. Series: Materials Science and Engineering **722** (2020) 012013 doi:10.1088/1757-899X/722/1/012013

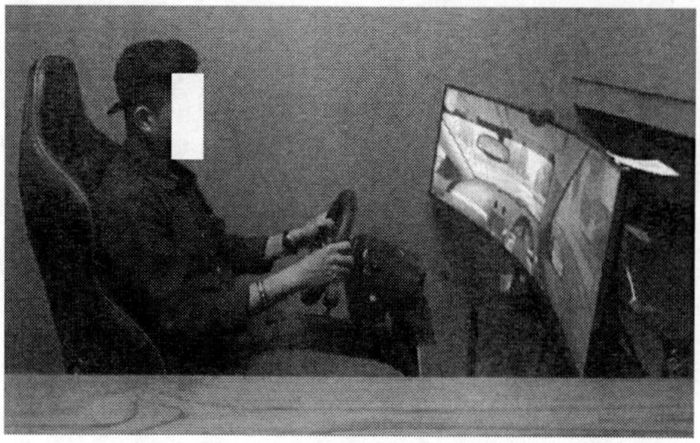

(a)

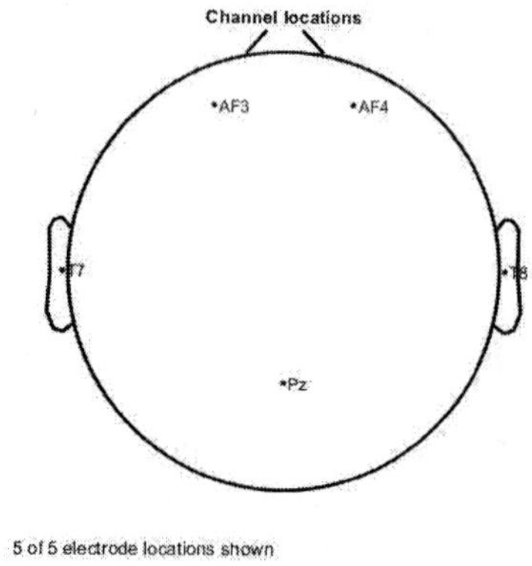

(b)

Figure 1. (a) Respondent when using a car simulator; (b) Emotive Insight channels location

2.3. Data processing method

EEG data obtained, then processed using the EEGLAB toolbox. EEGLAB is a toolbox and graphic user interface that running under MATLAB environment used for processing raw EEG data. EEGLAB functions include channel and information importing, data visualization (scrolling, scalp map, ERP image plots), pre-processing (artefact rejection, filtering, epoch selection and averaging) Independent Component Analysis (ICA), as well as STUDY menu to analyse multiple data file. EEGLAB STUDY menu also provide a statistical analysis to compare multiple conditions. EEGLAB allows performing statistical parametric tests (paired t-test, unpaired t-test, ANOVA) on ERPs, power spectra, ERPs and ITSs [16].

The processing begins by importing raw EEG data in the EEGLAB and filtering unwanted frequencies, followed by removing the artifacts. EEG data was imported in EDF format and the importing process used the guidelines provided by previous research [20]. Furthermore, the initial power spectrum for the five channels are calculated. Independent component analysis and principal component analysis are applied to be able to analyze the power spectrum of each frequency band [16].

3. Result and discussion

3.1. Results

EEG signal filtering used EEGLAB toolbox was performed by filtering the desired frequency limit between 0.1 Hz to 60 Hz. Artifacts, arise due to eye blinking and low conductivity channels, are subsequently removed. The EEGLAB toolbox displays the entire channel signal, and the artifact can be seen in plain view and trim off manually.

EEG data processing in four experimental scenarios, i.e., (1) complex road without Quran; (2) complex road with Quran; (3) simple road without Quran and (4) simple road with Quran, was carried out using Study menu in EEGLAB. The result of statistical analysis using EEGLAB shown in Figure 2a-b and Figure 3a-b.

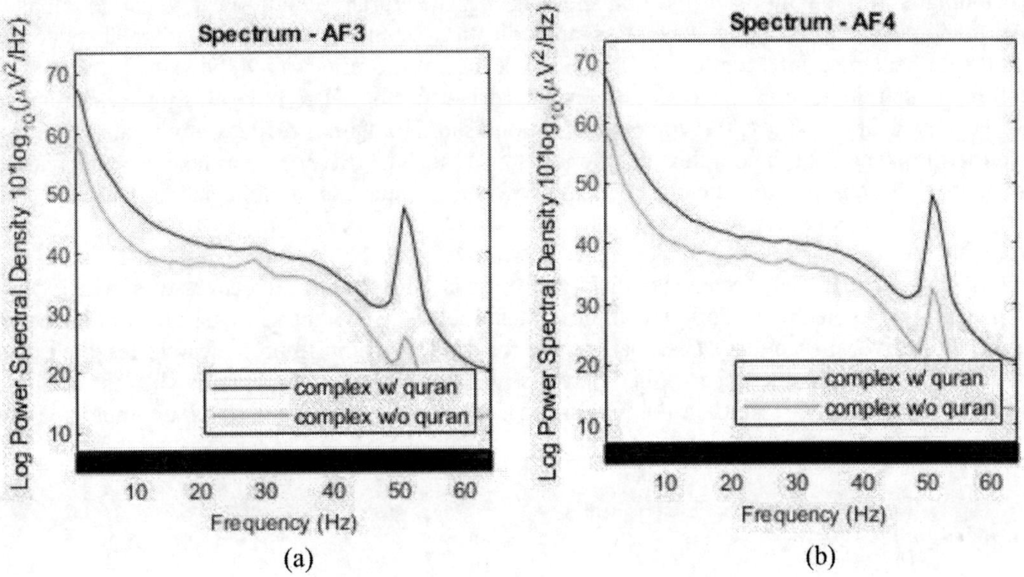

Figure 2. (a) AF3 power spectra – complex road with vs without Quran; (b) AF4 power spectra - complex road with vs without Quran;

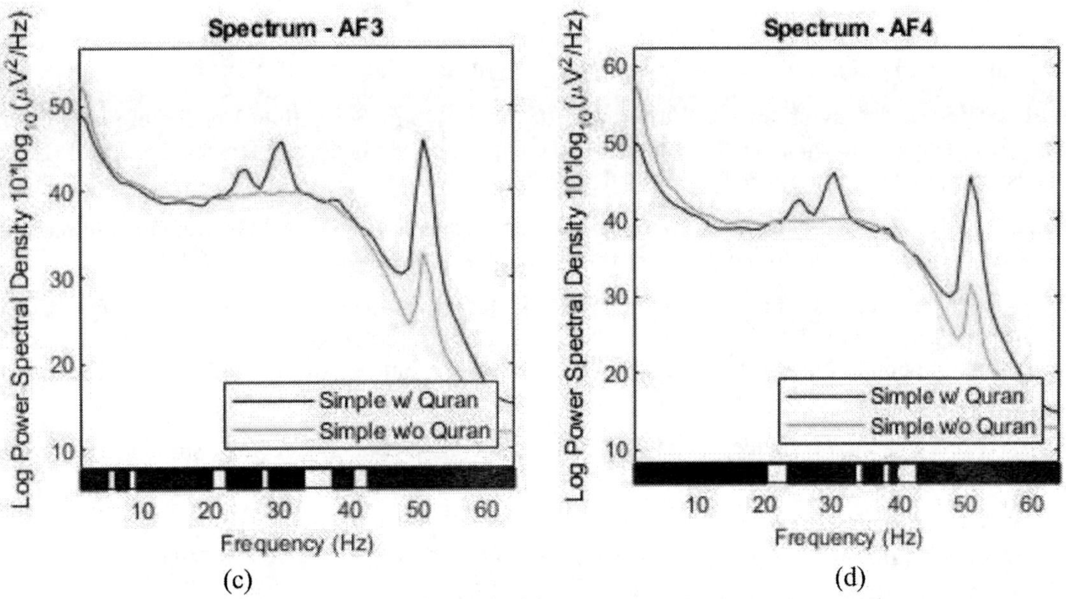

Figure 3. (a) AF3 power spectra – simple road with vs without Quran; and (b) AF4 power spectra – simple road with vs without Quran

3.2. Discussion

Figure 2a-b and Figure 3a-b shows the result of paired t-test in both scenarios: scenario 1 (driving without Quran recitation) and scenario 2 (driving with Quran recitation). The result indicates that there is a statistically significant difference in Alpha (8-13 Hz) dan Beta (13-30 Hz) wave (p-value 0,05). T-test statistic was performed in the EEGLAB toolbox. Statistically significant differences in EEGLAB are shown in black color in the frequency axis (if there is no statistically significant difference, the color on the frequency axis remains white).

This finding is in line with prior research showing that the Quran recitation can generate Alpha wave [13]. As mentioned before, alpha wave was an indication of alertness, focus and relaxed condition. Compared to other voice intervention, i.e., audiobook and music, this study shows a different finding. Research found that audiobooks have an influence on driver performance only on simple or monotonous road [8]. However, in this study, the voice intervention used, i.e., Quran recitation, was able to influence the driver performance both in complex and monotonous road. The driver performance was increased due to the increased of focus, relaxed and alertness, as shown in the increase of alpha and beta wave.

4. Conclusion

The results showed that there were statistically differences in alpha and beta waves when Quran recitation was given to the driver, both in complex and monotonous road conditions. This finding can be used as an alternative of secondary task that is save for drivers. In the future research, research using wireless EEG tools are possible to be applied in real driving environment, especially when the scenarios are safe for the driver. EEG results in the real driving environment can be used as a comparison data with the results of driving simulator.

5. Acknowledgment

This study was funded by the research grant from Direktorat Penelitian dan Pengabdian Masyarakat (DPPM) Universitas Islam Indonesia (UII). Authors would also like thank to all participants involved in this research.

6. References

[1] V. K. Lee, C. Champagne, and L. H. Francescutti, "Fatal distraction: cell phone use while driving," Can. Fam. Physician, vol. 59, no. 7, pp. 723–725, 2013.

[2] D. Haigney, R. Taylor, and S. Westermen, "Concurrent mobile (cellular) phone use and driving performance: task demand characteristics and compensatory processes," Transp. Res. Part F Traffic Psychol. Behav., vol. 3, no. 3, pp. 113–121, 2000.

[3] M. L. Courage, A. Bakhtiar, C. Fitzpatrick, S. Kenny, and K. Brandeau, "Growing up multitasking: The costs and benefits for cognitive development," Dev. Rev., vol. 35, pp. 5–41, 2015.

[4] J. F. A. Schwarz, M. Ingre, C. Fors, R. A. N. Kecklund, J. Taillard, and A. Anund, "Driver sleepiness and countermeasures In-car countermeasures open window and music revisited on the real road: popular but hardly effective against driver sleepiness," J. Sleep Res., vol. 21, pp. 595–599, 2012.

[5] V. Desai and M. A. Hague, "Vigilance monitoring for operator safety: A simulation study on highway driving," J. Safety Res., vol. 37, no. 2, pp. 139–147, 2006.

[6] A. Williamson, D. A. Lombardi, S. Folkard, J. Stutts, T. K. Courtney, and J. L. Connor, "The link between fatigue and safety," Accid. Anal. Prev., vol. 43, no. 2, pp. 498–515, 2011.

[7] D. Dinges, F. Park, K. Williams, K. Gillen, J. Powell, G. Ott, C. Aptowicz, and A. Park, "Cumulative sleepiness, mood disturbance, and psychomotor vigilance performance decrements during a week of sleep restricted to 4-5 hours per night," Sleep, vol. 20, no. 4, pp. 267–277, 1997.

[8] R. J. Nowosielski, L. M. Trick, and R. Toxopeus, "Good distractions: Testing the effects of listening to an audiobook on driving performance in simple and complex road environments," Accid. Anal. Prev., vol. 111, no. November 2017, pp. 202–209, 2018.

[9] N. Dibben and V. J. Williamson, "An exploratory survey of in-vehicle music listening," Psychol. Music, vol. 35, pp. 570–589, 2007.

[10] B. Ünal, L. Steg, and K. Epstude, "The influence of music on mental effort and driving performance," Accid. Anal. Prev., vol. 48, pp. 271–278, 2012.

[11] B. Ünal, D. De Waard, K. Epstude, and L. Steg, "Driving with music: Effects on arousal and performance," Transp. Res. Part F Psychol. Behav., vol. 21, pp. 52–65, 2013.

[12] W. Brodsky, "A performance analysis of In-Car Music engagement as an indication of driver distraction and risk," Transp. Res. Part F Psychol. Behav., vol. 55, pp. 210–218, 2018.

[13] M.S Shekha, AO. Hassan and SA. Othman, "Effect of Quran listening and music on Electroencephalogram brain waves," Egyp. J. Exp. Biol (Zool.) vol.9, pp. 119-121, 2013.

[14] JS. Kumar and P. Bhuvaneswari, "Analysis of EEG signals and its categorization – a study," Procedia Engineering, Vol. 38, pp. 2525-2536, 2012.

[15] N. Birbauner, T. Elbert, AG. Canavan, and B. Rockstroh, "Slow potentials of the cerebral cortex and behaviour," Physiol. vol. 70, pp. 1-41, 1990.

[16] Delorme and S. Makeig," EEGLAB: an open source toolbox for analysis of single-trial EEG dynamics including independent component analysis," Journal of Neuroscience Methods, vol. 134, pp. 9-21, 2004.

[17] P. W. Burgess and J. S. Simons, "Theories of Frontal Lobe Executive Function : Clinical Applications," in Effectiveness of Rehabilitation for Cognitive Deficits, Oxford: Oxford University Press, 2005, pp. 211–232.

[18] J. Perrier, S. Jongen, E. Vuurman, M. L. Bocca, and J. G. Ramaekers, "Driving performance and EEG fluctuations during on-the-road driving following sleep deprivation," Biol. Psychol., vol. 121, pp. 1–11, 2016.

[19] M. A. Puspasari, H. Iridiastadi, I. Z. Sutalaksana, and A. Sjafruddin, "Effect of driving duration on EEG fluctuations," Int. J. Technol., vol. 6, pp. 1089–1096, 2017.

[20] Heunis, "Export and Analysis of Emotiv Insight EEG Data via EEGLAB" 10.13140/RG.2.1.3081.4326, 2016.

IOP Conf. Series: Materials Science and Engineering **722** (2020) 012014 doi:10.1088/1757-899X/722/1/012014

Modified EOQ Model for Refrigerated Display's Shelf-Space Allocation Problem

D Satiti[*], A Rusdiansyah [**], R S Dewi[***]

Department of Industrial Engineering
Institut Teknologi Sepuluh Nopember
Jl. Raya ITS, Keputih, Kec. Sukolilo, Surabaya, Jawa Timur 60111

E-mail: [*]satiti.dyah92@gmail.com, [**]arusdian@ie.its.ac.id,
[***]ratna.sari.dewi80@gmail.com

Abstract. This study modifies the economic order quantity (EOQ) model for the problem of allocating refrigerated rack space in retail. The purpose of this study is to obtain the amount of space allocation for each product that can maximize retail profits. In the modified EOQ model, there are three components of storage costs, specifically storage costs, energy costs and quality costs. Energy costs and quality costs are adjusted to the ratio of cooled performance (i.e. Coefficient of Performance) and quality deterioration rate, respectively. This study adapts Water's algorithm for EOQ orders with constraints on storage space in determining the amount of space allocation for each product. The Water's algorithm uses an Additional Cost approach that is inherent in each unit of space used. The results show the cooling temperature is an important parameter in determining energy costs and quality costs. Through numerical examples, it is determined the right storage temperature and space allocation which gives maximum profit.

1. Introduction

Nowadays, grocery retails are not only selling packaged/processed food products, but also fresh food products. Freshness is a parameter for the quality of fresh food products. Quality of fresh food products can evaluate by three components, biological, chemical, and physical. In terms of biology, the growth of microorganisms, especially dangerous ones such as pathogenic bacteria, is one parameter in assessing the quality of fresh food products. In terms of chemistry, the nutrient content in the product can be lost during storage due to reactions with the environment. While in physical terms, changes in color and texture are the most recognizable parameters. Controlling the environment is one way to slow down the quality deterioration rate [1]–[5]. Thus, refrigerated facility become main handling equipment for storage fresh food products.

Retail has limited resources, which are divided into shop floor and shelf-space. Meanwhile, retailers sell various types of products and brands. Those two things are contradictory. Thus, space management becomes a quite critical retail operational decision.

The problem of shelf-space allocation (SSAP) is the development of EOQ with constraints on limited space. The objective of SSAP is profit maximization by determining the number of space used on each product until it is within acceptable limit [6]. Nogales and Gomez Suarez [7] mention that

SSAP is a challenging problem for retail industry as an increasing number of brands for one product type.

Some retail industries are trying to expand business opportunities by having their own brands. The decision to allocate space for private brands with national brands has been made by Nogales and Gomez Suarez [7] and Amrouche and Zaccour [8]. Spatial allocation decisions are not the sole concentration of retail, some researchers try to solve joint optimization between space allocation decisions with other operational decisions such as pricing decision [9], [10] and replenishment policies [11], [12]. The complexity of SSAP provides an opportunity for researchers to develop methods of solving including using a dynamic program approach [13]–[15]; heuristic and metaheuristic [16]–[21]; data mining approaches [22]–[24]; and game theory [25].

The best of our knowledge, SSAP that considers food products [26];[27];[28] as objects has not much discussed. Moreover, which combines the use of refrigerated facilities. Therefore, through this research, we modify the EOQ model to complete the SSAP for refrigerated facility. This study aims to get maximum retail's profit by determine the right number of refrigerated display's shelf space allocation for each product.

2. Related model

2.1. EOQ model

The inventory model that is still used today is the economic order quantity or EOQ model. By

knowing the demand rate D, RC ordering costs, and HC handling costs, the optimal number of Qo orders can be obtained. There are several assumptions that must be fulfilled so that the resulting order value is the optimal value, that is not permitted by stock out and backorder. Mathematically the EOQ model can be formulated as follows [29]:

$$Q_o=\left(\frac{2\times RC\times D}{HC}\right)^{1/2} \tag{1}$$

2.2. Quality deterioration model

Although it has been stored in cold conditions, the fruits and vegetables are still ripening until the end of their shelf life. Ripening makes the colour and texture change. These parameters (colour and texture) can indicate the fruit quality level. Ripening is a natural process that is inevitable. In general, a quality deterioration depends on the time t and temperature of storage T, and other environmental conditions (humidity, light intensity, etc.). Some quality deterioration models involve those two parameters [30], [31]. With the Arrhenius kinetic equation, the quality deterioration rate σ can be determined by:

$$\sigma=k_A e^{-\left(\frac{E_A}{RT_r}\right)} \tag{2}$$

With k_A is quality rate constants, E_A is the energy of activation for the reaction that controls quality loss, R is ideal gas constants, and T_r display's temperature.

2.3. Multi-temperature performance model

The performance of energy efforts in changing the air temperature in the cooling system is measurable. It is introduced by Rong, et al. [32] and commonly known as the Coefficient of Performance (COP). Furthermore, based on COP value, Zanoni and Zavanella [30] decides the storage requirement between entities in cold supply chain (producer, distributor, and retail). First, they determining the ratio of COP to the lower temperature and the higher temperature. Then, the ratio used as a comparison of costs as a form of energy needed to store at a certain temperature compared to the ideal temperature.

By adapting the model, the authors tried to synchronize the ideal storage temperature between products in one storage facility with consideration of the quality expected to be fulfilled. The energy needed between one product and another can be different. The performance of energy efforts between one product and another product will be formulated as follows:

$$COP_{cooling} = \frac{Q_{cold}}{Q_{hot} - Q_{cold}} = \frac{T_{cold}}{T_{hot} - T_{cold}} \tag{3}$$

$$\rho_{T_{ri}} = \frac{COP_{T_r}}{COP_{T_{ref_i}}} \tag{4}$$

with $COP_{cooling}$ is cooling system's performance measurement, Q_{cold} is the cool (low) temperature, Q_{hot} is the hot (high) temperature or ambient temperature, ρ_{Tri} is the COP ratio of item i consider facility's absolute temperature, T_{refi} is ideal storage temperature of item i.

3. Method

3.1. Modified EOQ model

The storage cost component in cold inventory for fresh food products is different from others. The cost of using refrigerated facilities can be considered through the calculation of energy costs (ec), while consideration of product quality levels is calculated through quality costs (qc). Both parameters are adjusted to ratio of COP, and quality deterioration rate, respectively. So, in this study, the handling cost is not only a single parameter (hc), but also quality cost and energy cost. Then, the total handling cost (HC) is:

$$HC = hc + \left(ec \times \rho_{T_r}\right) + (qc \times \sigma) \tag{5}$$

Based on Eq. (1), the optimal order quantity for single item by substituted HC with Eq. (5):

$$Q_o = \sqrt{\frac{2 \times RC \times D}{hc + \left(ec \times \rho_{T_r}\right) + (qc \times \sigma)}} \tag{6}$$

Then, the variable cost (VC_o) for equation (6) is:

$$VC_o = HC \times Q_o \tag{7}$$

If it is assumed that all the order quantity will sold, the profit (F_o) becomes:

$$F_o = SP \times Q_o - VC_o - UC \times Q_o - RC \tag{8}$$

3.2. Solving algorithm for modified EOQ model

IOP Conf. Series: Materials Science and Engineering **722** (2020) 012014 doi:10.1088/1757-899X/722/1/012014

Storage facilities have limited capacity. To save costs, one facility is used to store several types of products. The previous formula (EOQ) is no longer in accordance with this condition. Procedures of reducing the stock until it meets the capacity are needed. Water used one approach on the same kind

problem. Water puts an additional cost (AC) on space used (S_i). Notation i is respect to each item (i= 1, 2, 3,… n).

Actually, in practice this additional costs are usually not included in the calculation of total costs. By adding this cost, the equation (6) becomes equation (8) for each item:

$$Q_i = \left(\frac{2 \times RC_i \times D_i}{hc_i + \left(sc_i \times \rho T_{r_i}\right) + (qc_i \times \sigma_i) + (S_i \times AC)} \right)^{1/2} \tag{9}$$

Subject to:

$$\sum_{i=1}^{n} \frac{Q_i}{2} \times S_i \leq W \tag{10}$$

Equation (10) guarantees that the space allocated to each product does not exceed the display capacity.

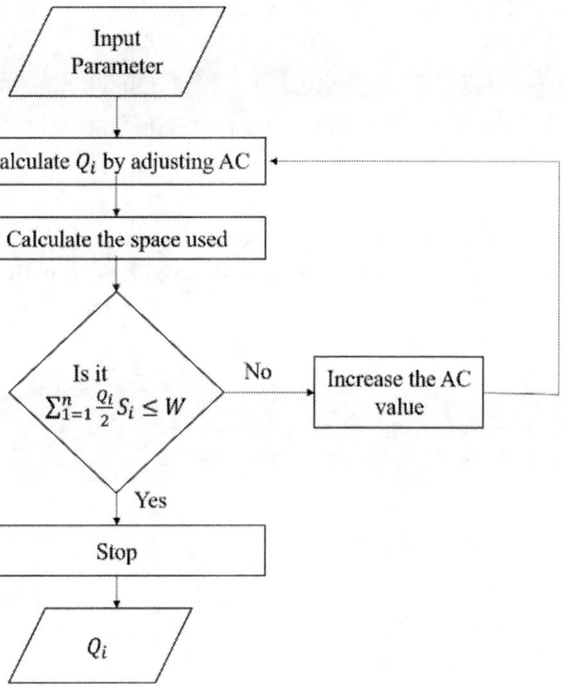

Figure 1. Additional cost on space algorithm (Water's algorithm)

To obtain Q_{oi}, adjust the AC value to achieve the acceptable proportion of space and not exceeding capacity's limit. The stages of completion can be seen in Figure 1. Then the average allocation space for each product in display's facility is:

$$AS_i = \frac{Q_i}{2} \times S_i \tag{11}$$

Then, the variable cost for Eq. (8) becomes:

$$VC_i = RC \times D_i/Q_i + HC_i \times Q_i \quad (12)$$

Thus, the retail's profit is:

$$F_i = SP \times Q_i - VC_i - UC \times Q_i \quad (13)$$

4. Results and discussion

4.1. Numerical example

Consider a retailer that has refrigerated display facilities to store their fresh fruit. It is assumed there are five variations of fresh fruit products sold by retailers. Each product has a different size (see Table 1). In this study, the five products are assumed to be divided into three product groups based on their storage temperature requirements, e.g. products with low temperatures (-1 °C - 3°C), products with high temperatures (7 °C -11°C), and mixed temperature products (-1 °C - 21 °C). Meanwhile, the temperature of the cooling facility for chilling needs is 2-8 °C.

Table 1. Product's attribute

Item	Space Needed (space/unit)	Demand Rate (unit)	Unit Cost (Rp/unit)	Selling Price (Rp/unit)
A	0.25	300	18,000.00	27,000.00
B	0.3	410	1,350.00	2,025.00
C	4	90	25,000.00	37,500.00
D	0.12	250	5,600.00	8,400.00
E	0.22	500	2,300.00	3,450.00

It is assumed that the holding cost is 20 percent of the unit cost, with an energy cost of Rp 15, - / unit and a quality cost of 5 percent of the unit cost. Each item is replenish independently, joint replenishments are not considered. Stock-outs and back orders are not permitted. (-1 °C – 21 °C).

4.2. Results and discussions

This study observe multi temperature products that must be storage in a single facility with single temperature set. Figure 2 show the average saving for high temperature product groups in different facility temperature set.

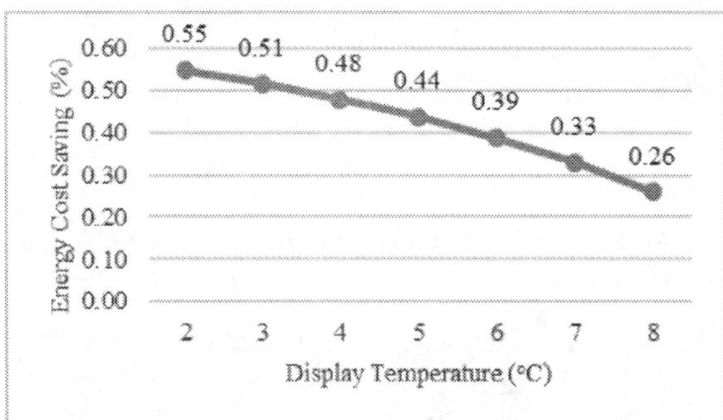

Figure 2. Energy cost saving on high temperature products

Table 2. Space allocation on mix temperature products

Using	A	B	C	D	E	Total Profit (Rp)
Qo	182	347	59	204	446	2,224,534.18
Qi	162	316	56	199	440	2,321,414.28

Table 3. Space allocation on low temperature products

Using	A	B	C	D	E	Total Profit (Rp)
Qo	182	345	58	204	445	2,204,932.64
Qi	163	315	56	199	439	2,321,894.19

If retail uses the economic order quantity for all items in an inventory, the resulting total stock might exceed the available capacity. By using the Water's algorithm that show in Figure 1, the number of each product can be reduced until it does not violate facility capacity limits. In the temperature mix product group, the temperature of the facility that provides the highest total profit is 3 °C, with allocations for each item as shown in Table II. As for the low and high temperature product groups, the temperature of the facilities that provide the highest total profit for each group is 5 °C and 6 °C, respectively. Comparison of the total profit generated from each storage temperature can be seen in Figure 3. It is known that there was an increase in total profit of 4,355% between the standard EOQ (Q_o) and revised EOQ (Q_i).

Table 4. Space allocation on high temperature products

Using	A	B	C	D	E	Total Profit (Rp)
Qo	182	347	59	205	448	2,236,882.92
Qi	162	315	56	199	441	2,326,069.71

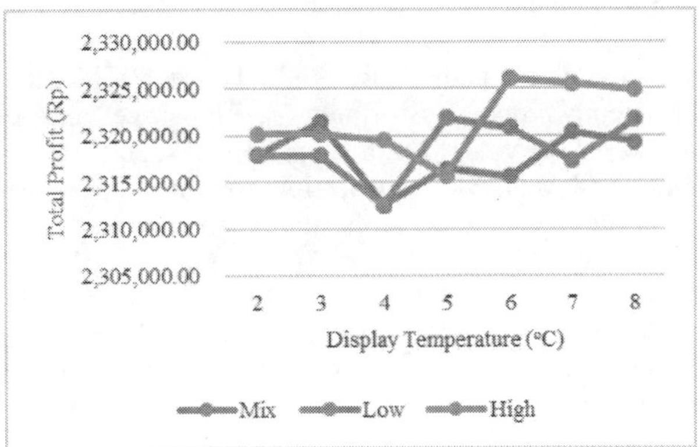

Figure 3. Total profit for each group of product

5. Conclusion

Among three product groups, the high temperature product group provides the highest profit value Rp 2,326,069.71, with the product composition i.e. A 162 units; B 315 units; C 56 units; D 199 units and E 441 units. Temperature has an important role in cost savings and retail's profit by determining the display shelf-space allocation number. Store the product at a lower temperature than the ideal temperature, giving an average energy cost savings of 55%. However, this needs to be reviewed in relation to the characteristics of fruit products that can experience chilling injuries if stored low temperature (extreme conditions).

References

[1] G. Romanazzi, S. M. Sanzani, Y. Bi, S. Tian, P. Gutiérrez Martínez, and N. Alkan, "Induced resistance to control postharvest decay of fruit and vegetables," *Postharvest Biol. Technol.*, vol. 122, pp. 82–94, 2016.

[2] S. Péneau, P. B. Brockhoff, F. Escher, and J. Nuessli, "A comprehensive approach to evaluate the freshness of strawberries and carrots," *Postharvest Biol. Technol.*, vol. 45, no. 1, pp. 20–29, 2007.

[3] C. Fagundes, K. Moraes, M. B. Pérez-Gago, L. Palou, M. Maraschin, and A. R. Monteiro, "Effect of active modified atmosphere and cold storage on the postharvest quality of cherry tomatoes," *Postharvest Biol. Technol.*, vol. 109, pp. 73–81, 2015.

[4] F. Nigro, A. Ippolito, V. Lattanzio, D. Di Venere, and M. Salerno, "Effect of ultraviolet-C light on postharvest decay of strawberry," *J. Plant Pathol.*, vol. 82, no. 1, pp. 29–37, 2000.

[5] J. Usall, A. Ippolito, M. Sisquella, and F. Neri, "Physical treatments to control postharvest diseases of fresh fruits and vegetables," *Postharvest Biol. Technol.*, vol. 122, pp. 30–40, Dec. 2016.

[6] T. Bianchi-Aguiar, E. Silva, L. Guimarães, M. A. Carravilla, and J. F. Oliveira, "Allocating products on shelves under merchandising rules: Multi-level product families with display directions," *Omega*, vol. 0, pp. 1–16, 2017.

[7] A. F. Nogales and M. Gomez Suarez, "Shelf space management of private labels: A case study in Spanish retailing," *J. Retail. Consum. Serv.*, vol. 12, no. 3, pp. 205–216, 2005.

[8] N. Amrouche and G. Zaccour, "Shelf-space allocation of national and private brands," *Eur. J. Oper. Res.*, vol. 180, no. 2, pp. 648–663, 2007.

[9] G. Martín-Herrán, S. Taboubi, and G. Zaccour, "The impact of manufacturers' wholesale prices on a retailer's shelf-space and pricing decisions," *Decis. Sci.*, vol. 37, no. 1, pp. 71–89, 2006.

[10] S. Y. Wang, G. J. Sheen, and Y. Yeh, "Pricing and shelf space decisions with non-symmetric market demand," *Int. J. Prod. Econ.*, vol. 169, pp. 233–239, 2015.

[11] M. A. Hariga, A. Al-Ahmari, and A. R. A. Mohamed, "A joint optimisation model for inventory replenishment, product assortment, shelf space and display area allocation decisions," *Eur. J. Oper. Res.*, vol. 181, no. 1, pp. 239–251, 2007.

[12] J. Zhao, Y.-W. Zhou, and M. I. M. Wahab, "Joint optimization models for shelf display and inventory control considering the impact of spatial relationship on demand," *Eur. J. Oper. Res.*, vol. 255, no. 3, pp. 797–808, 2016.

[13] H. K. Gajjar and G. K. Adil, "A dynamic programming heuristic for retail shelf space allocation problem," *Asia-Pacific J. Oper. Res.*, vol. 28, no. 2, pp. 183–199, 2011.

[14] F. S. Zufryden, "New computational results with a dynamic programming approach for product selection and supermarket shelf-space allocation," *J. Oper. Res. Soc.*, vol. 38, no. 2, pp. 201–203, 1987.

[15] F. S. Zufryden, "A dynamic programming approach for product selection and supermarket shelf-space allocation," *J. Oper. Res. Soc.*, vol. 37, no. 4, pp. 413–422, 1986.

[16] R. Bai, E. K. Burke, and G. Kendall, "Heuristic, meta-heuristic and hyper-heuristic approaches for fresh produce inventory control and shelf space allocation," *J. Oper. Res. Soc.*, vol. 59, no. 10, pp. 1387–1397, 2008.

[17] M. Castelli and L. Vanneschi, "Genetic algorithm with variable neighborhood search for the optimal allocation of goods in shop shelves," *Oper. Res. Lett.*, vol. 42, no. 5, pp. 355–360, 2014.

[18] H. K. Gajjar and G. K. Adil, "Heuristics for retail shelf space allocation problem with linear

profit function," *Int. J. Retail Distrib. Manag.*, vol. 39, no. 2, pp. 144–155, 2011.

[19] H. Hwang, B. Choi, and G. Lee, "A genetic algorithm approach to an integrated problem of shelf space design and item allocation," *Comput. Ind. Eng.*, vol. 56, no. 3, pp. 809–820, 2009.

[20] S. Raut, S. Swami, and M. P. Moholkar, "Heuristic and meta-heuristic approaches for multi-period shelf-space optimization: The case of motion picture retailing," *J. Oper. Res. Soc.*, vol. 60, no. 10, pp. 1335–1348, 2009.

[21] P. M. Reyes and G. V. Frazier, "Goal programming model for grocery shelf space allocation," *Eur. J. Oper. Res.*, vol. 181, no. 2, pp. 634–644, 2007.

[22] M.-C. Chen and C.-P. Lin, "A data mining approach to product assortment and shelf space allocation," *Expert Syst. Appl.*, vol. 32, no. 4, pp. 976–986, 2007.

[23] Y.-L. Chen, J.-M. Chen, and C.-W. Tung, "A data mining approach for retail knowledge discovery with consideration of the effect of shelf-space adjacency on sales," *Decis. Support Syst.*, vol. 42, no. 3, pp. 1503–1520, 2006.

[24] C.-Y. Tsai and S.-H. Huang, "A data mining approach to optimise shelf space allocation in consideration of customer purchase and moving behaviours," *Int. J. Prod. Res.*, vol. 53, no. 3, pp. 850–866, 2015.

[25] X. Li, S. Nukala, and S. Mohebbi, "Game theory methodology for optimizing retailers' pricing and shelf-space allocation decisions on competing substitutable products," *Int. J. Adv. Manuf. Technol.*, vol. 68, no. 1–4, pp. 375–389, 2013.

[26] R. Bai and G. Kendall, "A model for fresh produce shelf-space allocation and inventory management with freshness-condition-dependent demand," *INFORMS J. Comput.*, vol. 20, no. 1, pp. 78–85, 2008.

[27] T. Bianchi-Aguiar *et al.*, "Using analytics to enhance a food retailer's shelf-space management," *Interfaces (Providence).*, vol. 46, no. 5, pp. 424–444, 2016.

[28] U. Toft, L. L. Winkler, B. E. Mikkelsen, P. Bloch, and C. Glümer, "Discounts on fruit and vegetables combined with a space management intervention increased sales in supermarkets," *Eur. J. Clin. Nutr.*, vol. 71, no. 4, pp. 476–480, 2017.

[29] D. Waters, *Inventory Control and Management*. West Sussex, England: John Wiley & Sons, Ltd, 2003.

[30] S. Zanoni and L. Zavanella, "Chilled or frozen? Decision strategies for sustainable food supply chains," *Int. J. Prod. Econ.*, vol. 140, no. 2, pp. 731–736, 2012.

[31] X. Wang and D. Li, "A dynamic product quality evaluation based pricing model for perishable food supply chains," *Omega*, vol. 40, no. 6, pp. 906–917, 2012.

[32] A. Rong, R. Akkerman, and M. Grunow, "An optimization approach for managing fresh food quality throughout the supply chain," *Int. J. Prod. Econ.*, vol. 131, no. 1, pp. 421–429, May 2011.

IOP Conf. Series: Materials Science and Engineering **722** (2020) 012015 doi:10.1088/1757-899X/722/1/012015

Risk mitigation on product distribution and delay delivery : A case study in an Indonesian manufacturing company

E Kusrini [1], E Sugito [2], Z M Rahman [3], T N Setiawan [4], R P Hasibuan [5]

[1,2,3,4,5.] *Master of Industrial Engineering, Indonesian Islamic University of* Yogyakarta

E- mail: elisakusrini@uii.ac.id

Abstract. In a dynamic supply chain environment, problems that arise due to delays in the delivery of raw materials and distribution of products to consumers can pose risks and threats to the company. This study aims to identify and mitigate the risk of late delivery of goods. Risk management starts with identifying sources of risk, defining risks, creating risk maps and mitigating risks. a case study conducted at a manufacturing packaging company in Batam, Indonesia. The results showed that the biggest risks were supplier production capacity, supplier product quality, forecasting from consumers, daily finished good delivery schedule to consumers and loading and unloading processes at consumers. This research proposes mitigation to minimize the impact of short-term and long-term risks.

1. Introduction

Currently, the dynamics of competition efforts bring an impact on efforts that continuous so that goods can get to consumers at the right time and place. Many industrial companies are working to optimize the supply chain system to produce products and on-time delivery. The company seeks to work harder in order to delay in transport does not occur [1]. Respond to the level of production and the ever-changing technological innovation demands a conventional manufacturing company transformed into a more modern production system [2]. If the product does not arrive at the time of delivery in accordance with the agreed (delay delivery) by both sides then consumers will pick up the pace by giving sanction or fine [3]. Delivery time refers to the time customers wait between placing an order and receiving the order. As a consequence, the concept of the delivery time can be said to relate to the performance as perceived by customers [4, 5, 6, 7, 8].

Risk management can provide input or contribution is important for most types of management decisions and control, which is widely applied in various fields. Like the economy, insurance and manufacturing [9, 10, 11, 12]. The risk management process generally consists of 4 steps [13, 14]. That is as follows:

1) The first step is to identify the risks, to help improve the understanding of the current state against uncertainty in the future (the risk that may occur) to be able to receive an effective rule.
2) The second step is to evaluate the risk of identification with the possibility to associate events and identify the consequences.
3) The third step is to apply this action management risk, such as actions that are reactive (responding to risk identification who had previously occurred) or proactive (to reduce the likelihood of occurrence or the severity of the consequences and risk mitigation).
4) The fourth step is monitoring (monitoring) of risk on an ongoing basis in order to reduce the possibility of a major again.

This research takes a case study in one of the manufacturing companies in Indonesia, called PT X. PT.X engaged in the manufacture of cardboard packaging (packaging boxes) distributing its products in the area of Batam and Singapore. The distribution of the products is carried out on a daily basis based on consumer demand with the distance and the amount varies. Raw materials import process is carried out by the State of Singapore, Taiwan, Thailand, and the United States. Often the onset delay or a very late arrival hamper company production and delivery to the customer, graphics delivery delays can be seen in Figures 1 and 2. The delay caused by a late booking from the purchasing, production capacity of suppliers, transportation, delivery of materials, production process, trucking and other. This has resulted in the delivery of the product to the consumer becomes delays and raises fines or sanctions from the consumer. For it on-time delivery of badly needed in order for the company's operations are not disrupted and the lead time of the consumers can be reached, therefore companies need to conduct analysis for risk management from the perspective of a company in dealing with things that are not in want in distribution or delivery of goods heading to the consumer.

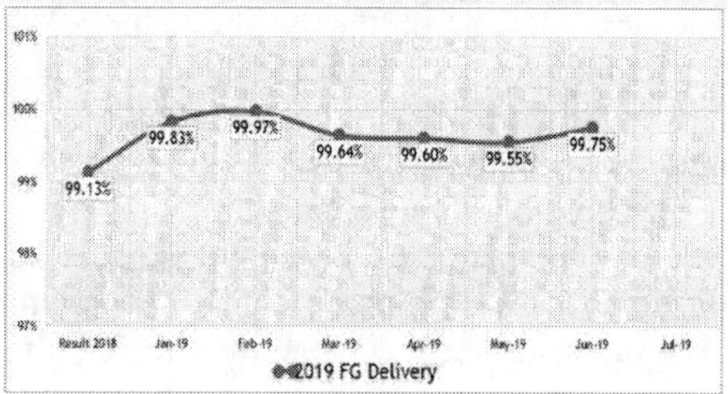

Figure 1. Finished Goods Delay Delivery Chart

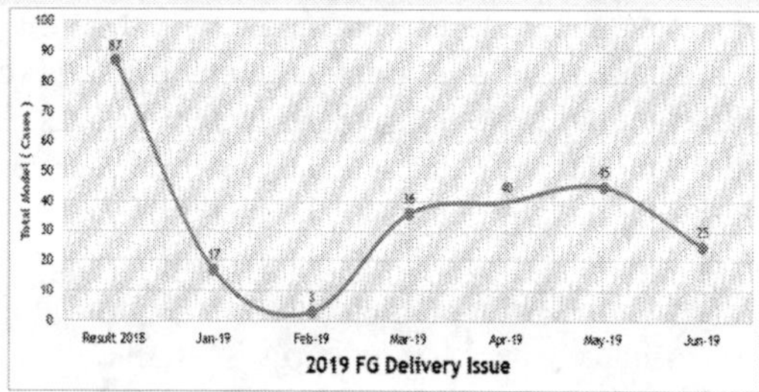

Figure 2. *Delivery Case Chart* Table

2. Literature Review

2.1. Risk management.

The risk is the likelihood of an event occurring and the impact on the success in achieving the strategic objectives of the organization [15]. The risk is the part that is inseparable from the decision-making process of the individual or organization in many situations. The risks cannot be eliminated. Organizations have to manage the factors that influenced the increase or decline the risk so that the organization can pursue strategic superiority with minimum fees [16]. Taking the risk seems to be the strategy that is available to resolve the risk [17]. Risk management in an organization contains the most important internal processes and strengthens resilience during period's prevention and which could not have been avoided in ensuring the security of the process [18].

Risk management is about identifying the things that are wrong, stop wrong, reduce the consequences when there are errors and recover after a few things wrong. This is important because there is always the possibility of error, which is why a Manager spends a lot of time to prevent things that are incompatible with minimize the possibility of failure and its effects [19]. Awareness about risk management in the supply chain to improve in those industries in Europe [20]. There are 4 categories of risk on corporate finance namely, hazard, operational and strategic risks. Risks of Finance comes from the strength of the market, assets or financial obligations. They occur in various forms, such as risk capital, investment risk, interest rate risk, credit risk, currency risk and market risk capital [21]. The risk of danger is called "pure risk" and its impact is always negative. They arise from natural or deliberate accidents caused by third parties. The consequences of these risks, for example, damage to property, personnel illnesses and disabilities [22, 23]. Operational risk arises from the failure that resulted from both the potential for human error and processes that can occur in the functional business units (e.g. related to product development, human resources and it) or in the management of

process [24]. Risk identification can be done by doing 1) identification of sources of risk, 2) Define the risks and consequences of encouraging risk identification) 3 4) to mitigate risk [25, 26].
There is a lot of research about managing risk. [27] Conducts research on risk management in the supply chain of fresh produce in Thailand. From the results of the research, there are nine known categories of risk in the supply chain of fresh produce that is weather risk, risk of demand, Finance risk, risk information, risk operational, risk policies, risk, price risk, and regulation.

2.2. Delay Delivery

Delivery is expected by customers is timely delivery (OTD) and delivery without damage. The delivery is delayed or late have a negative impact on the satisfaction of [28]. Delivery delays harm the next customer, a customer would not consider buying again because the company did not submit a time for the products ordered. Delivery delays lead to negative results and a lack of customer satisfaction [29].

2.3. Risk Mitigation

Mitigation is the action to reduce or minimize the potential negative impact of a disaster [30]. Risk is defined on the basis of the two points of view [31], namely: 1) point of view the results (output) where results cannot be predicted with certainty because it would be counter-productive. 2) point of view the process whereby the factor-factors that may affect the achievement of the goals, so unintended consequences. Mitigation in an enterprise is required to determine whether the company's risk in it can be fixed with good and right so that it does not risk impeding the activities of the company [32]. The base model introduced serves as the base for determining the magnitude of costs and risks. So as to provide the first phase of strategic decisions and could build a model to evaluate the risk mitigation options of the supply chain [33]. According to [34] to propose the theoretical decision framework that includes relevant elements derived from the literature on supply chain risk management, by providing quantitative decision support systems of the general framework the decision to select the risk mitigation strategy for the supply chain.

The research of [35] risk mitigation is done to minimize the impact of the environmental risks posed by the use of the game theory approach. [36] Conduct mitigation against supply chain risk and uncertainty. The results of this research to identify the type of supply chain flexibility appropriate to mitigate the risk of the three major risks of supply, the risk of the manufacturing process and the risk of shipment.

3. Methodology

Stages of research methods are as follows:

3.1. *Identification of the source of the risk.*

a. The identification process business
   Process business is material flow from supplier to the finished product into the hands of customers.
b. Identification of the risk
   Do the identification over the incident or potential incident in the case will affect the achievement of the objectives of the company or would potentially harm the company.

3.2. *Defining the risk.*

a. Identification of the point the number of cases of risk, using a table of researchers taking points or risk which is often the case and grouped based on areas.
b. Risk Measurement.
   Measurement (measurement) is performed to find out the level of risk by using the probability of the occurrence of the risk requirements.

3.3. *Risk Maps*

Are used to show the position of risk and determine priorities risk response. In the making of maps can be made with 2 models i.e.: inherent risk map and map of the residual risk.

IOP Conf. Series: Materials Science and Engineering **722** (2020) 012015 doi:10.1088/1757-899X/722/1/012015

3.4. *Risk mitigation*

Follow-up development to respond to the risk, this is a detailed and documented explanation of the strategic treatment of risk will be applied.

4. Discussion

4.1. *Identification of the source of the risk*

4.1.1. Identification Process business

Process business in PT. X showed in Figure 3. in the following.

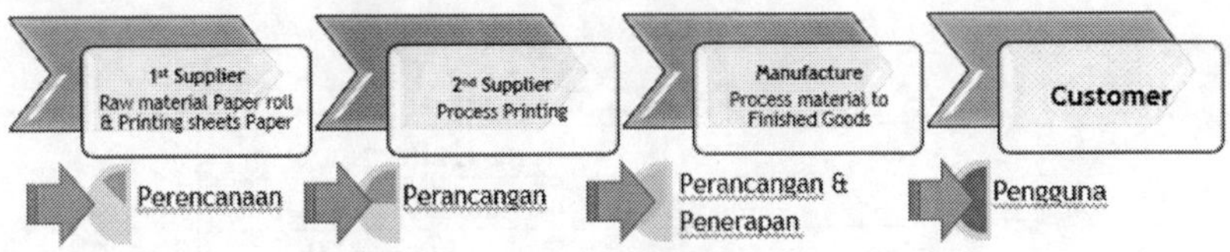

Figure 3. PT. X Business Process

4.1.2. Identification of Risk.

Identification risk for delay in distributing products either from the supplier or consumer to do with the analysis of fishbone diagrams [37], fishbone diagrams identification risk can be seen in Figure 4. Cases and the frequency with which led to delivery delays from suppliers and to consumers can be seen in Table 1 and 2.

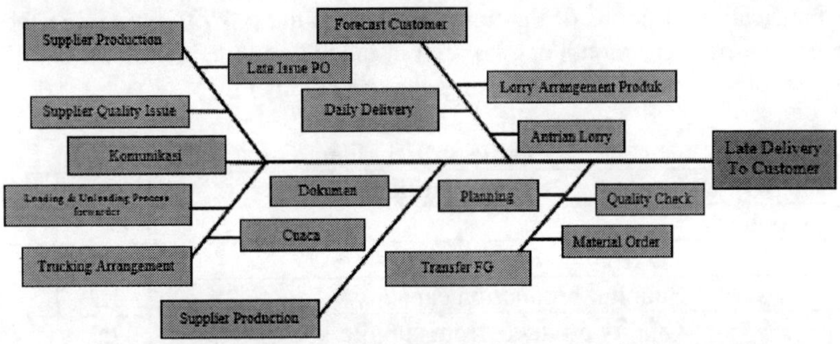

Figure 4. Risk Indicator (Fishbone Analysis)

Table 1. Risk Identification list

No	Risk event
Supplier	
1	PO from purchasing
2	Communication problem
3	Supplier production capacity
4	Product quality from suppliers
5	Shipping Documents
6	Unloading and loading goods by the forwarder
7	Weather
8	Truck Settings
Customer	
9	Forecast from consumers
10	Daily FG delivery schedule to consumers
11	Planning in production

No	Risk event
12	Ordering raw materials
13	Quality checking of material
14	Company production capacity
15	Transfer of finished goods to the warehouse
16	Lorry preparation for shipping to consumers
17	The process of loading and unloading on consumers
18	Lorry queues at consumers
19	The process of receiving goods documents

Tabel 2. Number of Delay

Month	Where Delays occur	
	Suppliers	Consumer
Jan 19	8	9
Feb 19	-	3
March 19	10	26
April 19	10	30
May 19	21	24
Jun 19	15	10
Total	64	102

The identification of points of the number of cases the risk of eliminating the risk derived from the risk of repeat cases every month can be seen in table 3 and the identification of risk impact can be seen in table 4, so it will be easier to determine the measurement.

Table 3. Risk Identification

No	Risk Indicator	Code	Case
Supplier			
1	PO from purchasing	L1	6
2	Supplier production capacity	L2	30
3	Quality products from suppliers	L3	20
4	Truck Settings	L4	8
Customer			
5	Forecast from consumers	L5	15
6	Daily FG delivery schedule to consumers	L6	30
7	Planning in production	L7	5
8	Ordering raw materials	L8	7
9	Transfer of finished goods to the warehouse	L9	15
10	The process of loading and unloading on consumers	L10	20
11	Lorry queues at consumers	L11	10

IOP Conf. Series: Materials Science and Engineering **722** (2020) 012015 doi:10.1088/1757-899X/722/1/012015

Tabel 4. Risk Impact Identification

No	Impact Indicator	Value
1	Consumers stop producing	30 USD / Hrs
2	Increased production costs	23 USD / Hrs
3	High maintenance costs for Production Machines	500 USD / Month
4	The number of products damaged in the production area	300 USD / Month
5	Consumer Confidence	-

4.2. ***Defining Risk***

4.2.1. Risk Point Identification

In measuring risk, the probability scale and risk impact scale data are needed, presented in tables 5 and 6.

Table 5. Risk Probability Scale

Scale	Criteria	Range	Explanation
(1)	Almost never happens	<20%	Max occurs 1 in 6 months
(2)	Can (might happen)	20–50%	Max occurs 5 times in 6 months
(3)	Rarely happening	50 -70%	Max occurs 10 times in 6 months
(4)	Often occur	70-90%	Max occurs 15 times in 6 months
(5)	Very often happens	90-100%	Occurred more than 20 times in 6 months

Table 6. Risk Impact Scale

Scale	Impact	Indicator
(1)	Small	Easy to complete, does not cause late, does not affect the overall distribution process
(2)	Very small	Does not cause delays when arriving at the customer
(3)	Is	Cause delay to the customer
(4)	Big	Causing delays, there are many complaints to the company
(5)	Very large	The source of the problem is very difficult to identify, causing delays, many complaints to the company, termination of employment

Source: Wiryani, et al., 2013

4.2.2. Risk Measurement

Risk Measurement Step for next was done by calculating each risk by multiplying the probability scale and the scale of the resulting impact. Risk measurement results can be seen in table 7.

IOP Conf. Series: Materials Science and Engineering **722** (2020) 012015 doi:10.1088/1757-899X/722/1/012015

Table 7. Measurement of Risk Points

No	Risk Indicator	Scale		Score
		Impact	Probability	
Supplier				
1	PO from purchasing	1	3	3
2	Supplier production capacity	4	5	20
3	Quality products from suppliers	4	5	20
4	Truck Settings	3	2	6
Customer				
5	Forecast from consumers	4	4	16
6	Daily FG shipping schedule to consumers	4	5	20
7	Planning in production	3	2	6
8	Ordering raw materials	2	2	4
9	Transfer of finished goods to the warehouse	2	4	8
10	The process of loading and unloading on consumers	4	4	16
11	Lorry queues at consumers	3	3	9

4.3. *Risk Map*

Risk maps are made based on the scale of impact and probability of each risk indicator. Green areas indicate a low-risk level, yellow indicates a moderate risk level, and red indicates a high-risk level. The risk map can be seen in Figure 5.

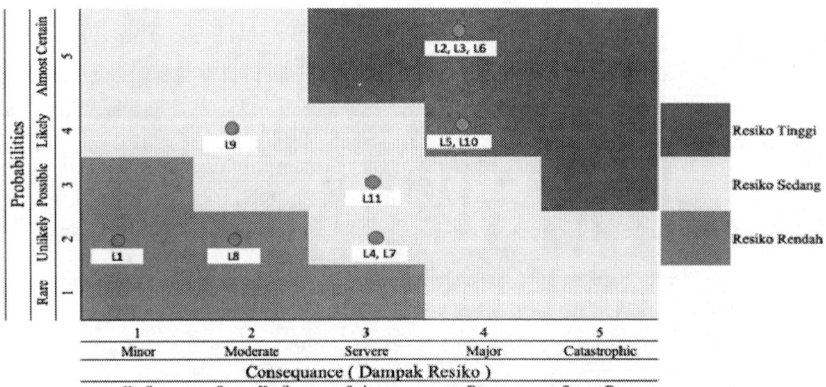

Figure 5. Current Risk Map

4.4. *Risk Mitigation.*

After the measurement and mapping, obtained the greatest impact from the risk of delay in distributing the products get to the consumers on a scaled score of 20 and 16. Risk indicators that are at the highest score become a priority for the company. High-risk indicators grouped into 8 tables and do a repair using the guidelines of risk control, then mapped back to an estimated decline of shifting the impact

IOP Conf. Series: Materials Science and Engineering **722** (2020) 012015 doi:10.1088/1757-899X/722/1/012015

and probability of the risk. Expected improvement occurs so that the risk can be decreased from high risk to medium risk and continue to do repairs in order to continue to decline to low risk.

Table 8. Risk Mitigation Control Guidelines

No	Risk Indicator	Risk Level	Risk Management Guidelines	4T strategy
1	Supplier production capacity	High	Transparency	Treat
			Request planning collaboration	Treat
2	Quality products from suppliers	High	Making quality standards	Treat
			Planning a business process together	Treat
3	Forecasting from consumers	High	Integrating system information	Treat
4	Daily FG shipping schedule to consumers	High	Perform data sharing for daily shipping schedules	Treat
5	The process of loading and unloading on consumers	High	Delivery time sharing so that no stuck vehicle in the customer	Treat

Risk mitigation strategies against the five indicators of risk that has a high level of risk are shown in table 8. The fifth indicator of risk mitigation can be done by doing some treatment. In particular, treatment is done in table 8 spelled out the company. Treatment was done in an effort to 1) information sharing both data or process lead time in finishing products; 2) conduct discussions with the buyer to discuss the problem of forecasting and daily delivery schedules, as well as the duration of the time unloading goods at the consumer's location; 3) Collaboration plan request (collaboration demand planning); 4) Monitoring the production process, specifically in the supplier due to a lot of the occurrence of defects during a process lasting;
5) Perform regular inspections for production machines and recorded for monitoring; 6) Customer service and planner team to be more active in following up on the production process and production lead time. On the research of supply chain risk and Uncertainty: a moderation role supply chain flexibility in risk mitigation risk three stated that the major risk of supply, the risk of manufacturing and delivery risk [38].

Table 9. Risk Scale after Mitigation

No	Risk Indicator	Scale		Score
		Impact	Probability	
Supplier				
1	PO from purchasing	1	3	3
2	Supplier production capacity	3	4	12
3	Quality products from suppliers	3	4	12
4	Truck Settings	3	2	6
Customer				
5	Forecast from consumers	3	3	9
6	Daily FG shipping schedule to consumers	3	4	12
7	Planning in production	3	2	6
8	Ordering raw materials	2	2	4
9	Transfer of finished goods to the warehouse	2	4	8
10	The process of loading and unloading on consumers	3	3	9
11	Lorry queues at consumers	3	3	9

After mitigation is carried out, the impact and probability of risk decreases, shown in table 9.

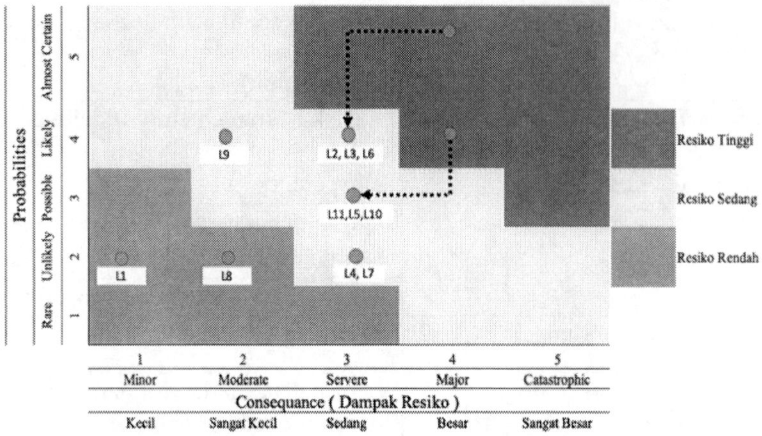

Figure 6. Estimation Risk Map after Mitigation Result

The risk mitigation map shows the five risk indicators that were originally in the red quadrant move to the yellow quadrant. The risk map for mitigation results can be seen in Figure 6.

5. Conclusion

Some conclusions that can be drawn from this study are as follows:

1. The risk that has the greatest impact and has an effect on delivery delay comes from supplier production capacity, product quality, daily delivery schedules from consumers, forecasts and loading and unloading processes at consumers.
2. Risk mitigation is done by improving the system to suppliers and consumers by sharing information using an integrated system so that lead-time can be more accurate, as well as collaborating on demand plans between suppliers and manufacturing companies. This is done continuously and monitored, whether there are changes and developments that occur in the future.

References

[1] Hajej. Z, Turki. S, and Rezg. N, "Modeling and Analysis for Sequentially Optimising Production, Maintenance and Delivery Activities Taking into Account Product Returns," International Journal of Production Research , 2015, 53 (15), pp. 4694-719.

[2] Agrawal. A, Minis. I, and Nagi. R, "Cycle time reduction by improved MRP-based production planning". International Journal of Production Research, 2000, vol. 38 (18), 4823–4841.

[3] Turki. S, and Rezg. N, "Optimization of Manufacturing Supply Chain with Stochastic Demand and Planned Delivery Time," International Journal of traffic and transportation Engineering, 2017, vol 5. Pp. 32-43.

[4] Parasuraman. A. A, "Delivering quality service: Balancing customer perceptions and expectations," Simon and Schuster.com, 1990.

[5] Chang. D.S, and Wang. T.H, "Consumer preferences for service recovery options after delivery delay with shopping online. Social Behavior and Personality," Journal: An International Journal, 2012, vol 40, pp. 1033-1044.

[6] Giannakis. M, and Papadopoulos. T, "Supply chain sustainability: a risk management approach," Int. J. Prod. Econ, 2016, 171, pp. 455-470.

[7] Harrison. A, and Van Hoek. R, "Logistics management and strategy- competing through the supply chain," 3rd Edn., UK: F.T. Prentice Hall, 2008.

[8] Stephen. O, Jonathan. A, and Kwabena. S.S, "Evaluation of the effect of lead time on quality service delivery in the bank industry in Kumasi metropolis of Ghana," Journal of art and humanities, 2015, Vol. 04 (07), pp. 29-44
[9] Ritchie. B, and Brindley. C, "Supply Chain Risk Management and Performance: A Guiding Framework for Future Development," International Journal of Operations & Production Management, 2007, vol. 27, pp. 303-322.
[10] Beske. P, Seuring. S, "Putting sustainability into supply chain management," Supply Chain Manag: Int. J. 2014, 19 (3), pp. 322-331.
[11] O'Rourke. D, "The science of sustainable supply chains," Science 344, 2014, (6188), pp. 1124-1127.
[12] Tseng. M.L, Islam. M.S, Karia. N, Fauzi. F.A, and Afrin. S, "A literature review on green supply chain management: trends and future challenges," Resour. Conserv. Recycle, 2019, 141, pp. 145-162.
[13] Giannakis. M, and Papadopoulos. T, "Supply chain sustainability: a risk management approach," Int. J. Prod. Econ, 2016, 171, pp. 455-470.
[14] Chan. F. T, "Performance measurement in a supply chain," The International Journal of Advanced Manufacturing Technology 21, 2003, vol (7), pp. 534-548.
[15] COSO and WBCS, "Enterprise Risk management, Applying Enterprise Risk Management to Environmental," Social and Governance-Related Risk, 2018, p-1.
[16] Borghesi, Antonio, and Gaudenzi Barbara, "Risk Management How to Assess, Transfer and Communicate Critical Ris," Springer-Verlag Italia, 2013.
[17] Hora. M, and Klassen. R.D, "Learning from others' misfortune: factors influencing knowledge acquisition to reduce operational risk," J. Oper. Manag, 2013, 31 (1), pp. 52-61.
[18] Katarina Buganova and Jana Simickova, "Risk management in traditional and agile project management," 13th International sientific conference on sustainable, moderen and safe transport, 2019, 40, pp. 986-993.
[19] Nigel. Slack, Alistair. Brandon, and Roberth. Johnston, "Seventh edition operation management," Myomlab, 2013, pp. 610
[20] Andre. S, Maximiliano. U, and Jan. CF, "A stochastic program to evaluate disruption mitigation investment in the supply chain". European journal of operation research, 2018, vol. 274, pp. 516-530.
[21] Priscila. F, Maria. C, and Chiara. V, "Risk management in SMEs: a systematic literature review and future directions," European Management Journal, 2019, pp. 0263-2373.
[22] Jorgensen. K, D Micheli. GJL, Mogre. R, and Perego A, "How to choose mitigation measures for supply chain risks," International Journal of Production Research, 2014, Vol. 52(1), pp. 117–29.
[23] Uijm. N. J, and Troen. H, "Accident prevention in SME using ORM," Safety Science, 2010, 48 (8), pp. 1036-1043.
[24] Chiara Verbano and Keren Venturini, "Development paths of the risk management approach, methods, and fields of application," Taylor & Francis online journal, 2014, Vol. 14 (5), pp. 519-550.
[25] Marcelino. Sadaba, S, Perez-Ezcurdia. A, and Lazcano. A. M. E, (2014). "Villanueva. Project risk management methodology for small firms," International Journal of Project Management, 2014, 32(2), pp. 327-340.
[26] Juttner. U, Peck. H. and Christopher. M, "Supply chain risk management: out-lining agenda for future research," Int. J.Logist.Manag, 2003, 6(4), pp. 197–210.
[27] Michael. D, Hugh. Medal, and Steven. A, "Proactive cost-effective identification and mitigation of supply chain delay risk in a low volume high-value supply chain using fault-tree analysis," Int. J. Production Economics Journal, 2016, Vol. 175, pp. 153-163.
[28] Raka. Chatchai, and Liangrokapart. Jirapan, "Supply Chain Risk Management: A Case Study in Thailand. In: Innovations and Strategies for Logistics and Supply Chains," Hamburg International Conference of Logistics, 2015.

[29] Ryan. G, and Valverde. M, "Waiting for service on the internet: defining the phenomenon and identifying the situations," Internet Research, 2005, 15, pp. 220-240.
[30] Chan. F. T, "Performance measurement in a supply chain," The International Journal of Advanced Manufacturing Technology 21, 2003, vol (7), pp. 534-548.
[31] Jokowinarno. D, "Mitigasi bencara tsunami di wilayah pesisir lampung," Jurnal Rekayasa, 2011, Vol.15 (1)
[32] Alijoyo. Antonius, "Enterprise risk management pendekatan praktis (edisi kedua)", Jakarta: Penerbit Ray Indonesia, 2006.
[33] Mensah. Peter, Merkuryev. Yuri, Klavins. Eric, and Manak. Sukhvir, "Supply chain risks analysis of a logging company: Conceptual Model," Journal Procedia Computer Science, 2017, vol.104 (207), pp. 313 – 320.
[34] Andre. S, Maximiliano. U, and Jan. CF, "A stochastic program to evaluate disruption mitigation investment in the supply chain". European journal of operation research, 2018, vol. 274, pp. 516-530.
[35] Gao. Yue, "A Game Theory Approach for Corporate Environmental Risk Mitigation," Journal Resources, Conservation & Recycling, 2018, 130, pp. 240-247.
[36] Sreedevi. R, and Saranga. Haritha, "Uncertainty and Supply Chain Risk: The Moderating Role of Supply Chain Flexibility in Risk Mitigation," International Journal of Production Economics , 2017, vol. 193, pp. 332-342.
[37] Tongyuan. L, Chao. Wu, and Lixiang. Duan, (2017). "Fishbone diagram and risk matrix anlaysis methode and its application in safety assesment of nantural gas spherical tank," Journal Cleaner Production, 2017, Vol. 174, pp. 296-304.
[38] Sreedevi. R, and Saranga. Haritha, "Uncertainty and Supply Chain Risk: The Moderating Role of Supply Chain Flexibility in Risk Mitigation," International Journal of Production Economics , 2017, vol. 193, pp. 332-342.

IOP Conf. Series: Materials Science and Engineering **722** (2020) 012016 doi:10.1088/1757-899X/722/1/012016

Redesign of wudhu facility for disabilities using integrated kansei method

A D Prabaswari, M A Fernanda, N P Jati, S Indrawati, R A Aldhiza

Department of Industrial Engineering, Universitas Islam Indonesia, Indonesia

E-mail: Atyanti.dyah@uii.ac.id

Abstract. The research is a wudhu place that is friendly to people with disabilities, especially wheelchair users. The purpose of this study is to find out and develop a design using integrated TRIZ, Kansei and anthropometry method. The first assessment wudhu facility using RULA method. The score of RULA is 5 which means changes required soon. TRIZ and Kansei method to defined solution the redesign wudhu facilities and anthropometry to define the dimension of design. The scope of research only redesigns, implementation not yet.

1. Introduction

Persons with disabilities have the same rights as others, but in reality, persons with disabilities based on research [1] bear the stigma of imperfection, making marginalized persons from social acceptance marginalized, and very far from fair, which is still there is a lot of discrimination related to the fulfillment of rights, education, public facilities such as transportation, places of worship, entertainment venues and the same position in public.

The design of the place of ablution is important for persons with disabilities because design errors can be fatal and the risk that can be caused increases as an accident when taking ablution water. Especially for persons with disabilities such as wheelchair users, improper design of ablution places such as height measurements that are unsuitable taps, then other supporting facilities such as putting feet and the distance that is too far between the floors used and the limiting floor can cause difficulties and potentially increase the danger and accidents for them, such as difficulties in accessing the ablution place due to limited space, then the difficulty in reaching the tap, the position of bowing during ablution water, the physical limitations that are the obstacles and fatal is falling from a wheelchair.

Based on these problems, it is certainly very necessary to design a place of ablution that is friendly to a disability, so that in designing the ablution place it can be done using an ergonomic approach and using the smart city concept approach. It is hoped that the proposed design improvements will be able to improve design not only in terms of aesthetic appearance of places of worship but are expected to be able to improve disability's friendly ability from places of worship, especially places for ablution. This research has only redesign proposal scope. This design implementation not yet.

2. Literature Review

2.1. RULA

Rapid Upper Limb Assessment (RULA) is a method developed in the field of ergonomics which a subjective observation that investigates and assesses the work position performed by the upper body but includes the lower body [2]. This method is used to take the value of work posture by taking posture samples from one work cycle that is considered to have a dangerous risk to the health of the worker, then held an assessment or scoring. After obtaining the results of the assessment, we can find out the posture of the worker is in accordance with ergonomic principles or not, if not then it is necessary to take corrective steps.

2.2. Theory of inventive problem solving (TRIZ)
TRIZ is a method developed by Altshuller in 1946 to solve contradictory problems based on logic and data [3]. There are three main steps in TRIZ, i.e. understand the problem and determine the objectives, determine the problem and formulate contradictive engineering using 39 contradiction tables, and formulate solutions with 40 inventive principles [4].

One example of a successful TRIZ is applied to an industrial problem. TRIZ was successfully used in designing an Ergonomic Milling Machine Control Knob to reduce occupational injuries and trauma disorders [5].

2.3. Kansei
Schutte (2002) says Kansei engineering is one of method to development product which as methodology for translating psychology human process to a current product or new concept design product [6].

2.4. Anthropometry
Anthropometric data is used to identify the physical size or dimensions of the workspace, work equipment, and clothing so that people can do their jobs properly. Anthropometric data is also used to avoid mismatching dimensions or sizes between equipment and products used with the size of the user or user [7].

Anthropometry will be widely used as an ergonomic consideration in the process of product design and work systems that will require human interaction. Anthropometric data that have been obtained will be widely applied, there are [8]:

- Designing a work area
- Designing work equipment such as machinery, equipment, tools and so on
- Designing consumptive products such as clothing, chairs, tables, and so on.
- Designing a physical work environment.

2.5. Percentile
Percentile is a value that shows a certain percentage of people who have a size in percentage or below a percentage [9]. For example, the 95th percentile shows that 95% of people are in or below value and can use the product. To determine anthropometric data, the normal distribution is used which can be formulated based on the average and standard deviation of the data obtained, percentiles can be determined if the data are normally distributed.

3. Research Method

3.1. Subjects
The subjects in this study were 11 people with disabilities who use wheelchairs in aged 25 to 50 years. The location for observation is located in Darussalam Mosque, Griya Perwita Wisata, Besi, Sukoharjo, Kec. Ngaglik, Kabupaten Sleman, Daerah Istimewa Yogyakarta 55581.

3.2. Method
The object of the research was to redesign the Darussalam mosque ablution facilities using the method of RULA, TRIZ, Kansei, and Anthropometry. This is intended so that persons with disabilities, especially wheelchair users can comfortably use ablution facilities without having to experience difficulties and of course to create safe facilities and reduce the risk of harm that exists. Based on the results of work postures analysis, contradictory problems in product re-design and inventive solutions are identified using TRIZ.

IOP Conf. Series: Materials Science and Engineering **722** (2020) 012016 doi:10.1088/1757-899X/722/1/012016

In its use, the Anthropometry method is used to adjust the shape of the place of ablution so that it is comfortably used by the user, while the use of the RULA method is used to keep the user's posture in a comfortable and safe.

4. Result and Discussion

4.1. RULA

The RULA method is used to investigate and assess work positions. The following is the RULA data operator that will be used as a measure of the proposed design of Wudhu place (Table 1).

Table 1. RULA

Respondent	Measured Angle				
	Neck	**Hips**	**Upper Arm**	**Elbow**	**Wrist**
1	119.76	220.32	333.37	229.09	228.03
2	55.82	447.94	115.72	662.55	440.18
3	99,48	229.86	335.7	84.6	118.4
4	338.93	331.02	52.2	52.2	214
5	88.47	443.38	44.81	75.52	38.23
6	118.56	444,71	47.89	15.06	15.01
7	223.27	339	40.33	82.75	32.82
8	337.17	663.67	44.97	33.59	37.57
9	22.18	330	22	54	62
10	119.98	33.2	30	37	32.96
11	222.57	447.56	21.82	71.69	10.83

It can be seen (Table 1) that the average value of Neck, hips, Upper Arm, Elbow, Wrist is 28.5;46.4;40.4;44.5;59.6;30.5 degree it can be implied that the posture is not good state. Respondents demonstrate their position when they will perform wudhu, especially in the bending section, based on this, it can be seen clearly the angle produced from Wudhu. The data will be processed with Ergofellow.

It can be seen (Table 1) the score of the RULA angle is 5, it means investigation and changes are required soon, so the wudhu place must be applying change soon.

This can happen because when doing work, the operator's body position is not in an ideal and comfortable position. Deviation of body posture that occurs, among others, is found in the back, the arm averages occur between 20-40 degrees. In addition, respondents also did not feel comfortable in the legs because respondents were required to bend to reach the legs.

4.2. TRIZ and Kansei

Based on the results of RULA, it can be identified that the main problem is that existing ablution facilities are not comfortable to use. So that the re-design process is needed in accordance with the dimensions of the user. But if the ablution facilities are redesigned, then the design will become more complex. In this case, there are two contradictory parameters, i.e .: Improving parameter: Productivity and Worsening parameter: Device complexity.

There are four alternative inventive solutions to the problem, i.e. equipotential (12), another dimension (17), intermediary (24) and mechanic substitution (28). In this case, the solution chosen is another dimension (17) using anthropometry.

The Kansei method is using users interview. The users are disability person who prays regularly in the location. The result of word Kansei method is flexible faucet. The faucet that can move on its own, elongate, and be shortened. Not only flexible faucet, but also water sensor that is useful for removing water when the user directs their body parts toward the faucet. This is useful to make it easier for persons with disabilities not to bend during ablution activities, especially on the legs so that users feel comfortable and avoid accidents when performing ablution activities.

The integrated TRIZ and Kansei are using inventive solutions from TRIZ and using Kansei words to supported TRIZ result.

4.3. Anthropometry

Researcher using anthropometry disability data with age range 20 to 50 years old. Table 2 is the anthropometric data from the measurements of the operator's body dimensions that will be used as a measure of the proposed wudhu.

Table 2. Anthropometry difable data

No	TBD	PPO	JT	TL	PTI	LI
1	22.5	40.5	67.5	47.4	24.5	9
2	19.8	35.3	78.5	40.8	23.5	10.4
3	23.4	40.5	79	47	24.6	9.4
4	27.8	45.6	74.5	45.5	23.2	9
5	22.1	37.6	87	46.1	22.3	9.7
6	22	29	75	44	25	11
7	22	50	73	41	22	12
8	31	37	76	45	27	11
9	29	37	58	38.5	24	12
10	16	37	57	41	22	9
11	18	42	97	45,5	26	10
12	22	43	83	41	23	11
13	20.3	41.6	68.4	48.1	24.5	10.3
14	18	38	67	46	24	11
15	23	37	72	45	23	10
16	25	32	93	51	23,5	11
17	21	44	71	45	23	10
18	24	45	69	45	24	6
19	19	41	70	43	26	13.5
20	24	48	74	45.2	23	11
21	23	44	76	45.4	25	12
22	16	44.5	89.2	41	22	11
23	24	36	78	50	23.4	10.5
24	21	34	74	48	25.5	11.5
25	19	41	72	43	23,3	10
26	20	41	80	44	18	10
27	22	42	77.5	48.5	24.2	10.8
28	23	46	79	48	25	11
29	18	33.5	67	45	22	10
30	21	40	67	40	23	9

The results of the measurement of the Anthropometry is Percentile, it can be calculated using the formula:

$$P = \bar{X} \pm Z_x . SB \tag{1}$$

Zx = Normal standard value, where (+) if P> 51, and (-) if P <50
SD = Standard Deviation
The selection of dimensions used is based on the design requirements for ablution sites. The dimensions used are as follows:

- Sitting shoulder height (TBD)

IOP Conf. Series: Materials Science and Engineering **722** (2020) 012016 doi:10.1088/1757-899X/722/1/012016

The choice of dimensions of TBD (Sitting Shoulder Height) is useful for the height of the tap so that in this dimension the percentile used is percentile 5. The reason for using the 5th percentile is for someone who has a low or small posture to be able to use the ablution tap comfortably. Then based on the percentile calculation the TBD value is 16.34 cm. This size is used as a high size tap that will be designed.

- Popliteal butt (PPO)

The choice of the PPO dimension (Popliteal Butt) is useful for determining the distance of the ablution place, so that in this dimension the percentile used is the 50th percentile. Then based on the percentile calculation, the PPO value is 40.1033 cm, this size is used as a high measure of the tap to be designed.

- Hand reach (JT)

The selection of the dimensions of JT (Hand Reach) is useful for determining the range of the user's hand with a tap, so that on this dimension the percentile used is percentile 5. The reason for using the 5th percentile is for someone who has a small posture to reach the tap comfortably and easily. Then based on the percentile calculation, the JT value is 60.55 cm. this size is used as the distance of the user's hand with the tap.

- Knee height

The selection of TL dimensions (Knee Height) is useful for determining the distance of the tap movement on the foot, so that in this dimension the percentile used is the 95th percentile. The reason for using the 95th percentile is that someone who has a large posture can use the tap comfortably. TL value is 49.7465 cm. this size is used to determine the distance of the tap movement.

- Foot length (PTI)

The selection of TL dimensions (Knee Height) is useful for determining the distance of the tap movement on the foot, so that in this dimension the percentile used is the 95th percentile. The reason for using the 95th percentile is that someone who has a large posture can use the tap comfortably. TL value is 49.7465 cm. this size is used to determine the distance of the tap movement.

- Foot width (LI)

The choice of LI dimension (Leg Width) is useful to determine the width of the footrest so that on this dimension the percentile used is the 95th percentile. The reason for using the 95th percentile is that someone who has a large posture can put his foot comfortably. 12.55 cm size is used to determine the width of the footrest.

4.4. Recommendation

The following is a recommended picture of the layout of the observation room based on TRIZ and Kansei method.

Figure 1. Product isometric view

The Product designed for disabled people using a wheelchair, with a moving tap so the user can wudhu without bent. Moving tap can move into four directions, Up, Bottom, Left and Right. It can be controlled with a blue box on the picture, it will give more capability to wudhu if the tap can move with adjusting our movement. In the bottom there are some sensors that can turn on the tap, then the tap will release the water if foot area on the above. The hole in the last picture is another waterway that can get out beside tap. The user doesn't need to wash his feet because it will automatically be washing.

5. Conclusion

The indicators needed to measure the form of ablution friendly ablution places with the smart city concept based on research are using human anthropometry, where dimensions are used namely sitting shoulder height (TBD), Popliteal Buttocks (PPO), Hand Reach (JT), Knee Height (TL), Foot Length (PTI) and Leg Width (LI). Each dimension has a purpose to determine the height of the tap at the top, to determine the distance of the foot to the place of ablution, then to determine the distance of the user's hand with the tap, then to determine the distance of the tap movement on the foot, then to determine footrest length and determine the width of the footrest. Then as a supporter of the redesign of the ablution place is done by taking the TRIZ method and the Kansei method by the voice of customer.

Results obtained based on the Anthropometry method in each dimension are sitting shoulder height (TBD), Popliteal Butt (PPO), Hand Reach (JT), Knee Height (TL), Footprint (PTI) and Leg Width (LI) are 16.34 cm, 40.1033 cm, 60.55 cm, 49.7465cm, 26,359 cm, and 12.55 cm respectively. Then, using the RULA (Rapid Upper Limb Assessment) method where the RULA method calculates the degree of slope of several parts of the user's body to be tested so that the researcher can find out what the RULA score obtained during the ablution process, whether or not a humped disability user is obtained. immediate repairs need to be done, so that design changes will be made to places with special disabilities for wheelchair users, especially wheelchair users

In designing the place of ablution proposal, the researcher uses the basis of Anthropometry and the standard size of the place of ablution in general which is obtained from the architect's data the size of the place of ablution. As for the innovations that we added based on user needs using the Kansei method, we found several innovations in the form of movable taps, self-moving taps, elongated taps, and long taps that can reduce size and water sensors that are useful for removing water when the user directs their body parts towards the water tap. The redesign is recommendation design that is useful to make it easier for persons with disabilities not to bend during ablution activities especially in the legs so that users feel comfortable and avoid accidents when performing wudhu activities.

Acknowledgment
Researchers thank the support of the Industrial Engineering Department, Faculty of Industrial Technology, Universitas Islam Indonesia.

References

[1] Prakosa P W 2011 *Social dimention disability in semin community: An approach social representation* Vol. 32 pp. 61-73
[2] Dockerell S et all 2012 *An Investigation of realibility of Rapid Upper Limb Assessment as a method of assessment of children's computing posture Applied Ergonomics* vol. 43 pp. 632-636
[3] Navas V G, Helena 2013 *TRIZ: Design Problem Solving with Systematic Innovation, Advances in Industrial Design Engineering* ISBN: 978-953-51-1016-3 InTech DOI: 10.5772/55979
[4] Yeoh T S, Yeoh T J, Song C L 2016 *TRIZ: Systematic Innovation in Manufacturing* First Fruit Publishing Malaysia
[5] Kiat Ng, P Siong Jee K 2016 *Design and Development of an Ergonomic Milling Machine Control Knob using TRIZ Principles American Journal of Applied Science* 13 (4): 451.458 DOI: 10.3844/ajassp.2016.451.458
[6] Rahmayani N, Yuniar, Desrianty A *2015 Loose Powder Design using Kansei Engineering Method Jurnal Teknik Industri Itenas* no. 04 vol. 03
[7] Bridger R P 1995 *Introduction to ergonomics* McGraw-Hill
[8] Wignjoesoebroto S 2003 *Ergonomi: a study of movement and time* Guna Widya Surabaya
[9] Tayyari S 1997 Occupational ergonomics principles and applications Great Britain

IOP Conf. Series: Materials Science and Engineering **722** (2020) 012017 doi:10.1088/1757-899X/722/1/012017

Spare part requirement and inventory policy for Rovema's 1 machine using Reliability Centered Spare (RCS) and Min-Max stock methods

C F Angelina[1], F T D Atmaji[2], B Santosa[3]

[1,2,3] Industrial Engineering Department, School of Industrial and System Engineering Telkom University, Bandung, Indonesia

E-mail:cfreskyangelina.student.telkomuniversity.ac.id,franstatas@telkomuniversity.ac.id, budisantosa@telkomuniversity.ac.id

Abstract. XYZ company is a manufacturing company engaged in the field of medicines, food, and natural products. One of the problems found in the machine delay, called "outstanding". An outstanding occurs is mostly caused by the unavailability of the spare parts. Based on the downtime losses data, the highest downtime, as well as unavailability spare parts, was owned by Rovema's machine, one of the packaging machines in food plant division. Therefore in this research, the Reliability Centered Spare (RCS) method was applied to calculate the optimal critical spare part policy using the Poisson process and Min-Max stock analysis. The analysis result shows that the critical component of Rovema's machine is steel band and brass insert are must be stored in a technical warehouse for one year ahead with optimal inventory quantity is 18 and 15 pieces with minimum stock is 10 pieces each

1. Introduction

Policies in managing inventory are very important in a company. Without a good inventory control system, the company cannot run their production smoothly, and it's difficult to fulfill customer demand. Inventory is one part of company spending that absorbs the largest cost. Over excess inventory causes high storage costs, but if there is a shortage of inventory there can be a risk of not being able to meet consumer demand. Spare part inventory is one of the critical aspects to support the whole inventory system in the manufacturing process.

The purpose of spare parts inventory management is to maintain the level of inventory of spare parts at an optimal level with a minimum inventory cost, determine when spare parts are ordered and determine the number of spare parts ordered. D. Louit et al, 2011 [1] present the spare parts inventory model used to determine the optimal stock size of the non-repairable and repairable component. Non-repairable parts are components that do not allow to be repaired physically, so the components must be replaced, meanwhile, repairable parts are damaged components that can be returned to their original function or operational state by repairing the components. The need for the number of critical spare parts that must be provided within a certain period using the Reliability Centered Spares (RCS) is analyzed in [2]-[3]. Meanwhile, [4] presented an integrated spare part inventory management approach for multi-

IOP Conf. Series: Materials Science and Engineering **722** (2020) 012017 doi:10.1088/1757-899X/722/1/012017

component systems subjected to random failures. Others researcher [5]–[7] used the deterministic deteriorating inventory (DDI) model, the stochastic deteriorating inventory (SDI), stocking policy, and dynamic predictive maintenance model for spare part analysis which follows the principle of inventory control level system.

In the manufacturing process, a spare parts inventory control must be concerned by the Engineering Maintenance Department to ensure the availability of each machine's spare part, especially for its critical and fast-moving item. A problem related to spare part inventory also faced by XYZ company, one of the manufacturing company engaged in medicine, food, and natural products. In the food plant division, there are several machines which used for the production process, however, in fact, the reliability of a machine can decrease over time, which can cause the machine downtime. Figure 2 shows a graph of the total downtime per machine in the food plant division.

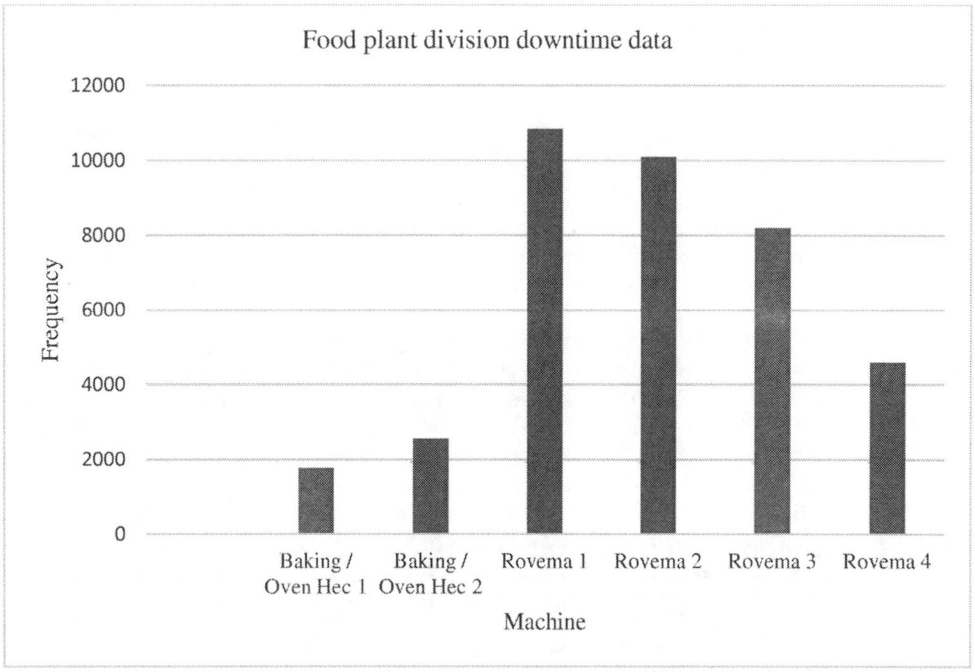

Figure 1. Downtime data of several machines in food plant division

Based on Figure 2, it can be seen that the highest downtime occurred on Rovema's 1 machine. Rovema's machine is one of the packaging machines which used for the final package of the food product.

Based on data from the company, there are several main factors causing machine downtime or in this term is called an "outstanding". An outstanding is the delay in machine repair which caused by several factors. Figure 2 shows the Pareto diagram for the outstanding of Rovema's machine in 2018.

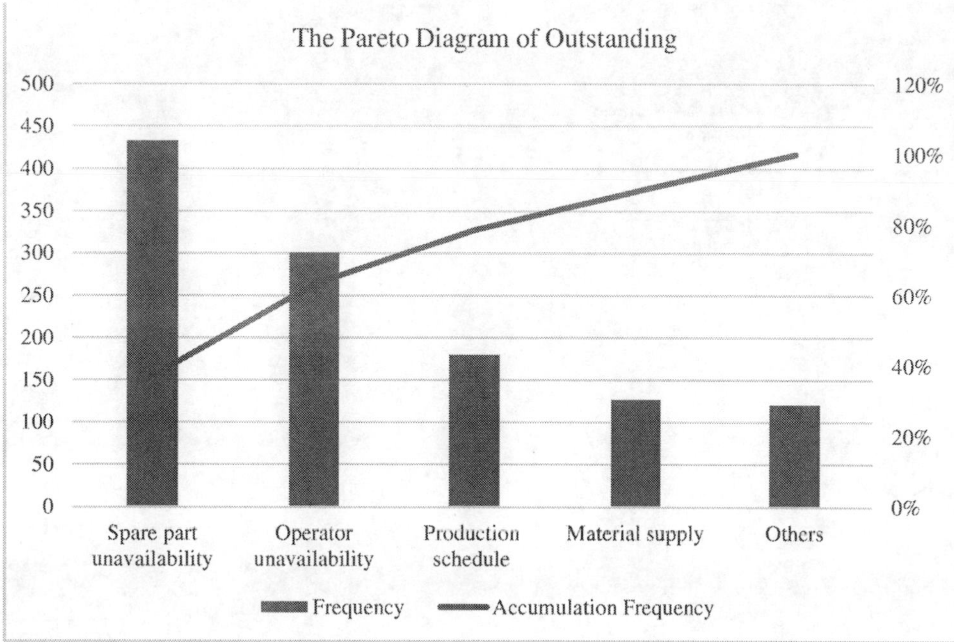

Figure 2. Pareto diagram of outstanding causes in 2018

From Figure 2, it can be seen that the outstanding cause consists of several factors such as spare part unavailability, operator unavailability, production schedule, material supply, and other causes. From the Pareto diagram above, it can be seen that the main cause of outstanding is the spare part unavailability. To minimize the problem of availability of spare parts above, the purpose of this study is to analyze, determine, and propose an optimum spare parts inventory stock policy, especially for critical components in the Rovema's 1 machine. The analysis is using Reliability Centered Spares (RCS) method with the Poisson process and Min-Max stock analysis. Therefore, the RCS method is used to determine the optimum number of spare parts requirements based on equipment needs and maintenance operations for a certain period, whereas, the Min-max stock is used to determine the minimum and maximum spare parts inventory.

2. Research Methodology

As stated in the introduction, this research was conducted in Food plant division, in the Rovema's machine, one of the packaging machine which mainly produces snack products. Figure 3. shows and explains the main steps of this research.

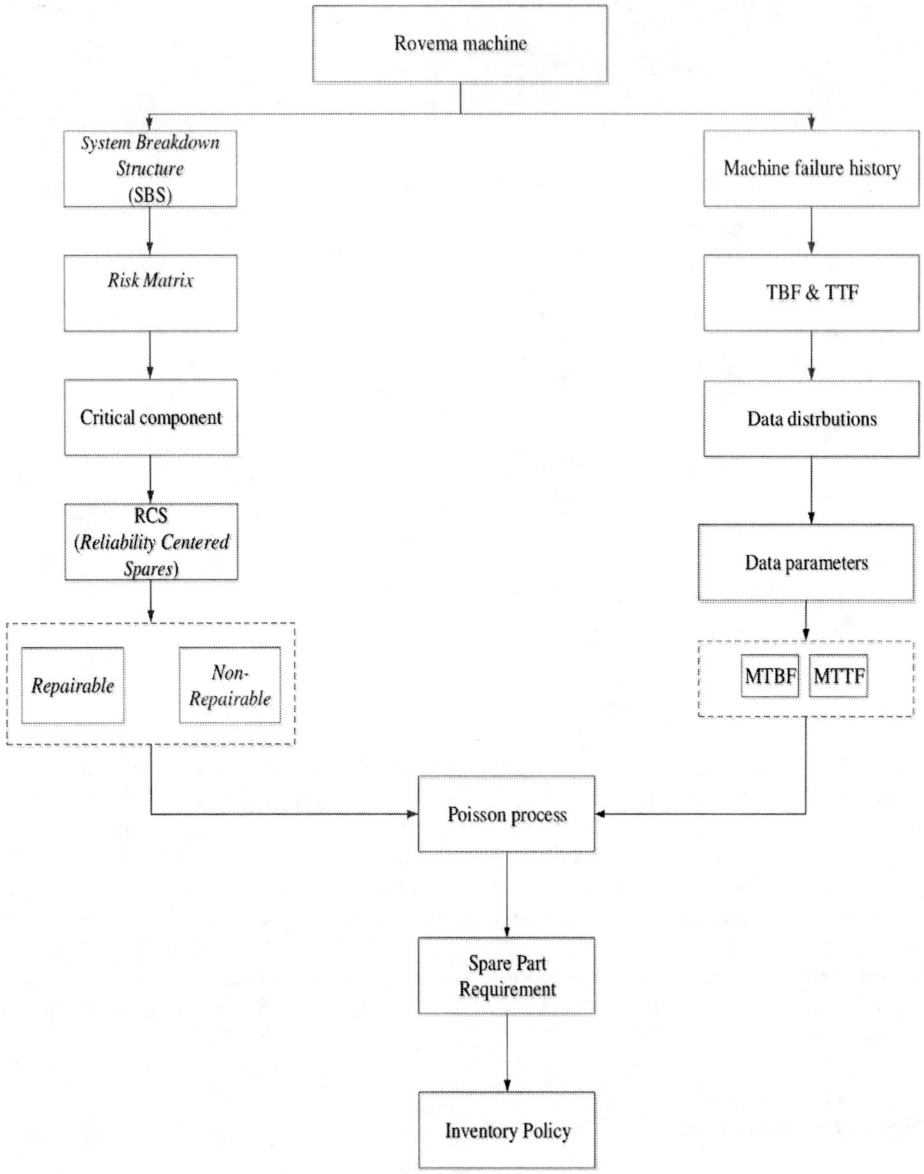

Figure 3. Research methodology

Based on Fig. 3, this research was conducted at Rovema's 1 machine. To identify the structure of a machine, a System Breakdown Structure (SBS) of the Rovema's 1 machine is performed to determine the subsystems and components. Then, perform a Risk Matrix based on the severity and likelihood levels to determine the critical subsystems and critical components. Critical components are classified into two, namely non-repairable or repairable. If the component is non-repairable, then the component must be replaced with a new component. In the meantime, if the component is repairable, then the component can be repaired or serviced to return it as a new one. From machine history data, we can find out TTF data and TBF data. TTF and TBF data were tested for distribution to determine the distribution of TTF and TBF data using software Minitab. Then, determine the parameters of TTF and TBF using software AvSim+ 9.0. The parameters obtained for each data distribution are used to calculate MTTF and MTBF values. Furthermore, spare parts requirements are calculated using the Poisson Process. After obtaining spare parts requirements, each critical spare part is determining inventory policy using the Min-Max Stock and Reorder Point method. Min-Max Stock is used to determining the minimum and maximum inventory of critical parts, meanwhile the Reorder Point is used to decide the limit of -reorder the spare part.

The first step in this research is to determine the component or spare parts of the Rovema's 1 machine which will be the focus and further analysis. System Breakdown Structure (SBS) must be carried out to facilitate the process of identifying parts of the machine. System Breakdown Structure (SBS) is a tool for analyzing, documenting and communicating machine structures at the system, subsystem, and component-level in detail. SBS classifies systems in equipment based on function. The existing functional system can be identified more easily and clearly. SBS is the first step that must be done to identify the components of a system to be further analyzed in [8]. SBS can also be used to identify the level at which failures occur and analyze their effects on systems, subsystems or related components. SBS provides a representation of parts of the equipment in the form of interrelated diagrams based on their functions. How to analyze SBS is to divide the machine into several main parts of the machine, such as electrical systems, mechanical systems, pneumatic systems, hydraulic systems, and machine support systems. From the main parts of the machine will be divided again into more detailed parts or sub-systems and components.

After analyzing the SBS, the next step is to determine the critical components or parts of the machine. The determination and selection of critical spare parts use the Risk Matrix. The Risk Matrix is a risk matrix that is used to determine the various levels of risk from several categories of hazard probabilities and the effects of these risks [9]. Risk assessment in the risk matrix can be seen from the level of severity or consequences and probability or likelihood [10]–[12] [13]. For the 5x5 risk matrix model, the severity level consists of insignificant, minor, moderate, major, and catastrophic categories. Meanwhile, the likelihood level consists of rare, impossible, possible, likely, and almost certain categories.

After the critical components (spare parts) have been selected, then the failure history data of each component is used to determine the Mean Time To Failure (MTTF), Mean Time Between Failure (MTBF). MTTF is the average interval of damage from distribution of damage, MTB is the time between failures of a component operating under normal conditions.

For the calculation of MTTF and MTBF use the following basic reliability formulas [14].

$$MTTF = \int_0^\infty t.f(t)dt \quad (1)$$

$$MTBF = \int_0^\infty R(t)dt \quad (2)$$

Which t is time, f (t) is time function and R (t) is reliability function.

There are several types of failure distribution functions based on reliability that are often used to analyze maintenance problems, for example, Exponential distribution, Normal distribution, and Weibull distribution. Failure data from critical components are analyzed using Minitab 17 and AvSim 9.0 software to determine the type of failure distribution. This software is one of the software to calculate and determine the type of distribution that represents the failure in critical components of the engine. The next step is processing and analyzing data using the Reliability Centered Spares (RCS) method. Reliability Centered Spares (RCS) is one method of managing spare parts analysis by considering several aspects such as maintenance needs of what is needed by the machine and the consequences that occur if parts are not available. This method allows maintenance activities without having to wait for the procurement of replaced components. RCS can be used to determine spare parts inventory levels based on equipment needs and maintenance operations [15]–[17]. In carrying out this RCS calculation the parts or components will be divided into components that can be repaired (repairable) and components that cannot be repaired (non-repairable). The calculation of these parts is based on the failure rate of each critical component using Poisson Proces [5], [18], [19]. Poisson Process is a technique to calculate the need for parts based on reliability. The demand for spare parts due to a component replacement or failure that occurs as a result of maintenance is an event that is described as a Poisson distribution that occurs based on events that occur within a certain interval. For components that cannot be repaired (non-repairable) use the following formula [20]–[23]:

$$\lambda t = (A \times N \times M \times T) / MTTF \quad (3)$$

As for components that can be repaired (repairable) using the formula:

$$\lambda t = (A \times N \times M \times T) / MTBF \quad (4)$$

Where λt is failure rate, A is a Number of Components, N is a number of units/machines used, T is an initial period, M is utility / operating machine, MTTF is Mean Time To Failure and MTBF is Mean Time Between Failure. In general, using the RCS analysis, it will help to ensure the availability of spare parts in conducting maintenance, determine the strategy of spare parts, and the number of spare parts needed in a certain period, for example in one year.

To support the RCS method, the Min-Max Stock and Reorder Points analysis also will be conducted. The Min-Max Stock method is a method in controlling spare parts inventory, which is determined by minimum stock and maximum stock of spare parts [24]. If it has reached the minimum, it is necessary to order spare parts to reach the maximum limit of spare parts. The application of the Min-Max Stock method will help the warehouse to determine how much minimum stock must be available to meet the production quantity capacity and the maximum stock of spare part in the warehouse. Inventory control using the Min-Max Stock method consists of several stages, namely determining safety stock, min stock, and max stock [25]. The following are the equations in the calculation using the Min-Max Stock method [26], [27]:

$$\text{Safety Stock} = (\text{Maximum Usage} - T) \times C \quad (5)$$

$$\text{Min Stock} = (T \times C) + S \quad (6)$$

$$\text{Max Stock} = 2 \times (T \times C) \quad (7)$$

Where Q is an average usage per certain period (unit), C is lead time (month), and S is safety stock (unit).

Reorder Points (ROP) is a method of inventory points or limits, where reorder will be made [28]. In calculating this reorder point is determined by the length of the lead time, average usage and safety stock [29][30]. Here's the equation in reordering point calculations:

$$ROP = (AU \times LT) + SS \quad (8)$$

Where the AU is an average usage per certain period (unit), LT is lead time (month), and SS is Safety stock (unit).

3. Result and Discussion

The result and of this research will be explained in several subchapters as follow.

3.1. System Breakdown Structure (SBS)

The results of the analysis System Breakdown Structure (SBS) for the Rovema's machine are described in Figure 4

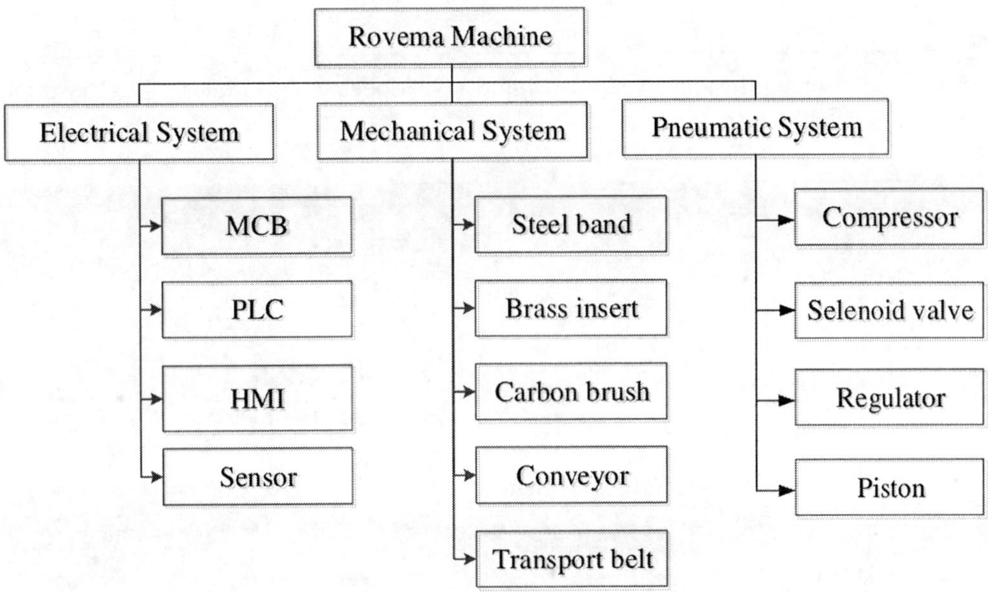

Figure 4. System Breakdown Structure

Figure 4 shown that Rovema's machine is divided into three main systems which are an electrical system, mechanical system, and pneumatic system. Each of the systems consists of several components. Following completed the System Breakdown Structure (SBS) of the Rovema machine system, the next step is selecting critical components that will be the main focus of this research.

IOP Conf. Series: Materials Science and Engineering **722** (2020) 012017 doi:10.1088/1757-899X/722/1/012017

3.2. Determination of Critical Subsystem and Component

Selection of critical subsystems and critical components using the Risk Matrix. The selection of subsystems and critical components is carried out to determine the subsystems and components that have a very high impact or risk for the whole machine system. In the risk matrix, it consists of two categories, which is severity and likelihood. In the severity category, there are three aspects used, specifically production, operational, and safety. The likelihood is the probability of damage or failure of subsystems and components.

Table 1. The Risk Matrix of the critical component

		Severity				
	Sub sistem	Insignificant (1)	Minor (2)	Moderate (3)	Major (4)	Catastrophic (5)
Likelihood	Almost Certain (5)				**Steel Band, Brass Insert**	
	Likely (4)					
	Possible (3)		**Carbon Brush**			
	Unlikely (2)		**Transport Belt**	**Conveyor**		
	Rare (1)					

Based on the results of the Risk Matrix in Table 1, it can be seen that the components of steel band and brass insert are included in the category of very high (red-colored) which means further action needs to be taken, carbon brushes and conveyors are included in the medium category (yellow colored) which means the risk can be tolerated, and the transport belt is categorized as low (green) which means the risk is acceptable or still safe. Consequently, the critical components that will be further investigated or as critical components in the mechanical sub-system are steel band and brass insert

IOP Conf. Series: Materials Science and Engineering **722** (2020) 012017 doi:10.1088/1757-899X/722/1/012017

3.3. Determination of TTF and TBF data distribution tests

Time To Failure (TTF) is the time of damage or failure after a system has been repaired until it is damaged again. Time Between Failure (TBF) is the time of failure or failure of a system where the system has been damaged again after experiencing previous damage. A distribution test is done to determine the distribution that represents TTF, and TBF data using the Anderson-Darling test. The distribution used in this distribution test is a Normal, Exponential, and Weibull distribution. Determining the distribution of TTF and TBF data using Minitab software. This data uses the significance level (α) which is 0.01. If P-Value < α then reject the initial hypothesis (H0). The chosen distribution is the distribution with the smallest Anderson-Darling (AD) value and the P-Value > α. Following Anderson-Darling's hypothesis:
H0: Component TTF / TBF data with Normal / Exponential / Weibull distribution

H1: Component TTF / TBF data are not Normal / Exponential / Weibull distribution

Table 2. Time To Failure (TTF) analysis result

Components	**Distribution**	**AD**	**P-Value**	**Selected Distribution**
Steel Band	Normal	1.208	< 0.005	Weibull distribution
	Exponential	3.287	< 0.003	
	Weibull	0.426	> 0.250	
Brass Insert	Normal	0.889	0.015	Weibull distribution
	Exponential	0.268	0.851	
	Weibull	0.238	> 0.250	

Based on table 2, the TTF data of Steel Band component has a Weibull distribution because it has the smallest AD value of 0.426 and the largest P-value of more than a significant level (α) which is 0.250. The TTF data of the Brass Insert component has a Weibull distribution because it has the smallest AD value of 0.238 and the largest P-value of more than a significant level (α) which is 0.250.

IOP Conf. Series: Materials Science and Engineering **722** (2020) 012017 doi:10.1088/1757-899X/722/1/012017

Table 3. Time Between Failure (TTF) analysis result

Components	Distribution	AD	P-Value	Selected Distribution
Steel Band	Normal	1.134	< 0.005	Weibull distribution
	Exponential	2.118	0.005	
	Weibull	0.323	> 0.250	
Brass Insert	Normal	0.851	0.019	Exponential distribution
	Exponential	0.253	0.873	
	Weibull	0.261	> 0.250	

Based on table 3, the TBF data of Steel Band component has a Weibull distribution because it has the smallest AD value of 0.323 and the largest P-value, where the P-value is more than the significant level (α) that is 0.250. Meanwhile, the TBF data of the Brass Insert component has an Exponential distribution because it has the smallest AD value of 0.253 and the largest P-value, where the P-value is more than the significant level (α) that is 0.250.

3.4. Determining the Value of Mean Time To Failure (MTTF) and Mean Time Between Failure (MTBF)

Mean Time To Failure (MTTF) is the average time interval of failure that occurs when the machine or component is repaired until the machine or component is a failure again. Whereas, Mean Time Between Failure (MTBF) is the average distance between failure.

Table 4. MTTF and MTBF analysis result

Components	Parameters		MTTF (hours)	Parameters		MTBF (hours)
Steel Band	η	1234.34	3546.87	η	1480.12	3583.12
	β	0.422761		β	0.45466	
Brass Insert	η	2326.85	3193.12	η	2539.51	2539.51
	β	0.647443		β		

IOP Conf. Series: Materials Science and Engineering **722** (2020) 012017 doi:10.1088/1757-899X/722/1/012017

Based on table 4, the MTTF value for Steel Band and Brass Insert components are 3546.87 hours and 3193.12 hours respectively, in the meantime the MTBF value for Steel Band and Brass Insert components are 3583.12 hours and 2539.12 hours.

3.5. Component Classification

After the MTTF and MTBF calculation, the classification of component is needed to specify of repairable or non-repairable. Non-repairable is a component that cannot be repaired, therefore that a component is must be replaced. Meanwhile a repairable is a component that can be repaired which mean that it can be used and operated again.

Table 5. Component classification

Components	Classification
Steel Band	Non-repairable
Brass Insert	Non-repairable

Based on Table 5, both of Steel Band and Brass Inserts including into non-repairable components, which means when it gets damaged, this component need to be replaced and can't be repaired again.

3.6. Determination of component and inventory policy

Determination of component requirements using the Reliability Centered Spares (RCS) method with Poisson analysis shows the final policy and quantity of critical component for 1 year ahead. The following is the calculation based on equation (3) and (4) of critical component requirements for the next year.

Table 6. The component policy and requirement

Components	Policy	Requirement (unit)
Steel Band	Store	18
Brass Insert	Store	11

Based on table 6, the final policy both of the Steel band and Brass Insert needs to be stored in a technical warehouse for the next one years with the number of components needed is 18 units and 11 units respectively. This policy is to minimize the downtime machine caused by the deficiency of a critical component.

3.7. Determination of Minimal and Maximal stock

To optimize inventory policy for the critical components of Rovema'1 machine, it's important to analyze the minimal and maximal (Min-Max) stock for this critical component using equation 5, 6 and 7. This analysis is important to minimize problem-related to the possibility of out of stock for the critical component, especially when the machine suddenly getting a breakdown. The Min-Max Stock method is a method used to determine the minimum and maximum inventory of spare parts with the aim of no shortage of spare parts and excess parts or accumulation in the warehouse. If there already reached the minimum level of stock, it will need to order spare parts or in general called reorder point. Reorder point is the point or boundary of the company having to order spare parts again. Moreover, safety stock is needed because there is a possibility that the use of component will change and suddenly increase and the ordered items arrive late. The following results determine the min-max inventory policy for each critical component.

Table 7. The min-max policy

Component	Min Stock	Max Stock
Steel Band	10	18
Brass Insert	10	15

Table 7 shows the min-max stock policy for steel band and brass insert which min-max stock for Steel Band is 10 and the maximum is 18 and for Brass Insert minimum is 10 and maximum is 15 pieces respectively. Based on the equation, the reorder point for this critical component is 10 pieces. This mean, after the remaining inventory reaching 10 pcs, the component must be ordered again to meet the optimal stock level.

4. Conclusion

This paper proposed of spare part and inventory analysis for steel band and brass insert as a critical component in Rovema's packaging machine. The Reliability Centered Spare (RCS) method was applied to calculate the optimal critical spare requirement using the Poisson process and Min-Max stock analysis. The RCS method using Poisson process with considering the failure rate of critical component is used to determine the optimal policy of inventory, meanwhile, the min-max stock analysis used to decide the quantity minimum and maximum critical spare part. The result shows that both of steel band and brass insert must be stored in certain quantity at the technical warehouse for one year ahead to minimizing the downtime which caused by unavailability spare part. Using this result analysis, the company can consider their current spare part inventory system to keep the Rovema's packaging machine run smoothly during the production process. For the next research, further analysis is needed for the others component, i.e. electrical or pneumatic system to increase the Rovema's machine reliability, availability and maintainability.

References

[1] D. Louit, R. Pascual, D. Banjevic, and A. K. S. Jardine, "Optimization models for critical spare parts inventories-a reliability approach," *J. Oper. Res. Soc.*, vol. 62, no. 6, pp. 992–1004, 2011.

[2] F. Tatas Dwi Atmaji and A. Agung Ngurah Nanda Utama Putra, "Kebijakan Persediaan Suku Cadang Di PT ABC Menggunakan Metode RCS (Reliability Centered Spares) Spare Part Inventory Policy at ABC Company Using RCS (Reliability Centered Spare) method," *J. Manaj. Ind. DAN LOGISTIK*, vol. 2, no. 1, pp. 84–94, 2018.

[3] Z. Ega, J. Alhilman, and F. T. D. Atmaji, "Perencanaan Pengelolaan Suku Cadang Pada Pompa Produk Menggunakan Reliability Centered Spares Pada Terminal BBM PT. XYZ," *J. Ind. Serv.*, vol. 4, no. 1, 2018.

[4] C. Diallo, D. Ait-Kadi, and A. Chelbi, "Integrated spare parts management," in *Handbook of Maintenance Management and Engineering*, 2009, pp. 191–222.

[5] Y. Jiang, M. Chen, and D. Zhou, "Joint optimization of preventive maintenance and inventory policies for multi-unit systems subject to deteriorating spare part inventory," *J. Manuf. Syst.*, 2015.

[6] A. B. M. Z. Kabir and A. S. Al-Olayan, "A stocking policy for spare part provisioning under age based preventive replacement," *Eur. J. Oper. Res.*, 1996.

[7] Q. Liu, M. Dong, and Y. Peng, "A dynamic predictive maintenance model considering spare parts inventory based on hidden semi-Markov model," *Proc. Inst. Mech. Eng. Part C J. Mech. Eng. Sci.*, 2013.

[8] S. Cavalieri, M. Garetti, M. MacChi, and R. Pinto, "A decision-making framework for managing maintenance spare parts," *Prod. Plan. Control*, 2008.

[9] L. S. Lipol and J. Haq, "Risk analysis method : FMEA / FMECA in the organizations.," *Int. J. Basic Appl. Sci.*, 2011.

[10] A. C. Cagliano, A. De Marco, S. Grimaldi, and C. Rafele, "An integrated approach to supply chain risk analysis," *J. Risk Res.*, 2012.

[11] S. Glöser, L. Tercero Espinoza, C. Gandenberger, and M. Faulstich, "Raw material criticality in the context of classical risk assessment," *Resour. Policy*, 2015.

[12] X. Ruan, Z. Yin, and D. M. Frangopol, "Risk Matrix Integrating Risk Attitudes Based on Utility Theory," *Risk Anal.*, 2015.

[13] D. Ristic, "A Tool for Risk Assessment," *Saf. Eng.*, vol. 3, no. 3, 2013.

[14] J. Moubray, "Introduction to Reliability-centered maintenance," in *Reliability-centered maintenance*, 2001, p. 426.

[15] S. O. Duffuaa and A. Raouf, *Planning and control of maintenance systems: Modelling and analysis*. 2015.

[16] D. Moore, "Reliability Centered Maintenance support tools," in *Failure Prevention: Implementation, Success Stories and Lessons Learned - Proceedings of the 2009 Conference of the Society for Machinery Failure Prevention Technology*, 2009.

[17] A. Harding, "Reliability, Maintenance and Logistic Support: A Life Cycle Approach," *J. Oper. Res. Soc.*, 2002.

[18] M. C. Van Der Heijden, E. M. Alvarez, and J. M. J. Schutten, "Inventory reduction in spare part networks by selective throughput time reduction," in *International Journal of Production Economics*, 2013.

[19] A. Molenaers, H. Baets, L. Pintelon, and G. Waeyenbergh, "Criticality classification of spare parts: A case study," in *International Journal of Production Economics*, 2012.

[20] H. G. H. Tiemessen and G. J. Van Houtum, "Reducing costs of repairable inventory supply systems via dynamic scheduling," in *International Journal of Production Economics*, 2013.

[21] J. P. J. Van Kooten and T. Tan, “The final order problem for repairable spare parts under condemnation,” *J. Oper. Res. Soc.*, 2009.
[22] H. Wong, G. J. Van Houtum, D. Cattrysse, and D. Van Oudheusden, “Multi-item spare parts systems with lateral transshipments and waiting time constraints,” in *European Journal of Operational Research*, 2006.
[23] W. D. Rustenburg, G. J. Van Houtum, and W. H. M. Zijm, “Spare parts management at complex technology-based organizations: An agenda for research,” *Int. J. Prod. Econ.*, 2001.
[24] A. P. Kinanthi, D. Herlina, and F. A. Mahardika, “Analisis Pengendalian Persediaan Bahan Baku Menggunakan Metode Min-Max (Studi Kasus PT.Djitoe Indonesia Tobacco),” *PERFORMA Media Ilm. Tek. Ind.*, vol. 15, no. 2, pp. 87–92, 2018.
[25] T. Iqbal, D. Aprizal, and M. Wali, “Aplikasi Manajemen Persediaan Barang Berbasis Economic Order Quantity (EOQ),” *J. JTIK (Jurnal Teknol. Inf. dan Komunikasi)*, vol. 1, no. 1, pp. 48–60, 2017.
[26] N. Agin, “A Min-Max Inventory Model,” *Manage. Sci.*, 1966.
[27] Y. J. Chen and Y. M. Chen, “A fundamental analysis-based method for stock market forecasting,” in *Proceedings of the 2013 International Conference on Intelligent Control and Information Processing, ICICIP 2013*, 2013.
[28] T. Lukmana and D. T. Y, “Penerapan Metode EOQ dan ROP,” *J. Tek. Inform. dan Sist. Inf.*, vol. 1, no. 3, pp. 271–279, 2015.
[29] G. G. Kencana, “Analisis Perencanaan dan Pengendalian Persediaan Obat Antibiotik di RSUD Cicalengka Tahun 2014,” vol. 3, pp. 42–52, 2015.
[30] A. Shtub and M. Simon, “Determination of reorder points for spare parts in a two-echelon inventory system: The case of non identical maintenance facilities,” *Eur. J. Oper. Res.*, 1994.

Observing knowledge of ergonomic product in Indonesia

G B Pratama, N Nurfitrisari, and A Widyanti*
Department of Industrial Engineering, Institut Teknologi Bandung, Bandung, Indonesia

*Corresponding author email: widyanti@ti.itb.ac.id

Abstract. A range of ergonomic products have started to be widely produced and used in Indonesia. Office supplies as well as computer peripherals appear to be the majority of product types which are ergonomic labeled by the manufacturer. However, how well is people's knowledge regarding the characteristics of ergonomic product and its benefit are still unidentified. Not to mention the intention to use ergonomic-labeled product after realizing the benefit. This research is aiming to observe people's level of awareness and intention to use the ergonomic-labeled products through a set of experiments. The investigation used several types of office chair, keyboard, and mouse representing the ergonomic product and standard product. Participants' productivity by using ergonomic and standard product were compared and analyzed. The intention to use ergonomic products before and after trial were also observed.
Keywords*. ergonomic product, intention to use, knowledge*

1. Introduction

Ergonomic product is designed to improve the usability of the product and to reduce the risk of injury which is possibly appeared during the product use. Ergonomics aspects on product development in industry is one of major strategies for prevention of work-related injuries and illnesses among employees at the moment of manufacture of new products [1]. In fact, many ergonomic products are having an increase in its production and utilization rate, because they are claimed to be able to give more benefit than any other standard products with the same function. The claims are supported by several relevant studies. Smith and Bayeh [2] for instance, showed that the ergonomic product, in this case a fully adjustable chair, utilized by call center team members in Midwest, United States, can improve the average productivity by 4.78%. Study by Cofco [3] mentioned that the probability of having injury, such as low back pain, carpal tunnel sydrome, sprained, hernia, etc. is higher when the product was not designed based on the ergonomics principles.

Not only for the customer, the ergonomic product can also bring benefit to the manufacturer of the product. The "ergonomic" label appears to be one of marketing strategies one can use to increase their business. The manufacturer can bring the ergonomic design as a competitive advantage in the market [4]. Some manufacturers even use ergonomic-product label explicitely to attract the customers. Not to mention, some are chasing the certification of ergonomic product through a credible institution like US Ergo in United States [5]. The certificate basically guarantees scientifically that the product has better comfort as well as minimum risk of injury for the user. However, this type of marketing strategy will be succeed only when the manufacturer could deliver the advantage of ergonomic product to the correct group of targeted customers.

It is important to deliver the ergonomic-labeled product to a group of target customers who aware of the existence and the benefit of the particular product. On the other hand, unfortunately, customer's level of awareness of ergonomic product in some developing countries, including Indonesia are still considered as low. In Indonesia particularly, a big problem of applying the ergonomics principles in a work system is the stakeholders' low level of awareness [6]. A study by Kamaroddin [7], involving 60 university students in South East Asia shows that 96.3% of the participants agreed to the importance of the application of ergonomics principle and design paticularly related to a computer-based work. However, it is also interesting to see that only 50% of participants in mentioned study were actually apply the ergonomics principles or use any ergonomic product for their work. Therefore this study aims to observe Indonesian people's level of awareness as well as their intention to use the ergonomic-labeled product..

2. Methods

This study involved 40 university students and office workers ranging from 18~53 years old; 52% female - 48% male; 60% student - 40% office workers. An experimental scheme were designed in this study to investigate the effect of user perception to the intention to use an ergonomic product. A set of activities was used to evaluate the effectiveness of particular work using a range of ergonomic-labeled products. The product used in this experiment were office chair, computer keyboard, and mouse. A set of questionnaire were also given to the participants in order to know their intention to use the product, before and after experimental activities were conducted.

Before the experiment were conducted, a questionnaire was given to each participant to measure their level of awareness on ergonomic product. First, the participants were asked whether they have any knowledge of human factors/ergonomics either from formal or nonformal education. Second, they needed to indicate whether they have known or aware of any ergonomic product exists in the market. Lastly, the participants were given three sets of different product (office chair, keyboard, and mouse) with only one actual ergonomic-labeled product in each set (Figure 1). Participants had to choose which product they perceived to be the ergonomic product and stated their reason. The correct answer for each set were determined based on existing patents of relevant ergonomic products.

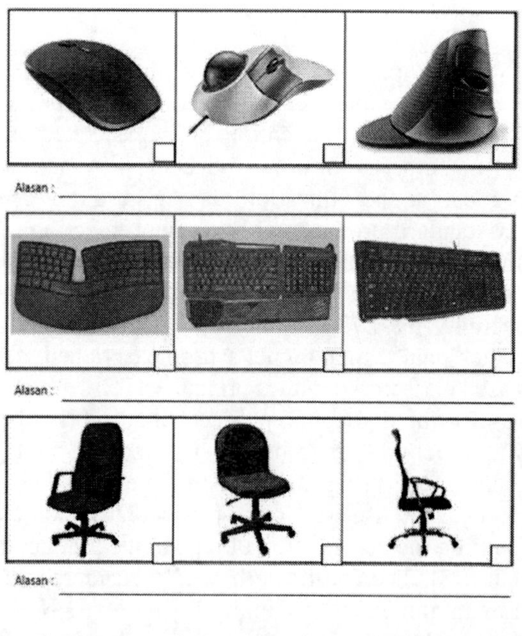

Figure 1. Set of products to be chosen by the participants

IOP Conf. Series: Materials Science and Engineering **722** (2020) 012018 doi:10.1088/1757-899X/722/1/012018

The experiment was divided into three parts of activity: reading a book on office chair, typing with keyboard, and highlighting digital text with mouse. In the first activity, participant was asked to read a novel book within 5 minutes on office chair. Each participant did two treatments using a standard office chair and an ergonomic-labeled office chair. In the end of each reading activity, the participant was asked the last two words on last page they have read so far in order to measure the reading speed of participant. Second activity involved a typing task with a standard keyboard and ergonomic-labeled keyboard as each treatment. The participant typed a set of english text using Mavis Beacon Teaches Typing Software [8] within five minutes. Before the task, participant was allowed to practice using mini games feature of the software for one minute. At the end of the task, the software recorded participant's typing speed. The third activity was highlighting digital text using a standard mouse and an ergonomic-labeled mouse. On each treatment, participant should highlight character "e" within a set of text within 5 minutes [9]. Participant was not allowed to use any navigation feature in the microsoft word. At the end of the task, the percentage of success rate was measured by counting the total character highlighted.

3. Results and discussion

The initial questionnaire were intended to identify participants' knowledge of ergonomics in general. The result shows that 70.98% of participants did have initial knowledge regarding human factors or ergonomics principles. Data also indicated that 56.66% of participants were aware that there are ergonomic-labeled products existed in the market. Afterwards, the participants were given three sets of different product figures (office chair, keyboard, and mouse) and they had to choose the product which they perceived as the ergonomic product. The percentage of participants answer is indicated in Table 1.

Table 1. Percentage of participants' answer on each product type.

Product	Answer	
	Correct	*False*
Office chair	32.75%	67.25%
Keyboard	31.43%	68.57%
Mouse	4.62%	95.38%

Participants were also mentioned their reason of choosing particular option on each product type. The reasons can be categorized based on the similar keyword stated in the questionnaire. The keywords represent participants' definition of ergonomic product. The percentage of appearance of each keyword is shown in Table 2 below. It is clear that the most frequent keywords which appear on three types of product is how the design follows the corresponding body parts' shape.

Table 2. Percentage of keywords' appearance.

Keywords	% Appearance		
	Office chair	*Keyboard*	*Mouse*
Comfortable	17.81%	16.67%	23.81%
Efficient	0.68%	0.00%	4.76%
Makes work easier	0.00%	7.25%	19.05%

IOP Conf. Series: Materials Science and Engineering **722** (2020) 012018 doi:10.1088/1757-899X/722/1/012018

Keywords	% Appearance		
	Office chair	*Keyboard*	*Mouse*
Unique	1.37%	3.62%	14.29%
Follows body shape	70.55%	61.59%	28.57%
Interesting design	5.48%	8.70%	9.52%
Simple	2.05%	2.17%	0.00%
Functional	0.68%	0.00%	0.00%
Adjustable	1.37%	0.00%	0.00%

The experiment data was processed using non-parametric statistics where the *median* differences were analyzed on each data group. The comparison of *median* value from each task is shown in Table 3 below.

Table 3. Median value of each experiment task.

Experimental measures	*Median*		Unit
	Ergonomic product	*Standard product*	
Reading effectiveness on office chair	195.5	212.5	*Words per minute*
Typing efficiency using keyboard	25	36	*Words per minute*
Highlighting effectiveness using mouse	87.8	98.89	%

The results shown in Table 3 indicates that user's productivity on activities associated with particular product was lower when utilizing the ergonomic product. Within the experiment, all participants have not used any ergonomic products which became the study's object. As matter of fact, the procedure of experiment was designed to assign the participants to have trial on each product. However, it is possible that the period of trial in the experiment was too short and incomparable to the frequency of usage of standard product they have been using all this time.

Mann-Whitney U test (α = .05) was then conducted to compare participant's intention to use ergonomic product, before and after experimental tasks. The recapitulation of Mann-Whitney U test result is indicated in Table 4. It is shown that only the keyboard and mouse product have a significant difference on each type of design (ergonomic & non-ergonomic).

Table 4. Mann-Whitney U test results on experiment tasks.

Data group	*p-value*	Note
Reading effectiveness on office chair	0.572	No significant difference
Typing efficiency using keyboard	< .001	Significantly different
Highlighting effectiveness using mouse	< .001	Significantly different

IOP Conf. Series: Materials Science and Engineering **722** (2020) 012018 doi:10.1088/1757-899X/722/1/012018

Before and after the experiment tasks, the participant was asked their intention to use an ergonomic product. The median of questionnaire results on users' intention to use ergonomic product before and after experiment shows equivalent number (*median* before = 6; *median* after = 6; *p-value* = 0.574). This result means that the trial did not actually affect participant's intention to use the ergonomic product after the trial in the experiment.

4. Conclusions

This study aimed to observe people's level of awareness and intention to use the ergonomic-labeled products. The questionnaire and experiment results showed that people's level of awareness of ergonomic product is still low, indicated by many participant choose the wrong product which is perceived as the ergonomic product. People's perception of the ergonomic product mostly defined by the design of the product which follows the corresponding body parts' shape.

The interesting part of this result is the fact that the effectiveness of every determined experimental measures were lower on the utilization of ergonomic product. This outcome may indicate that the unfamiliar design of product could affect significantly how the user behave and make their productivity decrease when compared to the use of the standard product.

5. References

[1] Broberg O 1997 Integrating ergonomics into the product development process *Int. J. of Industrial Ergonomics* **19** 317
[2] Smith M J and Bayeh A D 2003 Do ergonomics improvements increase computer workers' productivity?: An intervention study in a call centre *Ergonomics* **46** 1
[3] Cofco 2012 The Cost of Ergonomic Related Injuries in The Office. http://www.cofcogroup.com/the-cost-of-ergonomic-related-injuries-in-the-office/. Accessed on 19 August 2018
[4] Dul J and Neumann P 2009 Ergonomics contributions to company strategies *Appl. Ergonomics* **40** 745
[5] US-Ergo 2018 Product Certification. http://us-ergo.com/product-ergonomics/ergonomics-product-certification/. Accessed on 18 August 2018
[6] Sutardjo U S 2007 Ergonomics policy in Indonesia *J. Human Ergonomics* **36** 57
[7] Kamaroddin J H, Abbas W F, Aziz M A, Sakri N M, and Ariffin A 2010 *Proc. Int. Conf. on User Science and Engineering (Shah Alam)* (Malaysia: IEEE) p 296
[8] Rempel D, Barr A, Brafman D, and Young E 2007 The effect of six keyboard design on wrist and forearm postures *Appl. Ergonomics* **38** 293
[9] Schmid A B, Kubler P A, Johnston V, and Coppieters M W 2015 A vertical mouse and ergonomic mouse pads alter wrist position but do not reduce carpal tunnel pressure in patients with carpal tunnel syndrome *Appl. Ergonomics* **47** 151

Extraction optimization for antioxidant phenolic compounds in black grass jelly (Mesona palustris BL) using response surface methodology

H Hendratama, K Harismah and A M Fuadi
Master of Chemical Engineering, Graduate School, Unversitas Muhammadiyah Surakarta, Surakarta, Indonesia

E-mail: U200166001@student.ums.ac.id, kun.harismah@ums.ac.id, Ahmad.fuadi@ums.ac.id

Abstract. Black grass jelly is a traditional plant that is more widely used as medicine to prevent carcinogenesis, diabetes, and hypertension, to overcome diarhea, because it is antimutanogenetics, stabilizers, and immunomudulators, besides black grass jelly also contains antioxidants. This study examines changes of temperature and solvent ratio toward the composition of phenols, tannins, and flavonoids which are antioxidant constituents. Analysis of phenols and tannins uses folin quality reagents, while analysis of flavonoids uses the spectrophotometer method. The change of temperature and solvent ratio gives effect to change of phenol, tannins, and flavonoids amount. Optimal extraction of phenol, tannin and flavonoids using the response surface methodology (RSM) occur at a temperature range of 94 -100°C with a solven ratio of 23-24 ml/g.

1. Introduction

Black grass jelly (Mesona palustris BL) in the Java is known as janggelan, a traditional food that is used as folk medicine. In China and Taiwan, this plant is known as hasian-tsao (Mesona procumbens Hemsl), which is used to prevent heat illness, hypertension and diabetes [1,2]. As a local plant, black grass-jelly can be used as a functional food.

Black grass jelly is a plant of Lamisaceae family that has a height of 30-60 cm and grows at an altitude of 150-1800 above sea level. It has been owned by this plant containing bioactives which are beneficial to the body [3]. Besides having bioactive ingredients, the black grass jelly plant also has benefits to prevent carcinogenesis, antiosidative, antimutagenetic, hepatoprotector, antioxidants, immunomodulators, antidiarrhea, antidiabetic and antihypertensive [3-8]. Black grass jelly plan contains polyphenol compounds which function as anti-cancer substances and are easily digested gels [9,10].

Black grass jelly has a phenolic content [11]. Phenol is an antioxidant that can protect cells from oxidative stress, namely the imbalance between relative oxygen species and the antioxidant ability of the body's antioxidants. Oxidative stress can cause tissue damage in diabetics. Antioxidant treatment can have the effect of hyperglycemia [12]. Antioxidants are compounds as having one or more hydroxyl groups on the aromatic ring, where derivatives of these antioxidants contain more flavonoids, alkaloids, tannins, and other phenolic mixtures [13].

Besides containing bioactive compounds, black grass jelly leave also contains energy, carbohydrates, protein, fat, calcium, phosphorus, iron, vitamins, and water [14]. The main present content in black grass jelly is pectin, which is a low methoxy [15]. Pectin is a food fiber that is soluble in water, which can reduce blood fat and glycemic response [16].

Extract of black grass jelly uses 3 types of solvents, namely ethanol, ethyl acetate and water to determine the effect of the cytotoxic effects. The black grass jelly extraction was carried out for 2 hours with a solvent ratio of 20 mL/g. Then the extract is concentrated using an evaporator at a pressure of 34-36 kPa. The results of the study, black grass jelly extract has anticancer potential [17].

Based on several previous studies, black grass jelly has a bioactive composition and antioxidants. Specifically, the types of materials, bioactives, and antioxidants found in the extract, black grass jelly needs to be studied further. Therefore this study aims to study the effect of solvent and temperature ratio on the content of phenol, tannin and flavonoid compounds in black grass jelly extract, besides testing using RSM has not been done to study the effect of two types of independent variables on changes in phenol, tannin and flavonoid amount.

2. Material and Methods

2.1. Ingredients

The ingredients used are dried black grass jelly leaves obtained by black grass jelly farmers in Pakis Baru village, Nawangan District, Pacitan Regency, East Java, while the analysis material used are folin-ciocalteu (Merek), quercetin (Merek), gallic acid reagents . (Merek), aluminum chloride (Merek), Na_2CO_3, oxalic acid, sulfuric acid, sodium nitrite, and sodium hydroxide.

2.2. Extraction

The research was conducted in two stages, namely extraction and analysis of results. The ingredients of black grass jelly are dry, then dried and mashed to 40 mesh. Then, the material is extracted by using water for 120 minutes and temperatures 80, 90, 100°C. The remaining solids are squeezed for 3 minutes to increase the amount of comparison coming out of the material [18, 19]. Determination of optimal temperature is carried out at a solvent ratio of 24 mL/g. Then, the extraction is carried out to analyze the optimal solvent ratio, at optimal temperature.

2.3. Analysis

- Determination of total phenolic

The total phenolic from the extract was estimated using the folin-ciocalteu method with gallic acid standard [20]. The reaction mixture consists of 1 ml of extract diluted with 9 ml distilled water. 1 ml of folin-ciocalteu was treated to the mixture and 7% of sodium carbonate solutions was treated to the mixture, after 5 minutes. Standard solution of gallic acid (10, 20, 30, 40 and 50 μg/ml) were prepared,

IOP Conf. Series: Materials Science and Engineering **722** (2020) 012019 doi:10.1088/1757-899X/722/1/012019

then incubated for 90 minuts at room temperature [21]. Absorbace for test and standard solution were determined agains the reagent blank at 550 nm with a visible spectrophotometer.

- Determination of tannin

The tannins were determined by Folin-Ciocalteu method. About 0.1 ml of the sample extract was added to a volumetric flask (10 ml) containing 7.5 ml of distilled water and 0.5 ml of Folin- Ciocalteu phenol reagent, 1 mi of 35% sodium carbonate solution and dilute to 10 ml with distilled water. The mixture was shaken well and kept at room temperature for 30 min. a set of reference standard solutions of tannic acid (10, 20, 30, 40, 50 μg/ ml) were prepared in the same manner as described earlier. Absorbance for test and standard solutions were measured against the blank at 700 nm with an UV/ Visible spectrophotometer [21].

- Determination of flavonoid

Total flavonoid content was measured by the aluminium chloride colorimetric. The reaction mixture consists of 1 ml of extract and 4 ml of water. 0.30 ml of 5 % sodium nitrite was treated and after 5 minutes, 0.3 ml of 10 % aluminium chloride was mixed. After 5 minutes, 2 ml of 1 M Sodium hydroxide was treated and diluted to 10 ml with water. A set of reference standard solutions of quercetin (10, 20, 30, 40 and 50 μg/ml) were prepared in the same manner as described earlier. The absorbance for test and standard solutions were determined against the reagent blank at 510 nm with an UV/Visible spectrophotometer [22].

3. Result and Discussion

Extraction efficiency is determined by the five parameters issued are temperature, time, solvent ratio, size of material and type of solvent. These parameters can be either independent or interactive effects [23]. This study uses an independent variable solvent ratio (x) and temperature (z). While the interactive variables are analyzed are phenol (y_1), tannin (y_2), and flavonoids (y_3). Based on the equations from standard graph, the concentration of phenol, tannins and flavonoid obtained in the Table 1.

Table 1. Amount of phenol, tannin, and flavonoid toward Changes of temperature and solvent ratio

Solvent Ratio (mL/g)	Temperature (°C)	Phenol (mg GAE/mL)	Tannin (mg GAE/mL)	Flavonoid (mg Qc/mL)
18	100	0,06586	0,03957	0,14347
20	100	0,06390	0,03817	0,12109
22	100	0,06915	0,04113	0,11254
24	100	0,07502	0,05008	0,13479
24	90	0,05910	0,04442	0,12902
24	80	0,06203	0,03678	0,10193

3.1. Phenol

Based on Table 1, it can be seen that the phenol content increases at a ratio of 20-24 mL/g, while the decrease in phenol amount occurs at the solvent ratio of 18-20 mL/g. This increases the nature of phenol which is easily dissolved in water [24, 25], while the decrease in other effects of black grass jelly extraction. Increased temperature causes changes in the questionnaire to changes in phenol [25, 26]. This happened because the temperature on phenol causes two optimum extraction peaks. At temperatures more than 100°C there is an optimum global phenol extraction.

The regression test of order 2 polynomial was used to determine the changes in temperature and ratio of changes to phenol amounts. The temperature and solvent ratio have a significant effect on the probability of 0.009. While the temperature and solvent ratio did not give a significant effect, the square of the temperature and the ratio of solvents had a significant effect on changes in phenol amounts. The use of this regression model shows that changes in temperature and ratio have a very strong influence on changes in phenol levels, this is a multiple correlation value (R) which approaches 1. The value of R this regression model is 0.9932 , with the correlation between temperature and the number of seafarers on the phenol level is mathematically written as follows:

$$y_1=-0,0121z+0,005x^2+0,0001z^2-0,0002xz+0,4947 \quad (1)$$

Based on the mathematical equation, the correlation between phenol amount, temperature and solvent ratio was connected in 3 dimensions. The results of analyzed variables indicate that phenol amount increase with increasing amounts of solvents. This is in accordance with the properties of phenol compounds dissolved in water [26].

At 80°C, the increase of phenol amount from the changesof solvent ratio showed a less significant increase. While the temperature of 100°C the solvent ratio showed a significant change in changes in phenol amounts. This happens because the temperature and ratio of solvents have a strong influence on changes in phenol levels. This result shows that the maximum condition of phenol occurs at 100°C and the solvent ratio is 24 mL/g. While the minimum content occurs at 85°C and the solvent ratio is 19 mL/g. Unlike the maximum conditions that exist in maximum operating conditions, the minimum requirements at a point in Figure 1.

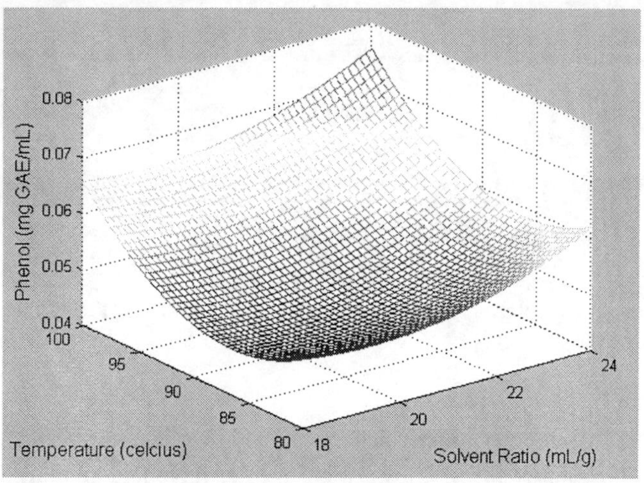

Figure 1. Correlation temperature and solvent ratio toward phenol amount

3.2. Tannin

Tannins is a group of constituents of antioxidants that dissolves in the water, but it dissolves in the water due to a bitter taste. Tannin is compound that decomposed at 2100°C. Based on the data obtained from Table 1 obtained from tanning levels increasing with temperature and solvent ratio [25, 27]. This happened because tannins dissolve easily in water and decomposes at high temperatures.

The regression method of order 2 polynomial is used to determine the correlation between temperature, and the ratio of solvents to tannin levels. This regression method is very significant because it provides a probability value of 0.00002, when compared with a linear regression model that gives a probability value of 0.075. Based on the regression method of order 2 polynomial, it was found that the temperature and ratio of the solvent simultaneously had a significant effect toward changes of tannin amount. While in parallel only the temperature and the solvent ratio have a significant effect. The temperature and ratio have a very strong influence toward changes of tannin because the R value of this regression model is 0.9994 with a constant correlation (R^2) of 0.9988. While the mathematical correlation of the regression model is written as follows.

Based on the equation of Figure 2 which represents a 3-dimensional correlation between temperature and the ratio of solvents toward tan amount. At 80°C, the effect of increasing the solvent ratio is not significantly increased on tannin amount, while at 100°C, there is a significant increase in the amount of the solvent ratio of 22-24 mL/g. This result shows that the maximum extraction conditions are 100° C and a solvent ratio of 24 mL/g.

$$y_2=0,0086z + 0,00065x^2 - 1,03E-05z^2 -0,00025xz - 0,471 \quad (2)$$

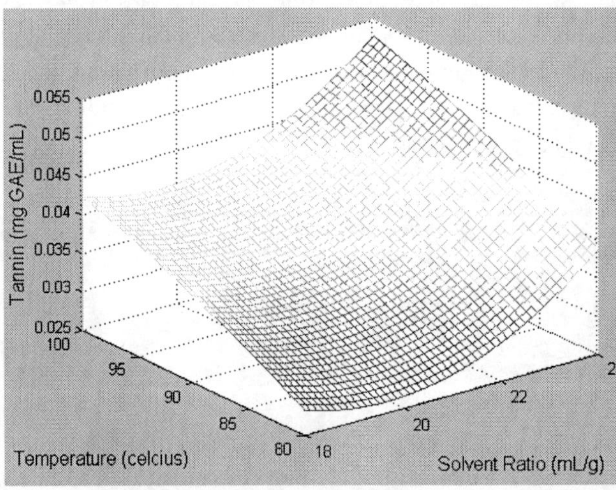

Figure 2. Correlation temperature and solvent ratio toward tannin amount

3.3. Flavonoid

Table 1 shows an increase in temperature which causes an increase in flavonoid amount [24, 25, 28]. The Increase of solvent ratio causes changes toward changes of flavonoid amount [24]. These results prove that not all antioxidants combinations are dissolved in water. From these results it can be seen that the increase in temperature causes an increase in flavonoid amount, but increases in the solvent ratio does not always increase flavonoid amount.

The regression method of order 2 polynomial is used to determine the correlation between temperature, and the ratio of solvent toward flavonoid amount. The temperature and the solvent ratio simultaneously have a very strong influence on flavonoids, this causes the R value = 0.9937. While the regression model shows a significant difference and the solvent ratio has a significant effect on the probability of 0.007. While the temperature and solvent ratio squared have a significant effect toward changes of flavonoid amount. From this regression model is obtained mathematic correlation between temperature and solvent ratio toward flavonoids amount as follows:

$$y_3 = 0{,}05z + 0{,}0028x^2 - 0{,}0001z^2 - 0{,}0012xz - 2{,}51984 \quad (3)$$

From this mathematical equation, it can be known that realistic between temperature and solvent ratio toward flavonoids the amount in a 3-dimensional plot The result, the maximum extraction conditions occur at 100°C and the solvent ratio is 18 mL/g. This result indicates that an increase of temperature causes an increase of amount, while the increase of solvents ratio does not cause an increase in flavonoid amount. Minimum flavonoid amount occurs at 80°C with a solvent ratio of 20-22 mL/g (Figure 3).

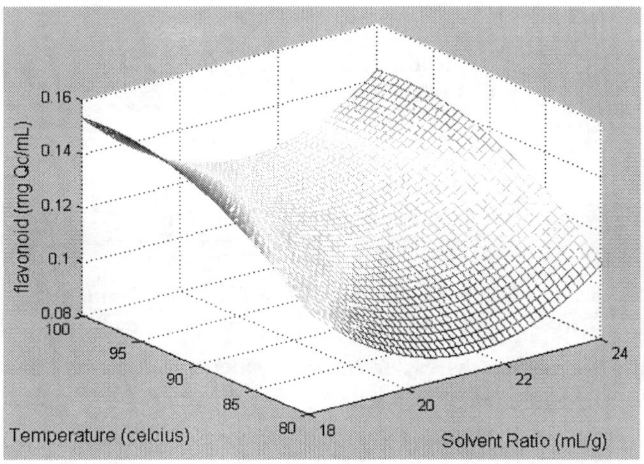

Figure 3. Correlation temperature and solvent ratio toward flavonoid amount

3.4. Analysis Regresion

Regression analysis was conducted to determine the accuracy of the use of regression model of order 2 polynomial toward the results of the analysis of the effect of temperature and solvent ratio of phenol, tannin, and flavonoid amount. The use of this regression model gives a value of R and R^2 above 0.98. Then, the standard error value of the regression model also has a relatively small value of less than 0.005.

IOP Conf. Series: Materials Science and Engineering **722** (2020) 012019 doi:10.1088/1757-899X/722/1/012019

Based on R value of square found in Table 2 shows that the magnitude of the independent variables and influences changes in interactive variables. The accuracy of the regression model is very accurate used to determine unknown data. Based on the adjusted square values, there are several other factors that influence the extraction of phenols, tannins, and flavonoids from black grass jelly.

Table 2. Regression statistic

Regression Statistic	
Phenol	
Multiple R	0,9931612
R Square	0,9863691
Adjusted R Square	0,9318459
Standard Error	0,0014726
Observation	6
Tannin	
Multiple R	0,999434
R Square	0,998868
Adjusted R Square	0,994340
Standard Error	0,000267
Observation	6
Flavonoid	
Multiple R	0,993714
R Square	0,987468
Adjusted R Square	0,937341
Standard Error	0,003792
Observation	6

The test model is shown in Table 3. The diversity of statistical models occurs significantly in the calculated less than 0.01 with the magnitude (F) shown in Table 3. From these results it can be used for the use of second-order polynomial regression models, which are used for comparison with temperature and ratio of solvents with phenols, tannins, and flavonoids.

Table 3. Analysis of variance (ANOVA)

	df	*SS*	*MS*	*F*	*Significance F*
		Phenol			
Regression	4	0,000156	3,92E-05	18,09	0,174331
Residual	1	2,17E-06	2,17E-06		
Total	5	0,000159			
		Tannin			

	df	*SS*	*MS*	*F*	*Significance F*
			Phenol		
Regression	4	0,000119	2,97E-05	220,59	0,050450
Residual	1	1,35E-07	1,35E-07		
Total	5	0,000119			
			Flavonoid		
Regression	4	0,001133	0,00028	19,699	0,167217
Residual	1	1,44E-05	1,44E-05		
Total	5	0,001147			

3.5. Optimization of Extraction Conditions

The optimization model used is the selection of the maximum area conditions with the contour model. The optimal condition is determined based on the conditions of the amount of phenol, tannin, and flavonoids at high concentrations, namely the conditions shown in red on the contour. Figure 4 shows the correlation between temperature and solvents ratio toward amount of phenol, tannins, and flavonoids. Based on Figure 4, the optimum conditions for extracting phenol, tannin, and flavonoids from black grass jelly occur in the temperature range of 94-100°C and the amount of solvent is 23-24 mL/g.

Based on the results of Figure 4, it is known that the distribution of maximal extraction conditions occurs in several variations of the solvent ratio. But in general all maximum conditions occur at 100°C. Thus, it can be seen that the increase in temperature causes an increase in the amount of phenol, tannins, and flavonoids. But the increase in the solvent ratio does not cause an increase in flavonoid levels. Nevertheless the determination of optimal extraction conditions by using this contour model does not show optimal conditions that produce an optimal condition. The optimal condition is chosen according to the maximal conditions which gave the condition of phenol amount more than 0.07 mg GAE/ mL, tannins more than 0.05 mg GAE/ mL, and flavonoid levels more than 0.123 mg Qc/mL

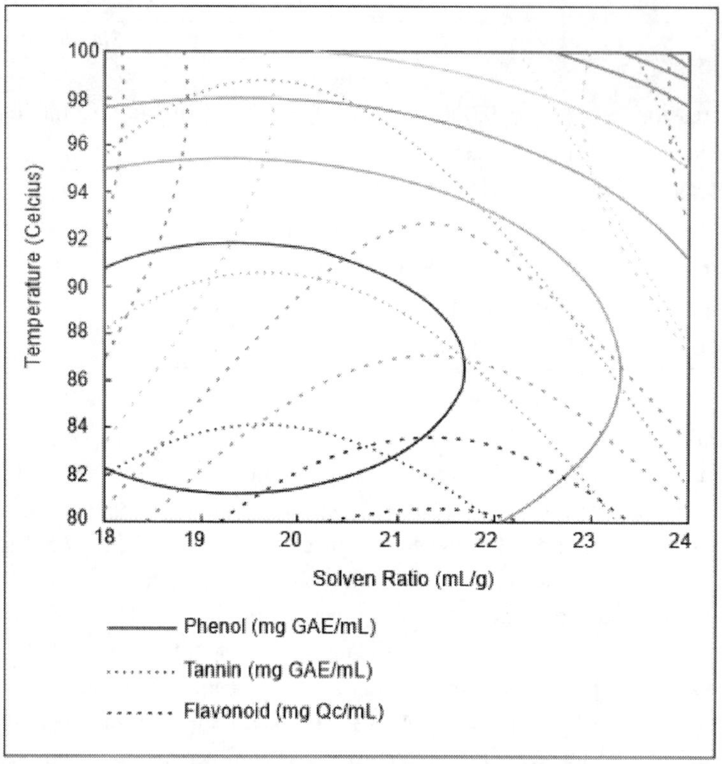

Figure 4. Contour plot optimal extraction conditions

4. Conclusion

Based on the Response Surface Methodology (RSM) analysis, it is known that the maximum amount of phenol and tannin occurs in the same conditions, but the maximum amount of flavonoids occurs at the solvent ratio of 18 mL/g. Overall, the increase in temperature causes increases in the amount of phenol, tannins, and flavonoids. With the contour model, it can be seen that the optimal extraction conditions occur in the temperature range of 94-100°C and the ratio of solvent ratios is 23-24 mL/ g, while the minimum extraction conditions occur in the temperature range 80 - 84°C with solvent ratio 19-21 mL/g.

5. Acknowledgment

Thanks for the Institute for Research and Community Service (LPPM) of the Muhammadiyah University of Surakarta, which has funded the research process.

References

[1] Harjadi D P, dan Bangun P Nuswantoro. 2002. *Purification og gel forming component extracted from janggelan (Mesona palustris BL) and characterization of the resulted gel*. Seminar PATPRI

[2] Ruhayat A. 2002. *Cincau hitam tanaman obat penyembuh*. Penebar Swadaya Indonesia

[3] Tri Dewanti Widyaningsih, dan B T F Sari. 2017. *International Journal of Chem Tech Reserc* **10** 45

[4] L S Lai, S T Chou, and W W Chao. 2001. *J. Agric. Food Chem.* **49** 96

[5] C G Yen, P D Duh, and Y L Hung. 2001. *J. Agric. Food Chem.* **49** 5000

[6] Tri Dewanti Widyaningsih, A P Sukardiman, Djoko and D Win. *J. Technol. Food Ind.* **23** 29

[7] C T Yeh, W H Huang, and G C Yen. 2008. *J. Nutr. Biochem*

[8] Tri Dewanti Widyaningsih, E Martanti, and D M Lukitasari. 2017. *Asian Journal of Pharmaceutial and Chlinical Research.* **10** 326

[9] S U Nudin, R K L Liu, G P Young, J C R Stangoulis, C T Christophersen, and C A Abbott. 2017. *Nutrients.* **9** 355

[10] J Lim, S Adisakwattana, and C J Henry, 2018. *Asia Pac J. Clin Nutr.* **27** 336

[11] T D Widiyaningsih, J Mahar, N Wijayanti, Rahmani, and M Najmuddin. 2013. *Food and Public health.* **3** 142

[12] R Amalia, and T D Widyaningsih.2014. *Jurnal pangan dan agroindustri.* **2** 28

[13] M Sayuti, dan R Yenrina. 2018. *Antioksidan alami dan sintetik.* Andalas University Press. Indonesia

[14] T D Widyaningsih, S D Wijayanti, and N Tisnaningrum. 2018. *International Journal of Phytomedicines and Related Industries.* **10** 185

[15] T D Widyaningsih, and P Adilaras. 2008. *Journal of food since and technology.* **10** 1390

[16] S A Moharib, and S A El-Batran. 2008. *Reserch Journal of Agricultural and Biological Science.* **4** 455

[17] T D Widyaningsih.2012. *ACPBEE Procedia.* **2** 110

[18] T D Widyaningsih, and R M Safitri. 2014. *J. Agroteknologi.* **8** 74

[19] B P Nuswantoro, and Haryadi. 2003. *Agritech.* **23** 28

[20] S Tabasum, S Khare, and K Jain. 2016. *Asian J. Pharm Clin. Res.* **9** 371

[21] C K Chandran, and G Indira.2016. *Journal of Medicinal Plants Studies.* **4** 282

[22] V D Tambe, and R S B Gambar. 2014. *Journal Of Pharmacognosy And Phytochemistry.* **9**

[23] Montgomery. 2001. *Design and analysis of experiments.* John Wiley & Sons. New york

[24] M Radojkovic, Z Zekovic, S Jokic, and S Vidovic. 2012. *Romanian Biotechnological Letters.* **17** 7295

[25] E N Fombang, and W R Saa. 2016. *Journal of Nutrition & Food Sciences.* **6** 1

[26] A Rajaeni, M Barzegar, Z Hamidi, and M A Sahari. 2010. *J. Agr. Sci. Teach.* **12** 605

[27] R Tabraki, A Safari, and A Y Faal. 2013. *Journaal of Applied Chemical Research.* **7** 67

[28] H N Rajha, N El Darra, Z Hobaika, N Baussetta, E Vorobiev, R G Maroun, and N Louk. 2014. *Food and nutrition sciences.* **5** 397

Usability Analysis of an Academic Faculty Website: A Case Study of the Selected University

H Soewardi, and M H Ramadhan

Industrial Engineering Department, Faculty of Industrial Technology
Islamic University of Indonesia, Yogyakarta-Indonesia

E-mail: hartomo@uii.ac.id; hafizia.ramadhan@gmail.com

Abstract. Website constitutes a pivotal element in an academic institution. It is useful to provide academic information for stakeholder requirements like final scores, academic calendar, scholarship, student competition, etc. This paper presents an investigation about the level of usability of the website at the selected faculty of the university in Indonesia based on 5 dimensions, namely learnability, efficiency, memorability, error, and satisfaction. It is noteworthy that the faculty should supply the information in detail and accurately. However, as a matter of fact, few of a faculty website in some certain universities sometimes experiencing matters for the user. Thus, it is crucial to do the analysis of the usability of the website at the faculty level. A qualitative study was conducted by distributing some Likert-scale questionnaire to stakeholders that consisted of students, academic staff, student's parents, and lecturers. The questionnaire contains 25 statements divided into five groups of elements by referring five dimensions above with each group having 5 questions or statements. The non-parametric statistic was conducted to test the hypothesis. The result of this study shows that the website of the selected faculty is classified into a moderate level with 68% of usability grade. Thus, this study recommends that the faculty improve its website in the layout of menu, content, add some features to ease login and increase usability level.

1. Introduction

Website constitutes a pivotal element in an academic institution. It is useful to provide academic information for stakeholder requirement like final scores, academic calendar, scholarship, student competition, etc. Meanwhile, website is also used as a media to promote university educational qualification and recognition [1] [2]. A good website should meet the user needs such as easy access and ease of getting information [3]. To find out whether a website is good or not, it is necessary to measure its usability level.

A preliminary study has been conducted to stakeholder of the website. It reveals that intensity level in using the selected academic website is 50% of the respondents occasionally accessing the website, 42% of respondents infrequently accessing the website and 7.5% of respondents oftentimes accessing the website. This finding pops up a question why the stakeholder does not frequently access such website. Thus, it is important to test the academic faculty website in the selected university about its usability level.

The objective of this study is to analyze the usability level of the academic website of the selected university as a case study based on its learnability, efficiency, error, memorability and satisfaction.

2. Research Method

2.1. Usability Analysis Method

Usability is a measure that describes the level of ease of use of some product with learnability, efficiency, error, memorability and satisfaction as the parameter [4] [5] [6]. Academic websites are said to be usable if the users feel satisfied after using the website; find them easy to learn the process for finding certain information in the website; users can find the information they want quickly and with a minimum error; and the contents of the website meet user's needs [7][8].

The testing was done by giving respondents a series of tasks that they must do on the website being tested. This test aims to form opinions from respondents about the particular website tohelp them when filling out the questionnaire. The test contains some tasks related to menus and features provided by the website. Once the respondent completed the tasks, they would be given a Likert-scale questionnaire. The data from the questionnaire would be used to measure the usability level of the website by calculating its mode value. The mode value is analysed to find the usability level of the website. To support the results of the analysis, interviews were conducted with respondents to find out their difficulties when using this website.

2.2. Survey

The survey was conducted by distributing questionnaires to stakeholders consisting of students, lecturer, student's parents and faculty's staff, and by interviewing them about their complaints related to the particular website. This survey aims to identify the intensity level of access for respondents to the faculty's website and to test the usability level of the website. The first survey contains questions about the intensity of accessing the website under study; what device they mostly used to access the website; and what information that they wanted to find on the website. The second survey is to test hypothesis by asking about the use of the website.

2.3. Apparatus

The tools used to support this experiment are:

- Likert-scale questionnaire
  Likert-Scale is a type of questionnaire that uses a number of statements to measure individual behaviour by responding to 5 choice points on each item statement. The choice is a strongly agree, an agree, a quite agree, a disagree, and a strongly disagree [9] [10]. The questionnaire consists of 25 statements divided into five group of elements, which can be seen on TABLE I.
- Laptop with a stable internet connection was required to do the test.
- Stopwatch was used to calculate loading and downloading time
- Stationary was used to write the results of interviews.
- Microsoft Excel 2013 software was used to process the data.

Table 1. List of the questionnaire statement

Elements	Statements	Elements	
Learnability	The texts on the website are easy to understand.	Errors	I rarely find errors that occur on the website.
	The information that I am looking for can be found easily based on the menu available on the main page of the website without using the search feature.		I always find the information I'm looking for on this website.
	The search feature can be found easily.		The search feature works well.

Elements	Statements	Elements	
	I don't need technical assistance when using this website.		All links on this website work well.
	The menu functions of this website are easy to understand.		The website provides procedures for correcting errors that I have encountered
Efficiency	The loading time needed to access this website is quite fast.	Errors	I rarely find errors that occur on the website.
	With the search feature, I can find the information I want quickly.		I always find the information I'm looking for on this website.
	Without the search feature, I can find the information I want quickly.		The search feature works well.
	The time needed to download a file from this website is fast.		All links on this website work well.
	This website can be opened at any time.		The website provides procedures for correcting errors that I have encountered
Memorability	I can memorize how to use this website easily.	Satisfaction	The appearance of this website is attractive.
	The symbols on this website are easy to remember		I will visit this website often.
	The functions of the menus of this website are easy to remember.		I feel comfortable with the menu layout of this website.
	The address of this website is easy to remember.		I feel comfortable with the colors of this website.
	I can remember the step that must be taken to get the information I want easily.		I am satisfied with the information I got from this website.

2.4. Design of Experiment

The usability testing was done in campus by assigning participant to complete the tasks. 30 subjects participated in this experiment. When conducting the test, the participant was allowed to tell their opinion of the website and the researcher recorded it. The tests were carried out using a laptop provided and with no time limit. Layout design of the test room is presented in Fig.1, while the homepage of the website under study can be seen in Fig. 2

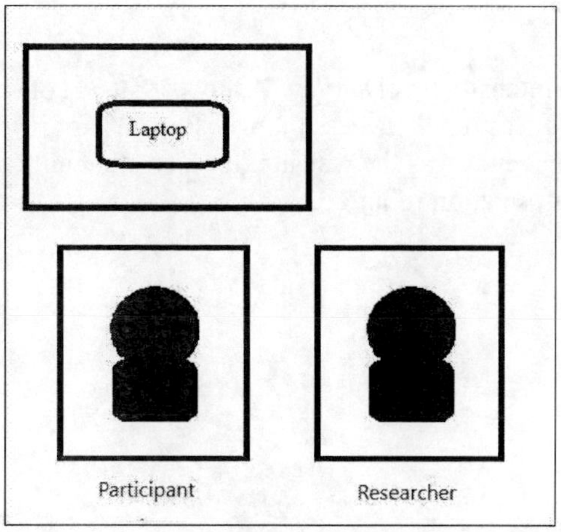

Figure 1. Layout of the usability testing room

Figure 2. Website's Homepage

2.5. *Task*
The followings tasks were to complete by the respondents:
- Find the information regarding the profile of this faculty (history, vision and mission, and organization structure).
- Find the information about this faculty study program and the facility.
- Find information about what innovations the faculty is doing.
- Find the information regarding the academic services such as test scores, academic calendar, exam schedule, submission of academic leave and thesis.
- Find information about this faculty partner.

2.6. *Procedure of experiment*
Participants were explained about the purpose of an experiment and a task done. Subsequently, participants were instructed to operate the computer and login to open the website and to complete the tasks. After that, the participants were asked to fill the Likert-scale questionnaire provided via Google Forms. Then, the participants were interviewed to find out the difficulties that they felt during the test.

2.7. *Statistical Analysis*
The data obtained using the Likert-scale questionnaire will be processed using Microsoft Excel software by calculating its mode value. Mode is the number or value that appears most often in a set of data, which can be used to define the meaning of those data [10] [11]. Thus an discriptive analysis is done.

3. Result and Discussion

3.1. Result of The Survey

Fig.3 shows the result of the intensity level survey. It shows that 7% of respondent frequently access the website, 50% of respondent occasionally access the website, 43% of respondent infrequently access the website, and no respondent never access the website. It means that almost no users are attracted to open this website gaining some information required.

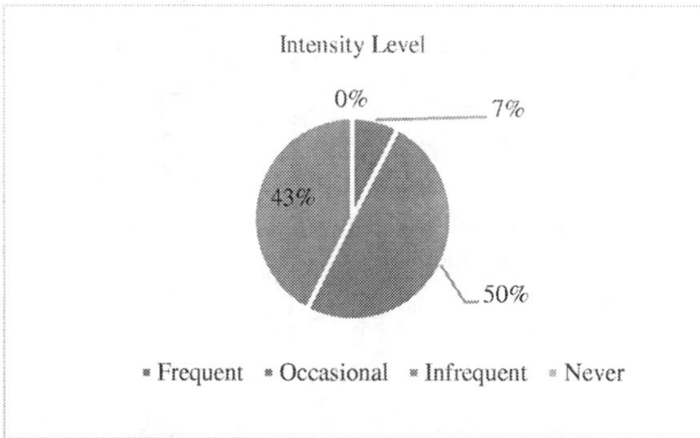

Figure 3. Result of intensity level survey

Table 2 presents the summary of interview, which contains complaints about the website. The first complaint is about the menu layout of test score portal and online mail because the menu is not on main menu of the website. Also, it does not appear when we firstly open the website and on the search feature which confused the respondents. The second complaint is about the fact that not all test scores are available on the menu. Some courses test score can be clicked and seen, while some cannot. The third complaint is related to the sixth complaint. Since the academic calendar is located on the announcement menu, it made it difficult for the users to find it without the search feature because the announcement menu cannot be clicked easily. For the fourth complaint, the respondent thought that the username and password for logging in are the same as another core feature of the university. As for the fifth complaint, there are 3 menus on the main menu that contains a blank page when opened. No information can be found on those 3 menus.

Table 2. Summary of interview

No	Complaint	Percentage of Respondent
1	Menu Layout of Test Score Portal and Online Mail	57% (17)
2	Some test score are not available.	100% (30)
3	It's difficult to find information such as academic calendar & exam schedule without using the Search feature.	80% (24)
4	Difficult to login into the Online Mail feature.	73% (22)
5	There are still 3 links containing empty content.	100% (30)
6	Announcement menu cannot be clicked.	100% (30)

3.2. Usability Analysis Result

Fig. 4 presents the mode value of usability attributes of this website. They are learnability, efficiency, memorability, error, and satisfaction. Learnability attribute has mode value of 4, which means the respondent agree with the statements related. The text and the menu of the website are easy to understand because they use a word that represent the function of the menus. For efficiency attribute, it has mode value of 3 or average. The loading and downloading time for the website is fast, but there is some information that cannot be found quickly without search feature, such as academic calendar.

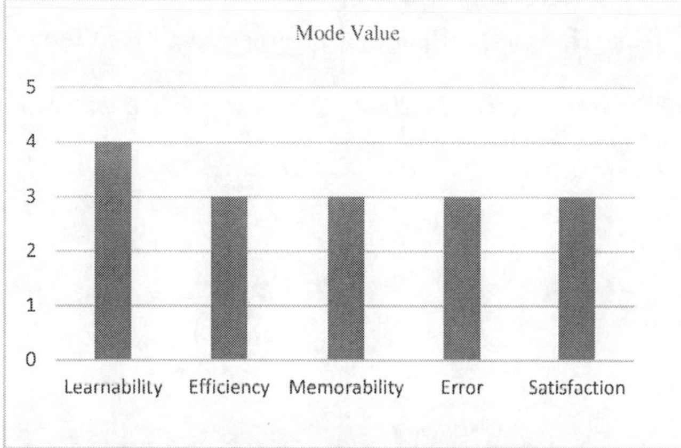

Figure 4. Mode value of usability attributes

Memorability attribute has mode value of 3, which means that the memorability level of this website is on average. Some information can be easily found, while some others are not such as exam schedule. Exam schedule information can be found on announcement menu, but the menu cannot be clicked and the respondent must find another way to reach the menu. For error attribute, it has mode value of 3 or average. There are still broken links or empty content on the website. Also, the search feature does not display all information on the website such as test score. As for satisfaction attribute, it has mode value of 3. It means that the respondent satisfaction level is still on average. Although the respondent like the interface of the website, they do not like the menu layout of score portal feature. Also, the respondents felt that the information on the website was incomplete.

Based on the explanation above, this website is not entirely easy to use and not all information is easy to find. Thus, the usability of this website is still on moderate level-3.

4. Conclusion

This study concluded that the usability level of this website is still on moderate level-3. Therefore, some features need improvements such as menu layout, empty content and incomplete information.

Acknowledgment

We wish to acknowledge Faculty of Industrial Technology Islamic University of Indonesia Yogyakarta Indonesia for supporting this project.

References

[1] Kothainayaki, S., Sivakumaren, K.S., Gopalakrishnan, S. 2012. User Preferences on University Websites: AStudy. Library philosophy and Practice, 788-795.
[2] Mentes S.A., Turan A.H. 2012. Assessing the usability of university websites: an empirical study on Namik Kemal University. TOJET. 11: (3)61-69
[3] Nielsen, J. 1999. *Designing Web Usability.* New Riders.

IOP Conf. Series: Materials Science and Engineering **722** (2020) 012020 doi:10.1088/1757-899X/722/1/012020

[4] Krug, S. 2006. *Don't Make Me Think! A Common Sense Approach to Web Usability (Second Edition)*. Berkeley: New Riders Publishing
[5] Dumas S.J, Redish. C.J.1999, A Practical guide to Usability Testing, John Wiley & Sons.
[6] Nielsen, J. 1994. *Usability Engineering*. Elsevier.
[7] Rubin, J. and Chisnell, D. (2008), Handbook of Usability Testing: How to Plan, Design, and Conduct Effective Tests, 2nd ed. Wiley: Indianapolis.
[8] Pant, A. 2015. Usability evaluation of an academic library website. The Electronic Library. 33:896 - 915
[9] Likert RA. 1932. Technique for the measurement of attitudes. Archives of Psychology, 140 pp: 1-55
[10] Boone Jr, Harry N, and Deborah A. Boone. 2012. Analyzing Likert Data. *Journal of Extension* 50 (2)
[11] Gravetter, F. J., & Wallnau, L. B. 2007. *Statistics for the behavioral sciences*. Australia: Thomson/Wadsworth.

IOP Conf. Series: Materials Science and Engineering **722** (2020) 012021 doi:10.1088/1757-899X/722/1/012021

NCTF Control Performance Analysis on Rehabilitation Robot

Herianto[1], A F Riyadi [2], and H Mastrisiswadi [3]
[1, 2] Department of Mechanical and Industrial Engineering, Universitas Gadjah Mada, Indonesia
[3] Department of Industrial Engineering, UPN "Veteran" Yogyakarta, Indonesia

E-mail: herianto@ugm.ac.id

Abstract. The need for rehabilitation robots is increasing, while the current price is too high. One reason for the high price of rehabilitation robots is the complicated controller design. For this reason, there is a need for a simpler controller design to be produced at an affordable price. In this study, the NCTF control system tested its performance on a rehabilitation robot prototype with a horizontal linear motion system. The tests are carried out using point to point and set point tracking commands. Point to point testing is carried out at a distance of 50 mm, 150 mm and 230 mm. Meanwhile, set point tracking testing was carried out on the sinusoidal wave with a 100 mm amplitude and a frequency of 0.25 Hz. Both tests were carried out by giving a load of 0 Kg, 0.82 Kg, and 1.64 Kg. The results of this study indicate that the NCTF control provides a good response. The steady-state error value for all point to point motion commands below 2%. As for set point tracking commands, the system can follow orders properly, but there is a 0.24-second delay.

1. Introduction

Stroke is one of the deadliest diseases [1]. Besides death, stroke sufferers often experience paralysis after an attack [1–5]. Therefore, regular physiotherapy is needed, but it requires a long time to be done. This process is conducted by moving paralysed limbs repeatedly [6–9]. Physiotherapy process requires a long time, intensive frequency, and high labour [4,10]. Therefore, robots are now being developed to rehabilitate post-stroke patients to help physiotherapists [10–15]. But to be able to do physiotherapy using a robot costs a lot.

The process of identifying robot rehabilitation needs has been carried out [4,16–18], one of which is low cost and easy to control. During this time, the rehabilitation robot control system is relatively sophisticated, so it requires a high cost to produce it. This control system has a direct impact on increasing the value of rehabilitation robot. To reduce these costs, a simpler and cheaper control system is needed but still reliable. One control system that meets these requirements is the nominal characteristic trajectory following (NCTF). The model on the NCTF control system is not based on complex mathematical calculations but is based on a simple open-loop experiment mechanism. But the control system performance is still unknown. In this research, the performance analysis of the NCTF control system will be carried out on a rehabilitation robot prototype with a horizontal linear motion system.

IOP Conf. Series: Materials Science and Engineering **722** (2020) 012021 doi:10.1088/1757-899X/722/1/012021

2. Methods

2.1. Rehabilitation robot prototype with a horizontal linear motion system

The Rehabilitation robot prototype in this study used a belt and pulley mechanism. This prototype has a carriage that can move linearly positively and negatively. Carriage is an eccentric load that can be added with the ballast to increase its weight. The belt connected to the carriage is pulled by a pulley that is rotated by a DC motor. The illustration of this robot prototype can be seen on Figure 1.

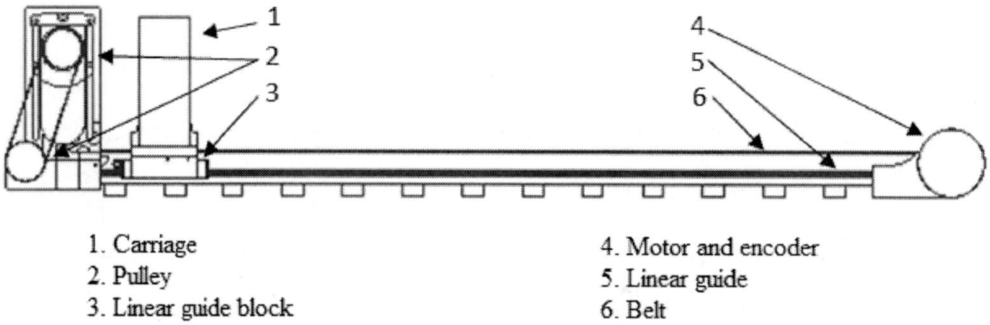

Figure 1. Robot prototype illustration

In addition to the robot prototype, system design is also needed. System design is done by configuring Arduino mega and myRio with a computer and calibrating the encoder. Mega Arduino is connected via MATLAB, while myRio via Labview. MATLAB and LabView are used as user interfaces to program, run, and retrieve data on the controller. Meanwhile, the encoder used to read the displacement of the controller needs to be calibrated to obtain precise results. Calibration is conducted by displaying all data actually when there is a displacement.

2.2. NCTF control system design

The design of the NCTF control system is divided into three stages: (1) moving the system in an open-loop together with calculating displacement and speed, (2) making the NCT system in accordance with the information of the loop-driven system, and (3) designing the PI compensator sought by trial and error method.

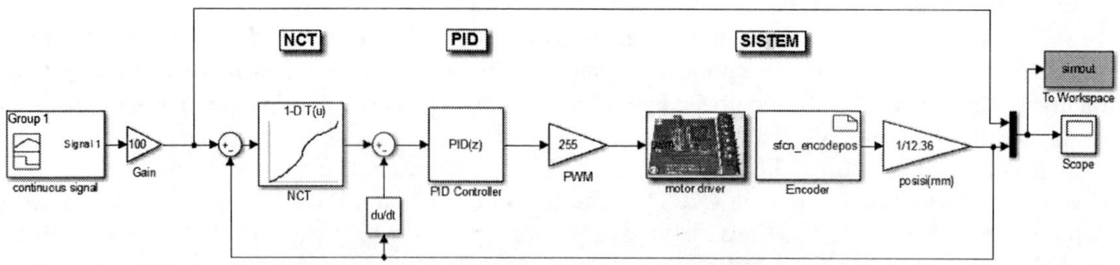

Figure 2 NCT program design using MATLAB

The NCT program design is carried out using MATLAB (Figure *2*). In designing NCT, stepwise signals are given to the system in running the control system in an open-loop together with calculating displacement and speed. The step input signal and object response graph can be seen in

IOP Conf. Series: Materials Science and Engineering **722** (2020) 012021 doi:10.1088/1757-899X/722/1/012021

Figure 3. The step input signal graph is shown in blue, the object response in the form of displacement is shown in red. Meanwhile, green indicates speed. The results of the NCT design are then the basis for determining the value during trial and error.

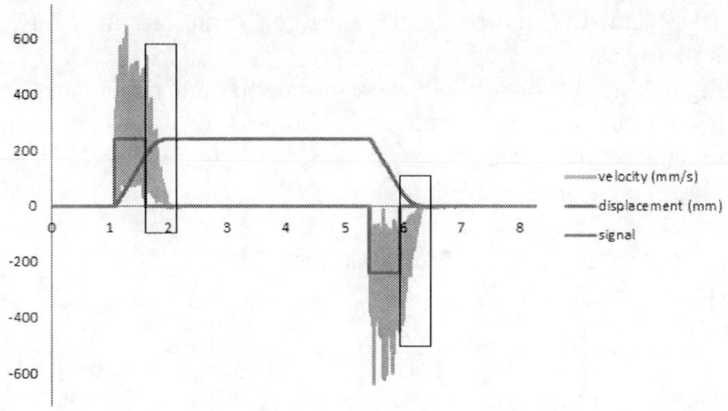

Figure 3 Step input signal and object responses graph

2.3. Experiment

Experiment for the NCTF control system on Rehabilitation robot prototype with a horizontal linear motion system carried out by a point to point and set point tracking commands. Experiment with the command point to point is done by ordering the system to move towards the position of 50 mm, 150 mm and 230 mm. Meanwhile, for experiments with set point tracking commands, the input command is a set point tracking command with a sinusoidal wave with a 100 mm amplitude and a frequency of 0.25 Hz. Both experiments were carried out using loads of 0 Kg, 0.82 Kg and 1.64 Kg. The data collection process in this study can be seen in Figure 4 and the system testing process on Figure 5.

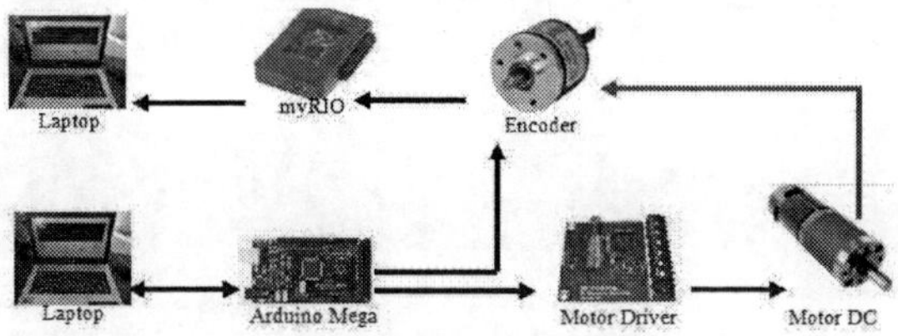

Figure 4. Data collection process

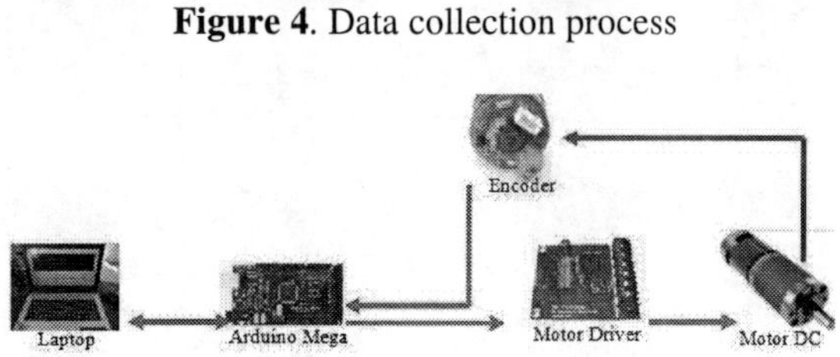

Figure 5. System testing process

3. Result and Analysis

3.1. NCTF performance evaluation (point to point system)

NCTF performance evaluation is done on point to point and set point tracking systems. The parameter for evaluating the performance of the NCTF control is the overshoot, rise time, settling time, and steady-state values.

At the point to point system testing is carried out at three different distances, that are 50 mm, 150 mm, 230 mm. The response of the system at three different distances is applied for load variations 0 Kg, 0.82 Kg and 1.64 Kg. After the experiment, the response system was obtained with two varieties of loading at each distance. The system response received from the experimental results at a distance of 50 mm can be seen in Figure 6 (a), at a distance of 150 mm *in* Figure 6 (b), and at a distance of 230 mm *in* Figure 6 (c). Meanwhile, the parameter data for variations in the two mass loads of each distance can be seen in Table 1.

Based on Figures and Tables, there were no significant differences in parameters for each distance due to differences in load. Parameter differences occur when changes in distance. The overshoot, rise time, and settling time values increase with increasing load. Meanwhile, the largest steady state error value at a distance of 150 mm

Table 1. Parameter data

Distance	Mass (Kg)	Overshoot (%)	Rise time (Second)	Settling time (Second)	Steady state error (mm)	Steady state error (%)
	0	17.771	0.42	0.56	0.0809	0.1618
50 mm	0.82	19.386	0.4	0.56	0.0809	0.1618
	1.64	24.272	0.4	1.72	0.0809	0.1618
	0	96.009	0.43	2.34	1.2136	0.8091
150 mm	0.82	86.839	0.42	2.42	1.0518	0.7012
	1.64	91.743	0.42	2.37	1.2136	0.8090
	0	363.125	0.58	3.21	0.5825	0.2533
230 mm	0.82	374.032	0.58	3.21	0.7443	0.3236
	1.64	374.604	0.58	3.19	0.3398	0.1477

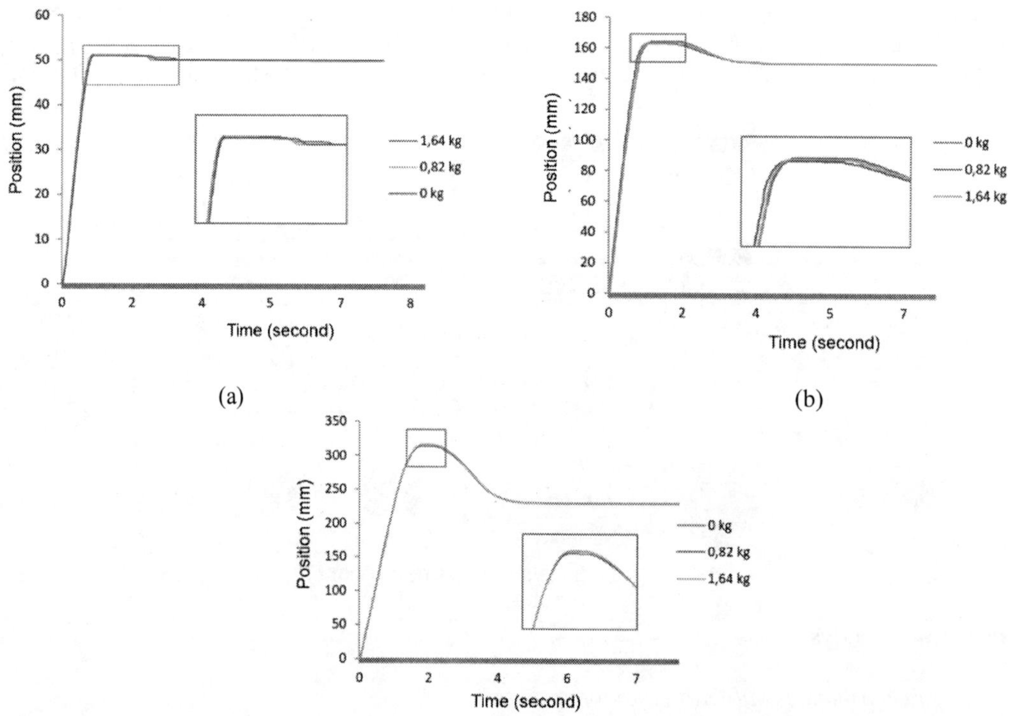

Figure 6. Point to point system response received at a distance of (a) 50 mm, (b) 150 mm, and (c) 230 mm

3.2. NCTF performance evaluation (Set point tracking system)

In a set point tracking system, the system is given a signal input in the form of a sinusoidal signal with an amplitude of 100 mm, a frequency of 0.25 Hz, and a trial time of 25 seconds. The response of the system is applied to variations in the load of 0 kg can be seen in Figure 7 (a), 0.82 Kg in Figure 7 (b) and 1.64 Kg in Figure 7 (c). Based on the Figure, it can be seen that the system can follow the set point tracking commands properly, but there is a delay of 0.24 seconds for each load. The delay occurs because of the controller limitations, so there is no significant difference in performance with changes in load.

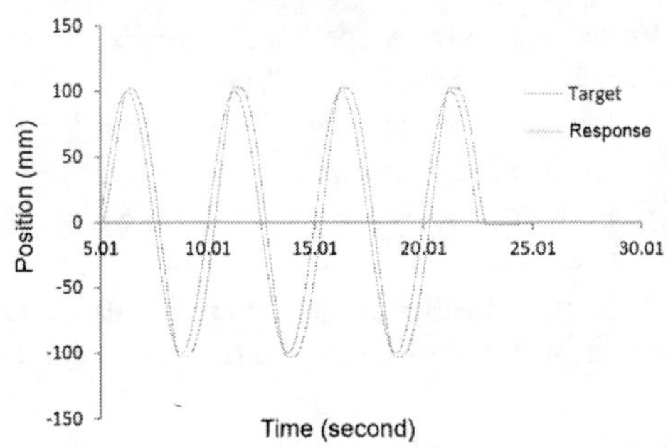

(a)

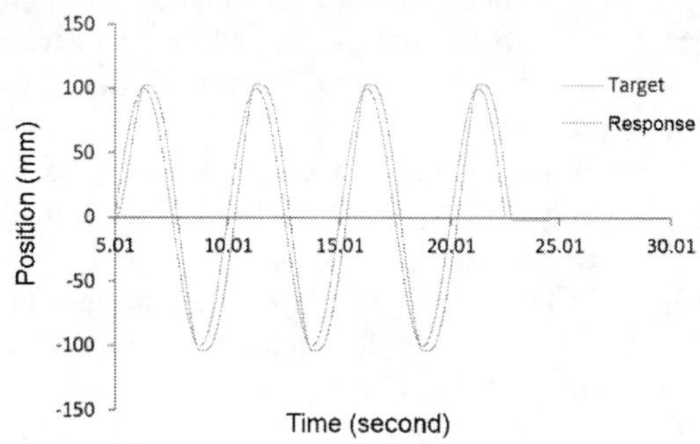

(b)

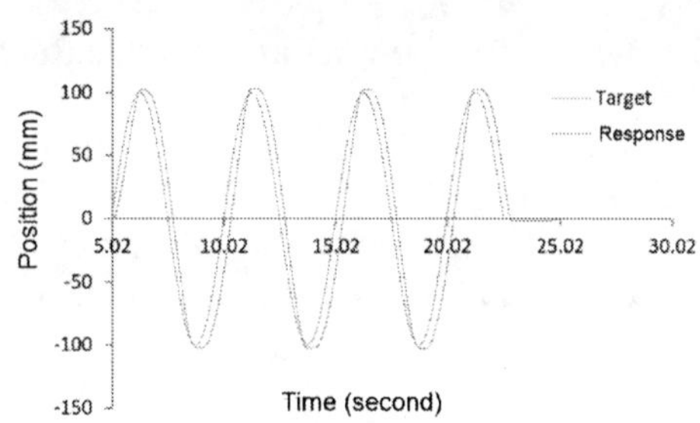

(c)

Figure 7. Set point tracking system response in the load of (a) 0 Kg, (b) 0.82 Kg, and (c)1.64 Kg

4. Conclusion

From this study it was concluded that the movement of linear rehabilitation robots using the NCTF system showed good results. At the point to point command, the steady-state error value is less than 2% for each distance and loading. Meanwhile, on the set point tracking command, the target movement and response are going well. It's just that there is a delay of 0.24 seconds for each load which is a limitation of the controller. Variation of loading only affects the point to point command, while for set point tracking commands no effect.

References

[1] Katan M and Luft A 2018 Global Burden of Stroke *Semin. Neurol.*
[2] Bright F A S, Kayes N M, McPherson K M and Worrall L E 2018 Engaging people experiencing communication disability in stroke rehabilitation: a qualitative study *Int. J. Lang. Commun. Disord.*
[3] Lallukka T, Ervasti J, Lundström E, Mittendorfer-Rutz E, Friberg E, Virtanen M and Alexanderson K 2018 Trends in diagnosis-specific work disability before and after stroke: A longitudinal population-based study in Sweden *J. Am. Heart Assoc.*
[4] Stephenson A and Stephens J 2018 An exploration of physiotherapists' experiences of robotic

therapy in upper limb rehabilitation within a stroke rehabilitation centre *Disabil. Rehabil. Assist. Technol.*

[5] Benjamin E J, Virani S S, Callaway C W, Chamberlain A M, Chang A R, Cheng S, Chiuve S E, Cushman M, Delling F N, Deo R, De Ferranti S D, Ferguson J F, Fornage M, Gillespie C, Isasi C R, Jiménez M C, Jordan L C, Judd S E, Lackland D, Lichtman J H, Lisabeth L, Liu S, Longenecker C T, Lutsey P L, MacKey J S, Matchar D B, Matsushita K, Mussolino M E, Nasir K, O'Flaherty M, Palaniappan L P, Pandey A, Pandey D K, Reeves M J, Ritchey M D, Rodriguez C J, Roth G A, Rosamond W D, Sampson U K A, Satou G M, Shah S H, Spartano N L, Tirschwell D L, Tsao C W, Voeks J H, Willey J Z, Wilkins J T, Wu J H Y, Alger H M, Wong S S and Muntner P 2018 Heart disease and stroke statistics - 2018 update: A report from the American Heart Association *Circulation*

[6] Wolf S L, Blanton S, Baer H, Breshears J and Butler A J 2002 Repetitive Task Practice: A Critical Review of Constraint-Induced Movement Therapy in Stroke *Neurologist*

[7] Feys H, De Weerdt W, Verbeke G, Steck G C, Capiau C, Kiekens C, Dejaeger E, Van Hoydonck G, Vermeersch G and Cras P 2004 Early and Repetitive Stimulation of the Arm Can Substantially Improve the Long-Term Outcome after Stroke: A 5-Year Follow-up Study of a Randomized Trial *Stroke*

[8] Woldag H, Stupka K and Hummelsheim H 2010 Repetitive training of complex hand and arm movements with shaping is beneficial for motor improvement in patients after stroke *J. Rehabil. Med.*

[9] Woldag H, Waldmann G, Heuschkel G and Hummelsheim H 2003 Is the repetitive training of complex hand and arm movements beneficial for motor recovery in stroke patients? *Clin. Rehabil.*

[10] Díaz I, Gil J J and Sánchez E 2011 Lower-Limb Robotic Rehabilitation: Literature Review and Challenges *J. Robot.*

[11] Boyard N, Christmann O, Rivette M, Kerbrat O and Richir S 2018 Support optimization for additive manufacturing: Application to FDM *Rapid Prototyp. J.*

[12] Ang K K, Chua K S G, Phua K S, Wang C, Chin Z Y, Kuah C W K, Low W and Guan C 2015 A Randomized Controlled Trial of EEG-Based Motor Imagery Brain-Computer Interface Robotic Rehabilitation for Stroke *Clin. EEG Neurosci.*

[13] Louie D R and Eng J J 2016 Powered robotic exoskeletons in post-stroke rehabilitation of gait: A scoping review *J. Neuroeng. Rehabil.*

[14] Polygerinos P, Wang Z, Galloway K C, Wood R J and Walsh C J 2015 Soft robotic glove for combined assistance and at-home rehabilitation *Robotics and Autonomous Systems*

[15] Kim M S, Kim S H, Noh S E, Bang H J and Lee K M 2019 Robotic-Assisted Shoulder Rehabilitation Therapy Effectively Improved Poststroke Hemiplegic Shoulder Pain: A Randomized Controlled Trial *Arch. Phys. Med. Rehabil.*

[16] Mastrisiswadi H and Herianto H 2015 Identifikasi kebutuhan konsumen robot rehabilitasi pasien pasca stroke dengan menggunakan metode Quality Function Deployment (QFD) *SEMINAR NASIONAL TEKNIK INDUSTRI UNIVERSITAS GADJAH MADA 2015* ed B M Sopha and T Wijayanto (Yogyakarta: Universitas Gadjah Mada) pp 27–36

[17] Mastrisiswadi H and Herianto H 2017 Analisis Kebutuhan Robot Rehabilitasi Pasien Pasca Stroke dengan Menggunakan Metode Kano *J. Ilm. Tek. Ind. Vol 15, Iss 2, Pp 151-156 VO - 15*

[18] Mastrisiswadi H and Herianto H 2017 Identifikasi Kepentingan Relatif Konsumen Terhadap Robot Rehabilitasi Pasien Pasca Stroke dengan Menggunakan Conjoint Analysis *J@ti Undip J. Tek. Ind.* **12** 35

IOP Conf. Series: Materials Science and Engineering **722** (2020) 012022 doi:10.1088/1757-899X/722/1/012022

Recycled Polypropylene Filament for 3D Printer: Extrusion Process Parameter Optimization

Herianto[1], S I Atsani [2], and H Mastrisiswadi [3]
[1, 2] Department of Mechanical and Industrial Engineering, Universitas Gadjah Mada, Indonesia
[3] Department of Industrial Engineering, UPN "Veteran" Yogyakarta, Indonesia

E-mail: herianto@ugm.ac.id

Abstract. The presence of recycled plastic materials for the 3D printing process expects that the production process will be more environmentally friendly and solve the problem of plastic waste globally. However, recycled plastic as a filament material in the 3D printing process is still limited. One of the ways to make recycled plastic filaments is by extrusion. This research aims to optimize the extrusion process parameters in the manufacture of recycled plastic filaments (Polypropylene) that have never been conducted before. The extrusion process was analysed using the Taguchi and ANOVA methods. The results of this study indicate that the spooler speed of 4 rpm, extrusion speed of 40 rpm, and extrusion temperature of 200°C in the parameter setting produce an average filament diameter of 1.6 mm. However, filaments made from Recycled Plastic (Polypropylene) have a rough and easily curved surface, so further research is needed.

1. Introduction

Plastics are used in various fields of life such as packaging, transportation, building materials, electronic components, and consumer products [1,2]. The high demand for plastic causes an increase in plastic production in the world [3]. However, most plastics are only used in a short time and then discarded. This plastic waste then contaminates almost all parts of the world and becomes a global problem [1,4–6].

To be able to deal with the increasing amount of plastic waste, various efforts have been made to be able to recycle. The use of recycled plastic is expected to reduce the amount of plastic waste in the world. One of the developing technologies and applications of plastic as its main material is Fused deposition modelling (FDM). In FDM, material in the form of plastic filament is heated to near the melting point and then extruded according to the existing design [7,8]. The use of filament made from recycled plastic has been done before [9,10], but it is still very limited. With the use of recycled plastic material on FDM technology, it is expected that an FDM process that is more environmentally friendly and can solve the problem of plastic waste globally is expected.

The process of making material in the form of plastic filament for FDM can be done using an extrusion process [11]. The extrusion process is carried out by an extruder machine which functions to push or force the material through the gap to get the product as an extrudate [11]. In this extrusion process, many factors can affect the quality of the product.

Research on optimal parameters in the extrusion process has been carried out by many researchers. [12] used the extrusion temperature and winding machine parameters for PLA material from corn. The results of this study stated that there was a match between extrusion temperature and extrusion speed and the diameter of the resulting filament. Besides, the winding machine influences the regularity of the filament. [13] conducted a study using the die temperature, roller puller speed, spindle speed, and inlet temperature as parameters for PCL-PLA composite materials. In his research, the results showed that the roller puller speed and spindle speed have a strong interaction as a factor in making filaments. Meanwhile, [14] conducted research using barrel and platform temperature parameters on PP material. In the study, the best temperature parameters were obtained to process PP between 210°C-250°C.

Until now, there have been no studies to find out about how to optimise the extrusion process on recycled plastic as FDM material. Although [14] uses PP material, the material used is not recycled material. Besides that, the parameters used are only limited to the temperature. In this study, other parameters will be added, namely spooler speed and extrusion speed. Therefore, this study aims to determine the optimal parameters in recycled plastic extrusion as FDM material so that it can be utilized in the production process.

2. Methods

2.1. Material

The material used in this study is Polypropylene (PP). PP has been widely used in human life, such as electric devices, automotive parts, and household appliances. Research on recycled PP material as FDM material has been carried out by [10]. From the research, it can be concluded that recycled PP can be used as an FDM material, except that it hasn't optimized the extrusion process.

2.2. Response variable

The response variable in this study is the filament diameter. Ideally, the filament must have a constant diameter along with the spool. The diameter of the commonly used filament in FDM is 1.75, with a tolerance standard of 0.05 mm. If the filament diameter is not constant, a severe problem will arise. If there is a decrease in diameter, the extrusion process will decrease the amount of material needed so that the filament fails to enter the hot end. If there is an increase in diameter, the extrusion process will experience an excess amount of material so that the extruder is not strong enough to attract the filament to enter the hot end.

2.3. Experimental design

The process of data retrieval is conducted by doing an extrusion process under the parameters to be optimised. The parameters in the extrusion process are spooler speed, extrusion speed and extrusion temperature. Spooler speed has two levels, 2 and 4 rpm. Extrusion speed also has two levels, 40 and 50 rpm while extrusion temperature has three levels, that are 180°C, 190°C, and 200°C.

2.4. Based on the parameters that have been determined, the experiment was compiled using L36 orthogonal array. The experimental design used in this study can be seen in

Table *1*. The extrusion process is carried out according to the design of the experiment that has been made. Filament results of extrusion for each element were then measured in diameter and analysed using the Taguchi and ANOVA methods. The use of the Taguchi method is used to determine the optimal level for each parameter, while the ANOVA is used to determine the relationship of parameters with the response variable. The results of processing these data are then tested for validity.

Table 1. Experimental design

Experiment	Parameter			Experiment	Parameter		
	Spooler speed (Rpm)	Extrusion speed (Rpm)	Extrusion temperature (°C)		Spooler speed (Rpm)	Extrusion speed (Rpm)	Extrusion temperature (°C)
1	2	40	180	19	4	40	180
2	2	40	190	20	4	40	190
3	2	40	200	21	4	40	200
4	2	40	180	22	4	40	180
5	2	40	190	23	4	40	190
6	2	40	200	24	4	40	200
7	2	40	180	25	4	40	180
8	2	40	190	26	4	40	190
9	2	40	200	27	4	40	200
10	2	50	180	28	4	50	180
11	2	50	190	29	4	50	190
12	2	50	200	30	4	50	200
13	2	50	180	31	4	50	180
14	2	50	190	32	4	50	190
15	2	50	200	33	4	50	200
16	2	50	180	34	4	50	180
17	2	50	190	35	4	50	190
18	2	50	200	36	4	50	200

3. Result and Analysis

3.1. Taguchi analysis

Before carrying out the Taguchi analysis, measurements were made on the diameter of the filament produced by each experiment. Filament measurements are carried out at three different points every 9 cm. Measurements are made using a digital Vernier calliper. The results of these measurements are then calculated as averages and standard deviations (Table 2).

Table 2. Experimental results

Experiment	Location			Average	Standard deviation	S/N Ratio
	1	2	3			
1	1.88	1.92	1.82	1.87	0.05	18.09

IOP Conf. Series: Materials Science and Engineering **722** (2020) 012022 doi:10.1088/1757-899X/722/1/012022

Experiment	Location			Average	Standard deviation	S/N Ratio
	1	2	3			
2	1.44	1.49	1.66	1.53	0.12	23.51
3	1.93	1.94	1.95	1.94	0.01	28.43
4	1.5	1.63	1.61	1.58	0.07	*
5	1.48	1.43	1.34	1.42	0.07	*
6	2.08	1.96	1.93	1.99	0.08	*
7	1.3	1.48	1.59	1.46	0.15	*
8	1.34	1.4	1.42	1.39	0.04	*
9	2.1	2.08	1.92	2.03	0.10	*
10	1.53	1.61	1.41	1.52	0.10	20.34
11	1.43	1.77	1.53	1.58	0.17	19.74
12	1.73	1.79	1.76	1.76	0.03	22.3
13	1.35	1.42	1.32	1.36	0.05	*
14	1.68	1.59	1.5	1.59	0.09	*
15	1.88	1.59	1.57	1.68	0.17	*
16	1.21	1.26	1.26	1.24	0.03	*
17	1.39	1.32	1.33	1.35	0.04	*
18	1.63	1.48	1.79	1.63	0.16	*
19	1	1.02	1.05	1.02	0.03	25.97
20	1.02	1.16	1.15	1.11	0.08	18.92
21	1.73	1.69	1.56	1.66	0.09	24.89
22	1.04	0.95	0.91	0.97	0.07	*
23	1.03	0.82	1.06	0.97	0.13	*
24	1.6	1.66	1.62	1.63	0.03	*
25	0.94	1	1.04	0.99	0.05	*
26	0.98	1.04	0.86	0.96	0.09	*
27	1.44	1.5	1.59	1.51	0.08	*
28	0.89	0.85	0.81	0.85	0.04	20.9
29	1.18	1.09	1.13	1.13	0.05	25.85
30	1.52	1.48	1.64	1.55	0.08	23.29
31	0.8	0.74	0.75	0.76	0.03	*
32	1.12	1.04	1.04	1.07	0.05	*
33	1.64	1.8	1.72	1.72	0.08	*
34	0.69	0.9	0.75	0.78	0.11	*
35	1.04	1.04	1.02	1.03	0.01	*
36	1.75	1.79	1.74	1.76	0.03	*

In this study, the variable response is the filament diameter. The ideal filament diameter is a diameter close to 1.75 mm. Therefore the "normal is best" condition in Taguchi is chosen. After measuring the diameter, the next step is to calculate the S / N ratio. The S / N ratio calculation in Taguchi is done by entering the response value and factors in the Minitab Software. The results of the calculation of S / N ratio can be seen in Table 2. After calculating the S / N ratio, the next step is to calculate the average signal to noise ratio for each parameter and level (Table 3). In Table 3, the largest delta value is in the C (Extrusion temperature) parameter and followed by the spooler speed and extrusion speed.

Table 3. Average signal to noise ratio for each parameter and level

Level	Parameter		
	A	B	C
1	22.07	23.3	21.32
2	23.3	22.07	22
3	-	-	24.73
Delta	1.24	1.23	3.4
Rank	2	3	1

IOP Conf. Series: Materials Science and Engineering **722** (2020) 012022 doi:10.1088/1757-899X/722/1/012022

The average signal to noise ratio data is then created in the form of scatter plots that can be seen on Figure 1. Based on Figure 1, the best diameter response from factor parameters is extrusion temperature 200°C (C3), spooler speed 4 rpm (A2), and extrusion speed 40 rpm (B1).

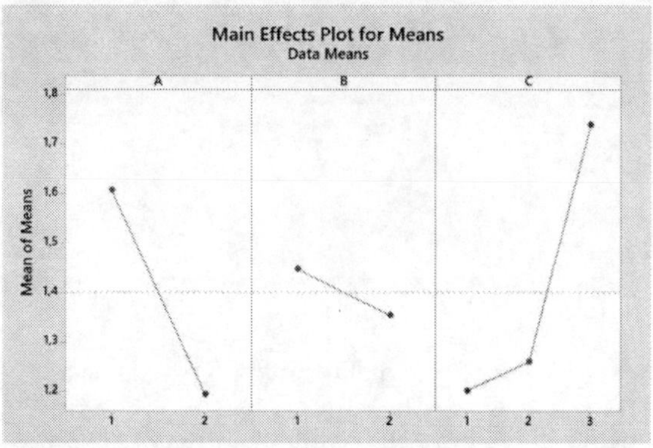

Figure 1. Scatter plot of average signal to noise ratio

3.2. ANOVA

ANOVA is used to determine the relationship between each parameter to the filament diameter. ANOVA can only be done if the data used is normally distributed. Before conducting ANOVA, a normality test was carried out on the average data (Figure 2) and the S/N ratio (Figure 3).

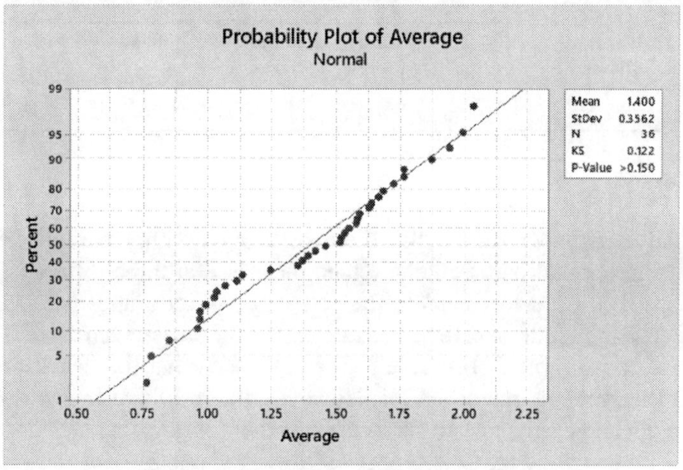

Figure 2. Normality test for average data

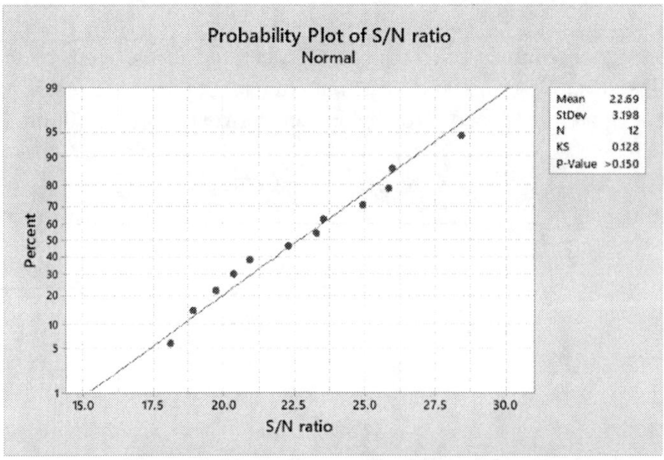

Figure 3. Normality test for S/N ratio data

ANOVA is done using the help of the Minitab software. The results of ANOVA can be seen in Table 4. In Table 4, it can be seen that the parameters A (spooler speed) and C (extrusion temperature) have a P-value of less than 0.05. It means that both settings are parameters that have a significant effect on filament diameter.

Table 4. ANOVA results

Source	DF	Adj SS	Adj MS	F-value	P-value
A	1	1.538	1.538	64.4	0
B	1	0.076	0.076	3.22	0.08
C	2	2.084	1.042	43.65	0

3.3. Validity test

Validity test is conducted by extruding the filament using optimal parameters. The results of the extrusion process using optimal parameters are then measured at three different points. The results of this measurement are then used to calculate the actual S / N ratio and mean. The actual results are compared with prediction results and then the rate of deviation is calculated. The calculation of the validity test can be seen in Table 5. In Table 5, it can be seen that the deviation of the S / N ratio is 5.7%. It shows that the accuracy rate is 94.3%. However, the results of the extrusion process with this parameter have a rough and easily curved surface, so further research is needed.

Table 5. The calculation of the validity test

Point	Diameter	Actual		Prediction		Error Level	
		S/N ratio	Mean	S/N ratio	Mean	S/N ratio	Mean
1	1.72						
2	1.65	27.44	1.65	25.96	1.39	5.7%	18.7%
3	1.58						

4. Conclusion

The extrusion process to produce filament as material on FDM can be done using recycled PP. Based on the Taguchi analysis, the optimal parameter for producing recycled PP is extrusion temperature 200°C (C3), spooler speed 4 rpm (A2), and extrusion speed 40 rpm (B1). Meanwhile, based on

IOP Conf. Series: Materials Science and Engineering **722** (2020) 012022 doi:10.1088/1757-899X/722/1/012022

ANOVA parameters that have a significant effect on filament diameter are spooler speed and extrusion temperature. The results of this study were tested for validation and obtained an S / N ratio deviation rate of 5.7%. Further research is still needed to be able to produce recycled PP filaments with better properties.

References

[1] Gourmelon G 2015 Global Plastic Production Rises, Recycling Lags *Vital Signs, Worldwatch Inst.*

[2] Lahimer C, Ayed M and Horrichfile N 2013 Characterization of plastic packaging additives: Food contact, stability and toxicity *Arab. J. Chem.*

[3] Velis C 2014 *Global recycling markets - plastic waste: A story for one player – China*

[4] MacArthur D E 2017 Beyond plastic waste *Science (80-.).*

[5] Singh N, Hui D, Singh R, Ahuja I P S, Feo L and Fraternali F 2017 Recycling of plastic solid waste: A state of art review and future applications *Compos. Part B Eng.*

[6] Domingues J, Marques T, Mateus A, Carreira P and Malça C 2017 An Additive Manufacturing Solution to Produce Big Green Parts from Tires and Recycled Plastics *Procedia Manuf.*

[7] Tlegenov Y, Wong Y S and Hong G S 2017 A dynamic model for nozzle clog monitoring in fused deposition modelling *Rapid Prototyp. J.*

[8] Geng P, Zhao J, Wu W, Ye W, Wang Y, Wang S and Zhang S 2019 Effects of extrusion speed and printing speed on the 3D printing stability of extruded PEEK filament *J. Manuf. Process.*

[9] Singh R, Singh H, Farina I, Colangelo F and Fraternali F 2019 On the additive manufacturing of an energy storage device from recycled material *Compos. Part B Eng.*

[10] Pickering K and Stoof D 2017 Sustainable Composite Fused Deposition Modelling Filament Using Post-Consumer Recycled Polypropylene *J. Compos. Sci.*

[11] Rauwendaal C 2014 *Polymer extrusion: Fifth edition*

[12] Mirón V, Ferrándiz S, Juárez D and Mengual A 2017 Manufacturing and characterization of 3D printer filament using tailoring materials *Procedia Manuf.*

[13] Haq R H A, Khairilhijra K R, Wahab M S, Sa'Ude N, Ibrahim M, Marwah O M F, Yusof M S, Rahman M N A, Ariffin A M T, Hassan M F, Yunos M Z and Adzila S 2017 PCL/PLA Polymer Composite Filament Fabrication using Full Factorial Design (DOE) for Fused Deposition Modelling *Journal of Physics: Conference Series*

[14] Kumar N, Jain P K, Tandon P and Pandey P M 2018 Experimental investigations on suitability of polypropylene (PP) and ethylene vinyl acetate (EVA) in additive manufacturing *Mater. Today Proc.* **5** 4118–27

IOP Conf. Series: Materials Science and Engineering **722** (2020) 012023 doi:10.1088/1757-899X/722/1/012023

Measurement of magnetic fields on electric train

H W Nugroho, M I Sudrajat, A N Bakti, Yudhistira and R H Arjadi[1]

[1]Research Center fot Testing Technology, PUSPIPTEK Serpong building 417, South Tangerang, Banten, Indonesia

E-mail: huto001@lipi.go.id

Abstract. Electric train is one of the most widely used transportation modes in Indonesia especially Jakarta. In general, all electrical and electronic based equipment when operated will produce electromagnetic fields. At present, the railway system uses electric power, both as to its driving force, as well as a supporting part (light system, air conditioner, etc.), and instead uses very large electric power. So surely the train system produces an electromagnetic field. Electromagnetic fields are a combination of invisible electric and magnetic fields of force. They are generated by natural phenomena, but also by human activities, mainly through the use of electricity. The magnetic field on the train generally occurs in areas close to the source. In this paper, already measured the magnetic field (near field) in the electric train in the frequency range 0.015 kHz up to 1 kHz. The measurement results show that the highest magnetic field is obtained when the train starts up at 333.76 nT. The magnetic field level is small when away from the electrical inverters. The average level of the magnetic field when the train is at its highest is 223.26 nT and during braking there is an increase of 323.93 nT.

1. Introduction

Today, travelling by train is the easiest, fastest way to explore another city. People using electric train is increasing year by year and become one of alternative public transportation that likely choose by peoples and society around Jakarta. Based on data from PT. Kereta Api Indonesia (KAI), the number of electric train passengers throughout 2018 reached 336.71 million passengers, up approximately 6% from previous year [1]. The amount is around 1.2 million/day [2].

In general, all electrical and electronic based equipment when operated will produce electromagentic fields [3]. Electromagnetic fields are present everywhere in our environment but are invisible to the human eye [4]. Electromagnetic fields are a combination of invisible electric and magnetic fields of force. More than 3 billion people across the world are exposed to electromagnetic fields every day [5]. At present, the railway system uses electric power, both as to its driving force, as well as a supporting part (light system, air conditioner, etc.), and instead uses very large electric power. In general, train infrastructure operates in a complex and non-homogeneous environment, where low-power electronic equipment must function in an environment with large voltages and currents such as on electric rail trains [6]. The environment of railway tracks is exposed to magnetic and electric fields from trains, the railway power supply systems, nearby electric power transmission lines [7]. In addition, according to research conducted by the International Commiccion on Non-Ionizing Radiation Protection (ICNIRP), the electromagnetic field can also affect human health, which means here are train passengers themselves [8]. In 1998, ICNIRP issued a guideline to limit radition emitted by trains [9]. The magnetic field on the train generally occurs in areas close to the sources. In

2011, Wuwus et al condusted a research of the measurement of the electromagnetic field from the electric train [10]. Therefore, the current study aims to measure the intensity of magnetic field in the electric trains of Jabodetabek in different conditions.

2. Basic Theory

2.1. Magnetic Field

Magnetic fields and electric fields are interrelated, and are both components of the electromagnetic force, one of the four fundamental forces of nature [11]. Magnetic fields are widely used throughout modern technology, particularly in electrical engineering and electromechanics. A magnetic field is an area which is under the influence of a magnetic charge. A magnetic field, like an electric field, is due to the polar nature of electric or manetic charges. Poles are oriented in north and south directions, and a magnetic fields strength depends on the strength of magnet that is creating it. Magnetic fields are a basic principle behind countless appliances and modern-day gadgets.

A magnetic field can be generated in two ways: either around a magnet or in the victim of moving electric field (when electric charge carries such as electrons move through space or within an electrical conductor). Magnetic and electric fields interact differently depending on the environment into which they are placed. An object placed in a magnetic field experiences a forces which depends on its orientation within the field. Properties of a magnetic field are rendered in electromagnetism, which is basis of technological advancements such as bullet trains, elevators, escalators, and catthodc-ray tubc televisions [12].

2.2. Electric Train

In an electric railway system, the train is provided by sliding contact from a supply line that is placed over the railway tracks. The electric train uses two types of electricity sources, namely a DC source of 600 V, 750 V, 1500 V, and 3000 V, while for an AC source of 15 kV (16.7 Hz) and 25 kV (50 Hz). A pantograph use to deliver electricity from the source above to the converter then forward to the motor (Figure 1) so that the train runs. The pantograph must always contact continuously with the source conductor, besides that the pantograph must be aerodynamic because it is used at relatively high speed continuously [13].

The current generally return to the substation via the rails, a separate return conductor, or via the earth. The large electrical plants in the network are constituted of sub-plants, which are essentially independent. Every single sub-plant consists essentially of overhead lines, buried cables, and rails. Two different substations that are equipped with static AC–DC conversion groups supply the sub-network.

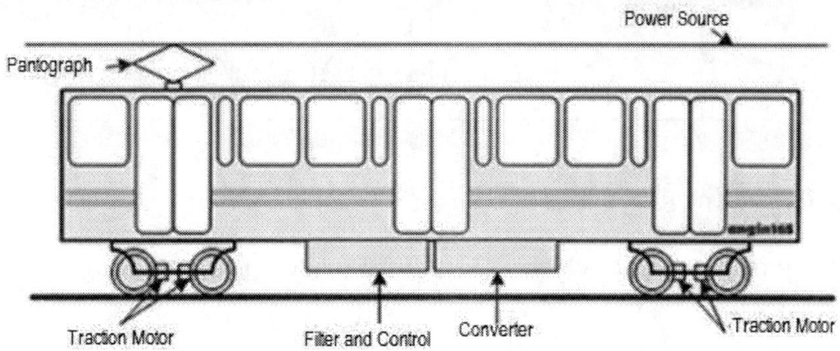

Figure 1. The main parts of the electric train

The return path of the current is constituted by the rails that are connected by means of cables to the negative pole of the supply. DC electrical motors controlled by choppers are employed for traction. In order to compensate for the voltage drop along the lines, several substations are used as line subway feeders. The power delivered by the substation is transmitted to the traction vehicle via a system of flexible suspension contact lines (overhead or centenary) with which a locomotive mounted articulated

IOP Conf. Series: Materials Science and Engineering **722** (2020) 012023 doi:10.1088/1757-899X/722/1/012023

device (pantograph) is brought into contact. On the traction vehicle, the power is regulated using choppers and then supplied to electric motors to control the movement. Auxiliary power that is lower than that which is supplied to the electric traction motors is also conditioned and regulated using static converters, inverters, and rectifiers. The rails ensure that the current return and the sources of magnetic fields on the train are often under the floors. [14].

3. Methodology

Ideally, the measurement is at the distance as close as possible to the source so that get the maximum measurement results. However, because the condition of the train is moving and the position of the motor is under the carriage, it is not possible to take the ideal near field measurement. The possible approach is to take measurements on the train floor area exactly where the motor is installed. Magnetic field measurements are carried out in the frequency range of 0.015 kHz-1 kHz. The antenna is placed 10 cm above the surface of the train floor. Setup measurements can be seen in Figure 2.

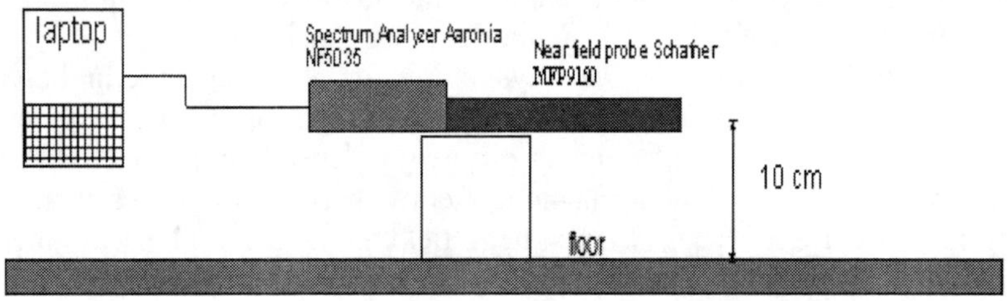

Figure 2. Measurement setup

The measurement conditions are as follows (Figure 3) :

a. When the electric inverter starts at point 1
b. When after the electric inverter starts up at point 2
c. When after the electric inverter starts up at point 4
d. When after the electric inverter starts up at point 5
e. When the first acceleration starts at point 5
f. When the train is running at point 5
g. When braking starts at point 5

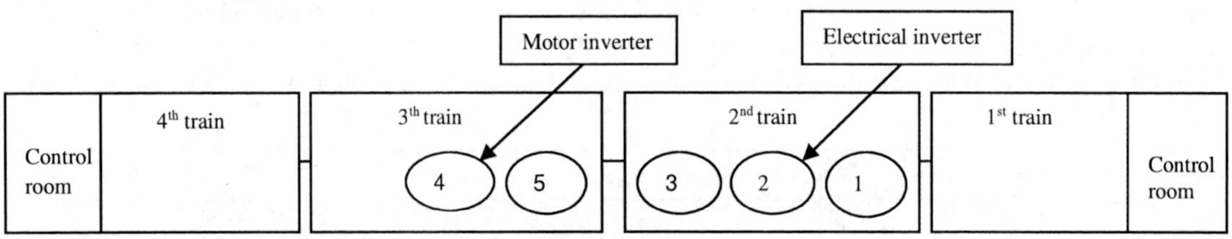

Figure 3. Measurement point

4. Results and Discussion

The results of near field magnetic field measurements are shown in the following graphs:

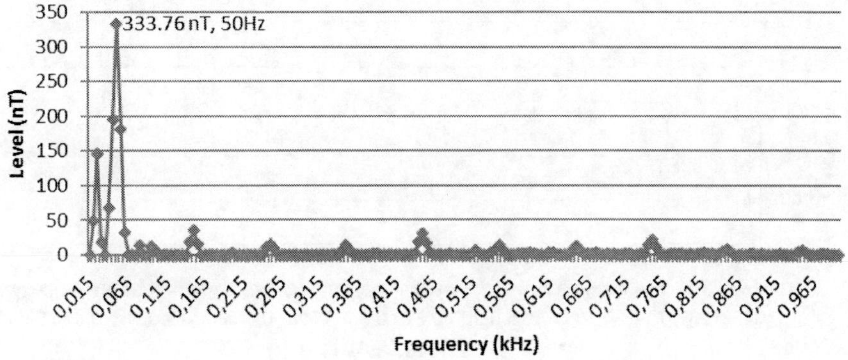

Figure 4. Measurements graph when the electric inverter starts at point 1 (0,015 kHz-1 kHz)

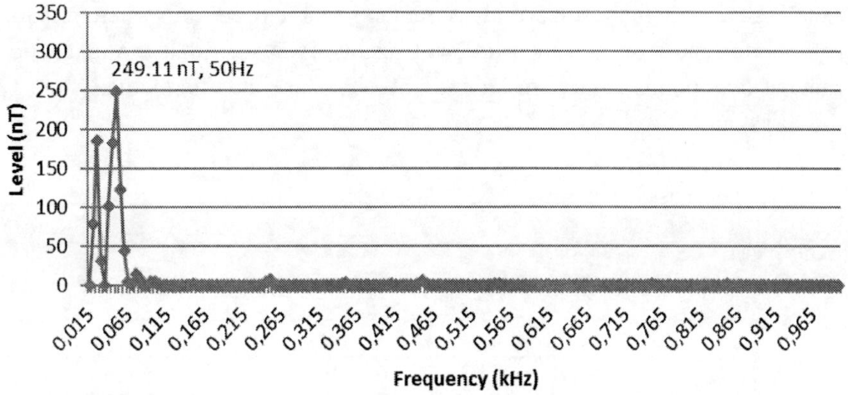

Figure 5. Measurements graph after the electric inverter starts at point 2 (0,015 kHz-1 kHz)

From the picture above it can be seen at the measurement point 1 when the electric inverter starts to turn on the highest value of the magnetic field reaching 333.76 nT which is at the frequency of 50 Hz. A few moments after the inverter startup is seen in point 2 there is a decrease in the magnetic field value to 249.11 nT at the same frequency. At this point, it should be the maximum magnetic field due to the electricity inverter. But because of the limited measurement time so that the initial data when the inverter starts to be completely turned on cannot be done at this point.

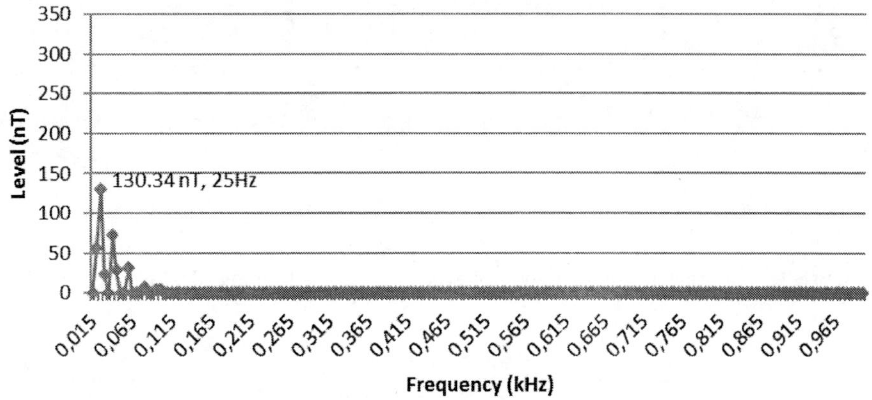

Figure 6. Measurements graph after the electric inverter starts up at point 4 (0,015 kHz-1 kHz)

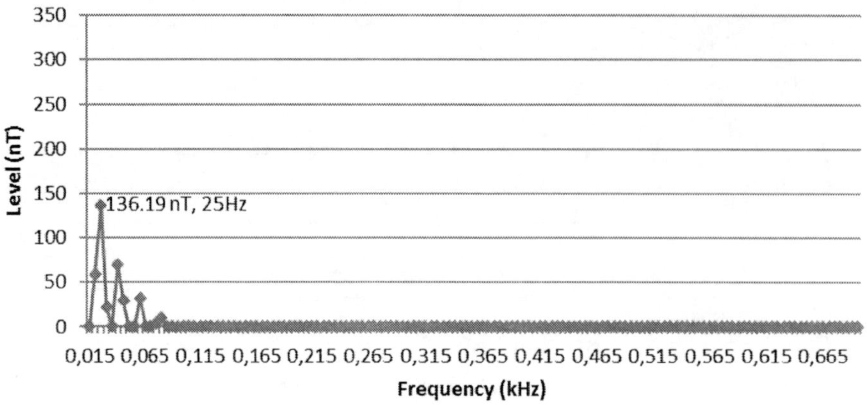

Figure 7. Measurements graph after the electric inverter starts up at point 5 (0,015 kHz-1 kHz)

Figure 6 and Figure 7 are the results of measurements on the third train where it is located far from the electricity inverter and from the measurement results, it is seen that the magnetic field level is almost the same at both points which are around 130 nT. This value is relatively much lower than at the start of the starts up.

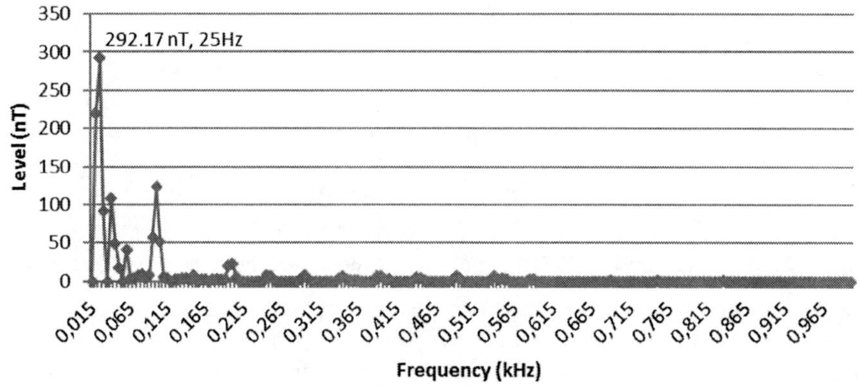

Figure 8. Measurements graph when the first acceleration starts at point 5 (0.015 kHz-1 kHz)

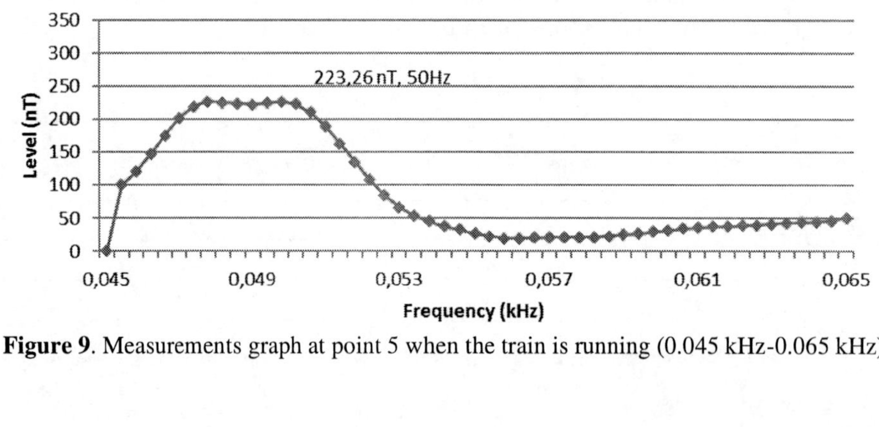

Figure 9. Measurements graph at point 5 when the train is running (0.045 kHz-0.065 kHz)

IOP Conf. Series: Materials Science and Engineering **722** (2020) 012023 doi:10.1088/1757-899X/722/1/012023

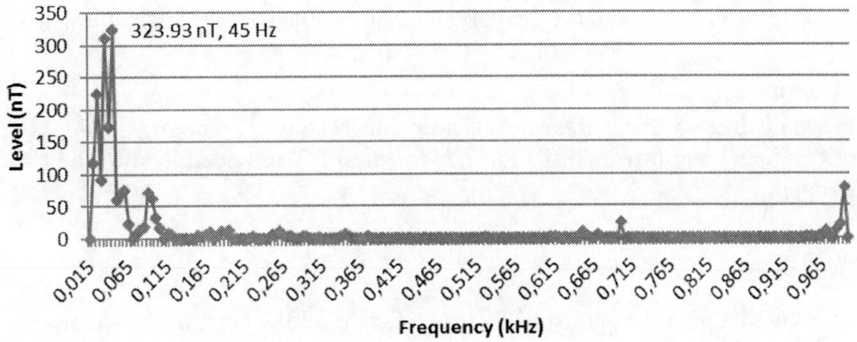

Figure 10. Measurements graph at point 5 when braking begins (0,015 kHz-1 kHz)

Measurement results (Figure 8-10) show that the highest field is generated during regenerative braking and maximum acceleration. According to measurements done in Switzerland by the Biel Technical University [15] the strength of the magnetic fields changes constantly as the car travels along, and is greatly dependent on the way the car is advancing or braking, the magnetic fields were between 0.1 and 3 μT. Figure 8 shows an increase in the magnetic field level at point 5 (above the inverter motor) when the first acceleration occurs or the train starts moving from a stationary position. Figure 9 is a graph of the average magnetic field measurement results measured at a point just above the motor inverter when the train runs. The graph shows that the largest magnetic field level at a frequency of 50 Hz with a magnetic field strength of 223.26 nT. From Figure 10, it can be seen when the train is braking, there is an increase in magnetic field level with a field strength of 323.93 nT.

If all measurement results are compared with the limit recommended by the ICNIRP which is only 100μT for passengers (general public) then the magnetic field level of the train is relatively low. For further application, the units in μT can be converted into units of dBμA/m using the following equation:

$$dB\mu A/m = 20 \log (\mu T/1.25) + 120 \qquad (1)$$

So that the results are as follows:

Table 1. Magnetic Field (dBμA/m)

Condition	Measurement point	Frequency (Hz)	Magnetic field (nT)	Magnetic field (dBμA/m)
Electric inverter starts	1	50	333.76	108,53
After electric inverter starts up	2	50	249.11	105,98
After electric inverter starts up	4&5	25	130	100,34
First acceleration	5	25	292.17	107,37
When driving	5	50	223.26	105,03
When breaking	5	45	323.93	108,27

5. Conclusion

In this study, magnetic field measurements have been carried out on the electric trains. The measurement results show the magnetic field level when measured near the electricity inverter is higher than when measured far from the electricity inverter. The highest level of the magnetic field is obtained when the train starts, which is equal to 333.76 nT. The average level of the magnetic field when the train driving is 223.26 nT and during braking an increase of 323.93 nT. The value of the magnetic field strength at the two farthest points from the electricity inverter looks almost the same, which is around 130 nT. This value is relatively much lower than at the start of the starts up.

Acknowledgments
The authors would like to thank Research Centre for Testing Technology, Indonesian Institute of Sciences for the equipment provided, PT. INKA for the EUT provided, and to the EMC Laboratory staff for the support.

6. References

[1] https://www.tribunnews.com/bisnis/2019/01/10/tahun-2018-penumpang-kereta-api-naik-delapan-persen
[2] https://metro.tempo.co/read/1161866/tahun-2020-menhub-targetkan-2-juta-penumpang-krl-per-hari/full&view=ok
[3] Kadarwati S, Wibowo P, Sudrajat M I and Arjadi R H 2010 *Radiasi elektromagnetik pada alat telekomunikasi militer* (Jakarta: LIPI press) p3
[4] WHO Electromagnetic Field " http://www.who.int/peh-emf/about/WhatisEMF/en/index3.html. 2009
[5] Fragopoulou A F, Koussoulakos S L, Margaritis L H 2010 Cranial and postcranial skeletal variations induced in mouse embryos by mobile phone radiation. Pathophysiology ;17:169–77
[6] Niska F 2008 *Measurements and Analysis of Electromagnetic Interferences in the Swedish Railway Systems* (Sweden: Lulea University of Technology)
[7] Al Dhalaan S M and Elirbawy M A May 2003 A quantitative study of the electromagnetic field coupling between the electric power transmission line and railway. *Proc. Canadian Conf. on Electrical and Computer Engineering*
[8] ICNRP Guidelines for limiting exposure to time-varying Electric, magnetic, and electromagnetic fields (up to 300 GHz) 1998 Published in: *Health Physics* 74 (4):494-522
[9] Hardell L, Koppel T, Carlberg M, Ahonen M, and Hedendahl L. Radiofrequency radiation at Stockholm Central Railway Station in Sweden and some medical aspects on public exposure to RF fields 2016 *Int J Oncol* 49(4): 1315–1324.
[10] Ardiatna W, Mandaris D, Andiani D and Arjadi R H 2010 Pengukuran Tingkat Emisi Radiasi Elektromagnetik Kereta Rel Listrik Terhadap Lingkungannya. *Jurnal Teknologi Indonesia* Vol 33 No 2
[11] https://en.wikipedia.org/wiki/Magnetic_field
[12] https://www.techopedia.com/definition/8211/magnetic-field
[13] https://konversi.wordpress.com/2010/04/23/kereta-rel-listrik/
[14] Halgamuge MN, Abeyrathne CD, and Mendis P. 2010. Measurement and Analysis of Electromagnetic Fields From Trams, Trains And Hybrid Cars." Radiation Protection Dosimetry 2010 *Radiation protection dosimetry* Vol. 141, No. 3, pp. 255–268
[15] Federal office of Public Health, "Cars, hybrid cars". https://www.bag.admin.ch/bag/en/home/gesund-leben/umwelt-und-gesundheit/strahlung-radioaktivitaet-schall/elektromagnetische-felder-emf-uv-laser-licht/emf.html

IOP Conf. Series: Materials Science and Engineering **722** (2020) 012024 doi:10.1088/1757-899X/722/1/012024

Applying the Theory of Planned Behavior in Functional Food Purchasing: a Young Consumers Perception

I G M Y Bakti , S Sumaedi, N J Astrini, T Rakhmawati, M Yarmen, and S Damayanti

Research Center for Testing Technology, Indonesian Institute of Sciences, South Tangerang, Indonesia

E-mail: iged010@lipi.go.id, siks002@lipi.go.id, nidy001@lipi.go.id, trir005@lipi.go.id, medi001@lipi.go.id, sihd001@lipi.go.id

Abstract. The 4.0 industrial era has impact not only on the industrial sector but also on the changes of people's lifestyle, including the healthy lifestyle changes. More specifically, one of the impacts of the 4.0 industrial era is the growth of functional food industry. This research aims to analyze the effect of attitude, subjective norm, and perceived behavioral control on young consumers' purchase intention on functional food. This research focused on young consumers because they are susceptible to consume a food that can cause non-communicable diseases. This study used a quantitative approach. A survey was used to collect data. Questionnaires were spread in a university in South Tangerang. This study obtained 123 students that participated in the survey. Multiple regressions analysis was performed to test the conceptual model and the proposed hypotheses. The findings showed that attitude and subjective norm influence the purchase intention of the young consumers to buy functional foods. In the other hand, this research also revealed that perceived behavioural control has no significant impact on repurchase intention.

1. Introduction

1.1. Background

We are entering the 4.0 industrial era. This era utilizes various intelligent manufacture technology in companies' value chain, such as the Internet of Things (IoT), cyber-physical systems (CPSs), cloud computing, big data analytics (BDA), and information and communications technology (ICT) [1]. The emergence of those technologies benefitted companies. They (1) drove down production costs; (2) cut time-to-market of new products; (3) enabled a custom mass production without significantly increasing overall production costs; (4) created a more friendly and flexible working environment, and (5) increased the efficiency of natural resource and energy use [2].

The effect of technology arose in the 4.0 industry affected not only affect companies, but also human life [1]. Specifically, the technology development in the 4.0 industrial era also changed people's lifestyle. One of the notable changes was a healthy lifestyle. People now are more aware of the importance of a healthy lifestyle [3]. In addition, this modern era has seen many new products including functional food products [4,5].

Generally, functional foods are food or beverage which can give health benefits beyond basic nutrition [6]. The consumption of functional food is a new way for consumers in expressing health [7].

The health benefits include reducing the risk of certain diseases, improving body condition, and even healing diseases [8].

In the last several years, the functional food industry has grown considerably [9]. Even in 2020, the market for this industry was projected to reach $192bn [3,10]. Even though the trend was significantly positive, the growth of this industry was not evenly spread. The industry tended to grow in developed countries, such as Japan, China, the US, and European countries [3]. For example, in the Asia-Pacific region, 78% of the total sales of functional foods came from the US, Canada, and Europe [11, 12]. On the other hands, in the developing county, this industry did not grow as expected even though developing countries have rich biodiversity [13]. One of the causes of this predicament was because functional food products have not been accepted. Based on that condition, one of the efforts that can be done to understand the consumers was to study the determinants of functional food consumers' purchase intention.

1.2. Research Gap and Objective

The research on functional food have been done for more than two decades [14]. Several researchers have studied the behavior of functional food consumer. Studies on the consumption of functional food have also been done in many countries, like Europe [15], Japan [16], Canada [17], Australia [18], Croatia [19], Mexico [12], Turkey [3], Lebanon [20], Iran [21], Malaysia [22], etc. On the other hands, many researchers also studied the antecedents of functional foods consumers' purchase intention [23-26].

From multiple consumers behavior theories, one of the most used was the theory of planned behavior (TPB). The theory explained that purchase intention was influenced by three determinants, which are attitude, subjective norm, and Perceived Behavioural Control (PBC). Several researchers who applied this theory in the context of functional food consumers' behavior. [27-29].

Even though there are several researches on functional foods that focused on TPB, there has not been a study that applied this theory to young consumers, especially in Indonesia. This research has become important because younger consumers have different behavior compared to the older generation. This was because they have a different experience and faced different condition. The differences have also been proven by some researchers [30-32].

This research focused on younger consumers because of two reasons. First, the younger generation was susceptible to an unhealthy lifestyle that would lead to the probability of the increased incidence of non-communicable diseases (NCD). Septiana et al. stated that the younger generation has a habit of consuming junk food [33]. Second, the younger generation was in a transition between living under their parents' roof and living independently.

Based on the above gap, this research aims to analyze the effect of attitude, subjective norm and PBC on younger consumer's purchase intention.

2. Literature Review

2.1. Functional food

The concept of functional food was first popularized in Japan in 1984 [8]. Even though the concept has been studied by researchers for some time, there was no single standardized definition. One of the most referred definition was one offered by Siró et al. [8]. Functional foods are foods that have been enriched with important and physiologically useful effects. Siró et al. [8] added that a product was called as a functional food when the food: (1) could improve the human health, condition in general (e.g., prebiotic or probiotic), (2) reduce the risk of certain diseases (e.g., reducing cholesterol or blood sugar level), or (3) foods that can cure diseases.

2.2. *Theory of planned behaviour (TPB)*

One of the most popular theories that explain consumers' behavior was TPB. This theory was first developed by Icek Ajzen in 1985 [34]. According to TPB, a consumer's real behavior can be predicted by their behavioral intention. In other words, the strength of their intentions will determine their real behavior. In the context of purchase intention, a stronger intention would lead to a higher probability of purchasing behavior. Aside from the importance of behavioral intention, TPB also explained that there were three important factors that affected behavioral intention, namely attitude, subjective norm, and PBC [34].

2.3. *Purchase Intention (behavioral intention)*

In the context of consumer behavior, one of the behavioral intention forms was repurchase intention [35]. Morinez et al. defined purchase intention as a situation when a consumer prefer a certain product [36]. According to Halim and Hameed, purchase intention was related to a plan to buy a certain product in the future [37]. In this study, we defined purchase intention as a condition when a young consumer has a plan to buy a functional food product in the near future.

2.4. *Attitude*

Attitude is an important factor in the product purchasing process. Generally, this factor is a factor that represented consumers' perception of a product. In TPB, attitude was defined as how far someone has a favorable or unfavorable evaluation on a product [34]. In the context of this study, attitude was viewed as young consumers' favorable or unfavorable evaluation on functional food products.

TPB explained that attitude can affect behavioral intention. It means that higher attitude will increase consumers' attention to buy a functional food product. Empirical evidence related to the effect of attitude on behavioral intention has been found by several researchers [35, 38-41]. In the context of functional food, that relationship applies [27, 41]. Given this, the first hypothesis of this study is proposed as follows:

H1. Attitude toward functional food influences purchase intention positively

2.5. *Subjective Norm*

Subjective norm was one of the factors that affect consumers' behavior. More specifically, the subjective norm was a factor that illustrated the effect of the social environment [42]. According to Ajzen, a subjective norm was a consumer's perception of peers' pressures to do or not to do something [34]. Based on the definitions, we defined subjective norm as young consumers' perception of functional food consumption based on the reference group that is considered important for them.

According to TPB, consumers would be willing to consume functional foods when there was a subjective norm to support the consumption of functional food. Several studies have proven the positive relationship of subjective norm and behavioral intention [35, 38-40, 43, 44]. In the context of functional food, O'conner and White have proven the effect of subjective norm on behavioral intention 28]. The second hypothesis of this study is formulated as follows:

H2. Subjective norm influences intention to purchase functional food positively

2.6. *Perceived Behavioral Control (PBC)*

One of the important factors that affect consumers' behavior was PBC. Ajzen defined PBC as "people's perception of the ease or difficulty of performing the behavior of interest" [34]. This factor was a factor that showed consumers' abilities to act. In other words, this factor was related to resource, opportunities, and barriers. Furthermore, bigger resources or opportunity to act or lower barriers would result in a stronger intention to act. More briefly, we defined PBC as young consumers' perception of the ease or difficulty of consuming functional food.

Several researchers have empirically proven that PBC had impacts on behavioral intention [35, 38, 41]. In the context of functional food, Some researchers have also studied likewise [27, 28]. The final hypothesis of this study is formulated as follows:

H3. PBC influences intention to purchase functional food positively

3. Methodology

3.1. Research design

This is a quantitative research. Data was gathered through a survey with a questionnaire as the instrument. A survey was done in April 2019. Questionnaires were spread in a university in South Tangerang.

3.2. Research object

Because functional foods have various categories, the object of this research only focused on the ones that prevent hypertension. This study chose this type of functional food because of three reasons. First, hypertension is the leading mortality cause in Indonesia [45]. Second, the number of hypertension incident was growing over the years [45]. Third, the younger generation tended to prefer junk food [46] and this habit increased the risk of hypertension.

3.3. Respondents

The respondents of this research are young consumers. In this study, we defined young consumer as undergraduate student. The respondents of this research were selected by using a convenience sampling technique. Questionnaires were distributed among regular students in a university in South Tangerang. This study obtained 123 students. The demographic profile can be seen in Table 1.

Table 1. Respondents'demographic profile

Characteristics	Category	%
Sex	• Male	76.4
	• Female	23.6
Age	• 18 Years old	7.6
	• 19 Years old	37.3
	• 20 Years old	24.6
	• 21 Years old	20.3
	• 22 Years old	5.9
	• 23 Years old	2.5
	• 24 Years old	0.8
	• 25 Years old	0.8
Residency status	• With family	65.9
	• Boarding house	24.4
	• Other	9.8
Location of resident	• The District of Tangerang	19.2
	• The City of South Tangerang	44.2
	• Other	36.7
Monthly allowance	• < Rp600,001	54.8
	• Rp600,000-Rp1,200,000	35.7
	• > Rp1,200,000	9.6
Source of monthly allowance	• Parents	71.9
	• Scholarship	0.8
	• Wage/salary	20.7
	• Other	6.6

3.4. *Questionnaire*

This research questionnaire was divided into two parts. The first one asked about the respondents' demographic profile, such as sex, age, residency status, resident location, allowance, and the source of their allowance. The second part is filled with questions related to the variables used in this research, namely attitude, subjective norm, PBC, and purchase intention.

Based on the previous research, those variables were categorized as latent variables. A latent variable was a variable that could not be measured directly but through several indicators. Those indicators were taken from previous studies [46, 38, 40]. The variables and indicators used in this research can be seen in Table 2. This study employed a five-point Likert scale to measure each indicator.

Table 2. The variables and indicators

Variables	Indicators
Attitude	AT1: I like functional food that can prevent hypertension AT2: A functional food that can prevent hypertension is good AT3: Consuming functional food that can prevent hypertension is a beneficial behavior for me AT4: It is interesting to consume functional food that can prevent hypertension
Subjective Norm	SN1: The people that are important for me agreed that I need to consume functional food that can prevent hypertension SN2: My family members want me to consume functional food that can prevent hypertension SN3: My friends suggest me to consume functional food that can prevent hypertension
PBC	PBC1: I believe that I can consume functional food that can prevent hypertension easily PBC2: I have a full control (not dependent on others) for consuming functional food that can prevent hypertension PBC3: My money allows me to buy functional food that can prevent hypertension PBC4: I can decide when I want to consume functional food that can prevent hypertension
Purchase Intention	PI1: I plan to consume functional food that can prevent hypertension in the future PI2: I expect to consume functional food that can prevent hypertension PI3: I want to try functional food that can prevent hypertension

3.5. *Data Analysis*

Because this is a quantitative research, this study employed several statistical analyses, which were factor analysis, Cronbach's Alpha analysis, and multiple regressions. The factor analysis was done to test the construct validity. The questionnaire was considered as valid if the factor analysis shows: (1) Kaiser Meyer Olkin (KMO) value $\geqslant 0.5$; (2) Measure of Sampling Adequacy (MSA) for each indicator ≥ 0.5; and (3) Factor loading value for each indicator $\geqslant 0.5$. Cronbach's Alpha was used to test the questionnaire's reliability. The questionnaire was deemed as reliable is the value for each variable is higher than 0.6. The multiple regressions were done to test the hypothesis. A hypothesis is accepted if the t-test for each variable is more than 1.96 (5% error margin). This research utilized SPSS

4. Result and discussion

4.1. *Validity Test*

Based on the factor analysis, this study found that the instrument was valid. The indicators were suitable to measure attitude, subjective norm, perceived behavioral control, and purchase intention. Based on the factor analysis of each variable, it was found that (1) Kaiser Meyer Olkin (KMO) value $\geqslant 0.5$; (2) Measure of Sampling

Adequacy (MSA) for each indicator ≥ 0.5; and (3) Factor loading value for each indicator ⩾ 0.5. The result can be seen in Table 3.

4.2. Reliability Test

Based on the Cronbach's Alpha analysis, the result showed all variables have alpha values above 0.6. This means the instrument of this study was reliable. In other word, the questionnaire was deemed as consistent in measuring those four variables. The results of reliability test can be seen in Table 4.

Table 3. The result of the validity test

Variables (Indicators)	KMO	% Eigenvalue	MSA	Factor Loading
Attitude	0.760*	67.475		
AT1			0.792	0.856
AT2			0.741	0.787
AT3			0.751	0.876
AT4			0.755	0.761
Subjective Norm	0.694*	74.766		
SN1			0.758	0.830
SN2			0.645	0.904
SN3			0.704	0.859
PBC	0.740*	54.952		
PBC1			0.788	0.704
PBC2			0.704	0.796
PBC3			0.804	0.668
PBC4			0.710	0.790
Purchase Intention	0.705*	75.196		
PI1			0.700	0.871
PI2			0.661	0.898
PI3			0.777	0.832

Table 4. The result of the reliability test

Variables	Cronbach Alpha
Attitude	0.839
Subjective Norm	0.831
PBC	0.724
Purchase Intention	0.833

4.3. Hypothesis Testing

The result of the hypothesis testing can be seen in Table 5. Hypothesis 1 (H1) and Hypothesis 2 (H2) were accepted. Attitude and subjective norm were positively and significantly affected repurchase intention in the context of a younger generation that consumed functional foods. Based on the multiple regressions, the t-tests of H1 and H2 showed values higher than 1.96 (alpha= 5%). The unstandardized coefficients beta for attitude was 0.466 and 0.222 for subjective norm. This research also shows that PBC has no significant impact on repurchase intention. The t-test was below 1.96 (alpha= 5%). Overall, the multiple regressions' R^2 was 38.8 percent. In other words, 38.8 percent of the variance in purchase intention was explained by the independent variables proposed in the model.

4.4. Discussion and Implications

Purchase intention was an important factor explaining why consumers were willing to purchased functional foods. Consumers would not want to buy functional food products because they had low purchase intentions. This research found that the younger generation's purchase intention to buy functional foods was significantly influenced by attitude and subjective norm. On the other hands, the result also stated that PBC did not significantly affect purchase intention.

The effect of attitude on purchase intention was supported by previous studies [27, 28, 35]. The effect of subjective norm on purchase intention was also backed by previous findings [35, 40]. The non-significant effect of PBC no purchase intention was not new. Several previous studies also found that purchase intention was not influenced by PBC [27, 28, 40, 44].

Table 5. The result of the hypothesis test

Independent Variables	Unstandardized Coefficients Beta	t-test	Sig.	Hypothesis
Attitude	0.466	5.429	0.000	Accepted
Subjective Norm	0.222	2.725	0.007	Accepted
PBC	0.048	0.624	0.534	Rejected

Based on that result, this research created several implications. First, it was important to stimulate consumers so that they have a positive attitude towards functional foods. The younger generation might not know the benefit of functional food. Therefore, the producers must educate young consumers on the health benefits of functional foods. This type of food does not only fill the belly but also provides health benefits. The image of functional food as a healthier choice must be embedded in the mind of the younger generation. Second, in promoting and educating consumers, producers could use important public figures to affect the younger generation's perception so that they would be willing to consume functional food. The public figure chosen must be the one that represented a healthy lifestyle. Furthermore, producers should position functional foods as family foods. Family members can influence each other in consuming functional foods.

5. Conclusion

This research has tested TPB in explaining the purchase intention of the young generation to buy functional foods. This study showed that the instrument to measure attitude, subjective norm, PBC and purchase intention was valid and reliable. In addition, this research has empirically proven that purchase intention was only influenced by attitude and subjective norm. Attitude has the most significant impact on purchase intention.

Even though this research has generated important findings on the young generation's behavior related to the consumption of functional foods, there were still several limitations. First, the sample was only taken from one location, in a university. The result of this research cannot be widely generalized to the young generation. Second, the object of this research was limited to functional foods that prevent hypertension. There was a chance of different results if the objects were altered. Third, the research only used variables offered by TPB while there are other theories that can explain consumers' behavior. The result did show that 61.2 percent of variance in purchase intention was explained by other independent variables. The limitations of this research can be used as considerations for future studies.

References

[1] Zhong R T Xu X Klotz E and Newman S T 2017 *Intelligent Manufacturing – Review* Intelligent Manufacturing in the Context of Industry 4.0: A Review **3 (5)** 616 – 630.

[2] Rojko A 2017 *International Journal of Interactive Mobile Technologies* Industry 4.0 Concept: Background and Overview **11 (5)** 77 – 90

[3] Gok I and Ulu E K 2018 *Nutrition and Food Science* Functional foods in Turkey: marketing, consumer awareness and regulatory aspects DOI: 10.1108/NFS-07-2018-0198

[4] Sarkar S 2019 *Nutrition and Food Science* Potentiality of probiotic yoghurt as a functional food – a review DOI: 10.1108/NFS-05-2018-0139

[5] Kraus A 2015 *British Food Journal* Factors influencing the decisions to buy and consume functional food **117 (6)** 1622-1636

[6] Bech-Larsen T and Grunert K G 2003 *Appetite* The perceived healthiness of functional foods. A conjoint study of Danish, Finnish and American consumers' perception of functional foods **40 (1)** 9 – 14.

[7] Badrie N Reid-Foster S Benny-Ollivierr C and Roberts H 2007 *Nutrition and Food Science* Exercise enthusiasts' perceptions and beliefs of functional foods in Trinidad, West Indies **37 (5)** 345 – 357

[8] Siró I Kápolna E Kápolna B Lugasi A 2008 *Appetite* Functional food. Product development, marketing and consumer acceptance—A review **51** 456–467

[9] La Barbera, F Amato M and Sannino G 2016 *British Food Journal* Understanding consumers' intention and behaviour towards functionalised food **118 (4)** 885-895

[10] Kaur N and Singh D P 2017 Appetite Deciphering the consumer behaviour facets of functional foods: a literature review **112** 167-187.

[11] Vicentini A Liberatore L and Mastrocola D 2016 *Italian Journal of Food Science* Functional foods: trends and development of global market **28 (2)** 338-351.

[12] Rojas-Rivas E Espinoza-Ortega A Thomé-Ortiz H and Moctezuma-Pérez S 2019 *British Food Journal* Consumers' perception of amaranth in Mexico **121(6)** 1190-1202

[13] Rezai G Teng P K Shamsudin M N Mohamed Z and Stanton J L 2017 *Journal of Agribusiness in Developing and Emerging Economies* Effect of perceptual differences on consumer purchase intention of natural functional food 7 (2) 153-173

[14] Brečić R Gorton M and Barjolle D 2014 *British Food Journal* Understanding variations in the consumption of functional foods – evidence from Croatia **116 (4)** 662 - 675

[15] Menrad K 2003 *Journal of Food Engineering* Market and marketing of functional food in Europe **56** 181–188

[16] Shimizu T 2003 *Nutrition Research Reviews* Health claims on functional foods: the Japanese regulations and an international comparison **16** 241–252

[17] Lu J. 2015 *British Food Journal* The effect of perceived carrier-ingredient fit on purchase intention of functional food moderated by nutrition knowledge and health claim **117 (7)** 1872 - 1885

[18] Williams P Ridges L Batterham M Ripper B Hung M.C 2008 *food policy* Australian consumer attitudes to health claim – food product compatibility for functional foods **33(6)** 640-643

[19] Markovina J Cacic J Kljusuric J G and Kovacic D 2011 *British Food Journal* Young consumers' perception of functional foods in Croatia **113(1)** 7 – 16

[20] Chammas R Jessy E-H. Mira F Reine M Christelle B-M *Nutrition and Food Science* Consumer knowledge and attitudes toward functional foods in Lebanon DOI: 10.1108/NFS-09-2018-0263

[21] Bazhan M Keshavarz-Mohammadi N and Hosseini H 2017 *British Food Journal* Consumers' awareness and perceptions regarding functional dairy products in Iran: a qualitative research **119 (2)** 253 - 266

[22] Hassan S.H. 2011 Managing conflicting values in functional food consumption: the Malaysian experience **113 (8)** 1045 - 1059

[23] Lu J 2015 *British Food Journal* The effect of perceived carrier-ingredient fit on purchase intention of functional food moderated by nutrition knowledge and health claim **117 (7)** 1872 – 1885

[24] Krutulyte R Grunert K G Scholderer J Lähteenmäki L Hagemann K S Elgaard P S Nielsen B and Graverholt J P 2011 *Food Quality and Preference* Perceived fit of different combinations of carriers and functional ingredients and its effect on purchase intention 22 (1) 11-16

[25] Huang L Bai L Zhang X Gong S 2018 Food *Quality and Preference* Re-understanding the antecedents of functional foods purchase: Mediating effect of purchase attitude and moderating effect of food neophobia **73** 266 - 275

[26] Ares G Besio M Gimenez A and Deliza R 2010 *Appetite* Relationship between involvement and functional milk desserts intention to purchase - Influence on attitude towards packaging characteristics **55 (2)** 298-304

[27] Patch C Williams P G and Tapsell L C 2005 *Journal of Nutrition Education and Behavior* Attitudes and intentions toward purchasing novel foods enriched with omega-3 fatty acids **37** 235-241

[28] Connor E L and White K M 2010 *Food Quality and Preference* Willingness to trial functional foods and vitamin supplements: The role of attitudes, subjective n orms, and dread of risks **21** 75-81

[29] Nolan-Clark D J Neale E P Probst Y C. Charlton K E and Tapsell L C 2011 *BMC Public Health* Consumers'salient beliefs regarding dairy products in the functional food era: a qualitative study using concepts from the theory of planned behaviour **11** 843

[30] Ares G and Gambaro A 2007 *Appetite* Influence of gender, age and motives underlying food choice on perceived healthiness and willingness to try functional foods **49** 148–158.

[31] Mullie P Guelinckx I Clarys P Degrave E. Hulens M and Vansant G 2009 *European Journal Clinical* Nutrition Cultural, socioeconomic and nutritional determinants of functional food consumption patterns **63 (11)** 1290 - 1296

[32] De Jong N Ocké M Branderhorst H and Friele R 2003 *British Journal of Nutrition,* Demographic and lifestyle characteristics of functional food consumers and dietary supplement users **89(2)** 273-281.

[33] Septiana P Nugroho F A Wilujeng C S 2018 *Jurnal Kedokteran Brawijaya* Konsumsi Junk food dan Serat pada Remaja Putri Overweight dan Obesitas yang Indekos **30 (1)** 61 - 67

[34] Ajzen I. 1991. *Organizational Behavior and Human Decision Processes* The theory of planned behavior **50** 179-211

[35] Alam S S and Sayuti S M 2011 *International Journal of Commerce and Managemen*t Applying the Theory of Planned Behavior (TPB) in halal food purchasing **21 (1)** 8-20

[36] Mirabi V Akbariyeh H and Tahmasebifard H 2015 J*ournal of Multidisciplinary Engineering Science and Technology* A study of factors affecting on customers purchase intention **2(1)** 267-273

[37] Halim W Z W and Hamed A B 2005 Consumer Purchase Intention at Traditional Restaurant and Fast Food Restaurant University Utara Malaysia

[38] Ajzen I 2015 *Rivista di Economia Agraria* Consumer attitudes and behavior: the theory of planned behavior applied to food consumption decisions **70(2)** 121–138

[39] Bakti IGMY Rakhmawati T Sumaedi S Widianti T Yarmen M and Astrini N J 2019 *World Conference on Transport Research - WCTR* Public transport users' WOM: an integration model of the theory of planned behavior, customer satisfaction theory, and personal norm theory May 2019.

[40] Sumaedi S Yarmen M Bakti IGMY Rakhmawati T Astrini N J and Widianti T 2016 *Management of Environmental Quality: An International Journal* The integrated model of theory planned behavior, value, and image for explaining public transport passengers' intention to reuse **27(2)** 124 – 135

[41] Conner M Norman P and Bell R 2002 *Health Psychology* The theory of planned behavior and healthy eating **21(2)** 194-201.

[42] Venkatesh V and Davis F D 2000 *Management Science* A theoretical extension of the technology acceptance model: for longitudinal field studies **46 (2)** 186 - 204

[43] Yarmen M Rakhmawati T Bakti IGMY Astrini N J Sumaedi S Widianti T 2016 *International Journal of Quality and Service Sciences* Investigating Patient Loyalty: An Integrated Framework for Trust, Subjective Norm, Image, and Perceived Risk (A Case Study in Depok, Indonesia) **8.(2)** 179 – 196.

[44] Sumaedi S Bakti I G M Y Rakhmawati T Astrini N J Yarmen M and Widianti T 2015 *Leadership in Health Services* Patient loyalty model **28(3)** 245-258

[45] Ministry of Health 2018 Hasil Utama Riskesdas Jakarta Kemenkes

[46] Bamberg S Ajzen I and Schmidt P 2003 *Basic and Applied Social Psychology* Choice of travel mode in the theory of planned behavior: the roles of past behavior, habit, and reasoned action **25(3)** 175-187.

Acknowledgments

This research is supported by National Incentive Program (INSINAS) 2019 from Indonesia's Ministry of Research and Higher Education. All Authors are main contributors.

IOP Conf. Series: Materials Science and Engineering **722** (2020) 012025 doi:10.1088/1757-899X/722/1/012025

Modified fuzzy Delphi to determine the consensus of government facility governance

I N Sutapa[1] and M Wullur[2]

[1] Industrial Engineering Department, Petra Christian University, Surabaya, Indonesia
[2] Management Department, Sam Ratulangi University, Manado, Indonesia

E-mail: mantapa@petra.ac.id

Abstract. The article discusses the development of governance indicators for work facility policies of local government. These indicators will be used as a policy to ensure that work facilities are durable, cost-effective in procurement and maintenance, are effective in use, and added value for end of use/life. Many researchers previously developed governance indicators with the Fuzzy Delphi method, but there were still gaps, namely the differences in the weight of expertise had not been accommodated in determining the final consensus of the importance of each indicator. In this article, the indicators are developed by expert, by considering the weight of their expertise, using a modified Fuzzy Delphi method. The final results of the study are four dimensions of government work facilities governance with twenty indicators that are important in improving the operational performance of local governments.

1. Introduction

Determination of expert consensus with traditional Delphi method requires several stages to get a convergent opinion, which is time-consuming and expensive [1]. Furthermore, the opinions of each expert on an indicator in the traditional Delphi method are generally expressed in one choice interest score on a Likert scale for the research using questionnaire [2]. The choice of interests represented in a score point is not representative, because humans in judging something is not 'black' and 'white', if in one point assessed 'black' and then shifted slightly 'white' directly, there is a slow degradation from black to white [3]. Based on the two reasons mentioned above, namely to reduce the steps and overcome the lack of a single point assessment, the Delphi method modification is needed. To overcome those problems, some researchers have used the concept of fuzzy sets, such as [4, 5, 6, 7, and 8].

Delphi decision model with existing fuzzy approaches has weaknesses, decision aggregation is determined based on the average value of the opinions of all experts. According to Kamonpatana *et al.* [7] and Quyen [8] each expert is considered to have the same level of importance. Meanwhile, according to Liu *et al.* [9] expert opinions must be distinguished based on differences in the level of importance of each expert, due to differences in educational qualifications, work experience, and level of authority. Need different weights for each expert opinion in determining the aggregate final assessment of an indicator. The experts who have higher weight have a bigger voice, more to hear. For this reason, in this article, a modified Fuzzy Delphi method concerning different expert weights was developed.

2. Formulation of modified fuzzy Delphi

The selection of work facility governance policies, ranging from procurement policies to end-of-life/use, requires a collective consensus of various stakeholders. In this study, collective decision choices were formulated using the modified Fuzzy Delphi method through focus group discussion (FGD). The purpose of using fuzzy concepts is to accommodate the uncertainty of human choice boundaries regarding the choice of policy importance [3-7]. This can reduce the decision-making process in the traditional Delphi method which takes a lot of time and money. Consensus in the traditional Delphi method generally cannot be reached in one stage, since the choice of stakeholders is expressed in a narrow or crisp choice [1, 2, and 8].

Collective decision by several interested parties to get a consensus on work facility governance policies in the local government with modified Fuzzy Delphi, starting from the identification of various dimensions of governance policies to policy indicators through literature review, field studies and input from stakeholders when FGD is performed.

Furthermore, for the model formulation, the policy dimension is denoted by PI, the ith policy indicator of jth dimension is denoted by PI_{ij}, and the kth expert or interested party is denoted by E_k. The importance level of each indicator is expressed in five Likert scale [5, 7], namely *EI* (extremely important), *VI* (very important), *RI* (relatively important), *SI* (somewhat important), and *NI* (not important) with the domain of all levels in $[0, 1]$ and the degree of membership μ of each point is expressed in $[0, 1]$, where the degree of membership $\mu = 0$ means not being at a level of importance, $\mu = 1$ means definitely in a level of importance, and a score between 0 and 1 means being at an importance level with membership degrees $(0, 1)$. Then the membership function of each level of importance of policy indicators can be formulated as a triangular fuzzy membership function with a domain $[0, 1]$, where the midpoint of a triangle function has a membership value of one, then the membership value is degraded to zero to the left and right as far as 0.25 [3-7].

Membership function of fuzzy set μ_{PI} of the ith indicator importance level of jth dimension who is evaluated by k^{th} expert with membership value is one at middle point M, i.e. $\mu_{PI_{ijk}}(M)=1$ can be formulated as triangular function TRF, with lower limit L, upper limit U, and middle point M:

$$\mu_{MF_{ijk}}(x) = TRF(L_{ijk}, M_{ijk}, U_{ijk}) \quad (1)$$

$$= TRF(M - 0.25, M, M + 0.25) \quad (2)$$

$$= \begin{cases} 4(x-M)+1, & M - 0.25 \le x \le M \\ -4(x-M)+1, & M < x \le M + 0.25 \end{cases} \quad (3)$$

The triangular membership function of all values of the importance levels of *NI, SI, RI, VI,* and *EI* can be described as follows (Fig. 1):

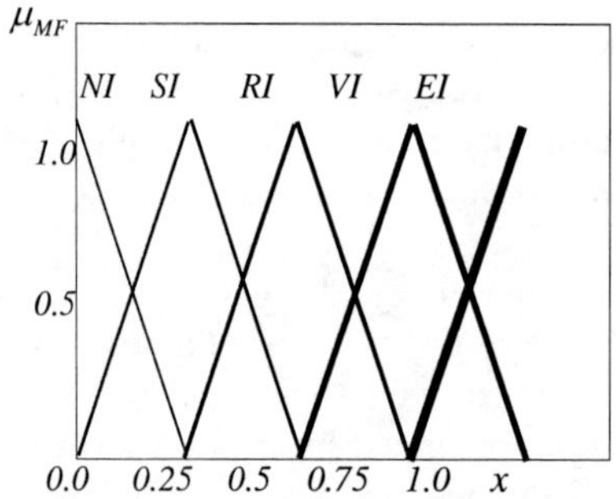

Figure 1. Triangular membership function of indicators importance level

Furthermore, if the weight of the kth expert is denoted by WE_k where $\sum_k WE_k = 1$, the weight of the kth expert is determined based on the level of knowledge, skills, and experience, then the fuzzy set of

IOP Conf. Series: Materials Science and Engineering **722** (2020) 012025 doi:10.1088/1757-899X/722/1/012025

the consensus of all experts in determining the importance level of a policy indicator PI_{ij} can be formulated as union U of all fuzzy set of the expert, i.e.:

$$\mu_{PI_{ij}}(x) = \bigcup_k \mu_{PI_{ijk}}(x) = TRF(L_{ij}, M_{ij}, U_{ij}) \quad \forall i, j \tag{4}$$

Where

$$L_{ij} = \min_k \{L_{ijk}\} \quad \forall i, j \tag{5}$$

$$U_{ij} = \max_k \{U_{ijk}\} \quad \forall i, j \tag{6}$$

$$M_{ij} = \sum_k WE_k M_{ijk} \quad \forall i, j \tag{7}$$

Equations (5) and (6) illustrate the left L_{ij} and right U_{ij} boundaries of the importance level of all experts related to the evaluation of the ith policy indicator on the jth dimension. The left and right boundaries are the value with the greatest scope, meaning that the left boundary of the assessment of all experts is taken the smallest (*min*) and the right boundary taken from the largest expert assessment (*max*). These left and right boundaries correspond to Liu [5], Kamonpatana *et al.* [7], and Quyen [8]. These limits (5) and (6) aim to reduce the decision-making process that is repeated in the Delphi method. Meanwhile, equation (7) is the midpoint M_{ij} of the importance level of all experts, namely the highest importance membership value of *1*, its value is determined as the weighted average of all the highest importance assessments of all experts. Where the weight of each expert WE_K is taken into account Liu *et al.* [9], it means that experts with greater weight have greater votes and vice versa, this is by the rules of decision making in local government in general. Equation (5, 6, and 7) is a modification of the existing equation by researchers [4-7]. Specifically, in equation (7) the authors setting the arithmetic mean with different weights for experts, whereas researchers [4, 6, and 8] used arithmetic mean, and researchers [5, 7] with geometric mean, all of which assume each expert has the same weight.

Finally, the conclusion of the importance level of the ith policy indicator in the jth policy dimension by all *PI* experts can be formulated as a center of gravity CoG as determined by Kamonpatana *et al.* [7], which is the defuzzification of equation (4-7), i.e.:

$$PI_{ij} = \frac{\int_0^1 x\mu_{MF_{ij}}(x)dx}{\int_0^1 \mu_{MF_{ij}}(x)dx} \tag{8}$$

PI policy indicators in the jth policy dimension can be used as a consensus by all experts, considered as important policy indicators in the governance of work facilities if they meet the following threshold values (*TV*):

$$PI^*_{ij} > TV_j \quad \forall j \tag{9}$$

Where the threshold value is the first quartile Q_1 of all the indicators of importance in each dimension j^{th} as mentioned by Quyen [8], i.e.:

$$TV_j = Q_1^j = \min_j \{PI_{ij}\} + 0.25 range_j \{PI_{ij}\} \tag{10}$$

The level of importance of policy indicators *PI* that meet the threshold value (9) can be classified into high importance (PI>0.75), medium importance ($0.51 \le PI \le 0.75$), and low importance (PI<0.51). Furthermore, the level of consensus of all experts for each indicator can be formulated as a coefficient of quartile value (CQV):

$$CQV_i = \frac{Q_3 - Q_1}{Q_3 + Q_1} = \frac{0.5 range_k \{M_{ijk}\}}{2\min_k \{M_{ijk}\} + range_k \{M_{ijk}\}} \quad \forall i, j \tag{11}$$

According to Quyen [8] consensus can be divided into high consensus if $CQV \le 0.15$, medium consensus, if $0.15 < CQV \le 0.20$, and low consensus, if $CQV > 0.20$.

3. Formulation of modified fuzzy Delphi

As a case study, the following is the application of modified fuzzy Delphi in determining the level of importance and consensus level of work facility governance policy decision in a local government in Indonesia. Dimension of policy decisions include decisions on procurement, usage, and maintenance of work facilities, as well as recapturing the value of end of use/life of them [10, 11]. Based on the dimensions of the governance policy, the following are each policy indicators derived from a review of some previous literature.

3.1. Procurement policy of work facility

The procurement policy of an organization has a strategic position to maintain the stability of the operational process. Government procurement division aims to supply facilities which are ready and reliable to use by the user with low on price and maintenance cost. There are many ways to procure material or facilities [12, 13, and 14]. Purchase (abbreviated as policy indicator *Proc1*). Meeting the needs of work facilities using the institution handing the amount of money to the seller to obtain work facilities by the agreement of both parties. Receipt of a grant (*Proc2*). Meeting the needs of work facilities by accepting voluntary gifts from other parties. Grants can be received from the government (national or regional) and the private sector. Rentals (*Proc3*). Meeting the needs of work facilities by temporarily utilizing the property of other parties for the benefit of the institution and then paying it based on the lease agreement. Borrowing (*Proc4*). Fulfillment of work facilities by utilizing other parties' goods for the interests of institutions voluntarily by the loan agreement. Exchange (*Proc5*). Fulfillment of work facilities by exchanging goods owned by institutions with goods owned by other parties, meanwhile exchanged work facilities must be work facilities that are no longer useful for the institution. Reconditioning/repairing (*Proc6*). Reconditioning or repairing is a means of meeting the needs of work facilities that have been damaged. Repairs can be done by replacing damaged parts so that damaged work facilities can be reused as they should.

3.2. Usage policy of work facility

All facilities must be used properly and properly so that conditions remain excellent and can be used in the long run. Some of the results of previous studies [11, 15, and 16] suggest various ways of using work facilities. Installation of facilities according to the manual book and operation of facilities by work instructions (*Usage1*). All facilities are placed in the area needed in sufficient quantities, well-organized, always clean, and ready to use (*Usage2*). Raw materials or spare parts needed to work or to operate facilities are available and adequate (*Usage3*). And, there is visual control or warning system of irregularities or abnormalities (damage, run out of material, errors) (*Usage4*).

3.3. Maintenance policy of work facility

Maintenance of work facilities is very important to keep the facilities in top condition, ready to use and can be used for a long time. There are various ways to maintain the work facilities [17, 18, and 19]. Care facilities are carried out regularly following a predetermined maintenance schedule regardless of the condition of the facility (*Maint1*). Facility maintenance is carried out by looking at the current condition (decline in performance), without waiting for the maintenance schedule or waiting until the facility is damaged (*Maint2*). Facility maintenance is carried out with repairs due to damaged or non-functioning facilities (*Maint3*). Facility maintenance is carried out by estimating the conditions or workload in the future (*Maint4*).

3.4. Recapture value of end of use/life

Handling of work facilities that are end of use/end of life or damaged ones requires a big cost. To reduce handling costs and/or recapture the remaining value, the institution must be clever. If not, then the handling costs will increase. Several ways can be done to reduce costs or to recapture value [20,

21]. Outdated facilities sold or auctioned openly (*EndUL1*). Facilities that have expired are exchanged for facilities needed by the institution (*EndUL2*). Facilities that have expired are moved to places (within the same institution) that need them more (*EndUL3*). Facilities that have expired service are given to other institutions that need them (*EndUL4*). Facilities that have expired have been reused or recycled (*EndUL5*). Facilities that have been used up have been destroyed/burned (*EndUL6*).

3.5. *Model implementation on a case study*

This case study was carried out at a local government that has a variety of work facilities, such as personal computers, laptops, printers, photocopiers, LCDs, etc. Seven respondents are willing to participate in this study. The seven respondents were called experts because they had very adequate knowledge and work experience in the procurement, use, maintenance, or management of finished goods. Data in Table 1 shows the level of knowledge ($Know_k$) and work experience ($Expr_k$) of each expert (E_k, $k=1, 2,...,7$), with a total weight (WE_k) calculated from 40% knowledge and 60% experience.

In Table 2, by using equations (5)-(7) the fuzzy set of the importance level of each policy indicator is calculated, the lower limit of *L*, the middle value of *M*, and the upper limit of *U*, namely the boundaries of the combined fuzzy triangular membership function of the judgment of the seven experts. Furthermore, using equation (8) the centre of gravity of each fuzzy set of importance is calculated for each policy indicator *PI.* Equation (9) to calculate threshold value *TV* for each dimension policy, as a criterion for determining a *PI* in a dimension whether it is important or not important. Equation (10) was used to calculate the consensus level CQV for each policy indicator *PI* from all experts.

Table 1. Expert weight based on knowledge and experience

Expert	E1	E2	E3	E4	E5	E6	E7
$Know_k$	3	1	3	2	3	3	2
$Expr_k$	15	10	20	10	5	18	9
WE_k	0.17	0.11	0.22	0.12	0.07	0.20	0.11

Table 2. Fuzzy membership function l-m-u, level of importance pi, and level of consensus cqv

Policy		*L*	*M*	*U*	*PI*	*CQV*
Procurement	Proc1	0.25	0.59	1.00	0.61	0.20
	Proc2	0.50	0.89	1.00	0.80	0.14
	Proc3	0.50	0.92	1.00	0.81	0.14
	Proc4	0.00	0.47	1.00	0.49	0.14
	Proc5	0.00	0.35	0.75	0.37	0.33
	Proc6	0.50	0.88	1.00	0.79	0.14
Usage	Usage1	0.25	0.76	1.00	0.67	0.08
	Usage2	0.50	0.89	1.00	0.80	0.14
	Usage3	0.00	0.58	1.00	0.53	0.11
	Usage4	0.50	0.88	1.00	0.79	0.14
Mainte-nance	Maint1	0.25	0.69	1.00	0.65	0.09
	Maint2	0.50	0.89	1.00	0.80	0.14
	Maint3	0.00	0.37	0.75	0.37	0.33
	Maint4	0.50	0.85	1.00	0.78	0.14
End of use/life	EndUL1	0.25	0.78	1.00	0.68	0.08
	EndUL2	0.25	0.55	1.00	0.60	0.11
	EndUL3	0.00	0.41	1.00	0.47	0.14
	EndUL4	0.00	0.52	1.00	0.51	0.20
	EndUL5	0.25	0.83	1.00	0.69	0.14
	EndUL6	0.00	0.36	0.75	0.37	0.20

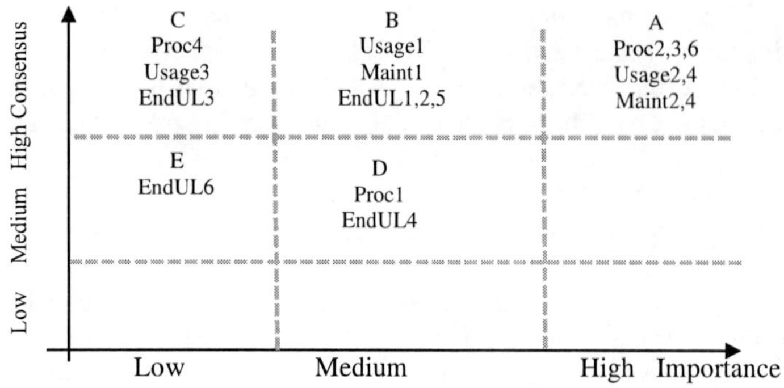

Figure 2. Importance and consensus level of each policy indicator

Threshold values *TV* for procurement, usage, maintenance, and end of use/life dimension are 0.52, 0.63, 0.58, and 0.48. Based on Table 2 and Figure 2 for procurement dimension, it can be explained that from six indicators (with TV_1=0.52), four indicators agreed to be very important by all experts, namely purchase, receive a grant, rental, and recondition/repair. While the remaining indicators are less important, i.e. loans and exchange work facilities. Meanwhile, 4 policy indicators have a high level of consensus, namely receipt of grants, leases, loans, and reconditioning/repairs. These indicators are supported by all experts. While the other two indicators received varied support, namely the purchasing indicator was in moderate consensus, and the exchange indicator received very low consensus. The same explanation for indicators in other dimensions.

In general, calculation result of modified fuzzy Delphi are all experts have high consensus for 15/20 indicators or 75% (A, B, and D groups), 15% medium consensus, and only 2/20 or 10% are low consensus, in deciding the importance level of work facility governance policy indicators government.

4. Conclusion

Modified fuzzy Delphi was quite successful in capturing the uncertainty of the approval limits of experts regarding the importance of a policy indicator. This model also succeeded in reducing the decision-making process to reach a consensus that was repeated several times. In this case, the study can produce a high consensus level of 90%, only 10% is low, indicating a decision-making process that does not need to be repeated much. This model still has weaknesses related to the fuzzy membership function which is limited as triangular, changes in the degradation of the level of importance are considered symmetrical and linear, which in reality is not the case. Future studies are expected to capture the degradation of non-linear membership functions by developing non-linear fuzzy membership functions.

References

[1] Le Pira M, Inturri G, Ignaccolo M and Pluchino A 2017 *Trans. Res. Proc.* **25** pp 3725–35
[2] Musa H D, Yacob M R, Abdullah A M and Ishak M D 2015 *Proc. Environ. Sc.* **30** pp 244–9
[3] Hsueh S-L 2015 *Renew. & Sustain. Energy Rev.* **49** pp 1286–95
[4] Habibia A, Jahantigh F F and Sarafrazic A 2015 *As. J. of Res. in Bus. Ec. & Man.* 5 **2** pp 130-143
[5] Liu W-K 2013 *Int. J. of e-Ed. e-Bus., e-Man. & e-Learn.* 3 **4**
[6] Pham T Y, Ma H M and Yeo G T 2017 *As. J. of Ship. & Log.* 33 **4** pp 211-9
[7] Kamonpatana K, Anuntavoranich P, Varodompun J, Sunakorn P, Thaitakoo D and Wattanapreechanon K 2015 *Kasetsart J.* (Soc. Sci) **36** pp 554–567
[8] Quyen D T N 2014 *Proc. Soc. & Behav. Sc.* **141** pp 828–833
[9] Liu X, Xu Y, Montes R and Herrera F 2019 *Inf. Sc.* **505** pp 215–232

[10] Hanis M H, Trigunarsyah B and Susilawati C 2010 *Proc. of 1st Makassar Int. Conf. on Civil Eng.* Indonesia
[11] Lima E S and Costa A P C S 2019 *Rel. Eng. & Sys. Safe.*, **190** 106523 pp. 1-11
[12] Filho W L, Skouloudis A, Brandli L L, Salvia A L and Rayman-Bacchus L 2019 *J. of Cleaner Prod.* **231** pp 1267-1280
[13] Bakir S, Khan S, Ahsan K and Rahman S 2018 *J. of Environ. Man.* **225** pp 325-335
[14] Patrucco A S, Luzzini D, Ronchi S and Essig M 2017 *Pub. Money & Man.* 37 **4** pp 269-276
[15] Leming-Lee T, Polancich S and Pilon B 2019 *Nursing Clinics of North Am.* 54 **1** pp 53-79
[16] Antony J, Rodgers B A and Cudney E A *Int. J. of Qual. & Rel. Man.* 34 **9** pp 1402-11
[17] Sani S I A, Mohammed A H, Misnan M S and Awang M 2012 *Proc.-Soc. & Behav. Sc.* **65** pp 827–832
[18] Ighravwe D E and Oke S A 2019 *J. of Build. Eng.* **24** 100753 pp 1-18
[19] Dickerson D E and Ackerman P J 2016 *Proc. Eng.* **145** pp 685–692
[20] Farahani S, Otieno W and Barah M 2019 *J. of Cleaner Prod.* **224** pp 25-39
[21] Agrawal S, Singh R K and Murtaza Q 2015 *Res. Conservation & Recycling* **97** pp 76–92

IOP Conf. Series: Materials Science and Engineering **722** (2020) 012026 doi:10.1088/1757-899X/722/1/012026

Job Safety Analysis in Production Floor of PT BD Based on Semi Quantitative AS/NZS4360 Method

I D Widodo[1], and N R A Dwinanda[1]

[1] Industrial Engineering Department, Universitas Islam Indonesia, Yogyakarta, Indonesia

E-mail: imamdjati@uii.ac.id

Abstract: Every industry must pay attention to the job safety of each employee by implementing good job safety effort to create a comfortable, healthy and prosperous work environment. It is related to the relationship of labor with work equipment, materials and processes, workplace foundations and ways of doing the work. This research was done in PT BD by implementing Job Safety analysis based on AS / NZS 4360. The analysis indentifies 6 jobs that should be improved, namely, (1) Lifting the core (2) cutting mother core (3) The process of removing the email layer (4) Placing the results of the cutting core to the decoiler (5) Taking the cutting results for semi stacking and (6) Taking cutting results for semi wound. Some recomended actions are proposed such as improving the tool, developing SOP, better controlling procedure and improving operator (training).

1. Introduction

Every industry must pay attention to the job safety of each employee by some safety efforts. Work safety effort generally aim to protect the operator from accidents and protect the machine from damage, both during operation and during repair and maintenance activities. There are two biggest things that cause workplace accidents, such as unsafe worker behavior and unsafe environmental conditions. The frequency of work accidents and the occurrence of contamination of the product is very large if it does not pay attention to the two causes.

There are differences in the level and magnitude of workplace accidents between developed and developing countries. The rate of work accidents in developing countries reaches three to four times more fatal than developed countries [1]. Work safety is related to machines, working tools, materials, processes, workplace foundations, and the environment, as well as ways of doing work. Work safety targets all workplaces [2]. Therefore, companies should implement occupational safety and health efforts in the company as an effort to create a comfortable, healthy and prosperous work environment. This research was conducted in factory plan 1 PT BD. During a period of four years (2011-2015) the frequency of work accidents is 27% - 29%. The company's biggest zero accident achievement was 6697 man hours.

The Job Safety Analysis (JSA) method will be used to analyze and reduce the risk of work accidents. This method is a method to look for risks and potentials that can be generated and overcome them against the possibility of workplace accidents that can occur by analyzing operator activities. Benefits of JSA are formalisation of work, accountability, worker participation, organisational learning, awareness, and loss prevention [3]. JSA was also applied in some fields such as construction[4], automation [5], mining [6] and shipyards [7].

2. Basic Theory

A. Occupational Health and Safety

Occupational Safety and Health can be described philosophically and scientifically [8]. Philosophically, it is a thought and effort to guarantee the integrity and perfection of both the physical and spiritual workforce, the work and culture towards a just and prosperous society. Whereas scientifically, occupational safety and health is a science and its application in an effort to prevent the possibility of accidents and diseases caused by work.

According to Dainur [9], Occupational Safety and Health is "safety related to the relationship of labor with work equipment, materials and processing processes, workplace foundations and ways of doing the work". Sutrisno in Moenir [10] stated that work safety is a condition in the workplace / environment that can guarantee the maximum safety and health of people who are in the area / place, both the employee and non-employee of the work organization. Work safety is safety related to equipment, workplaces and the environment, as well as ways of doing work. Another opinion also expressed by Taslimin [11] is that occupational safety and health involve all elements involved in work activities involving subjects (people who do work), objects (material), ie objects or items being worked on, tools - tools used in the work and involve the environment.

From some of the definitions and concepts above, the researcher concludes that application and work safety is a way to apply oneself or organize oneself on a job in order to work safely and healthily both physically and spiritually related to the work process and work environment.

The aim of Occupational Safety and Health is essentially to protect workers from work-related accidents. Sutrisno and Ruswandi [12] stated that the purpose of occupational safety and health is to achieve employee safety at work and after work.

Suma'mur [13] stated that the aim of Occupational Health Safety is to protect operators for their safety rights in doing work for the welfare of life and increase production and work productivity, to ensure the safety of others who are in the workplace environment and the source of production is maintained and used efficiently.

B. Risk management

Risk management must be part of the organizational culture. Organizations or companies that manage risk effectively and efficiently are more likely to achieve their objectives and do so at a lower overall cost (AS / NZS 4360, 2004) [14]. The risk management process can be seen from the standards set by AS / NZS 4360: 2004, the flow of the process can be seen from Figure 1. In identifying risks in occupational safety and health, the JSA (Job Safety Analysis) technique is used, ie before the activity starts, it is necessary to conduct a risk analysis study to find out what the magnitude and potential hazards are caused during the activity [15].

3. Methodology

The object used in this study is the operator and operator activities in the core section relating to Occupational Safety and Health, namely the hazards and risks contained in the operator's work process. In this study, the data used are safety data, the possibility of accidents that occur, the consequences, and the risks that exist in the core section. The data collection methods used in this study are observation and interviews with expert informants/experts. This data includes data on the type of danger, the type of injury caused, the risk of danger, and susceptibility when the injury occurs.

The study was conducted in several stages:

1. At this first stage, JSA is used to look for risks of work accidents that might occur, what might be the cause and control efforts that have been made by the company in minimizing and preventing work accidents.

IOP Conf. Series: Materials Science and Engineering **722** (2020) 012026 doi:10.1088/1757-899X/722/1/012026

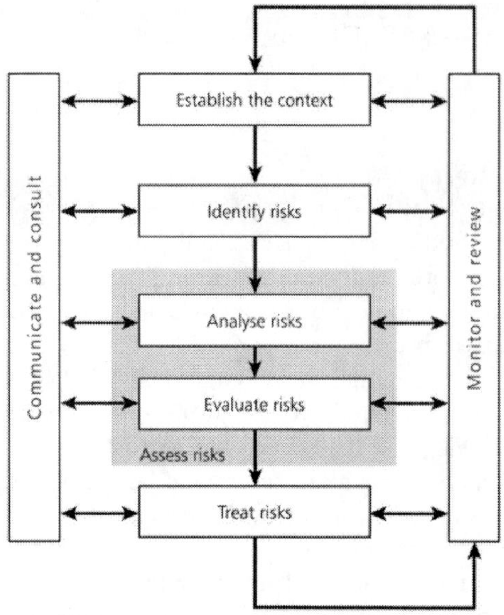

Figure 1. Risk management overview

2. A risk analysis is performed using a semi-quantitative analysis method based on AS/NZS 4360: 2004 risk management standards to determine the consequences, exposure, and likelihood of work accidents on the operator. From the results of the semi-quantitative analysis, then the data is processed to find out how much risk the operator has using the formula:
Scoring is done as an interpretation of the value of risk (Value of Risk = Consequences (C) × Exposure (E) × Probability (L)). Based on the results and discussion of risk analysis, the authors can then provide recommendations for companies in managing safety and health for workers with the principle of hierarchy of control.
3. The conclusions and recommendations stage contains the conclusions and suggestions of the authors of the company in accordance with the focus of the issues raised based on the OHS risk analysis of the operator.

4. Results And Discussion

The method for hazard identification is carried out using Job Safety Analysis (JSA). In the implementation of JSA, the authors take the data directly by observing the Core section of the plant production department I PT. BD and interviews with the operators concerned. Observations are recorded on the Job Safety Analysis form based on AS / NZS 4360 (2004) Risk Management. Safety risks were analyzed based on 13 identified work details (Table 1). The risk identification and score determination is obtained through direct observation at the work location and interviews with experts which in this case are the operator and the head of the team at the core section. Risk analysis is carried out using a semi-quantitative risk analysis method which includes identification of risks, determining the value of consequences (consequences), likelihood, and exposure of each stage of work in the core section which is then used to determine the level of risk based on AS / NZS standards 4360: 2004, which will be described in Table 1. Based on the score, Six jobs/work need to be analysed. They are (1) cutting the mother core (2) Lifting the core (3) The process of removing the email layer (4) Placing the results of the cutting core to the decoiler (5) Taking the cutting results for semy stacking and (6) Taking cutting results for semi wound. Some identified conditions and facts of job/work that have high are:

1. The work that has the highest risk value is the Core Cutting operator with details of the work cutting the mother core as indicated by a risk value of 300. This indicates that the work is included in the priority 1 category, so it is necessary to handle risk until the risk value is more acceptable. Based on September's work accident data, the core cutting operator suffered a hand accident wedged by a

sharing core machine. The victim's finger entered the machine and was cut 1cm. Barrier to avoid being exposed to the blade that is on the sharing core machine looks loose so that the victim's finger can easily enter. In addition it is needed an improvement on the engine pedal which sometimes moves on its own. This incident occurred on 18 September at 01.00 (night shift). During the incident, the operator was cutting the cover seat, with a thickness of 0.6 mm with a maximum cutting thickness of 5 mm.

Table 1. Accident Risk Assessment

Job/work	**Risk**	*Basic Level**			**Risk Value**	**Risk Level**
		C	**E**	**L**		
Lifting the core	Scratched on the blunt end of the core and crushed by the core	5	6	6	180	*Priority 1*
Cutting the mother core	Cut off the core sharing machine, exposed to engine noise	5	6	10	300	*Priority 1*
Placing the results of the cutting core to the decoiler	Hit by Core	5	3	2	30	*Priority 3*
Taking the cutting results for semy stacking	Scratched edge blunt	5	3	3	45	*Priority 3*
Stacking process for *silicone steel*	*Hit by silicone steel*	5	2	0.5	5	*Acceptable*
Placing *stacking core* result into *decoiler*	Hit by silicone steel, awkward posture	5	2	1	10	*Acceptable*
Taking the cutting results for semy wound	Scratched edge blunt	5	3	3	45	*Priority 3*
The process of wound / making a pile of silicone steel (prisma)	*Hit by silicone steel*	5	2	0.5	5	*Acceptable*
Placement of the wound core results to the decoiler	Hit by silicone steel, awkward posture	5	2	1	10	*Acceptable*
The process of removing the email	Contact with flammable chemicals	15	2	3	90	*Substantial*
Assemble cables between high / low voltage coil	Scratched cable wire assembly	5	3	1	15	*Acceptable*
Do assembly with tic core welding	Exposed to ultraviolet light and sparks	5	2	1	10	*Acceptable*
Perform copper welding and connector welding	Contact with hot objects and smoke	5	2	1	10	*Acceptable*

* AS/NZS 4360 Standart

The next day, the operator can re-enter so that a restart for zero accident can be prevented. Accidents occur due to several factors, namely the unsafe action factor of the operator who is less focused on his work, besides that it is necessary to do engine rejuvenation by repairing machines that are no longer safe to use.

2. Other work that has a high risk analysis value is the Core Cutting operator with details of the work lifting the mother core with a score of 180. The score is included in the priority 1 category, which means that it requires control as soon as possible. The process of lifting the mother core requires special expertise for the operator in running the crane, so it takes seriousness in lifting the core. Based on the analysis results, core cutting operators have the highest potential danger in each of their activities. Accidents can occur due to the work of core cutting operators are still limited in the number of PPE safety shoes and helmets so that some operators are seen not to use them. This undisciplined operator can cause work accidents.
3. The next job with a high risk analysis value is the HV&LV Wiring operator with details of the work eliminating the email layer with a score of 90. The score falls into the substantial category which means it requires corrective action. In the process of removing the enamel layer on the wire is carried out by the operator in direct contact with sulfuric acid chemicals through a high temperature combustion process. Direct contact with chemicals and exposure to high temperatures can endanger the operator if the operator does not use Personal Protective Equipment (PPE) correctly. Masks also

need to be used so that exposure to heat does not affect the face and does not suck chemicals, but from the workers themselves are still not disciplined in using masks because they feel their work is safe and fine. In addition, it also requires leather gloves for operators to avoid being exposed to hazardous chemicals.

4. Other jobs have risk analysis values that fall into the priority 3 and acceptable categories. Operators sometimes are seen not using complete PPE so that they can cause work accidents. The company does not need to take any corrective actions other than providing continuous supervision so as not to increase the risk of work accidents that may occur in the future.

The last stage of risk assessment is about developing work accident analysis solutions as structured tools based on risk events that have been obtained from the risk identification stage. Risk assessment actions are carried out according to risk events that require improvement to reduce work safety risks, starting from the highest to the lowest risk value. Actions of control over existing hazards must be carried out in accordance with the principle of control hierarchy according to AS / NZS 4360: 2004 as in Table 2.

Table 2. Risk control

Job/work	Risk	Risk Value	Risk Level	Current Conditi on	Causes	Control Description
Cutting the mother core	Cut off the core sharing machine, exposed to engine noise	300	*Priority 1*	Wear gloves, safety shoes, safety helmet	*Unsafe act, the operator is out of focus, there is noise from the cutting machine*	***Engineering:*** Repairing the loose blade barrier on the sharing machine to make it safer ***Administrative:*** Supervision of SOP implementation, operator occupational health safety training, installing warning signs ***PPE*** Additional PPE is the ear plug
Lifting the core	Scratched on the blunt end of the core and crushed by the core	180	*Priority 1*	Wear gloves, safety shoes, safety helmet	Do not use PPE and the appropriate tools	**Administratif :** Supervision of SOP implementation, material handling training, installing warning signs
The process of removing the email	Contact with flammable chemicals	90	*Substantial*	Wear gloves, safety shoes, safety helmet	Do not use PPE gloves	**Training:** Job safety analysis training on a regular basis, work procedure training

In Table 2, Some risk control descriptions are recommended based on semi Quantitative AS / NZS 4360: 2004. The focus of researchers in analyzing corrective actions is only for the risk level that is higher than 20. Meanwhile, at acceptable level (risk level <20), corrective action is not necessary and the company can only carry out continuous supervision so the risk of work accidents might not occur in the future .

The application of occupational health and safety on the production floor is actually quite good. The Occupational Health and Safety Organizing Team has installed warning signs that are installed in every corner of the factory. However, one of the main factors is the frequent occurrence of work accidents at PT. BD originated from the operator error itself in the form of unsafe act. Companies need a scheduling for controlling or periodic monitoring of operator performance to avoid human errors. In addition, regular training is needed to increase the expertise and insight of operators on the importance of work safety. In this case, overall the company needs to further promote order and availability of PPE for operators, especially in the core section. From the head of Committee for the Development of Occupational Safety and Health at PT. BD also should appeal and explain the importance of occupational health safety implementation and the use of PPE for operator safety.

5. Conclusion
The following are the conclusions of the research carried out:
1. The main danger with priority risk level 1 contained in the core section is that the operator can be cut off by the core sharing machine and exposed to engine noise in the process of cutting the mother core, and scratched by the blunt end of the core during the core lifting process, both can be said to be a main hazard because it has the highest risk value of 180-350.
2. One job/activity (process of removing the email layer) has in substantial risk. The recommendation is that beside wear gloves, safety shoes, safety helmet, operators should be trained by Job safety analysis training on a regular basis, work procedure.

Reference
[1] Asl M M 2017 Uni. J. of Man. **5(7)**, 355-363
[2] Santoso G 2004 *Manajemen Keselamatan dan Kesehatan Kerja* (Jakarta: Prestasi Pustaka)
[3] Albrechtsen E Solberg I Svensli E 2019 Safety Science **113** 425–437
[4] Zhang S Boukamp F Teize J 2015 Autom. Constr **52** 29–41.
[5] Gopinath V Johansen K 2016 *Proc. CIRP* 44 199–203
[6] Morrish C 2017 Int. J. Min. Sci. Technol **27** (4) 635–664
[7] Zheng W Shuai J Shan K 2017 Saf. Sci. 93 9–15.
[8] Situmorang, C. 2003 *Mengikuti Prosedur Menjaja Keselamatan Dan Kesehatan Kerja* www.google.com/http://modul.tedebandung.com/ industry (October, 2018)
[9] Dainur 1995 *Materi-materi Pokok Ilmu Kesehatan Masyarakat* (Jakarta: Widya).
[10] Moenir 1993 *Mengikuti Prosedur Menjaga Kesehatan dan Keselamatan Kerja* (Jakarta: Departemen Pendidikan Nasional).
[11] Taslimin H A 1993 *Keselamatan Dan Kesehatan Kerja* (Jakarta: Erlangga)
[12] Sutrisno and Ruswandi R 2007 *Prosedur Keamaanan, Keselelamatan dan Kesehatan Kerja* (Jakarta: Yudistira)
[13] Suma'mur 1981 *Keselamatan Kerja dan Pencegahan Kecelakaan.*(Jakarta: CV Haji Masagung)
[14] Australian/New Zealand Standart. 2004 *OHS Risk Management Handbook* (Australia : Standarts Australia International Ltd.)
[15] Ramli S 2010 *Pedoman Praktis Manajemen Risiko Dalam Perspektif K3* (Jakarta : Dian Rakyat)

IOP Conf. Series: Materials Science and Engineering **722** (2020) 012027 doi:10.1088/1757-899X/722/1/012027

Characteristics of Brake Lining Cessna Caravan Type 208B

I Suroso
Aeronautical Programs, Aerospace College in Yogyakarta

E-mail: indreswari.suroso@gmail.com

Abstract. The brake system is a braking system located on the landing gear on an airplane. Brakes are one of the most important components of safety, and performance in the airplane. This study uses the observation method, the interview method with the engineer and the experimental method at The Material Laboratory, Polytechnic Ceper, Klaten. The purpose of this study is to determine the function, characteristics, and effect of chemical compositions on the characteristics of the brake lining on the Cessna Caravan Type 208B aircraft. The results of the brake lining as, and characteristic of brake lining are reducing speed and stopping the aircraft wheels. The characteristics of the brake lining as follows result in a hardness value of 205 VHN and the chemical composition consist of Fe 95,9%; Cu 0.8%; Carbon 0,33%; 0,235% Si; Mn 0,468%; Al 0,235%; Mo 0.151%; and Cr 0,772%. This brake lining materials are included in medium carbon steel with chromium alloy. The brake lining material is steel medium carbon which has hard, resistance wear, durable use life. The influence of chemical composition on characteristics of brake lining on this aircraft is the higher the composition carbon the resistance to wear.

1. Introduction

The brake is a tool to slow or stop the wheel. The brake system is a braking system located on the landing gear on an airplane. Brakes are one of the most important components of safety and performance in the airplane. The purpose of this study is to determine the function of brake lining; knowing the brake lining hardness, knowing the chemical composition of the brake lining; and to find out the correlation between material hardness and chemical composition.

Characterization of wear surface squealing material with external silica sand particles. The tire model and brake performance are very useful for controlling friction force, anti-slip on aircraft braking system is very important [1]. The target declaration rate is compared with the declaration rate which is calculated to determine the effectiveness of the aircraft wheel braking system [2]. Cu alloys which have high anti-wear are combined with composite C and C/SiC to obtain stable friction coefficient, low wear rate and high material hardness, namely 1441 HVN [3]. Brake canvas which has a microstructure in the form of Cu, sheet steel with microstructures in the form of ferrite and pearlite [4]. The results of measuring the friction coefficient on the aircraft wheels using the anti-lock braking [5]. Banana skin particles can be used a replacement for asbestos in the manufacture of brake pads in the braking system [6]. The wear or tribology test obtained by weight reduction is 0,01 and 0,02 gram for each pressure 1.31; 1.44; 1.58 bars the duration 30, 60, 90 and 120 seconds. The brake canvas has good wear resistance [7]. By adding silicon, the Al-Primer microstructure is getting smaller and there is no dislocation [8]. Increased variation percentage of Titanium can increase aircraft material type Boeing 737-500. With the addition of Ti-V-6A1-4V by recycling recycled pistons, the strength increases by 5% and is more useful. Thus, the brake characteristics are better [9].

30% composition of corncob powder; 30% brass; 20% MgO; 20% polyester resin with a coefficient of friction 0.395 has the optimal brake pad characteristics over the Nissan brake linings [10]. The results of castings from used aluminum from former brake linings have a high hardness when dyeing fast at low temperatures and if the temperature is too much porosity [11]. The results of castings from used aluminum the results of testing the production of brake pads with the highest alloy material is the composition of 20% ash; 50% polyester receipt; 0% iron powder [12]. Numerical simulation of CFD as a groove method to suppress lift force on aircraft tires, the smaller the groove the greater the compression force [13].

Brake lining with a composition of 20% bamboo powder; 15% MgO; 15% Al and 25% epoxy and non-asbestos resins have the best characteristics like federal [14]. The more disk brake ventilation holes, the temperature on the disk brake surface is lower but the brake distance is far away [15]. This study aims to the braking system with ABS frequency 10 Hz has the best system to reduce vibration [16]. This study aims to the braking system with fuzzy logic capable of braking even though high speed [17]. The critical condition of the engine during take-off, landing requires good brakes and meets flying standards [18]. Composite aircraft brake base material with added 15-25% clay is very good for anti-wear [19]. The rate of wear of composite brakes is affected by Cu modification to reduce the brake wear rate [20]. This study investigated the wear surface of the brake lining, with non-asbestos and low metal materials. It can be seen Figure 1 [21]. Sintered brake pads have the characteristics of lighter and stronger C/C-SiC composite disc brakes than the classic metal brake. Can be seen from mechanical energy in Figure 2 [22].

Figure 1. Brake Disk

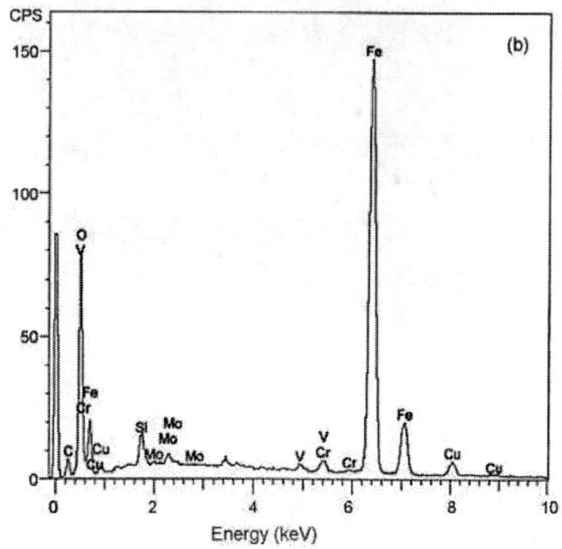

Figure 2. EDX spectra characteristics of brake lining from mechanical energy

IOP Conf. Series: Materials Science and Engineering **722** (2020) 012027 doi:10.1088/1757-899X/722/1/012027

The Function of Brake Lining

Brake linings are composed of a relatively soft but tough and heat resistant material with a high coefficient of dynamic friction and ideally an identical coefficient of static friction. Typically mounted to a solid metal backing using high-temperature adhesives or rivets. The complete assembly including lining and backing is then often called a brake pad or brake shoe. Brake lining as shown in Figure 3.

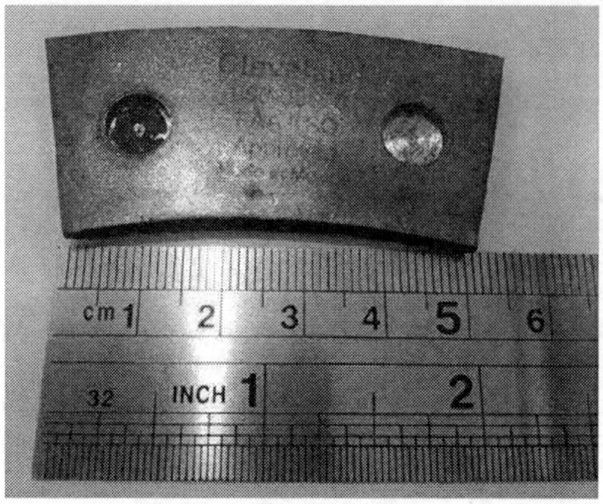

Figure 3. Brake Lining

Brake System

The brake system is part of the aircraft system that is used when the aircraft lands. The component of the braking system is normal braking, alternate braking, and parking braking. The brake system and wheel can be seen Figure 4.

Figure 4. Brake system

Cessna Caravan Type 208B can be seen in Figure 5.

Figure 5. Cessna Caravan Type 208B

2. Methods

This study used a chemical composition test equipment with the Atomic Absorption Spectrometer and Vickers hardness testing. This research was conducted at Polytechnic Ceper, Klaten. Research material is brake lining in Cessna Caravan Type 208B. The flowchart research can be seen in Figure 6.

Hardness test:

a) Preparation specimen, take the size of the test sample where the sample can be tested three times, flatten the sample specimen.
b) Turn the handwheel to the left so that the piece press 2 cm, install the penetrator, set the load by turning the wrench located. The results show in a standard sample and analyze the test object.
c) Sample testing, preparing test object, put the sample stand that the sample surface is perpendicular to indenter axis, rotate the hand wheel, read the designated number, indent at least 5 times for example and test completed.

Chemical Composition Test:

a) Preparation specimen.
b) Standardization of tool.
c) Specimen analysis.

3. Results and discussion

Chemical composition testing on Cessna Caravan Aircraft Brake Lining can be concluded that the element iron (Fe) of 95.9% and chromium (Cr) of 0.772% is the dominant element because it includes elements Fe-Cr alloy. This results in the component having heat-resistant properties which are higher, corrosion-resistant, wear-resistant and higher hardness Chromium (Cr) is one element that can increase the hardness of steel carbon, resistant to wear and resistant to corrosion. Formation of Brake Lining material with a mixture of very Fe-Cr elements suitable because it has the appropriate properties of the brake lining function. The brake lining function that is to reduce or even stop an object's rotational speed. The following characteristic of brake lining:

Hardness Vickers

Material hardness is tested using the Vickers hardness method that is by pressing the specimen or specimen with a diamond indenter pyramid-shaped with a rectangular base and a large angle from the surface facing 136°. In the Vickers Brake Lining test using sample point, load (P) is 200-gram force and loading time is 15 seconds. The result of hardness Vickers and the characteristics of the brake lining as follows result in a hardness value of 205 VHN can be seen in Table 1 and the hardness Vickers equipment can be seen in Figure 7.

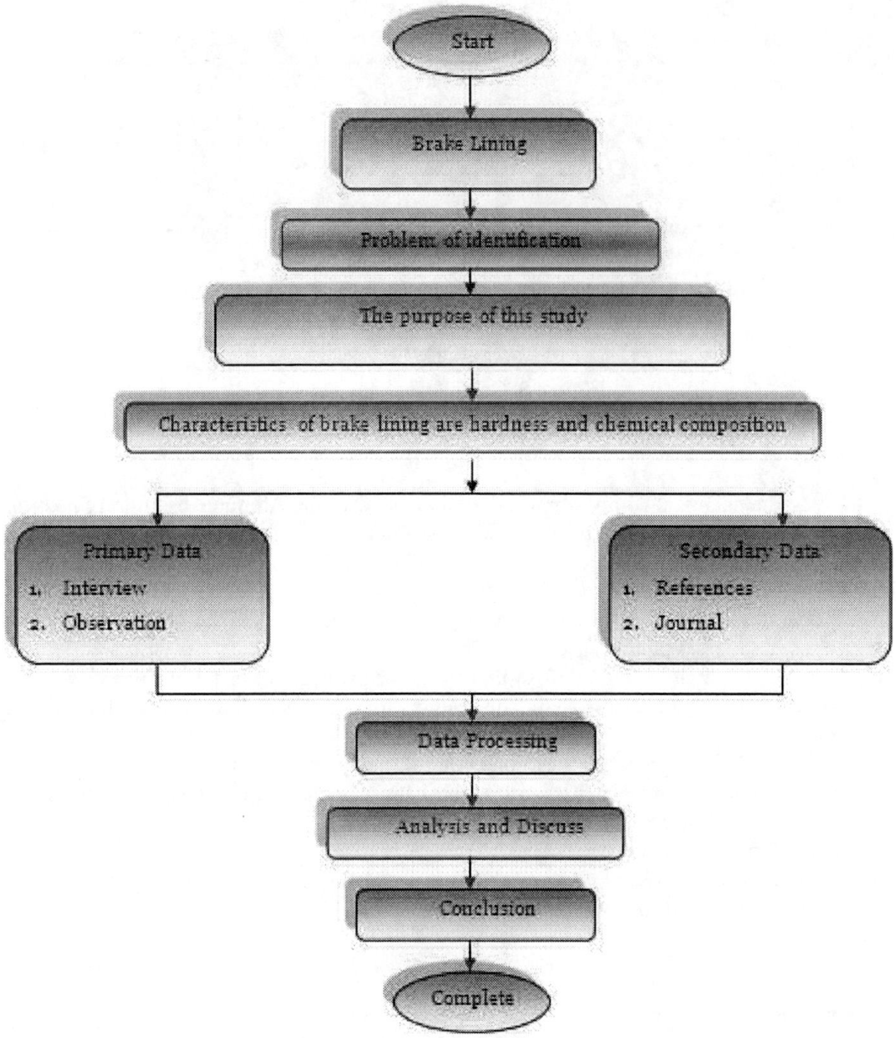

Figure 6. Flowchart

Table 1. The hardness Vickers of brake lining [Source: Polytechnic Ceper, Klaten]

Point	Hardness Vickers (VHN)
Point 1	204, 77
Point 2	205,58
Point 3	205,66
Average	205, 34

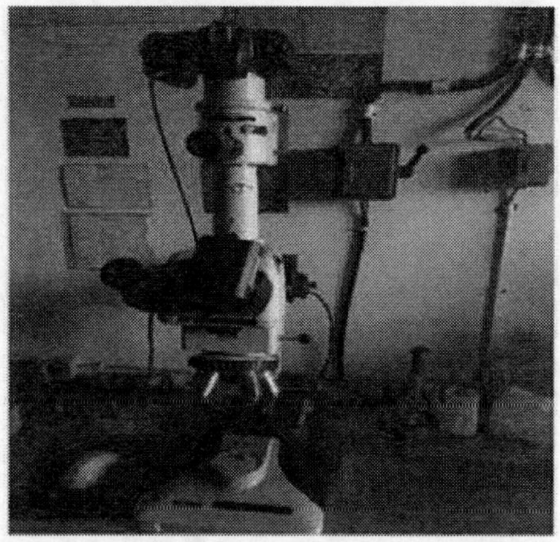

Figure 7. Hardness Vickers [Source: Polytechnic Ceper, Klaten]

These brake lining materials are included in medium carbon steel with chromium alloy. The influence of chemical composition on characteristics of brake lining on this aircraft in the Carbon content with the chemical composition of medium carbon steel, the harder the material and the higher the wear resistance, the more durable the brake lining usage. The correlation between the chemical composition and the hardness of the material as well as the brake lining includes medium carbon steel which has a hardness of 205.34 VHN so that the harder the brake lining material is, the more durable the material uses.

Chemical Composition

The results of testing the chemical composition of the Brake Lining are classified as medium carbon steel because it has a mixture of elements Iron (Fe) and carbon of 0.333% affect the strength and hardness. But there is a mixture of other elements that affect the strength of the Brake Lining properties as following:

a) Iron (Fe): Brake Lining consists of 95.9% Fe. The function of Fe due to material strength, tenacity, and hardness.
b) Chromium (Cr): Brake Lining consists of 0.772% Cr. It has a very big influence to improve toughness, ability to be hardened and resistant to high temperature. The most important use of chromium is increased corrosion resistance.
c) Mangan (Mn): Brake lining consist of 0.468% Mn. This element can be increased strength, hardness, resistance to corrosion, and experiencing strengthening in cold formation.
d) Carbon (C): Brake lining consists of 0,333% C. The function of carbon to increase hardness and tensile strength.
e) Silicon (Si): Brake lining consists of 0,235% Si. The function of Si to increase hardness, strength, weariness, the resistance of corrosion.
f) Aluminum (Al): Brake lining consists of 0,235% Al. The function of Al to increase the resistance of corrosion.
g) Molybdenum (Mo): Brake lining consist of 0,235% Mo. Mo is usually used to be mixed with other metals. This mixture itself will have different properties from the first metal element, molybdenum is usually mixed often with steel to increase strength, toughness, durability against wear and corrosion and the ability to harden steel.

The results of the brake lining as, and characteristic of brake lining are reducing speed and stopping the aircraft wheels. The chemical composition consists of Fe 95,9%; Cu 0.8%; Carbon 0,33%; 0,235% Si; Mn 0,468%; Al 0,235%; Mo 0.151%; and Cr 0,772% can be seen on Table 2 and Chemical Composition Testing Equipment can be seen in Figure 8.

Table 2. The Result of chemical composition brake lining [Source: Polytechnic Ceper, Klaten]

No	Element	Percentage (%)
1	Al	95,9
2	Cu	0,8
3	C	0,33
4	Si	0,235
5	Mn	0,468
6	Al	0,235
7	Mo	0,151
8	Cr	0,772

Figure 8. Optical Emission Spectroscopy [Source: Polytechnic Ceper, Klaten]

4. Conclusion

Brake lining serves to hold the wheel rotation so that braking occurs. The characteristics of the brake lining as follows result in a hardness value of 205 VHN and the chemical composition consists of Fe 95,9%; Cu 0.8%; Carbon 0,33%; 0,235% Si; Mn 0,468%; Al 0,235%; Mo 0.151%; and Cr 0,772%. The correlation between the chemical composition and the hardness of the material as well as the brake lining includes medium carbon steel which has a hardness of 205.34 VHN so that the harder the brake lining material is, the more durable the material uses.

Acknowledgment
The author acknowledgment for academic of Aerospace Technological College in Yogyakarta and prospective promoter Prof. Triyono and Ass. Prof Agung T. Wijayanto.

References

[1] Yi, M. S 2017 Experimental Research on Braking Characteristics of Aircraft ABS Brake System with Ground Conditions, Journal of the Korean Society for Aviation and Aeronautics. 25 2 18-24

[2] McKeown, S. L., Kleinschmidt, P. P., & Laidman, J. A 2016 U.S. Patent No. 9,296,488, Washington, DC: U.S, Patent and Trademark Office

[3] Li, Z., Liu, Y. Z., Zhang, B. G., Lu, Y. H., Li, Y., & Xiao, P 2016 Microstructure and tribological characteristics of needled C/C–SiC brake composites fabricated by simultaneous infiltration of molten Si and Cu. Tribology International. 93 220-228

[4] Irmawan E 2017 Kerusakan Nose Wheel Steering pada Pesawat Boeing 737-200 Series, Jurnal Teknika. 4 1

[5] Li, Z., Xiao, P., Zhang, B. G., Li, Y., & Lu, Y. H 2016 Preparation and tribological properties of C/C–SiC brake composites modified by in situ grown carbon nanofibers. Ceramics International. 41 9 11733-11740

[6] Idris, U. D., Aigbodion, V. S., Abubakar, I. J., & Nwoye, C. I 2015 Eco-friendly asbestos free brake-pad: Using banana peels. Journal of King Saud University-Engineering Sciences. 27 2 185-192

[7] Putra, N. A. D, Pengaruh Beban Dan Waktu Kontak Antara Rotor Disk Dan Kanvas Rem (Lining Pad) Pada Steel Brake Pesawat Terbang. Skripsi Program Studi Teknik Mesin.

[8] Salam, A. N., & Bayuseno, A. P 2016 Pengaruh Penambahan Unsur Silikon Material Model Shaft Propeller Berbahan Dasar Al6063 Terhadap Sifat Fisis Dan Mekanis Hasil Pengecoran Hpdc, Jurnal Teknik Mesin 4 2 138-144

[9] Sutrisno, S 2018 Pengaruh Penguat Serbuk Ti-6Al-4V terhadap Kekuatan pada Matrix Al-Si pada Fuel Intake Pesawat Terbang Boeng 737-500 (Doctoral dissertation, University of Muhammadiyah Malang).

[10] Wicaksono, R. B 2015 Kaji Eksperimental Performansi Pengereman Kampas Rem Serat Bonggol Jagung Sebagai Bahan Alternatif Kampas Rem Mobil

[11] Rahman, S., & Siswanto 2019 Pengaruh Temperatur Tuang Dengan Jenis Material Paduan Al Rongsok (Kampas Rem Panci) Terhadap Porositas, Kekerasan Dan Struktur Mikro Hasil Pengecoran Evaporative. sjme KINEMATIKA. 4 1 87-94

[12] Hadi, Q., & Zamheri, A 2017 Pengaruh Fraksi Volume Penguat Abu Terbang, Serbuk Besi Dan Matrik Resin Terhadap Keausan Dan Kekerasan Untuk Bahan Kampas Rem. AUSTENIT. 9 1 25-32

[13] Wuwung, V., Anggreyni, N., Hitoyo, V. M., & Bintoro, C 2017 Justifikasi Cfd Kedalaman Groove Ban Pada Proses Perawatan Harian Pesawat B737-800 Akibat Hydroplaning (B737-800 Tire Groove Depth Cfd Justification On Its Daily Maintenance Process Due To Hydroplaning). Jurnal Teknologi Dirgantara. 15 1 29-44

[14] Siallagan, R. P. Studi Eksperimental Performansi Kampas Rem Berbahan Serbuk Bambu, Aluminium, Magnesium pada Kendaraan Satria FU, 2018

[15] Albana, M. H., & Putra, Y 2017 Variasi Jumlah Lubang Ventilasi Disc Brake serta Pengaruhnya terhadap Jarak Pengereman dan Temperatur Permukaan Disc. JURNAL INTEGRASI. 9 2 125-128

[16] Juwana, W. E 2018 Rancang bangun sistem rem Anti-Lock Brake System (ABS) dengan penambahan komponen vibrator solenoid. Jurnal Teknik Mesin Indonesia 11 2 83-88

[17] Munandar, A., & Aria, M 2016 Sistem Pengereman Otomatis Menggunakan Logika Fuzzy Berbasis Mikrokontroler. Telekontran. 4 1 1-17

[18] Syamsuar, S 2019 Analisis Data Uji Prestasi Terbang Field Performance pada pesawat udara CN235. Warta Penelitian Perhubungan 25 5 337-343

[19] Agbeleye, A. A., Esezobor, D. E., Balogun, S. A., Agunsoye, J. O., Solis, J., & Neville, A 2017 Tribological properties of aluminium-clay composites for brake disc rotor applications. Journal of King Saud University-Science

[20] Li, Z., Liu, Y. Z., Zhang, B. G., Lu, Y. H., Li, Y., & Xiao, P 2016 Microstructure and tribological characteristics of needled C/C–SiC brake composites fabricated by simultaneous infiltration of molten Si and Cu. Tribology International 93 220-228

[21] Neis, P. D., Ferreira, N. F., Fekete, G., Matozo, L. T., & Masotti, D 2017 Towards a better understanding of the structures existing on the surface of brake pads, Tribology International. 105 135-147

[22] Li, G., & Yan, Q 2015 Comparison of friction and wear behavior between C/C, C/C-SiC and metallic composite materials. Tribology Letters. 6 0 1278-285

IOP Conf. Series: Materials Science and Engineering **722** (2020) 012028 doi:10.1088/1757-899X/722/1/012028

Adoption of 3D Printing in Indonesia and Prediction of Its Application in 2025

Ismianti[1], Herianto[2]
[1] Department of Industrial Engineering, Universitas Pembangunan Nasional Veteran Yogyakarta, Indonesia
[2] Department of Mechanical and Industrial Engineering, Universitas Gadjah Mada, Indonesia

E-mail: ismianti@upnyk.ac.id

Abstract. Additive manufacturing, known as 3D printing (3DP) is currently growing rapidly in many countries. The applications of 3DP may vary in each country, depending on the country's 3DP development and local needs. This study aims to find out the adoption of 3DP in Indonesia and the prediction of its application in the future. Data were collected either by interviews or questionnaires. Data retrieval was done 3 times. First, data were collected from society with 109 respondents. The second data comes from 3DP user with 40 respondents, and the third data comes from the expert with 15 respondents. This study used a pairwise comparison method to predict the application of 3DP in the future. The results of this study indicate that 67% of respondents in society have known about 3DP. The number of 3DP users was increasing every year, a considerable increase in the last 3 years. From the pairwise comparison results, it was found that in the health sector, the highest prediction is in the fabrication of medical equipment. In the industrial sector, the highest prediction is fabrication for electronic and robotic. Then in the sociocultural sector, the highest prediction is education.

1. Introduction

Additive manufacturing (AM) technology is currently growing rapidly in various country. Additive manufacturing or commonly referred to 3D printing has an important role not only in various fields of business but also in the daily lives of consumers[1]. Various countries such as the United States, New Zealand, Australia, Britain, Germany, Singapore, and China invest tremendous value for the developments of additive manufacturing[2]. In these countries, additive manufacturing is growing rapidly for various productions. However, the use of additive manufacturing in each country can be different. The use of AM in each country can vary depending on the development of additive manufacturing in the country and also depends on the local needs of the country[3]. Research on AM's use is still growing in various countries. More research on the use of AM and the impact of this use is still needed for future development[4].

In addition to the great opportunities of this technological development, there is still uncertainty and speculation of this technology in the future[5]. This uncertainty can come from many things such as the development of new technologies, changes in demand, government policies, changed in economic conditions, and so on. Research on the prediction of this technology is still rare. Existing research is predicting AM technology in 2030 but with experts in European and South African respondents. Further research is needed on the prediction of this AM technology in the future in Asian countries[5]. This study would like to see how large the use of 3D printing in Indonesia now and prediction the future of its use. From this research, the use of 3DP in Indonesia and predictions of use in what fields will be optimally potential in Indonesia with the local character of the Indonesian people can be seen.

2. Review References

Additive manufacturing is a technology that has long been researched and developed. There have been many studies on this technology. However, this additive manufacturing still has strengths and weaknesses. Research on additive manufacturing is still needed so that additive manufacturing can overcome its weaknesses[6]. Many researches on 3D printing have been carried out, but more research is needed on the adoption of 3D printing and the impact of adopting this technology to find out more about the consequences of using this technology[7].

One of the studies on the adoption of 3D printing was carried out by Kianian et al.[2]examined the adoption of 3D printing technology in Sweden with respondents being users in companies, universities, and research institutions. The research would like to see variations in the use of 3D printing in the country and the cause of the variations. Another study conducted by Schniederjans[7] tried to see the adoption of 3D printing in the manufacturing sector. In this study investigated factors that influence the speed of adoption of 3D printing in companies. However, the speed of adoption and the impact of the use of 3D printing will differ depending on the socio-economic conditions of the country[3]. In the condition of developed and developing countries, the implications of 3D printing can show the opposite pattern [3].

A lot of research on 3D printing prediction is going forward. Various studies have been conducted to predict how 3D printing technology will be in the future. McKinsey Global Institute [8]tried to provide a review of 3D printing predictions in 2025. Gebler et al.[3] then used this data to see the sustainability of these 3D printing products in the future. The 3D printing application areas seen are consumer goods, aviation industry, automotive industry, health components, and equipment. Other study conducted by Jiang et al.[5], in the study, predicted the implications of 3D printing for industry, policy, and future research. This research was conducted with expert respondents from countries in Europe and Africa using the Real-Time Delphi method. One of the weaknesses of the research is that the respondents are not yet representative, one of them is the absence of representatives of Asian countries in it[5].

3. Research Methodology

Objects in this study were divided into 3 groups. The first group was the community to find out how much people's understanding of 3DP. The second group was 3D printing users in educational institutions, companies, businesses, and individuals in Indonesia. The third group was experts in the fields of 3D printing, government, manufacturing, health, arts, and education. Data collection was performed 3 times. The stages of this research can be seen in Figure 1.

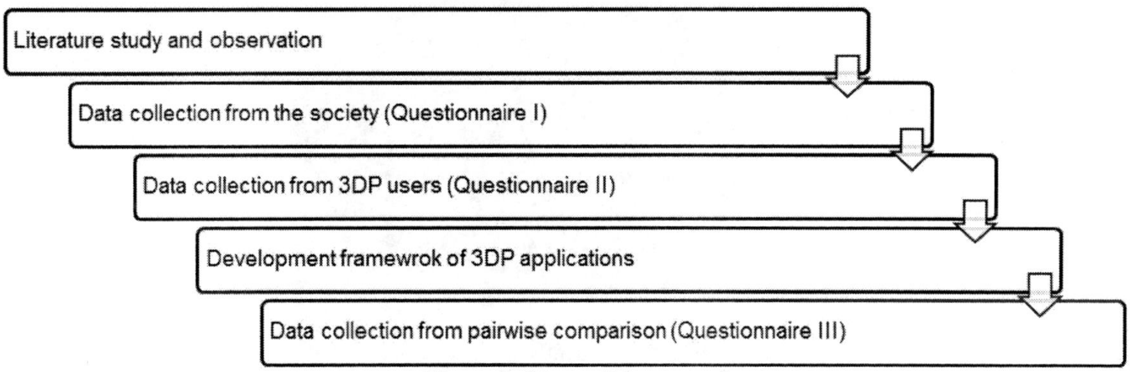

Figure 1. Stages of the research

The following is an explanation of each of the stages in this research.

1. Literature study and observation
   Literature studies and observations were made to find out things related to the use of 3DP in Indonesia and other countries through journals, books, social media, articles, websites, and previous research.
2. Data collection from society (questionnaire I)
   The first data collection was done online with non-probability sampling method (snowball sampling).
3. Data collection from 3DP users (questionnaire II)
   The second data collection was done online and offline with nonprobability sampling (purposive sampling method). Data collection at this stage was carried out mainly through the Facebook group "3D Printer Indonesia", which consist of about 5800 members from various region in Indonesia.
4. Development framework of 3DP applications
   Data on various types of 3DP applications were used as factors and subfactors then AHP hierarchy structure was made from these factors. This structure was used to pairwise comparison data collection.
5. Data Collection from the pairwise comparison (questionnaire III)
   Data collection at this stage was done online and offline with expert respondents.
6. Data processing and factor weighting
   Data processing and factor weighting were done with software Expert Choice 11. After all of the data was consistent, then weighting each factor and subfactor was carried out.
7. Prediction sequence of 3DP applications
   From the weight obtained, then the order of priority prediction of 3DP usage in 2025 was obtained.

4. Result and Discussion

4.1. Indonesian People's Knowledge of 3D Printing

Respondents in the first questionnaire were the general population of 109 respondents with respondent's age range from 20 to 38 years (25.65 ± 4.18). These respondents were spread from various regions in Indonesia. From the 109 respondents, 73 people (67%) know about 3D printing, 22 people (20%) do not know, and 14 people (13%) might know. The respondent who knows about 3DP mostly gets information about 3DP from the internet, campus, school, friends, television, and community. With so many people who know information about 3DP technology, this will be the catalyst of 3DP adoption in Indonesia. Figure 2 is the distribution of respondent's knowledge about 3DP.

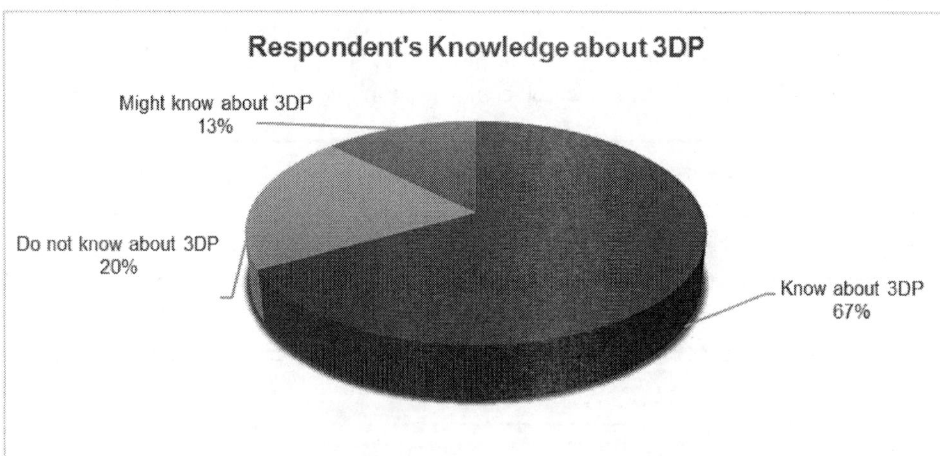

Figure 2. Respondent's knowledge about 3DP

From 109 respondents, 100% of respondents did not have 3DP. From the 73 respondents who knew about 3DP, only 43 respondents (59%) were interested in having 3DP, others were not interested or not yet decided. From the 43 interested respondents, 22 people (51%) were interested in the price of 3DP below Rp. 5 million, 15 people (35%) were interested in printers at the price of Rp. 5 million - Rp. 10 million, and 6 people (14%) were interested in printers with prices>Rp. 10 million.

4.2. Adoption of 3D Printing in Indonesia

Respondents in the second questionnaire were 40 people with 39 men and 1 woman. The age range of respondent was from 16 to 49 years, with an average age of 28.48 ± 7.82 years. Respondents of 3DP users were spread in various provinces in Indonesia, with most users still concentrated in Java (75%). Figure 3 shows the distribution of 3DP users respondents in this research. Most of respondents were in Yogyakarta because data collection in Yogyakarta was searched offline and online and in other cities data collection was only searched offline.

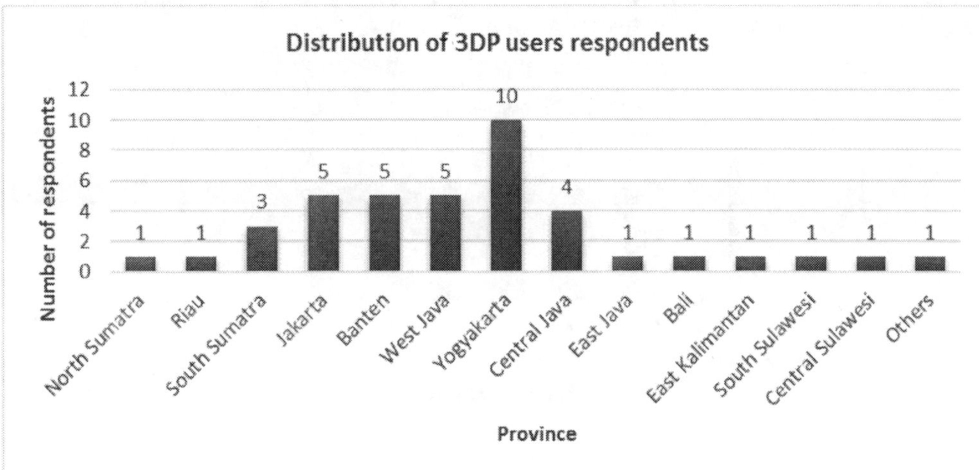

Figure 3. Distribution of 3DP users respondents

Most respondents of 3DP users worked as private employees and 3DP entrepreneurs. 3DP users had a monthly income ranging from less than Rp. 1,500,000 to more than Rp. 10,000,000.Most respondents were direct owners of 3DP (65%), while 32% use the company's 3DP. The type of 3DP that most respondents have (86%) is FDM (Fused Deposition Modeling). FDM printers were affordable in the community, materials were easy to get and easy to use. In addition to FDM, several users who pursue 3DP also have SLA (Stereolithography), and SLS (Selective Laser Sintering).

The highest percentage of 3DP prices owned by respondents in the price range of Rp. 5 million to Rp. 10 million with a percentage of 51% while 31% of respondents have a 3DP with a price range of <Rp. 5 million. For the duration of knowing 3DP, most users (80%) know 3DP in the last three years. Figure 4 shows the duration of the respondents knowing 3DP.

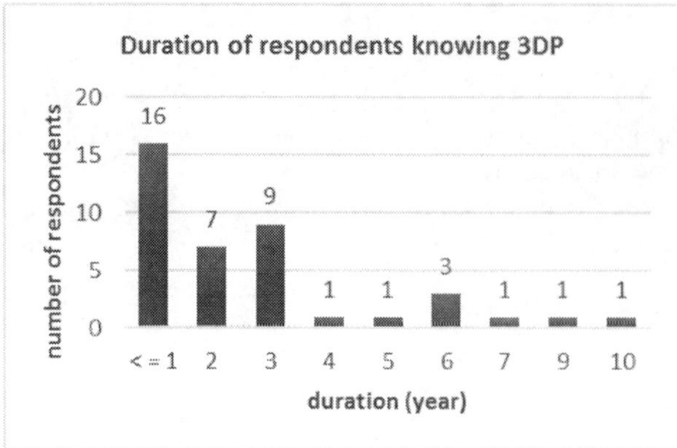

Figure 4. The duration of respondents knowing 3DP

Ownership of 3DP was increasing year by year. In the last two years, the increase in ownership of respondents can reach 137.5 %%. This shows that 3DP was increasingly popular in the community. Figure 5 shows the duration of 3DP ownership of respondents.

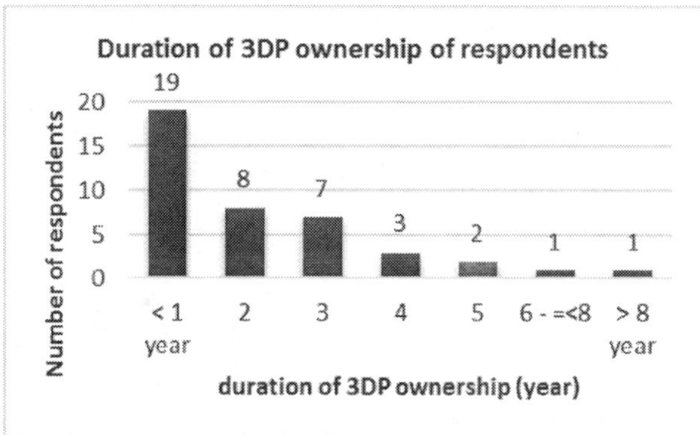

Figure 5. Duration of 3DP ownership of respondents

The use of 3DP in the user respondents was mostly in the fields of electronics and robotics, toys and the creation of miniature art objects. Table 1 shows the 3DP application by respondents.

4.3. Development the Use of 3DP Framework

After taking data on 3DP users, the next step was the selection and identification of the expert. From the results of literature studies, respondent data, and interviews obtained various kinds of 3DP usage. These various kinds of 3DP usage were divided into three sectors, health, industrial, and sociocultural [9]. In the health sector, users use 3D printing to help the healing process, for example replacing organs, assistive devices and medicines. In the health sector, the use of 3DP was divided into more detail based on the characteristics of its use into three sub-sectors, bio-printing, medical equipment, and medicine.

In the industrial sector, 3DP was used for economic and business activities in both small and large companies. In this industrial sector, the use of 3DP was divided into seven sub-sectors, fashion, construction and architectural, military, food printing, automotive and aerospace, electronics, and robotics, and toys.

IOP Conf. Series: Materials Science and Engineering **722** (2020) 012028 doi:10.1088/1757-899X/722/1/012028

Table 1. 3DP application by respondents

Field	Number of users
electronic and robotics	28
toys	27
miniature	24
construction and architecture	19
education	19
manufacture	18
households	15
automotive and aerospace	11
historical and cultural heritages	10
3D selfies	9
military	7
fashion	5
jewelry	5
medical equipment	5
environmental objects	1
musical instrument	1

The use of 3DP in the third sector was sociocultural. With the growing development of 3DP, the use of this machine also expanded to personal, especially hobbies. This personal use covers various aspects of social, art, culture, environment, and education. The use of the sociocultural sector was divided into eight subsectors, miniature art objects, jewelry, 3D selfish, musical instruments, household, education, environmental objects, then historical and cultural heritage. Existing sectors and sub-sectors are then formed in the structure into factors and sub-factors. In Indonesia, not all of these fields have developed. This is because of various factors such as local potential, government, society and technological progress[10]. Figure 6 shows the sector and sub-sector of 3DP applications.

4.4. Data Collection and Processing of Pairwise Comparison

Ratings through pairwise comparison (1-9) conducted by the expert. Experts were asked to provide an assessment of 3DP usage predictions in 2025 totalling 15 people from various backgrounds such as academia, businessmen, government, industry, and health who lived in various cities in Indonesia. Table 2 shows the respondents of pairwise comparison. Data from pairwise comparison processed by software Expert Choice 11.

4.5. Prediction the Use of 3D Printing in 2025 in Indonesia

After the data from each respondent is combined, then the weight of each sub-factor is obtained. This weight was then sorted and made a priority sequence. Table 3 shows combined pairwise comparison data from all respondent in all sector.

In the comparison of level 1 factors, it was found that the industrial sector was predicted to be the sector with the highest use compared to other sectors, followed by the sociocultural sector and then health. An overview of the priority order can be seen in Figure 7.

IOP Conf. Series: Materials Science and Engineering **722** (2020) 012028 doi:10.1088/1757-899X/722/1/012028

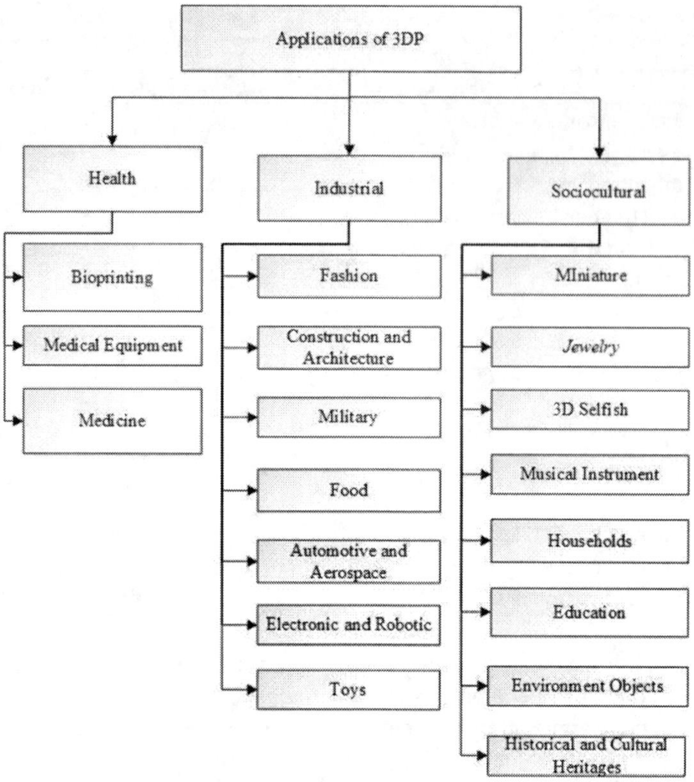

Figure 6. Applications of 3DP divided into sector and sub-sector

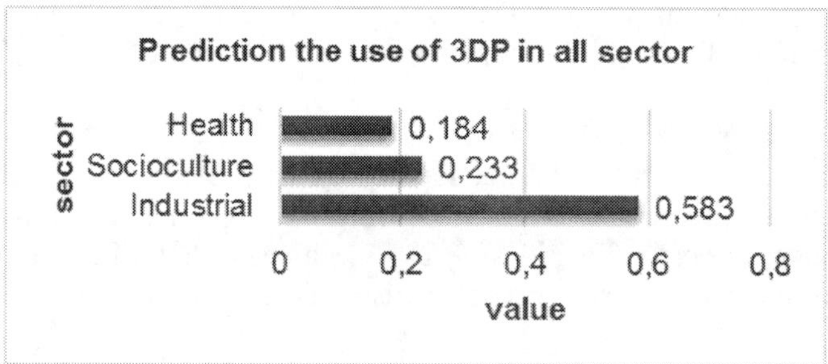

Figure 7. Prediction the use of 3DP in all sector (inconsistency = 0,00006)

4.5.1. Predictions in the Health Sector

In the health sector, the prediction of the largest 3DP use was medical equipment. Medical equipment was also already widely manufactured by several companies 3DP in Indonesia and also by individuals. This medical equipment was predicted to continue increase in use until 2025. Some example of the medical equipment is prosthetic, dental equipment, assistive devices and others. Figure 8 shows prediction the use of 3DP in health sector.

4.5.2. Predictions in the Industrial Sector

In the industrial sector, the most predicted fields for use were electronics and robotics. In this field, 3DP was usually used for making drones, 3DP replication, manufacturing robots, and various other electronic devices. The increasing need for automation made this electronic and robotics field more widely used. Even today, electronic and robotics were also ranked first in their use. Figure 9 shows prediction the use of 3DP in the industrial sector.

Table 2. Respondents of pairwise comparison

Fields	Profession	Domicile
3D Printing	Co-founder of 3D printing maker and medical devices company	Bandung
	Founder of 3DP maker company	Tangerang
	Engineer in 3D printing company	Yogyakarta
	CEO 3DP maker and services company	Tangerang
	CEO 3DP maker and services company	Malang
Manufacture	RND Engineer in an automotive manufacturing company	Jakarta
	Engineer FMCG company who use 3DP	Tangerang
	Senior engineer aerospace company	Bandung
	RND in an automotive manufacturing company	Tangerang
	Technology material and process staff in the aerospace company	Bandung
Government	Researcher in Industrial Minister	Yogyakarta
Academics	Robotics and automation lecturer in a state university	Yogyakarta
	Design lecturer in private university	Jakarta
Health	Dentist and lecturer in a state university	Yogyakarta
	Doctor and lecturer	Wonosobo

Table 3. Combined paiwise comparison data

	Health	Industrial	Sociocultural
Health	1,000	0,318	0,782
Industrial	3,145	1,000	2,520
Sociocultural	1,278	0,397	1,000

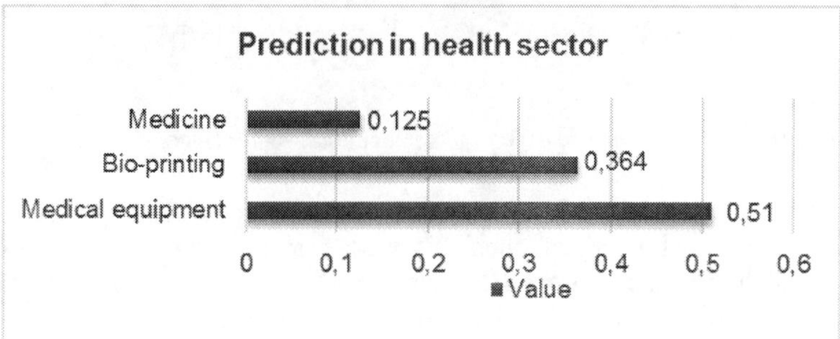

Figure 8. Prediction the use of 3DP in health sector (inconsistency= 0,00048)

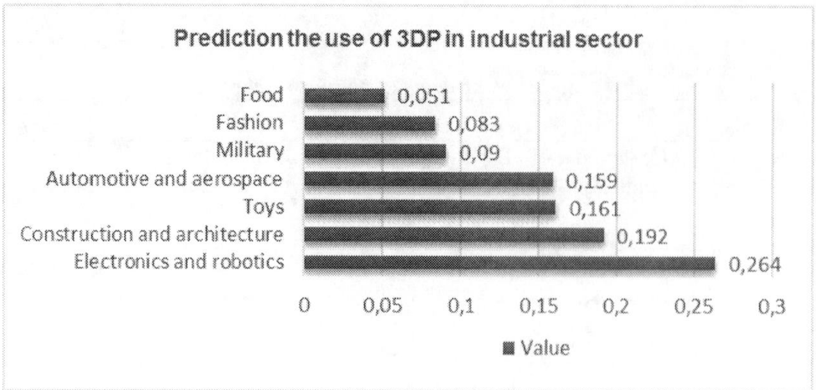

Figure 9. Prediction the use of 3DP in health sector (inconsistency= 0,00356)

4.5.3. Predictions in the Sociocultural Sector
In the sociocultural sector, education was the highest prediction priority for the use of 3DP in the future. At present it has been seen that in the education sector there have been many formal and informal institutions that began holding 3DP. From the results of weighting, the highest priority sequence after education was 3D selfish. 3D selfish were currently not very high in use in Indonesian society, but with the tendency of the Indonesian people in the selfish, 3D selfish were predicted to grow rapidly in the future after the education factor. Figure 10 shows prediction the use of 3DP in sociocultural sector.

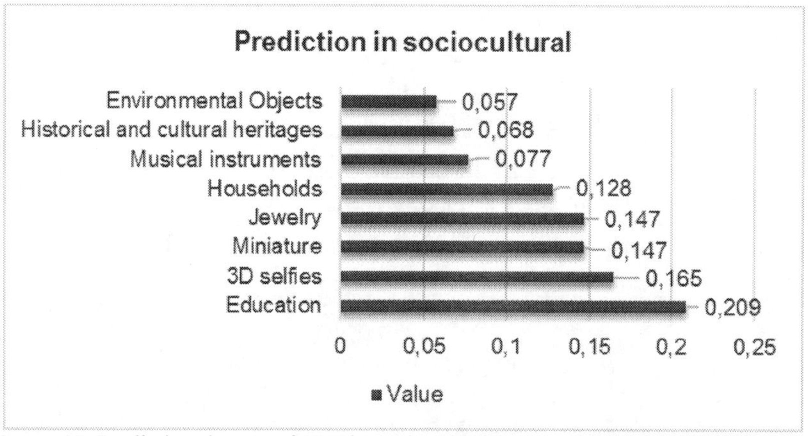

Figure 10. Prediction the use of 3DP in sociocultural sector (inconsistency = 0,00386)

4.6. Prediction the Use of 3DP in All Sector
After prediction in each sector gained, then this value can be used to get the order value for all subsector. The prediction sequence for 3DP usage in all sector can be seen in Figure 11 and the framework prediction based on each sector can be seen in Figure 12.

4.7. Factors that Encourage the Development of 3DP Usage in Indonesia
Some of the things that encourage the development of 3DP use in Indonesia are the introduction of 3DP in the community, the increasingly mature of 3DP technology, the ease of getting 3DP (marketplace), affordable 3DP prices, growth of local companies that sell 3DP, and the community's need for 3DP technology. With the 3DP, items that must initially be printed in conventional can be printed with 3DP at a lower cost, faster work, better results, and safer.

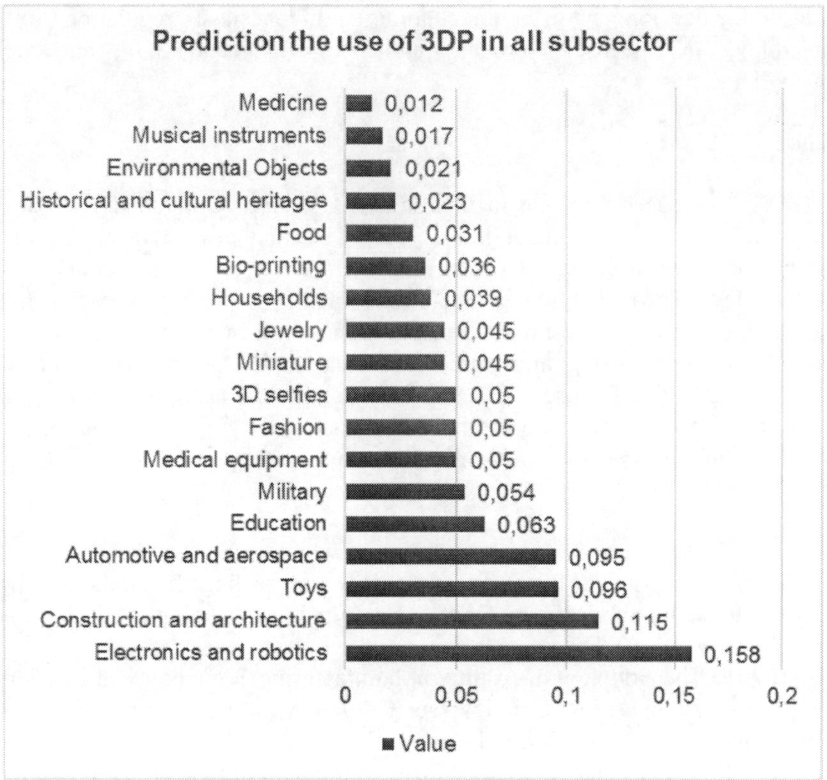

Figure 11. Prediction the use of 3DP in all subsector

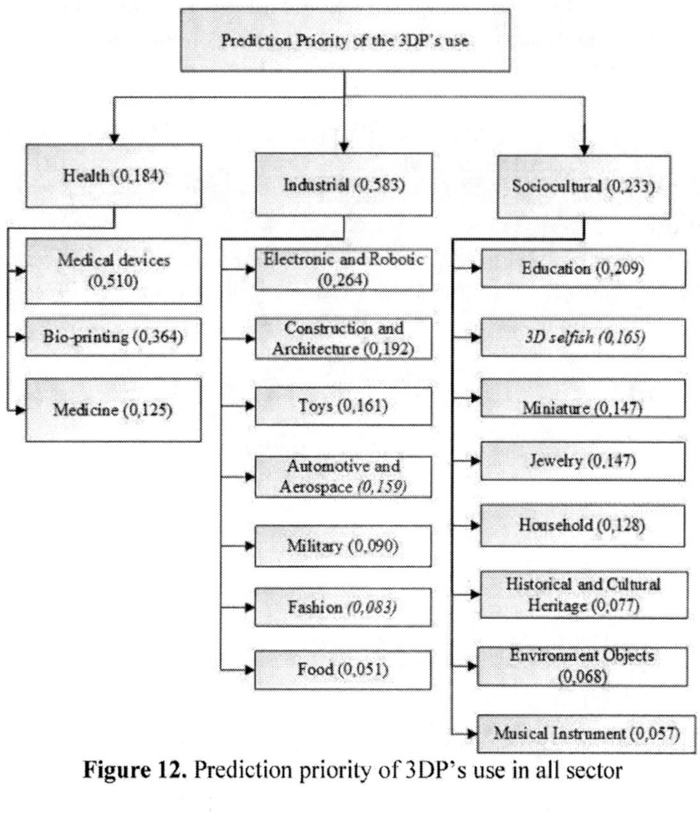

Figure 12. Prediction priority of 3DP's use in all sector

IOP Conf. Series: Materials Science and Engineering **722** (2020) 012028 doi:10.1088/1757-899X/722/1/012028

4.8. Factors that Inhibit the Development of 3DP Usage in Indonesia

Some obstacles to the development of 3DP technology in Indonesia are regulation (tax), infrastructure, government policy, SNI license, material, public knowledge of 3DP, and lack of research collaboration in 3DP fields.

5. Conclusion

Based on research that has been done in this study, it can be concluded that the use of 3DP in Indonesia increasingly widespread in the last 3 years. The number of 3DP users was increasing, the high increase has occurred in the last 3 years. 3DP users were spread throughout Indonesia with the most users still on Java. The most types of machines owned by 3DP users were FDM. The use of 3DP in Indonesia was increasingly widespread with the highest use in the manufacture of electronic and robotic, making toys, and making miniature art objects. By using pairwise comparison, obtained prediction priority order of 3DP's use in 2025 in Indonesia. In the health sector, the highest prediction is fabrication in medical equipment. In the industrial sector, the highest prediction is fabrication for electronic and robotic. Then in the sociocultural sector, the highest prediction is in the education sector, followed by 3D selfish.

References

[1] Rayna T and Striukova L 2015 Technological Forecasting & Social Change From rapid prototyping to home fabrication : How 3D printing is changing business model innovation *Technol. Forecast. Soc. Chang.* 1–11

[2] Kianian B 2015 The Adoption of Additive Manufacturing Technology in Sweden **2** 152–8

[3] Gebler M, Uiterkamp A J M S and Visser C 2014 A global sustainability perspective on 3D printing technologies *Energy Policy* 1–10

[4] Khajavi S H, Partanen J and Holmstro J 2014 Computers in Industry Additive manufacturing in the spare parts supply chain **65** 50–63

[5] Jiang R, Kleer R and Piller F T 2017 Technological Forecasting & Social Change Predicting the future of additive manufacturing : A Delphi study on economic and societal implications of 3D printing for 2030 *Technol. Forecast. Soc. Chang.* **117** 84–97

[6] Wong K V and Hernandez A 2012 A Review of Additive Manufacturing **2012**

[7] Schniederjans D G 2017 Adoption of 3D-printing technologies in manufacturing: A survey analysis *Int. J. Prod. Econ.* **183** 287–98

[8] Anon 2013 Disruptive technologies : Advances that will transform life , business , and the global economy

[9] Taufik M, Jain P K and Jabalpur M 2016 Additive Manufacturing : Current Scenario

[10] Ismianti and Herianto 2018 Framework Prediksi Penggunaan 3D Printing di Indonesia pada Tahun 2030 *Seminar Nasional IENACO 2018* pp 546–53

IOP Conf. Series: Materials Science and Engineering **722** (2020) 012029 doi:10.1088/1757-899X/722/1/012029

Application of ZnO nanoparticles for Producing Antibacterial Batik

I N Eskani[1*], A Haerudin[1], J Setiawan[1], D W Lestari[1], Isnaini[1] and W Astuti[2]

[1]Center for Handicraft and Batik, Ministry of Industries, 7 Kusumanegara Rd, Yogyakarta, Indonesia

[2]Research Unit for Mineral Technology, Indonesian Institute of Sciences, Ir. Sutami Rd. 15, Tanjung Bintang, South Lampung, Indonesia

*E-mail: hana.eskani@gmail.com

Abstract. The government of Indonesia has launched a program of 'Making Indonesia 4.0' since April 4th 2018. The program is an integrated roadmap to implement a number of strategic action to enter the 4.0 Industrial era. One of the strategic initiatives is the development of functional clothing by producing a cloth that has multifunctional properties, such as antibacterial clothing. Batik, a traditional cloth is originally from Indonesia. It is fabricated by applying hot wax onto a fabric to create a pattern before colouring process which is then removed in the final stage. Since people use batik in daily life for clothing and household appliances, it is worth to produce antibacterial batik. This research aimed to make antibacterial batik by applying zinc oxide (ZnO) nanoparticles to the material of batik i.e. cotton. Colouring process was conducted by synthetic dyes of indigosol. To see the effect of ZnO application, the antibacterial activity, the colour fastness to washing, the colour strength (K/S) and surface morphology of batik have been examined. The results showed that the batik exhibited antibacterial activity due to the incorporation of ZnO onto the fabric. The colour strength was also enhanced. To maintain the colour fastness to washing, the ZnO should be applied to the fabric before the batik process.

1. Introduction

Industry 4.0 refers to a new phase in the Industrial Revolution that emphasizes on interconnectivity, automation, machine learning, and real-time data. Indonesia has launched a program of 'Making Indonesia 4.0' since April 4th 2018. The program is an integrated roadmap to implement a number of strategic action to enter the 4.0 Industrial era. Five industrial sectors that have been selected for Making Indonesia 4.0 are: Food and beverages, textile and apparel, automotive, electronics and chemicals. In the textile and apparel sectors, Indonesia is targeted to become a leader of functional clothing producer in 2030. Functional clothing exhibits multi-functional properties, including hydrophobicity, enhanced thermal stability, antibacterial, UV-protection, anti-wetting and laundering performance [1].

In recent years, nanotechnology has become one of the most important and exciting fields in chemistry, physics, biology and engineering [1]. Nanotechnology deals with particles at the atomic level (1–100 nm). Structures on this scale have been shown to have unique and novel functional properties. Nano materials in various forms such as metal nanoparticles, nano composites and metal oxides have been used for functional finishing of textiles, such as UV protection, conductive, water repellent, antibacterial, and deodorizing functionalization of textiles [2]. They can be applied onto textile surfaces by different means like in situ synthesis, spraying and pad-dry-cure method. One of the most studied nanoparticles is zinc oxide (ZnO). Zinc oxide (ZnO) nanoparticles are preferred because of their unique properties such as electrical, photo catalytic, optical, and antibacterial [3]. The antibacterial properties of ZnO are excellent because they are heat-resistant and stable in a harsh conditions [4]. ZnO exhibits significant antimicrobial activities when particle size is reduced to the nano meter range. The nano-sized ZnO can interact with bacterial surface where it enters inside the cell, and subsequently exhibits distinct bactericidal mechanisms [5]. Different methods have been adopted for investigation of antibacterial activity. The agar diffusion method is the most frequently used method and has been standardized as an official method for detecting bacteriostatic activity by the American Association of Textile Chemists and Colourists (ATCC).

Batik is a traditional Indonesian cloth that have been recognized by UNESCO as the Intangible Cultural Heritage of Humanity on October 2, 2009. According to SNI 0239: 2014, the term batik is a handicraft as a result of resist dyeing technique using a hot wax that attached onto media (textile or non-textile) using the batik stamp (called a canting) [6]. The fabrics commonly used for batik are silk and cotton. Cotton fabric has several advantages including high absorption, smooth and comfortable to wear [7]. However, they certainly lack easy-care, dimensional stability, antibacterial, UV-blocking and self-cleaning properties [8]. Application of ZnO nanoparticles on cotton fabrics is expected to improve batik performance, especially providing antibacterial properties. The process of batik fabrication is carried out by attaching a hot wax onto fabric, which acts as a color barrier using a batik canting. Then the dyeing process may be carried out by natural dyes or synthetic dyes. Furthermore, the wax is removed by boiling the cloth in water. The batik cloth is then washed thoroughly and dried in the shade place [9].

The application of zinc oxide onto the fabric is not only able to provide antibacterial properties but also improve its coloring quality. These have been investigated by a number of researchers [8] [10]. Cotton fabrics that have previously been carboxymethylated and then applied using zinc oxide and titanium isopropoxide have good antimicrobial properties and color fastness [8]. Tavanaie (2017) applies ZnO-TiO2-PET (Polyethylene Tereptalat) to the fabric and has excellent color absorption and color fastness [10]. To the best of author knowledge, there has been no research yet about application of ZnO nanoparticles for producing antibacterial batik, therefore this research was conducted to apply ZnO nanoparticles for producing antibacterial batik with good coloring quality.

2. Experimental

2.1. Materials and Equipments

Batik wax, cotton fabric, synthetic dye of Indigosol, aluminium sulphate ($Al(SO_4)_2 \cdot 12H_2O$), non-ionic detergents (TRO), soda ash ($NaCO_3$), natrium nitrit ($NaNO_2$), and HCl were purchased from Prawoto Shop, Yogyakarta. ZnO nanoparticles powder (average particle size of 67 nm) was obtained from Sigma Aldrich. Analytical balance (Fujitsu), oven (Memmert), magnetic stirrer (SCILOGEX MS7-H550-Pro) and standard laboratory glassware were used for the experimentation. The experiments were carried out at the laboratory of Centre for Craft and Batik Yogyakarta.

2.2. Methods

2.2.1. Making the batik. The batik processes includes an application of the batik wax onto fabric, dyeing process and removal of the wax. Initially, the white cotton fabric was immersed in a non-ionic detergent (TRO) solution (0.1%) in order to clean it up. To apply the batik wax onto fabric, the melted wax was applied onto the fabric by batik stamp to create desired pattern. The dyeing process was carried out by the following procedures: the indigosol dye solution was prepared by dissolving 5 g/l of dyes with 10 g/l of $NaNO_2$. Dyeing of the fabric was conducted by immersing the fabric in the indigosol solution for 5 minutes and then it was dried under sun light for 10 minutes. The dyeing process was repeated twice. Then the fabric was immersed in the acid solution (15 cm^3/l of HCl) and it was neutralized with soda ash solution (1 gr/l of Na_2CO_3). At last the fabric was rinsed thoroughly in tap water and exposed to the sun to get dried. To remove the wax, fabrics was immersed in the boiling water until the wax was removed. Soda ash (2 gr/l) was added in the water to help removing the wax.

2.2.2. Application of ZnO nanoparticles onto fabric. ZnO nanoparticles was applied onto the fabric by the pad-dry-cure method. To investigate the effect of ZnO concentration, temperature and order of the ZnO application, the research was carried out at a concentration of ZnO of 1% and 2%, temperature of 25°C and 80°C and the coating of ZnO conducted before and after the batik process. Initially, the fabric was immersed in a non-ionic detergent (TRO) solution (0.1%) in order to clean it up. Thereafter ZnO was applied to the fabric using the pad-dry-cure method. When it was applied before the batik process, white cotton fabric (330 mm x 350 mm) was dipped into ZnO solution (1% and 2%) at the temperature variation of 25°C and 80°C. The cotton fabric was then padded at room temperature to reduce moisture content. Afterward the fabric was dried at 80°C for 5 minutes. The dipping-padding-drying processes were repeated 3 times. Then the fabric was cured at 140°C for 3 minutes. The batik process was then subsequently carried out. When ZnO was applied after the batik process, the batik process was carried out first and then followed by the application of ZnO using the same procedures.

2.3. Characterizations

2.3.1. Antibacterial Activity. Antibacterial activity of the ZnO coated fabric was examined against Gram-positive bacteria Staphylococcus aureus and Gram-negative bacteria Eschericia coli using agar diffusion method. The bacteria were prepared in nutrient broth medium. Test organisms diluted x100 from a 3 hour culture, were wiped on the surface of Isosensitest agar plates. The batik of 15 mm diameter were then slightly pressed on to the surface of one plate and the control fabric was onto another plate. The incubation of the plates were carried out at 37°C for 18 – 24 hours. The antibacterial activity of batik was indicated by the inhibition zone diameter (mm) compared to the control fabric [7].

2.3.2. Colour Strength. Color strength (K/S value). Color strength of the batik were evaluated using a UV-vis Spectrometer. The reflectance (R) of the samples were measured so the K/S value were calculated using equation 1. The K/S value indicates the color intensity of dyed samples [8].

$$K/S = \frac{(1-R)^2}{2R} \quad (1)$$

K: absorption coefficient, S: scattering coefficient, and R: reflectance at complete capacity

2.3.3. Color fastness to washing. Washing fastness was carried out in accordance to ISO 105-C10:2006 standard. The change in color and stains of the adjacent batik were assessed by grey scale.

3. Result and Discussion

Antibacterial activity of white fabric and batik with ZnO treatment have been examined using the agar plate method. The results are shown in Figure 1 and 2. Figure 1 shows inhibition zone around the sample fabric of white fabric and batik with ZnO treatment against E.coli bacteria compared to antibacterial drug (Chloramphenicol) and batik without ZnO treatment. Figure 2 shows inhibition zone of the same sample against S. aureus bacteria. As can be seen in Figure 1, ZnO application affect the ability of the fabric to inhibit the growth of E. coli bacteria, since the value of inhibition area of fabric without ZnO is zero. The inhibition area of coated fabric is approximately 60% of inhibition zone of the antibacterial drug (Chloramphenicol). The treated white fabric exhibit slightly greater antibacterial effect than the treated batik. This indicates that the batik process reduce slightly the antibacterial activity against E. coli of the treated fabric. The order of ZnO treatment (before and after the batik process), the difference of ZnO concentration and temperature did not significantly affect the antibacterial activity against E. coli bacteria.

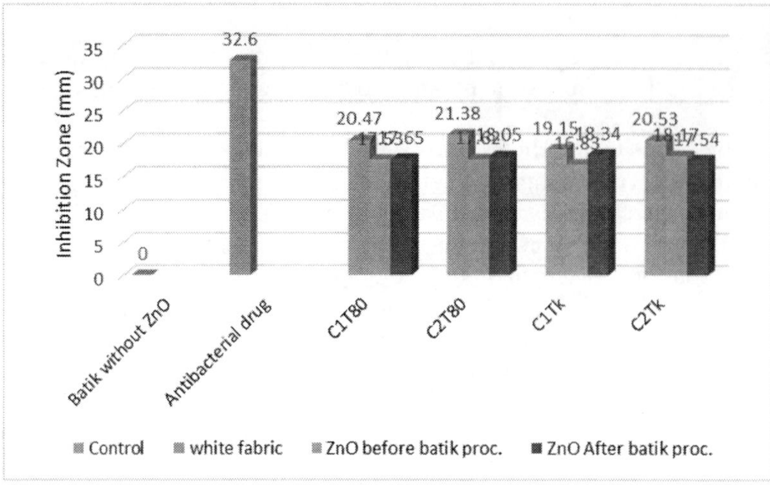

Figure 1. Antibacterial activity against E. coli of batik with ZnO treatment and the control, C1 : ZnO concentration 1%, C2 : ZnO concentration 2%, Tk : temperature of ZnO application at 25°C, T80 : temperature of ZnO application at 80°C

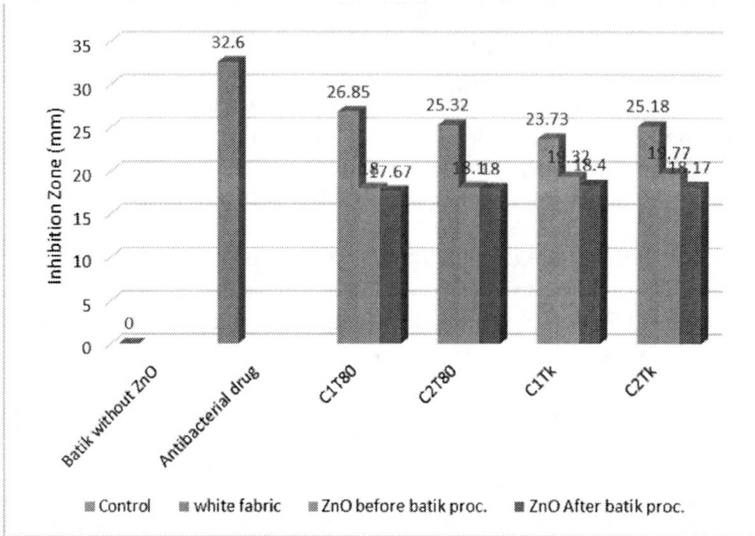

Figure 2. Antibacterial activity against S. aureus of batik with ZnO treatment and the control, C1 : ZnO concentration 1%, C2 : ZnO concentration 2%, Tk : temperature of ZnO application at 25°C, T80 : temperature of ZnO application at 80°C

The same trend of antibacterial activity is shown by the sample fabric against S. aureus as can be seen in Figure 2. The inhibition area of coated fabric against S. aureus bacteria is approximately 60% - 75% of inhibition zone of the antibacterial drug (Chloramphenicol), slightly greater than the inhibition zone against E. coli bacteria. This is in line with the study of some researchers which state that gram-negative E. coli bacteria are more resistant to the antibacterial activity of ZnO compared to gram-positive S. aureus bacteria [4][12][9]. In this case the effect of temperature, concentration of ZnO solution and the order of ZnO treatment are also insignificant. Even though the effect of the ZnO application order is insignificant, there was a slight decrease of antibacterial activity against both E. coli and S. aureus when ZnO was applied before the batik process. This is due to the batik process, mainly dyeing and wax removal process. The dyeing and wax removal process caused the ZnO separated from the fabric thus providing less antibacterial activity. The mechanism of antibacterial activity of ZnO nanoparticles was studied by

Mishra which states that ZnO nanoparticles can generate Reactive Oxygen Species (ROS) that damage bacterial membranes, DNA, and mitochondria, resulting in the death of bacterial cells [13].

The inhibition area due to the effect of ZnO application were photographed and shown in Figure 3 and 4. In Figure 3b, the white cotton cloth applied by ZnO shows the zone of inhibition around the fabric. Whereas the zone of inhibition not seen in Figure 3a, indicates no antibacterial activity on the untreated white fabric. This result corresponds with Rajendran's study on producing antimicrobial cotton fabric using ZnO nanoparticles [11]. Figure 4b also shows the zone of inhibition of the batik fabric applied by ZnO whereas there is no inhibition area around the untreated batik fabric in Figure 4a. It can be seen that the zone of inhibition of treated batik is smaller than the zone of inhibition of white fabric applied by ZnO, indicates that the process of batik causes a decrease in antibacterial activity of the fabric, due to the reduction of ZnO during the batik process.

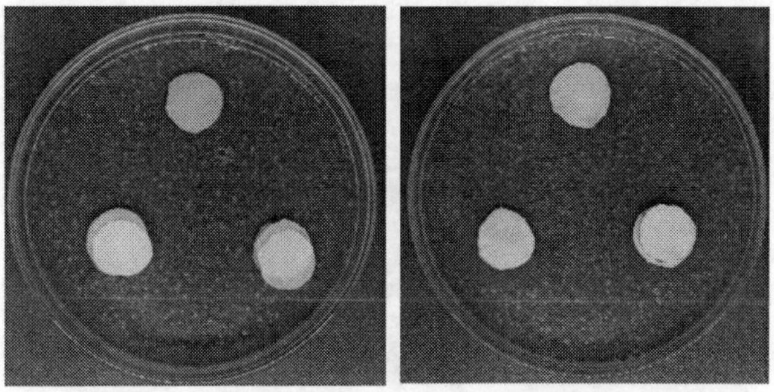

(a) (b)

Figure 3. The inhibition area of (a) untreated white fabrics and (b) white fabrics treated with 2% ZnO at temperature of 25°C

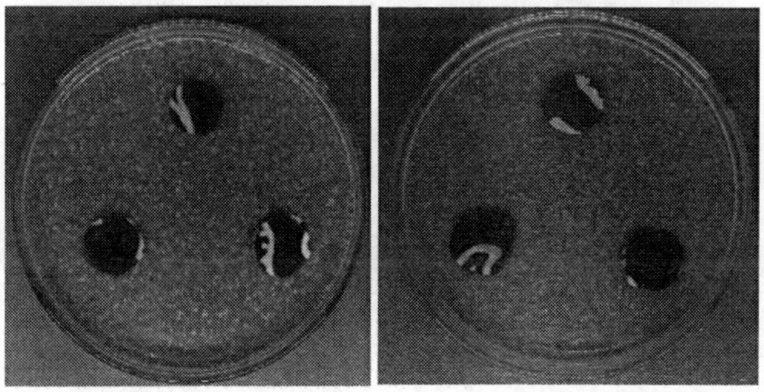

(a) (b)

Figure 4. The inhibition area of (a) untreated batik fabric and (b) batik fabric treated with 2% ZnO at temperature of 25°C

The assessment of surface morphology by Scanning Electron Microscope (SEM) shows the presence of ZnO nanoparticles on the fabric. Figure 5 shows that the ZnO nanoparticles were imparted on the white fabric before being applied with wax (batik process). Figure 6 shows the presence of ZnO nanoparticles on the fabric after the batik process. It can be seen that ZnO in the fabric after the batik process is less than in the white fabric. This shows the ZnO loss during the batik process. But from the assessment of antibacterial activity previously discussed, although there is a reduction of ZnO during the batik process, the batik fabric still exhibit antibacterial activity. This indicates that ZnO nanoparticles is suitable for producing antibacterial batik, and the proof of Dimapilis's study which states that the antibacterial properties of ZnO is excellent because they are heat-resistant and stable in a harsh conditions [4].

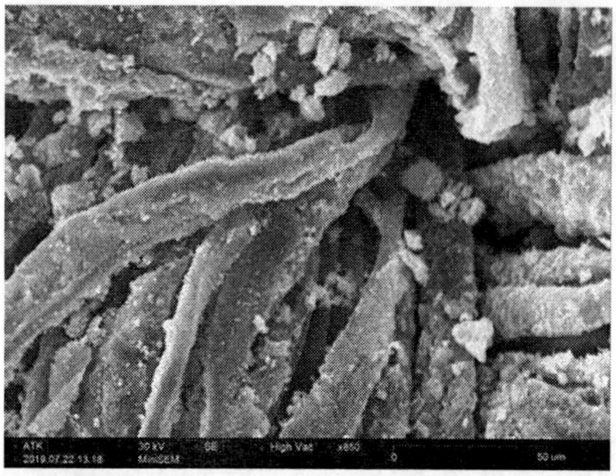

Figure 5. Morphology of ZnO-treated white fabrics before the batik process

Figure 6. Morphology of ZnO-treated batik fabric

The assessment of colour strength and washing fastness of ZnO-treated batik fabric is seen in Table 1. The colour strength is shown by the K/S value that indicates the colour intensity of the dyed fabric, calculated using equation 1. In all ZnO-treated batik fabrics gives a K/S value that are greater than the batik fabric without ZnO application. The higher K/S value indicates a high intensity of the colour. The higher *K/S* values of ZnO-treated batik fabric indicate that the presence of ZnO nanoparticles increases the dye affinity towards the fabric[12]. When ZnO was applied after the batik process, the K/S value is higher, but the colour fastness to washing value is lower than untreated batik fabric. Meanwhile, when ZnO was applied before the batik process, the value of colour fastness to washing is the same as untreated batik fabric and the K/S values are still higher than untreated batik fabric. So to increase the K/S value and to maintain colour fastness to washing, ZnO application should be carried out before the batik process.

Table 1. Color strength and color fastness to washing of ZnO-treated batik fabric dyed with indigosol

Samples	Assesment	
	Colour Strength	Colour fastness to washing

Application of ZnO before batik process:		
C1T80[b]	1.10	4-5
C2T80[b]	1.03	4-5
C1Tk[b]	1.05	4-5
C2Tk[b]	0.86	4-5
Application of ZnO after batik process:		
C1T80[a]	4.09	4
C2T80[a]	3.50	4
C1Tk[a]	3.20	4
C2Tk[a]	4.30	4
Synthetic dye batik without ZnO	0.51	4-5
White fabric	0	

C1 : ZnO concentration 1%
C2 : ZnO concentration 2%
Tk : temperature of ZnO application at 25°C
T80 : temperature of ZnO application at 80°C
[a] : application of ZnO after batik process
[b] : application of ZnO before batik process

4. Conclusions

The ZnO application can provide antibacterial activity of synthetic dye batik against E. coli and S. aureus. The antibacterial activity of treated batik fabric against S. aureus is greater than that against E.coli. The order of ZnO treatment (before and after the batik process), the difference of ZnO concentration and temperature did not significantly affect the antibacterial activity of batik fabric against E. coli and S. aureus bacteria. However, there was a slight decrease of antibacterial activity against both E. coli and S. aureus when ZnO was applied before the batik process. This is due to the batik process, mainly dyeing and wax removal process that caused the ZnO separated from the fabric thus providing less antibacterial activity. The ZnO application also enhance color strength that indicate the presence of ZnO nanoparticles increases the dye affinity towards the fabrics. To maintain the value of color fastness to washing, the ZnO application should be carried out before the batik process.

5. Acknowledgment

The author greatly acknowledge the financial support from INSINAS, Ministry of Research, technology and Higher Education Republic of Indonesia.

References

[1] M. Fouda, "Antibacterial Modification of Textiles using nanotechnology," 2012.
[2] S. Islam, M. Shabbir, and F. Mohammad, "Insights into the Functional Finishing of Textile Materials Using Nanotechnology," *Text. Cloth. Sustain.*, pp. 97–115, 2017.
[3] O. E. Novarini and T. Wahyudi, "Sintesis Nanopartikel Seng Oksida (ZnO) Menggunakan Surfaktan Sebagai Stabilisator dan Aplikasinya pada Pembautan Tekstil Antibakteri," *Arena Tekst.*, vol. 26, no. 2, pp. 81–87, 2011.
[4] E. A. Dimapilis, C. Hsu, R. Marie, O. Mendoza, and M. Lu, "Zinc oxide nanoparticles for

water disinfection," *Sustain. Environ. Res.*, vol. 28, no. 2, pp. 47–56, 2018.
[5] A. Sirelkhatim, S. Mahmud, and A. Seeni, "Review on Zinc Oxide Nanoparticles : Antibacterial Activity and Toxicity Mechanism," *Nano-Micro Lett.*, 2015.
[6] BSN, *Batik - Istilah dan Definisi*. Jakarta, Indonesia: Badan Standarisasi Nasional, 2014.
[7] A. K. Yetisen, H. Qu, A. Manbachi, H. Butt, and M. R. Dokmeci, "Nanotechnology in Textiles," 2016.
[8] F. A. Mohamed, H. M. Ibrahim, E. A. El-kharadly, and E. A. El-alfy, "Improving Dye ability and antimicrobial properties of Cotton Fabric," *J. Appl. Pharm. Sci.*, vol. 6, no. 02, pp. 119–123, 2016.
[9] S. Ristiani, "Pengembangan Teknik Smock pada Batik Untuk Meningkatkan Daya Saing Produk Fesyen," Yogyakarta, 2017.
[10] M. A. Tavanaie, "Production of cationic dyeable poly (ethylene terephthalate) fibers via nanotechnology," no. May, pp. 1–9, 2017.
[11] R. Rajendran, C. Balakumar, H. Mohammed Ahammed, S. Jayakumar, and K. Vaideki, "Use of zinc oxide nano particles for production of antimicrobial textiles," *Int. J. Eng. Sci. Tecnol.*, vol. 2, no. 1, pp. 202–208, 2010.
[12] P. Kanade and B. Patel, "Copper nano - sol loaded woven fabrics : structure and color characterization," *Fash. Text.*, 2017.

IOP Conf. Series: Materials Science and Engineering **722** (2020) 012030 doi:10.1088/1757-899X/722/1/012030

Development of Chitosan/TiO2 Nanocomposite for Multifunctional Sunscreen Application

J Petrick[1], M Ibadurrohman[1], and Slamet[1,2]

[1] Department of Chemical Engineering, Faculty of Engineering, Universitas Indonesia, Depok 16424, Indonesia.

Corresponding Author : slamet@che.ui.ac.id

Abstract. Several problems that often occur on the skin are acne and sunburn due to UV exposure. Acne is caused by the activity of bacteria dwelling on the surface of skin. This study has successfully synthesized several multifunctional cream based on chitosan loading in the chitosan/TiO2 nanocomposite. Multifunctional sunscreen defined by anti-bacterial and sunscreens performance with photocatalytic technology. Multifunctional sunscreen is made by Chitosan/TiO2 nanocomposite as active ingredient. Nanocomposite which has been synthesized is characterized by FT-IR, UV-Vis DRS, and SEM-EDX. To make Multifunctional sunscreen, the nanocomposite chitosan/TiO2 as active ingredient is mixed with water and carboxymethyl cellulose (CMC) as natural emulsifier. The result of testing the disinfection of bacteria *Escheria coli* by nanocomposite chitosan/TiO2 showed the bacteria can be eliminated up to 99.7% within 2 hours. As sunscreens, creams have the ability to moderate protection under UV rays with SPF values from 21.30 to 22.77.

1. Introduction

Indonesia as one of the tropical country in southeast asia, has a quite high sun exposure around 1800-2000 kWh/m3 per year. Continous exposure of UV light contained in sunrays can do serious damage to the skin, or people might call it sunburn. On critical condition is that extensive exposure of ultraviolet radiation can cause skin cancer.

Sunscreen generally has two types based on the protection properties offered, namely physical sunscreen and chemical sunscreen. Physical sunscreen functions is to drive UVA and UVB rays out of the skin by reflecting it on the surface of the skin. Physical sunscreen protection mechanism can work immediately after being applied on the skin and to obtain maximum protection, the application of physical sunscreen must be evenly distributed to all parts of the body but often leaves white spots after use. This encourages the creation of new sunscreen performance that leaves no white files but is able to work as a sunscreen. Chemical sunscreen can answer this problem, chemical sunscreens work by filtering and absorbing UV light and turning it into heat energy released through the skin. This causes a chemical sunscreen to be applied 20 minutes before being in the sun, because the content takes time to absorb into the skin. In the chemical sunscreen the utilization of the content of titanium oxide / titania (TiO2) and zinc oxide (ZnO2) as UV filters is very often found in the community. On the other hand, titania and zinc oxide can release ROS (Reactive Oxygen Species) in the form of radical hydroxyl compound, into the skin so that it can potentially cause irritation, itching, and redness.

Chitosan is a compound synthezied by de-acetylation of shrimp skin and able to reduce ROS compounds to Glucosamine and Chito-oligosaccharides which do not cause irritation, itching and redness in the skin (Chang, 2001). Chitosan also has anti-bacterial properties (Goy R.C, 2009). Chitosan itself comes from the skin of crustacean living things and is often found in the fishing industry. The export value of the fisheries industry in the shrimp sector increased by 34% from 2008 (Fronthea Swastawati, 2008) made an increase in shrimp waste because the byproducts in the form of head, skin, and tail of shrimp were around 30% -50% of initial weight. It can be concluded that shrimp skin has a lot of availability and its use in Indonesia as a processed material from waste is still minimal. Therefore, the key objective of this study is by making a composite of titania and chitosan to multifunctional sunscreens that do not cause irritation, itching, and redness when applied to the skin while being anti-bacterial.

2. Experimental

2.1. Materials
Aquades, Chitosan Analytical Grade (De-acetylation degree >75%), Acetic Acid (CH3COOH), Nitric Acid (HNO3), TiO2 Evonik P25, and Methylene Blue..

2.2. Methods
Titania / Chitosan composite synthesis using Wet Impregnation Method. The first step is by making a 5% acetic acid solution of 10 mL, then adding HNO_3 solution until it reaches pH 3. Chitosan compounds are added to the mixture as much as loading variations (0%; 0.5%; 1%; 5%; and 10%). Next adding Titania to the chitosan solution and stirred manually, after that sonication for 30 minute reqired. The solution then separated using centrifuge and washed with distilled water until pH 7 was obtained from Titania / Chitosan solution. Furthermore, the composite was dried using a magnetic strirrer for 2 hours at 800C. The dried composites were then crushed using mortar to obtain a fine catalyst powder.

3. Result and Discussion

3.1. Characteristic of Nanocomposite
The FTIR spectrum of several Chitosan/TiO2 nanocomposite (% Chitosan based on weight) is shown in Figure 1. To see whether a bond between chitosan and TiO2 has occurred, functional group peak that shows the presence of chitosan is needed, according to Shahida (2016) the presence of chitosan can be seen in the 1500 nm wave number peak which indicates the presence of NH- bonds in the chitosan chain, then at wave number 3400 nm which indicates the presence of an -OH bond. The presence of a peak at wave number 1177 nm shows the stretching and bending of the CH3 group, Ti-O-C bond, or Ti-OH (Kavitha, et al. 2013).

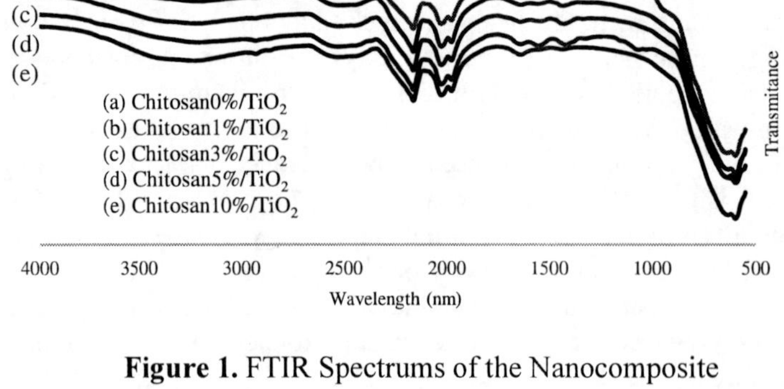

Figure 1. FTIR Spectrums of the Nanocomposite

For 1% chitosan loading, 3% chitosan, 5% chitosan, and 10% chitosan have similar and identical wave number ranges found in the four samples, namely in the wavelength region 1637 - 1440 nm and 3211 - 3227 nm. For wave number 3211 - 3227 nm shows the vibration of the hydroxyl group (-OH) which shows the occurrence of bonding from the chitosan hydrogen group to TiO_2 (Fajriati, 2014). The spectrum of wave numbers 1637 - 1440 nm also shows the presence of H+ atomic bonds from amino groups (NH2) from chitosan structures that make hydrogen bonds to titania groups (Fajriati, 2014). Therefore, it can be concluded that the possible bond between chitosan and titania is a hydrogen bond.

The results in table. 1 show that the response of Chitosan/TiO2 nanocomposite is better to receive photon energy compared to Chitosan 0%/TiO2. The reduction in gap energy by chitosan showed that chitosan/TiO2 nanocomposite would be more easily activated by photon energy than normal TiO2. After acquiring the band gap energy from table. 1, Kubelka-Munk analysis can be used to determine the light wavelength needed by the nanocomposite to be activated. The result can be seen in table. 1, the energy gap values of chitosan/TiO2 nanocomposite are decreasing as the addition of chitosan loading which initially, chitosan0%/TiO2 has a gap energy value 3,17 to 3.08 on chitosan10%/TiO2 nanocomposite. Even though the reduction in gap energy is not too significant, still when applying in daily activities to chitosan/TiO2 sunscreen, it can help activate TiO2 under sun exposure. This can occur because chitosan / TiO2 nanocomposite can be responsive with wavelengths up to 402.53 cm-1. Chitosan here is acting as if it was dopant for TiO2, when metal dopant ussually attracting the electrons which produced when photocatalysis occurred, chitosan is reacting with hole (h+) produced by photocatalysis so that the recombination for electrons and hole can be minimized, so the band gap of TiO2 are reduced[7].

Table 1. Band Gap Energy and Activation Wavelength

Sample	Band Gap Energy (eV)	Wavelength (nm)
Chitosan0%/TiO2	3.17	391.10
Chitosan1%/TiO2	3.13	396.10
Chitosan3%/TiO2	3.11	398.70
Chitosan5%/TiO2	3.09	401.53
Chitosan10%/TiO2	3.08	402.53

Characterization of SEM (Scanning Electron Microscopy) was carried out to determine the morphology and structure. The nanocomposite characterized by SEM / EDX was loading Chitosan3%/TiO2. The results of which can be seen in Fig. 3. In Fig. 3 (a) it can be seen that one of the matrix formations of chitosan / TiO2 nanocomposite. White powder covering the aggregate surface is TiO2 nanoparticles, the aggregate is estimated to be chitosan. This is reinforced by the next image (b) with a 5 times magnification at the same point, where the EDX results show the part that is not covered in white powder (B), the C atom is detected more than the Ti atom while the part covered by white powder (A) shows an increase in Ti atoms compared to C atoms. As seen in Table. 2 for comparison.

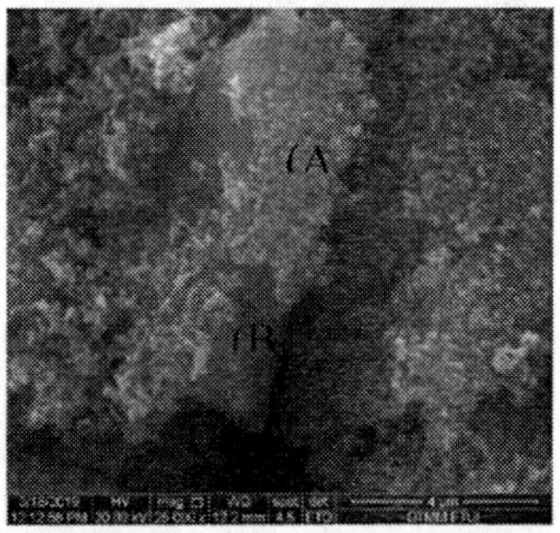

Figure 2. Morphology of Chitosan 3%/TiO2 at 25.000x

IOP Conf. Series: Materials Science and Engineering **722** (2020) 012030 doi:10.1088/1757-899X/722/1/012030

Table 2. EDX Result on Sample Chitosan 3%

Elements	(A)(%Wt)	(B)(%Wt)
Carbon (C)	6.06	13.74
Oxygen (O)	29.38	35.16
Titanium (Ti)	64.58	51.10

3.2. Performance of Nanocomposite

3.2.1. Decolorization of Methylene Blue. Decolorization of MB testing was carried out to see the ability of Chitosan/TiO2 nanocomposites to reduce the activity of ·OH radical compounds to degrade colors as a representation of human skin which can be irritated by radical compounds · OH. This test uses five samples, namely chitosan loading 0%, loading 1%, loading 3% loading 5%, and loading 10%. Methylene blue is used as a colored organic compound, the test is carried out using a photoreactor for one hours per sample by taking data in 0, 15, 30, and 60 minutes. Percentage of color degradation is then seen through the value of methylene absorbance resulting from degradation compared to the absorbance early methylene blue.

On Figure 5(a) and Figure 5(b) it can be seen that the degradation of methylene blue is best done by nanocomposites with 0% chitosan loading or 93.5% TiO2 within 1 hour, the mechanism that occurs in this event is photocatalysis, where when the energy of a photon excites TiO2 produces an electron pair and hole. Holes that are positively charged oxidize hydroxyl compounds or water compounds, then produce OH (· OH) radicals. This hydroxyl compound can later degrade methylene blue into other compounds. But according to Zuo (2014) methylene blue is not directly degraded to CO2 and H2O but through a series of reactions, where when methylene blue encounters electrons from TiO2 it will become methylene blue radicals which can be degraded by · OH radicals thus losing methylene blue constituent elements such as C, N, S, to become CO2 and H2O. The loading variation of 1% to 10% has decreased due to the presence of chitosan compounds which are able to reduce OH radical formation through the mechanism of binding of hydrogen peroxide compounds to glucosamine compounds and chito-ogliosaccarides (Chang, et al., 2001). Decreasing the percentage of degradation of methylene blue along with increasing chitosan loading proves that the more chitosan compounds, OH radical activity can be reduced. The mechanism of suppressing OH radical activity as follows:

$$(GlcN)_m - (GlcN)_n + OH \bullet \rightarrow (GlcN \bullet)_m - (GlcN)_n + H_2O \quad (1)$$

$$(GlcN \bullet)_m - (GlcN)_n + H_2O \rightarrow (GlcN) + (GlcN) \quad (2)$$

Table 3. MB Decolorization percentage at 1 hours of each nanocomposite

Chitosan/TiO2	%A/A0 when 60 Minutes
0% (TiO2)	93,5%
1%	92,6%
3%	91,1%
5%	84,6%
10%	77,0%

3.2.2. Antibacterial Activity. In Figure 6 it can be seen that samples with chitosan loading have the ability to disinfect bacteria more effectively than samples that are not given chitosan (0%). This proves that the addition of chitosan is able to improve the disinfection ability of TiO2 bacteria. This disinfection process can occur due to two things, the first by the photocatalytic activity of TiO2 itself and the antibacterial properties possessed by chitosan. The disinfection process through photocatalytic activity occurs because TiO2 is activated by photon energy in this case in the form of UV light and then photocatalytic activity results in · OH radicals or can also be referred to as ROS (Reactive Oxygen Species) compounds, then ROS compounds react with bacterial cell membranes so that causing bacterial cells to leak causing the molecular molecules inside the cell to come out, making the cell unable to move again and then die. This degradation process continues until the cell nucleus makes the bacteria break down completely into CO2 and H2O. (Sangchay et al., 2013) (Tsai et al., 2012).

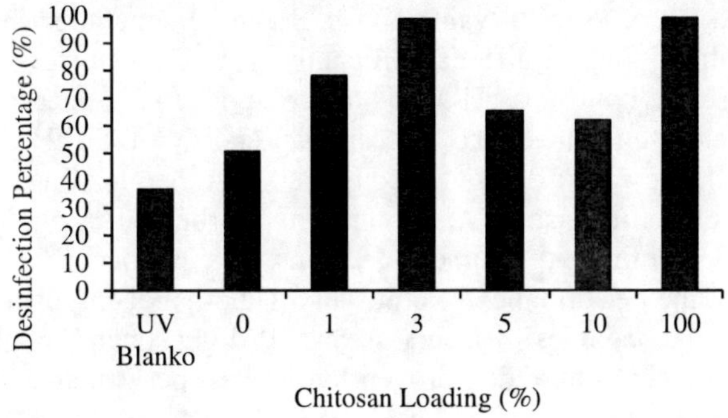

Figure 3. Morphology of Chitosan 3%/TiO2 at 25.000x

Chitosan also helps the bacterial disinfection process, besides helping to facilitate the activation of TiO2 by UV light, chitosan also has antibacterial properties that are able to simultaneously disinfect bacteria along with photocatalytic activity. According to Goy, R.C (2009) the mechanism of the antibacterial chitosan involves the interaction between chitosan molecules that are positively charged with a negatively charged bacterial membrane. The interaction makes two results. The internal osmotic imbalance of bacterial cells is due to changes in the permeability properties of bacterial cell membranes thereby inhibiting bacterial cell growth. Leakage of electrolyte fluid contained in bacterial cells due to peptidoglycan hydrolysis in bacterial cell membranes. This model has been evaluated by Raafat et, al (2008). using an electron microscope to see the ultrastructural changes of bacterial cells when they are exposed to positively charged chitosan. He found the presence of chitosan molecules still attached to the membrane of dead bacterial cells. Therefore, with minimal light or no light, chitosan / TiO2 nanocomposites can still disinfect bacteria by the presence of chitosan. It can be seen also from the results of 100% chitosan loading without TiO2 proving that chitosan indeed has antibacterial properties.

3.2.3. Sun Protection Factor Performance. Sun Protection Factor or SPF testing is used as a parameter to determine how well synthesized sunscreens can work in absorbing UV light. The higher the SPF value, the more effective it is to protect the skin from the adverse effects of UV light (Dutra et al., 2004). The SPF value can be interpreted as the amount of UV energy needed to achieve. MED (Minimal Erythemal Dose) on skin given sunscreen compared to the amount of UV energy needed to reach MED on sun-protected skin (FDA, 2001). In this test, the method used is the Mansur method, where using the value of absorbance of sunscreen between wavelength ranges of 290 nm - 320 nm (UV-B) at intervals of 5 nm then processed into SPF values using the existing equations. To use the Mansur method, the value of Correction Factor (CF) is obtained from commercial sunscreens that have known SPF values. In this study, the correction factor value used refers to research from Taqqiyah, Shofi (2012) that has been predetermined to facilitate calculations.

Table 4. SPF Value on Each Nanocomposite

Chitosan/TiO2	SPF Value
0%	21.30
1%	21.66
3%	21.77
5%	21.42
10%	21.46

Based on the calculation of the SPF value above, it can be concluded that the addition of chitosan loading does not make the SPF value differ significantly. The SPF value generated is consistently in the range 21,3-21,4. In this case chitosan / TiO2 ¬ nanocomposite is in accordance with previous results where the addition of chitosan can reduce the bandgap energy of TiO2 when viewed from Figure 7 chitosan0%/TiO2 has the lowest SPF value compared to other loading variations, this proves the addition of chitosan is able to increase TiO2 response in absorbing UV / photon light. Chitosan3%/TiO2 has the highest SPF value compared to other variations, this can occur Chitosan3%/TiO2 is the most effective composition in the performance of chitosan / TiO2 nanocomposite because the least shading effect on chitosan / TiO2 nanocomposite occurs, so that UV light / energy of photons entering the surface of TiO2 is not blocked by chitosan and is proven by the best performance of bacterial disinfection in these nanocomposites.

4. Summary

From the experimental data we can conclude that: Wet Impregnation Method is proven to be one of the methodology of synthezising Chitosan/TiO2 nanocomposite. In FT-IR result emphasize that under the certain range of wave number the most significant bonds between TiO2 and Chitosan are Amino group attached from Chitosan and O from TiO2[5]. Under UV-Vis spectroscopy test, implied that Chitosan also reduce the band gap energy from TiO2 making TiO2 more active when irradiated with certain type of light[7]. Antibacterial activity also generated by Chitosan/TiO2 nanocomposite until the loading of 3% chitosan in the composite. SPF value which generated are classified in high protection of sunlight class by Indonesian Government (SNI 164-399-1996). Based on those two performance test the most optimal chitosan loading on TiO2 is loading with a concentration of 3% chitosan / TiO2.

References

[1] Ezzedine K, Guinot C, Mauger E, Pistone T, Rafii N, Receveur M et al. Expatriates in High–UV Index and Tropical Countries: Sun Exposure and Protection Behavior in 9,416 French Adults. Journal of Travel Medicine. 2007;14(2):85-91.

[2] Harper J. An update on the pathogenesis and management of acne vulgaris. Journal of the American Academy of Dermatology. 2004;51(1):36-38.

[3] Goy R, Britto D, Assis O. A review of the antimicrobial activity of chitosan. Polímeros. 2009;19(3):241-247.

[4] Yasmeen S, Kabiraz M, Saha B, Qadir M, Gafur M, Masum S. Chromium (VI) Ions Removal from Tannery Effluent using Chitosan-Microcrystalline Cellulose Composite as Adsorbent. International Research Journal of Pure and Applied Chemistry. 2016;10(4):1-14.

[5] Fajriati I, Mudasir M, Wahyuni E. Photocatalytic Decolorization Study of Methyl Orange by TiO2–Chitosan Nanocomposites. Indonesian Journal of Chemistry. 2014;14(3):209-218.

[6] Gurkan Y, Kasapbasi E, Cinar Z. Enhanced solar photocatalytic activity of TiO2 by selenium(IV) ion-doping: Characterization and DFT modeling of the surface. Chemical Engineering Journal. 2013;214:34-44.

[7] Sangchay W, Lek S, Kooptarnond K. The Photocatalytic and Antibacterial Activity of Cu-Doped TiO2 Thin Films. Walailak Journal of Science and Technology. 2013;10. 19-27. 10.2004/wjst.v10i1.530.

[8] Tsai C, Kuo T, Hsi H. Fabrication of Al-Doped TiO2Visible-Light Photocatalyst for Low-

Concentration Mercury Removal. International Journal of Photoenergy. 2012;2012:1-8..
[9] Raafat D, von Bargen K, Haas A, Sahl H. Insights into the Mode of Action of Chitosan as an Antibacterial Compound. Applied and Environmental Microbiology. 2008;74(12):3764-3773.
[10] Chang K, Tai M, Cheng F. Kinetics and Products of the Degradation of Chitosan by Hydrogen Peroxide. Journal of Agricultural and Food Chemistry. 2001;49(10):4845-4851.

IOP Conf. Series: Materials Science and Engineering **722** (2020) 012031 doi:10.1088/1757-899X/722/1/012031

Design of SCADA for Protection System of Uncoiled DC Motor's Temperature using PLC ABB AC 800PEC Based on Wonderware Intouch

J T Putra, A Diantoro, U Yusmaniar Oktiawati, A Surriani, A B Pradana, L Subekti

Department of Electrical and Informatics Engineering, Universitas Gadjah Mada, Yogyakarta, Indonesia

E-mail: jimmytrioputra@ugm.ac.id

Abstract. Continuous Tandem Cold Mill (CTCM) operates many types of DC motor for various processes. The continuity of the DC motor is very important to ensure the automation process are working perfectly and safe. One of the abnormal condition when operating DC motor is overheating at Field winding and Field bearing. Therefore, this research is needed to avoid further damage in the DC motor by designing and implementing protection system of temperature based on PLC (Programmable Logic Controller) ABB AC 800PEC and PT100 sensor. SCADA (Supervisory Control and Data Acquisition) system was implemented so the operator can supervise the motors. When temperature is increasing and higher than the allowed maximum temperature, the system should generate Fast Stop signal to stop every process that have been running at entry section CTCM. The protection system was implemented successfully and generated Fast Stop signal when the Field winding temperature was higher than 100-degree Celsius or when the Field bearing temperature was higher than 75-degree Celsius.

1. Introduction

PT. Krakatau Steel is the largest integrated steel producer in Indonesia and produces various steel product. As a state-controlled company engaged in steel making and rolling mill industry, PT. Krakatau Steel has 7 production facilities, and one of them called CRM (Cold Rolling Mill) that produces cold rolled coil.

CTCM (Continuous Tandem Cold Mill) is one of seven production lines in CRM plant which reduce the thickness of strip. The entire process in CTCM are consist of many mechanical equipment driven by DC motor such as uncoiled motor, tension reel, bridle and main drives. This research focused on uncoiled DC motor. DC motor has several heating issues when it is operated. Increasing temperature of DC motor reduces its lifespan, insulation life, physical damage and eventually causes motor failure. Therefore, it is important to develop temperature monitoring and protection system for DC motor as a preventive action to maintain the continuity of production line and minimize the risk.

The temperature of uncoiled DC motor obtained by RTD (Resistance Temperature Detectors) sensor placed in bearing and inductor. When temperature is increasing and higher than the allowed maximum temperature, it can generate Fast Stop signal to stop every process that have been running at entry section CTCM. Fast Stop Signal will be forwarded to DC Drive ABB DCS 800 through AA11 controller module.

The type of RTD used in this research is PT100, this sensor has a high accuracy and linearity. This sensor has been tested to protect temperature of induction motor. The system was built without SCADA and used 2 wire PT100 and shows that PT100 has linear characteristic towards temperature changes and has a high sensitivity [18]. Other research about temperature monitoring also performed by Ponjavi [21]. Meanwhile, previous research to monitor temperature also used PT100 [22].

SCADA is an industrial control system that help operator and engineer to monitoring and control the plant [15, 20, 23,24]. SCADA is implemented as protection system of temperature controlled with PLC ABB AC 800PEC. PLC is control module consist of Central Processing Unit (CPU), I/O module and Programming Device. PLC has become industrial standard for control system because of its reliability [13]. PLC could be used to monitor and control line as reported by Kaur and Bansal [23]. HMI is designed with integrated development environment called Wonderware Intouch and data acquisition system is developed with IbaFop4 module.

In the previous research, PLC ABB AC800 PEC was used to test the performance of ABB OPC server compared to other communication protocol [17]. According to that research, the time required to transmit data through OPC is not more than 45ms with 30% CPU load. In the other hand, the controller of ABB x800 system have 25 μS cycle time that is fast enough to handle complex automation process [1]. It is obviously clear that this protection system needs faster controller to make sure Fast Stop signal is generated as immediate as possible. This research used newest controller from ABB programmed by Function Block Diagram that is easier to maintain than any other PLC programming language such as Ladder and Structured Text [6].

Controller module AC 800PEC is also can be programmed by MATLAB as mentioned in previous research [16]. This system consists of an I/O module and single Field Communication Interface (FCI) to simulate PLC with soft controller.

Wonderware Intouch has commonly used by engineer in industrial field such as packaging automation system [19]. In this case Wonderware Intouch is used to develop HMI for DC motor temperature system. Wonderware Intouch has better graphic library compared to other SCADA software such as Citect SCADA and WinCC [19]. This advantages make Wonderware Intouch become a standard SCADA software for industrial purposes [19].

The design of protection system on DC Motor based SCADA has been carried out on the research. In addition, HMI based in Wonderware Intouch and ABB OPC was also made to facilitate the operator in monitoring the temperature that occurs in DC Motor. Analysis of the resistance characteristics on the 3 wire PT100 sensor to temperature changes is also described.

2. Methodology

The whole SCADA system has been designed using PLC ABB AC800 PEC as a controller and I/O module as a Remote Terminal Unit (RTU). Temperature data is obtained from PT100 sensor, a RTD sensor that commonly become industrial standard for measuring temperature. Analogue value from PT100 will be converted to digital using internal ADC in Analogue Input Module 830. Digital signal from AI830 will be transmitted to Controller through Profibus. In the controller data will be processed due to program logic which has been made using Function Block Diagram. HMI in this case is necessary to display value of temperature in real time and control several configuration regarding temperature protection system. IbaDAQ is a data acquisition module, it can be used to saved temperature data with high sampling rate, stored in hard drive and make it accessible for assessment purpose or checking if there is a problem in the future. Function block of the system shown in Figure 1.

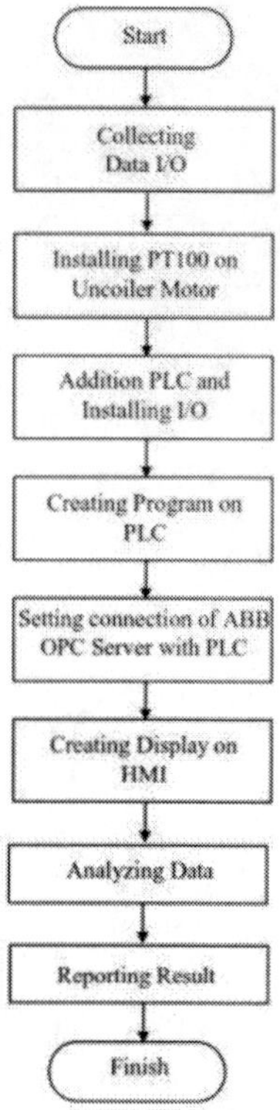

Figure 1. Research methodology

PT100 sensor refers to Wheatstone bridge connected to RTD, from equation (1) shows that wire A, B and C need to be the same length to make Wheatstone bridge balanced.

$$\frac{R_1}{R_2} = \frac{R_A + R_{RTD}}{R_B + R_3} \tag{1}$$

When each branches have the same potential difference, the output will be zero and ADC convert it to digital value equal to 0°C.

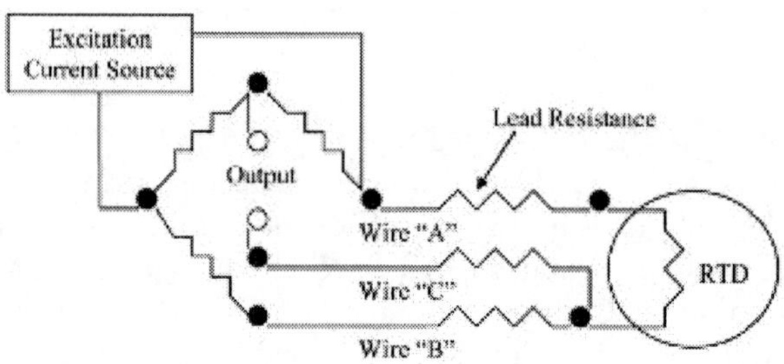

Figure 2. Wheatstone bridge connected with RTD

The resistance of PT100 obtained from equation (2).

$$R = 0.385T + 100 \tag{2}$$

where R is the resistance of PT100, T is temperature measured by PT100 and 100 is a constant that actually represent the resistance when Wheatstone bridge is in balance condition. On the other word, when temperature is 0°C then the R as output of Wheatstone bridge is equal to 100 Ω.

Meanwhile, the power dissipated in the sensor causes its temperature to rise above its ambient temperature by equation (3).

$$\Delta T = (I^2 R_{RTD})\theta \tag{3}$$

Where ΔT = the change in temperature due to internal power dissipation; I = excitation current; RRTD = the value of the RTD in ohms; and θ = the self-heating effect in °C/mW.

3. Design of PLC

ABB AC 800PEC is a PLC controller module from ABB designed to handle complex automation process. It has dual core CPU, 1 GB SDDR3 RAM and communication port. This PLC is modular which has been separated into piece of modules depend on the functionality [4]. This module can be added or reduced as needed, so this make AC 800x system has high flexibility to make automation system run efficiently.

This system consists of three controller module, CI854 as a Master FCI (Field Communication Interface), CI801 as a Slave FCI, AI830 as an Analogue Input Module designed to read RTD sensors and TU 810 (Termination Unit) as a connector [6]. AI830 is placed in cabinet RA05, CI801 can handle up to 12 I/O device [7]. Data transmitted from I/O module to communication interface CI801 through module bus [3]. CI801 is connected with other slave FCI 801 through PROFIBUS cable in series [5]. The last FCI will be connected to Master FCI CI854 through FROFIBUS cable in series and connected to other CI854 as well. The last CI854 will be connected to controller module through CEXBUS cable to process the data from sensor. Data will be calculated and the result will be transmitted from AA15 to AA12 and AA12 to AA11 through fibre optic cable. The system architecture shown in Figure 3.

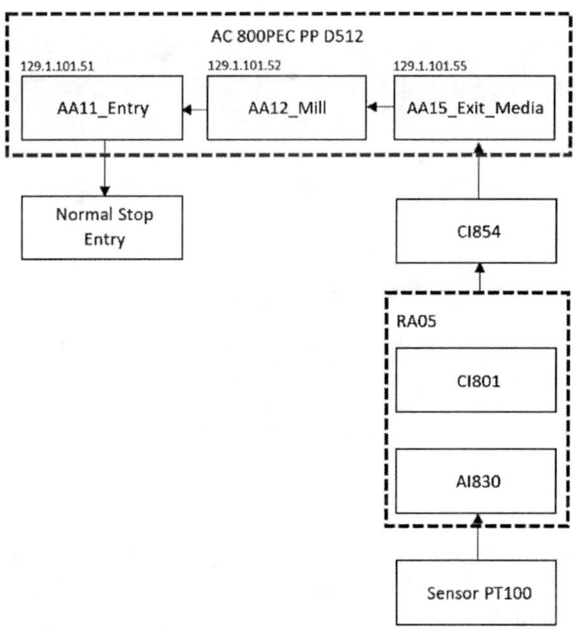

Figure 3. PLC ABB AC 800x system architecture

IOP Conf. Series: Materials Science and Engineering **722** (2020) 012031 doi:10.1088/1757-899X/722/1/012031

4. Design of PLC Programme

PLC Program is developed using FBD (function block diagram) of integrated development system provided by ABB called Compact Control Builder. Program in Compact Control Builder separated in control module that will be connected to task-time [2]. This system consists of four main program as following details.

4.1. Main Program

The main purpose of this program is to determine motor status by checking the temperature sensor measurement. Fault signal can be classified into four different conditions, each condition generate depend on minimum bearing and inductor temperature. Bearing temperature alarm will be generated if bearing temperature is higher than 60°C. Bearing temperature trip will be generated if bearing temperature is higher than 75°C. Inductor temperature alarm will be generated if field winding temperature is higher than 90°C. Inductor temperature alarm will be generated if field winding temperature is higher than 100°C. Flowchart of main program can be seen on Figure 4.

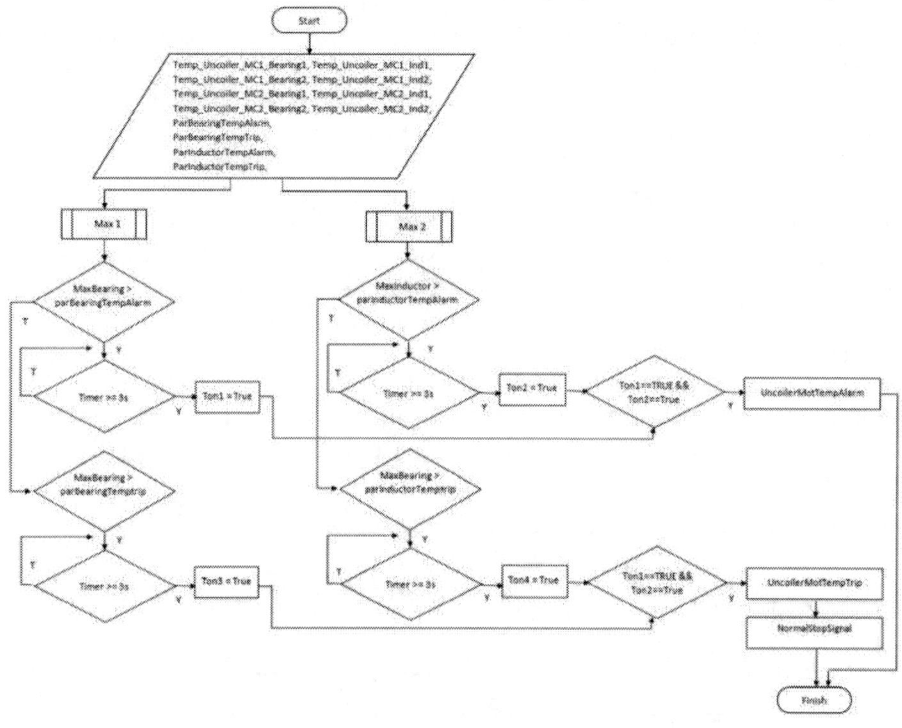

Figure 4. Main Program Flowchart

4.2. Selector Program

Selector program is made to implement force I/O when sensor gained false sensing that might be different with actual temperature. False sensing occurred when PT100 cannot measure temperature of DC motor accurately. This condition implies that fast stop entry executed for wrong reason. Forcing the I/O is necessary to avoid this condition by passing the measurement signal through selector program.

4.3. HMI Program

Data measurement from PT100 sensor will be transmitted to HMI server using OPC. One of the common problem in industrial field is the diversity of communication protocol developed by the manufacturer. This protocol is used to communicate as read or write from PLC to computer or

IOP Conf. Series: Materials Science and Engineering **722** (2020) 012031 doi:10.1088/1757-899X/722/1/012031

otherwise [12]. Diversity of protocol communication causes decrease capability some of PLC with the system. Therefore, a universal protocol called OPC (Ole for Process Control) have been made to accommodate all of PLC, make it able to communicate between PLC to PLC or PLC to computer despite there are from different manufacturer [8]. OPC was developed by a foundation and distributed as a open source project [9]. OPC is also implement COM and DCOM technology which allows transfer data between computer to computer using TCP/IP architecture [11].

4.4. Data Acquisition Program

Data acquisition is a part of SCADA that have been playing important role to make system can be monitored and integrated efficiently. Data measurement from PT100 sensor will be transmitted to Iba Data Acquisition server using FPGA library included in Compact Control Builder. Data type AA15ComIbaReal is declared on FPGA library to receive real value of temperature signal and transmit them to the server. This data obtained simultaneously and can be monitored in real time.

4.5. Data Transfer Program

Fast stop signal generated from main program in AA15 controller module need to be transmitted to AA11. AA11 controller module is designed to handle every mechanical and electrical equipment for all automation sequence in entry section. Uncoiled DC motor is located in entry section, therefore fast stop signal will be executed in AA11 controller module, but unfortunately there is no direct fibre optic communication module from AA15 to AA11 and that why it must be transmitted to AA12 first.

Data transfer from one controller to another controller require two function block called BctoDint and DintToBc. BcToDint placed in transmitted controller that will convert binary coded data to double integer and same process will be implemented in receiver controller. DintToBc will convert double integer data to binary coded, thus the data will be returned to the original bit.

5. Design of HMI

Human Machine Interface (HMI) is used to monitor control process of the plant and warning system for operator when fault occurred eventually. HMI layout and program has been created with Wonderware InTouch. Designing the HMI is conducted with several steps, started with identifying process, listing parameter tag name, determining the logic flowchart and designing the layout.

This system is consisting of three window application, there are main window showed text based temperature data, main window 1 showed real time trend based temperature data and setting windows showed slider to set minimum temperature fault. The layout can be seen respectively on Figure 5.

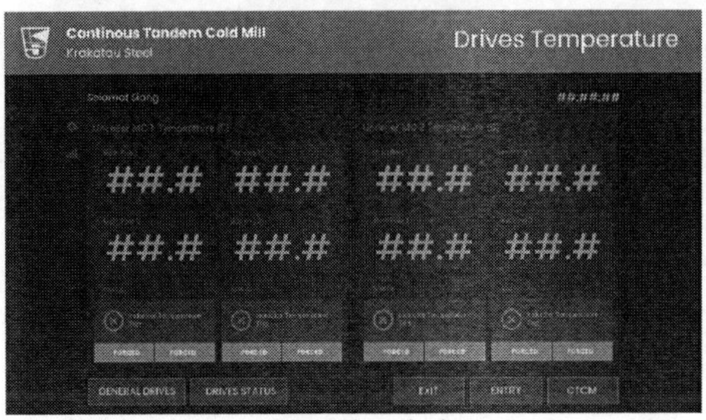

(a) Main window

IOP Conf. Series: Materials Science and Engineering **722** (2020) 012031 doi:10.1088/1757-899X/722/1/012031

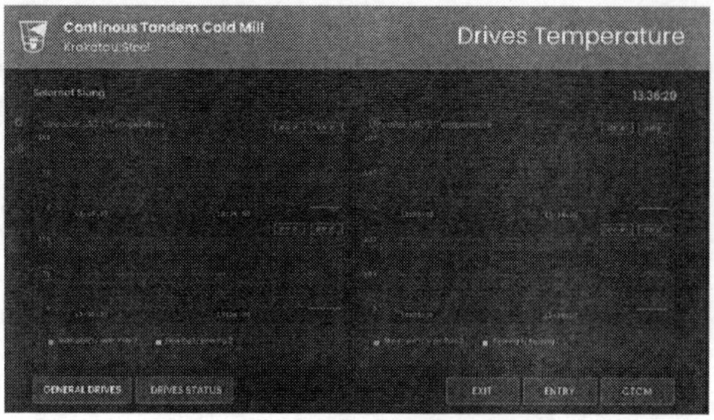

(b) Main window 1

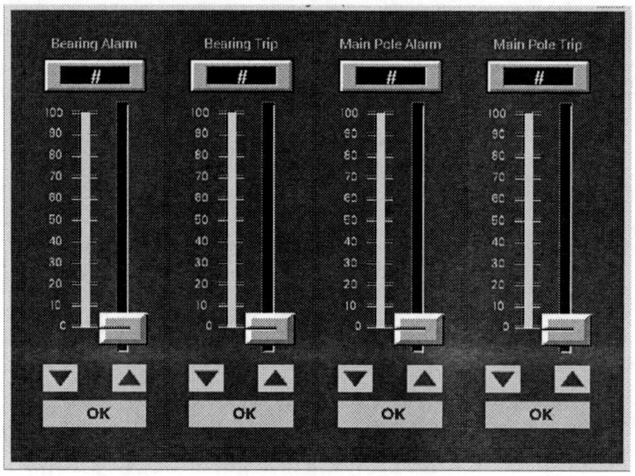

(c) Window setting

Figure 5. Layout Application

6. PT100 Sensor

The are eight sensor placed in bearing and inductor of uncoiled DC motor. Uncoiled DC motor is mechanical equipment to hold coil and maintain the strip tension between uncoiled DC motor and bridle. Uncoiled DC motor consist of 2 DC motor called MC1 (Motor Control1) and MC2 (Motor Control2). The sensor placed in the following order that can be seen on Table 1.

Table 1. PT100 Sensor Placement

Name	Sensor Placement
Sensor 1	MC1 Bearing 1
Sensor 2	MC1 Bearing 2
Sensor 3	MC1 Inductor 1
Sensor 4	MC1 Inductor 2
Sensor 5	MC2 Bearing 1
Sensor 6	MC2 Bearing 2
Sensor 7	MC2 Inductor 1
Sensor 8	MC2 Inductor 2

IOP Conf. Series: Materials Science and Engineering **722** (2020) 012031 doi:10.1088/1757-899X/722/1/012031

The important thing that need to be considered during commissioning is the length of extension cable. It must have the same length for each cable because RTD sensor implement Wheatstone bridge to compensate resistance from extension cable [14].

7. System Testing

System testing is conducted to ensure the entire system is successfully implemented without problem and run perfectly and safe. The testing process divided into 4 sections.

7.1. Testing PLC Program

PLC program has been tested with ABB soft controller which is made for simulation purposes only. Input A1830 is given with different value for each condition such as Bearing Temperature Alarm, Bearing Temperature Trip, Inductor Temperature Alarm and Inductor Temperature Trip. The result of this experiment showed on Table 2.

Table 2. Testing Result PLC Program

Sensor	Input AI830	Temp Limit (Alarm/Fast Stop)	Output PLC
MC1 Bearing 1	61	60/75	Alarm
MC1 Bearing 2	62		Alarm
MC1 Inductor 1	92	90/100	Alarm
MC1 Inductor 2	94		Alarm
MC2 Bearing 1	61	60/75	Alarm
MC2 Bearing 2	63		Alarm
MC2 Inductor 1	94	90/100	Alarm
MC2 Inductor 2	95		Alarm
MC1 Bearing 1	76	60/75	Fast Stop
MC1 Bearing 2	79		Fast Stop
MC1 Inductor 1	101	90/100	Fast Stop
MC1 Inductor 2	105		Fast Stop
MC2 Bearing 1	78	60/75	Fast Stop
MC2 Bearing 2	77		Fast Stop
MC2 Inductor 1	102	90/100	Fast Stop
MC2 Inductor 2	110		Fast Stop
MC1 Bearing 1	34	60/75	Normal
MC1 Bearing 2	33		Normal
MC1 Inductor 1	37	90/100	Normal
MC1 Inductor 2	36		Normal
MC2 Bearing 1	35	60/75	Normal
MC2 Bearing 2	38		Normal
MC2 Inductor 1	34	90/100	Normal
MC2 Inductor 2	32		Normal

The response of Function Block Diagram for Bearing Temperature Alarm and the whole testing experiment can be seen in Table 3.

Table 3. FBD Response

Condition	Alarm	Trip	Normal
Bearing Alarm	**true**	**false**	**false**
Bearing Trip	**true**	**true**	**false**
Inductor Alarm	**true**	**false**	**false**
Inductor Trip	**true**	**true**	**false**

IOP Conf. Series: Materials Science and Engineering **722** (2020) 012031 doi:10.1088/1757-899X/722/1/012031

7.2. Testing HMI Application

Testing HMI application conducted with three different condition, there are when normal condition, when signal temperature alarm is set to be true, when signal temperature trip is set to be true, when input signal AI830 forcing with HMI and when minimum temperature changed from setting window. The result of this experiment can be seen respectively on Figure 6.

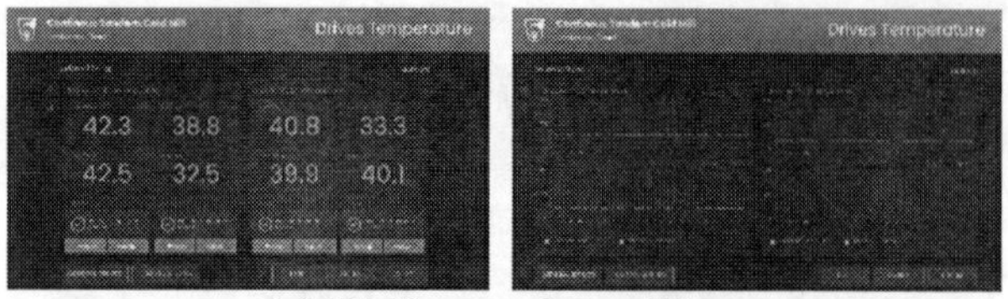

(a) HMI normal condition

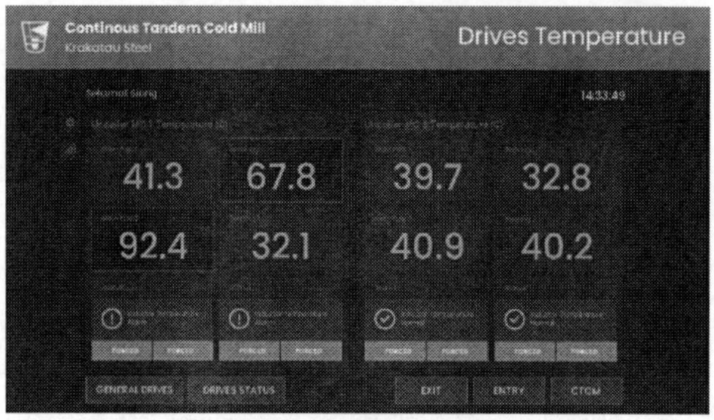

(b) HMI temperature alarm

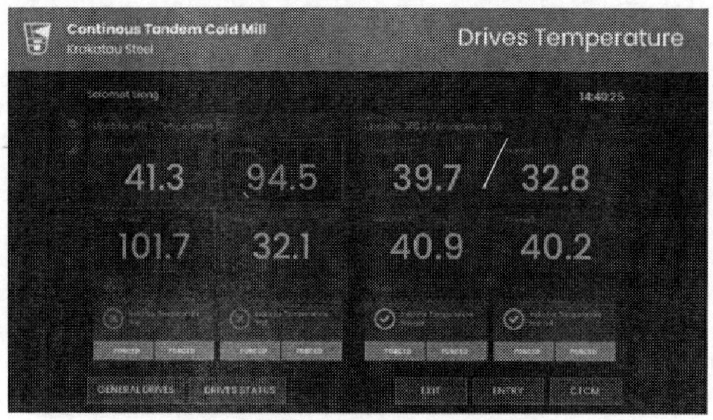

(c) HMI temperature trip

Figure 6. Result window

Force mode is implemented to bypass the input signal received in AI830 with standard value for this case, it set by 30°C. Thus when false sensing occurred or PT100 itself has been broken, it might be decrease the ability of measuring physical value accurately and then fast stop will be executed for wrong condition. Production line need to be operate immediately as quick as possible after stop and false sensing has been confirmed, therefore forcing the I/O through HMI is necessary. To ensure that the system is working properly Bearing1 MC1 set to 95.6 Inductor2 MC1 set to 93.5 and Inductor 2 set 95.1, it will generate both fault signal. Force button is pressed and the result can be seen on Figure 7.

IOP Conf. Series: Materials Science and Engineering **722** (2020) 012031 doi:10.1088/1757-899X/722/1/012031

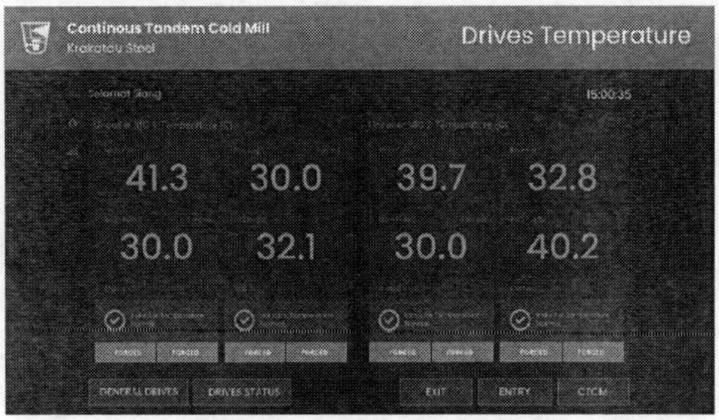

Figure 7. HMI force mode

The last one is testing window setting. Operator have ability to change the minimum temperature for bearing nor inductor. For this case bearing alarm decreases by 40°C and inductor alarm decreases by 40°C as well. Based on Figure 8, it can be seen that the HMI show alarm for MC 1 Main Pole 1 and MC2 Bearing 2 even though the temperature is normal.

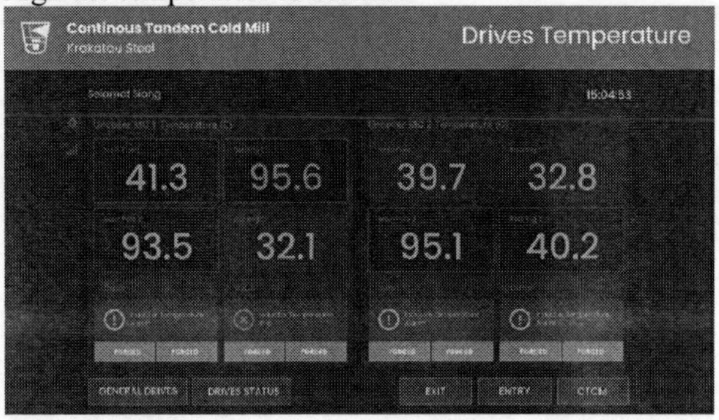

Figure 8. HMI response when setting parameter changed

7.3. Data Acquisition Testing

Data Acquisition is one of the most important part in SCADA system. Data acquisition help engineer to analyse anomaly and data that have been measured by sensor. Data is gained from data acquisition module called IbaFob41, this module runs 24 hours a day and gather data from sensor continuously. Every 15 minutes, all data will be stored in hard drive and can be accessed for assessment purposes. However, during data acquisition process, engineer has ability to monitor the sensor real-time. The graph can be seen on Figure 9 for MC1 and Figure 10 for MC2.

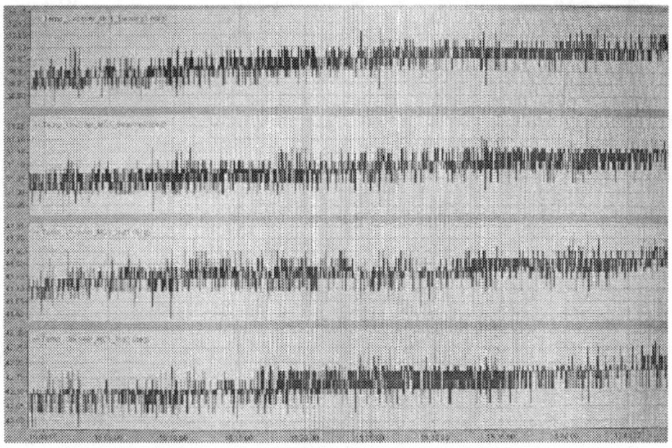

Figure 9. MC1 sensor monitoring

IOP Conf. Series: Materials Science and Engineering **722** (2020) 012031 doi:10.1088/1757-899X/722/1/012031

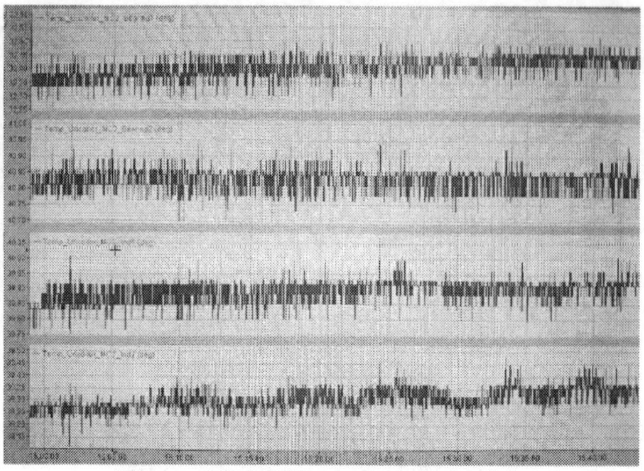

Figure 10. MC2 sensor monitoring

7.4. PT100 Sensor Testing

PT100 sensors that have been installed on uncoiled DC motor need to be tested to make sure the sensors are working properly. Extreme Industrial Environment condition need to take into account during commissioning. So, if extension cable of the sensor interferes with any other signal, it would be decreases the sensor accuracy. The testing conducted by checking the resistant value over the terminal AI830 and compared with calculation result.

This equation is used for calculate R_{meass} shows in Table 4. From Table 4, it can be confirmed that commissioning PT100 is successfully installed due to small error below 5% obtained from this experiment.

Table 4. PT100 Sensor Testing

Sensor	Temp (°C)	R_{meas} (Ω)	R_{calc} (Ω)	Error (%)
MC1Bearing1	38.6	115.8	114.86	0.81
MC1Bearing2	31.5	114.5	112.13	2.07
MC1Inductor1	41.7	121.3	116.05	4.32
MC1Inductor2	42.1	120.6	116.21	3.64
MC2Bearing1	31.3	114.5	112.05	2.14
MC2Bearing2	40.1	118.4	115.44	2.50
MC2Inductor1	39.6	119.6	115.25	3.64
MC2Inductor2	38.9	117.3	114.98	1.98

The error of sensor measurement in Table IV can be occurred in several reasons. There are the wires are not equal in length, wire have resistance and if one of the wires is unequal to the other will implies to unbalanced Wheatstone bridge that have been explained by equation (1) in previous section and self-heating of the sensor. Therefore, PT100 with three wires should have the same length in single wire, if one of them have different length this will increase measurement error.

AI830 have 10mA excitation current to the PT100 sensor and based on the equation (3), an PT100 with $\theta = 0.05$°C/mW, and a resistance value of 100 Ω at 0°C using a 10 mA excitation current will have a drift error of 0.5°C. To reduce error caused by sift heating, it can be done by reducing the excitation current for example by 100 μA that will be effectively lower this error by 10000. However, the sensor output will be reduced by 100. The PT100 with $\alpha = 0.385$ Ω/°C will have 100mV of output scale and if the excitation current reduced to 100 μA will have 1 mV of output scale. This will become problem because we know that industrial environment has a lot of interference signal and any other noise. So choosing an ADC with 10mA excitation current will be fine as long as the temperature error not exceed by 5%.

Table IV showed difference of inductor and bearing temperature. This is normal because the heat that arises in the winding is a square function of the current, therefore a slight increase of armature current will result a large increase of heat. Uncoiled DC motor have 50 A of current when its operated in nominal speed. Therefore, each of DC motor must be equipped with ventilation system. Ventilation system consist of an induction motor placed on the top side of DC motor. Without ventilation system, the temperature may arise above 90 °C. On the other hand, bearing of DC motor has a collier consist of grease lubricated in the bearing that will prevents the harsh metal-to-metal contact between the rotating element and races that reduces friction and thus improves performance and reduces temperature as well. That's why the temperature in bearing relatively lower than inductor temperature.

8. Conclusion

The design of SCADA based temperature protection system for uncoiled DC motor was successfully implemented due to several testing and experiment. Fast Stop Entry will be executed when bearing temperature increase and higher than 75°C or when inductor temperature higher than 100°C, Alarm will occur when bearing temperature increase and higher than 60°C or when inductor temperature higher than 90°C. HMI and data acquisition as part of SCADA has successfully implemented due to several test and experiment that showed data acquisition to HMI and IbaPDA running properly.
PT100 sensor is used for measuring temperature of uncoiled DC motor has a good accuracy and the data indicate that error measurement below 5%.

References

[1] ABB. System 800xA 5.1, Product Catalogue (FP4 included). URL: http: / / www05. abb. com / global / scot / scot354. nsf / verity display / a7dc1ca53bd960e2c1257b9d003d63a0 / $file / 3BSE062937 _ E _ en _ System_800xA_5.1_Product_Catalog_(FP4_included).pdf.
[2] ABB (2006). IndustrialITCompactControlBuilderAC800MVersion5.0-Getting Started Introduction and Installation.
[3] ABB (2010). AC800M Communication Protocols.
[4] ABB (2012). System 800xA Network Configuration - System Version 5.1.
[5] ABB (2013b). Field bus and Network Communications. URL: http://www.abb. se/product/us/9AAC115770.aspx.
[6] ABB (2013). Control IT AC 800PEC Hardware Guide.
[7] ABB (2013). S800 I/O Module and Termination Unit
[8] OPC Overview Version 1, OPC Task Force 1998
[9] OPC Data Access 2.0 Technical Overview, Al Chrisholm, 1998
[10] Asrom, Karl J-Wittenmark, Bjorn (1997), Computer Controlled System, Prentice Hall.
[11] Davidsson, Peter-Hansson, Fredrik (2000), Communication between Advent Control Builder and MATLAB/SIMULINK, Master thesis at the Department of Automatic Control, Lund Institute of Technology.
[12] Neskovic, Nelica (2000), OLE for process control in Advent Soft Controller as communication interface for simulation, Master thesis at Lund Institute of Technology.
[13] Olsson, Gustaf- Piani, Gianguido (1998), Computer Systems for Automation and Control, Second Edition, Prentice Hall
[14] Suryatmo, F. 1974. Electrical Measuring Technique. Bandung: Bina Aksara
[15] Bailey, David and Edwin Wright. 2003. Practical SCADA for Industry. Great Britain, India.
[16] Victor Anderson, Tommi Nylander (2014). Simulation of Process Control Network Traffic. Department of Automatic Control, Lund Institute of Technology.
[17] Mathias Larsson (2000). Modelling and Control in MATLAB for ABB's Control Builder.

Department of Automatic Control Lund Institute of Technology.
[18] Aleksejs Gedzurs, 2015. Temperature Protection Method of Induction Motor. Latvia University of Agriculture.
[19] Hilmi Fathoni dkk, 2015. Design of Supervisory Control and Data Acquisition (SCADA) for Automatic Process of Packaging Station in PT. Perkebunan Nusantara VIII Rancabali
[20] Nicola M, Nicola C, Ţă MDU, Ianu DS Ţ, Hurezeanu I. System for Monitoring of Hot Spot Temperature of Power Transformer Windings Using Fiber Optic Sensors , Kalman Filter and SCADA Integration. 2018;99–104.
[21] Ponjavi MM. Online Temperature Monitoring , Fault Detection , and a Novel Heat Run Test of a Water-Cooled Rotor of a Hydrogenerator. 2013;28(3):698–706.
[22] Svelto C, Galzeraiio G, Suardi G, Bava E. Compact aind Accurate Digital Thermometer Based on an Anderson ' s Loop and a Pt-100 Sensor. 1999;1619–23.
[23] Kaur A, Bansal D. Monitoring and controlling of continue furnace line using PLC and SCADA. 2016 5th Int Conf Wirel Networks Embed Syst. IEEE; 2016;(V):1–5.
[24] Basuki A. Integration of Online Transformer Condition Monitoring into SCADA Systems. 2018 53rd Int Univ Power Eng Conf. IEEE; 2018;1–6.

IOP Conf. Series: Materials Science and Engineering **722** (2020) 012032 doi:10.1088/1757-899X/722/1/012032

Textile Wastewater Treatment using Advanced Oxidation Process (AOP)

S S Hutagalung[1], I Muchlis[1], K Khotimah[1]

[1]Research Center For Testing Technology, Indonesian Institute of Sciences, Kawasan Puspiptek Gd 417, Tangerang Selatan, Indonesia

E-mail: khusnulkhotimah@lipi.go.id, sutr007@lipi.go.id, imamulmuklis@lipi.go.id

Abstract. Wastewater from printing and dyeing industries is often colorful, contains reactive dye residues and chemicals, and requires proper treatment before being released into the environment. The treatment of textile wastes with biotechnology through the general use of high inorganic and organic concentrations is difficult for further process. Therefore, many industries make use of the advanced oxidation process (AOP) because of its ability to eliminate non-degradable organic components and avoid residual deposits removal. a new design of synthetic waste treatment was developed by installing the AOP on the existing wastewater treatment process (WWTP). However, the textile industry faces significant environmental problems because of its vast water consumption. This makes the resulting wastewater from their activities characterized by intense color, a large number of suspended solids, very unpredictable pH and high COD concentrations. Therefore, there is a need to implement some control measures in the future in order to minimize these waste problems. Experiments show that the pollutants can be physically removed by using the AOP process and COD decrease 51.7 % from 16.167 mg/L into 8.364 mg/L.
Keywords— textile, waste, paint and sludge, treatment, AOP

1. Introduction

1.1. Background of the study

Textile industry is a leading industry in most countries such as China, Singapore, England, Bangladesh, Italy, Turkey, Indonesia and many more other countries in the world. Unfortunately, during production activities, several hazards that potentially damage the environment arise as a result of waste produced. Textile wastewater is characterized by the high content of dyes, salt, high COD of additives, suspended solids (SS) and fluctuating pH values. C. Sahunin, et al. explained that the textile industry consumes around 21-377 m^3 of water per tons of textile produced, thus generating a large amount of wastewater from various stages of dyeing and refining processes (1). In agreement with this, S. Eswaramoorthi K. Dhanapal revealed that treating wastewater is a severe problem for several reasons including high total dissolved solids (TDS) from wastewater, Cu, Zn, etc.; types of organic dyes which cannot be broken down by waste bacteria; and the presence of dissolved silica free of chlorine (2). One of the major environmental concerns is the removal of dyes from textile waste because it has been reported that only 87% of its content can be decomposed (3).

The primary reason color could be seen in the textile industry waste is because of the use of large amounts of dyes during the dyeing process of textile manufacturing (4). Yonar et al. explained that

IOP Conf. Series: Materials Science and Engineering **722** (2020) 012032 doi:10.1088/1757-899X/722/1/012032

inefficient dyeing methods often produce significant coloring residues in the final hydrolysed or irregular dye effluent (5).
Even though the production of waste has been minimized, the remaining wastes must be processed into several forms before being discharged into the river or sea. The reduction has significant effects because it helps to reduce operating costs, the risk of responsibility and the requirement for waste treatment. It also helps to improve the efficiency of the production process as well as the protection of health and the environment.
There are several textile wastewater treatment methods, including electro-oxidation, bio-treatment, photochemistry and membrane processes. The use of nano-filtration has been observed to be less energy intensive compared to reverse osmosis and, as a result of this, it is mostly used in processing several industrial wastes (6).
Currently, AOP has a particular advantage over conventional alternative treatments because of its ability to eliminate non-degradable organic components and avoid residual deposits removal. It is based on the high oxidation strength of free radicals, notably hydroxyl, and highly reactive oxidations (7).
This paper presents the implementation of AOP in treating textile wastes. For this purpose, a new design of synthetic waste treatment was developed by installing the AOP on the existing plant. AOP was developed by combining ozone and ultraviolet in order to safely dispose of processed liquid waste and meet the environmental quality standard requirements set by the Ministry of Environment.
The X corporate is currently using and operating a WWTP in combination with chemical and biological methods. The plant is also equipped with sludge treatment facilities, including belt press and incinerator. The textile wastes are processed from massive accumulated sludge waste chemically, especially through PAC coagulant used in paint removal, and biologically through sewage sludge from the active basin residual.
The production process of X corporate is quite diverse and tailored based on the needs of consumers in terms of fabric type and desired color. As a result, compound liquid waste is produced. Therefore, a large amount of PAC coagulants is needed in order to remove the dyes contained in the waste making the organization produce $5m^3$ sludge residual daily.

1.2. Purpose of the study
The purpose of this study was to overcome pollutants from liquid wastes of X corporate containing organic and inorganic compounds that have the potential of becoming toxic to the environment using AOP system. Some text.

2. Materials and Methods

2.1. Materials
AOP system with a processing capacity of 1 m^3/day, COD influent 5,000 mg/L, a process pressure of 1 atm, process temperature at 20-40 ^{o}C, the power consumption of 1 kWh, and voltage source of 220 Volts 1 Phase 50 Hz was used for field trials.
The AOP unit divided into five parts which have specification and function, as seen in Table 1.

Table 1. Parts of AOP unit.

No	Part	Specification	Function
1	Inlet pump set	Multi-stage/Centrifugal/Submersible 1 piece with a capacity of 40 L/minute, head 15 m, and power at 220V, 1 phase, 50Hz	drive the waste for the treatment process

No	Part	Specification	Function
2	oxidizing section	Fiberglass material, capacity of 8 SCFH	decompose chemical compounds and kill microorganisms in the waste
3	oxidation chamber	SS material, 1000 L capacity and atmospheric pressure	absorbing micropollutants which have not been deposited in the previous process
4	filtration tank room	SS material, 500 L capacity, and atmospheric pressure	absorbing micropollutants which have not been deposited in the previous process
5	pH adjusted tank	SS material, 500 L capacity	controlling pH

2.2. Methods

The hazardous waste treatment activities ranged from regulatory restrictions to savings in chemical recycling costs. Most organizations make use of waste treatment industries in order to achieve environmental security. Unfortunately, there is no specific processing method suitable for all types of waste treatment or which can be applied universally. Therefore, wastewater treatment can be conducted through several physical, chemical and biological processes depending on the pollution load.

The method used in this research was ozone/UV (O3/UV) because the photolysis of ozone in water with UV radiation in the range of 200 to 280 nm might cause hydrogen peroxide. However, the hydroxyl radicals produced by hydrogen peroxide under UV radiation and/or ozone can be seen in the following equation:

$$O_3 + hv + H_2O \rightarrow H_2O_2 + O_2 \quad (1)$$

$$H_2O_2 + hv \rightarrow 2(\cdot OH) \quad (2)$$

$$2O_3 + H_2O \rightarrow 2(\cdot OH) + 3O_2 \quad (3)$$

From low-pressure mercury vapor lamps, all types of UV light sources can be used for this process. As a result of the fact that the O_3/UV does not have the same limits as the H_2O_2/UV process, low-pressure UV mercury vapor is the most commonly used source of radiation for this process. It is, however, important to state that many variables such as pH, temperature, entry, turbidity, UV intensity, spectral properties of light and types of pollutants affect the system efficiency.

Ozonation is a phenomenon of simultaneous mass transfer while ozone is a well-known potent oxidizing agent widely used for the treatment of water and wastewater, theoretically and practically. It has high efficiency at high pH values such that at a high pH (> 11.0), it reacts almost indiscriminately with everything that is in the organic and inorganic compounds of the reaction. It also reacts with wastewater compounds in two different ways, direct and indirect radical chain reactions. They are simultaneously continuous; therefore, their reaction kinetics are dependent on the characteristic of the treated wastewater.

Simplified ozone reaction mechanism at high pH is as shown below:

$$3O_3 + H_2O \rightarrow 2(\cdot OH) + 4O_2 \quad (4)$$

Ozon (O_3) is an allotrope of oxygen (O_2)

Textile dyeing and industrial finishing are one of the main contributors to the quantity and chemical composition of wastes disposed. Like many other industrial wastes, textile industry waste varies significantly in quantity and composition (8). It contains organic and inorganic chemicals such as equipment, carriers, surfactants, precipitates, rating agents, etc.

This is the reason it is mostly characterized by high COD (400-3000 mg/L), BOD5 (200-2000 mg/L), total solids (1,000 to 10,000 mg/L), suspended solids (100-1000 mg/L, TKN) (~ 10-100 mg/L), total phosphorus (~ 5-70 v) conductivity (1,000-15,000 Ms/cm) and pH (~ 5-10 usually basic). Another important problem of waste from the textile industry is the color (9). Without proper processing, the color of the discarded dye can remain in the environment for a long time, and this has become a major challenge due to strict environmental regulations.

In order to eliminate discarded dye remain in the environment using AOP, it is necessary to have a processing flow chart for textile waste. The treatment is often conducted with a single aerobic sludge or the combination of chemical coagulation and flocculation of sludge. However, aerobic biological treatments have been observed only have the ability to remove paint from the textile waste while the chemical coagulation and flocculation are also not effective in removing reactive dyes. Therefore, dyes and chemicals used in the textile industry have severe inhibitory effects on aquatic ecosystems and visible water pollution (9).

It is, however, important to point out that there are other alternative methods of eliminating textile waste pollution and they include various combinations of physical, chemical and biological treatments and color removal methods. Moreover, they cannot be applied effectively to all dyes and also not cost effective. Therefore, the use of AOP reduces non-biodegradable organic wastes in textile dyeing (10). This ability can be established on the report that ozonation is very effective in removing paint, COD, and toxicity from textile waste.

The main advantage of ozonation is that there is no chemicals addition needed to water or wastewater. As a consequence, it is possible to apply ozonation in textile waste without adjusting pH and adding chemicals. However, the process has several disadvantages, such as inefficient production capacity of the corona destroyer generator (CCD) (2-4%). This can be minimized by using an efficient ozone generator such as the electrochemical ozone generator.

The oxidation process of hydroxyl radicals has shown that toxic organic compounds in wastewater can be destroyed because of the radicals' high oxidation potential and several production possibilities. They can be made by combining ozone with ultraviolet light; ozone, hydrogen peroxide with ultraviolet light; hydrogen peroxide with iron or iron ions and by photocatalysis which involves combining semiconductors with UV radiation and oxygen molecules. The schematic diagram of the textile waste treatment flow rate with AOP and CA is shown in Fig. 1(11).

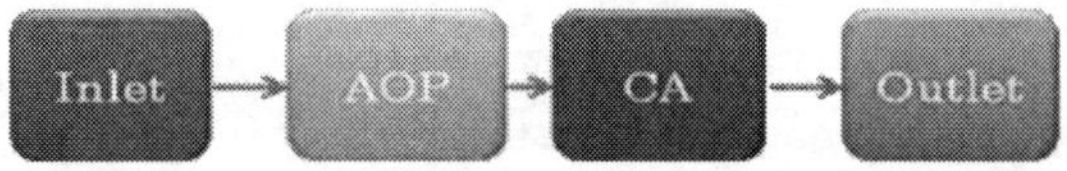

Figure 1. The flow of textile waste treatment with AOP and CA.

The basic concept of the system created is the AOP, with ozone and ultraviolet as the main component, which is further combined with activated charcoal in the last stage to produce a phylophone. The process of AOP produces a potent but relatively non-selective hydroxyl radical in sufficient quantities and is used to oxidize waste chemicals (12). This is possible because hydroxyl radical (OH) has the highest oxidation potential after fluorine (potential oxidation, E0: 2.8 eV compared to normal hydrogen electrode (NHE) while the most potent oxidant, fluorine has E_0: 3.06 V (11) and cannot be used for wastewater treatment due to its high toxicity). For this reason, hydroxyl radicals including AOP have attracted the attention of several scientists and technology developers. The understanding of ozone oxidation has led to the expansion of its applications in drinking water and wastewater treatment industries.

The waste treatment was divided into primary, secondary and tertiary processing as represented by the AOP prototype system and activated carbon in the image. The main purpose of the secondary processing is to remove more BOD than the simple sedimentation and also to save a lot of oil and phenol. During further processing, dissolved organic compounds, colloids, and colors in wastewater are removed or reduced and organic matter stabilized.

3. Experiment Results and Discussion

The first step included the observation of the equipment requirements and field conditions in order to obtain the proposed field trial design as shown in Fig. 2(13).

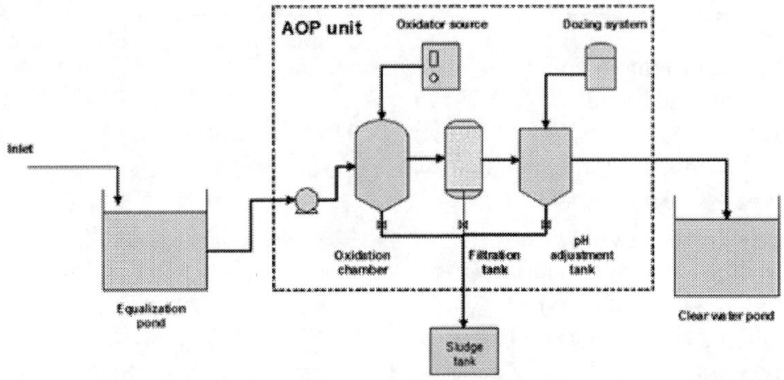

Figure 2. The layout of field trial design.

The next has to do with the installation of the AOP unit prototype in stages as shown in Fig. 3. It started with the trial preparation at X corporate then followed by the research stage I (model A) conducted through activated sludge ozonation method, stage II (model B) through the use of medium-color Ex-Jet Dyeing T/R with Ozone method and UV, stage III (model C) using Ex desizing scouring non color with the Ozone method, stage IV (model D) with spunpolly Ozone and UV method, stage V (model E) treated sludge from coagulation with Ozone method, stage VI (model F) with C26 (Finish Resin) with Ozone and UV methods, and stage VII (model G) conducted with T/R solid color discharge by the Ozone method. These can be seen in the WWTP Field Experiment Design Layout in Fig. 2 and field tests in Fig. 3(13). AOP prototype system with active carbon.

Figure 3. Field trial at X Corporate.

In the field trial, the prototype unit of liquid waste treatment, including ozone and UV was partially operated according to the request of X corporate and coagulation and filtration were not used. As shown in Fig. 4, wastewater stored in the re-circulated container by pumping through the AOP (ozone and UV) process for less is returned to the original reservoir. The circulation was conducted for 1 hour for each sample tested. As stated in Table II, the trial was conducted on seven types of liquid waste produced by the corporate and the experiments and analysis were conducted during the trial process.

Table 2. Results of field trials

No	Sample	pH	COD	SS	Color
1	Activated sludge	7	15,693	9,225	heavy
	After process (homogenous)	7	15,504	5,100	
	After process (sediment)	7	7,956	687	
2	ex jet dyeing medium color T / R (ozone + UV)	7.9	2,346	162.5	heavy
	after the process of 10 minutes PAC was deposited	5.6	476	8	light
	After process (30 minutes)	6.9	5,916		
	After process (60 minutes)	7.2	5,022		
3	Ex Desizing non-color scouring (ozone)	7.1	28,254	162.5	moderate
	After process (60 minutes)	7	23,511		
4	Spunpolly (weight reduce) (Ozone + UV)	8.6	16,167	75	light
	After process	7	8,364	87.5	light
5	Mud from coagulation	5.9	7,089	9,075	heavy
	After process	5.7	7,293	8,850	heavy
6	C 26 (Resin finish) (Ozone + UV)	6.1	28,458	1,600	light
	After process	5.2	28,050	1,237	light
7	Concentrated color waste	7.9	1,224	62.5	heavy
	After process (homogenous)	7.2	1,648	46	decreased
	After process (sediment)	7.2	1,360	50	light

The results showed that the maximum capacity for the activated sludge (ozone) samples was 5 m^3/day. This shows that the sludge is decreasing the maximum capacity for weight reduces (ozone + UV) (ozone) at 110 m^3/day as well as the CSB. It was also found that the maximum capacity for ozone concentrated color samples was 350 m^3/day which helped in reducing the color.

The result also showed that the activated sludge was well processed with AOP. This can be seen from the decrease in COD and SS that occurs after the process (Table II). Fig. 4 also shows that the pollutants contained in the activated sludge can be removed.

While liquid waste is produced in the APS process, it is possible to reduce the COD content. It was observed in this research, there was a significant decrease of 51.73 % from its original value from 16.167 mg/L to 8.364 mg/L after the process. This value is slightly higher than stated by Balasubramanyan et al. which is said that COD reduction usually not exceeding 50 % (14). Reduction value can be made continuous through the increment of the ozone dose in the operation process.

Figure 4. SS decline in activated sludge.

In the seventh experiment, a test was conducted to remove the paint in the liquid waste and a significant color reduction was observed as shown in Fig. 5. This shows that AOP is also very suitable for removing colors contained in the liquid waste. However, this is not accompanied by a decrease in COD. Further research is needed to find out the exact cause of the phenomenon that occurs

Experiment result also shows that the degradation of chemical compounds must be accompanied by a decrease in COD in correlation with the decrease in sludge content and color. This can be associated with the final result of the color reduction test after one hour of processing through the AOP prototype unit as shown in Fig. 6.

Figure 5. Color drop.

To get the expected results, a processing capacity of up to 20 m^3/day, COD 25,000 mg/L (coloring), 1 atm printing process, process temperature at 20-40 °C, 5 kWh power consumption, 220 volts 1 phase voltage source 50 Hz are required for all pH values.

From the test results and through the calculation of the required requirements, the specifications needed to make the design of textile wastewater treatment can be divided into two parts, civil parts, and the AOP units. In the Civil Division, the AOP and control room are further divided into two parts. The AOP room functions as a storage area for the AOP units with dimensions of 4m x 3m x 4m while the control room serves as a storage area for the oxidation unit.

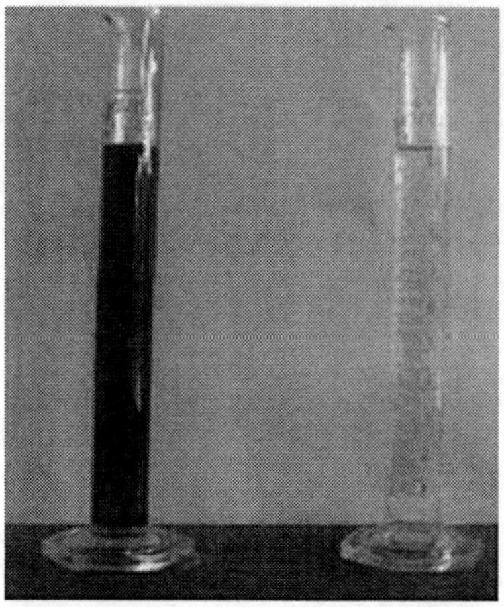

(a) Before processing (b) After processing

Figure 6. Color reduction tested in the laboratory.

As the literature clearly showed, ozonation is very effective in removing paint, COD, and toxicity from textile waste. However, it also important to understand its limitations before it is applied in any other industry.

O_3/UV is the most effective method in changing dye colors when compared to the use of UV oxidation or ozonation only. Although ozone can be used to increase organic degradation through its hydroxyl radical, UV-light also has a strong ability to absorb dyes and in producing minimal amounts of free radicals (HO•) required to degrade dyes. Therefore, the same color removal efficiency can be expected from both O_3 and O_3/UV.

4. Conclusion

To optimize the processing and possible reuse of textile industry waste, the flow must always be considered separately. It is also essential to investigate all aspects involved in reducing emissions and waste products because it does not only improve environmental performance but also leads to significant savings.

From the results of several tests conducted and observation of the colored, smelly and cloudy synthetic textile processing, it can be concluded that the pollutants can be physically removed by using the AOP process. The experiment result shows that COD reduction is 51.73%. This result shows that the content of chemical compounds contained in the wastewater can be effectively removed through the oxidation. Therefore, the AOP process is a promising alternative to traditional processing methods for textile waste treatment.

5. Acknowledgment

The author shows gratitude to the Centre for Calibration, Instrumentation, and Metrology - Indonesian Institute of Sciences (LIPI) for providing the opportunity to conduct this research in their laboratory. PT. ISTEM, Tangerang is also appreciated for its collaboration in providing research trial locations.

References

[1] C. Sahunin, J. Kaewboran & M. Hunsom. (2006). Treatment of textile dyeing wastewater by photo oxidation using UV/ H_2O_2/Fe^{+2} reagents. Science Asia, 32.

[2] S. Eswaramoorthi K. Dhanapal & D.S. Chauhan. (2008). Advances in textile waste water treatment: the case for UV ozonation and membrane bioreactor for common effluent treatment plants in Tirupur. Tamil Nadu, India, Environmental Technology Awareness Series, pp.1-17.

[3] U. Pagga & D. Brown. (1986). The degradation of dyestuffs: part II Behaviour of dyestuffs in

aerobic biodegradation tests. Chemosphere, vol. 15, pp. 479-491.
[4] Georgiou, D., Melidis, P., Aivasidis, A. & Gimouhopoulos, K., (2002). Degradation of AzoReactive Dyes by Ultraviolet Radiation in The Presence of Hydrogen Peroxide. Dyes and Pigments.Vol.52, pp. 69–78, ISSN 0143-7208.
[5] Yonar, T., Yonar, G.K., Kestioglu, K. & Azbar, N., (2005). Decolorisation of Textile Effluent Using Homogeneous Photochemical Oxidation Processes. Color. Technol. Vol.121, pp. 258-264, ISSN 1472-3581.
[6] Meier, J., T. Melin, & L.H. Eilers. (2002). Nanofiltration and adsorption on powdered adsorbent as process combination for the treatment of severely contaminated waste water. Desalination, 146:361-366.
[7] Kestioglu, K.; Yonar, T. & Azbar, N. (2005). Feasibility of Physico-Chemical Treatment and Advanced Oxidation Processes (AOPs) as a Means of Pretreatment of Olive Mill Effluent (OME). Process Biochemistry, Vol.40, pp. 2409-2416, ISSN 1359-5113.
[8] Corriera. V.M.; Stephenson. T. & Judd, S.J. (1994). Characterisation of textile wastewaters-A review. Environ Technol, Vol.15, pp. 91.7 -919. ISSN 0959-3330.
[9] Willmott, N.J.; Guthrie, J. & Nelson, G. (1998). The Biotechnology Approach to Color Removal from Textile Effluent. J. of the Soc. of Dyers and Colorists, Vol.114, pp. 38-41, ISSN 1478-4408.
[10] Azbar, N.; Kestioğlu, K. & Yonar, T. (2005). Application of Advanced Oxidation Processes (AOPs) to Wastewater Treatment. Case Studies: Decolorization of Textile Effluents, Detoxification of Olive Mill Effluent, Treatment of Domestic Wastewater, Ed. A.R. BURK, Water Pollution: New Research, pp. 99-118, New York, Nova Science Publishers, ISBN-1-59454-393-3.
[11] Sutrisno Salomo Hutagalung., Anto Tri Sugiarto, A.T., & Veny Luvita.(2010). Aplikasi Metode Advanced Oxidation Processes untuk mengolah Limbah Resin Cair. Jurnal Teknologi Pengelolaan Limbah Volume 13 Desember Nomor 2, hlm. 18-25.
[12] Gogate, P.R. & Pandit, A.B. (2004a). A Review of Imperative Technologies for Wastewater Treatment I: Oxidation Technologies at Ambient Conditions. Advances in Environmental Research, Vol.8, pp. 501-551, ISSN 1093-0191.
[13] Anto Tri Sugiarto, Sutrisno Salomo (2010). Laporan uji coba limbah cair di PT. ISTEM.
[14] Balasubramanyan, Shriram & S, Kanmani. (2014). Ozonation of Textile Dyeing Wastewater - A Review. Journal of Institue of Public Health Engineers. 2014-15. 2014.

IOP Conf. Series: Materials Science and Engineering **722** (2020) 012033 doi:10.1088/1757-899X/722/1/012033

An Anthropometric Study of Head for Designing Ergonomic Helmet for Acehnese Adults

M Andriani[1], A Widyanti[2], Yassierli[3], R Pramanda[4], Suheri[5]

Dept. of Industrial Engineering Samudra University
Faculty of Industrial Technology Institut Teknologi Bandung
Faculty of Industrial Technology Institut Teknologi Bandung.
Dept. of Industrial Engineering Samudra University
Dept. of mechanical Engineering Samudra University

E-mail: meri_tind@unsam.ac.id, widyanti@ti.itb.ac.id, yassierli@ti.itb.ac.id

Abstract. One of the fundamental applications of ergonomics is the use of an anthropometric approach for product design and evaluation. Redesigning ergonomic helmets using anthropometric methods is the aim of the study. In this study, participants consisted of 102 women and 102 men who were involved voluntarily. Based on the preliminary study, we found the following head dimensions considered important for helmet design, including: Head breadth, Menton to top of the head, Diametric menton to back of the head, Nose to top of head, Pronasale to back of head, Face breadth, and Eye to back of head. Anthropometric data are presented in this paper, and proposed helmet size design data is discussed.

1. Introduction

Motorcycles are among common mode of transportations by people with low-moderate incomes in developing countries including Indonesia [1]. Although traveling on a motorcycle is associated with a higher risk of death or injury [2], motorcycles are still among the most preferred transportation mode by people [3]. This condition is supported by the fact that the prices of motorcycles are affordable, coupled with the advantage in dealing with traffic jam conditions.

Unfortunately, the high use of the motorcycle is accompanied by high rates of motorcycle accidents [4]. The World Health Organization reports that nearly 60% of deaths caused by road accidents involve motorcyclists, especially in low and middle-income countries [5]. Most motorcycle accidents occur at young ages. Young ages range from fifteen and nineteen years [6]. The severity of motorcycle accident varies from minor, moderate to severe injuries. The death of motorcycle riders from the results of previous studies due to head injury [7]. In this case, efforts have been made to minimize deaths from motorcycle accidents, including, one of which is to ride a motorcycle must use a helmet.

Safety helmets, although not the best protective equipment but protect motorcycle users from injury and can prevent about 32% of head injuries during a motorcycle accident [8]. Although regulations have been implemented in several countries, there seems a reluctance to use helmets while driving for several reasons, including discomfort in wearing a helmet [9].

Indonesia is developing countries that face the problem of high motorcycle use and high motorcycle accidents [10]. The high number of fatality accidents seems due to the lack of use of helmets. This problem is often associated with helmet size. There is no standard helmet size for Indonesian yet. The current helmet size is determined by helmet manufacturers with lack support of

IOP Conf. Series: Materials Science and Engineering **722** (2020) 012033 doi:10.1088/1757-899X/722/1/012033

the anthropometric study, resulting in varying size helmet size is simply determined based on head circumference. Considering that head anthropometry is not only related to head circumference, but it also is not surprising that helmet size, in general, is not suitable for users.

Anthropometry is defined as the application of ergonomics to measurements people's body [11]. Measurement of the people's body [12] is key in determining the product size. For helmet design, the anthropometric head is used as a basis for determining size. Given Indonesia's anthropometric data there are ethnic differences. [13], in this study, we focused on Acehnese adults, because Aceh was a province that suffered from high motorcycle accidents and head injuries, the purpose of the study is to measure Acehnese anthropometry and to use the data for redesigning helmet sizes. The more fit the helmet to the motorcycle riders, the more will they use the helmet, and the risk of fatality injury will be minimized.

2. Methods

A. Participant

Participants are Acehnese adults (ranging from 18 to 22 years) consisted of 102 women and 102 men who were involved voluntarily in this study and the participants were randomly selected.

B. Tools

Head anthropometry measurement was conducted using Martin anthropometer, due to its flexibility.

C. Procedure

The measurements were conducted by a research assistant who had been trained in using body martin instruments and was also proficient in anthropometric measurements. Following [14], seven dimensions of head anthropometry were measured, including Head breadth. Menton to top of the head, Diametric menton to back of the head, Nose to top of the head, Pronasale to back of the head, Face breadth, and Eye to back of the head as can be seen in Table 1.

Table 1. Anthropometry for Helmet Design

No	Dimension	Picture	Associated helm dimensions
1	Head breadth		Helmet width
2	Menton to top of the head		Helmet height
3	Diametric menton to back of the head		Helmet diameter
4	Nose to top of the head	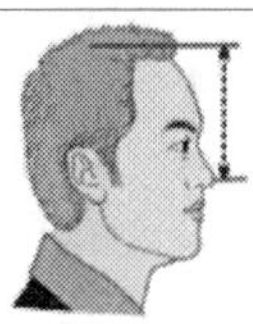	Glass helmet height

5	Pronasale to eye (The value of "Pronasale to back of the head" minus the value of "Eye to back of the head")	 	Helmet glass length
6	Face breadth		The width of the front of the helmet
7	Eye to back of the head	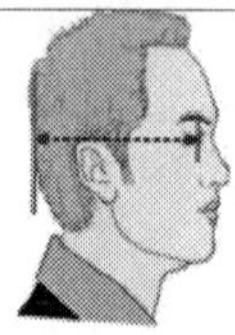	Helmet length

3. Result

The anthropometric head dimensions for Acehnese adult's samples are shown in table 2 and Table 3, consecutively for men and women.

Table 2. Head Antropometry For Men

No	Dimensions	Mean (cm)	σ (cm)	Percentiles 5 (cm)	Percentiles 50 (cm)	Percentiles 1 95 (cm)
1	Head breadth	17.52	2.99	12.61	17.52	22.43
2	Menton to top of the head	22.55	3.47	16.86	22.55	28.24
3	Diametric menton to back of the head	25.66	2.16	22.11	25.66	29.21
4	Nose to top of the head	15.77	1.84	12.76	15.77	18.78
5	Pronasale to eye	19.66	2.46	15.62	19.66	23.70
6	Face breadth	16.34	2.14	12.84	16.34	19.85
7	Eye to back of the head	16.91	2.61	12.63	16.91	21.18

Table 3. Head Antropometry For Women

No	Dimensions	Mean (cm)	σ (cm)	Percentiles 5 (cm)	Percentiles 50 (cm)	Percentiles 1 95 (cm)
1	Head breadth	17.30	2.94	12.48	17.30	22.13

IOP Conf. Series: Materials Science and Engineering **722** (2020) 012033 doi:10.1088/1757-899X/722/1/012033

2	Menton to top of the head	22.24	1.57	19.66	22.24	24.81
3	Diametric menton to back of the head	25.53	1.79	22.59	25.53	28.47
4	Nose to top of the head	16.23	2.04	12.88	16.23	19.58
5	Pronasale to eye	19.80	2.08	16.39	19.80	23.21
6	Face breadth	16.28	1.99	13.02	16.28	19.55
7	Eye to back of the head	17.68	2.82	13.05	17.68	22.30

As expected, there were statistical differences in dimensions between men and women, in which four of the seven dimensions of male dimensions are larger than women.

4. Discussion

The aim of the study was to redesign helmets in accordance with the anthropometric dimensions of Acehnese adults. Our preliminary study showed the inconvenience of using helmets because of their unfit size to users.

There are several methods for determining the standard size of a product, the method often used is based on percentiles [15]. In this case, the 5th percentile is used as a reference for size S, 50th percentile for size M, 95th percentile for size L.

This study has some boundary that needs to be considered. First, the number of samples is quite low considering the Acehnese adult population (i.e. 5.19 million, BPS, 2019). Further research including more samples is needed to strengthen the results. Second, the measurement conducted manually may have an issue related to lack of reliability, so that a more sophisticated and reliable measurement method is needed

Despite its limitations, this research can be seen as the first step in the measurement of head anthropometry in Indonesia, focused on Aceh. The ergonomic helmet size design based on the anthropometry measures is shown in Figure 1.

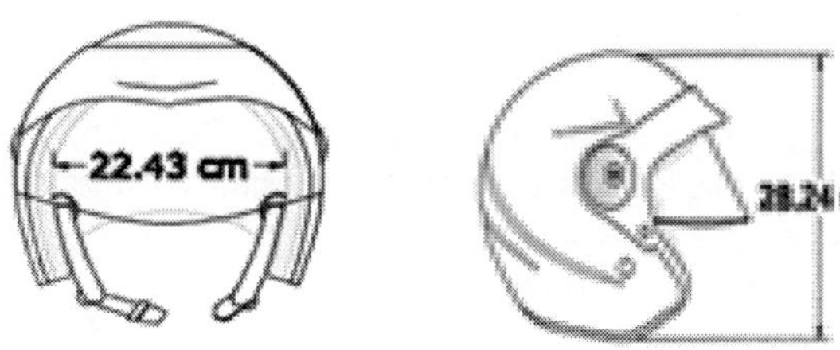

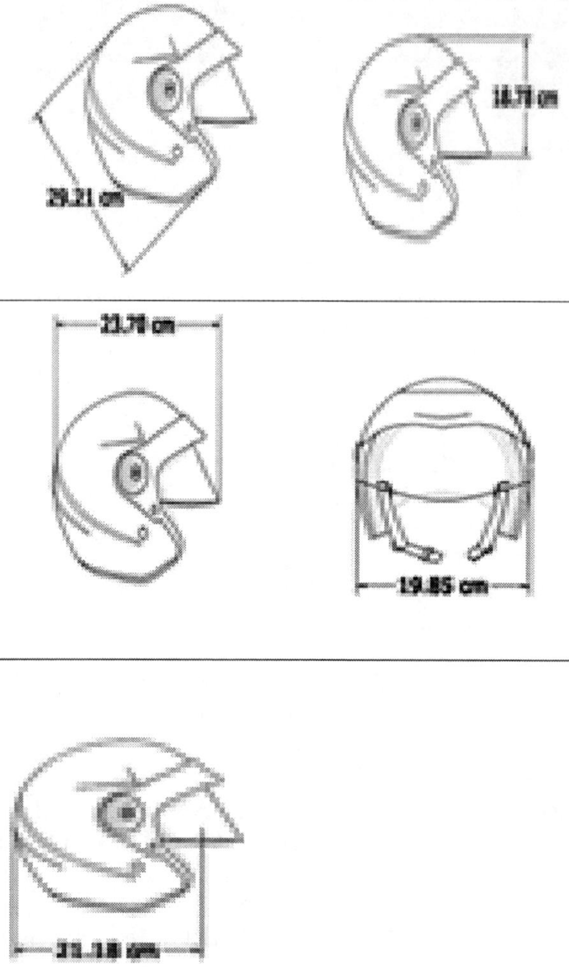

Figure 1 The proposed helmet size

5. Conclusion

There are seven head dimensions considered to be important for helmet design, namely: Head breadth, Menton to top of the head, Diametric menton to back of the head, Nose to top of the head, Pronasale to back of the head, Face breadth, and Eye to back of the head. Data of each dimension is presented, and the size of the helmet is proposed.

Acknowledgment

The authors thank the Ministry of Research and Technology of the Republic of Indonesia for funding in this study. to back of the head, Face breadth, and Eye to back of the head. Data of each dimension is presented, and the size of the helmet is proposed.

References

[1] M. Ali, M. M. S. Saeed, M. M. Ali, and N. Haidar, "Determinants of helmet use behaviour among employed motorcycle riders in Yazd, Iran based on theory of planned behaviour," *Injury*, vol. 42, no. 9, pp. 864–869, 2011.

IOP Conf. Series: Materials Science and Engineering **722** (2020) 012033 doi:10.1088/1757-899X/722/1/012033

[2] T. A. Trinh and T. P. L. Le, "Motorcycle helmet usage among children passengers: Role of parents as promoter," *Procedia Eng.*, vol. 142, pp. 10–17, 2016.

[3] V. Ratanavaraha and S. Jomnonkwao, "Community participation and behavioral changes of helmet use in Thailand," *Transp. Policy*, vol. 25, pp. 111–118, 2013.

[4] N. Yoshida and T. Koyanagi, "Empirical analysis of hazard perception and driving behaviors among high school and college students on motorcycles in Phnom Penh, Cambodia," *IATSS Res.*, vol. 42, no. 4, pp. 171–179, 2018.

[5] N. S. A. Sukor, A. K. M. Tarigan, and S. Fujii, "Analysis of correlations between psychological factors and self-reported behavior of motorcyclists in Malaysia, depending on self-reported usage of different types of motorcycle facility," *Transp. Res. Part F Traffic Psychol. Behav.*, vol. 46, pp. 509–523, 2017.

[6] T. Champahom, S. Jomnonkwao, T. Satiennam, N. Suesat, and V. Ratanavaraha, "Modeling of safety helmet use intention among students in urban and rural Thailand based on the theory of planned behavior and Locus of Control," *Soc. Sci. J.*, 2019.

[7] S. H. Keng, "Helmet use and motorcycle fatalities in Taiwan," *Accid. Anal. Prev.*, vol. 37, no. 2, pp. 349–355, 2005.

[8] S. Wadhwaniya *et al.*, "A comparison of observed and self-reported helmet use and associated factors among motorcyclists in Hyderabad city, India," *Public Health*, vol. 144, pp. S62–S69, 2017.

[9] M. Papadakaki *et al.*, "Barriers and facilitators of helmet use in a Greek sample of motorcycle riders: Which evidence?," *Transp. Res. Part F Traffic Psychol. Behav.*, vol. 18, pp. 189–198, 2013.

[10] H. H. Purba, H. Prayogo, R. Wibowo, Y. Pradipta, and S. Aisyah, "Increasing the Thermal Comfort, Ergonomics and Safety of Helmet by Using of Quality Function Deployment Method: A Case Study in Indonesia," vol. 4, no. 7, pp. 184–192, 2017.

[11] A. O. Adeodu, I. A. Daniyan, and O. O. Adaramola, "Anthropometry Evaluation Using Analytical Hierarchical Process and Independent Test Method for Hospital Bed Design : A Case Study in Nigeria," *J. Comput. Biosci. Eng.*, vol. 1, no. 1, pp. 1–10, 2014.

[12] M. Andriani, M. T. Hasan, N. Nazaruddin, and N. Ninafahriana, "Application of Anthropometry to Overcome Musculoskeletal Problems," *J. Phys. Conf. Ser.*, vol. 1114, no. 1, 2018.

[13] I. Z. Sutalaksana and A. Widyanti, "Anthropometry approach in workplace redesign in Indonesian Sundanese roof tile industries," *Int. J. Ind. Ergon.*, vol. 53, pp. 299–305, 2016.

[14] M. Wyganowska-Świątkowska, I. Kowalkowska, K. Mehr, and M. Dabrowski, "An anthropometric analysis of the head and face in vocal students," *Folia Phoniatr. Logop.*, vol. 65, no. 3, pp. 136–142, 2014.

[15] M. Vasu and A. Mital, "Table 1 . Stature values of males and females for the chosen percentiles in cm . Stature percentile selection," no. 2, pp. 304–306.

A literature review on reducing carbon emission from supply chain system: drivers, barriers, performance indicators, and practices

M F Ibrahim[1*], M M Putri [1] and D M Utama[2]
[1]Department of Logistics Engineering, Universitas Internasional Semen Indonesia, Gresik, Indonesia
[2]Department of Industrial Engineering, Universitas Muhammadiyah Malang, Malang, Indonesia

*E-mail: faisalibrahim.ie@gmail.com

Abstract. This study aims to review the literature in reducing carbon emission from supply chain system for the past few years and presenting drivers, barrier, performance indicator, and practice. The issue of reducing carbon emission become frequently discussed since unplanned and irresponsible actions by industries are potential threats to sustainability. Companies are now trying to minimize environmental impacts by integrating environmental concerns into their supply chain operations. In the aim of reducing carbon emissions, the industry is faced with different drivers, barriers, and performance indicator. With a variety of problems faced, the practices taken to reduce carbon emissions are very diverse. Thirty papers were collected from previous studies and linked to discussion points. The result showed that good coordination with various media in the supply chain system would be able to achieve common goals in reducing gas emissions.

1. Introduction

Supply chain management is strategically positioned for sustainability, performance, and organizational sustainability [1]. Furthermore, the supply chain is an essential branch of operations management, and it has an essential impact on the environment, including emissions, pollutions, and the health hazard of the community. However, unconsciously, inefficient supply chain management activities from upstream to downstream can cause excessive carbon gas emissions.

The increase of global warming and changing biodiversity has brought the world's sustainability towards immediate danger. People from different areas, including researchers, academicians, practitioners, and scientists got all together to suggest ways to maintain the environmental sustainability.

According to King and Lenox [2] claimed that unplanned and irresponsible actions by industries are potential threats to sustainability. Companies are now trying to minimize environmental impacts by integrating environmental concerns into their supply chain operations. According to Sarkis and Dou [3], the integration of supply chain elements with corporate environmental management that referred to as “Green Supply Chain.”

In today's globalized supply chains, environmental issues are of critical importance. During the past decade, carbon emissions and pollution associated with economic development have caused severe issues such as the greenhouse effect, abnormal climate, and environmental degradation [4]. Based on this background, the handling of gas emission in the company must be better and carefully planned. Many previous studies have tried to find strategies and scenarios in minimizing gas emissions.

This paper presents a review of reducing carbon emission problem, which classifies the study in terms of driver, barrier, and performance indicator. The remainder of this paper is organized as follows: Section 2, is for presenting the literature review. Section 3 is for drivers and barriers and performance indicators in Section 4. Section 5 is for presenting the practice. Section 6, which is, for the conclusion.

2. Literature Review

Tapia Granados and Spash [5] suggest that annual costs in emissions are highly correlated with economic growth. In California, the cap-and-trade scheme has been implemented since 2013, and it did not have a significant impact because the energy supply was minimal. When making decisions about cap-and-trade allowance allocations or carbon tax designs, policymakers should consider that industries may be concentrated and thus already limit output below socially optimal levels [6]. Carbon cap-and-trade regulation and green technology investment is an effective method of reducing carbon emissions however carbon cap-and-trade regulation and green technology investment is an effective method of reducing carbon emissions Bai, et al. [7].

Waltho, et al. [8] and Elhedhli and Merrick [9] try to build network design with the aim to reduce carbon emission. Four policies are recommended by Waltho, et al. [8] that successfully reduce emissions effectively without increasing high costs, i.e. carbon offset, carbon cap, carbon tax, and cap-and-trade. Cap-and-trade is the most preferred policy because it has high flexibility. Peng, et al. [10] utilize structural decomposition analysis and structural path decomposition analysis method to identify air pollution emission. Adjusting the energy consumption structure as soon as make some guiding consumption will reduce emissions gradually.

According to Valadkhani, et al. [11], the country's economic conditions are an essential part of the transition to renewable energy. They study energy consumption portfolio by utilizing data from 79 countries from 1965 to 2017 and investigate the contribution of consumption of primary energy (i.e., coal, gas, oil) to changes in CO2 emission per capita. The result of the study suggests that only high-income countries are eligible to switch to renewable energy. On the other hand, in low-income countries, natural gas remains a potential solution.

Du, et al. [12] utilize a system dynamic method to investigates carbon reduction technology policies to the level of economic growth. According to the study, emissions could be reduced by promoting the development of carbon reduction technology. One of the policies that can be implemented is promoting the use of renewable energy in related industries so that it can reduce carbon emissions. Recovery supply chain model with consideration of safety stock and carbon emission has been developed by Darom, et al. [13]. In this model, safety stock is used as an essential mitigation strategy in overcoming supply disruptions, and the cost of carbon emissions is also considered for its relationship with the environment.

Tseng and Hung [14] developed a strategic decision-making model considering social costs and operational cost costs incurred due to carbon dioxide emissions during the process in the supply chain system. Evaluation is carried out through several different scenarios with case studies on industrial apparels. The results of the study suggested that the government make strict regulations on companies to bear the social costs of carbon emissions produced. This needs to be done because, from the experimental results, it can be concluded that the higher the level of social costs can reduce carbon emissions effectively.

According to Wang, et al. [15], the retailer can achieve the goal of jointly reducing carbon emission with the manufacturer. In their study, numerical experiment and analysis successfully were done to get the optimal cost-sharing rate and the optimal premium rate of the wholesale price. Incentive mechanism policy from the government to work collaboratively among supply chain member to reduce carbon emission has been investigated by Ding, et al. [4]. The results show that collaboration between supply chain members plays a crucial role in improving their environmental performance. Ji, et al. [16] has been compared two emission reduction strategies, i.e., production strategy and joint emission reduction strategy, the result is joined emission reduction strategy is more profitable for manufacturer and retailer.

In recent years, the business and scientists communities have a significant interest to discuss the danger of gas emissions that always increases [17]. According to Chen, et al. [17], the main goal of emission reduction will not succeed if companies often choose actions to reduce emissions without coordination with other business partners. This kind of event results in an increase in emissions for other partners which will not cause emissions reductions in the supply chain system

Daryanto, et al. [18], Zissis, et al. [19], Peng, et al. [20], Bai, et al. [21], Liu [22], Xu, et al. [23], Ding, et al. [24], [25] in their research they found that integration in the supply chain system can be an essential way to reduce total costs and carbon emissions. Contracts are needed that are designed in such a way as to create the coordination of the entire supply chain system to make perfect coordination between players in the supply chain [26]. Coordination between players in the supply chain system has succeeded in creating optimal costs, especially in transportation costs and carbon emissions [27]. Improving performance and obtaining the benefits of an overall supply chain system will be achieved if the supply chain can be appropriately coordinated [28].

According to Zissis, et al. [19], just the right environmentally friendly policies can save the total cost of the supply chain system. They suggest a quantity discount mechanism to coordinate players along the supply chain. Daryanto, et al. [18] develop three-echelon supply chain models to determines the optimal number and size of the delivery. Delivery starts from supplier to third-party logistic (3PL) and from 3PL to buyers to reduce carbon emissions. Zhou and Ye [26] analyzed shared emission reduction strategies and contract design in multiple channel supply chains under a low carbon environment. Cooperative advertising contract and emission reduction cost-sharing contract successfully to make more supply chain improvements.

Utilizing quantity discount contracts as a medium of coordination between players in the supply chain system has been done by Peng, et al. [20]. The result show, that quantity discount contracts, and revenue sharing contract can coordinate the low-carbon supply chain perfectly with some several combinations.

According to Dash Wu, et al. [29], the win-win solution achieved when manufacturer invests more in reducing emissions. They design a contract to coordinate the supply chain under optimal order quantity, the optimal wholesale price of the manufacturer, and optimal carbon emission level to be the focus for exploration.

Bai, et al. [21] analyze coordination effect on emission reduction, revenue-and investment-sharing (RIS) contract successfully coordinating the supply chain system perfectly. RIS can make supplier and manufacturer gain more profit and not negatively affected by carbon cap. In other studies Liu [22] and Xu, et al. [30] also succeeded in creating a low-carbon supply chain through revenue sharing contract.

Cost-sharing and wholesale price contract is some contract that can coordinate supply chain Xu, et al. [23]. They focus on the problem of coordinating supply chain systems under cap-and-trade regulation. With proper coordination, carbon emissions can be reduced without affecting the profit of each player in the supply chain system. One example, although the retailer shares the manufacturer's total cost, the manufacturer may increase the wholesale price, which indicates that the manufacturer can charge a high price for the low-carbon product.

Ding, et al. [24] developed an optimal pricing strategy model for environmentally friendly supply chain system. In this supply chain model, government and manufacturing have mutually integrated with strong coordination relationships. The results of this study suggest that the performance of an environmentally friendly supply chain system can be improved by the existence of good cooperation between manufacturers and suppliers, regardless of the role of the government.

Table 1. Paper Classification According to Source of Drivers, Barriers and Performance Indicator

No.	Authors	Drivers Source		Barriers Source		Performance Indicator			Practices
		IN	EX	IN	EX	EC	EN	OP	
1.	Tapia Granados and Spash (2019)		X		X	X			Policies (must be balanced with renewable material and energy supply)
2.	Valadkhani, Nguyen, and Bowden (2019)	X		X		X			Renewables energy policies (only eligible to high-income countries)
3.	Di Filippo, Karpman, and DeShazo (2019)		X	X		X			Carbon-pricing policies (limit output and optimal level)
4.	Du et al. (2019)	X		X			X		Low carbon development policy (technology need)
5.	Chen, Benjaafar, and Elomri (2019)	X		X	X				Penalties in the decentralized supply chain (buyer-vendor supply chain framework)
6.	Daryanto, Wee, and Astanti (2019)	X			X	X	X		Optimizes the number of deliveries and delivery size (SCM contract)
7.	Bai, Gong, Jin, and Xu (2019)	X		X	X	X			Carbon cap-and-trade regulation and green technology investment
8.	Liu (2019)		X			X	X		Revenue sharing contract, pricing, and targeted advertisement policy
9.	Zissis, Saharidis, Aktas, and Ioannou (2018)	X			X	X			Supply chain coordination with quantity discount
10.	J. Peng, Zhang, Xie, and Liu (2018)	X			X		X		Coordinate the relationship between household consumption by adjusting energy consumption structure
11.	Darom, Hishamuddin, Ramli, and Mat Nopiah (2018)	X				X			a recovery model with safety stock as the mitigation strategy to respond to environmental issues
12.	Zhou and Ye (2018)	X			X	X			Shared emission reduction strategies and contract design under a low-carbon environment
13.	H. Peng, Pang, and Cong (2018)		X		X		X		quantity discount contracts as a medium of coordination between players in the supply chain system
14.	Dash Wu, Yang, and Olson (2018)	X			X	X			contract to coordinate supply chain under optimal order quantity, the optimal wholesale price of the manufacturer, and optimal carbon emission level
15.	Bai, Xu, and Zhang (2018)	X	X		X	X			coordination with emission reduction, revenue-and investment-sharing (RIS) contract
16.	L. Xu, Wang, and Zhao (2018)	X	X		X	X			Revenue sharing contract
17.	Aljazzar, Gurtu, and Jaber (2018)		X		X	X			Delay in payment to improve economic performance
18.	X. Xu, He, Xu, and Zhang (2017)		X	X	X	X			Coordinate by Wholesale price and cost-sharing contract
19.	Ji, Zhang, and Yang (2017)	X		X		X			Production strategy and join emission reduction strategy
20.	Yang, Zhang, and Ji (2017)	X	X	X	X	X			Profit-sharing contracts
21.	Bai, Xu, Xu, and Wang (2016)	X	X		X	X			Revised revenue sharing contract
22.	Wang, Zhao, and He (2016)	X		X	X		X		Retailer and manufacturer join carbon reduction contract
23.	Ding, Zhao, An, and Tang (2016)		X	X	X	X			Incentive mechanism policy
24.	Zhou, Bao, Chen, and Xu (2016)		X	X	X	X			co-op advertising contract and the co-op advertising & emission reduction cost-sharing contracts
25.	Sarkar, Ganguly, Sarkar, and Pareek (2016)	X		X		X			Coordination between players
26.	Ding, Zhao, An, Xu, and Liu (2015)	X			X	X			Coordination with optimal pricing strategies
27.	Tseng and Hung (2014)		X	X		X			Government regulation, level of social costs effectively reduce carbon emissions
28.	Zhang, Wang, and Ren (2014)	X	X	X	X	X			Pricing and coordination strategy
29.	Liu, Kasturiratne, and Moizer (2012)		X	X	X	X	X		Multi-dimensional integration strategy
30.	Elhedhli and Merrick (2012)		X	X	X	X	X		Network design

Zhang, et al. [31], has investigated the issue of pricing and coordination of players in the green supply chain system. System performance in cooperative games is better than non-cooperative games.

Deteriorating item in inventory management is the challenging issues [28]. Research conducted by [28] proposes a revised revenue-sharing (RRS) contract to coordinate the decentralized system. They constructed two models with multivariate decisions for both decentralized and centralized systems. Then proposed the use of the RS contract and RRS contract to coordinate the system. The result show RSS contract successfully coordinates the supply chain system perfectly and give a higher profit than RS contract.

Yang, et al. [32] has analyzed two competitive supply chains under the cap and trade scheme. The win-win solution between manufacturers and retailers could be achieved effectively with profit-sharing contracts provided by retailers, while also benefiting the environment with a higher level of emission reduction. Members in the supply chain system could accept the cap and trade mechanism only if consumers had strong low carbon preferences Ji, et al. [16].

According to studies by Zhou, et al. [33] Co-op advertising and emission reduction cost-sharing contracts can achieve channel coordination and achieve win-win situations under certain conditions. On another research by Aljazzar, et al. [34], delay in payment has successfully improved the economic performance of the supply chain system and environmental performance.

3. Source of Drivers and Barriers to Reducing Carbon Emission from Supply Chain System

To reduce carbon emission from a supply chain system, there are various drivers and barriers in the process of implementation. Drivers in this study are factors that can encourage companies to reduce carbon emissions, for example, the target of company image or competitive pressure. While the barrier is factors that inhibit some process from reducing carbon emission. According to Tseng, et al. [35], drivers on reducing carbon emission can be classified into internal and external sources.

A. Internal Source of Drivers on Reducing Carbon Emission

The drivers come from internal organizational, for example, company green image, the requirement of ISO 14000, internal environmental awareness, resource efficiency by way of reducing waste.

B. External Source of Drivers on Reducing Carbon Emission

The drivers come from external organizational, for example, a regulator from government, a regulator from an environmental group, awareness-raising of business partners, competitive pressure because of competitors, requirements of business partners.

On the other hand, barriers to reducing carbon emission are also classified into internal and external sources.

A. Internal Source of Barriers to Reducing Carbon Emission

The barriers come from internal organizational, for example, lack of environmental knowledge, lack of environmental awareness, new system switching cost, environmentally friendly design costs, financial problem, lack of top management involvement, lack of collaboration between departments, afraid of failure, lack and high cost of eco-technology.

B. External Source of Barriers to Reducing Carbon Emission

The barriers come from external organizational, for example, lack of government support, lack of eco-friendly training, no eco-friendly appreciation from partners or government, lack of awareness between partners, market uncertainty, lack of commitment between partners, lack of support from related parties.

4. Performance Indicator

Performance Indicator is a unit of measurement that can describe how successful the implementation of policies in reducing carbon emissions. In general, economic factors are the most frequently used performance indicators. According to Huang, et al. [36] refined by Tseng, et al. [35], in general, performance indicators are divided into three major groups. Performance can be seen through environmental performance, economic performance, and operational performance.

A. Economic Performance Indicator

Performance valuation indicators related to economic impacts, for example, minimizing costs, new market opportunities, profit margin, sales increase, market share, net income, cost of goods sold (COGS), overall business performance, cost to the customer.

B. Environmental Performance Indicator

Performance valuation indicators related to environmental impacts, for example environmental performance, waste minimization, environmental impact analysis, pollution level.

C. Operational Performance Indicator

Performance valuation indicators related to operational impacts, for example efficiency improvement, quality improvement, productivity improvement, delivery accuracy and speed, level of flexibility, operational performance, minimize lead time.

5. Practice

In the literature, this review found a variety of practices that had been submitted by previous researchers. Various policies from the government have a good impact on reducing gas emissions. These policies include renewables energy, carbon-pricing, low carbon development, and incentive mechanism policy. To produce the best output, these policies must be supported by all stakeholders. According to Tapia Granados and Spash [5], policies that have been made will run smoothly if balanced with the availability of renewable material and energy supply. Thus, the policy to reduce gas emissions will not succeed if it is not balanced with the supply of renewable energy. Renewables energy policies were only eligible for high-income countries [11]. Supporting technology is very important for the success of policy [12].

Green supply chains will be challenging to achieve in the decentralized supply chain. According to Chen, et al. [17], the use of penalties in the decentralization supply chain is quite influential on low-carbon policies. Coordination can be the best way to achieve common goals in a supply chain system. In the literature review, several media in the coordinated contract were found for coordinating the supply chain and achieve the goal of reducing gas emissions. The coordination media include quantity discount, energy consumption structure, optimal safety stock, optimal order quantity, optimal wholesale price, optimal carbon emission level, revenue-and-investment sharing, pricing, delay in payment, production strategy, deliveries, deliveries size, and profit-sharing.

Zissis, et al. [19] and Peng, et al. [20] use quantity discount as a medium in coordinating the supply chain to reduce carbon emission. Energy consumption structure successfully used by Peng, et al. [10]. Darom, et al. [13] succeeded in using safety stock with consideration of emissions. According to past research optimal order quantity [29], optimal wholesale price [29];[23], optimal carbon emission level [29] under a coordinated supply chain can reduce carbon emission effectively. In another side, revenue sharing is used by [22] [30] [28] and investment sharing used by [21] to coordinate the supply chain to reduce carbon emission. Other than that, Liu [22], Ding, et al. [24], and Zhang, et al. [31] utilize a pricing strategy to coordinated supply chain and reduce carbon emission. Moreover, coordination has also been carried out with other media, among others delay in payment [34], production strategy [16], deliveries [18], deliveries size [18], and profit sharing [32].

Based on the practices that have been carried out in previous studies, it can be concluded that coordinated supply chains can reduce gas emissions more effectively.

Through proper coordination with any media, the common goal of becoming a green supply chain will be easier to achieve.

6. Conclusion

Reduce carbon emission has recently become a company goal. The emergence of concern about carbon emissions can come from internal and external companies. Many companies are directly aware of the awareness of reducing carbon emissions. However, it is not uncommon for companies that actually do not have awareness but must do so because of government policies related to carbon emissions. In this paper, 30 papers relating to the practice of reducing carbon emissions were reviewed. The author tries to explore drivers, barriers, performance indicators, and practices in reducing carbon emissions from each paper.

Some papers suggest that drivers of reducing gas emissions originate from the company's internal awareness. On the other hand, some papers also say that consciousness occurs due to coercion or the terms requested by external parties. Regarding performance indicators, most papers use economic value as a performance indicator

Most of the papers suggest that coordination must be done with various media. With proper coordination in the supply chain system, a common goal will be achieved in reducing gas emissions. For future research work, it is essential to combine approaches that can more effectively and efficiently coordinate every player in the supply chain system to reducing carbon emission. As we know, based on this literature review coordination is very important in achieving the common goal of reducing carbon emissions.

This review concludes that there is a relationship between approaches in the aim of reducing carbon emissions related to drivers, barriers, performance indicators, and practices.

7. References

[1] H. Reefke and D. Sundaram, "Key themes and research opportunities in sustainable supply chain management – identification and evaluation," *Omega,* vol. 66, pp. 195-211, 2017/01/01/ 2017.

[2] A. A. King and M. J. Lenox, "Industry Self-Regulation without Sanctions: The Chemical Industry's Responsible Care Program," *The Academy of Management Journal,* vol. 43, no. 4, pp. 698-716, 2000.

[3] J. Sarkis and Y. Dou, *Green Supply Chain Management: A Concise Introduction*. Taylor & Francis, 2017.

[4] H. Ding, Q. Zhao, Z. An, and O. Tang, "Collaborative mechanism of a sustainable supply chain with environmental constraints and carbon caps," *International Journal of Production Economics,* vol. 181, pp. 191-207, 2016/11/01/ 2016.

[5] J. A. Tapia Granados and C. L. Spash, "Policies to reduce CO2 emissions: Fallacies and evidence from the United States and California," *Environmental Science & Policy,* vol. 94, pp. 262-266, 2019/04/01/ 2019.

[6] J. Di Filippo, J. Karpman, and J. R. DeShazo, "The impacts of policies to reduce CO2 emissions within the concrete supply chain," *Cement and Concrete Composites,* vol. 101, pp. 67-82, 2019/08/01/ 2019.

[7] Q. Bai, Y. Gong, M. Jin, and X. Xu, "Effects of carbon emission reduction on supply chain coordination with vendor-managed deteriorating product inventory," *International Journal of Production Economics,* vol. 208, pp. 83-99, 2019/02/01/ 2019.

[8] C. Waltho, S. Elhedhli, and F. Gzara, "Green supply chain network design: A review focused on policy adoption and emission quantification," *International Journal of Production Economics,* vol. 208, pp. 305-318, 2019/02/01/ 2019.

[9] S. Elhedhli and R. Merrick, "Green supply chain network design to reduce carbon emissions," *Transportation Research Part D: Transport and Environment,* vol. 17, no. 5, pp. 370-379, 2012/07/01/ 2012.

[10] J. Peng, Y. Zhang, R. Xie, and Y. Liu, "Analysis of driving factors on China's air pollution emissions from the view of critical supply chains," *Journal of Cleaner Production,* vol. 203, pp. 197-209, 2018/12/01/ 2018.

[11] A. Valadkhani, J. Nguyen, and M. Bowden, "Pathways to reduce CO2 emissions as countries proceed through stages of economic development," *Energy Policy,* vol. 129, pp. 268-278, 2019/06/01/ 2019.

[12] Q. Du, L. Shao, J. Zhou, N. Huang, T. Bao, and C. Hao, "Dynamics and scenarios of carbon emissions in China's construction industry," *Sustainable Cities and Society,* 2019/04/30/ 2019.

[13] N. A. Darom, H. Hishamuddin, R. Ramli, and Z. Mat Nopiah, "An inventory model of supply chain disruption recovery with safety stock and carbon emission consideration," *Journal of Cleaner Production,* vol. 197, pp. 1011-1021, 2018/10/01/ 2018.

[14] S.-C. Tseng and S.-W. Hung, "A strategic decision-making model considering the social costs of carbon dioxide emissions for sustainable supply chain management," *Journal of Environmental Management,* vol. 133, pp. 315-322, 2014/01/15/ 2014.

[15] Q. Wang, D. Zhao, and L. He, "Contracting emission reduction for supply chains considering market low-carbon preference," *Journal of Cleaner Production,* vol. 120, pp. 72-84, 2016/05/01/ 2016.

[16] J. Ji, Z. Zhang, and L. Yang, "Carbon emission reduction decisions in the retail-/dual-channel supply chain with consumers' preference," *Journal of Cleaner Production,* vol. 141, pp. 852-867, 2017/01/10/ 2017.

[17] X. Chen, S. Benjaafar, and A. Elomri, "On the effectiveness of emission penalties in decentralized supply chains," *European Journal of Operational Research,* vol. 274, no. 3, pp. 1155-1167, 2019/05/01/ 2019.

[18] Y. Daryanto, H. M. Wee, and R. D. Astanti, "Three-echelon supply chain model considering carbon emission and item deterioration," *Transportation Research Part E: Logistics and Transportation Review,* vol. 122, pp. 368-383, 2019/02/01/ 2019.

[19] D. Zissis, G. K. D. Saharidis, E. Aktas, and G. Ioannou, "Emission reduction via supply chain coordination," *Transportation Research Part D: Transport and Environment,* vol. 62, pp. 36-46, 2018/07/01/ 2018.

[20] H. Peng, T. Pang, and J. Cong, "Coordination contracts for a supply chain with yield uncertainty and low-carbon preference," *Journal of Cleaner Production,* vol. 205, pp. 291-302, 2018/12/20/ 2018.

[21] Q. Bai, J. Xu, and Y. Zhang, "Emission reduction decision and coordination of a make-to-order supply chain with two products under cap-and-trade regulation," *Computers & Industrial Engineering,* vol. 119, pp. 131-145, 2018/05/01/ 2018.

[22] P. Liu, "Pricing policies and coordination of low-carbon supply chain considering targeted advertisement and carbon emission reduction costs in the big data environment," *Journal of Cleaner Production,* vol. 210, pp. 343-357, 2019/02/10/ 2019.

[23] X. Xu, P. He, H. Xu, and Q. Zhang, "Supply chain coordination with green technology under cap-and-trade regulation," *International Journal of Production Economics,* vol. 183, pp. 433-442, 2017/01/01/ 2017.

[24] H. Ding, Q. Zhao, Z. An, J. Xu, and Q. Liu, "Pricing strategy of environmental sustainable supply chain with internalizing externalities," *International Journal of Production Economics,* vol. 170, pp. 563-575, 2015/12/01/ 2015.

[25] S. Liu, D. Kasturiratne, and J. Moizer, "A hub-and-spoke model for multi-dimensional integration of green marketing and sustainable supply chain management," *Industrial Marketing Management,* vol. 41, no. 4, pp. 581-588, 2012/05/01/ 2012.

[26] Y. Zhou and X. Ye, "Differential game model of joint emission reduction strategies and contract design in a dual-channel supply chain," *Journal of Cleaner Production,* vol. 190, pp. 592-607, 2018/07/20/ 2018.

[27] B. Sarkar, B. Ganguly, M. Sarkar, and S. Pareek, "Effect of variable transportation and carbon emission in a three-echelon supply chain model," *Transportation Research Part E: Logistics and Transportation Review,* vol. 91, pp. 112-128, 2016/07/01/ 2016.

[28] Q. Bai, X. Xu, J. Xu, and D. Wang, "Coordinating a supply chain for deteriorating items with multi-factor-dependent demand over a finite planning horizon," *Applied Mathematical Modelling,* vol. 40, no. 21, pp. 9342-9361, 2016/11/01/ 2016.

[29] D. Dash Wu, L. Yang, and D. L. Olson, "Green supply chain management under capital constraint," *International Journal of Production Economics,* 2018/09/17/ 2018.

[30] L. Xu, C. Wang, and J. Zhao, "Decision and coordination in the dual-channel supply chain considering cap-and-trade regulation," *Journal of Cleaner Production,* vol. 197, pp. 551-561, 2018/10/01/ 2018.

[31] C.-T. Zhang, H.-X. Wang, and M.-L. Ren, "Research on pricing and coordination strategy of green supply chain under hybrid production mode," *Computers & Industrial Engineering,* vol. 72, pp. 24-31, 2014/06/01/ 2014.

[32] L. Yang, Q. Zhang, and J. Ji, "Pricing and carbon emission reduction decisions in supply chains with vertical and horizontal cooperation," *International Journal of Production Economics,* vol. 191, pp. 286-297, 2017/09/01/ 2017.

[33] Y. Zhou, M. Bao, X. Chen, and X. Xu, "Co-op advertising and emission reduction cost sharing contracts and coordination in low-carbon supply chain based on fairness concerns," *Journal of Cleaner Production,* vol. 133, pp. 402-413, 2016/10/01/ 2016.

[34] S. M. Aljazzar, A. Gurtu, and M. Y. Jaber, "Delay-in-payments - A strategy to reduce carbon emissions from supply chains," *Journal of Cleaner Production,* vol. 170, pp. 636-644, 2018/01/01/ 2018.

[35] M.-L. Tseng, M. S. Islam, N. Karia, F. A. Fauzi, and S. Afrin, "A literature review on green supply chain management: Trends and future challenges," *Resources, Conservation and Recycling,* vol. 141, pp. 145-162, 2019/02/01/ 2019.

[36] Y.-C. Huang, C.-H. Huang, and M.-L. Yang, "Drivers of green supply chain initiatives and performance: Evidence from the electrical and electronics industries in Taiwan," *International Journal of Physical Distribution & Logistics Management,* vol. 47, no. 9, pp. 796-819, 2017/10/02 2017.

Acknowledgement
This research is supported and financed by the Ministry of Research, Technology, and Higher Education of the Republic of Indonesia under the PDP scheme.

IOP Conf. Series: Materials Science and Engineering **722** (2020) 012035 doi:10.1088/1757-899X/722/1/012035

Corn husk Fibers Reinforced Polyester Composites: Tensile Strength Properties, Water Absorption Behavior, and Morphology

N H Sari[1], S Suteja[2]

[1,2] Mechanical Engineering Department, Faculty of Engineering, University of Mataram, Mataram, Indonesia

Corresponding author e–mail: n.herlinasari@unram.ac.id

Abstract. The effects of fiber content and immersion time in the water on the tensile, morphology and water absorption properties of composites made from corn husk fiber/polyester have been studied. Composite made with a variety of different fiber contents namely: 20%, 25%, 30%, 40%, 50% and 60% respectively. All composite specimens were immersed in water for 24 h and 72 h. The effects of fiber content and time of immersion of composites in water have been determined by examining the nature of tensile strength, water absorption behavior, morphology. The results demonstrated that after soaked in water for 24 h and 72 h, the water absorption properties of the composites increased with increasing fiber content. The tensile strength and modulus of elasticity of composites tend to increase from 20% to 30% fiber content after immersed for 24 h, and then decrease with increasing fiber and soaking time because the interface between fiber and polyester becomes weak. These results suggest that corn husk fiber composites could have the potential to use as decking, siding, and exterior windows.

1. Introduction

Composites made from natural fibers and thermoset resins are increasingly developing in the polymer industry, specifically as a substitute for wood for outdoor products under wet environments. The advantages of hydrophobic resins have been to protect natural fibers and increase the durability of the final product; therefore they are successfully used in structures such as decking, docks, and exterior windows, etc [1] that are directly in contact with water.

Although natural fiber composites are widely used in many industries, long–term performance and durability are still not comprehensively understood. The fibers and adhesives are inevitable from changes and hostile environmental conditions. Water diffusion in composites and polymer adhesives is considered as one of the main reliability problems for the performance of composites.

Composites made from corn fiber are increasingly interesting to study and their properties still need to be developed. Some researchers have reported the best properties of corn husk fiber composites (CHF) with a polymer matrix. CHF composites with polyester matrices have a sound absorption coefficient of 0.8–0.9 at a frequency of 2 kHz. They also reported the tensile and Young's moduli of CHF composites around 18.81 MPa–25.73 MPa [2]. The ability to absorb sound from CHF-polypropylene composites is superior to jute–polypropylene [3]. Cornhusk fiber plastic composites had the highest flexural and tensile strengths of 46.10 MPa and 26.58 MPa respectively [4]. 5% CHF composites showed deformability than 0%–8% CHF with low methoxy pectin (LMP) films [5]. From this previous study, it was agreed that other properties associated with corn husk fiber composites are still very limited.

As materials to be applied to a structure under a wet environment, the absorbed moisture will cause changes in the polymer microstructure, and degradation in their mechanical, thermo-physical, and chemical characteristics [6–8, 2]. The effect of moisture or water exposition on mechanical, morphology, and water absorption of composites is very important to be studied and to explain the performance of composites in wet environments.

Therefore, this study aims to explore the properties of corn husk fiber composites in water immersion. The effects of the CHF content on water absorption behaviour, tensile strength properties, and morphology were investigated.

2. Materials and Methods

2.1. Materials.

Cornhusk has been obtained from the Pagesangan market, Mataram, Indonesia (see in **Fig 1a**). Corn husk selected on the outside; to maintain uniformity. The average length and width of corn husk is 13.5 cm–15.2 cm. The polyester resin (PE) has a density of 1.2 g/cm^3, the tensile strength and a tensile modulus of 8.8 kg/mm^2 and 500 kg/mm^2, respectively, and elongation of 2.3%.

2.2. Extraction of fibers

They are immersed in fresh water for 10 days to undergo decay (**Fig. 1b**). The fibers were taken using a wooden comb with teeth diameter of 0.02 mm; to maintain fiber uniformity, and dried under the sun's heat (**Fig. 1c**).

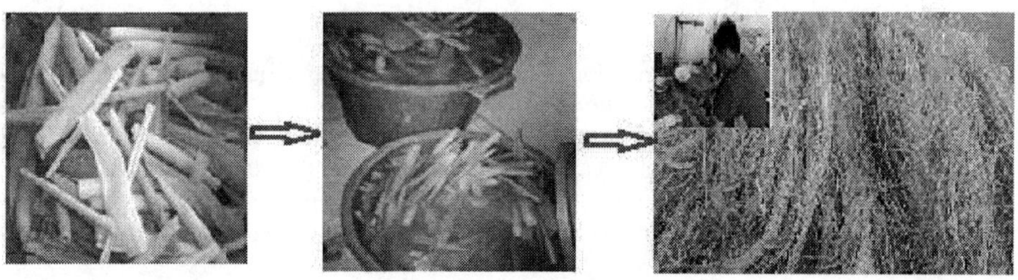

Figure 1. Material preparation, a. Corn husk, b. Extration of CHFs, and c. CHF raw.

2.3. Alkaly treatment of fibers

The prepared CHFs were immersed in NaOH 8% for 2 h (**Fig. 2a**). They are washed and rinsed with fresh water and repeated three times, and then followed by drying under the sun (**Fig. 2b**), then dry CHFs (**Fig. 2c)** are stored in a platics storage box.

Figure 2. Alkaly treatment process, a. CHFs immersion in NaOH8% for 2 h, b. The drying process of CHFs, and c. CHFs after alkaly treatment.

2.4. Preparation of composites

CHF with a fiber length of 4 cm was prepared. A mixture of polyester and catalyst is poured into a mold that filled CHFs with different volume fraction (see in **Table 1**). Then the mold is closed and pressed at 5MPa for 4 min (temperature 175°C), followed by cooling at ambient temperature. The composite is removed from the mold and ready for testing.

Table 1. Composition of CHF and polyester of different composites

Code of specimens	Corn husk fibers (% vol)	Polyester (% vol)
NC20	20	80
NC25	25	75
NC30	30	70
NC40	40	60
NC50	50	50
NC60	60	40

All of the test specimens are given the carrying out by soaking them in water for 24 h and 72 h. They are lifted and wiped using blotting paper. In total there were 54 test specimens with repetition 3 times for each test parameter tests.

2.5. Characterization

2.5.1. Water absorption and Swelling tests. Water absorption tests have been measured according to the international standard ASTMD570 [9]. The water absorption (WA) was calculated using expression 1 [10]:

$$WA,\% = \frac{(N_1 - N_2)}{N_1} x100 \qquad (1)$$

Where, N_1 and N_2 show the dry weight (g), and the weight after time t (g).

Measurement of the percentage swellability evaluated using equation 2 [8]:

$$Swellability,\% = \frac{(x - y)}{y} x100 \qquad (2)$$

Where, x and y are composite volumes after and before immersion.

2.5.2. Tensile strength test. The specimen prepared was according to ASTMD3039 standard [11] used a Tensilon RTG–1310 that operated at a speed and load cell of of 5 mm/min and 5 kN respectively.

2.5.3. Flexural test. The CHF composites were conducted by the three–point bending method on a UTM and follow ASTMD790 standards [12]

2.5.4. Scanning electronic microscopy, SEM. In this test, the fracture surfaces of the specimen were characterized by SEM Inspect–S50type at 18mA and 10 kV.

3. Results and Discussion

3.1. Water absorption and swelling analysis

The water absorption capacity of composite as display in **Fig. 3**. The nature of water uptake increases with an increasing amount of fiber content and immersion time of the composite. During 24 h – 72 h period, the polyester resin demonstrated negligible water uptake and the CHF induced significant water uptake. Maximum water uptake is obtained from polyester composites with a volume fraction of 60% CHF (NC60) of 5.62% at 72 h immersion.

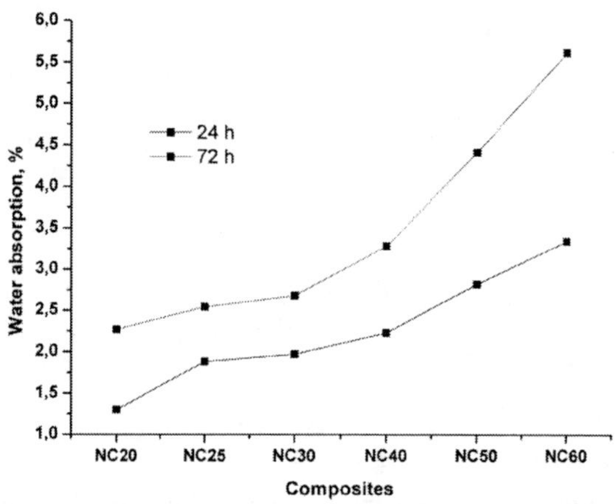

Figure 3. Water absorption of corn husk fiber/polyester composites.

A possible reason for this behavior might be because CHF shows tendency to absorb water higher than polyester (hydrophobic). The presence of lumens, defects, fissures at the interface, hydrogen bonds in fibers, and micro crevices in the matrix can cause the composite to absorb water [13,14]. Hence, the water uptake increases with more CHFs content.

Conversely, composites with low CHF content have better interface adhesion which reduces the interface width between fibers and reduces water uptake through this part to the interior of the composite [15]. It was noted that fiber adhesion/strong interface can help reduce water hygroscopicity, reduce penetration, hence avoiding deterioration in the mechanical performance of composites [16–18]. This also answers the reason why the ability to absorb water from NC20 is lower than other samples. This result has been confirmed by mechanical test results.

Typical swelling data for all composites displayed in **Fig. 4**, which shows that CHF/polyester swelling increases with increased water absorption, and thus the rate of swelling changes increases with immersion time. The effect of CHF on the polyester ratio on swelling thickness can also be explained by the difference in water uptake between CHF- polyester (see discussion on composite water absorption). Thickness swelling is affected by water uptake and change due to the same mechanism as water uptake.

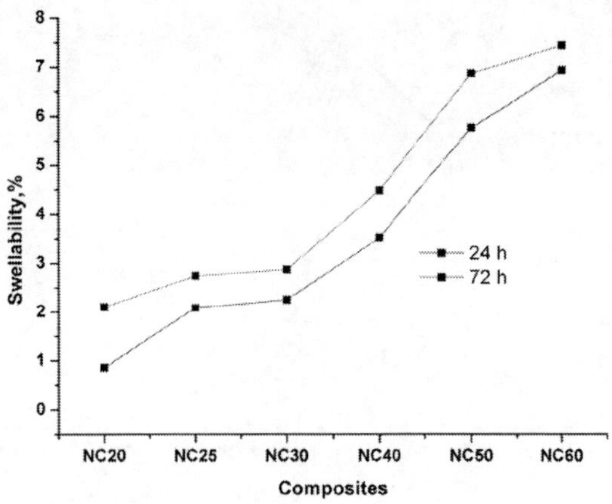

Figure 4. Swelling of polyester/corn husk fiber composites.

3.2. Tensile strength analysis

From **Fig. 5a** shows that the strength of composites with a 20 – 30% fo fiber content (NC20, NC25, and NC30) at the 24 h immersion stage tends to increase due to the strong bonding interface between polyester and CHF, and after being soaked in water for 72 h, the strength of the composite tends to decrease with the increasing number of CHFs. This is indicated that the amount of water absorbed in the composite has caused the interface bond between CHF and resin to be quite weak; as a result, the tensile and modulus of elasticity (MOE) of the composite tend to decrease with longer immersion time. A sharp decrease in tensile strength and Young's modulus values was also seen in composites with 40 – 60% fiber content (see NC40, NC50, and NC60 specimens). This drastic decrease is indicated that when CHFs content was increased, the matrix is no longer evenly distributed and many CHFs overlap one another, resulting in bad bond at the interface, causing the composite strength to be small. The same tendency behaviour is also found in the modulus of elasticity of the composite (**Fig. 5b**). The Young's moduli demonstrated a gradual increase, its value increased up to 30% CHF content then decreased; it is attributed to the flow of polyester which increased the bond strength and the composite strength.

Furthermore, the NC20, NC25, and NC30 composites have increasing strain values (seen in **Fig. 5c**), this means that there is an opposite response to the large tensile load received, which is indicated by the effect of internal shifts at the atomic level in the composite material so that the composite increases in length thus the strain produced to be high. Conversely, the low strain value is due to the opposite response given by the composite to the small tensile load received.

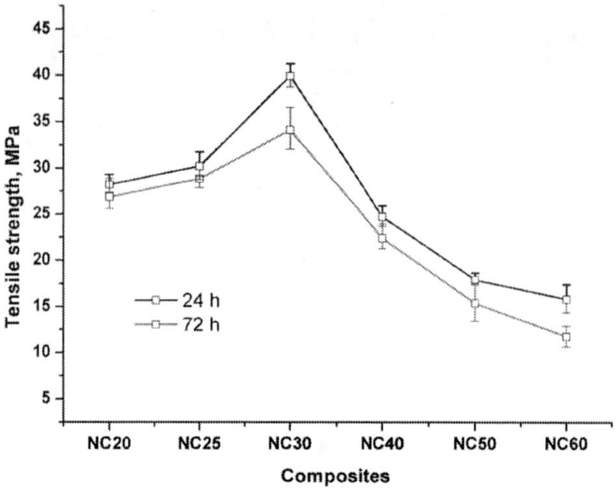

(a)

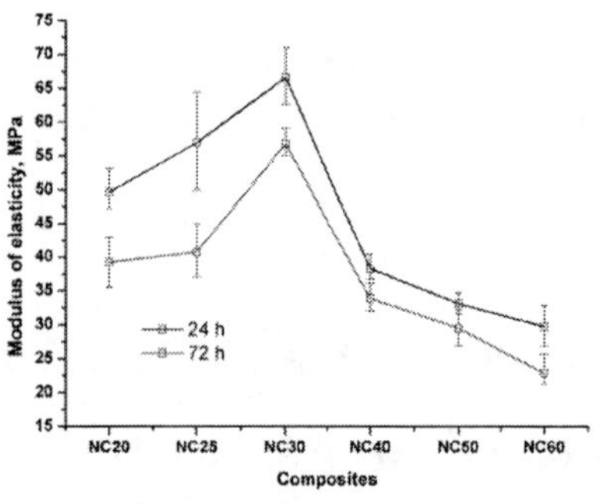

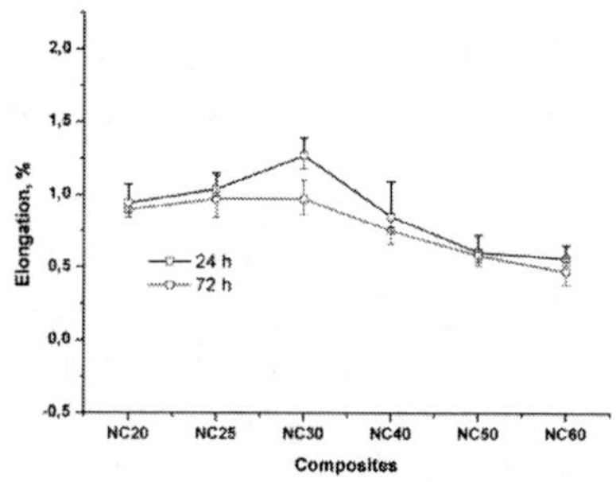

(b). **(c)**

Figure 5. a. Strength, b. Young's moduli, and elongation of tensile of corn husk fiber/polyester composite

3.3. Flexural strength analysis

Fig. 6 shows the flexural strength of the composite which tends to be the same as the tensile strength. From Fig. 6 it is known that the average flexural strength value of the composite after immersed in water for 24 h varies from 36.495 MPa to 35.650 MPa, and after immersed for 72 h the flexural strength varies from 30.301 MPa to 32.370 MPa. The flexural strength trend is seen to increase with increasing CHF content from 20% to 25%, and subsequently decreasing. Maximum flexural strength is obtained in the composite after soaking 24 h with 25% fiber content and 75% PE resin of 42.739 MPa. The detected increase in term of tensile and bending behaviour was related to interfacial bonding between the CHF-polyester, and the modification of single corn husk fibers. After 72 h of soaking, the flexural strength of the composite is known to decrease. For example, NC25 with a 25% fiber content have a higher flexural strength value (42.194 MPa) when compared to NC50 specimen which have a flexural strength of only around 36.192 MPa. This decrease was maybe due to the higher level of brittleness of the incorporation of CHF into the PE.

IOP Conf. Series: Materials Science and Engineering **722** (2020) 012035 doi:10.1088/1757-899X/722/1/012035

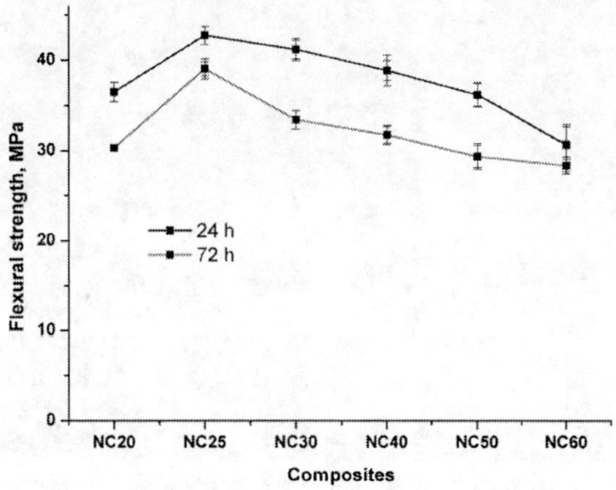

Figure 6. Flexural strength of CHF/polyester composites cornhusk fiber composites.

3.4. SEM

Morphology of the fractured surface of specimen composite in tensile is shown in **Figs. 7** and **8.** After immersed for 24 h (seen in **Figs 7a**, **7b**, and **7c**), it was observed that the composite display the interfacial bonding between the CHFs – PE was high and strong. Localized bunch of CHFs is shown, which indicates the good dispersion of CHFs within the polyester, and the fracture occurred at the CHFs itself. This shows that the stress was well propagated between CHFs–polyester, resulting in enhanced flexural and tensile strength in response to stress. The composite with higher fibers content (seen in **Figs. 7d–f**) appears to be dominated by fibers breakage. The interfacial fracture accompanied by cross–section damage of the CHFs, resulting in decreased tensile strength.

Figs. 8a, **8b**, and **8c** shows a crack running through the CHF, and this an indication of the lack of stress–transfer from polyester to CHFs. **Figs 8d**, **8e**, and **8f**, it was found that composite had a damage area interface between CHF and PE is loose. The interfacial fracture is demonstrated by CHF cross–section damage, resulting in decreased tensile, and flexural strength.

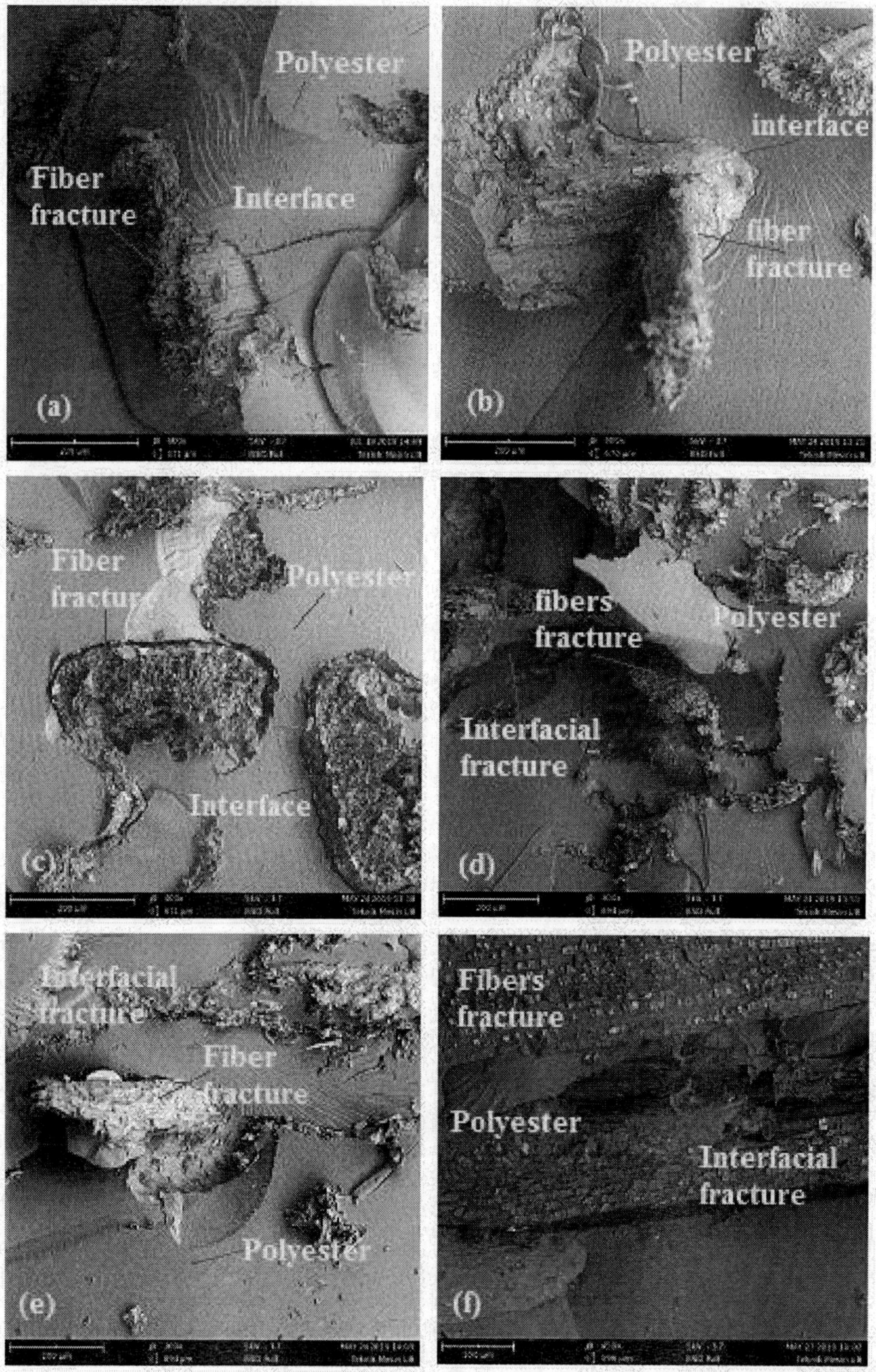

Figure 7. SEM photos, (a) NC20, (b). NC25, c. NC 30, d. NC40, e. NC50, and f. NC60 after water immersed for 24 h.

Figure 8. SEM images, (a) NC20, (b). NC 25, c. NC 30, d. NC40, e. NC50, and f. NC60 after water immersed for 72 h.

4. Conclusion

An experimental investigation of the behavior of tensile strength, morphology and water absorption from CHF–based composites under the water environment was carried out. The water uptake and swelling properties of the composites increase with an increasing amount of fiber content and soaking time. Consequently, the tensile and bending strength of the composite to decreased. The maximum tensile strength, and young's moduly are obtained from composites with 30% fiber content (NC30) after 24 h water immersed, and then decreases. SEM images display the interfacial fracture accompanied by cross–section damage of the CHFs. Composites based CHF are suitable as an alternative material for decking, siding, exterior windows, and doors.

Acknowledgments
Authors would like to thank the Directorate General of Strengthening Research and Development, Ministry of Research, Technology and Higher Education, Republic of Indonesia for the research financial support under the research scheme of fundamental research of National competitive. Contract number: 182/SP2H/LT/DRPM/2019).

References

[1] Jiang X, Kolstein H, and Bijlaard F S K 2013 Moisture diffusion in glass–fiber–reinforced polymer composite bridge under hot/wet environment. *Composites: Part B* **45** 407–416. http://dx.doi.org/10.1016/j.compositesb.2012.04.067
[2] Sari N H, Wardana I N G, Irawan Y S, and Siswanto E 2017 Corn Husk Fiber–Polyester Composites as Sound Absorber: Nonacoustical and Acoustical Properties. *Advances in Acoustics and Vibration* **2017**. https://doi.org/10.1155/2017/4319389.
[3] Huda S, and Yang Y 2008 A novel approach of manufacturing light–weight composites with polypropylene web and mechanically split Cornhusk. *Industrial Crops and Products* **30** 17–23. doi:10.1016/j.indcrop.2008.12.007.
[4] Luo Z, Li P, Cai D, Chen Q, Qin P, Tan T, and Cao H 2016 Comparison of performances of corn fiber plastic composites made from different parts of corn stalk. *Industrial Crops and Products* **95** 521–527. http://dx.doi.org/10.1016/j.indcrop.2016.11.005.
[5] Bernhardt D C, Perez C D, Fissore E N, De'Nobili M D, and Rojas A M 2017 Pectin–based composite film: Effect of corn husk fiber concentration on their properties. *Carbohydrate Polymers* **164** 13–22. http://dx.doi.org/doi:10.1016/j.carbpol.2017.01.031.
[6] Levy R L, Fanter D L, and Summers C J 1979 Spectroscopic evidence for mechanochemical effects of moisture in epoxy–resins. *Journal Application Polymer Science* **24**(7)1643–64.
[7] Mikols W J, Seferis J C, Apicella A, and Nicolais L 1982 Evaluation of structural–changes in epoxy systems by moisture sorption–desorption and dynamic mechanical studies. *Polymer Composite* **3**(3) 118–24.
[8] Youssef A M, Gendy A E, and Kamel S 2014 Evaluation of corn husk fibers reinforced recycled low density polyethylene composites. *Materials Chemistry and Physics* 1–8.
[9] ASTM D570–98 2002 Standard test method for water absorption of plastics: *Annual book of ASTM Standards.* West Conshohocken PA.
[10] Saenghirunwattana P, Noomhorma A, and Rungsardthong V 2014 Mechanical properties of soy protein based "green" composites reinforced with surface modified cornhusk fiber. *Industrial Crops and Products*, **60** 144–150.
[11] Sari N H, Sanjay M R, Arpitha G R, Pruncu C I, and Siengchin S 2019 Synthesis and properties of pandanwangi fiber reinforced polyethylene composites: Evaluation of dicumyl peroxide (DCP) effect. *Composites Communications* **15** 53–57.
[12] ASTMD790 2010 Standard test methods for flexural properties of unreinforced and reinforced plastics and electrical insulating materials. ASTM Standards International.
[13] Stokke D D, and Gardner D J 2003 Fundamental aspects of wood as a component of thermoplastic composites. *Journal of Vinyl & Additive Technology* **9**(2) 96–104.
[14] Sari N H, Wardana I N G, Irawan Y S, and Siswanto E 2018 Characterization of the Chemical, Physical, and Mechanical Properties of NaOH–treated Natural Cellulosic Fibers from Corn Husks. *Journal of Natural Fibers* **15**(4) 545–558. doi: 10.1080/15440478.2017.1349707.
[15] Toro P, Quijada R, Murillo O, and Yazdani–Pedram M 2005 Study of the morphology and mechanical properties of polypropylene composites with silica or rice–husk. *Polymer International* **54** 730–734.
[16] Dobreva T, Benavente R, Perea J M, Perez E, Avella M, García M, and Bogoeva–Gaceva G 2009 Effect of different thermal treatments on the mechanical performance of Poly (l– lactic acid) based eco–composites. *Journal of Applied Polymer* **116** 1088–1098 https://doi.org/10.1002/app.31584

IOP Conf. Series: Materials Science and Engineering **722** (2020) 012035 doi:10.1088/1757-899X/722/1/012035

[17] Machado J S, Santos S, Pinho F F S, Luís F, Alves A, Simões R, and Rodrigues J C 2016 Impact of high moisture conditions on the high moisture conditions on the serviceability performance of wood plastic composite decks. *Materials & Design* **103** 122–131.
[18] Kuciel S, Jakubowska P, and Kuzniar P 2014 A study on the mechanical properties and the influence of water uptake and temperature on biocomposites based on polyethylene from renewable sources. *Composites: Part B* **64** 72–77.

Online Consumer Segmentation Study Based on Factors Affecting E-Commerce Selection

N B Puspitasari, S N Widyo Pramono, D I Rinawati, and F Fidiyanti

Industrial Engineering, Faculty of Engineering, Diponegoro University, Semarang - Indonesia

E-mail: niabudipuspitasari@gmail.com[1] ; nwp.susatyo@gmail.com[2] ; dyah.ika@gmail.com[3] ; finsafidi@gmail.com[4]

Abstract. Indonesia has considerable e-commerce growth potential supported by the increasing number of internet users. The rapid growth of e-commerce is characterized by the increasing number of e-commerce resulting in the increasingly intense competition among e-commerce. The purpose of this study is to do online buyer segmentation based on the attributes of the website. The survey using an online questionnaire was given to 338 online buyers in Indonesia. Factor analysis is used to verify the construct model used using Confirmatory Factor Analysis (CFA). Finally, segmentation analysis is applied to identify groups of respondents and create their profiles. Based on segmentation analysis, the study identified five segments, namely, assurance seekers, secure seekers, convenience seekers, convenience and secure seekers, and social connectivity seekers. Each of the five segments varies greatly in the selection of websites for purchases and describes the differences in their online shopping orientation. Companies need to develop a plan to consider the unique characteristics of the online market segment to attract and retain users. The findings of this study can be used as guidelines for the development of a multi-dimensional strategic framework by online retail companies.

1. Introduction

Indonesia has considerable e-commerce growth potential supported by the increasing number of consumer online. The research entitled "The Opportunity of Indonesia" initiated by TEMASEK and Google, discussed that the growth of Indonesian e-commerce has increased. In 2015, Indonesia has 18 million online consumers and is predicted to increase to 119 million online consumers by 2020.Along with the increasing number of consumer online, businessmen and traders have begun to switch from offline sales to online, especially through e-commerce platforms. Based on the results of the Economic Census in 2016, total e-commerce in Indonesia reached 26.2 million, an increase of about 17% during 2006 to 2016. This resulted in the increasingly intense competition between e-commerce.

E-commerce business models can generally be categorized in different models one of which is B2C (Business to Consumer). In B2C model, a customer can view a product and order the same. In this model, in addition to business organizations, online shopping websites are useful and are growing day by day at an unprecedented rate and the volume of transactions in e-marketplaces is expanding [1].

To attract the final consumer online in using e-commerce sites, e-commerce companies need special strategies. One strategy that can be used is to segment the final consumer online. Online consumer Segmentation is one of the strategies that can be utilized by the marketing department by grouping heterogeneous consumers (having many differences) into homogeneous groups (having many similarities) [2]. Segmentation is very important to do, this is in accordance with the statement of [3], stating that increasing competition, more informed and educated customers, and changing demands, urges companies to pay attention to segmentation issues. Segmentation can be the right tool for identifying unmet customer needs.

One final consumer online Segmentation that can be done is segmentation based on factors that influence the end consumer online in choosing an e - commerce by using the website attribute variable. This variable is used to determine the characteristics of each segment in choosing an e-commerce based on the attributes of an e-commerce website. From the results of existing research, consumer online Segmentation has been carried out. Some of them are consumer online Segmentation in India based on consumer shopping behavior [4], consumer online Segmentation based on web portals attributes [5] and Greenfield consumer online Segmentation based on shopping motivation and e-attribute store [6].

In addition to segmenting online consumers, another strategy that can be done is to learn and understand the factors that influence the end consumer online in choosing an e-commerce based on the attributes of the website. This strategy is needed to improve the quality of e-commerce so that it can attract end consumer online interest in using e-commerce.

Based on a number of things that have been described, final consumer online segmentation based on psychographic factors, namely the factors that influence the final consumer online in choosing e-commerce is the topic of this research. The purpose of this study is to do online buyer segmentation based on the attributes of the website. This research was conducted to provide information and advice to e-commerce business people in Indonesia to be able to compete and maintain their existence in the midst of increasingly e-commerce competition.

2. Literature Review

2.1. E-commerce

E-commerce (Electronic Commerce) is an activity of buying and selling goods and services as well as the transmission of funds or data through electronic networks, especially the internet. Types of e-commerce include B2B (Business-to-Business), B2C (Business-to-Consumer), C2B (Consumer-to-Business), C2C (Consumer-to-Consumer), G2G (Government-to -Government), G2B (Government-to-Business), G2C (Government-to-Consumer) and C2G (Consumer-to-Government) [7]. Business-to-Consumer (B2C) is e-commerce with transactions between business people / companies and final customers or in other words B2C is retail trade that occurs online. Whereas Consumer-to-Consumer (C2C) is an e-commerce with electronic transaction products and services between two final customers.

2.2. Factor Analysis

Factor analysis aims to minimize and rearrange very many variables into meaningful categories based on the same variance to get the underlying concepts and help understand interpretations [8]. There are two main factors analysis techniques, namely Exploratory Factor Analysis (EFA) and Confirmatory Factor Analysis (CFA). CFA tries to confirm hypotheses and uses path analysis diagrams to represent variables and factors, while EFA tries to reduce data from origin variables to new variables or smaller factors and find relationships between new variables formed [9].

2.3. Segment Analysis

Segmentation analysis is an analysis used to group a number of objects into groups based on certain characteristics, so that members in the group have homogeneous characteristics and heterogeneous characteristics with other groups. The method in clustering is divided into 3, namely [10]:

a. Hierarchy Method

This method classifies two or more objects that have the closest similar. Then the process is continued with another object that has the second closest similarity. And so on until all objects enter the group. The hierarchy method can be used when there is no prior knowledge of how many groups should be formed. In the hierarchical method, there are two basic types:

- Agglomerative (Centering)

In the agglomerative method, each object is considered as a group then similar groups are combined into a new group and so on. In agglomerative there are 5 methods, namely single linkage, complete linkage, average linkage, ward's method and centroid method.

- Divisive (Spread)
  The divisive method starts from a large group consisting of all objects, then objects with the highest value of dissimilarity separated and so on.

b. Non-Hierarchy Method
Non-hierarchical methods begin by determining the number of groups desired and then the grouping process is done by placing objects into groups at once so that certain groups are formed. This method is commonly called k-means and is more suitable for large samples (n> 100). Non-hierarchical methods have three approaches, namely sequential threshold, parallel threshold and optimization.

c. Two Stage Method
This method identifies groupings by executing pre-grouping first, then carrying out hierarchical methods. In the first step of the procedure, the data is pre-grouped into many small sub-clusters. Then, the sub-cluster is combined into the desired number of clusters. The two stage method can handle very large data and scale and ordinal data in the same model.

3. Research Method

3.1. Data Collection Technique

Data collected is primary data obtained by survey techniques through online questionnaires. Sampling was done by the convenience sampling technique. The characteristics of respondents needed are Indonesian people aged at least 17 years and having made transactions at least 5 times on e-commerce types B2C and C2C. The questionnaire consists of 2 parts, namely the respondent's personal data and statements regarding the factors that influence the final consumer online in choosing e-commerce. The factors used are derived from previous research conducted by [11]. Respondents were asked to provide an assessment regarding the statement given in accordance with their opinions / perceptions using the five-point Likert scale. Based on research conducted by [11], this research uses the attributes of a website to determine the factors that influence online consumers in choosing e-commerce. In this study there are 7 variables, namely the design and quality of the website (DKW); security and privacy (KP); reliability and trust (KK); the popularity of the website (KW); accessibility (A); compatible with social media (KMS); search engine optimization (OMP).

Design and quality of the website (DKW) are divided into five indicators, namely visually attractive appearance (DKW1), user-friendly interface with easy navigation (DKW2), complete content and interactive mechanisms (DKW3), Regular information updates (DKW4), and Load all text and graphics quickly (DKW5). Security and privacy (KP) are divided into three indicators, namely provides error-free transactions (KP1), Protect personal information (KP2), Provides various payment mechanisms (KP3). Reality and trust (KK) are divided into three indicators, namely send orders to the right place and time (KK1), notify about order status (KK2), exchange and replacement policy (KK3). The popularity of the website (KW) is divided into five indicators, namely Offer good service (KW1), appropriate website image (KW2), offering quality products and services (KW3), the website offers original products (KW4), good relationship with users (KW5).

Accessibility (A) is divided into four indicators, namely, providing services 24 hours per day (A1), provide in-depth information about the product (A2), products can be ordered easily and quickly (A3), Product categorization for easy shopping (A4). Compatible with social media (KMS) is divided into four indicators, namely keep up with the latest trends (KMS1), interactive descriptions and reviews (KMS2), benefits of sweepstakes and promotions (KMS3), enables customers to share information (KMS4). Search engine optimization (OMP) is divided into three indicators, namely only provide relevant information (OMP1), minimum effort to find the product needed (OMP2), provides links related to products (OMP3).

3.2. Data Processing Techniques

This study uses factor analysis with the Confirmatory Factor Analysis (CFA) method using AMOS software and Segmentation analysis using the k-means method using SPSS software. The CFA method was chosen to provide valid information by testing or confirming a priori theory [12]. Segmentation analysis is done to obtain the final consumer Segment or online group based on the factors that influence the selection of e-commerce. The k-means method was chosen because it has high efficiency in the clustering process.

4. Results and Discussion

4.1. Normality and Outlier Test

Before carrying out factor analysis, a normality test was carried out to determine whether the data were normally distributed multivariate. The data are normally distributed multivariate if it has a critical ratio value between -2.58 to _2.58. From the results of AMOS software processing, the value of the critical ratio is 27,373, which means that the data is not normally multivariate. In order for data to be normally distributed multivariate, outlier elimination is carried out by deleting data that have a value of expensive d-squared distance> chi square value (52,260). From the results of AMOS software processing, there are 34 outlier data and eliminated so that the data is normally multivariate with a critical ratio of 2.57. In the next stage 304 data are used.

Table 1. Recapitulation of Stage I Factor Analysis

No.	Information			Estimate	C.R.	P	Loading Factor
1	DKW5	<---	DKW	1,000			0,555
2	DKW4	<---	DKW	1,039	6,973	***	0,536
3	DKW3	<---	DKW	0,931	7,411	***	0,647
4	DKW2	<---	DKW	1,020	7,355	***	0,648
5	DKW1	<---	DKW	1,062	6,377	***	0,517
6	KP3	<---	KP	1,000			0,703
7	KP2	<---	KP	1,055	10,430	***	0,719
8	KP1	<---	KP	1,000	6,942	***	0,458
9	KK3	<---	KK	1,000			0,661
10	KK2	<---	KK	0,796	10,247	***	0,680
11	KK1	<---	KK	0,714	10,204	***	0,706
12	KW5	<---	KW	1,000			0,651
13	KW4	<---	KW	0,895	8,016	***	0,510
14	KW3	<---	KW	0,867	11,195	***	0,754
15	KW2	<---	KW	0,846	7,966	***	0,512
16	KW1	<---	KW	0,900	10,557	***	0,706
17	A4	<---	A	1,000			0,691
18	A3	<---	A	1,031	12,152	***	0,776
19	A2	<---	A	1,161	12,373	***	0,784
20	A1	<---	A	1,111	8,427	***	0,521
21	KMS4	<---	KMS	1,000			0,590
22	KMS3	<---	KMS	0,918	6,635	***	0,465
23	KMS2	<---	KMS	0,925	8,681	***	0,735
24	KMS1	<---	KMS	0,893	6,976	***	0,504

25	OMP3	<---	OMP	1,000			0,561
26	OMP2	<---	OMP	1,004	8,921	***	0,804
27	OMP1	<---	OMP	1,091	8,001	***	0,609

4.2. Factor Analysis

The model structure that contains 7 factors regarding the factors that influence online consumers in choosing e-commerce is tested for their strength and validity using CFA with Analysis of Moment Structures (AMOS) software. Evaluation is done to check the adequacy of the parameters and the overall model.

Table 2. Recapitulation of Stage II Factor Analysis

No.	**Information**			**Estimate**	**C.R.**	**P**	**Loading Factor**
1	DKW5	<---	DKW	1,000			0,555
2	DKW4	<---	DKW	1,036	6,964	***	0,535
3	DKW3	<---	DKW	0,934	7,410	***	0,650
4	DKW2	<---	DKW	1,016	7,338	***	0,646
5	DKW1	<---	DKW	1,055	6,347	***	0,514
6	KP3	<---	KP	1,000			0,715
7	KP2	<---	KP	0,998	10,429	***	0,692
8	KK3	<---	KK	1,000			0,657
9	KK2	<---	KK	0,802	10,168	***	0,681
10	KK1	<---	KK	0,722	10,163	***	0,708
11	KW5	<---	KW	1,000			0,651
12	KW4	<---	KW	0,892	7,991	***	0,508
13	KW3	<---	KW	0,866	11,186	***	0,754
14	KW2	<---	KW	0,844	7,942	***	0,511
15	KW1	<---	KW	0,903	10,592	***	0,708
16	A4	<---	A	1,000			0,691
17	A3	<---	A	1,032	12,208	***	0,777
18	A2	<---	A	1,160	12,388	***	0,783
19	A1	<---	A	1,108	8,423	***	0,520
20	KMS4	<---	KMS	1,000			0,573
21	KMS2	<---	KMS	0,972	8,359	***	0,750
22	KMS1	<---	KMS	0,884	6,644	***	0,484
23	OMP3	<---	OMP	1,000			0,560
24	OMP2	<---	OMP	1,008	8,895	***	0,806
25	OMP1	<---	OMP	1,089	7,971	***	0,607

4.2.1. Factor Analysis: Stage 1

In stage I factor analysis, all indicators have a critical ratio of> 1.96 but there are 2 indicators that have a loading factor value of <0.5 (KP1 and KMS3) so that both indicators are eliminated and a new construct model is tested.

IOP Conf. Series: Materials Science and Engineering **722** (2020) 012036 doi:10.1088/1757-899X/722/1/012036

4.2.2. *Factor Analysis: Stage 2*

In stage II factor analysis, all indicators have a critical ratio of> 1.96 but there are 1 indicator that has a loading factor value of <0.5 (KMS1) so that the indicator is eliminated and a new construct model is tested.

4.2.3. *Factor Analysis: Stage 3*

In stage III factor analysis, all indicators have a critical value ratio of> 1.96 and all indicators have a value of loading factor> 0.5 (KP1 and KMS3). The Goodness-of-fit index used Root mean square error of approximation (RMSEA = 0.057), Goodness-of-fit (GFI = 0.889), Adjusted Goodness of Fit Index (AGFI = 0.856), Normed Fit Indices (NFI = 0, 85), Comparative Fit Index (CFI = 0.918), Incremental Fit Indices (IFI) and Tucker-Lewis Index (TLI = 0.902). Because the goodness-of-fit value is more than the recommended index, which is good fit if> 0.9 and marginal fit if> 0.8; the model can be said to be quite fit and validated.

Table 3. Recapitulation of Stage III Factor Analysis

No.	Information			Estimate	C.R.	P	Loading Factor
1	DKW5	<---	DKW	1,000			0,555
2	DKW4	<---	DKW	1,029	6,926	***	0,531
3	DKW3	<---	DKW	0,937	7,410	***	0,652
4	DKW2	<---	DKW	1,021	7,345	***	0,649
5	DKW1	<---	DKW	1,052	6,328	***	0,512
6	KP3	<---	KP	1,000			0,713
7	KP2	<---	KP	1,002	10,413	***	0,694
8	KK3	<---	KK	1,000			0,656
9	KK2	<---	KK	0,803	10,162	***	0,681
10	KK1	<---	KK	0,723	10,182	***	0,709
11	KW5	<---	KW	1,000			0,651
12	KW4	<---	KW	0,888	7,953	***	0,506
13	KW3	<---	KW	0,867	11,192	***	0,754
14	KW2	<---	KW	0,844	7,943	***	0,511
15	KW1	<---	KW	0,905	10,608	***	0,709
16	A4	<---	A	1,000			0,688
17	A3	<---	A	1,038	12,169	***	0,778
18	A2	<---	A	1,167	12,354	***	0,785
19	A1	<---	A	1,109	8,386	***	0,519
20	KMS4	<---	KMS	1,000			0,561
21	KMS2	<---	KMS	0,962	8,154	***	0,726
22	OMP3	<---	OMP	1,000			0,556
23	OMP2	<---	OMP	1,020	8,876	***	0,810
24	OMP1	<---	OMP	1,093	7,923	***	0,605

4.3. Segmentation Analysis

Segmentation analysis is used to Segment respondents based on their choice of website. Segmentation analysis in this study uses non-hierarchical techniques (k-means) by determining the optimum number of Segments using the elbow method. In the elbow method, the optimal number of Segments is determined by minimizing the total number of squares between points and centroids. Each data point is the distance between the data point and the center of mass of the corresponding Segment. Furthermore, the total number of squared distances between points and centroids is calculated.

Table 4. Recapitulation of Intra-Cluster Numbers

No.	k (Number of Segments)	Intra Cluster	%change
1	2	336,526	
2	3	301,082	10,53%
3	4	288,730	4,10%
4	5	276,638	4,19%
5	6	270,639	2,17%
6	7	265,604	1,86%
7	8	261,421	1,58%
8	9	249,444	1,58%
9	10	246,361	1,24%

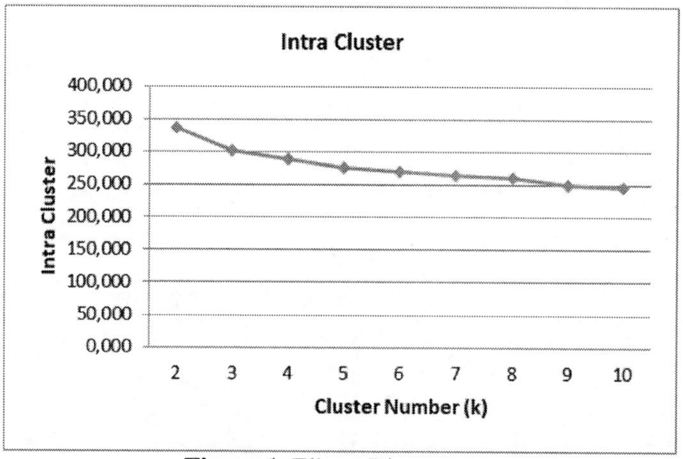

Figure 1. Elbow Diagram

As can be seen in Figure 1, the total number of squared distances between the corresponding points and centroids (intra-cluster variations) decreases with increasing number of Segments. However, there is a drastic decrease from k = 2 to k = 3. In other words, the total number of squared distances when k = 2 and k = 3 are 336,526, and 301,082, respectively, with a decrease of about 10.53%. This is an indication that there are several levels to group consumer online into several Segments.

However, outside k = 5, the total number of squares between the points and centroids that are suitable is not much different. When k = 5 and k = 6, the total number of squared distances is 276,638 and 270,639, respectively. Therefore, there is not much variation between k = 5 and k = 6. This gives an indication of the optimal number of Segments to classify online consumer as 5 Segments. So that the optimum number of Segments is 5 Segments.

Table 5. K-Means Analysis Results

Variable	Cluster					F-value	*p*
	1	*2*	*3*	*4*	*5*		
DKW	4,43	4,02	4,23	3,76	4,43	22,36	0,000
KP	4,04	4,28	4,13	3,64	4,48	23,33	0,000
KK	4,70	3,91	4,21	3,60	4,62	89,09	0,000
KW	3,98	3,94	4,13	3,54	4,44	38,03	0,000
A	3,97	3,90	4,10	3,40	4,54	67,78	0,000
KMS	3,61	3,71	4,22	3,45	4,75	112,09	0,000
OMP	4,27	3,79	4,18	3,57	4,67	54,19	0,000
Number of members	51	48	121	21	63		
Percentage of members	16,78%	15,79	39,80	6,91	20,72		

The formed Segment has its own characteristics according to the variables that affect the online consumer in choosing e-commerce and based on demographic factors. The following is an explanation of each Segment:

1. Segment 1

   This Segment has 51 members. The demographic characteristics of respondents in this Segment were dominated by female respondents (64.71%), had the last high school education and were equivalent (47.06%), worked as students (50.00%), the majority were 17-24 years old (54,90%) and has an income of Rp. 2,600,000 - Rp. 6,000,000 (50.98%). Most of the respondents in this Segment came from Central Java (33.33%). Variables that have the highest ranking in influencing members in Segment 1 in choosing e-commerce is reliability and trustworthiness, website design and quality as well as search engine optimization. This shows that members in Segment 1 choose e-commerce that has the best ability in fulfilling services to consumers and e-commerce that has high integrity. Online consumers in this Segment tend to be careful and afraid of errors in the transaction process so that reliability and trust will greatly affect online consumers in this Segment to choose e-commerce. In addition, online consumers in this Segment also pay attention to the convenience of using e-commerce both from website design and the ease of searching for the desired product. This Segment is 'assurance seeker'. This Segment has characteristics similar to those of [5] namely the selection of e-commerce is influenced by personal guarantees.

2. Segment 2

   This Segment has a total of 48 respondents. The demographic characteristics of respondents in this Segment were dominated by female respondents (62.50%), the majority were 17-24 years old (47.92%), single status (68.75%) and the majority were students (35.42 %). Most of the respondents in this Segment came from East Java (33.33%) with their last education, undergraduate (47.92%) and had an average income of IDR 2,600,000 - IDR 6,000,000 (58.33%). Variables that have the highest ranking in influencing members in Segment 2 in choosing e-commerce are security and privacy, website design and quality and the popularity of the website. The results of this study are consistent with the results of research conducted by [13] and [14] that security and privacy are important aspects of e-commerce selection. This shows that members in Segment 2 choose e-commerce that has the ability to prevent fraud. Members in this Segment tend to be very careful in providing information to e-commerce. In addition, online consumers in this Segment also pay close attention to the popularity of e-commerce because of the notion that popular e-commerce has good quality. This Segment is a 'secure seeker'. This Segment has characteristics similar to those of [14] namely the selection of e-commerce is influenced by the security of personal and financial data.

3. Segment 3
   This Segment has the most number of members, 121 respondents. The demographic characteristics of the respondents in this Segment were dominated by female respondents (58.68%), the majority were 17-24 years old (52.07%), single status (68.75%) and the majority were students (41,32 %). Most of the respondents in this Segment came from West Java (20.66%) with the last education of senior high school and the equivalent (42.15%). Variables that have the highest ranking in influencing members in Segment 3 in choosing e-commerce are a website design and quality, reliability and trust and compatibility with social media. These results are consistent with [15] which provide corroborating evidence that website design, in the form of usefulness and availability of information, is a key factor for e-commerce. This shows that members in Segment 3 choose e-commerce, which has an attractive interface appearance and complete content and information provided. Members in this Segment tend to find convenience in e-commerce use such as easy navigation and information retrieval. This Segment is 'convenience seeker'. This Segment has almost the same characteristics as the findings of [5], namely the selection of e-commerce is influenced by the convenience of transactions.
4. Segment 4
   Segment 4 has the least number of members, namely 21 respondents. Demographic characteristics of respondents in this Segment were dominated by respondents of male sex (71.43%), the majority were 17-24 years old (61.90%), single status (71.43%). Respondents in this Segment mostly came from Central Java (23.81%), had the last education S1 (42.86%) and the majority were students (42.86%). Variables that have the highest ranking in influencing members in Segment 3 in choosing e-commerce are website quality design, security and privacy and reliability and trustworthiness. This shows that members in Segment 4 choose e-commerce that provides convenience in transactions and has the ability to prevent fraud. In addition, online consumers in this Segment are also very concerned about the reliability of e-commerce in providing services. This Segment is a combination of the previous 3 Segments with the name 'convenience and secure seeker'. This Segment has characteristics similar to those of [14], namely the selection of e-commerce is influenced by ease of use and security / privacy.
5. Segment 5
   Segment 5 has 63 members. The demographic characteristics of respondents in this Segment were dominated by female respondents (57.14%), the majority were 17-24 years old (46.03%), single status (61.90%). Most of the respondents in this Segment came from Central Java (25.40%), worked as private employees (34.92%) and had an average income of Rp2,600,000-Rp6,000,000 (52.38%). Variables that have the highest ranking in influencing members in Segment 5 in choosing e-commerce are compatible with social media, reliability and trust, and search engine optimization. This shows that members in Segment 5 choose e-commerce that has a relationship with social media. Members in this Segment tend to look for e-commerce that can be directly connected with social media so as to facilitate online consumers in sharing information. This Segment is 'social connectivity seeker'. This Segment has different characteristics from previous studies.

Based on the results of Segmentation, there are differences with existing studies. This research provides 5 online segmentations as we can see at Picture 1. The research conducted by [5] has 4 online consumer Segments based on website attributes. The four Segments are traditional shopper, benefit seekers, convenience and assurance seekers and information and price seekers. Although the number of Segments formed is different, there are several Segments that have characteristic similarities, namely assurance seeker and convenience seeker. The difference between the results of this study and the research of [5] shows that there are differences in preferences between online consumer Indonesia and online consumer India.

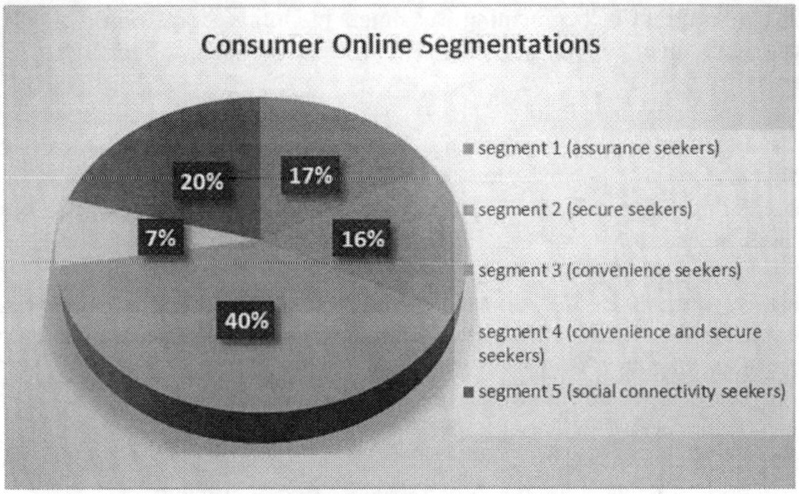

Figure 2. Consumer Online Segmentations

4.4. Analysis of factors affecting segmentation

All variables can distinguish Segments that are formed significantly (significant value <0.05). The variable that has the largest Chi-Square value is the variable compatibility with social media of 112,093. A greater Chi-Square value indicates that the variable has a greater influence in distinguishing the formed Segment. While the variable that has the smallest Chi-Square value is the website quality design variable of 22.360, so that the website quality design variable has a smaller influence in distinguishing the formed Segment. Based on the crosstab test, all demographic factors have a significance value of> 0.05, which means that all demographic factors are not significant in distinguishing consumer online Segments.

5. Conclusion

In this study, there are 24 indicators that influence online consumers in choosing e-commerce. The twenty-four indicators relate to website attributes such as website design and quality, security and privacy, reliability and trust, website popularity, accessibility, compatibility with social media and search engine optimization. There are 5 consumer online segments that are formed based on the factors that influence the final consumer online in choosing an e-commerce by using the website attribute variable. The five Segments include assurance seekers, secure searchers, convenience seekers, convenience and secure searchers and social connectivity seekers. Based on demographic factors, all factors were not statistically significant in distinguishing formed Segments. While based on the ANOVA test, all variables are significant in distinguishing the formed Segments. Based on the results of Segmentation, this study can be used as a reference to improve the quality of e-commerce websites in order to attract and target the Indonesian consumer online Segment based on the factors that influence the end consumer online in choosing e-commerce.

Some recommendations that can be considered to improve the quality of web sites include Segment 1 (assurance seekers) can consider reliability and trust, Segment 2 (secure seekers) can consider security and privacy factors, Segment 3 (convenience seekers) can consider the design and quality factors web, Segment 4 (convenience and secure seekers) can consider web design and quality factors as well as security and privacy while Segment 5 (social connectivity seekers) can consider factors of compatibility with social media.

IOP Conf. Series: Materials Science and Engineering **722** (2020) 012036 doi:10.1088/1757-899X/722/1/012036

References

[1] Maadi M, Maadi M and Javidnia M 2016 Identification of factors influencing building initial trust in e-commerce *Iranian Journal of Management Studies (IJMS)* **9** 483-503

[2] Hong T and Kim E 2012 Segmenting Customers in Online Stores Based on Factors That Affect the Customer's Intention to Purchase *Expert Systems with Applications: An International Journal* 39 2127-2131

[3] Lancaster G and Reynolds P 2002 *Marketing Made Simple* Oxford UK: Elsevier

[4] Jain N and Ahuja V 2014 Segmenting Online Consumers Using K-Means Cluster Analysis *International Journal Logistics Economics and Globalisation* 6 161-178

[5] Prashar S, Vijay T S and Parsad C 2016 Segmenting Online Shoppers: A Study of Consumers' Web Portal Selection Motivations for E-shopping *Asian Academy of Management Journal* 21 27-46

[6] Ganesh J, Reynolds K E , Luckett M and Pomirleanu N 2010 Online Shopper Motivations, and e-Store Attributes: An Examination of Online Patronage Behavior and Shopper Typologies *Journal of Retailing* 86 106-115

[7] Bhat S A, Kansana K and Majid J 2016 Green Marketing: Sustainable Economy, Environment & Society- Concept & Challenges *ResearchGate*

[8] Rummel R 1970 *Applied Factor Analysis* Evanston: Northwestern University Press

[9] Child D 2006 *Essentials of Factor Analysis* New York: Continuum

[10] Gore P A 2000 *Handbook of Applied Multivariate Statistics and Mathematical Modelling* Illinois: Academic Press

[11] Katta R M and Patro C S 2017 Influence of Web Attributes on Consumer Purchase Intentions *International Journal of Sociotechnology and Knowledge Development* 9 1-16

[12] Stevens J 1996 *Applied Multivariate Statistics for the Social Sciences* New Jersey: Lawrence Erlbaum Associates, Publishers

[13] Bhatnagar A and Ghose S A 2004 Latent Class Segmentation Analysis of E-shoppers *Journal of Business Research* 57 758-767

[14] Bressolles G 2008 Segmenting Online Customers Based on Electronic Service Quality Dimensions: An Application to Wine Websites *4th Int. Conf.of the Academy of Wine Business Research (Siena)* p 1-17

[15] Demangeot C and Broderick A 2010 Consumer Perceptions of Online Shopping Environments: A Gestalt Approach *Psychology & Marketing* **27** 117-140

Innovativeness, Proactiveness, and Risk-taking: Corporate Entrepreneurship of Indonesian SMEs

N J Astrini, T Rakhmawati, S Sumaedi, I G M Y Bakti, M Yarmen and S Damayanti

Indonesian Institute of Sciences, PUSPIPTEK Area 417, Banten, Indonesia 15314

E-mail: nidy001@lipi.go.id

Abstract. Corporate entrepreneurship (CE) consists of three dimensions, which are innovativeness, proactiveness, and risk-taking. It has an essential role in ensuring SMEs willingness to adopt technology in the 4.0 industrial era. The aim of this study is twofold. First, to discover the pattern CE (innovativeness, proactiveness, and risk-taking) intensity on four SMEs and second, to identify the likely causes of that pattern from a strategic management perspective. This research employed a case study design with four Indonesian SMEs as the subjects. The data was gathered through guided interviews and observations. Dyadic data was collected to avoid bias. The answers were then put into a scale from one to seven by the informants and combined using a simple mean. A pattern was inferred from the scale. The results suggest that all four SMEs have a similar low-medium range of CE intensity. From the strategic management standpoint, this was likely due to the fact the SMEs have not conducted proper environmental scanning in terms of looking for new technology or embraced the involvement of low-level employees in their planning process.

1. Introduction

The use of technology in the 4.0 industrial era is rampant. Businesses of all sectors and sizes were vying to utilize technology to improve their overall businesses. However, unlike large companies, SMEs in Indonesia have a very limited resource. They neither have large capital to acquire advanced technology, a pool of highly qualified human resource to create or operate the latest technology, nor unlimited access to information on the most up-to-date technology[1]. Those reasons have created considerable constraints for SMEs to adopt new technology. So, even though the tech exists and the pressure to adopt the tech was felt [2] Indonesian SMEs might not know about it or unwilling to obtain it.

One of the determinants of an SME willingness to adopt new technology was something called Corporate Entrepreneurship intensity[3–5]. The definition of corporate entrepreneurship had transformed, but the one underlying notion of it – adapt or die – hold still from the time when corporate entrepreneurship was only associated with big investments made by corporate executive [6,7] until the day when single act of an employee in finding a new way to do his or her job was considered as an entrepreneurial behavior [8,9]. Corporate entrepreneurship has three underlying dimensions, namely innovativeness, proactiveness, and risk-taking, which are essential in gaining competitive advantage. Now, corporate entrepreneurship was seen as efforts to identify and exploit opportunities in radical or incremental ways in order to grow and survive. In the context of the 4.0 industrial era, it means exploiting technology to grow and survive in the industry.

Based on the above argument, it was important to explore the CE intensity of Indonesian SMEs and investigate the probable cause behind it from the strategic management standpoint because previous studies have proven the relationship between CE and strategic management practices [10,11] The result of this study can be used to identify ways to boost CE intensity so that the SMEs are more proactive, innovative, and willing to take risk in adopting new technology.

There are many studies on CE or entrepreneurship as an individual trait [12,13]. However, they mainly focused on one dimension (i.e., innovativeness) such as [2,4,5] or on the technical details of CE as an enabler to technology adoption [14]. Previous studies were also mostly quantitative and did not provide in-depth insight into CE. This study aims to investigate the level of corporate entrepreneurship intensity in Indonesian SMS and compare the strategic management practices done by an SME with the highest level of intensity and the rest of the research subjects.

2. Literature Review

2.1. Corporate Entrepreneurship

Entrepreneurship was generally associated with new venture creation [15], but when an organization grows, entrepreneurial spirit belongs not only to a single person or the founder but also to the entire corporation. Even though corporate entrepreneurship was defined as formal and informal activities related with the product, process, and marketing innovation in order to increase company's competitive position and financial performance[16], there was no single accepted consensus on the definition of corporate entrepreneurship (CE) [17]. According to Ferreira [18], there are three ways to see CE; CE perceived as a form of new venture creation by an existing organization[6,19], as the trigger of organizational transformation and integration, and as the force changing the competition map of an industry.

CE has three dimensions, which are innovativeness, proactiveness, and risk-taking[20–22]. Innovation can arise in many forms, new product development, or process improvement [23]. According to Barringer [20], a company's innovativeness can be measured through the company's investment in R&D or simple process improvement. This commitment can also be seen from the number of new products or services. A company that conducted dramatic changes was deemed to have high CE intensity. Proactiveness, on the other hand, refers to the willingness to pursue ideas from opportunities presented along the way. This dimension requires a company to be future-oriented and focus on formulating the next steps to face potential changes and competition. Proactiveness helps companies create a competitive advantage as a first-mover [24]. The last dimension is risk-taking. This dimension has been the part of CE literature since the Cantillon era (1734). Companies with high CE intensity would prefer being aggressive to maximize profits while companies with low-intensity CE would often 'wait and see' to minimize risk[25–27]. The dimensions of CE are listed in Table I.

2.2. Strategic Management Practices

CE can naturally occur in a company [28], or it can be the result of management initiative [29]. Studies show 93% of academicians argued that entrepreneurship could be taught through formal education or training [30] However, it will be better for a company to encourage entrepreneurial behaviour by providing a supportive environment [31]. Management practices are believed to have an important role in shaping corporate entrepreneurship by providing a supportive environment [26,32–34].

From previous studies on CE, it was found that strategic management practices affected CE intensity. According to Brown et al. [35], strategic management practices will influence entrepreneurial behaviour, commitment to opportunities, and company's growth orientation. Strategic management practices will promote entrepreneurial behaviour even though that behaviour could pose a certain risk to the company [8,36,37].

Table 1. The dimensions of corporate entrepreneurship

Dimensions	Explanations	References
Innovativeness	a. Describes how strong companies' encouragement towards R&D and innovation. b. Gauges the number of new product lines or innovative process within the company. c. The nature of product or process changes (i.e. incremental or radical)	[21,22,38–45]
pro-activeness	a. Tendency to try being the first in initiating new business concepts or launching new products. b. Aggressiveness in dealing with competitors	
risk-taking	a. Willingness (or unwillingness) to engage in high-risk high-return projects. b. Explores and exploits available opportunity using numerous means.	

The components of strategic management process included formulating company's mission and vision, analyzing internal and external environment, formulating and choosing a strategy, and implementing a strategy which also deals with the evaluation, control, and change management[37]. Barringer and Bluedorn [18,20,46] stated that environmental scanning, planning, and control are the components of strategic management practices that influence CEI. The strategic management practices explored in this study was limited to the scanning, planning, and controlling processes.

In order to keep their technology adoption relevant to the external and internal environment, SMEs need to conduct environmental scanning. It will help them recognize opportunities arose from current technology and prepare to counter a potential threat. It can also be facilitated through flexible planning with reasonable planning horizon, and deep locus of planning. Third strategic management practice related to CE intensity is the control strategy. Financial control is the simplest control available, especially for small firms, while strategic control is more complex because it takes non-financial measures into accounts, such as consumer, internal business process, and innovation improvement. It is believed that bigger emphasize on strategic control will support corporate entrepreneurship.

3. Methodology

3.1. Research Subjects

This is a case study involving four SMEs in Indonesia. The first one (A) was a drop-off laundry service in Bogor. The business has one workshop and four outlets across the city. It employed 48 workers divided into four major workgroups, which were washers, pressers, delivery and administration. The second (B) SME was an outdoor-event organizer based in Bogor. The company has a relatively small number of permanent employees which consisted of administration and marketing. The instructors for their events were mainly freelancers. The third (C) SME was a Muslim apparel company that focused on manufacturing dresses. Their employees were divided into production-line workers, delivery, and administration. The fourth subject (D) was a hand-embroidered apparel company with around twenty employees. Three subjects have two levels of hierarchy, which were top management (i.e. business owners) and middle managers. Only company A has three levels of hierarchy (i.e., director, general manager, and middle manager).

The subjects were chosen based on the purpose of the study, which were to investigate the level of CE intensity. One fundamental prerequisite must be met, the research subjects must show basic entrepreneurial behaviours (innovative, proactive, and willing to take risks). They should also be willing to be tailed by the research team for at least a year. To select the research subjects, the research team conducted a workshop and offered technical guidance to implement an organizational development intervention namely ISO 9001:2015. This intervention was deemed as risky, time- and resource-consuming, but if it was executed correctly, would generate a significant result for an organization [47,48]. The research subjects were then selected based on who were willing to join this 'risky program'. Companies A, B, C, and D were the ones who proactively sought the opportunity to join. Thus, they were considered as research subjects that could best inform the research question [49].

3.2. Research Design

This is qualitative research. This research design was chosen because the study on CE intensity of Indonesian SMEs was limited and qualitative research is needed to provide an in-depth understanding [40]. The research process was divided into several stages. First, a literature review was done to identify the dimensions of CE and strategic management practices that relate to the intensity of CE. Literature were gathered from scientific databases (e.g., Emerald, JSTOR, and ScienceDirect). This process resulted in three CE dimensions and three categories of strategic management practices. Second, this study created an instrument to help with the interviews. The instrument has three parts. The first part assessed the CE intensity of companies, the second part explored the strategic management practices, and the third one asked about the number of employees and annual revenue. The part of the instrument can be seen in Table II. The third was data collection. The data were gathered through interviews and observation. To avoid bias, two measures were taken. First, data were collected from two informants in each SMEs. The answers from two informants were synthesized and this study ensured their convergence. Second, the answers were put into one to seven scale and further validated by the informants. The fourth step was data analysis. After all informants validated the CE intensity of their business, their answers were combined using a simple arithmetic mean and presented in Figure 1.

Table 2. Corporate entrepreneurship intensity scale

Low Innovativeness	1	2	3	4	5	6	7	**Highly Innovative**
Selling products that were readily available; have been tested before								Conducting research, innovating, and being the pioneer in technology adoption
There was no new product or services								There are many new products or services
Changes were usually small								Changes were usually dramatic
Low Proactiveness	1	2	3	4	5	6	7	**Highly proactive**
Usually reacting to competitors' moves								Usually making the first moves which then being responded by competitors
Very seldom becoming the first small business that introduced new product, service, or technology								Very often becoming the first small business that introduced new product, service, or technology
Usually avoid competition								Usually positioning the business to compete
Mild risk-taker	1	2	3	4	5	6	7	**Avid risk-taker**
Taking low-risk project with certain and low-expected-results								Taking high-risk project with unpredictable, but potentially high results
In uncertain condition, the business tended to be careful, wait-and-see in order to minimize costs								In uncertain condition, the business tended to be aggressive to maximize profit
In terms of changes in business environment, the best action was to be careful in exploring the situation								In terms of changes in business environment, the best action was to engage in various action to achieve the target.

IOP Conf. Series: Materials Science and Engineering **722** (2020) 012037 doi:10.1088/1757-899X/722/1/012037

4. Results and Discussions

4.1. The General Level of Corporate Entrepreneurship Intensity Among Four SMEs

The CE intensity of SMEs in Indonesia is in medium-low range (see Fig 1). According to Barringer and Bluedorn [20] innovativeness of a company can be measured through the size of investment provided by the company to conduct the research, development, and creation of new products even though it might not be applicable for gaining commercial interests. This commitment can also be seen from the quantity of new products or services issued by the company within a certain timeframe. Companies that make dramatic changes are considered to have a higher intensity than the one with minor changes. In terms of innovativeness, SME B, C, and D often introduced new products or services. However, the new products or services offered by them were not radically different with their previous products or services. They relied on products that have been tested before. Put in the context of technology adoption, they did not want to adopt a technology that drastically changed their operation.

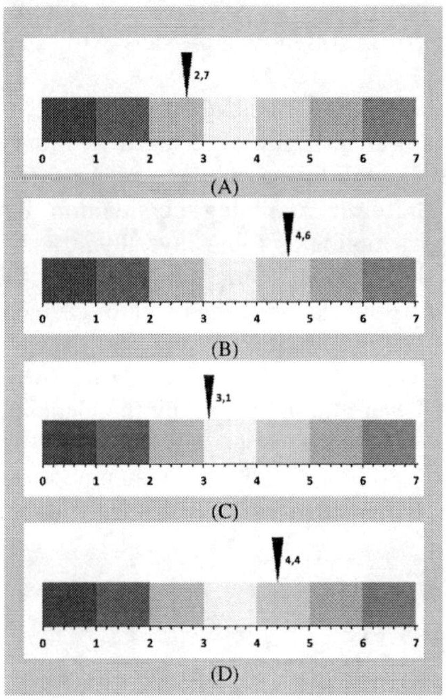

Figure 1. CE intensity of four SMEs

Proactiveness refers to the willingness to develop ideas from the opportunities found through research and trend analysis. This dimension requires companies to have future orientation; focus in formulating what to do to deal with change and competition in the future [32,50]. Proactiveness assists companies in creating competitive advantage by putting competitors in the position of 'must respond to first-mover initiatives'. As we know, a first-mover advantage refers to the gains for being the first company to introduce a new product or service, establish brand identity, enter new markets, or adopt a new operation [24,51]. The level of proactiveness can be seen from the behaviour of firms in the face of future competition. Companies that initiate the first action and being the first to introduce new products or to adopt new ways are considered to have a higher entrepreneurial intensity than companies that only provide a response to competitor actions, very rarely be the first to introduce products or new ways, and prefer a posture of live and let-live [22].

In terms of proactiveness, all SMEs were engaged in a push and pull with their competitors. They did not always become the first-mover but were not totally reliant on competitors' moves in deciding their next step. They adopted the stance of live and let live instead of trying to kill their competitors. Put into a technology adoption context, except for SME C, the rest of the subject did not proactively seek new technology to boost their performance. Technology adoption was considered only when the owners have previous experience with the technology or when there was a problem in the company (i.e. adopting ICT-based financial system to curtail fraud in the workplace).

How a company deals with risks that would occur is also a dimension that describes the intensity of corporate entrepreneurship. Courage to take risk has become an integral part in the entrepreneurship literature. In a business environment that is unstable and dynamic as today, risk management is a vital component in the strategic management and entrepreneurship consideration [52]. The risks which are well managed will create opportunities for the company. Organizations that do not take risks in a dynamic environment will lose their market share and unable to maintain position in the industry when faced with a more aggressive competitor [53]. Companies with high CEI will have a clear understanding of business and financial risks associated with entrepreneurial activity. Companies that prefer to do small projects with definitive results, taking the posture 'wait and see' to minimize risk, and too careful in dealing with the environment has a lower CEI than the companies that behave aggressively in the face of uncertainty with the objective to maximize potential. Companies with high CEI are generally willing to carry out high-risk projects with chances of getting a greater return [21,22,25,26,54].

Risk-taking is the dimension with lowest range point in this study. SMEs tended to choose low-risk projects with low and stable return. When they were faced with uncertainty, they tried to be careful, wait and see to minimize the risk of losing money. When there were changes in the business environment, they chose to be careful in examining the situation. Putting the technology adoption in context, SMEs saw a technology adoption as a big investment and the failure to fully utilize it will cripple their business since they did not have a large capital. That was why they tend to avoid adopting new technology. The subsections below elaborate the result of each company.

4.1.1. Company A

The informants form Company A were the Director and the General Manager. As mentioned before, to generate the final score, the arithmetic mean was used. For example, to obtain the result for Innovativeness, the answers for each indicator were summed and divided with the number of indicators as seen in equation 1. The subsequent results for Company B, C, and D were generated using the same equation.

$$\overline{x_1} = \sum \frac{x_i}{n} = \frac{2.5 + 2.7 + 2.8}{3}$$

$$\overline{x_1} = 2.7 \tag{1}$$

Table 3 shows the summary result for Company A. Company A was the least entrepreneurial compared to the other three companies especially in terms of Innovativeness. They relied on minuscule operational improvement based on the emergence of problems in the company. The owner and the managers were highly proactive when it came to knowledge improvement, but they tended to be extremely careful in a turbulent environment. They emphasized more on avoiding the risk of large cost rather than focusing on a potential high return from high-risk opportunity.

Table 3. Summary result of company A

Low Innovativeness (1)	$\overline{x_1}$ =2.5	(7) Highly Innovative
Low Proactiveness (1)	$\overline{x_2}$=2.7	(7) Highly proactive
Mild risk-taker (1)	$\overline{x_3}$=2.8	(7) Avid risk-taker
CE Intensity	2.7	

IOP Conf. Series: Materials Science and Engineering **722** (2020) 012037 doi:10.1088/1757-899X/722/1/012037

4.1.2. Company B

The informants for Company B were one of the owners and one of the staffs. Company B was the most entrepreneurial company among four SMEs (see Table 4). Company B focused on growth and expansion. They pride themselves on creating a variety of new services to cover several target markets. When this study was conducted, Company B was just an event-organizer of outdoor activities. They catered, mostly, to the needs of other organizations. However, they have a vision to develop their own outdoor activities venue, which can be enjoyed by people of all ages and to realize this vision, they continuously expanded their services. For example, they constructed a customized rooftop arena in malls to target children and test their business concepts. The profits were retained to build their equity.

Table 4. Summary result of company B

Low Innovativeness (1)	$\overline{x_1}$ =4.0	(7) Highly Innovative
Low Proactiveness (1)	$\overline{x_2}$=5.5	(7) Highly proactive
Mild risk-taker (1)	$\overline{x_3}$=4.3	(7) Avid risk-taker
CE Intensity	4.6	

4.1.3. Company C

The informants for Company C were the owners of the company. The highlight of their answers was their tendency to avoid risk (see Table 5). Minimizing potential cost in the future was one of their most important concerns. This might be because they operated under a slim profit margin and they were almost crippled by previous defect-related costs. It was essential for them to avoid higher cost, so that they could still enjoy a healthy cash-flow. In terms of proactiveness, Company C projected their production and price based on their big competitors in the market. Nevertheless, they were passionate to learn about new production technology and marketing strategy even though they did not necessarily implement them.

Table 5. Summary result of company C

Low Innovativeness (1)	$\overline{x_1}$ =3.7	(7) Highly Innovative
Low Proactiveness (1)	$\overline{x_2}$=3.7	(7) Highly proactive
Mild risk-taker (1)	$\overline{x_3}$=1.8	(7) Avid risk-taker
CE Intensity	3.2	

4.1.4. Company D

The informants for Company D were the owner and one of the production staffs. Company D's CE intensity was slightly below Company B (see Table 6). They were not as proactive or willing to take risk as Company B, but they invested in R&D activities. They invested on equipment to design new clothing patterns and models and they also realized the importance of analysing feedback from their distributors, resellers, and consumers. In addition, they also routinely assess their production and storage log to find ways to reduce production and distribution time. However, similar to the three other companies, they also tended to shy away from risk and chose to minimize cost in a turbulent condition.

Table 6. Summary result of company C

Low Innovativeness (1)	$\overline{x_1}$ =5.0	(7) Highly Innovative
Low Proactiveness (1)	$\overline{x_2}$=4.2	(7) Highly proactive
Mild risk-taker (1)	$\overline{x_3}$=4.0	(7) Avid risk-taker
CE Intensity	4.4	

4.2. Strategic Management Practices: Lessons from Company B
In terms of strategic management practices, it was not about whether a practice was adopted, but about the emphasize the SMEs given to each practice. As mentioned in the literature review, this study limited the type of strategic management practices to three, which are environmental scanning (method and frequency), planning (locus), and control (strategic and financial).

4.2.1. Environmental Scanning: Barringer and Bluedorn [20] suggested that environmental scanning refers to the managerial activity in learning events and trends within the organization. Hambrick also mentioned that environmental scanning is the basic unit of analysis that can facilitate the ability to recognize external environment. Organization's external environment includes knowledge and information about competitors, consumers, government regulations, changes in the macro economy, and issues, new trends [55] and new technology.

This study found that Company A, C, and D did not proactively scan their environment especially on the emergence of new technology that can boost their performance and the latest regulations regarding their business. They mostly concerned themselves with competitors and market, but not new technology to manufacture their products or deliver their services or regulations that might hamper their businesses. Company B, on the other hand, constantly tried to find an 'easier way' to do business. They adopted e-marketing and proactively collected information on the latest regulations, so that they could win tenders from government institutions.

4.2.2. Planning. Action plan is desired by companies and without a plan, the company will be swayed by environmental influences. Planning helps ensure that all work done by members of the organization is coordinated, not leading to different directions, and includes future predictions as a factor to consider. With a plan, the work will be coordinated, the employee will be facilitated to conduct long-term thinking, the company will have better control, and employees tend to give commitments to keep the business going according to plan. However, planning also has several drawbacks, among others: the uncertainty of forecasting, the use of assumptions which are prone to changes, stiffness, and sometimes, plans tend to inhibit change.

Ideally, to get a proactive, innovative, and realistic plan, companies should have deep locus of planning. Deep locus of planning indicates that all employees of every level of hierarchy within a company involved in the planning process. Deep locus of planning is demonstrated in the form of top-level managers desire to facilitate and support the entrepreneurial behaviour in firms; in line with the commitment of top-level managers to tolerate failure, provide freedom from excessive monitoring, and delegate authority and responsibility to middle and lower-level managers [7,10].

Comparing Company B with the rest of the subjects, this study found that Company A, C, and D rarely encouraged the involvement of employees in their planning process. This probably because their employees were not highly educated. However, the non-involvement of employees means that SMEs are losing valuable insights that might be relevant in terms of technology adoption. There was always a probability that the low-level employees were exposed to a low-level or a simple technology that can be useful for the SMEs.

4.2.3. Control. Controlling employees influences what they actually do [56,57]. Purpose of the control system is to ensure that business strategy really brings the organization to the right paths in achieving predetermined goals and targets. In the context of this study, if the company wants to stimulate the entrepreneurial intensity of firms, then it should use a control system that supports innovativeness, proactiveness, and risk-taking.

There are two types of controls that are relevant to the issue of entrepreneurship, namely financial control and strategic control. In most companies, both controls are generally used together. Financial control is based on financial criteria such as net income, sales, and ROE [58,59]. Financial control is a concept that is clear and unambiguous because of the discipline in the implementation process. Financial control is an easier instrument to use than strategic control. It is often used as a means of controlling tools and corporate performance measurement. However, financial control does not trigger entrepreneurial behaviour in the company because the standard used in the financial controls are very stiff and has been determined from the start so often do not provide room for companies to innovate. In contrast, strategic control could pay tribute to the creativity and the use of the opportunity through innovation.

This study found that Company A, C, and D preferred to measure their performance using a financial control rather than strategic control because it was easy to conduct. For them, the financial bottom line was the most important thing. However, this might be the reason why their CE intensity and consequently, their willingness to adopt new technology was low. They think that adopting technology is a big investment and for a short period, it would hurt their balance; making their financial performance looks poor. Company B, on the other hand, focused on growth. They were inclined to sacrifice short-term profit to expand.

5. Conclusion

For SMEs to be willing to adopt new technology in the 4.0 industrial era, they need to have a high CE intensity which existence can flourish under the support of strategic management practices. This research tried to capture the CE intensity of Indonesian SMEs and the underlying strategic management practices (i.e., scanning, planning, and control). This study found that (i) the CE intensity of SMEs in Indonesia is in the medium-low range. They were not proactive, innovative, and brave enough to adopt new technology and (ii) There are strategic management practices that can be learned from a company with high level of CE intensity.

The first implication was that managers or owners need to do a more intensive environmental scanning to identify new technology, involve their employees in planning process to capture more insights, and rely on both financial and strategic control in order to foster a more entrepreneurial thinking. The second implication is related to government support. Indonesian SMEs have a very limited resource in terms of R&D and it would be too risky for them to invest on R&D activities heavily. The government should invest in risk-sharing schemes with SMEs in terms of R&D like what have been done by Ireland [60] and Malaysia [61].

This study has limitations. First, this is qualitative research to explore CE intensity in four Indonesian SMEs. The results could not be generalized to other SMEs. Future research should aim to validate this CE and Strategic Management Practices pattern using a quantitative study specifically for Indonesian SMEs. Second, this study only considered the CE intensity as the impact of strategic management practices. Likely, there are other variables that might affect the CE intensity. Future research should explore other frameworks that can complement strategic management practices so that CE intensity can be better explained.

References

[1] Bank Indonesia 2015 *Profil Bisnis Usaha Mikro, Kecil dan Menengah (UMKM)* (Jakarta)
[2] Eze S C, Chinedu-Eze V C, Bello A O, Inegbedion H, Nwanji T and Asamu F 2019 Mobile marketing technology adoption in Service SMEs: a multi-perspective framework *J. Sci. Technol. Policy Manag.* **ahead-of-p**
[3] Han J and Park C 2017 Case study on adoption of new technology for innovation: Perspective of institutional and corporate entrepreneurship *Asia Pacific J. Innov. Entrep.* **11** 144–58
[4] Abebe M 2014 Electronic commerce adoption, entrepreneurial orientation and small-and medium-sized enterprises (SME) performance *J. Small Bus. Enterp. Dev.* **21** 100–16
[5] Alos-Simo L, Verdu-Jover A J and Gomez-Gras J-M 2017 How transformational leadership facilitates e-business adoption *Ind. Manag. Data Syst.* **117** 382–97
[6] Burgelman R A 1983 Corporate entrepreneurship and strategic management: Insights from a

process study *Manage. Sci.* **29** 1349–455
[7] Kuratko D F, Montagno R V. and Hornsby J S 1990 Developing an intrapreneurial assessment instrument for an effective corporate entrepreneurial environment *Strateg. Manag. J.* **11** 49–58
[8] Shemi A P and Procter C 2018 E-commerce and entrepreneurship in SMEs: case of myBot *J. Small Bus. Enterp. Dev.* **25** 501–20
[9] Urban B and Wood E 2017 The innovating firm as corporate entrepreneurship *Eur. J. Innov. Manag.* **20** 534–56
[10] Kuratko D F, Hornsby J S and Hayton J 2015 Corporate entrepreneurship: the innovative challange for a new global economic reality *Small Bus. Econ.* **45** 245–53
[11] Zahra S A 2015 Corporate entrepreneurship as knowledge creation and conversion: the role of entrepreneurial hubs *Small Bus. Econ.* **44** 727–35
[12] Burkholder P, Stam A M C E and Hulsink W 2017 A case of academic intrapreneurship *Acad. Manag. Proc.* **2016** 17743
[13] Gawke J C, Gorgievski M J and Bakker A B 2019 Measuring intrapreneurship at the individual level: Development and validation of the Employee Intrapreneurship Scale (EIS) *Eur. Manag. J.* **ahead-of-p**
[14] Kurnia S, Choudrie J, Mahbubur R M and Alzougool B 2015 E-commerce technology adoption: A Malaysian grocery SME retail sector study *J. Bus. Res.* **68** 1906–18
[15] Duobienė J 2008 The role or organizational culture in sustaining corporate entrepreneurship *Econ. Manag.* 350
[16] McCrea E and Betts S C 2013 Failing to learn from failure: An exploratory study of corporate entrepreneurship outcomes *Electron. Bus. J.* **12** 345–57
[17] Gautam V and Verma V 1997 Corporate entrepreneurship: Changing perspectives *J. Entrep.* **6** 233–44
[18] Ferreira J 2002 Corporate entrepreneurship: A strategic and structural perspective *International Council for Small Business 47th World Conference* p 20
[19] Kuratko D F and Goldsby M G 2004 Corporate entrepreneurs or rogue middle managers? A framework for ethical corporate entrepreneurship *J. Bus. Ethics* **55** 13–30
[20] Barringer B R and Bluedorn A C 1999 The relationship between corporate entrepreneurship and strategic management *Strateg. Manag. J.* **20** 421–44
[21] Grünhagen M, Wollan M L, Dada O and Watson A 2014 The moderating influence of HR operational autonomy on the entrepreneurial orientation-performance link in franchise system *Int. Entrep. Manag. J.* **10**
[22] Messersmith J G and Wales W J 2011 Entrepreneurial orientation and performance in young firms: The role of human resource management *Int. Small Bus. J. Res. Entrep.* **31** 115–36
[23] Lumpkin G T and Dess G G 2001 Linking two dimensions of entrepreneurial orientation to firm performance: The moderating role of environment and industry life cycle *J. Bus. Ventur.* **16** 429–51
[24] Ferrier W J, Smith K G and Grimm C M 2017 The role of competitive action in market share erosion and industry dethronement: A study of industry leaders and challengers *Acad. Manag. J.* **42** 372–88
[25] Miller D and Friesen P H 1982 Innovation in conservative and entrepreneurial firms: Two model of strategic momentum *Strateg. Manag. J.* **3** 1–25
[26] Covin J G and Slevin D P 1988 The influence of organizational structure on the utility of an entrepreneurial top management style *J. Manag. Stud.* **25** 217–34
[27] Clarke J E and Liesch P W 2017 Wait-and-see strategy: Risk management in the internationalization process model *J. Int. Bus. Stud.* **48** 923–40
[28] Sundbo J 2003 *The Theory of Innovation: Entrepreneurs, Technology and Strategy* (Cheltenham: Edward Elgar Publishing)
[29] Kenny M and Mutjaba B G 2007 Understanding corporate entrepreneurship and development: A practitioner view of organizational intrapreneurship *J. Appl. Manag. Entrep.* **12** 73–88

[30] Erkkilä K 2000 *Entrepreneurial Education: Mapping the Debates in the United States, the United Kingdom, and Finland* ed S Bruchey (New York: Garland Publishing, Inc)
[31] Antoncic B and Hisrich R D 2004 Corporate entrepreneurship contingencies and wealth creation *J. Manag. Dev.* **23** 518–50
[32] Miller D 1983 The correlates of entrepreneurship in three types of firms *Manage. Sci.* **29** 757–868
[33] Zahra S A 1991 Predictors and financial outcomes of corporate entrepreneurship *J. Bus. Ventur.* **6** 259–85
[34] Stone D L, Deadrick D L, Lukaszcwski K M and Johnson R 2015 The influence of technology on the future of human resource management *Hum. Resour. Manag. Rev.* **25** 216–31
[35] Brown T E, Davidsson P and Wiklund J 2001 An operationalization of Stevenson's conceptualization of entrepreneurship as opportunity-based firm behavior *Strateg. Manag. J.* **22** 953–68
[36] Hisrich R D 2013 *International Entrepreneurship: Starting, Developing, and Managing a Global Venture* ed L C Shaw, T Accomazzo, K Guarino, A Virding and T Bigelow (Thousand Oaks: Sage Publications, Inc.)
[37] Yunis M, El-Kassar A-N and Tarhini A 2017 Impact of ICT-based innovations on organizational performance *J. Enterp. Inf. Manag.* **30** 122–41
[38] Hayton J C 2003 Strategic human capital management in SMEs: An empirical study of entrepreneurial performance *Hum. Resour. Manage.* **42** 375–91
[39] Hayton J C 2005 Promoting corporate entrepreneurship through human resource management practices: A review of empirical research *Hum. Resour. Manag. Rev.* **15** 21–41
[40] Hayton J C and Kelley D J 2006 A competency-based framework for promoting corporae entrepreneurship *Hum. Resour. Manag. Rev.* **45** 407–27
[41] Kaya N 2006 The impact of human resource management practices and corporate entrepreneurship on firm performance: Evidence from Turkish firms *Int. J. Hum. Resour. Manag.* **17** 2074–90
[42] Morris M H and Jones F F 1993 Human resource management practices and corporate entrepreneurship: An empirical assessment from the USA *Int. J. Hum. Resour. Manag.* **4** 873–96
[43] Morris M H and Jones F F 1995 Relationships among environmental turbulence, human resource management, and corporate entrepreneurship *J. Bus. Entrep.* **7** 61
[44] Mustafa M, Richards J J and Ramos H M 2013 High performance human resource practices and corporate entrepreneurship: The mediating effect of middle managers knowledge collecting and donating behavior *Asian Acad. Manag. J.* **18** 17–36
[45] Montoro-Sanchez A and Soriano D R 2011 Human resource management and corporate entrepreneurship *Int. J. Manpow.* **32** 6–13
[46] Hill C W, Jones G R and Schilling M A 2015 *Strategic Management Theory* (Stamford: CENGAGE Learning)
[47] Heras-Saizarbitoria I and Boiral O 2013 ISO 9001 and ISO 14001: Towards a Research Agenda on Management System Standards *Int. J. Manag. Rev.* **15** 47–65
[48] Kafetzopoulos D P, Psomas E L and Gotzamani K D 2015 The impact of quality management systems on the performance of manufacturing firms *Int. J. Qual. Reliab. Manag.* **32** 381–99
[49] Sargeant J 2012 Qualitative research part II: Participants, analysis, and quality assurance *J. Grad. Med. Educ.* **4** 1–3
[50] Goldsby M G, Kreiser P M, Kuratko D F, Bishop J W and Hornsby J S 2018 Social proactiveness and innovation: The impact of stakeholder salience on corporate entrepreneurship *J. Small Bus. Strateg.* **28** 1–15
[51] Hirose K, Lee S-H and Matsumura T 2017 Environmental corporate social responsibility: A note on the first-mover advantage under price competition *Econ. Bull.* **37** 214–21
[52] Harris L C and Ogbonna E 2006 Initiating strategic planning *J. Bus. Res.* **59** 100–11

IOP Conf. Series: Materials Science and Engineering **722** (2020) 012037 doi:10.1088/1757-899X/722/1/012037

[53] Freel M S 2005 Patterns of innovation and skills in small firms *Technovation* **25** 123–34
[54] Tang G and Wei L-Q 2015 How effective human resource management promotes corporate entrepreneurship: evidence from China *Int. J. Hum. Resour. Manag.* **26** 1586–601
[55] Hay M and Williamson P 1997 Good strategy: the view from below *Long Range Plann.* **30** 651–64
[56] Hussey D 1997 *The Innovation Challenge* (Wiley)
[57] Biron M, Farndale E and Paauwe J 2011 Performance management effectiveness: Lessons from world-leading firms *Int. J. Hum. Resour. Manag.* **22** 1294–311
[58] Hitt M A, Hoskisson Ro E and Ireland R D 1990 Merger and acquisitions and managerial commitment to innovation in M-form firms *Strateg. Manag. J.* **11** 29–47
[59] Ismyrlis V and Moschidis O 2015 The effects of ISO 9001 certification on the performance of Greek companies: A multidimensional statistical analysis *TQM J.* **27** 150–62
[60] Hewitt-Dundas N and Roper S 2010 Output additionality of public support for innovation: Evidence for Irish manufacturing plants *Eur. Plan. Stud.* **18** 107–22
[61] Ismail R and Othman N A 2014 The effectiveness of government-support programmes toward business growth *J. Technol. Manag. Technopreneush.* **2** 41–52

IOP Conf. Series: Materials Science and Engineering **722** (2020) 012038 doi:10.1088/1757-899X/722/1/012038

Applying eye tracker technology for analyzing user interface design of online transportation applications (case study: grab application)

D S Dewi[1], N Ilmi[2], and R S Dewi[3]

[1,2,3] Industrial Enginnering Department, Institut Teknologi Sepuluh Nopember Surabaya, Indonesia

E-mail: dyahsd@gmail.com, nurulilmi136@gmail.com,ratna@ie.its.ac.id

Abstract. In the past few years, online transportation has gain popularity in Indonesia. Many passengers prefer to choose online transportation for traveling due to several reasons including accessibility and price. In response to this, several online transportations have been operating in Indonesia, the most popular one is Grab. Despite its popularity, little is known about the usability of its interface design in respect to specific groups of user especially for elderly. In order to provide information on how to better design online transportation application especially for elderly, this paper aims to analyze interface design of Grab application from the elderly point of view. For this purpose, an experiment was conducted by using the eye tracker. This technology has applied for different purposes including design and marketing research purposes. The experiment was conducted by involving 10 elderly participants with ages ranging from 55-69 years. The eye tracker data used in this study is a fixation number, fixation duration, saccade number, and saccade duration. The result of study reveals that gender influence the fixation number, fixation duration, saccade number and saccade duration. Eye condition also influences fixation number and fixation duration but not for saccade number and saccade duration. The highest fixation number occurred in the area of interest (AOI) 2 and the largest fixation duration is also in the area of interest (AOI) 2.

1. Introduction

The development of modern technology has influenced the way transportation business is operated in Indonesia. Nowadays, online transportation has become a choice that is more popular for traveling rather than conventional ones. This because the applications (apps) are easy to install on a smartphone and provide various offers. Grab is one of online transportation that operates in Indonesia. Grab, formerly known as Grab Taxi, is a company that originally established in Singapore that serves as transportation applications provider which are now available in six countries in Southeast Asia, including Malaysia, Singapore, Thailand, Vietnam, Indonesia, and the Philippines. The high population in Indonesia together with the increasing demand for transportation makes Indonesia become the potential market for the online transportation provider. One potential segment of markets for online transportation is elderly.

The population of elderly in Indonesia is increasing year by year. The percentage of elderly growth reaches 8.97% of the total population in Indonesia. Elderly according to the Republic Indonesia Ministry of Health (2006) is an advanced preparatory period that shows the maturity of 55-year-old souls and above. Physically, the process of aging is the gradual disappearance of the tissue's ability to repair itself or replace and maintain its normal function so that it cannot withstand the infection and repair the damage suffered [1]. Nowadays, the elderly are required to be more independent. Data from Indonesia Statistics revealed that nearly half of Indonesia's elderly (47.92%) still actively working at their age and 61.70% of elderly make the elderly as the head of the household. One of requirements for the elderly to be able to work and to be independence includes ability to use smartphone for acquiring their need such as transportation. Indonesian Internet Service Provider Association (2017) shows that the elderly is one of the active users of internet in Indonesia by percentage 15 - 72% [2].

However, most of design of online applications currently does not consider elderly users' limitations [1]. The current smartphone application interface design is not meet the needs of the elderly. This is supported by research conducted in South Korea regarding the design of interface design on smartphones for the elderly, which found that there are some difficulties experienced by the elderly users when interacting with interface devices on smartphones. The highest difficulty experienced by older people is the length of time they need to learn and understand the new icons that exist on an interface device [2]. Due to this reason, it is necessary to improve the interface design of applications used by elderly since in the future elder people will depend on such application for living.

Study on user interface design usually uses interview methods or questionnaires. While these methods are easy to apply, however, these approaches have some limitations such as the subjective nature of the result [3]. Different method is needed that could provide information on the behavior and attention of the elderly in using an application. In this case, the eye tracker technology could provide such benefit. Eye tracker is a sensor technology that allows the device to know precisely where the user's eyes focused [4]. Few studies have been conducted for applying this technology, for example, a study that uses eye tracker to investigate consumer behavior in choosing a product by considering the type of product being reviewed [5]. But, there is a lack of publication that analyzes online transportation interface design for the elderly by using an eye tracker. Therefore, in this study, the eye tracker experiment is conducted to focus on the elderly while using the Grab application. Tobii Pro Lab Eye Tracker is used for the experiment. It provides a comprehensive platform for recording and analysis of eye gaze data, which helps in the interpretation of human behavior, consumer responses, and psychology [6].

2. Methodology

Eye-tracking experiments were carried out in the Work System Design and Ergonomic Laboratory of ITS. In this experiment, elderly participants were given the task to order GrabCar services installed on smartphones. The eye tracker is placed in front of the smartphone to track the user's eye movements while working on the task of ordering GrabCar services. Before the experiment began, a calibration process is carried out to measure the level of precision and accuracy of the eyes for all elderly participants. Calibration results must have a minimum value of 0.5 ($<= 0.5$) so that the resulting data can be used [6]. Because elderly participants have limited vision, this process must be carried out repeatedly to get the right accuracy and precision.

2.1 Participants

Ten elderly participated in the experiment, these include six women and four men with an age range from 55 to 69 years. These participants are elderly who live in the city of Surabaya. They were divided into three groups based on their eye conditions, which are elderly with farsightedness, elderly with a combination of farsightedness and cylindrical eyes and elderly with normal eyes. Participants consisted of 6 elderly (1 men and 5 women) who wore glasses with nearsightedness, 3 elderly (2 men and 1 women) with a combination of farsightedness and cylindrical eyes, and 1 elderly man who didn't wear glasses (normal eyes).

2.2 Eye Tracker

Eye tracker is a sensor technology that allows the device to know precisely where the user's eyes are focused. Eye tracker technology is a technique for measuring physiological responses to visual stimuli that has the advantage that responses can be recorded in real-time [7]. The device determines the presence, attention, focus, sleepiness, awareness to other mental states. The information can be used to gain insight into consumer behavior or to design revolutionary new user interfaces on various devices. Eye tracker technology has many uses in obtaining product information, such as what information influences consumer decisions, whether visual attention varies based on level of importance, familiarity, or external pressure (time, purpose, etc.) and product design or label on which packaging is the most visually appealing [8]. Researchers can also focus on certain format choices (such as font style or size, color, image, number of products, etc.) or an advertising content (including branding, value-added information, and promotional messages).

The Tobii eye-tracker software comprises three modules and a Project Overview. It consisting of the design module, the record module, and the analysis module. Design module is how to outline the way the stimuli are presented to study participants. This includes what the stimuli look like on the screen, in which order they get displayed, as well as how to proceed to the next stimuli. The record module is where you make recordings in Tobii eye tracker. It enables you to select which Timeline to use for which recording and lets you control the recording process, in general. Then analyze module is necessary when you want to conclude the collected data. The analyze module provides analysis tools that enable you to visualize and export the data you have collected [6].

There are currently four kinds of projects consisting of screen projects which are supposed to be used when stimuli presentation and data collection are done using a screen-based eye tracker. Glasses projects which are supposed to be used when data has been collected using Tobii Pro Glasses 2. Scene Camera projects, when using an external video camera to record events in the real world and VR 360 projects, when analyzing 360-degree images and videos using Tobii Pro VR Integration. In this paper, the experiment used a scene camera project because the eye-tracking process while participants are doing the task will be recorded using an external camera. Eye tracker is also designed to work with other software commonly used for recording and analyzing data. Result data from eye tracker can be enabling data exports in standardized formats, for example for Microsoft Excel, Matlab, and SPSS. Result data from eye tracker enabling data exports in standardized formats, for example for Microsoft Excel, Matlab, and SPSS. In this study, the data used to be processed is fixation number, fixation duration, saccade number, and saccade duration. During the eye-tracking process, the data collected will be very much because the eye tracker records eye movement data every 45 milliseconds. So the data needs to be filtered and tested using the Anova Test.

2.3 Experimental Design
Tasks were given to all participants. This test is done by displaying the Grab interface application with a participation scenario to order GrabCar (car services). The main tasks are to order car to a specific destination. The task doesnot include the payment process. The tasks carried out in the eye tracker experiment is described in Figure 1. The first step is briefing with the elderly participants. This briefing includes requests that participants keep their seating positions during the test, how the calibration process takes place, and the main task of this research is to ask elderly participants to order GrabCar services in the Grab application. The explanation should be done one by one because the elderly have limitations in receiving new information. Then the calibration process is assessing the level of accuracy and precision of the user's pupils. As for the process, participants were asked to see the three red dots on the mobile screen (dots on the top left, middle and bottom right). The calibration process showed in Figure 2. To calibrate the user's students, they are asked to look at three points on the mobile screen, then if the accuracy and accuracy numbers reach 0.5, it can proceed to the main experimental stage of assessing the design of the Grab application.

2.4 Experimental Variables
2.4.1 Fixation Number
Fixation is a situation where the user's pupils focus on one point of the object [9]. Fixation number is the number of fixations that are formed when the user's eyes are focused on an object. Fixation number is calculated by tracking the number of fixations in the Area of Interest (AOI) area so that you can find out which area of interest is the most fixation.

2.4.2 Fixation Duration

Fixation duration is the time needed when a fixation occurs or when the user's eyes see an object [5].

2.4.3 Saccade Number
Saccade is data when the pupil moves from one object to another [10]. The Saccade number is the amount of saccade formed when participants see the interface tested using an eye tracker.

2.4.4 Saccade Duration
Saccade duration is the time needed when the saccade occurs or eye movement from one object to another object.

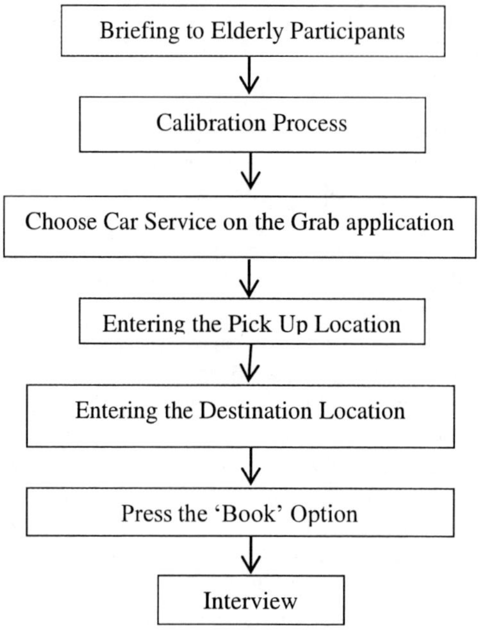

Figure 1. Procedure of Experiment

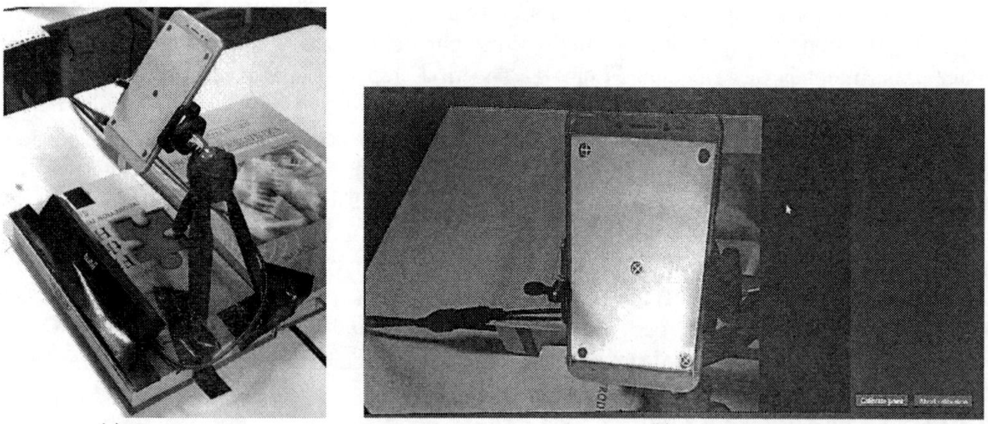

(a) (b)

Figure 2. Calibration Process. Figure (2a) is calibration point on the mobile phone screen and Figure (2b) is the calibration points that appear on the computer. The points used are dots on the top left, middle and bottom right.

IOP Conf. Series: Materials Science and Engineering **722** (2020) 012038 doi:10.1088/1757-899X/722/1/012038

2.5 Apparatus and Stimuli

Eye Tracker used in this study is Tobii Pro Lab Full Edition (x64). The Samsung Galaxy Grand Prime SM-G530 smartphone was also installed near to the eye tracker device. The smartphone is placed on a Gorilla Mini Camera Tripod to be easily seen by elderly users. The Grab version 5.51.0 application was installed in the smartphone. HP All-in-One 22-c0035d computer is also installed to see the calibration results. During the experiment, the Logitech C922 webcam recorded a screen display that integrated with Tobii's eye tracker, to record the point of fixation and eye movement. Participants sat at 50 cm in front of the Tobii Eye Tracker. The distance between a smartphone and a Logitech camera is 80 cm. The elevation angle between the eyes of the elderly participant and the smartphone is 40 degrees while the elevation angle between the eyes of the elderly participant and the eye tracker is 45 degrees. Figure 3 shows the experiment layout of this study. It presents the apparatus of the experiment with the relative position to the participant.

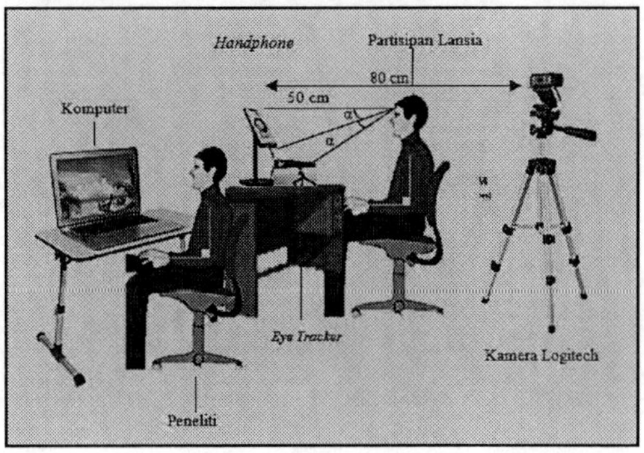

Figure 3. Illustration of experimental layout

3. Data Collection and Analysis

3.1 Data Screening Process

The result of eye tracker testing with 10 respondents produced 18,404 data. This data is taken by exporting data from the eye tracker that records the eye position of the respondent every 45 milliseconds. Since the recorded data is huge then it needs to be filtered.. The screening process showed in Figure 4. This eye record data consists of data fixation, saccade, unclassified, and eyes not found. Then unclassified data and eyes not found are removed because it is invalid data so that the data from the eye tracker becomes 7318 data. In this study, the results of the eye tracker data will be filtered according to the Area of Interest (AOI). Furthermore, the data filtration process is carried out in accordance with predetermined variables so that the data to be processed becomes as much as 1984 data.

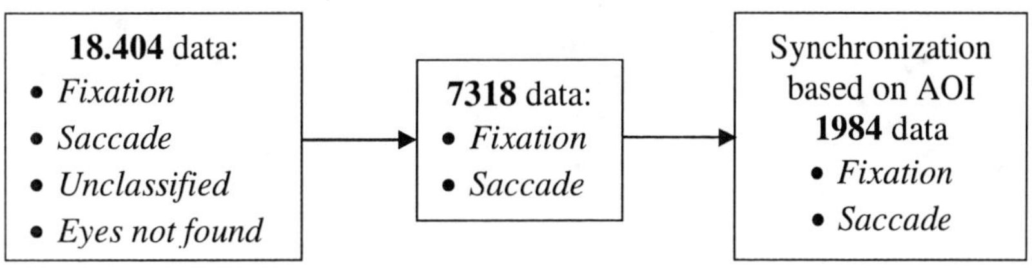

Figure 4. Eye tracker data screening process

3.2 Gaze plot

Gaze plots are lines that indicate saccade movement that occurs when the eye moves focus from one object to another [11]. The difference in color in the gaze plot display shows the number of participants who did the testing. While the numbers written on the circle gaze plots indicate the order in which the user's eye movements occur. In the Figure 5 showed 10 different colors are indicating 10 people tested participants.

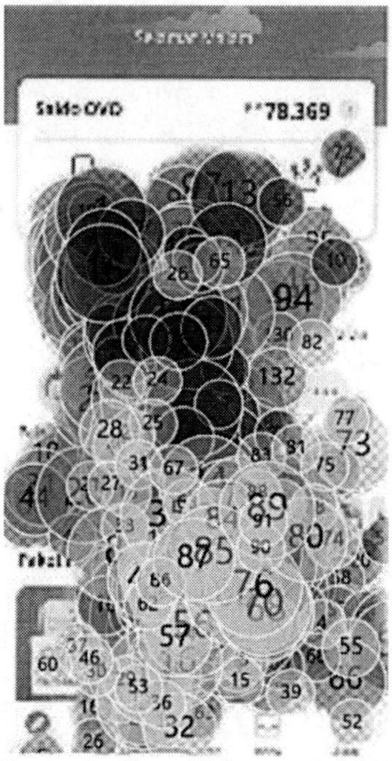

Figure 5. The Gaze plot showed the movement of the user's eye in the Grab application interface.

3.3 Heatmaps

Heat maps are color maps used to illustrate the amount of fixation that occurs when participants use an eye tracker [12]. Heat maps mark certain areas affected by participants' eye focus stimuli with certain color indicators. It is showed in Figure 6, the color indicators are red means the highest number of fixations and the longest duration of fixation and green means a small amount of fixation and short duration of fixation [13].

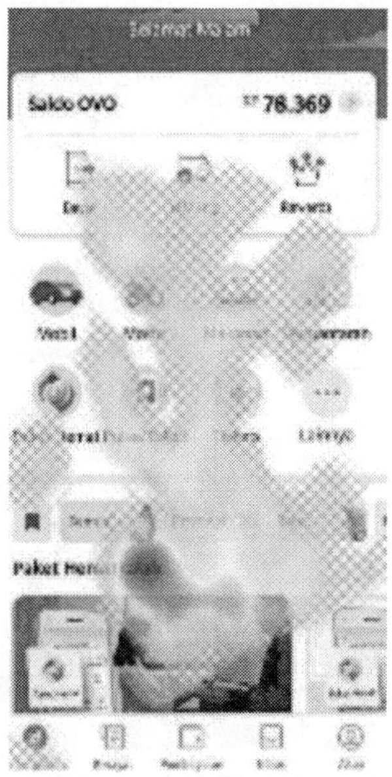

Figure 6. Heatmaps

3.4 Area of Interest (AOI)

Area of Interest is an area that is limited to being used as a focus of vision on the interface to be examined in testing [14]. The interface in the Grab application will be divided in three areas of interest that it is shown in Figure 7. The selection of three Area of Interest (AOI) is because it refers to the design of an existing Grab application, where they classify menus on the application interface based on the similarity of functions. The first area is the payment area which consists of the pay menu, refill menu, gift menu, and OVO balance. These menus are very important to be accessed in using the Grab application related to payment matters. The second menu is a core menu that contains services that can be used by Grab users, such as ordering cars, motorbikes, food, delivery, savings packages, credit, tickets, and others. The third area is the menu area consisting of the homepage, user purchase history, payment, inbox, and account profile menu.

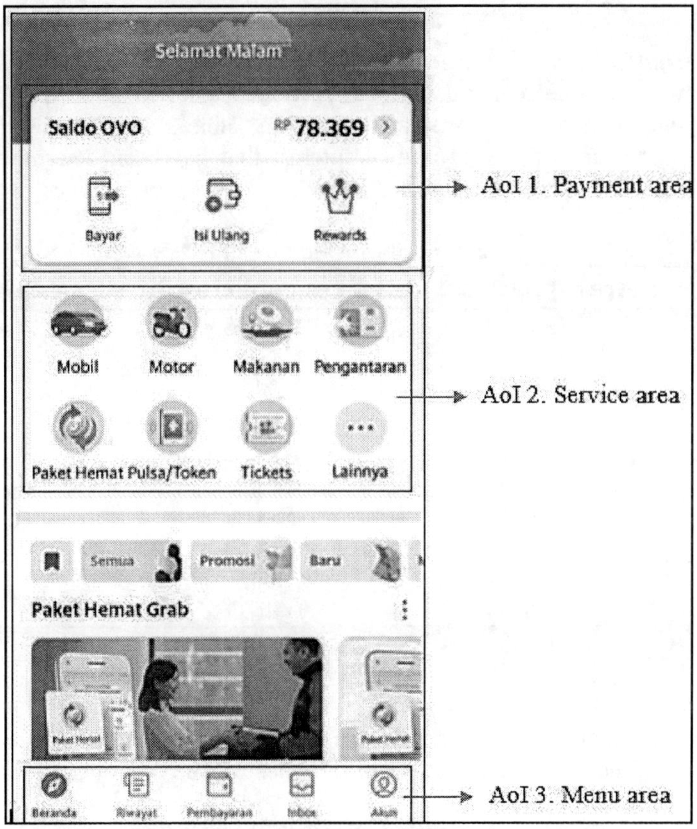

Figure 7. Area of Interest (AOI) in Grab Application

3.4.1 Fixation Number Data

Fixation number or number of fixations that are formed when the user's eyes are focused on an object. The fixation number is calculated by tracking the number of fixations in the Area of Interest (AOI) area so that it can be seen which area of interest has the most formed fixation. In the Table 1 showed the distribution of fixation data that is formed in each Area of Interest that has been determined. Then the biggest fixation number is in the area of interest 2, which is consist of services in the Grab application interface.

Table 1. Fixation Number in Every AOI

Participants	**Area of Interest 1**	**Area of Interest 2**	**Area of Interest 3**
1	56	60	21
2	45	67	34
3	52	74	40
4	49	68	37
5	38	51	26
6	33	46	24
7	41	63	30
8	35	58	32
9	39	61	28
10	30	68	32
Total	**418**	**616**	**304**

3.4.2 Fixation duration Data
Fixation duration is the time required when a fixation occurs. Fixation duration data can also be traced based on the location of the Area of Interest. In Table 2 showed the distribution of fixation duration data that is formed in each Area of Interest that has been determined. Then the biggest fixation duration is in the area of interest 2, which is consist of services in the Grab application interface.

Table 2. Fixation Duration in Every AOI

Participants	**Area of Interest 1**	**Area of Interest 2**	**Area of Interest 3**
1	353.9107	352.0833	383.1905
2	324.22	471.63	268.14
3	593.5	344	372
4	361.52	416.78	291.8
5	405.29	479.08	126.5
6	248.68	283.97	246.71
7	519.91	584.23	483
8	456.41	523.59	423.93
9	515.91	1094.25	379.04
10	386.6364	446.4	329.64
Average	**416.60**	**499.60**	**330.40**

3.4.3 Saccade Number Data
Saccade numbers are many saccades that are formed when participants view the interface tested using an eye tracker. The data on the number of saccades will be described in Table 3 below.

Table 3. Saccade Number

Participants	***Saccade Number***
1	88
2	76
3	82
4	69
5	64
6	50
7	67
8	61
9	46
Total	**646**

3.4.4 Saccade duration Data

Saccade duration is the time required when a saccade or eye movement occurs from one object to another. The saccade duration data for 10 participants are described in Table 4 below.

Table 4. Saccade Duration

Participants	Saccade Number
1	88
2	76
3	82
4	69
5	64
6	50
7	67
8	61
9	46
Total	**646**
Average	**64.6**

4. Result

In this section, the results of the eye tracker experiment with One Way Anova test revealed that the fixation number, fixation duration, saccade number, and saccade duration have a significant influence on the gender of the elderly. Also, fixation number and fixation duration have a significant influence on the eye condition of the elderly but saccade number and saccade duration have no significant influence on the eye condition of the elderly. To find out this, the One Way Anova test was carried out.

4.1 Normality Test

The eye tracker results are processed by statistical testing, namely the One Way Anova test by first performing the Shapiro-Wilk normality test and Kolmogorov-Smirnov to ensure that the distribution of fixation duration data is normally distributed. Data can be said to be normally distributed if the significance value (sig) in the Tests of Normality table is greater than 0.05. It shown in Table 5 the significance value for farsighted is 0.214 and the significance value for farsighted and cylindrical eyes is 0.722, it is mean greater than 0.05. So it can be concluded that the distribution of fixation duration data is normally distributed.

Table 5. Test of Normality

	Eye conditions	Kolmogorov-Smirnov[a]			Shapiro-Wilk		
		Statistic	df	Sig.	Statistic	df	Sig.
Fixation_duration	Farsighted	.237	6	.200*	.867	6	.214
	Farsighted and cylindrical eyes	.233	3	.	.979	3	.722

4.2 Homogeneity Test

Homogeneity tests are also needed to fulfill the assumptions on the One Way Anova test that will be carried out. Data is considered homogeneous if the significance value (sig) in the Test of Homogeneity of Variances table is greater than 0.05. It shown in Table 6 the significance value is

IOP Conf. Series: Materials Science and Engineering **722** (2020) 012038 doi:10.1088/1757-899X/722/1/012038

0.176 and it is mean greater than 0.05. So it can be concluded that the distribution of fixation duration data is homogeneous.

Table 6. Homogeneity Test

Fixation duration

Levene Statistic	df1	df2	Sig.
2.208	1	8	.176

The aggregate result of the relationship will be described in Table 7.

Table 7. One Way Anova Result

No	Variables	One Way Anova Result	
		There is a significant difference	There is no significant difference
1	Relationship between participants gender with fixation number	✓	
2	Relationship between participants gender with fixation duration	✓	
3	Relationship between participants gender with saccade number	✓	
4	Relationship between participants gender with saccade duration	✓	
5	Relationship between participants eye condition with fixation number	✓	
6	Relationship between participants eye condition and fixation duration	✓	
7	Relationship between participants eye condition and saccade number		✓
8	Relationship between participants eye condition and saccade duration		✓

4.3 Saccade Analysis

The Saccade number is the amount of saccade formed when participants see the interface tested using an eye tracker [15]. To find out the saccade number can only be known based on the results of the eye tracker in each participant. The average value of saccade number for 10 participants was 64.6 saccade which occurred with an average saccade duration of 50.56 milliseconds. Whereas to see the saccade movement on the Grab interface, it can be seen on the video recording carried out during the test. Figure 8 shows the saccade movement at the beginning process of seeing the Grab interface, it can be seen on the video recording carried out during the test.

1. Saccade movement at the beginning when participant looking at the Grab interface

The movement of the eyes of the elderly when they first see the Grab application interface is to look at the top of the interface that says "Good Morning". In Figure 8 showed the saccade movement in the beginning interface of Grab application. This means that such greeting words are quite interesting to the attention of the elderly. Then the saccade formed in the payment ad area which is also at the top of the interface. This advertisement also distracts the attention of the elderly when they first see this Grab display.

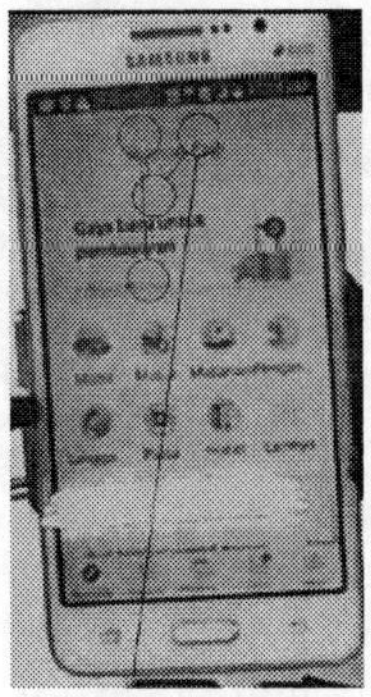

Figure 8. Saccade movement in beginning interface of Grab application

2. Saccade movement when order GrabCar services

Before starting to order, the focus movement of the elderly eye leads to the ads in the middle area of the lower part of the Grab interface like showed in Figure 9. This indicates that advertisements placed in the area are quite capable of distracting the elderly.

Figure 9. Saccade movement when order the GrabCar

3. Saccade movement when the elderly enter the location to order the GrabCar Service

After choosing the GrabCar service, the elderly are required to enter the pickup location point and also the destination location point. In Figure 10 showed that the saccade movement is right at the bottom of the place to enter the location. Then the elderly can add the 'Message' option to order GrabCar after entering their location. It also seen the saccade occurring in the pickup location map to the destination location. This section is enough to make the elderly experience confusion in seeing the map. Because several saccades form in the area.

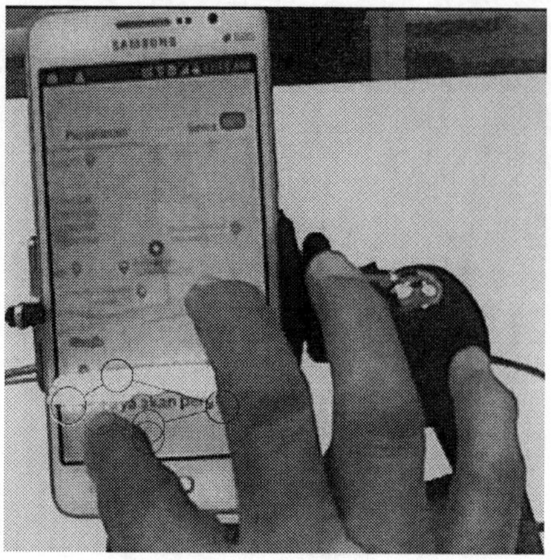

Figure 10. Saccade movement when the elderly enter the location

5. Conclusion

This study used four eye-tracking data variables, which are the fixation number, fixation duration, saccade number, and saccade duration. These four variables are associated with the elderly gender factor (men and women) and the eye condition of the elderly using the Anova test. The result is known that the relationship of the fixation number, fixation duration, saccades number, and saccade duration with the elderly gender has a significant influence. While variables of the fixation number and fixation duration, related to the eye condition of parents (nearsightedness, farsightedness near the combination of cylindrical eyes, and normal eyes) is known to have a significant effect. On the other hand, the number of saccade variables and the duration of the saccade is associated with the eye condition of the elderly (nearsightedness, nearsightedness combination of cylindrical eyes, and normal eyes) are known to have no significant effect. Also based on saccade analysis shows that elderly users experience confusion in using the existing Grab application interface. So that the existing Grab application design needs to be improved to meet the limitations of elderly users.

References

[1] J. Balata, Z. Mikovec, and T. Slavicek, "KoalaPhone : touchscreen mobile phone UI for active seniors KoalaPhone : Touchscreen Mobile Phone UI for Active Seniors," no. December, 2015.

[2] W. Fa and K. C. Soo, "Research on Smart Phone Interfaces for the Elderly Based on Ergonomics Research on Smart Phone Interfaces for the Elderly Based on Ergonomics," J. Phys. Conf. Ser., p. 9, 2018.

[3] P. Loyola, K. Muñoz, A. Couve, P. Maldonado, J. D. Velásquez, and G. Martinez, "Combining eye tracking and pupillary dilation analysis to identify Website Key Objects," Neurocomputing, vol. 168, pp. 179–189, 2015.

[4] H. M. Salman, "Usability Evaluation of the Smartphone User Interface in Supporting Elderly Users From Experts ' Perspective," no. April, 2018.

[5] J. Luan, Z. Yao, F. T. Zhao, and H. Liu, "Search product and experience product online reviews: An eye-tracking study on consumers' review search behavior," Comput. Human Behav., vol. 65, pp. 420–430, 2016.

[6] T. Pro, "Tobii Pro Lab User's Manual," 2017.

[7] M. Hassan, "Research design in eye-tracking experiments By Mohamed Hassan Submitted to Professor Sylwester Białowąs Marketing intelligence February 2017," Mark. Intell., no. February, p. 15, 2017.

[8] G. Ares, A. N. A. Giménez, F. Bruzzone, L. Vidal, L. Antúnez, and A. Maiche, "Consumer Visual Processing Of Food Labels : Results From An Eye-Tracking Study," Vol. 28, Pp. 138–153, 2013.

[9] Y. T. Huang, The female gaze: Content composition and slot position in personalized banner

ads, and how they influence visual attention in online shoppers, vol. 82. Elsevier B.V., 2018.

[10] K. Tzafilkou and N. Protogeros, Diagnosing user perception and acceptance using eye tracking in web-based end-user development, vol. 72. Elsevier B.V., 2017.

[11] Y. Herawati, S. Halim, and C. Tesavrita, "Evaluasi Website Rakuten Indonesia dengan Eyetracking Usability Testing," J. Rekayasa Sist. Ind., vol. 5, no. 1, p. 60, 2017.

[12] O. Špakov, "Visualization of Eye Gaze Data using Heat Maps," vol. 2, no. 2, 2007.

[13] G. Drusch, J. M. C. Bastien, S. Paris, G. Drusch, J. M. C. Bastien, and S. Paris, "Analysing eye-tracking data : From scanpaths and heatmaps to the dynamic visualisation of areas of interest To cite this version : HAL Id : hal-01223743 Analysing eye-tracking data : From scanpaths and heatmaps to the dynamic visualisation of areas of interest," 2015.

[14] M. Alonso Dos Santos, F. Calabuig Moreno, and M. Sánchez Franco, "Congruence and placement in sponsorship: An eye-tracking application," Physiol. Behav., vol. 200, no. May, pp. 159–165, 2019.

[15] L. Larsson, M. Nyström, R. Andersson, and M. Stridh, "Biomedical Signal Processing and Control Detection of fixations and smooth pursuit movements in high-speed eye-tracking data," Biomed. Signal Process. Control, vol. 18, pp. 145–152, 2015.

Increasing performance of straight pattern of rubber tyre on asphalt track due to sulfur composition

P I Purboputro[1,*], A S Darmawan[1,*], M A Hendrawan[1]

[1]Department of Mechanical Engineering, Faculty of Engineering, Universitas Muhammadiyah Surakarta, Jl. Ahmad Yani, Tromol Pos 1 Pabelan, Surakarta 57162, Indonesia

*Corresponding author: Pramuko.Ilmu@ums.ac.id, pip272@ums.ac.id

Abstract. The motion of a ground vehicle is primarily determined by friction force transferred from road via tires. Therefore, the effect of the compound composition on the hardness and wear resistance is investigated. The grip coefficients on the track in wet and dry conditions are also investigated. In this research, the three combinations of black carbon and sulfur composition namely compound 1, compound 2 and compound 3 are provided. The hardness, wear resistance and grip coefficient are then tested. The results of the tests are also compared to the existing compound material which is already on the market. The shore A hardness of compound 1, compound 2 and compound 3 are 75, 75, and 77 respectively. The wear rate of compound 1, compound 2 and compound 3 in dry asphalt condition are 19.49 mm^3/sec, 18.64 mm^3/sec and 13.42 mm^3/sec respectively. The wear rate of compound 1, compound 2 and compound 3 in wet asphalt condition are 6.457 mm^3/sec, 5.933 mm^3/sec and 4.133 mm^3/sec respectively. The grip coefficient of compound 1, compound 2 and compound 3 in dry asphalt condition are 0.756, 0.751, and 0.725 respectively. The grip coefficient of compound 1, compound 2 and compound 3 in wet asphalt condition are 0.702, 0.693, and 0.685 respectively. The compound material already on the market has shore A hardness of 71, wear rate in dry asphalt condition of 11.40 mm^3/sec, wear rate in wet asphalt condition of 2.800 mm^3/sec, grip coefficient in dry asphalt condition of 0.773 and grip coefficient in wet asphalt condition of 0.727. The shore A hardness and wear rate of compound 1, compound 2 and compound 3 are higher than the shore A hardness and wear rate of the compound material already on the market. While the grip coefficient of compound 1, compound 2 and compound 3 are lower than the grip coefficient of the compound material already on the market.

1. Introduction

Globally, motor vehicles are the manufactured product of the highest value. They roll on rubber tires that transfer friction force and wear out over the vehicle's lifetime. Vehicle tires are made of rubber because it is flexible and elastic. In addition, the rubber material does not quickly absorb heat. Tires are made of vulcanized rubber and various reinforcing materials. The most commonly used rubber matrix is the co-polymer styrene-butadiene (SBR) or a blend of natural rubber and SBR. In addition to the rubber compound, tires contain carbon black which is used to strengthen the rubber and aid abrasion resistance. Sulfur is also added in order to vulcanize the rubber compounds, transforming them into highly elastic material. Textile or steel fibers, usually in the form of a cord, used to provide the reinforcing strength or tensile component in tires [1-3].

Quality of product plays important role in manufacturing [4]. Tire is extensively developed in automotive industry nowadays. The development includes the creation of various types, physical understanding and modelling [5]. As the development of tires exist at many tire manufacturers, they are competing to produce quality tires both in terms of material mix, the tire models, performance and comfort during use. Factor composition is the most direct influence on the quality of tire to be produced.

Car tires release wear particles through mechanical abrasion. Wear and tear from tires significantly contributes to the flow of micro plastics pollution into the environment. The emissions from car tires are substantially higher than those of other sources of micro plastics, e.g., airplane tires, artificial turf, brake wear and road markings [6]. Pollution of the environment with plastics is recognized as a serious global threat because it can negatively affect human health, aquatic organisms, as well as the economy [7-9].

There are many factors that affect tire grip coefficient such as the vertical force of the tires against the asphalt, the coefficient of friction between the surfaces intersect, pattern, the air pressure in the tire and rubber types, road characteristics, the type of road surface, as well as tire temperature or the road temperature itself. Traction grip can be improved by improving the coefficient of friction between the tires with the road surface and increasing the vertical force. The road surface is a constant magnitude that no bias is changed, therefore the composition and coefficient of friction play the important role to improve the quality of the tire compound.

Coefficient of friction tire compound is a function of temperature. The best conditions are usually achieved compound tires at temperatures between 85 °C to 100 °C. Quality compound also depends on the type of rubber. The harder compound grip quality usually decreases, but resistance to the wear increases. Ideal conditions of course if the tire manufacturer can make a hard compound but having a good grip.

In this research natural rubber is mixed with other sup materials with some combination of the composition to make the outer layer of the tire. The objectives of this research is to study the effect of the compound composition on the hardness, wear rate and grip coefficient on the asphalt track in wet and dry conditions. The frictional behaviour of tires on any surface such as wet and dry condition is important for vehicle safety and control [10]. Hardness, one of mechanical property that important to consider, is a measure of a material's resistance to localized plastic deformation. Hardness is an indicator of wear resistance [11, 12].

2. Materials and Methods

In this study, material used in this research were RSS (Rubber Smoke Sheet), SBR (Styrene Butadiene Rubber), black carbon, Sulfur, Paraffinic oil, Stearic acid, Paraffin wax, MBTS, Resin epoxy, and ZnO.

In the early stages of making the compound, the rubber was heated using a heater, after the rubber becomes softer, then do the mixing ingredients one by one and rolled.

The process was to mix the ingredients with different compositions, namely the composition of compound 1 made of 70 phr RSS, 3 phr Sulphur and 50 phr Black Carbon, composition 2 made of 70 phr RSS, 3.5 phr Sulphur and 55 phr Black Carbon, while composition 3 made of 70 phr RSS, 4 phr Sulphur and 60 phr Black Carbon.

The mixing of compositions 1, 2 and 3 was conducted by using roll machine. Then the vulcanization process was performed to determine the time and temperature of vulcanization.

The next process was Shore A hardness testing with SNI standard 0778-2009 point 6.22, and testing grip with ASTM D 2047-99

The composition of composite can be seen in table 1.

Table 1. Composition of Composite

No	Materials	phr Compound 1	phr Compound 2	phr Compound 3
1	RSS	70	70	70
2	SBR	30	30	30
3	Black Carbon	50	55	60
4	Parafin Oil	6	6	6
5	ZnO	4	4	4
6	Stearic Acid	2	2	2
7	Parafin Wax	0.5	0.5	0.5
8	MBTS	1	1	1
9	Epoxy	2	2	2
10	Sulfur	3	3.5	4

2.1. Methods

This Material used in this research were RSS (Rubber Smoke Sheet), SBR (Styrene Butadiene Rubber), black carbon, Sulfur, Paraffinic oil, Stearic acid, Paraffin wax, MBTS, Resin epoxy, and ZnO.

In the early stages of making the compound, the rubber was heated using a heater, after the rubber becomes softer, then do the mixing ingredients one by one and rolled.

The process was to mix the ingredients with different compositions, namely the composition of compound 1 made of 70 phr RSS, 3 phr Sulphur and 50 phr Black Carbon, composition 2 made of 70 phr RSS, 3.5 phr Sulphur and 55 phr Black Carbon, while composition 3 made of 70 phr RSS, 4 phr Sulphur and 60 phr Black Carbon.

The mixing of compositions 1, 2 and 3 was conducted by using roll machine. Then the vulcanization process was performed to determine the time and temperature of vulcanization.

The next process was Shore A hardness testing with SNI standard 0778-2009 point 6.22, and testing grip with ASTM D 2047-99

3. Results and Discussion

3.1 The result of hardness testing

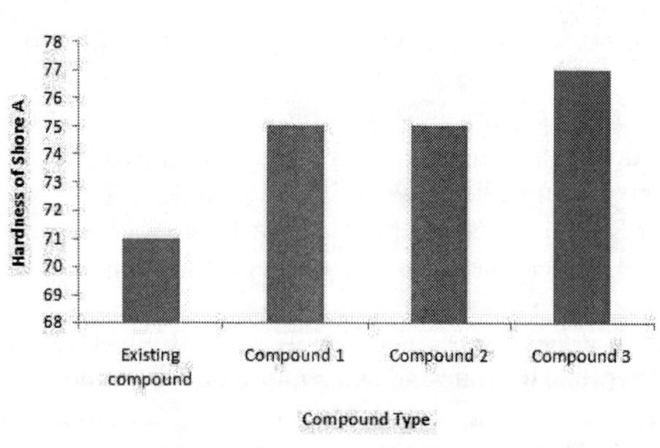

Figure 1. Histogram of the Relationship between Compound on Shore A hardness value

IOP Conf. Series: Materials Science and Engineering **722** (2020) 012039 doi:10.1088/1757-899X/722/1/012039

After mixing processes, the composition of compounds can be seen in Table 1linings made from the composites are shown in figure 1. Figure 1 shows relationship between the types of compound on the Shore A hardness. The hardness of existing compound, compound 1, compound 2 and compound 3 were 71, 75, 75 and 77 respectively. The more sulphur the higher the hardness of compound. The compounds also have higher hardness than the existing compound. These aluminium powders are filtered with mesh 40, 50 and 60. The larger the mesh values the smaller the size of aluminum powder.

3.2 The result of wear testing on dry asphalt

As illustrated in Figure 2, in the wear on the dry track, the existing compound had the lowest wear value of 11.40 mm^3/sec. In the same test, the highest wear value of 19.49 mm3/sec was occurred at compound 1. While compound 2 had wear value of 18.64 mm^3/sec and compound 3 had a value of 13.42 mm^3/sec at dry condition. The wear rate of compound 1, compound 2 and compound 3 higher than the existing compound.

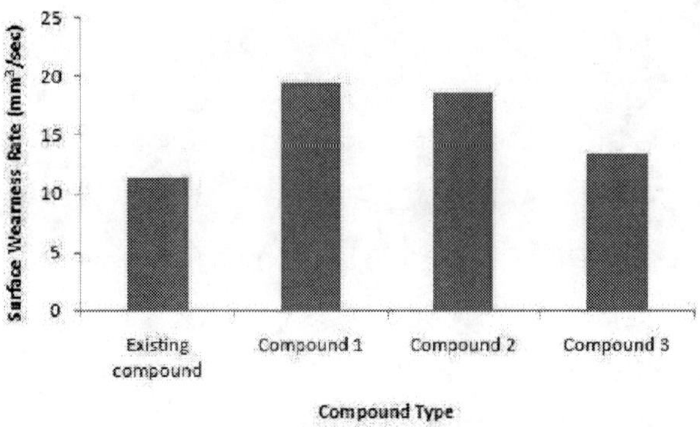

Figure 2. The relationship between the types of compound against the wear rate in dry conditions

3.3 The result of wear testing on wet asphalt

In the wear test on the wet track as can be seen in Figure 3, the existing compound had wear rate of 2,800 mm^3/sec. While the compound 1 had high wear value of 6.467 mm^3/sec. Furthermore, compound 2 and compound 3 had a wear value of 5,933 mm^3/sec and 4.133 mm^3/sec respectively.

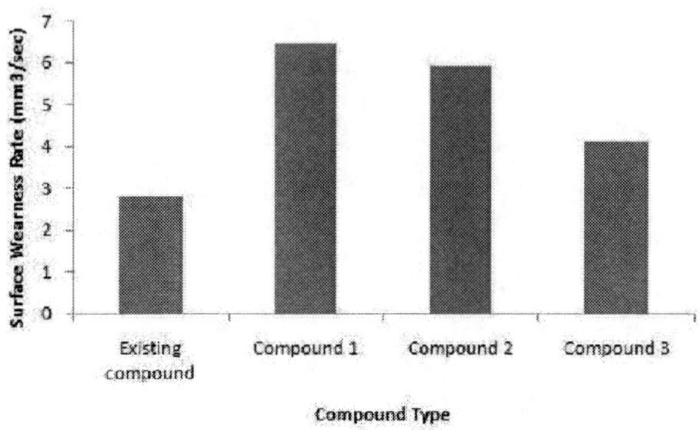

Figure 3. Relationship between the types of compound against the wear rate in wet conditions

The grip coefficient test results on the dry track were greater than the grip coefficient that occurs on the wet track conditions.

3.4 The result of wear testing on dry asphalt

In Figure 4 shows that the grip coefficient of the existing compound is higher than the grip coefficient of compound 1, compound 2, and compound 3. The figure also illustrate that the higher the composition of sulfur the lower the grip coefficient. The grip coefficient of the existing compound, compound 1, compound 2, and compound 3 were 0.773, 0.756, 0.751 and 0.725 respectively.

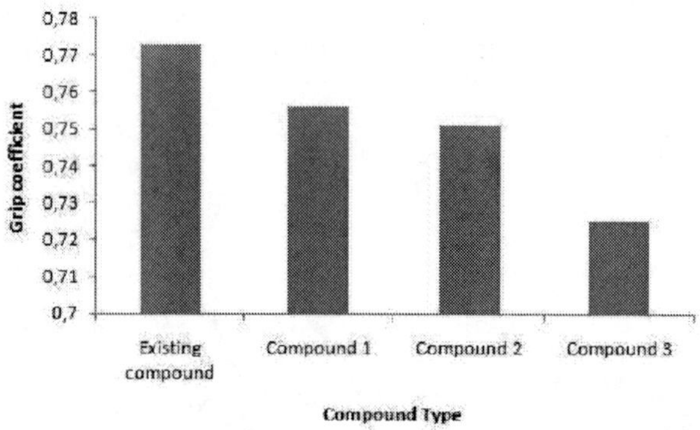

Figure 4. Grip coefficient of straight pattern on dry asphalt

3.5 The result of grip coefficient testing on wet asphalt

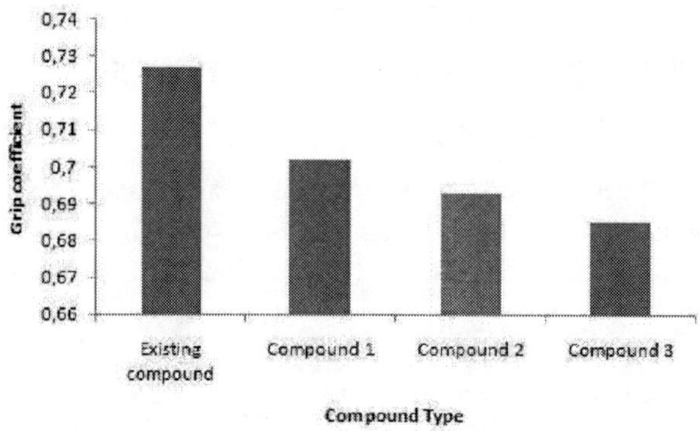

Figure 5. The grip coefficient on wet asphalt track

From Figure 5. The grip coefficient on wet asphalt track condition has 0.727 for existing compound. Compound 1 , compound 2, and compound 3 had 0.702, 0.693, and 0.685, respectively. The The grip coefficient of compound 1, compound 2, and compound 3 are lower than the grip coefficient of existing compound. Compound 1,

4. Conclusion

In this study, the hardness of compound 1, compound 2 and compound 3 were higher than the existing compound. However, the wear rate was higher, and the grip coefficient was lower.

Acknowledgment

The authors gratefully acknowledge financial support from Education Ministry of Indonesia. The authors also would like to thank Mechanical Engineering Laboratory of Universitas Muhammadiyah Surakarta for all specimens' preparation.

References

[1] Amari, T., Themelis, N. J., and Wemick, I. K.: 'Resource recovery from used rubber tires', *Resources Policy*, 1999, 25, (2), pp. 179-188.

[2] Okel, T. A., and Rueby, J. A.: 'Silica morphology and functionality: Addressing winter tire performance', *Rubber World*, 2016, 253, pp. 21-52.

[3] Sharifzadeh, M., Akbari, A., Timpone, F., and Daryani, R.: 'Vehicle tyre/road interaction modeling and identification of its parameters using real-time trustregion methods', *IFAC-PapersOnLine*, 2016, 49, (3), pp. 111-116.

[4] Darmawan, A. S., Anggono, A. D., and Hamid, A.: 'Die design optimization on sheet metal forming with considering the phenomenon of springback to improve product quality, *MATEC Web of Conferences*, 2018, 154, 01105.

[5] Heinrich, G., and Klüppel, M.: 'Rubber friction, tread deformation and tire traction', *Wear*, 2008, 265, pp. 1052-1060.

[6] Kole, P. J., Löhr, A. J., Van Belleghem, F. G. A. J., and Ragas, A. M. J.: 'Wear and Tear of Tyres: A Stealthy Source of Microplastics in the Environment', *Int. J. Environ. Res. Public Health*, 2017, 14, (10), 1265.

[7] Wagner, M., Scherer, C., Alvarez- Muñoz,D., Brennholt, N., Bourrain, V., Buchinger, S., Fries, E., Grosbois, C., Klasmeier, J., Marti, T., Rodriguez-Mozaz, S., Urbatzka, Vethaak, A D., WintherNielsen, M., and Reifferscheid, G.: 'Microplastics in freshwater ecosystems: What we know and what we need to know', *Environ. Sci. Eur.*, 2014, 26: 12.

[8] Chae, Y., and An, Y. J.: 'Effects of micro- and nano plastics on aquatic ecosystems: Current research trends and perspectives', *Mar Pollut Bull.*, 2017, 124, (2), pp. 624-632.

[9] Nakki, P., Setala, O., and Lehtiniemi, M.: 'Bioturbation transports secondary microplastics to deeper layers in soft marine sediments of the northern Baltic Sea', *Mar Pollut Bull.*, 2017, 119, pp. 255-261.

[10] Ella, S., Formagne, P. Y., Koutsos, V., and Blackford, J. R.: 'Investigation of rubber friction on snow for tyres', *Tribology International*, 2013, 59, pp. 292-301.

[11] Darmawan, A. S., Siswanto, W. A., and Sujitno, T.: 'The Influence of Plasma Nitrocarburizing Process Temperature to Commercially Pure Titanium Surface Hardness', *Applied Mechanics and Materials*, 2013, 315, pp. 700-704.

[12] Darmawan, A. S., Siswanto, W. A., and Sujitno, T.: 'Comparison of Commercially Pure Titanium Surface Hardness Improvement by Plasma Nitrocarburizing and Ion Implantation', *Advanced Materials Research*, 2013, 789, pp. 347-351.

RTS Threshold Variation Study on Contention-based IEEE 802.11n EDCA for Real-time Services

I A Herdiniamy[1], E F Cahyadi[1,2], E S Nugraha[1], P Yuliantoro[1], and M S Hwang[2,3]

[1] Faculty of Telecommunication and Electrical Engineering, Institut Teknologi Telkom Purwokerto, Purwokerto 53147, Indonesia
[2] Department of Computer Science and Information Engineering, Asia University, Taichung 41354, Taiwan (R.O.C.)
[3] Department of Medical Research, China Medical University Hospital, China Medical University, Taichung 40402, Taiwan (R.O.C.)

E-mail: ekofajarcahyadi@ittelkom-pwt.ac.id

Abstract. As a "wide-deployed" wireless local area networks (WLANs) standard, the IEEE 802.11n was designed to support a high data transmission rate (up to 600 Mbps) and maintain a compatibility with the previous versions. Due to a substantial demand of real-time data services, such as VoIP and video conference, an enhanced quality of service (QoS) mechanism become crucially important. The IEEE 802.11n standard inherits the enhanced distributed channel access (EDCA) mechanism from the IEEE 802.11e, as the MAC layer improvement. It classifies the traffic into four different classes of data, called access categories (ACs). However, EDCA still suffers from a data packet collision, caused by hidden nodes problems. The packet loss ratio and end-to-end delay increase along with the escalation of real-time traffic. To address the issue, we conduct a simulation in *Riverbed Modeler 17.5* environment with a 128, 256, 512, and 1,024 Bytes of request to send (RTS) threshold variation value to overcome a collision caused by hidden nodes problem. The result shows that the 512 Bytes of RTS threshold value gives the best QoS output among them all.

1. Introduction

The telecommunication traffic through the wireless local area networks (WLANs) and mobile devices will account for more than 63 % of total world IP traffic, while wired devices managed for 37 % by 2021 [1]. As the latest widely used WLANs technology, IEEE 802.11n [2] standard already gave us a very high data rates at both physical (PHY) and MAC layer by implementing the latest advances in wireless communications technology. The standard operates at 2.4 GHz and 5 GHz of bands frequency, which employed multi-input multi-output (MIMO) antennas for producing maximum throughput up to 600 Mbps [3]. However, IEEE 802.11n WLAN standard still inherit the same roots from its predecessor [4], which initially designed merely for data traffic applications, and not considering a quality of service (QoS) requirements.

The "legacy" distributed coordination function (DCF) and point coordination function (PCF) MAC schemes mechanism, that was implemented prior to IEEE 802.11n, no longer satisfy the demand of real-time applications [5]. Both of DCF and PCF MAC schemes do not support any prioritization of data traffic [6]. In 2005, IEEE 802.11e was deployed to support quality of service (QoS) over WLANs [7]. Its MAC employs a channel access function called the

IOP Conf. Series: Materials Science and Engineering **722** (2020) 012040 doi:10.1088/1757-899X/722/1/012040

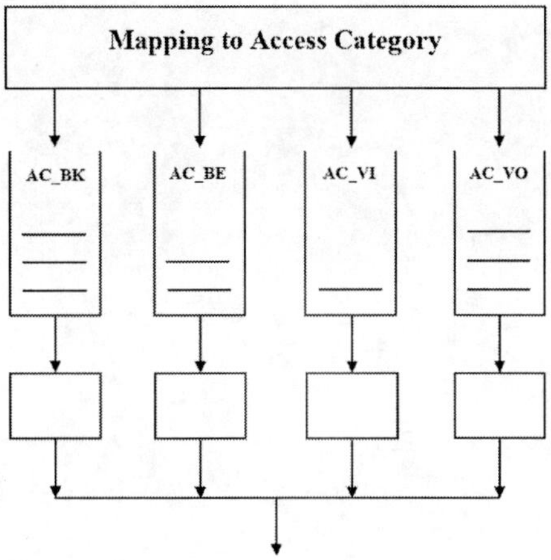

Figure 1. EDCA reference implementation model.

hybrid coordination function (HCF), which includes contention-based channel access, called enhanced distributed channel access (EDCA), and contention-free centrally controlled channel access, called HCF controlled channel access (HCCA).

DCF and PCF are soon replaced by HCF, with a new access types to ensure higher bandwidth. It also reduces latency of high priority network traffic, and thus, provides high throughput performance. As an extension of DCF, EDCA provides an enhanced QoS based on the priority classes. It classifies the traffic into four different classes of data called access categories (ACs), as depicted in Fig. 1. EDCA mechanism still inherit the basic principle of CSMA/CA. However, the channel access parameters are defined per ACs, called arbitration inter-frame space (AIFS), instead of distributed inter-frame space (DIFS) that employed in DCF [8]. As shown in Table 1, the value of AIFS (AIFSN) varies to every different ACs. The highest priority AC got the smallest value of AIFSN, and vice versa [8, 9]. Since EDCA give the higher priorities to real-time traffic, *e.g.,* voice and video, so it is likely suitable for such transmission. However, EDCA still suffers from a data packet collision, ultimately caused by the hidden nodes problems. This phenomenon can potentially affect any nodes for cannot hear each other even though they are an adjacent WLANs.

Table 1. Default EDCA access categories.

Priority	AC	CWmin	CWmax	AIFSN	TXOP Limit	Designation
Lowest	BK	31	1023	7	0	Background
	BE	31	1023	3	0	Best Effort
	VI	15	31	2	3.008 ms	Video
Highest	VO	7	15	2	1.504 ms	Voice

The employment of request to send (RTS) and clear to send (CTS) mechanism can overcome a collision caused by hidden nodes problem. As the blind nodes cannot receive any packets from the other nodes, it will lead to collisions and packet loss [10, 11, 12]. Fig. 2 shows a hidden node scenario consists of three nodes, A, B, and C. In this scenario, node B can receive packets

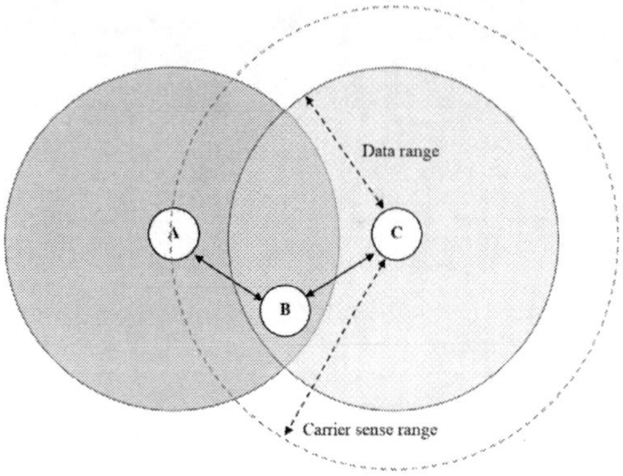

Figure 2. Collision and carrier sense range between two overlapping WLANs.

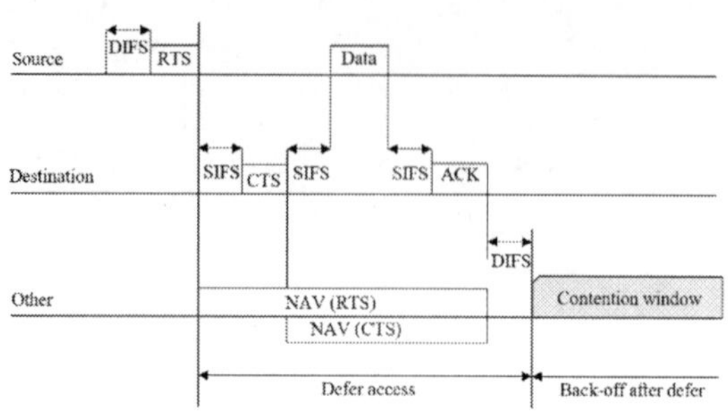

Figure 3. RTS/CTS mechanism in DCF.

from both nodes A and C, since its located in the data range of both nodes. However, node A is hidden from node C and vice versa. In this case, if both nodes A and C send packets to node B at the same time, then it would cause a data packet collision at node B. At this point, RTS/CTS protocol will allow the access points (APs) to regulate the utilization of the air interface media, so the stations (STAs) could see the state of the other stations that will process the packet transmission at the same time. Fig. 3 shows how the protocol works. The source node broadcasts an RTS packet. When the destination node receives the RTS packet, it will waits for the short inter-frame space (SIFS) time interval, then send a CTS packet to the source. When the source node transmits the data, the destination should send an ACK packet. Meanwhile, the network allocation vector (NAV) is used by any stations to determine channel access priority and future traffic on the medium, based on the duration information that announced in RTS/CTS frames before the actual data exchange [13]. The employment of the RTS/CTS mechanism could be done by setting the threshold value.

For a better understanding, the rest of this paper is organized as follows. In section 2, we cover the related works. It is followed by the proposed method in section 3, and results of the simulation is discussed in section 4. Finally, the conclusions are presented in section 5.

2. Related Works

Several works related to RTS/CTS mechanism in IEEE 802.11n have been published in recent years [13, 14, 15, 16]. Singh *et al.* [13] discuss the performance of the RTS/CTS mechanism in video conference service with RTS threshold variations of 128, 256, 512, and 1,024 Bytes. There are five deployed scenarios with 10, 20, 30, 40, and 50 of users density to observe QoS parameters *e.g.,* throughput, media access delay, retransmission attempts, and end-to-end delay. The results show that 512 Bytes of RTS threshold value gives the best overall output.

In 2015, Maraj *et al.* [14] discuss the effectiveness of the RTS/CTS mechanism to improve the efficiency in the heavy traffic condition in IEEE 802.11g/n with 20 workstations. The observed output parameters are data dropped and media access delay. The result shows the employment of the RTS/CTS mechanism in IEEE 802.11g can reduce the data drop and improve the network performance better than IEEE 802.11n.

In 2014, Khadrah *et al.* [15] compared the DCF and EDCA protocol performance to support QoS requirements, *e.g.,* end-to-end delay, packet loss, and jitter in voice and video services. They found a collision was detected on the EDCA access method when carrying out the real-time traffic transmission process. The collision arises along with the traffic escalation, causing

IOP Conf. Series: Materials Science and Engineering **722** (2020) 012040 doi:10.1088/1757-899X/722/1/012040

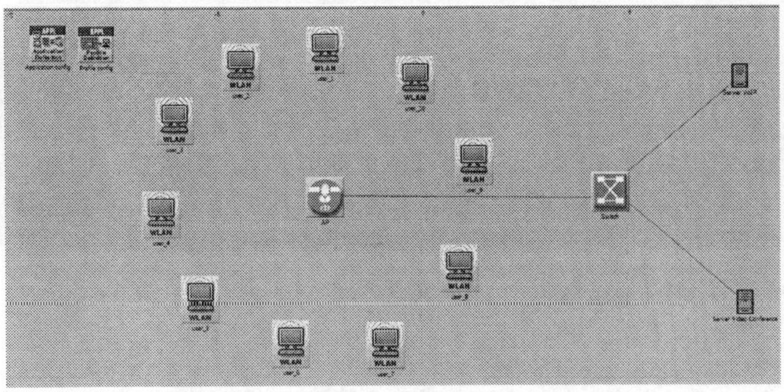

Figure 4. The network topology.

a delay and packet loss, even though, it still be tolerable in a specific small amount of traffic.

To address the issue, we employ an RTS/CTS mechanism with a 128, 256, 512, and 1,025 Bytes of RTS threshold variation value (as has been done in [13]), to overcome a collision caused by hidden node problem. Since we are only focused on studying the impact of those RTS threshold variations, towards its effect on data packets collision in EDCA (as happened in [15]), we only deployed ten nodes of the user. The key parameters in this study are packet loss ratio, delay, and throughput.

3. Proposed Method

This study was simulated in *Riverbed Modeler 17.5* environment. We uses a model to analyze the performance of several RTS threshold variation values in the IEEE 802.11n WLAN standard. In this section, we will elaborate how the simulation is conducted.

3.1. System Design

Fig. 4 shows the deployed networks topology simulated for this project. The center node is the AP, surrounded by 10 nodes of users, while the servers (VoIP and video conference servers) are located outside the circle. A switch connects both servers and access point through 100BaseT (100 Mbps over twisted pair) links. Two another objects located on the top left of the workspace are *Application* and *Profile Definitions* (to organize the services, its values, and parameter configurations).

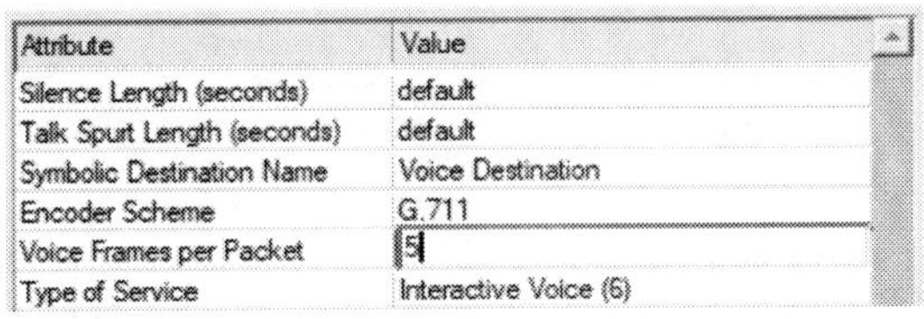

Attribute	Value
Silence Length (seconds)	default
Talk Spurt Length (seconds)	default
Symbolic Destination Name	Voice Destination
Encoder Scheme	G.711
Voice Frames per Packet	5
Type of Service	Interactive Voice (6)

Figure 5. VoIP application definition.

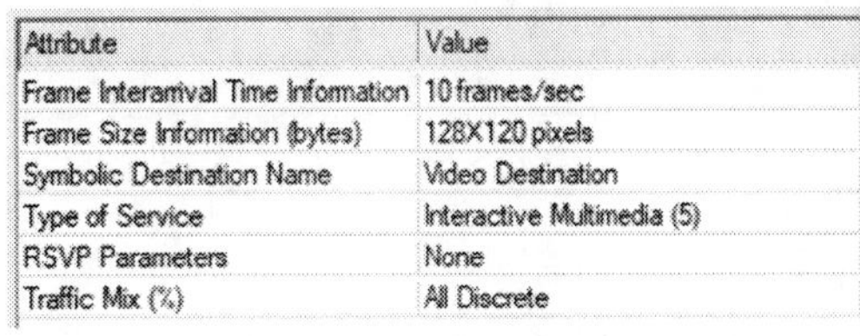

Attribute	Value
Frame Interarrival Time Information	10 frames/sec
Frame Size Information (bytes)	128X120 pixels
Symbolic Destination Name	Video Destination
Type of Service	Interactive Multimedia (5)
RSVP Parameters	None
Traffic Mix (%)	All Discrete

Figure 6. Video conference application definition.

The *Application Definition* has a role as a declaration node for several employed services, as depicted in Fig. 5 and 6. In Fig. 5, we could see the VoIP service configurations uses a G.711 encoder scheme, which has 160 Bytes of payloads (1 packet = 160 Bytes). Therefore, we set the "voice frames per packet" value to 5, since each voice frame has a size of 32 Bytes [17]. The type of service (ToS) value is set to interactive voice (6), which indicates the AC category

IOP Conf. Series: Materials Science and Engineering **722** (2020) 012040 doi:10.1088/1757-899X/722/1/012040

according to voice service on the EDCA access method. Meanwhile, on video conference service configuration, we use low resolution frame (frame inter-arrival time is 10 frames/sec), with frame size information for 128 × 120 pixels, which has a frame equivalent to 17,280 Bytes of payloads, as shown in Fig. 6. The value of ToS is set to interactive multimedia (5) which indicates that video conference has a lower priority compared to VoIP in EDCA access methods.

After the services has been declared in the *Application Definition*, they must be set in the *Profile Definition* node. Table 2 shows the *Profile Definition* table configurations. The simulation is started at 70th second, shown by the accumulation of start time (Constant(60)) and start time offset (Constant(10)).

Table 2. Parameters of the packet simulation.

Parameter	VoIP	Video Conference
Profile name	VoIP	Video conference
Application name	VoIP application	Video conference application
Start time offset (sec)	Constant(10)	Constant(10)
Start time (sec)	Constant(60)	Constant(60)
Duration (sec)	End of profile	End of profile
Repeatability	Once at start time	Once at start time
Operation mode	Simultaneous	Simultaneous

Table 3. Parameters of the packet simulation.

Parameter	Value
Wireless LAN MAC address	Auto assigned
Physical characteristics	HT PHY 2.4 GHz 802.11n
Data rates	65 Mbps (base) / 600 Mbps (max)
RTS threshold	Varying (128, 256, 512, 1024 Bytes)
Packet-reception threshold	-95
Long retry limit	4
Max. receive lifetime	0.5 sec
Buffer size	256,000 Kbps
Large packet processing	Drop

The RTS mechanism configurations in the AP are summarized in Table 3. We employ an RTS threshold values variations of 128, 256, 512, and 1,024 Bytes. The AP works in IEEE 802.11n WLAN standard with 2.4 GHz of frequency bands and a data rate of 65 Mbps / 600 Mbps. The RTS threshold range aims to determine the maximum size of the data packet. If the sent packet is higher than the RTS threshold value, then the station will send the RTS packet to the intended wireless node.

IOP Conf. Series: Materials Science and Engineering **722** (2020) 012040 doi:10.1088/1757-899X/722/1/012040

3.2. Scenarios

The performance parameters tested in this simulation are the packet loss ratio, delay, and throughput. These three variables are then simulated towards RTS threshold variations. The detailed scenarios of this work are summarized in Table 4.

Table 4. Scenarios.

Scenario	RTS Threshold (Bytes)	Services	QoS Parameters
1	128		
2	256	VoIP &	Packet loss, delay,
3	512	Video conference	& throughput
4	1,024		

4. Results and Discussion

In this section, we evaluate the effect of RTS threshold variation on VoIP and video conference services based on the arrangement in Table 4.

4.1. Packet Loss Ratio

As depicted in Fig. 7, VoIP service has an obvious lower average packet loss ratio compared to video conference. Since video conference got a lower priority compared to VoIP, it must wait for being transmitted, resulting in a higher packet loss ratio compared to VoIP. From all of the deployed scenarios, only in 512 Bytes of RTS threshold value employment has the lowest packet loss ratio for both services, as its traffic sent and received shown in Fig. 8. Hence, it shows that the 512 Bytes RTS threshold value gives a better effect to downturn the packet loss by preventing the occurrence of a collision. However, the overall results of the packet loss ratio in all scenarios are categorized good based on TIPHON standard [18].

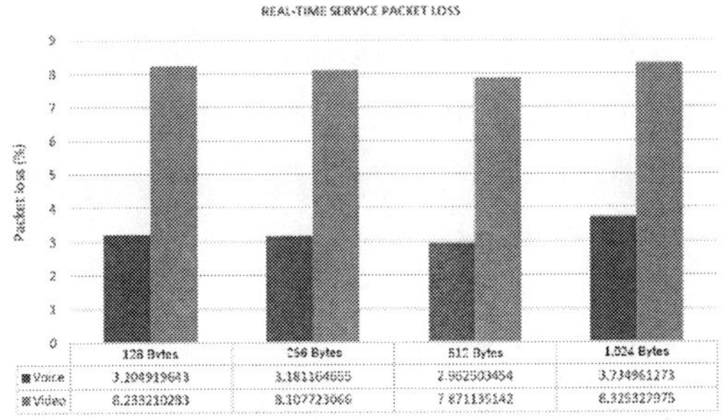

	128 Bytes	256 Bytes	512 Bytes	1,024 Bytes
Voice	3.204919643	3.181164655	2.962503454	3.734961273
Video	8.233210283	8.107723066	7.871135142	8.325327975

Figure 7. Packet loss ratio of real-time services.

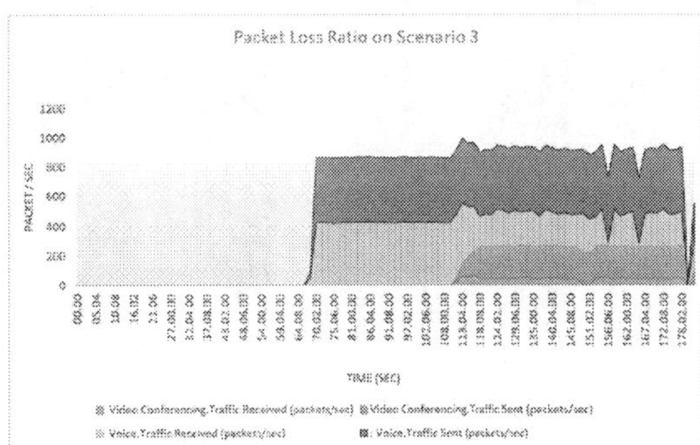

Figure 8. Packet loss ratio on 512 Bytes RTS threshold.

4.2. Delay

Fig. 9 shows that the most considerable delay occurred in video conference service compared to the VoIP. Since the video conference has more data packet to transmit i.e., image and sound, it requires a longer delivery time than VoIP service, which makes it vulnerable to being influenced by delay. From all of the deployed scenarios, the 512 Bytes of RTS threshold value employment in scenario 3, has the lowest delay for both services, as its fluctuation along the simulation time is shown in Fig. 10. The overall results of the delay in all scenarios are categorized good based on TIPHON standard [18].

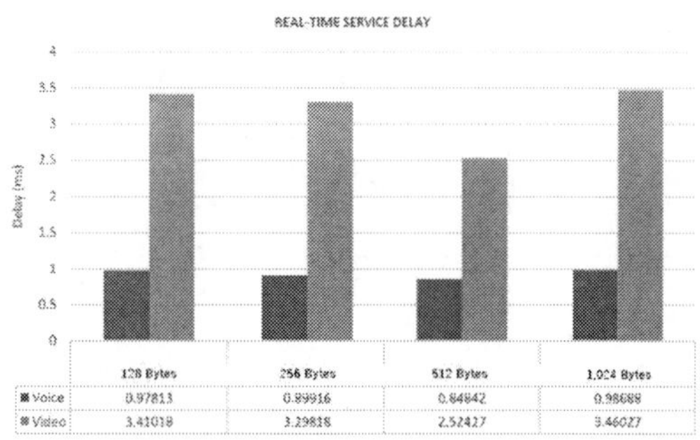

Figure 9. Delay of real-time services.

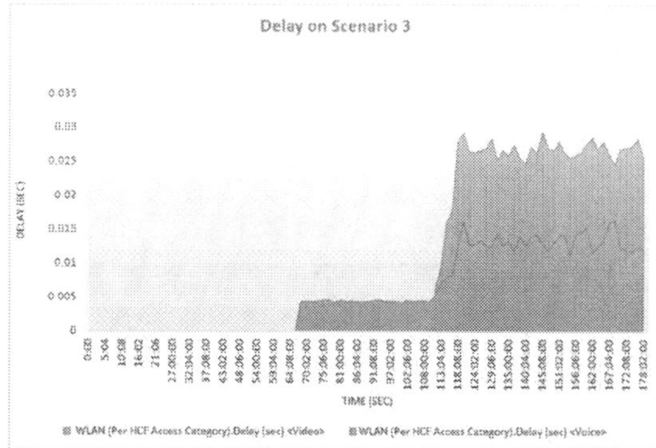

Figure 10. Delay on 512 Bytes RTS threshold.

4.3. Throughput

Throughput has an inverse relation towards the packet loss ratio parameter. Since the packet loss ratio in VoIP services is much smaller than the video conference, so it gains a better throughput, as depicted in Fig. 11. From all of the deployed scenarios, the 512 Bytes of RTS threshold value employment in scenario 3, has the highest throughput for both services, as its fluctuation along the simulation time is shown in Fig. 12. The overall results of the throughput in all scenarios are categorized good based on TIPHON standard [18].

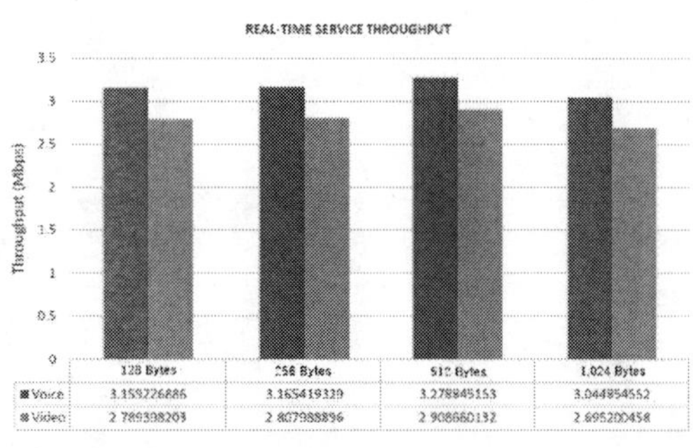

Figure 11. Throughput of real-time services.

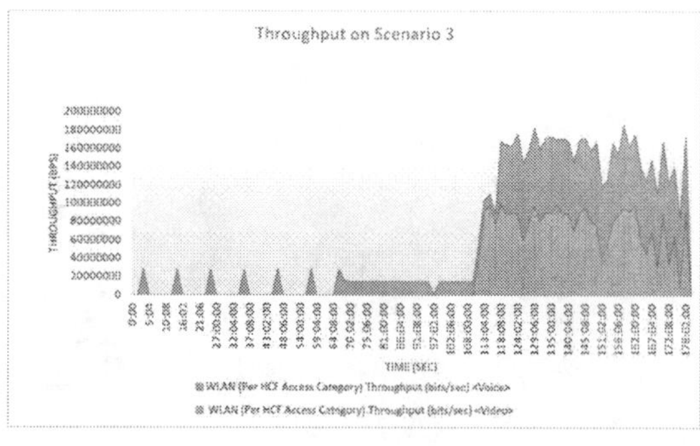

Figure 12. Throughput on 512 Bytes RTS threshold.

4.4. Discussion

From all four examined scenarios, the detailed packet loss ratio, delay, and throughput that affected by RTS threshold value variation are summarized in Table 5. The escalation of the

IOP Conf. Series: Materials Science and Engineering **722** (2020) 012040 doi:10.1088/1757-899X/722/1/012040

RTS threshold value slowly improves the network performance in real-time services, although it is not significant. The best RTS threshold value performance is shown in scenario 3, with 512 Bytes. However, not every increment will give a good impact on the network performance, as shown by scenario 4 with 1,024 Bytes. Overall performance of the output parameter calculation from all scenarios are excellent, as set by TIPHON standard [18].

The employment of the RTS/CTS mechanism can guarantee the transmission of data packets through the network without experiencing a collision. However, if there is no hidden nodes issue found in the network, the RTS/CTS mechanism instead will cause an overhead, that will resulting in degradation of throughput. Therefore, when applying a 1,024 Bytes of RTS threshold value in scenario 4, the threshold value will only cause a network overhead, which adversely affects the output parameters. Those value will disrupt the traffic delivery process, resulting in increased delay and packet loss, and decreasing throughput value. According to [19], if the network already experienced an eminent performance, marked by the lowering of the packet loss ratio and delay, followed by the increase of throughput, then upgrading the RTS threshold value is not necessary.

Table 5. The performance summary.

AC	Scenario	Delay (ms)	Packet Loss Ratio (%)	Throughput (Mbps)
Voice	1	0.97813	3.20492	3.15923
	2	0.89916	3.18112	3.16542
	3	0.84842	2.96250	3.27884
	4	0.98688	3.72496	3.04485
	Average	0.92815	3.26837	3.16209
Video	1	3.41018	8.23321	2.78940
	2	3.29818	8.10772	2.80799
	3	2.52427	7.87113	2.90866
	4	3.46027	8.32533	2.69520
	Average	3.17322	8.13435	2.80031

5. Conclusion

In this paper, we have simulated the RTS threshold value variation for 128, 256, 512, and 1,024 Bytes in *Riverbed Modeler 17.5* simulator environment. Based on the results and discussion in section 4, the employment of 512 Bytes of RTS threshold value gives the best QoS output, due to the active RTS/CTS frame in reducing the number of collisions. The improvement of network performance characterized by the reduction of delay in the video conference and VoIP services for 2.52427 ms and 0.84842 ms, respectively. Packet loss ratio increases for 7.87114 % in video conference service, and 2.96250 % in VoIP. Meanwhile, the throughput value in video conference and VoIP also escalates for 2.90866 Mbps and 3.27885 Mbps, respectively. On the other hand, the worst scenario resulted by 1,024 Bytes RTS threshold value. This thing happened due to the occurrence of the networks overhead, which lead to quality degradation. The delay in video services escalates for 3.46027 ms, while in VoIP rises for 0.9281475 ms. The packet loss ratio for video conference and VoIP services also increase for 8.32533 % and 3.73496 %, respectively — meanwhile, the throughput of VoIP decline for 3.27885 Mbps, and 2.90866 Mbps for video conference.

Acknowledgement
The authors would like to thank Lembaga Penelitian dan Pengabdian Masyarakat (LPPM), Institut Teknologi Telkom Purwokerto for funding this project.

References
[1] Cisco 2017 *The Zettabyte Era: Trends and Analysis* white paper.
[2] IEEE 2009 *IEEE 802.11n-2009.*
[3] Perahia E, Stacey R 2008 *Next Generation Wireless LANs 802.11n and 802.11ac* Cambridge University Press.
[4] IEEE 1999 *IEEE Std 802.11-1999.*
[5] Abichar Z, Chang J M 2013 *IEEE Trans. on Mobile Computing* **12** p 304-17.
[6] Karmarkar R, Chattopadhyay S, Chakraborty S 2017 *IEEE Comm. Surveys & Tutorials* **19** p 2050-91.
[7] IEEE 2005 *ANSI/IEEE Std 802.11e-2005.*
[8] Tinnirello I, Choi S 2005 *6th IEEE Int. Symp. on a World of Wireless Mobile and Multimedia Networks.*
[9] Sotenga P, Chuku P, Olwal T 2015 *Int. Conf. on Computing, Comm. and Security (ICCCS.*
[10] Boroumand L, Khokhar R H, Bakhtiar L A, Pourvahab M 2012 *Smart Computing Review* **2** p 95-110.
[11] Khalife H, Malouch N 2006 *2006 IEEE Int. Conf. on Comm.* p 3947-52.
[12] Krishna P V, Iyengar N C S N 2008 *Int. J. Network and Distribution Systems* **1** p 52-70.
[13] Singh H, Sharma V 2015 *2015 5th Int. Conf. on Adv. Comp. & Comm. Tech.* p 421-4.
[14] Maraj D, Maraj A 2015 *57th Int. Symp. ELMAR-2015* p 129-32.
[15] Khadrah A A, Zakaria A, Othman M 2014 *J. of Theoretical Applied Information Tech. (JATIT)* **64** p 261-6.
[16] Vijay B T, Malarkodi B 2016 *Procedia Computer Science* **89** p 17-26.
[17] Cahyadi E F, Nusantara H 2014 *7th Electrical Power, Electronics, Comm., Controls and Informatics Seminar (EECCIS)* p 105-10.
[18] ETSI 1998 *European Telecommunications Standards Institut* p 1-35.
[19] Geier J 2010 *Designing and Deploying 802.11n Wireless Networks* Cisco Press, USA p 1-27.

IOP Conf. Series: Materials Science and Engineering **722** (2020) 012041 doi:10.1088/1757-899X/722/1/012041

Applying Kaizen in Quality for Reducing Dent Defect per Unit

P A Wicaksono[1], and R Rozaq[1]

[1]Department of Industrial Engineering, Diponegoro University, Semarang, Indonesia

E-mail: purnawan@ft.undip.ac.id

Abstract. Continuous improvement (Kaizen) is considered as a main concern and prerequisite for competitiveness in today's manufacturing. One form of continuous improvement is in the field of product quality. The continuous quality improvement is the key to long-term success and high performance of the organization. A case study was conducted in a manufacturing company. In the year 2019, company has not yet reached the target in particular on cylinder head part. One of the defects that still occur is dent. Research carried out using Toyota Business Practices (TBP) approach. Toyota Business Practices is a method for applying kaizen in daily work. This study aims to reduce the number of defect using Toyota Business Practices. By applying Toyota Business Practices, the results of the improvement made were obtained improving the position of shutter table, installing slipper stand, and improving the design of slipper stopper. With these improvements, the average dent defect per unit reduced by 4.35 %. Furthermore, it will get the new standard to ensure the quality of product in the future.

1. Introduction

Many manufacturing and some service businesses today are using lean management principles and practices as means to improve business processes, which in turn improves productivity and competitiveness including quality, and delivers greater value to end-use customers [1]-[5]. Several studies have exhibited that companies could dramatically improve their performance by embracing the lean production approach [6], and preferred to many other conventional production models, including the mass production model, because of its ability to produce high-quality and diverse products to meet the needs of various customers [7]. Therefore, with the aim satisfy their customers, companies have to pursue their quality in products.

As in most countries within the region, the automotive industry was identified as a target for government support at an early stage of Indonesia's industrialization [8]. Total installed capacity of car production in Indonesia has increased to 2,208,131 units per year in 2017 [9], and car sales from the factory reached 553,757 units in January-June 2018, the value increased by 20,251 units compared to the period January-June 2017 [10]. The increasing pressure and demands from the stakeholders are the reason leads towards the continuous quality improvement of the management system to become more flexible, effective, and competitive in the industry [11]. The continuous quality improvement is the key to long-term success and high performance of the organization [12]. Therefore, it is very important for the car manufacturing industry to achieve zero defects with continuous improvement (Kaizen). Kaizen was introduced by Imai Masaki in 1986 and used by Toyota to improve their performance for waste reduction as a form of implementing lean manufacturing. It has become one of the important tools for getting improvement in any field i.e. production, process, quality, and maintenance in the manufacturing industry [13], [14].

A case study was conducted in engine production plant of "Company X" in Indonesia. Company X is a car automotive company which is engaged as a manufacturer and exporter of vehicles and parts. Zero defect is a matter that is highly prioritized by Company X in carrying out its production process. The good quality of the product can be achieved by practicing zero defects [15]. However, based on data of Defect per Unit (DpU) on November 2018-January 2019, the average defect per unit for cylinder head part is 7.33 %. The data shows that the production quality of Company X for cylinder head part is still very far from the target of 0.05 % (zero defect). Defect occurs on exhaust manifold has 6.74 % and intake manifold has 0.59 %. Defect that occurs in the cylinder head part is dent. Dent is a condition where the body is dented a certain point. In addition to having an effect on product aesthetics, large quantities of dent can affect production efficiency because it can cause high cycle time for rework. Rework is making corrections or scrapping work that has already been done before it will be sent to customer [15]. Those aforementioned problems have motivated us to develop suitable strategies for the purpose of reduce the number of defects per unit for cylinder head part in Company X.

In this research, we attempted to apply TBP to improve product quality in one of manufacturing company. Productivity and quality improvement are combined purposely to gain a competitive advantage for most manufacturing industries management [15]. An increase in quality does not only reduce cost, but also improves productivity by eliminating any rework and unnecessary inspections [16]. Toyota Business Practices (TBP) is a method to clearly apply kaizen in daily work developed by Toyota. We use TBP because it is the systematic pattern of work processes that integrates the wisdom of all members in pursuing continuous growth and satisfaction carried out continuously in eight steps and can be applied in various fields. Several studies confirmed that TBP can help businesses to achieve best results from existing problems [17], [18]. The direct contribution from this research are categorized into two categories, namely practical contribution and theoretical contribution. The practical contribution is the way improvements are made to reduce dent defect per unit for the Company X and can be emulated by other companies, and the theoretical contribution is the results of improvements made in the form of a reduction in the number of dent defect per unit as a further policy consideration for Company X to create the best quality products that are sustainable for the company. We provide the novelty of this research in the form of improvements made to reduce dent defects per unit. The repairs follow Karakuri Kaizen where the repairs do not use energy or large resources but instead use gravity, lever and cam utility, as well as inertia to move and transfer goods.

The rest of this paper is organized as follows. The second section contains the design of the research carried out to achieve the research objectives. The third section then demonstrates the case study result of applying Toyota Business Practices. Conclusion are finally drawn in the last section.

2. Researh Design

2.1. Research Methodology

The objective of this research is to reduce the number of defects per unit for cylinder head part. The object of the research is engine production plant of Company X, a car automotive company which is engaged as a manufacturer and exporter of vehicles and parts which is located in Jakarta, Indonesia. The study was conducted during the period January 2019-April 2019. Data collection was carried out through interviews, observation in the field, and using secondary data obtained from the company. Secondary data collected is data of defect per unit, and production output for cylinder head part at engine production plant.

2.2. Toyota Business Practices

Toyota Business Practices (TBP) broadly applies the problem-solving philosophy to the entire enterprise [19]. In this study, data has been collected will be processed to help obtain information that is useful in solving problem. Data of defect per unit, and production output for cylinder head part were processed so that problems can be clarified by describing current condition. It aims to make the problem clearer. At this stage the problem is described in the form of a gap between the current situation and the ideal situation which is the final destination. The second stage is breakdown the problem. It was used to see the problem become smaller so that the focus of research can be done to solve the problem that still has a major impact on the quality of the cylinder head part. The third stage is set a target as a goal achieved from solving the problem. Set a target was carried out to show the current condition with the ideal condition that were targeted at that time. The fourth stage is analyze the root cause. The results of observation are written in notes and investigated by asking directly to people who understand better or directly related to the object of observation. Root cause analysis is carried out on all aspects, i.e. man, machine, method, material, and environment. Root cause analysis aims to find the source of the problem that is dent defect on the exhaust manifold. The fifth stage is develop countermeasures as a form of planning to overcome the problems. The sixth stage is see countermeasures through as a form of the implementation of countermeasures according to the plan that has been carried out in the fifth stage. The seventh stage is monitor both results and processes. At this stage, total results achieved and the effect of activities in solving the problem are seen. The activities carried out when evaluating results and processes, i.e. : (1) check results using the same measuring instrument (tool, unit, time period, etc); (2) evaluation of side effects that are not expected either in the form of quality, cost, delivery, safety, etc; (3) summarize the benefits of improvement; (4) if the handling results are not satisfactory, check the work plan again (stage 5). The run chart can be used at the evaluation stage to be able to see more clearly the movement of defect number after repairs were made. The last stage is standardize successful processess. If the evaluation results show that the plan is achieved, then standardization of the system is made so that they can be permanently resumed in the production process, vice versa. In conducting standardization, there are three processes that need to be done, i.e. : (1) making new standards (standardization); (2) sharing new standards (yokoten); (3) starting next kaizen (improvement).

3. Case Study Result

3.1 Clarify The Problem

In accordance with Figure 1, there is a gap of 3.06 % between the target and the actual condition.

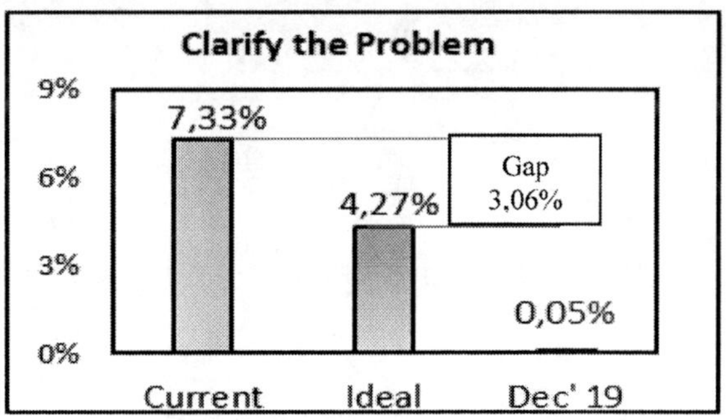

Figure 1. Clarify the Problem

3.2 Breakdown The Problem

To get a deeper understanding, the problem is broken down by dividing dent defect according to the area as shown in Figure 2. Due to greater dent defect occuring on the exhaust manifold, repairs focused on the exhaust manifold area.

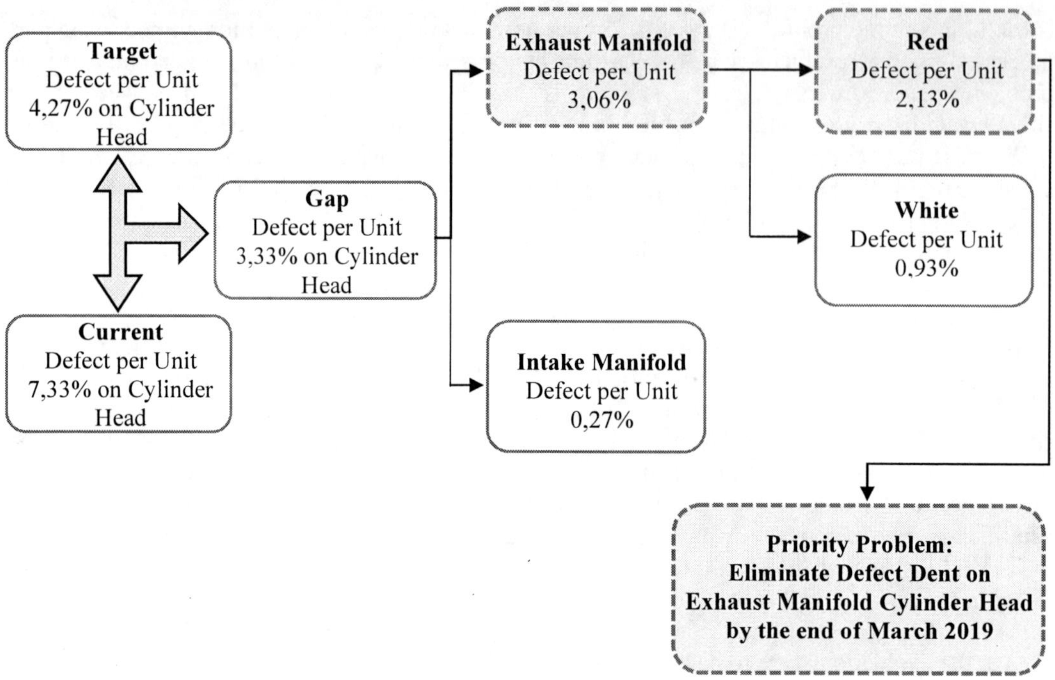

Figure 2. Breakdown the Problem

3.3 Set a Target

The next step is set a target to be achieved through the improvement that will be made. The target of reduce dent defect per unit to be achieved is 4.27 % until the end of March 2019.

3.4 Analyze the Root Cause

The next planning stage is to find the root cause of the problem through root cause analysis and get the results as shown in Figure 3. There are three root problems found are as follows: (1) The distance of the

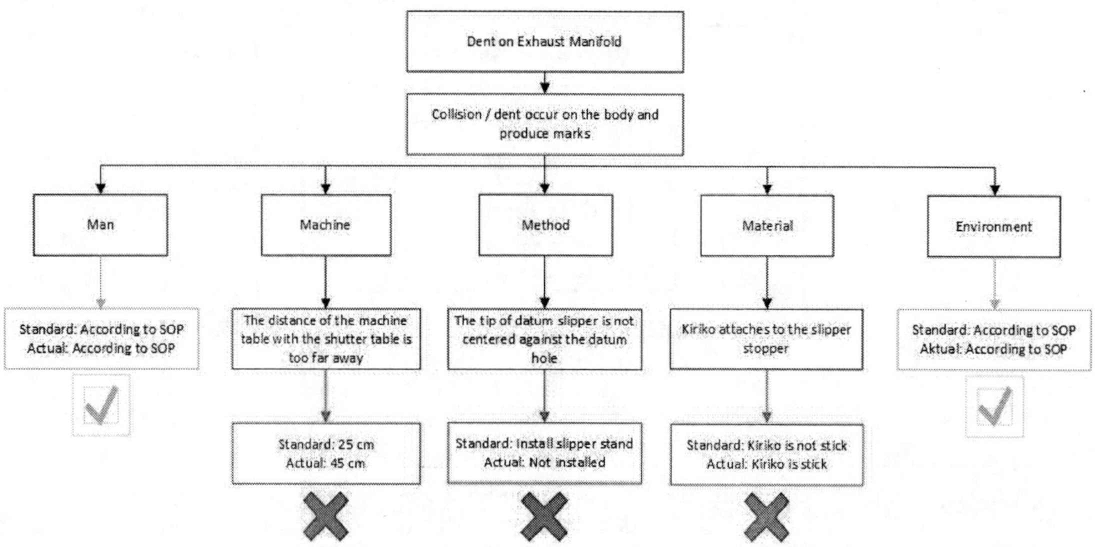

Figure 3. Root Cause Analysis

machine table with the shutter table is too far away. Shutter table is a table that is used as a cylinder head pedestal before entering the machine. The condition of the table is too far away from the machining table, causing collisions during loading and unloading. The loading activity is when the cylinder head is inserted into the machine for the machining process, while the unloading activity is when the cylinder head is pulled out of the machine to enter the next machining process; (2) The tip of datum slipper is not centered against the datum hole. Slipper is a tool or jig that is installed on the exhaust manifold which functions as a stand for the cylinder head or the base of the cylinder head during the machining process. However, the installation of the slipper so far is not right on the datum exhaust manifold holc, so that the movement of the cylinder head during the machinig process causes dent between the slipper and the exhaust manifold; (3) Kiriko (cut glass) attaches to the slipper stopper. The current slipper can make the cylinder head move when the slipper is installed, so that kiriko can stick to the exhaust manifold area. Kiriko which sticks long during the machining process can leave marks on the exhaust manifold body. This can happen, because according to the slipper function that is as the base of the cylinder head so that it does not directly come into contact with the machine part during the postal displacement does not work perfectly.

3.5 Develop Countermeasures

After finding the root cause of quality problems that occur, the next step is to plan improvement as shown in Table 1.

Table 1. Develop Countermeasures

Problem	Root Cause	Temporary Countermeasures	Fixed Countermeasures
Dent on Exhaust Manifold	The distance of the machine table with the shutter table is too far away	Improve the position of shutter table	Improve machine management system
	The tip of datum slipper is not centered against the datum hole	Install slipper stand	Install slipper stand at each post that needed it
	Kiriko attaches to the slipper stopper	Improve the design of slipper stopper	Improve the effectiveness of kiriko cleansing

3.6 See Countermeasures Through

Improving efforts are carried out according to the plans that have been made as follows:

1. Improve the position of shutter table

The design between the machine table and the shutter table is 45 cm. This distance is shortened by 20 cm to 25 cm. With a distance of 25 cm, the cylinder head's range of motion is smaller and the possibility of impact can be minimized as shown in Figure 4.

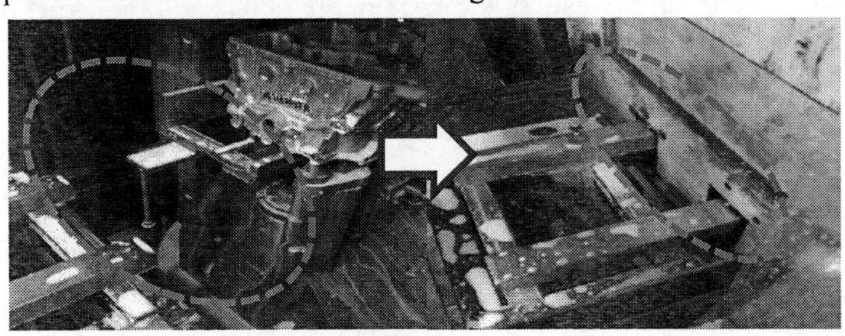

Figure 4. Improve the Position of Shutter Table

2. Install stand slipper

The installation of the slipper stand is carried out in the last machining process before final inspection is carried out. The slipper stand is installed with a size of 78.5 cm x 9.5 cm x 1 cm. With a height has 1 cm at slipper stand, it can help operators to position the slipper more precisely to the datum hole, so the dent between slipper and the body of exhaust manifold can be minimized as shown in Figure 5.

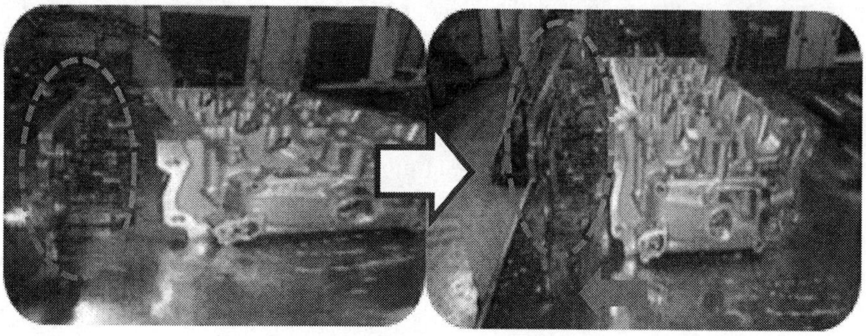

Figure 5. Install Stand Slipper

3. Improve the design of slipper stopper

The current slipper is undergoing design changes because the old design can still make the cylinder head move when the slipper is installed so that kiriko can stick on exhaust manifold. The slipper is redesigned by creating a multilevel design on the slipper face stopper and installing additional datum as a "dead point" and clamping when the slipper is installed. This addition can support the cylinder head more strongly and make kiriko not in direct contact with the body of exhaust manifold as shown in Figure 6.

Figure 6. Improve the Design of Slipper Stopper

3.7 *Monitor Both Results and Processes*

After improvement efforts, an evaluation is carried out by looking at the target using the run chart as shown in Figure 7. Before repairs are carried out, the average dent defect per unit on exhaust manifold is 6.74% and after repairing the average dent defect per unit the exhaust manifold reduced by 4.35% to 2.39%.

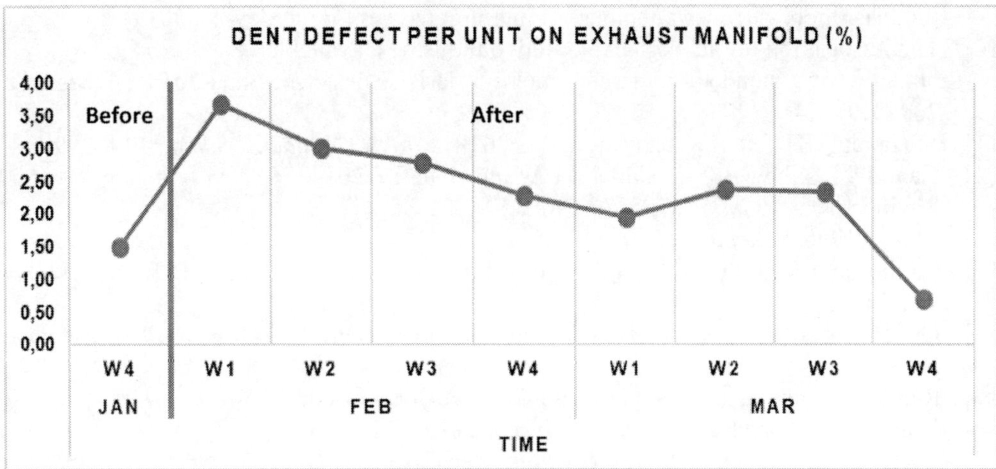

Figure 7. Run Chart

3.8 *Standardize Successful Processes*

Based on the results of the evaluation that has been carried out, standardization of the repair process has been established so that the engine production plant can proceed properly. Improvements set to standard are as follows: (1) Change the position of the shutter table and machine table. Standardization takes the form of rules for the work elements of the machining process; (2) Installation of stand slipper. Standardization is in the form of rules on the work elements of the slipper installation process, because with the addition of a tool that is a stand slipper, the working elements that have been used in this post have changed; (3) Improve slipper stopper design. Standardization in the form of making basic rules or rules that must be considered in the process of moving the cylinder head using a slipper to effectively reduce kiriko attached on cylinder head.

4. Conclusion

The study employed TBP to reduce the number of dent defects per unit for cylinder head part. A case study has been performed in Company X, located in Indonesia. By improving the position of shutter table, installing slipper stand, and improving the design of slipper stopper; the average dent defect per unit on exhaust manifold reduced by 4.35 % to 2.39 %. The study also creates new standards to maintain product quality sustainability. The method that has been performed can be beneficial for directors of the company to solve existing problems. Through such method, we can use the results as a reference in the improvement of manufacturing company in the future.

References

[1] Bowen D and Youngdahl W 1998 *Int. J. Serv. Ind. Manag.* **9** 207-25
[2] Goland A, Hall J, and Devereaux C 1998 *The McKinsey Quart.* **4** 59-66
[3] Lean Enterprise Institute, available: https://www.lean.org/Index.cfm?PageNotFound=https://www.lean.org:443/%20Community/Rgistered/SuccessStory.cfm?Success%20StoryId%20%C2%BC%205 (Accessed on May 26, 2019)
[4] Swank C K 2003 *Harvard Business Rev.* **81** 123-9
[5] Shingo, available: https://shingo.org/BusPrize/recipients/current.htm (Accessed on May 26, 2019)
[6] Womack J P and Jones D T 1996 *Harvard Business Rev.* 140-152
[7] Genaidy A M and Karwowski W 2003 *Human Fact. Ergonom. Manufact.* **13** 317-330
[8] Okamoto Y and Sjöholm F 1999 *Protection and the Dynamics of Productivity Growth: The Case of Automotive Industries in Indonesia* (Strockholm: The European Institute Studies)

[9] Saragih F A, available: https://otomotif.kompas.com/read/2018/01/17/122310015/kapasitas-produksi-mobil-indonesia-tembus-22-juta-unit (Accessed on May 26, 2019)
[10] The Association of Indonesia Automotive Industries, available: https://www.gaikindo.or.id/deretan-mobil-terlaris-semester-pertama-2018/ (Accessed on May 26, 2019)
[11] Muzaimi H, Hamid S R, and Chew B C 2018 *J. Adv. Manufact. Technol.* **12** 87-99
[12] Hamid S R, Chew B C, Abdullah M A, and Hanafi Z 2014 *J. Technol. Manag. Bus* **1** 113-133
[13] Ravinder K 2015 *Int. J. Mod. Eng. Res.* **5** 43-51
[14] Lyu J J 1996 *J. Manufact Syst.* **15** 125
[15] Soufhwee A R, Mahmood W H W, and Abdullah M I H C 2018 *J. Adv. Manufact. Technol.* **12** 343-355
[16] Gunasekaran A, Korukonda A R, Virtanen I, and Yli-Olli P 1994 *Int. J. Prod. Econom.* **36** 169-183
[17] Radite P, Priadythama I, and Fahma F 2015 *Performa* **14** 157-162
[18] Wirahadi K G and Rahardjo J 2015 *Jurnal Titra* **3** 211-214
[19] Liker J K and Ogden T N 2011 *Toyota Under Fire: Lessons for Turning Crisis Into Opportunity* (New York: McGraw-Hill Companies Inc)

IOP Conf. Series: Materials Science and Engineering **722** (2020) 012042 doi:10.1088/1757-899X/722/1/012042

The Influence of Blue Light in Maintaining Alertness in Tropical Country: A Preliminary Study

K T Rahma[1], S A Salma[1], A Widyanti[1], Suprijanto[2]

[1]Deparement of Industrial Engineering, Bandung Institute of Technology, Bandung, Indonesia
[2]Departement of Physical Engineering, Bandung Institute of Technology, Bandung, Indonesia

Corresponding Author: sasheilaamalia@gmail.com

Abstract. Several studies have discussed the influence of blue light on cognitive activity. This study aims to determine the effectiveness of blue light in increasing alertness at night in tropical countries. Twelve healthy young males joined in this study. All participants were asked to do a monotonous activity that is reading activity. The experiment was a within-subject design with the independent variable is lighting condition (normal light and blue light) and duration of exposure (30 minutes and 60 minutes). Electroencephalography (EEG) signals were recorded continuously during the experiment. Alertness was measured based on theta, alpha, and beta activity. The result is the light condition not significantly affected the theta ($F(1,11) = 0.608$, $\rho = 0.452$), alpha ($F(1,11) = 1.561$, $\rho = 0.237$), and beta ($F(1,11) = 0.608$, $\rho = 0.700$) activity. This shows that blue light is not effective in increasing alertness at night in the tropical country both in short and long duration of exposure.

1. Introduction

Blue light is light with wavelengths of natural light between 446 and 483 nm with an optimum wavelength of 468 nm [1]. Several studies have discussed the influence of blue light on cognitive activity, especially at night where the melatonin hormone plays a role in regulating the wake and sleep cycles to the maximum, resulting in increased drowsiness and decreased alertness.

Vigilance is a cognitive process related to the ability to maintain a continuous focus of attention when viewing signals over a long period [2]. Factors that can reduce alertness are the monotonous or low workload. The level of alertness can be known through physiological changes in the body, one of which is by using measurements of brain wave activity. Brain waves are measured using an electroencephalograph (EEG) device. Theta and alpha brain waves, with band frequency 4 Hz – 8 Hz and 8 Hz - 13 Hz, were associated with alertness [3]. Moreover, beta brain wave, with band frequency 13 Hz - 30 Hz, was associated with alertness [4].

One way to maintain vigilance is by providing the exposure of blue light, as in some studies that have been done in nontropical countries. A study explained that blue light for 6.5 hours was able to increase alertness at night by significantly reducing theta activity [5]. A study explained that blue light for 6 hours proved to be effective in increasing alertness at night by being able to reduce theta activity, but did not significantly affect alpha activity [6]. A study explained that blue light for 1 hour was able to increase alertness at night by significantly reducing alpha activity and increasing beta activity [7]. A study explained that blue light for 30 minutes can increase alertness at night by significantly reducing alpha activity [8].

In tropical countries, 60 minutes of blue light can increase alertness at night during monotonous activities, as explained in the research of Lisdiani and Yassierli [9]. Research of blue light in tropical countries is still rarely studied. Besides, the exposure of long duration blue light can make uncomfortable in eyesight [10] and cause retinal damage [11]. Thus, to reduce the negative impact of blue light, this study will discuss the exposure of blue light with a short duration of exposure, is there a difference with the exposure of long duration blue light as in the research Lisdiani and Yassierli [9].

2. Method

2.1. Participants
Twelve healthy young males with a mean (±SD) age of 21 years (±0.56) participated in this study. Participants were directed to sleep before 11.00 pm in the night before the experiment. Participants were informed to not consuming caffeine and cigarettes.

2.2. Procedure
The experiment was a within-subject design, conducted to evaluate the effects of two different lighting conditions that are normal room lighting (white light) and normal room lighting with a blue light enriched as additional lighting. Besides that, the experiment was conducted to evaluate the effects of duration with short duration (30 minutes) and long duration (60 minutes) of light exposure.
The experiment was held in a simulation room from 18:00 until 21:00. Before the experiment begins, the participant was given a brief explanation of the experiment. After that, EEG electrodes were placed in the participant head skin.

The experiment was started with 5 minutes of baseline condition (dark condition). After that, all participants were instructed to do a monotonous activity that is reading activity. The EEG was measured from beginning to the end of the experiment, including 5 minutes of the baseline condition. In total, the participants took 35 minutes (for 30 minutes duration of lighting exposure) and 65 minutes (for 60 minutes duration of lighting exposure) to complete the experiment. Each trial was carried out on different days, so participants had adequate rest and were able to reduce the carry over effect. The procedure of the experiment can be seen in Figure 1.

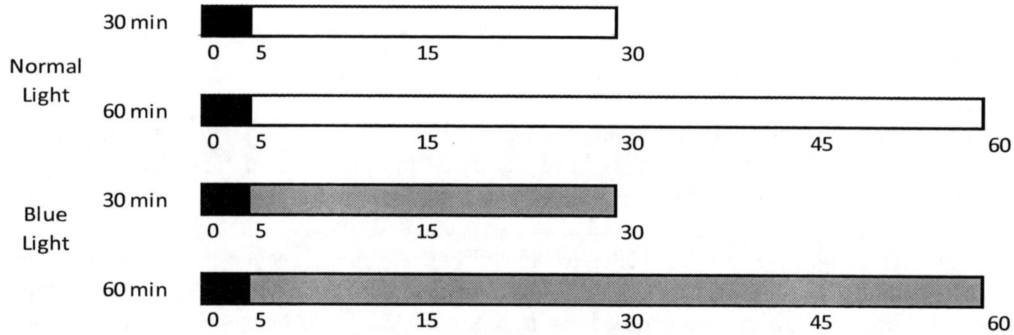

Figure 1. Procedure Experiment

2.3. Measure
Blue light apparatus, Philips goLITE BLU Light Therapy Device HF3332 7 × 7 inches, was used as the illumination source. EEG was measured using a Muse Brain Sensing Headband, with 7 electrodes (4 Channels and 3 References). Muse Headband Sensing can be seen in Figure 2. The electrodes were placed at frontal left (AF7), frontal right (AF8), temporal left (TP9), temporal right (TP10) and FPz, Fp1, Fp2 (as reference sensors). All data are transmitted via Bluetooth to a smartphone running the application that is Muse Monitor.

IOP Conf. Series: Materials Science and Engineering **722** (2020) 012042 doi:10.1088/1757-899X/722/1/012042

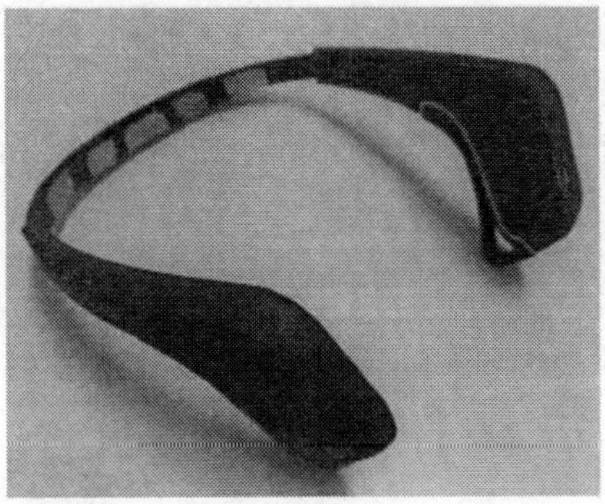

Figure 2. Muse Headband Sensing

2.4. Data Analysis

Based on Phipps-Nelson et al. [6] and Figueiro et al. [7], only EEG theta (4 Hz-8Hz), alpha (8Hz – 13 Hz), and beta (13-30 Hz) activity were examined. EEG absolute band power was recorded for each channel (frontal left, frontal right, temporal left, and temporal right). EEG absolute band power was transformed into relative band power before being normalized to the power obtained during the baseline. The formula of relative band power is [12]:

$$S_r = \frac{10^{S_{abs}}}{10^{\alpha_{abs}} + 10^{\beta_{abs}} + 10^{\gamma_{abs}} + 10^{\theta_{abs}} + 10^{\delta_{abs}}} \qquad (1)$$

in which S_r is the relative value of frequency range (delta, theta, alpha, beta, gamma) being calculated and S_{abs} is the absolute value of range frequency.

All data were analyzed using ANOVA repeated measure, with "lighting condition" (white and blue light enriched) and "duration" (30 and 60 minutes). All analyses were performed using SPSS ver. 20

3. Result

Theta, alpha, and beta activity as a function of light condition and duration can be seen in Figure 3, 4, and 5. The result of significance analysis can be seen in Table 1. The light condition not significantly affected the theta ($F (1, 11) = 0.608$, $\rho = 0.452$), alpha ($F (1, 11) = 1.561$, $\rho = 0.237$), and beta ($F (1, 11) = 0.608$, $\rho = 0.700$) activity. The duration significantly affected the theta ($F (1,11) = 51.378$, $\rho < 0.05$), alpha ($F (1, 11) = 18.378$, $\rho < 0.05$), and beta ($F (1, 11) = 51.378$, $\rho < 0.05$) activity. And, there is no interaction between light condition and duration to the theta ($F (1, 11) = 3.056$, $\rho = 0.520$), alpha ($F (1, 11) = 0.441$, $\rho = 0.237$), and beta ($F (1, 11) = 0.3.056$, $\rho = 0.503$) activity.

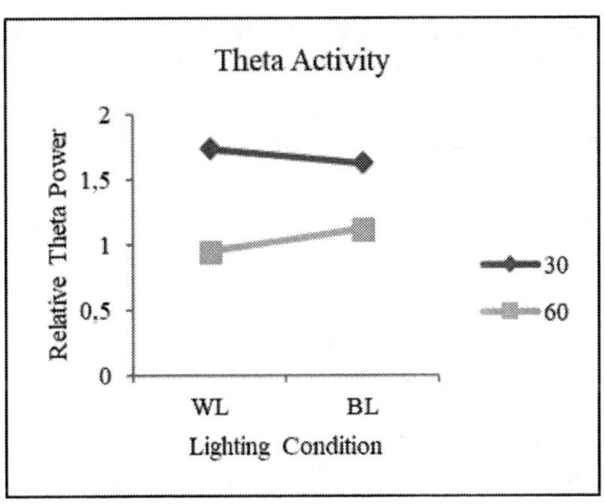

(a)

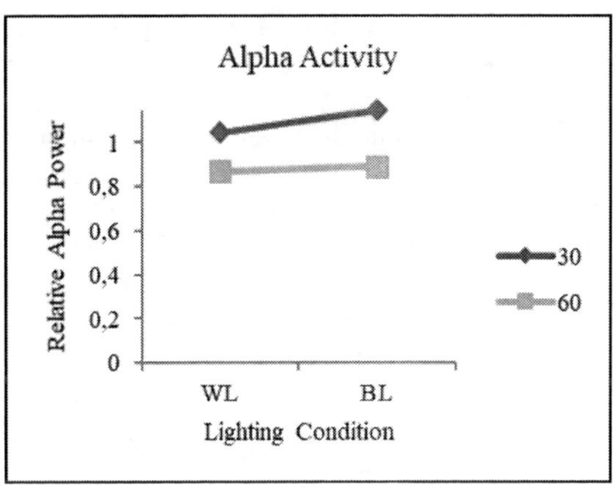

(b)

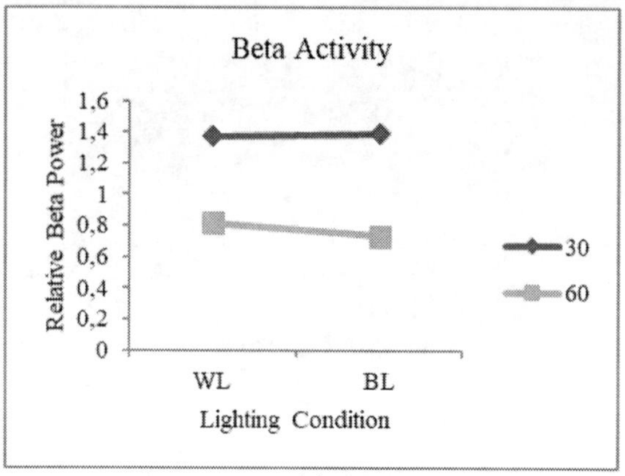

(c)

Figure 3. (a) Theta (b) Alpha (c) Beta activity as a function of lighting condition and duration

Table 1. The Result of Significant Analysis

Brainwave Activity	Factor	Sign	Conclusion
Theta	Illumination condition	0.452	Accept H0
	Duration	0.000	Reject H0
	Illumination condition*Duration	0.108	Accept H0
Alpha	Illumination condition	0.237	Accept H0
	Duration	0.001	Reject H0
	Illumination condition*Duration	0.520	Accept H0
Beta	Illumination condition	0.700	Accept H0
	Duration	0.001	Reject H0
	Illumination condition*Duration	0.533	Accept H0

4. Discussion

This study aimed to observe the effectiveness of blue light in a tropical country compared with a subtropical country. Besides that, this observed study will discuss the exposure of blue light with a short duration of exposure, is there a difference with the exposure of long duration blue light. The results show that blue light is not effective in increasing alertness in the tropical country seen from the wave activity of theta, alpha, and beta at night both in short and long duration of exposure. Blue light tends to reduce theta activity during 30 minutes of exposure, but the results are not significant. Blue light is not significantly in reducing alpha activity. It is in line with Phipps-Nelson et al. [6] that is blue light did not significantly affect alpha activity. Blue light tends to increase beta activity during 30 minutes of exposure, but the results are not significant.

The result is contrary to Lisdiani and Yassierli [9] that discuss 60 minutes of blue light with the ability to increase alertness at night. This difference in results is due to the research of Lisdiani and Yassierli [9] using the comparison value $(\theta + \alpha) / \beta$ as an alertness parameter.

This study contradicts the research of Iskra-Golec et al. [8] in the subtropical country which explains that blue light for 30 minutes can increase alertness at night by significantly reducing alpha activity. This result also contradicts the research of Figueiro et al. [7] which blue light for 1 hour was able to increase alertness at night by significantly increasing beta activity. This result can happen because in tropical country with high exposure of sun light (which is a main source of blue light) throughout the year, the high exposure of sun light can affect the degeneration of macula in retina that was the most sensitive part of eye to light at the shorter wavelengths including blue light [13]. Therefore, the high exposure of sunlight can reduce the sensitivity of blue light.

This study has several limitations worth noting. First, only one measurement was conducted that is brainwave activity. Further research should be conducted with different measurements, such as using task measurement and subjective measurement. Second, the respondents are limited. Enlarging the number of samples and enlarging the sample to include not only a student sample might enrich the result

5. Conclusion
This study shows that blue light is not effective in increasing alertness at night in the tropical country both in short and long duration of exposure.

6. Acknowledgement
Authors thank RISTEKDIKTI for funding this research under the scheme of Hibah Kompetensi fiscal year 2018-2019.

7. References
[1] Chellappa S L, Steiner R, Blattner P, Oelhafen P, Gotz T, and Cajochen C 2011 Non-Visual Effects of Light on Melatonin, Alertness and Cognitive Performance: Can Blue-Enriched Light Keep Us Alert? *PLoS One* **6** e16429

[2] Davies D R and Parasuraman R *1982 The Physiopsychology of Vigilance* (New York: Academic Press).

[3] de Waard D 1996 *The Measurement of Drivers' Mental Workload* (Den Haag: The Traffic Research Centre VSC).

[4] Surangsrirat D and Intarapanich A 2015 Analysis of the meditation brainwave from consumer EEG device *Proceedings of the IEEE SoutheastCon*.

[5] Lockley S W, Evans E E, Scheer F A J L, Brainard G C, Czeisler C A, and Aeschbach D 2006 Short-wavelength effectiveness for the direct effects of light on alertness, vigilance, and the waking electroencephalogram in humans *Sleep* **29** 161.

[6] Phipps-Nelson J, Redman J R, Schlangen L J M, and Rajaratnam S W M 2009 Blue light exposure reduces objectivemeasures of sleepiness during prolonged nighttime performance testing *Chronobiology International* **26** 891.

[7] Figueiro M G and Rea M S 2011Sleep opportunities and periodic light exposures: Impact on biomarkers, performance and sleepiness *Lighting Res Technol* **43** 349.

[8] Iskra-Golec I, Golanka K, Wyczesany M, Smith L, Siemiginowska P, and Wqtroba J 2017 Daytime Effect of Monochromatic Blue Light on EEG Activity Depends on Duration and Timing of Exposure in Young Men *Adv in Cogn Psy* **13** 241.

[9] Lisdiani N I and Yassierli 2012 Blue Light Exposure Improves Awareness during Monotonous Activities at Night *IEEE: 2012 Southeast Asian Network of Ergonomics Societies Conference (SEANES)*.

[10] Laufer L M A, Lang L D A, and Nemeth E M S 2009 Psychophsyological effect of coloured lighting on older adults *Lighting Res Tech* **41** 371.

[11] Revell V L, Arendt J, Fogg L F, and Skene D J 2008 Alerting effects of light are sensitive to very short wavelength *Neuroscience Letters* **399** 96.

[12] Wiechert G, Triff M, Liu Z, Yin Z, Zhao S, Zhong Z, Zhaou R, and Lingras P 2016 Identifying Users and Activities with Cognitive Signal Processing from a Wearable Headband *IEEE 15th International Conference on Cognitive Informatics & Cognitive Computing* 129.

[13] Tomany S C, Wang J J, van Leeuwen R, Klein R, Mitchell P, Vingerling J R, ..., and de Jong P T V M 2004 Risk factors for incident age-related macular degeneration *Ophthalmology* **111** 1280

IOP Conf. Series: Materials Science and Engineering **722** (2020) 012043 doi:10.1088/1757-899X/722/1/012043

Identify the Relevant Pages of Book to be Indexed Using Naive Bayes Classification Method

S Christina[1], and D Ronaldo[2]

[1,2]Informatic Departement, Faculty of Engineering, University of Palangka Raya, Palangka Raya, Indonesia

E-mail: sherly.christina.upr@gmail.com

Abstract. The back-of-book index is a component that is often found in non-fiction books. The back-of-book index contains important terms and page numbers where the terms appear in the book, it is useful for helping readers find certain term directly, without having to search on every page. Usually, the back-of-book index is compiled by the author or a professional indexer, although currently there are several automatic indexing applications available. To determine a term on a page worth to be indexed requires knowledge and expertise of the author whom better understand the book's context. Hence, generating a good back-of-book index is a task that requires great effort, knowledge and cost. Therefore, the back-of-book index is very possible containing indexed terms that refer to irrelevant page numbers due to human error or the weakness of the indexing application. This study aims to identify the relevance of the pages to be indexed in the back-of-book index, using Naive Bayes Classification. The testing result shows that the approach with Naive Bayes Classification produces an average precision value of 74.02 percent and 100 percent for the recall value. The average precision value that more than 50 percent indicates that the naive bayes classification approach is capable to identify the relevant and irrelevant pages.

1. Introduction

The Non-fiction books are usually equipped with a list of indexed terms to help readers find important terms. The following is an example of the indexed terms in a book entitled "Big Data, Data Mining and Machine Learning".

binary state, 225
binning, 199
Bluetooth, 6
boosting, 124–125

The list of indexed terms that usually called the back-of-book index, contains terms along with page numbers that indicate where the terms appear in the book. The list of indexed terms facilitates the reader to look for important words directly without having to open every page in the book [1]. Therefore, the back-of-book index is an important component of a book, thus much research has been done to find the optimal indexing method [2] [3] [4].

The list of indexed terms is usually created by the author or a professional book indexer that provided by the publisher. To compile an excellent list of indexed terms, the author must identify all concepts that are important to be indexed, then the author must scan each page of the book, and decide which pages are relevant to be indexed in concern to the context of the book.

Therefore, there are many software applications available to help the authors and the indexing experts to generate a list of indexed terms from a book [5], but although a software application can produce a set of index lists, there is still a problem regarding the relevance of an indexed page. The page numbers which referenced in the index list must lead the reader to the text which containing indexed terms and containing information which is relevant to the indexed terms. Donald and Anna Cleveland defined that a good index will guide the reader to find the information needed with no hurdles and no irrelevant material [6].

In this study, we aim to identify the pages which containing the indexed term that relevant to be indexed using Naive Bayes classification (NBC) method. The NBC method is implemented to identify the pages that are relevant and not relevant to be indexed. In this study, the data sets are the texts containing the indexed terms and indexed terms from the book's page, along with the values of semantic relatedness and cosine similarity between the text and the indexed term. To our knowledge, no research has been conducted to identify the relevant pages from a back-of-book index, by processing the semantic relatedness and cosine similarity in the data set and identifying the relevance of book pages using NBC method.

2. The Relevant Page and Irrelevant Page

The following is example of relevant and irrelevant page to be indexed, in the book titled C Programming for Arduino there is a term "double" indexed on page 57.

Page 57: double - It generally stores float values with a precision two times greater than the float value

The term "double" found on page 57 has a definition that related to data type in C programming. The term "double" on page 57 is closely related to the context of the book, therefore the term "double" on page 57 is important and relevant to be indexed. Then on page 49, also found the term "double" which is not referred by in the index list.

Page 49: Imagine that you have followed carefully the *Blink250ms* project, everything is wired correctly, you double-checked that, and the code seems okay too, but it doesn't work.

The term "double" that appears in the text on page 49 does not have an important meaning related to the context of the book in C Programming for Arduino, thus page 49 is not relevant to be referred by in the index list.

A good back-of-book index must contain page number that refers to the page that indeed contain the indexed term that is relevant to the book's context, because the terms defined in the index list represent the subject or concepts conveyed by the book [7]. Nevertheless, to produce a good back-of-book index, it requires accuracy, creativity and knowledge from the authors or book indexing experts [7] [8]. Therefore, indexing the terms is a task that requires expertise and knowledge, human error or the weakness of the indexing application can produce the irrelevant list of indexed terms.

3. Methodology

In the methodology section we briefly explain the methods implemented in this study.

3.1. Semantic Relatedness

The semantic relatedness approach [9] is a method to measure the semantic relatedness between text (S) containing the indexed term and the indexed terms (T) itself. In our previous studies [10][11], the semantic relatedness approach show a reliable performance that has been evaluated using Kappa values [12]. The following are the stages in the semantic relatedness measurement approach:

- The text mining stage is the step of text preparation for data sets, such as doing tokenization process, removing the stop words, and stemming the text (S) and indexed term (T).

- The speech tagging stage is the next stage after the text mining stage. The speech tagging stage defines the part of speech of S and T, such as noun, verb or adverb of each word in S and T.
- The Word Sense Disambiguation stage, is the stage to find senses of each words in S and T. The word sense disambiguation stage in this study use the wordnet lexical database.
- Create semantic similarity matrix M in mxn dimension, where M [i,j] is the semantic relatedness value between the most suitable sense of the word at the position i of T with the most suitable sense of the word at the position j of S.
- Calculate the semantic relatedness with the heuristic method, as in equation (1).

$$\text{Similarity}_{(T,S)} = \frac{\sum \text{maxR} + \sum \text{maxC}}{\text{number of rows and columns}} \quad (1)$$

Where maxR is the maximum value of each row, and maxC is the maximum value of each column.

3.2. Cosine Similarity

The cosine Similarity method [13][14] is a syntactic approach to measure the relatedness of S and T. The value of cosine similarity (C) is the term-frequency vector of T and S, as in equation (2).

$$C_{(T,S)} = \frac{T.S}{|T||S|} \quad (2)$$

The distance between vectors T and vector S is an indication of the relatedness between the indexed term and the text on the page which number has been indexed.

We use the cosine similarity approach in this study because the number of the occurrences of the indexed term and the number of the words in the text (S), also determines the relevance of a page to be indexed, although in previous studies [15], the testing result on the cosine cimilarity approach shows the Kappa value that is lower than the semantic relatedness approach.

3.3. Data Set

In this study, we obtained the data sets from several electronic books with different knowledge domains, such as social-politics, economy-business, religion, computer science, education, general and biology. From each e-book, we took texts containing indexed terms in the back-of-book index (S_1). Then we took other texts containing the same indexed terms from other pages that are not indexed in the back-of-the-book index (S_2). In this study, S_1 is considered as a relevant page to be indexed, because it is a page that is referenced in the back-of-book index. Whereas S_2 is considered an irrelevant page to be indexed because it is not referenced in the back-of-book index.

S_1 and S_2 are measured in relation to the indexed term (T) using the semantic relatedness and cosine similarity methods. The following is example of indexed term “double” from the "C Programming for Arduino" book.

T: double
S_1 (page 57): double - It generally stores float values with a precision two times greater than the float value.
Semantic Relatedness Value (T,S_1) = 0,52
Cosine Similarity Value (T,S_1) = 0,25

S_2 (page *49*): Imagine that you have followed carefully the *Blink250ms* project, everything is wired correctly, you double-checked that, and the code seems okay too, but it doesn't work.
Semantic Relatedness Value (T,S_2) = 0,45
Cosine Similarity Value (T,S_2) = 0,3

The number of the data sets in this study are 1664 data, consisting of S_1 and S_2 along with its features, that are semantic relatedness values and cosine similarity values.

IOP Conf. Series: Materials Science and Engineering **722** (2020) 012043 doi:10.1088/1757-899X/722/1/012043

3.4. Naive Bayes Classification

The Naive Bayes Classification (NBC) method [16] [17] is used as an approach to identify the relevant and irrelevant pages to be indexed. The NBC method has been proven effective and reliable to produce high accuracy in several studies [18] [19] [20].

On NBC, the hypothesis is a class label which is the target of classification, while evidence is a feature which is input into the classification model. If X is an input vector that contains features and C is the class label, then NBC is written as P(C|X). P(C|X) means the probability of class C is obtained after X features are observed. P(C|X) is also called the posterior probability for C, while P(C) is called the prior probability.

In the training phase, we build a model during the learning process which is done on P(C|X) for each combination of X and C based on the information obtained from the training data. Then the model is used to classify the testing data. Hence, the testing data that has the X_i features can be classified to find the value of C by maximizing the value of P(X|C) from the testing data. The NBC method in this study is shown by equation (3).

$$P(C|X) = \frac{P(C)\pi_{i=1}^{q} P(X_i|C)}{P(X)} \tag{3}$$

Where C is a class that affirms the relevance of S, C consists of Relevant Class (C_1) and Not Relevant Class (C_2). X is the attribute vector set of S_1 and S_2 consist of semantic relatedness (X_1) and cosine similarity (X_2) values.

P(C|X) is the probability of data with vector X in class C. P(C) is the initial probability of class C. $\pi_{i=1}^{q} P(X_i|C)$ is the independent probability class C of all X attributes. The value of P(X) is always fixed, therefore, to identify the relevant class, only calculate $P(C)\pi_{i=1}^{q} P(X_i|C)$ and only choose the largest value to be the chosen class as a result of predicting the relevant page. Then the independent probability $\pi_{i=1}^{q} P(X_i|C)$ is the effect of all the features of the data on each class C.

3.5. Method to Identify the Relevant and Irrelevant Page

Figure 1 shows the method for identifying the relevant and irrelevant page. In this study a number of the training data (X,C) have been provided as the input data for constructing the identifying model. The model is a black box that accepts input, then does reasoning on the input data and provides answer as the output. The training process for building models uses the NBC algorithm. The model is used to identify the relevance class from the test data (X,?), thus the true relevant class C(X|C) is known.

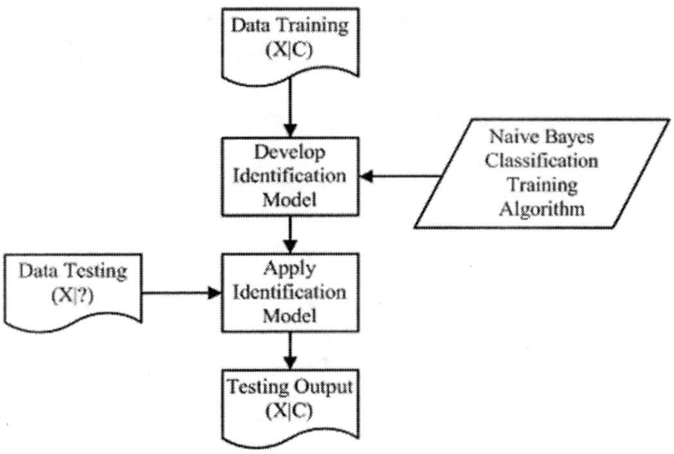

Figure 1. The method to identify the relevant and irrelevant class.

4. Experiment

In the experiment section, we explain the testing scenario in our study. In the testing scenario, the dataset is divided into training data and testing data. The amount of training data and testing data is divided in the equal proportions for each book domain, as shown in table 1.

Table 1. Data training and data testing.

Book Domain	Data Set	Data Training	Data Testing
Social Politics	434	217	217
Economy Business	286	143	143
Religion	122	61	61
Computer Science	346	173	173
Education	84	42	42
General	144	72	72
Biology	248	124	124

We created an identifier model that tested on each book domain and the input data for the identifier model are the training data from the related domain. The model is used to identify the relevant class of the testing data for each book domain. Furthermore, we evaluated the performance on NBC using the identification results, by calculating the precision and recall value of the model outputs.

5. Result and Discussion

In the result and discussion section we explain the results of our experiment. In every testing scenario, the NBC approach works very satisfactory in finding all data training and data testing, it is indicated by the recall value of 100% from every test result.

Table 2 shows the precision values of the NBC approach in identifying the relevant and irrelevant pages to be indexed. The precision average value from each book domain is more than 50 percent, namely 74.02%. The average precision that exceeds the 50% accuracy, indicates that the NBC approach result in reliable accuracy and can contribute to identifying the relevance of a page for constructing the excellent back-of-book index.

Table 2. Precision value of each book domain.

Domain	Precision Value
Social Politics	69,59
Economy Business	59,44
Religion	88,52
Computer Science	78,61
Education	66,67
General	81,94
Biology	73,39
Average Precision Value	74,02

Subsequently the graph in figure 2 shows the precision values of each book domain. It can be seen the accuracy of the NBC approach in each book domain. The NBC approach yields high precision values in the religion, general and computer science domains. In the books from the biology, social politics, education, and economy-business domains the accuracy of the NBC approach is also quite high, more than 50%.

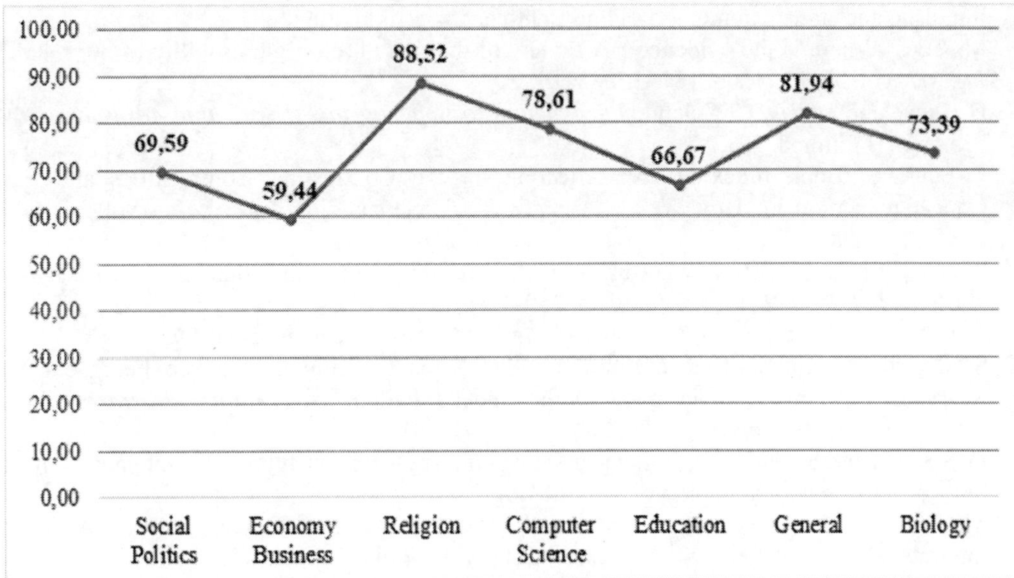

Figure 2. Precision values of each book domain.

However figure 2 also shows that the NBC approach produces the lowest precision value in the economy-business domain. It is because of the semantic relatedness value feature in the economic-business dataset. In the economy-business books, many S_1 data have low semantic relatedness values that are even lower than S_2. Whereas S_1 is the text from the pages which number is indexed in the back-of-book index, so it should have a higher semantic relationship with the indexed term. This anomaly occurs because of many terms such as product names, companies or trademarks from economy-business books that are not found in the WordNet database. Although in general the NCB can identify the relevance of pages in economy-business books, the problem of semantic relatedness in the economy-business books dataset needs to be solved by providing a more complete lexical database.

6. Conclusion

The results of this study indicate that Naive Bayes Classification in our model can be used to identify the relevant and irrelevant pages to be indexed, hence our model could contribute in creating the excellent back-of-book index. The accuracy of the Naive Bayes Classification results, is influenced by the reliability of the identification model, the more training data being trained, the higher the accuracy level of the Naive Bayes Classification approach [18].

Therefore, for future studies, it is necessary to analyze more data sets and apply the other machine learning approach beside Naive Bayes Classification method, in order to find the more reliable methods for identifying the relevant and irrelevant pages for the back-of-book index. The availability of a lexical database that has a broader scope of knowledge also needed for improving the accuracy result.

References

[1] M. Coe, "What do readers expect from book indexes and how do they use them? An exploratory user study," *Indexer*, vol. 33, no. 3, pp. 90–101, 2015.

[2] C. Gottschalg-duque and H. Lobin, "Ontology Extraction for Index Generation," *Electron. Publ.*, no. June, pp. 111–120, 2004.

[3] O. Medelyan, "Thesaurus-based index term extraction," in *In: The 6th Agricultural Ontology Service (AOS) Workshop*, 2005, pp. 1122–1129.

[4] T. Lahtinen, "Automatic indexing : an approach using an index term corpus and combining linguistic and statistical methods Timo Lahtinen," Helsinki, 2000.

[5] M. Coe, "Where is the evidence ? A review of the literature on the usability of book indexes," *Indexer*, vol. 32, no. 4, pp. 161–168, 2014.

[6] D. Cleveland and A. Cleveland, *Introduction to Indexing and Abstracting 4th Edition*, 4th ed. Libraries Unlimited, 2013.

[7] I. Fourier, "Book Indexing: A Reflection on the Contrasting Complexities and Ease of Conceptualization and How We Can Deepen Our Understanding.," *Mousaion*, vol. 26, no. 1, pp. 111–125, 2008.

[8] Z. Wu and C. L. Giles, "Measuring Term Informativeness in Context," in *Proceedings of NAACL-HLT 2013*, 2013, no. June, pp. 259–269.

[9] T. Dao, N. and T. Simpson, "Measuring Similarity between sentences," 2005.

[10] S. Christina, "Kinerja Cosine Similarity dan Semantic Similarity dalam Pengidentifikasian Relevansi Nomor Halaman pada Daftar Indeks Istilah," in *Seminar Nasional Teknologi Informasi dan Komunikasi 2014 (SENTIKA 2014)*, 2014.

[11] D. Siahaan and S. Christina, Using semantic similarity for identifying relevant page numbers for indexed term of textual book, vol. 516. 2015.

[12] M. Banerjee, M. Capozzoli, L. Mcsweeney, and D. Sinha, "Beyond kappa : A review of interrater agreement measures *," *Can. J. Stat.*, vol. 27, no. 1, pp. 3–23, 1999.

[13] A. Singhal, "Modern information retrieval: A brief overview," *IEEE Data Eng. Bull.*, pp. 1–9, 2001.

[14] A. C. A. B, A. Deraman, M. Masila, and A. Jalil, "An Ensemble Similarity Model for Short Text Retrieval An Ensemble Similarity Model for Short," in *17th International Conference on Computational Science and Its Applications*, 2017, no. March 2018.

[15] S. Christina and E. D. Oktaviyani, "Identifying the relevant page numbers that referred by the back-of-book index using syntactic similarity and semantic similarity," in *Proceedings of the 2nd International Conference on Informatics and Computing, ICIC 2017*, 2018, vol. 2018–January.

[16] N. Friedman, "Bayesian Network Classifiers," *Mach. Learn.*, vol. 29, no. 2–3, pp. 131–163, 1997.

[17] J. Dean, *Big Data, Data Mining,and Machine Learning*. Hoboken, New Jersey: John Wiley & Sons, Inc., 2014.

[18] B. Liu, E. Blasch, Y. Chen, D. Shen, and G. Chen, "Scalable Sentiment Classification for Big Data Analysis Using Na¨ıve Bayes Classifier.pdf," in *Big Data, 2013 IEEE International Conference*, 2013.

[19] I. Rish, "An empirical study of the naive Bayes classifier," in *Proceedings of IJCAI-2001 workshop on Empirical Methods in AI*, 2001, pp. 41–46.

[20] V. Rizqiyani, A. Mulwinda, and D. Mahadji, "Klasifikasi Judul Buku dengan Algoritma Naive Bayes dan Pencarian Buku pada Perpustakaan Jurusan Teknik Elektro," *J. Tek. Elektro*, vol. 9, no. 2, pp. 60–65, 2017.

IOP Conf. Series: Materials Science and Engineering **722** (2020) 012044 doi:10.1088/1757-899X/722/1/012044

Lean Concept Development in Fast Food Industry Using Integration of Six Sigma and TRIZ Method

S Indrawati, A 'Azzam, E Adrianto, S Miranda, and A D Prabaswari
Industrial Engineering, Universitas Islam Indonesia, Yogyakarta, Indonesia

E-mails: sriindrawati@uii.ac.id, abdullah.azzam@uii.ac.id

Abstract. The service industry plays an important role to the development of Indonesian economy. Recently, the fast food industry has a good growth due to customers preferences. This significant contribution and industrial growth are not accompanied by the quality of industrial performance. This condition lead to a weak competitiveness, especially to face of the ASEAN Economic Community (AEC). The fast food industry needs to develop a specific strategy to improve the service system. This study uses a lean approach to improve the service system in a fast food restaurant. The method used is the integration between six sigma and TRIZ (Theory of Inventive Problem Solving) to get a creative problem-solving innovation in improving service industry performance. Based on the research conducted, it can be identified that the waste occurs in fast food industry is the service delay due to the process of giving flavor with a total time of 527 seconds. Improvement strategy to reduce the waste is to change the process of giving flavor (parameter changes) and implementation of 5S in the work area (enriched atmosphere). There is a time improvement of 90% by implementing the strategies that have been obtained.

1. Introduction

In Indonesia, the service industry sector has a considerable potential development. This industrial sector has a significant contribution to the Indonesian economy, i.e. gross domestic bruto increased by 10% in 2010-2012 [1]. In addition, the service industry is able to reduce poverty in Indonesia because 52% of the workforce in this sector are women [2].
One of service industry that has a good growth is the fast food restaurant. As many as 80% of Indonesian citizen prefer to use fast food restaurant services compared to other restaurant categories [3]. This is due to a variety of unique food menu offers, fast in serving and offers an online purchases and other promotional packages.

The growth of fast food restaurant is not accompanied by some improvement strategies. Service system improvement is one strategy that should be done. With better performance, an industry can produce products with more competitive value. The improvement process can be done by business process mapping to determine the improvement priorities. Lean concepts can be developed to improve the service industry performance. This concept uses a systematic approach to identify and eliminate waste and various non-value-added activities. Lean is originally a manufacturing concept that used to improve the quality of production through efficiency by identifying and minimizing seven types of production waste, i.e. overproduction, transportation, waiting, unnecessary processes, inventory, movement and product defects [4].

The success of lean in manufacturing has triggered researchers and industry practitioners to adopt lean concepts in service industry. Implementation of lean service can increase the competitive advantage of a service industry by reducing process variability and waste of business activities [5]. Lean principles is implemented in a health care industry to improve the service performance and quality [6]. Lean implementation in health care industry in Europe, Australia, America and United Kingdom also shows a positive contribution to the industry, i.e. 50% of industries has an increased productivity and cost efficiency, service quality, patient and employee safety and financial benefits [7].

The application of lean 3P concepts (production, preparation, process) has been successfully used to rearrange the layout of hospital facilities with good effectiveness [8].

In implementing the lean service, several researchers have conducted studies on adjusting the types of waste and lean tools. Identification of appropriate lean tools for the service industry has been carried out in the mid-scale hotel industry in Europe [9]. There are five value stream mapping tools (VALSAT) that able to detect and eliminate waste in the hospitality service industry value chain, i.e. process activity mapping, quality filter mapping, demand amplification mapping, decision point analysis and physical structure [9]. Several other lean tools can also be used, i.e. supply chain optimization, ABC analysis, pareto diagrams, XYZ analysis, ideas management¸ Kaizen, visual management, poka yoke, PDCA, supplier management, six sigma and 5S [10].

Six sigma with the DMAIC cycle (define, measure, analyze, improve, control) can also being used as a tool for solving lean implementation problems. With six sigma, the process management, continuous improvement and business process reengineering can be done [11]. The integration of value stream mapping (VSM) with DMAIC is successfully done to reduce waste in service industry [12]. The Six Sigma method can be integrated with Theory of Inventive Problem Solving (TRIZ) to create a problem-solving innovation in improving service industry performance. The integration of TRIZ with other methods such as kansei engineering and SERVQUAL is successfully implemented to improve the quality of hotel and restaurant services [13].

2. Literature Review

2.1. Lean Concept

Lean is a continuous improvement concept to eliminate waste and increase the value added of products or services [14]. This concept can be applied to manufacturing or service companies. Implementation of this concept is based on five main principles, i.e [15]:

- Specify Value; determines the value of a product or service based on customer perspective.
- Identify Whole Value Stream; identify the business process, starting from the process of designing, ordering and making products based on the whole value stream to identify nonvalue added process.
- Flow; conduct activities that can create a value without interference, rework, backflow, waiting.
- Pulled; knowing the important activities that used to make what customer want.
- Perfection; trying to achieve perfection by eliminating waste continuously. *Formatting author names*

2.2. Six Sigma

Six Sigma methodology uses statistical tools to identify several important factors in improving the process quality [16]. There are five stages called DMAIC cycles, i.e.:

- Define; at this stage the problem statement is made based on existing facts, focused on what will be observed and compiled, not estimates or assumptions. After that, the objectives will be determined.
- Measure; the problem validation process is carried out.
- Analyze; an analysis of the problem root causes is carried out. Look at the process and data to identify possible causes, find the cause that is estimated and validate it.
- Improve; identify some potential solutions and select the most appropriate solution.
- Control; an evaluation of some potential problem after the solution is implemented and sets standards to maintain performance effectiveness

IOP Conf. Series: Materials Science and Engineering **722** (2020) 012044 doi:10.1088/1757-899X/722/1/012044

2.3. Theory of Inventive Problem Solving (TRIZ)
TRIZ is a method invented by Genrich Altshuller in 1940. Therefore, TRIZ comes from the Russian language, namely the Theory Resheniya Izobreatatelskikh Zadatch or in English known as Theory of Inventive Problem Solving. TRIZ is a strong methodology for finding solutions to a contradictive problem [17]. There are 11 types of TRIZ tools, i.e. function analysis, cause and effect chain analysis, perception mapping, ideality, S-curve analysis, trends of engineering system evolution, trimming, transfer features, function-oriented search, 9-windows and engineering contradiction. The entire tool can be used simultaneously or independently to get innovative solutions to a problem in a systematic and structured manner [17].

3. Method

Research on the integration of Six Sigma and TRIZ methods in lean concepts development for a fast food restaurant industry consists of five main stages, i.e.:

- Define; identification of business processes, process time, customer needs, number of workers and the number of customers in a fast food restaurant are used to develop the current value stream map. The function of this map is to provide an integrated information of current service conditions.
- Measure; identification of waste and performance measurements from the current process.
- Analyze; the integration process of six sigma and TRIZ is done to identify and analysis the causes of waste and contradictions that occurs.
- Improve; contradiction that already found then minimized or eliminated using 40 principles of invention.
- Control; a performance measurement process is estimated.

4. Result and Discussion

The service industry used in this research is a fast food restaurant that offers fried chicken-based foods with eight flavors. The main raw material is chicken meat from a supplier. There are four main processes carried out in this industry, i.e. the process of shoving chicken, frying, giving sauce according to customer orders and serving. Potential Customer from this industry are 2-25 years of age with professional dominance as students.

The detailed process consists of 18 activities on 6 workstations. The production time data has been tested for adequacy with N for all samples of 10. Based on the current value stream mapping, the delivery of 2000 pcs of raw materials or 50 packs to the store is done every day as shown in Figure 1. The cooking process capacity is one pack or 40 pieces of chicken. The cooking process begins in the process of chicken combing with three types of activities, i.e. the process of making flour, the process of dyeing chicken into the water and mixing the chicken with flour.

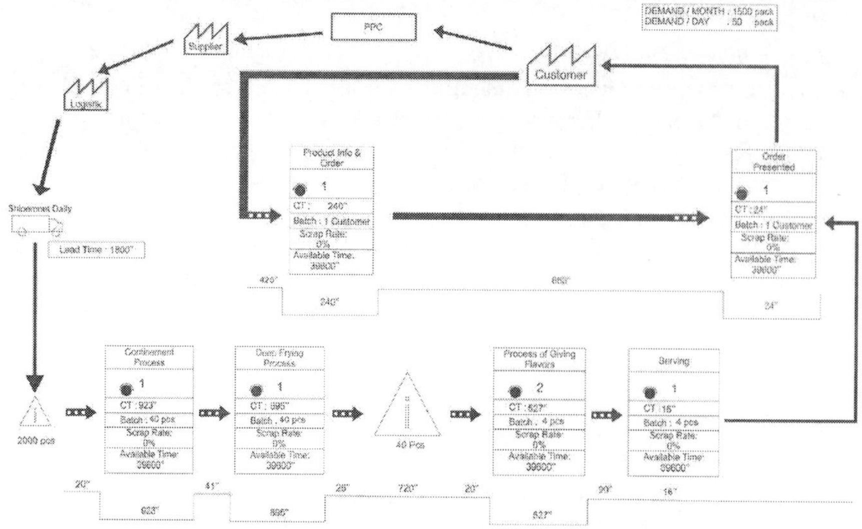

Figure 1. Current VSM of Fast Food Industry

The processing time is 742 seconds. Next is the frying process where in this process there are two types of activities, i.e. frying and draining by a worker within 895 seconds. When the chicken cooking process is complete, then work in process waits until there is a customer order.

The arrival of customer with an average of 4 people starts with the queue process to get product information and ordering services. Then customer will go to the cashier to make payments. The process of this customer arrival has a cycle time of 665 seconds. After the order process, the customer will wait for the order to be delivered. In this process, production activities at giving flavor station with cycle time of 527 seconds.

Based on the current VSM, customer waiting time is 10 minutes. The customer waiting time will cause a waiting time for the next order. To reduce customer waiting time, a process activities mapping (PAM) is used to identify the types of activities based on the categories of value added (VA), non-value added (NVA) and necessary nonvalue added (NNVA) activities. There are two types of PAM, i.e. production and service process. Based on current PAM in the production section, there are 70.7% of activities included as VA, 5% NVA and 23.9% NNVA. While in the service section, there are 25% of activities included as VA, 37.5% NVA and 37.5% NNVA.

The biggest NVA activity in the service process is the customer waiting time at 653 seconds. The production activities done in that time are the process of giving flavors and serving. That production activities time must be reduced to gain a better customer satisfaction. However, if the flavoring process takes over a longer time, the spices can absorb better even though the waste of time and energy becomes larger. There is a contradiction in this case, if one parameter is improved, the other parameters will decrease in performance. The TRIZ method can be used to solve these problems with three main stages, i.e.:

a. System parameters: There are two parameters identified from the fast food restaurant case, i.e.
   - Improving parameters: Result or Output Produced
   - Worsening parameters: Waste of Energy
b. Contradiction matrix: In the contradiction matrix for business and management [17], the identified system parameters have the following code, i.e.
   - Improving parameters: Result or Output Produced
   - Worsening parameters: Waste of Energy
c. Inventive Principles: There are three alternative solutions that can be implemented, i.e.
   - Prior action (10)
   - Parameter changes (35)
   - Strong Oxidants / enriched atmosphere / accelerated oxidants (38)

Based on the TRIZ method, the process improvement that can be done is to change the giving flavor process by pouring sauce over the chicken according to the standard (parameters changes) and the implementation of 5S in the work area (enriched atmosphere). With these improvements, the process of giving a flavor only take 48 seconds. So that the time reduction is 90%.

5. Conclusion

Based on the results of research on the integration of six sigma and TRIZ methods on lean service, it can be concluded that:

- The type of waste that occurs in the fast food industry is service delay. This waste is caused by the process of giving flavor with a time of 527 seconds.
- Improvement strategy to reduce waste that occurs in the fast food industry is to change the process (parameter changes) and implementation of 5S in the work area (enriched atmosphere). There is a time reduction of 90% by implementing the strategies.

6. Acknowledgment

Researchers thank to the support of Industrial Engineering Department, Faculty of Industrial Technology, Universitas Islam Indonesia

References

[1] Pangestu, M.E 2016 CEO talk the walk: the servicea sector as a driver of change: indonesia experience in the ASEAN context https://ugm.ac.id/id/berita/11510-industri.jasa.berpotensi.besar.terhadap.peningkatan.ekonomi.indonesia accessed on 15 April 2018.

[2] Findley, C 2016 CEO talk the walk: the servicea sector as a driver of change: indonesia experience in the ASEAN context https://ugm.ac.id/id/berita/11510-industri.jasa.berpotensi.besar.terhadap.peningkatan.ekonomi.indonesia accessed on 15 April 2018.

[3] Majalah SWA. 2016 Survei mastercard: restoran cepat saji masih favorit masyarakat Indonesia https://swa.co.id/swa/trends/business-research/survei-mastercard-restoran-cepat-saji-masih-menjadi-tempat-favorit-bagi-masyarakat-di-indonesia accessed on 15 April 2018.

[4] Martin, W 2007 Quality models: Selecting the best to deliver results The Physician Executive 33 pp. 24–31.

[5] Lopeza, E.A., Requenaa, I.G., Loberaa, A.S 2015 Lean service: reassessment of lean manufacturing for service activities. Procedia Engineering Vol. 132 pp. 23-30. doi: 10.1016/j.proeng.2015.12.463.

[6] Spagnol, G.S., Min, L.L., Newbold, D 2013 lean principles in healthcare: an overview of challenges and improvements Proceeding of 6th IFAC Conference on Management and Control of Production and Logistics September 11-13 Fortaleza Brazil pp. 229-234.

[7] D'Andreamatteoa, A., Legab, F., Sargiacomoa, M 2015 Lean in healthcare: a comprehensive review Health Policy Vol. 119 pp. 1197–1209 http://dx.doi.org/10.1016/j.healthpol.2015.02.002.

[8] Hicks, C., Mc Govern, T., Prior, G., Smith, I 2015 Applying lean principles to the design of healthcare facilities, Int. J. Production Economics Vol. 170 pp. 677–686 http://dx.doi.org/10.1016/j.ijpe.2015.05.029.

[9] Vlachos, I., Bogdanovic, A 2013 Lean thinking in the european hotel industry Tourism Management Vol. 36 pp. 354-363 http://dx.doi.org/10.1016/j.tourman.2012.10.007.

[10] Rauch, E., Damian, A., Holzner, P., & Matt, D. T 2016 Lean hospitality-application of lean management methods in the hotel sector Procedia CIRP 41 614–619 https://doi.org/10.1016/j.procir.2016.01.019

[11] Snee, R.D 2008 W. Edwards Deming's making another world a holistic approach to performance improvement and the role of statistics American Statistical Association Vol. 62 No. 3.
[12] Alpasa, F., Fitria, L 2014 Penerapan konsep lean service dan DMAIC untuk mengurangi waktu tunggu pelayanan Reka Integra Vol. 02 No.03 pp. 108-117.
[13] Suseno, A., Hartono, M., Surjani, R.M 2013 Aplikasi integrasi kansei engineering dan metode TRIZ pada layanan Villa Nunia, Bali. Calyptra Jurnal Ilmiah Mahasiswa Universitas Surabaya Vol.2 No.1 pp. 1-19.
[14] Mrugalska, B., Wyrwicka, M.K 2017 Towards lean production in industry 4.0 Procedia Engineering Vol. 182 pp. 466 – 473 doi: 10.1016/j.proeng.2017.03.135.
[15] Hines, P. and Taylor, D 2000 Going Lean Lean enterprise research centre Cardiff Business School Cardiff UK pp 3-43.
[16] Gaspersz, V 2007 Lean Six Sigma for Manufacturing and Service Industries Gramedia Pustaka Utama Jakarta.
[17] Yeoh, T.S 2014 Systematic Innovation in Business and Management Firstfruits Sdn. Bhd. Malaysia.

IOP Conf. Series: Materials Science and Engineering **722** (2020) 012045 doi:10.1088/1757-899X/722/1/012045

Performance Improvement in Aerospace Production Through Lean Manufacturing Implementation

A Az Zahra[1], S Indrawati[2], J Sulistio[3]
Industrial Engineering, Universitas Islam Indonesia, Yogyakarta, Indonesia

Corresponding author: sriindrawati@uii.ac.id

Abstract. The high demand in aerospace industry require on going improvement to meet its costumer. Prototyping is a form of effort to ensure the quality of airplanes. However, several problems rose in this phase but two of them are significant in disrupting the production system. They are the spare parts tardiness and unnecessary motion leading a high lead time. The objective of this article is to develop lean action plan to reduce lead time in airplane prototyping. The method applied in this paper is divided into three major phases. Firstly, evaluate current system performance using Value Stream Mapping (VSM). Secondly, determine the dominant waste by cooperating waste finding checklist. Finally, develop a lean action plan. Most waste occurred in production activities. Waste of motion is the highest among other waste with the score of 44 followed by waiting with the score of 24. The third highest waste is excess processing with the score of 2. The causes of these wastes are lack of facilities such as machines and tools and inefficient operators in picking and mixing sealant. The lean action plans are to add machines and tools for labelling, adding sealant room and to change the production plant lay out. By doing so, the lead time can be reduced up to 0,25 hours.

1. Introduction

Manufacturing industries involves wide spectrum of parts to its production process. It opens a possibility of occurring wastes which compromised lead times and at the end, the value to its costumer. Wastes are all activities requires time, space and resource but did not add any value to the product in costumer's perspective [1]. The purpose of this article is to develop lean action plan to reduce lead time. The object of this study is an airplane industry, particularly the prototyping process. It shows symptoms of waste happened in that process. Therefore, lean manufacturing is the proper approach to address that problems. It can be applied to identify wastes along the production stream. Lean manufacturing is powerful to eliminate those waste, improve quality and reduce lead time [2]. In addition, it can increase the output by eliminating any form of waste in production line [3]). Cooperating VSM in Lean Manufacturing implementation proofed to reduce production cost, shorten the lead time and increasing profit margin [4].

VSM is a proper tool to enhance the quality of information generating such as time, resources and information flows from parts to finish goods [5]. VSM and Value Stream Analysis Tools (VALSAT) manage to reduce waste in furniture industry by properly identify the significant waste which is inventory (19%) and determine the right tool to cooperate which is process activity mapping (PAM) [6]. VSM and Valsat also used in concrete industry [7]. Together VSM and Valsat powerful enough to identify wastes in ship manufacture industry and able to minimize the production time, not to mention improve production efficiency up to 91% [8].

IOP Conf. Series: Materials Science and Engineering **722** (2020) 012045 doi:10.1088/1757-899X/722/1/012045

2. Basic theory

2.1. Lean Manufacturing

Lean Manufacturing is an approach to help factory to identify and eliminate production waste, improve product quality and reduce production cost [9]. There are at least seven types of production waste:

- Waste of overproduction. In some occasion, some finished products couldn't make their way to costumer. Those products are called waste of overproduction. One of the reasons this waste occur is earlier production timeline or producing larger amount of product than costumer needs.
- Waste of motion. The movement of employees in working on products is a necessity. However, if there is a movement that does not provide added value to the product, it can be categorized as waste. Those movements include searching, selecting or stacking components, tools and so on.
- Transportation waste. In a well-designed system, workplaces and storage areas are close together so that material transfer is close. The equipment is placed where it is used, and material is transferred to the process as needed.
- Processing waste. Nonvalue added processes must be eliminated. In some cases, changes in product design often cause a reduction in some parts of the final product. Waste of processing can take the form of unnecessary processes or carry out inefficient processing.
- Waste time. This type of waste can be divided into two groups, namely waiting time and queuing time. Waiting time occurs when a part has finished, but other parts that will be assembled with it have not been completed. Queuing time occurs when a part has finished, but cannot continue because the next machine is still doing another job
- Defective product. This waste arises from producing defective products or components or requires repair. Repairing or reworking, scraping, producing replacement goods, and inspections, means that additional handling, time and effort are considered as waste.
- Unnecessary Inventory. This waste arises from excessive inventory. Expenditures due to waste include warehouse costs, costs due to obsolete products, and damaged products.

2.2. Waste Finding Checklist

Waste finding checklist is used to measure waste production in each production activity [10]. The measurement process is carried out using a form as shown in Figure 1.

No	Process/ work cell	Defect	Overproduction	Waiting	Not utilizing employees	Transportation	Inventory	Motion	Excess processing	Waste magnitude total	Improvement ranking
1											
n											

Figure 1. Waste finding checklist form

3. Research method

This article study one of the airplane parts which called elevator and use its 2017 production data. VSM applied to identify inefficiencies within manufacturing industry, specifically in production process [9]. VSM is lean manufacturing tool to analyse the future state of a production process. Beside VSM, this study also applied Waste Finding Checklist (WFC) to identify wastes. After wastes are identified, Fishbone Diagram is applied to find its root cause. Subsequently, lean action plan developed using 5W+1H.

4. Result and discussion

Value Stream Mapping (VSM) is used to determine the company's production floor's initial conditions. The elevator production process begins with the Tail Unit Assy N219 stage and ends with the final inspection stage as shown in Figure 2. In total there are 37 activities that occur during the elevator production process where the cycle time is 46,040 hours and lead time of 60,918 hours thus creating waiting time for 2 days 12 hours. In addition, the takt time is 14 days/elevator or 89.6 hours. This shows that takt time is below the cycle time so that it can be ascertained that the production is running slow.

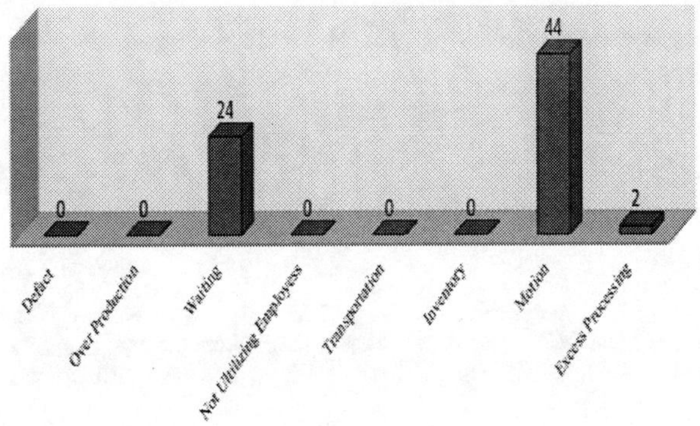

Figure 2. Waste in Aerospace Production

The overall processes lead to 2 days and 12 hours of waiting time. In respect to that time, the next step is to group the activities into Value Added (VA), Non-Value Added (NVA) and Necessary Non-Value Added (NNVA). Hence, the cause of waiting times can be identified. The results are as follow; Value Added is 44,520 hours; Non-Value Added is 0,418 hours; and Necessary Non-Value Added is 49,520 hours. Waste Finding Checklist (WFC) is cooperated to find the wastes in the elevator production. A series of discussion with experts are performed to fill the WFC filled. From the WFC the waste of motion has the highest value of 44, then waste of waiting is in the second place with a score of 24. The complete result is shown in Figure 3. Thus, the waste motion will be elaborated on the cause of the problem using fishbone diagram.

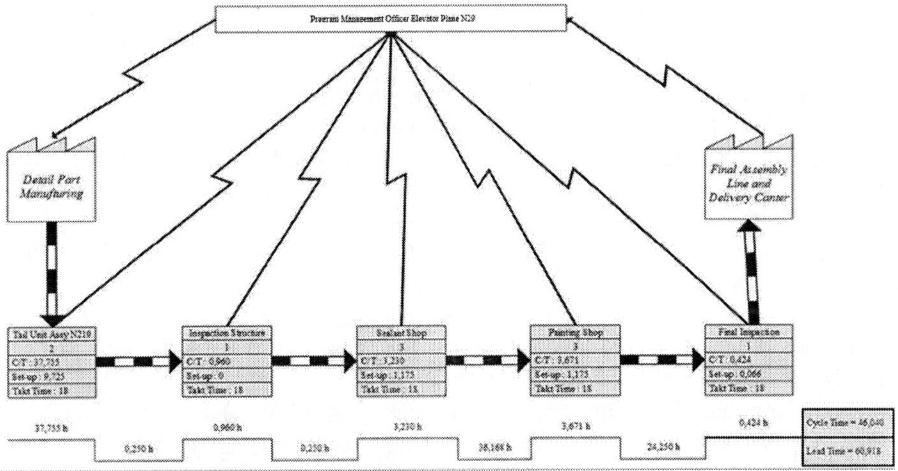

Figure 3. Current Value Stream Mapping

After all types of wastes are identified, Fishbone Diagram is applied to determine the root cause of the problem. Waste motion caused by 18 production process activities which caused by lots of unnecessary movements or excess movement which can cause additional time in the production process. Improvement analysis focus on waste of motion in several production process activities. The result is a program to change the layout to make JIG elevator closer. Another program it to add sealant space to the elevator production process location so that the operator does not have to bring the sealant too far from the elevator production process room. The last program is to restrict the operators to take and mix sealant. In addition, the labelling of the machine or tools is done in order to facilitate the operator in working. Based on those improvements' program, it can reduce cycle time and lead time. Figure 4 shows the Future Value Stream Mapping in accordance to those program.

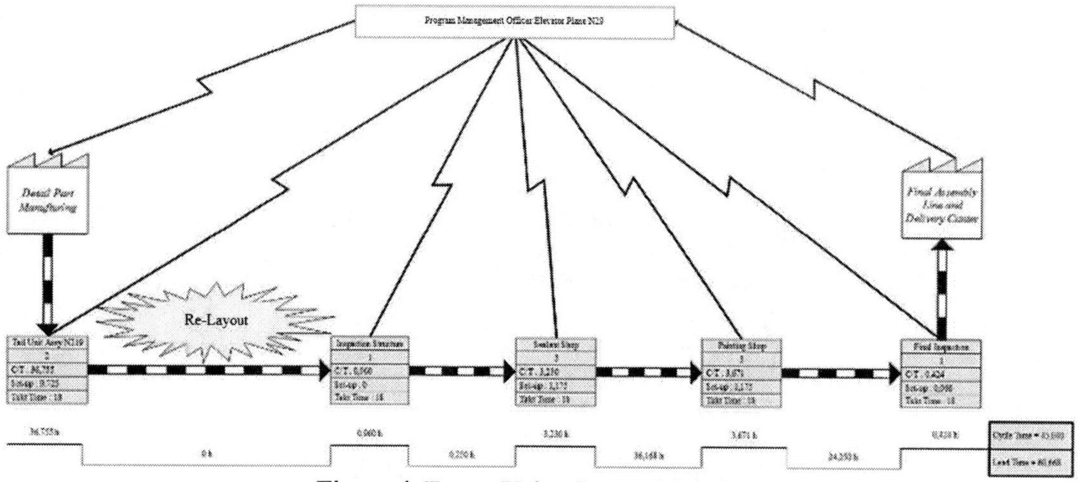

Figure 4. Future Value Stream Mapping

5. Conclusion

The results of WFC show that the greatest waste is waste of motion with a value of 44, followed by waste waiting with a value of 24. The activities that experience the highest waste and therefor the focus of improvement are activities that have a sealant process. The improvement programs are to add machines and/tools, change lay out, add sealant room in production process, and restriction for the operators. It reduces cycle time to 45,040 hours and lead time to 60,668 hours.

References

[1] Charron, R., Harrington, H., Voehl, F., Wiggin, H. 2015 The Lean Management System Handbook. U.S: CRC Press, Taylor & Francis Group.

[2] Abdulmalek, F., Jayant, R. 2007 Analyzing the benefits of lean manufacturing and value stream mapping via simulation: a process sector case study International Journal Production Economic 107 pp 223-236.

[3] Gaspersz, V. Lean Six Sigma For Manufacturing and Service Industries. Jakarta: PT. Gramedia Pustaka Utama, 2011.

[4] Jakfar, A., Setiawan, W., Masudin, I. 2015 Reducing waste dengan menggunakan pendekatan lean manufacturing Spektrum Industri Vol. 13 No. 1 pp. 1-114.

[5] Patel, N., Chauhan, N., Trivedi, P. 2015 Benefits of value stream mapping as a lean tool implementation manufacturing industries: a review IJIRST Vol.1 Issue 8.

[6] H Arif, D., Vanany, I 2016 Implementasi lean manufacturing dengan metode value stream mapping pada PT X.

[7] Setyastuti, Y., Dewi, S., Suharyanto, A. 2017 Peningkatan produktivitas pada proses produksi pracetak dengan penerapan metode lean construction untuk eliminasi waste Rekayasa Sipil, Vol. 11 No. 3.
[8] Tebiary, A., Suastika, I., Ma'ruf, B. 2017 Analisis non value added activity pada proses produksi kapal dengan pendekatan value stream mapping Jurnal Wave Vol. 11 No. 1 pp. 22-23.
[9] Salunke, S., Hebbar 2014 Value stream mapping: a continuous improvment tool for reduction in total lead time. International Journal of Current Engineering Vol.2 pp 931-934.
[10] Gaspersz, V 2007 Lean Six Sigma For Manufacturing And Service Industries Jakarta: PT. Gramedia Pustaka Utama.

Acknowledgments
Researchers thank to the support of Industrial Engineering Department, Faculty of Industrial Technology, Universitas Islam Indonesia.

IOP Conf. Series: Materials Science and Engineering **722** (2020) 012046 doi:10.1088/1757-899X/722/1/012046

Production Capacity Optimization with Rough Cut Capacity Planning (RCCP)

M Sugarindra, and R Nurdiansyah

Department of Industrial Engineering, Universitas Islam Indonesia, Yogyakarta, Indonesia

Corresponding author: sugarindra@uii.ac.id

Abstract. One of the factors supporting the company's success in achieving its goals is the ability to meet demand and provide satisfaction to customers. A musical instrument industry makes products with a make to order system, so that it requires the accuracy and smoothness of the production process to meet the demand in accordance with the specified time. Based on data on average total daily production of Spray and Sanding Balikan Factory 2, it known that there is still often an imbalance between the actual productions of the planned production at regular work time. This study aims to analyse and evaluate the capacity of production resources and facilities. The Rough Cut Capacity Planning (RCPP) method is a method that can show a comparison between the actual capacity available and the capacity needed. Based on the research results, workload rearrangement must be carried out to distribute labor and increase work hours 90 minutes and 120 minutes so that demand can be fulfilled.

1. Introduction

One of the factors supporting the success of a company in achieving its objectives is the ability to meet demand and provide satisfaction to customers. Successful planning and manufacturing control requires effective capacity planning to be able to meet the established production schedule [1]. The industry that is the object of research is a musical instrument industry that produces Upright Piano (UP) and Grand Piano (GP) pianos that apply make to order production systems, so that it requires the accuracy and smoothness of the production process to meet demand in accordance with the specified time.

In general, it can be described that the piano production process is wood working which is responsible for processing the abrasive material into cabinet - painting, which has the duty to provide coloring to the cabinet and assembly, which is to assemble the cabinet into a piano music instrument. Not achieving the production target in the painting department will have a major effect on the delay in sending the piano to the customer. One of the work stations under the painting department is the Spray and Sanding Balikan Factory 2. Based on the average total daily production for the last 6 months, it is known that there is still an imbalance between the actual production and production plan at regular work time. The imbalance between the actual production and the production plan causes the Spray and Sanding Balikan Factory 2 work stations to increase daily work hours (over time) and accept process delegation for some cabinets from other work processes, causing overload.

Production capacity planning needs to be balanced with available production capabilities [2], booth human resource capacity [3] [4] [5] [6] and the machinery capacity used [7] [8]. In dealing with these problems, integration between the production plan and the capacity plan needs to be done [9].

IOP Conf. Series: Materials Science and Engineering **722** (2020) 012046 doi:10.1088/1757-899X/722/1/012046

Therefore, it is necessary to conduct an analysis and evaluation of the capacity of production resources and facilities available in the Spray and Sanding Balikan Factory 2 production line so that they are suitable and can support the planned master production schedule. To analyze these problems the Rough Cut Capacity Planning (RCCP) method is applied. RCCP is an analysis to test the availability of production facility capacity to meet the established master production schedule [10]. RCCP is second in the priority-capacity planning hierarchy that plays a role in developing Master Production Schedule (MPS). RCCP validates the MPS that has been made.
RCCP method can show a comparison between the actual available capacity and the required capacity, so that the company is able to make the right production capacity planning to meet consumer demand. RCCP method can also be used to determine whether a production schedule requires overtime work, sub-contracts, as well as the addition of facilities or labor to meet demands on time.

This study aims to balance between the actual productions capacities available against the production targets at the Spray and Sanding Balikan Factory 2 work stations so that the spray department is able to meet the needs according to the production target.

2. Method

The research object is a music industry with the main material are wood. Focus of this research is to analyze the balance between the actual production capability of the company's production targets in the coming period in the "Spray Working Group" and "Sanding Balikan" Department of Painting Factory 2.

Primary data obtained in the field which includes direct observations and interviews, data Secondary data is obtained from company records to conduct capacity planning, including: data on total demand, work center cycle time, number of days and hours of work, number of workers, and others. Secondary data is also taken from the research literacy of various parties that have a correlation with the.

Figure 1. Step of research

3. Result and Discussion

3.1. *Production process*

Research focuses on the spray and sanding section of the Painting Factory 2 department. The production process in this section is generally as follows:

- Spray Undercoat, serves as a base coating so that the main paint can adhere properly
- Sanding Balikan is the process of sanding on the remnants of paint. The sanding and reverse process is carried out on the part after going through the process of painting. The process of reversing sanding goes through three stages: sanding machine belt sander on the surface, free hand sander on the rib part, and cleaning.
- Topcoat Spray is the main paint for piano parts that use a type of enamel paint.
- Waiting Room, is the process of drying freshly painted parts that lasts for 1.5 hours.
- Seasoning Room is a drying process after the waiting room process with a temperature of 40^0 C and lasts 2 hours.

3.2. Work Groups & Time Working

Groups are divided based on the similarity of tasks and workload on each operator. The list of working groups and direct operators in the Spray section is presented in the following Table 1:

Table 1. List of work groups and operators

No	Work Group	Operator
1	Painting Booth 1	Operator 1 - 4
2	Painting Booth 2	Operator 5
3	Painting Booth 3	Operator 6
4	Painting Booth 4	Operator 7 - 9
5	Painting Booth 5	Operator 10 - 13
6	Painting Booth 6	Operator 14
7	Painting Booth 7	Operator 15

Time hours in industry is Monday to Friday with daily work schedules 07.00 - 16.00 with a total rest period of 60 minutes. The number of normal working hours is assumed to be the same as 460 minutes. The data is used as input in calculating RCCP.

3.3. Cycle Time & Allowance

Data on standard time for piano products:

Table 2. Working cycle time data on piano cabinet each working group

No	Work Group	Total time (Minute)
1	Painting Booth 1	120,39
2	Painting Booth 2	117,35
3	Painting Booth 3	21,95
4	Painting Booth 4	673,14
5	Painting Booth 5	1240,78
6	Painting Booth 6	4,90
7	Painting Booth 7	1623,93

The standard time data in table 2 are the accumulation of standard time for working the piano cabinets which are the responsibility of each working group.

Table 3. Data of work sampling in working group spray

No	Work Group	Work Margin (%)
1	Painting Booth 1	15%
2	Painting Booth 2	28%
3	Painting Booth 3	20%
4	Painting Booth 4	37%
5	Painting Booth 5	17%
6	Painting Booth 6	30%
7	Painting Booth 7	30%

IOP Conf. Series: Materials Science and Engineering **722** (2020) 012046 doi:10.1088/1757-899X/722/1/012046

Allowances are all activities that are outside the main work activity group. The allowance level is represented in the form of a percentage ratio of the work sampling observations.

3.4. Capacity

Data taken is the demand data for piano products in the "period 195" which began in April 2018 until March 2019, which is 2939 consisting of various types of cabinets. Data on demand for piano products in the 195 period will then be used as input to predict the daily production of pianos in the coming period that is "period 196".

Data processing is done forecasting to predict demand for piano products in the period 196 (April 2019-March 2020) with product demand data in the period 195 (April 2018-March 2019) as initial input. Based on the results of the calculation of the three methods (time series analysis is Moving Average, Single Exponential Smoothing, and Liner Regression) forecasting used is known that forecasting with the dominant linear regression method produces the smallest forecasting error value. Based on these results, the forecasting method chosen is forecasting using linear regression method. By using demand data for selected forecasting products for the period 196 (April 2019-March 2020), the required capacity is calculated for each working group.

Capacity available for spray working groups in the Spray and Sanding Balikan Factory 2 section, capacity requirement and count of comparison between both for each work group are presented in the following table 4:

Table 4. Data of work sampling in working group spray

No	Work Group	Capacity Available (Hour/Day)	Capacity Requirement (Hour/Day)	± Capacity (Hour/Day)
1	Painting Booth 1	26,07	40,21	-14,14
2	Painting Booth 2	5,52	1,92	+3,60
3	Painting Booth 3	6,13	7,48	-1,35
4	Painting Booth 4	14,49	5,43	+9,06
5	Painting Booth 5	25,45	20,72	+4,73
6	Painting Booth 6	5,37	3,91	+1,46
7	Painting Booth 7	5,37	3,40	+1,97

Based on data in table 4, can be seen that there is an imbalance in capacity. The capacity needed, the working group Painting Booth 1 experienced a shortage of 14.14 hours per day, and Painting Booth 3 at 1.35 hours per day.

Operator workload is presented in Figure 2. Based on workload calculations, it can be seen that the operator workload is uneven, seen in operators 1–4 (Painting Booth 1) having the highest workload. To make this workload evenly needed distribution of workload to other operators who have a category and similarity of work processes. The working group that has similar expertise and work process is operator 5 (Painting Booth 2). Time requirements for operators 7-9 (Painting Booth 4).

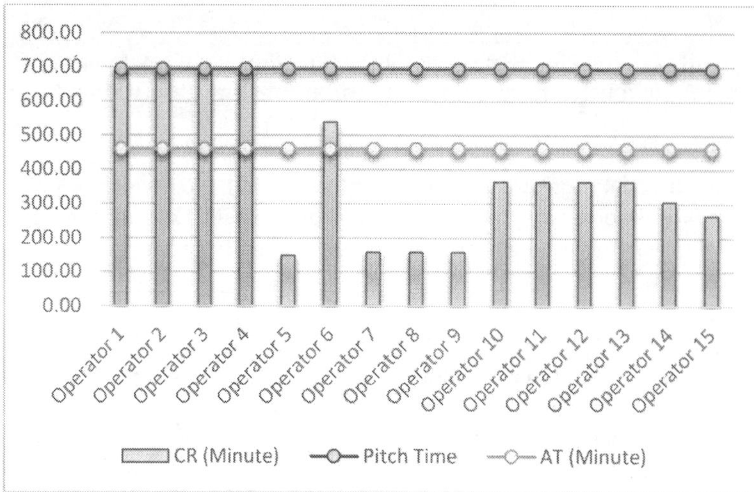

Figure 2. Operator Workloads in Spray Working Groups

1 operator on painting booth 4 will be moved to painting booth 2 to help work on painting booth 2. With this scenario the division of workload on each operator in the Spray Booth 1 working group will experience changes as shown in the following table 5.

Table 5. Data of after addition overtime

No	Work Group	Man	Capacity Available (Hour/Day)	Capacity Requirement (Hour/Day)	± Capacity (Hour/Day)
1	Painting Booth 1	4	26,07	30,33	-4,27
2	Painting Booth 2	2	11,04	10,99	+0,05
3	Painting Booth 3	1	6,13	7,48	-1,35
4	Painting Booth 4	2	9,66	3,69	+3,69
5	Painting Booth 5	4	25,45	20,72	+4,73
6	Painting Booth 6	1	5,37	3,91	+1,46
7	Painting Booth 7	1	5,37	3,40	+1,97

By distributing workloads, the results obtained there are still capacity shortages in the Painting Booth 1 and Painting Booth 3 work groups as much as 4.27 hours / day and 1.35 hours / day. Workload distribution in Painting Booth 1 and Painting Booth 3 can no longer be done, so the other strategy is by increasing are work hours. The data with the scenario of adding work hours is shown in the following table 6.

Table 6. Overtime

No	Work Group	Man	Capacity Available (Hour/Day)	Capacity Requirement (Hour/Day)	± Capacity (Hour/Day)
1	Painting Booth 1	4	1,5	-4,27	+0,83
2	Painting Booth 2	1	2,0	-1,35	+0,25

Based on the data in table 6 the capacities in PB 1 and PB 3 have been fulfilled, so the comparison between the available capacity and the required capacity after all the proposed improvements is as follows on table 7.

Table 7. Data after addition overtime

No	Work Group	Man	Capacity Available (Hour/Day)	Capacity Requirement (Hour/Day)	± Capacity (Hour/Day)
1	Painting Booth 1	4	31,17	30,33	+0,83
2	Painting Booth 2	2	11,50	10,99	+0,51
3	Painting Booth 3	1	7,73	7,48	+0,25
4	Painting Booth 4	2	9,66	5,97	+3,69
5	Painting Booth 5	4	25,45	20,72	+4,73
6	Painting Booth 6	1	5,37	3,91	+1,46
7	Painting Booth 7	1	5,37	3,40	+1,97

By redistributing excess workloads, moving operators from one group to another, and adding overtime hours of 90 minutes and 120 minutes each day, all of the existing capacity deficiencies, booth in Painting Booth 1 and Painting Booth 3 can be overcome.

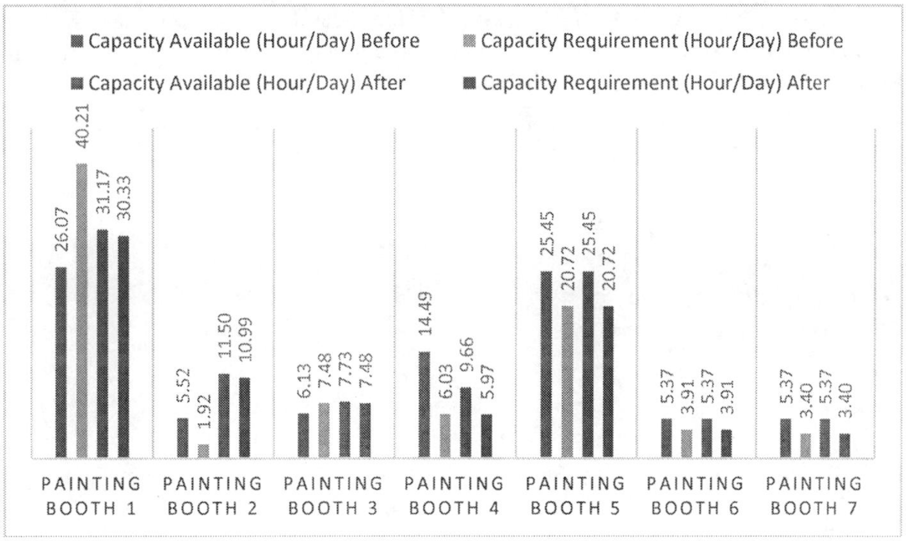

Figure 3. Capacity comparison after repair.

4. Conclusion

Comparative analysis of available production capacity against the capacity needed for the level of demand in the period 196 (April 2019-March 2020) there is still an imbalance in capacity. Based on the results of the analysis, it is necessary to optimize the capacity planning in each working group by managing and increasing the available capacity.

In order to balance between the actual production capacity and the planned production target in the period 196 (April 2019-March 2020) redistributing loads and overtime are redistributed.

References

[1] Gasperz, V., 1988. Production Planning and Inventory Control. Jakarta: Gramedia Pustaka Utama
[2] Aji, D. K., 2014. Perencanaan Kapasitas Produksi Untuk Memenuhi Permintaan Konsumen Dengan Menggunakan Metode Rough Cut Capacity Planning (RCCP), Semarang: Teknik Industri, Fakultas Teknik, Universitas Dian Nuswantoro.
[3] Kaouthar, C., Robert, P., Pierre, B. & Alain, H., 2015. A Time Driven RCCP Model With Two Levels Of Planning and A Reactive Planning Approach for Tactical Project Planning. Procedia Computer Science, Issue 64, pp. 257-264.
[4] Sirait, M. E., Sinulingga, S. & Ishak, A., 2013. Perencanaan Kebutuhan Kapasitas (Rough Cut Capacity Planning) Industri Pengolahan Peralatan Rumah Tangga Di PT. X. e- Jurnal Teknik Industri FT USU, 2(2), pp. 28-34.
[5] Hudori, M., 2016. Cable Clamp Production Capacity Planning Using Rough Cut Capacity Planning (RCCP) Method (A Case Study In PT Fajar Cahaya Cemerlang). Bekasi, Citra Widya Edukasi Polytechnic of Palm Oil, pp. 87-93.
[6] Baydoun, G. et al., 2016. A Rough-Cut Capacity Planning Model With Overlapping. OR Spectrum, Volume 38, pp. 335-364.
[7] Faisal, M. M. R., Suryadhini, P. P. & Juliani, W., 2016. Forecasting dan Analisis Perencanaan Kapasitas Produksi dengan Bill Of Labor Approach Pada Proyek Engine CT7 PT. XYZ. Bandung, e-Proceeding of Engineering.
[8] Kurniasih, D., Tarliah, D. T. & Suri, P. S., 2015. Perencanaan Kapasitas Produksi Untuk Box Kemasan Aluminium Foill, Bandung: Universitas Pasundan
[9] Chen, J. C., Chen, T.-L. & Harianto, H., 2016. Capacity Planning For Packaging Industry. Journal of Manufacturing Systems, Issue 42, pp. 153-169
[10] Christianty, E. M. A., Setyanto, N. W. & Hamdala, I., 2014. Optimasi Kapasitas Produksi Dalam Penyusunan Jadwal Induk Produksi Menggunakan Integer Linear Programming (ILP). Jurnal Rekayasa dan Manajemen Sistem Industri , 2(6), pp. 1147-115

IOP Conf. Series: Materials Science and Engineering **722** (2020) 012047 doi:10.1088/1757-899X/722/1/012047

Escalation proposal of supply chain management performance of oil and gas upstream industry in PT. XYZ by using SCOR model approach

T Immawan, and I Nugraha

[1,2]Faculty of Industrial Technology, Department of Industrial Engineering, Universitas Islam Indonesia, Yogyakarta, Indonesia

Corresponding author: 985220101@uii.ac.id

Abstract. Oil and gas industry in Indonesia is one of the industries with the largest contribution for country's foreign exchange. As recorded in 2014, 12% of country's budget is from oil and gas industry and able to provide a significant increase in national economic growth. But in the other hand, supply chain performance in this industry shows inconsistencies in each year, especially in PT. XYZ. As one of the largest oil and gas companies in East Kalimantan, Indonesia. As the result, company's targets are missed oftentimes and have not been able to achieve competitive advantage. Therefore, the purpose of this study is to propose performance improvements in the supply chain of PT. XYZ. This study uses the model of Supply Chain Operations Reference (SCOR) 11.0 which focuses on the attributes of reliability, Analytical Hierarchy Process (AHP), and fishbone in collecting and processing research data. The result shows that the supply chain performance of PT. XYZ. which focused on the reliability attribute is categorized as good, starting from 2012 to 2017. There is 1 indicator that need to be improved and get further control, that is the indicator of percentage of supplier meeting environmental metrics/criteria and fill rate. One means to overcome the problem is by using big data analysis approach, because the approach can analyze big number of data in the Company, whether it's structured, semi-structured, or unstructured.

1. Introduction

In Indonesia, oil and gas industry is one of the industries with the largest contribution for country's foreign exchange. As recorded in 2014, 12% of country's budget was sourced from oil and gas industry and able to provide a significant increase in national economic growth [1]. With rapid changes and the increasing of complexity in business processes, the industry needs to be guided in order to quickly and accurately provide the final product according to customer needs. The supply chain of this industry is relatively long and involves a lot of other companies from upstream to downstream. As a result, inconsistencies in the performance of this industry continue to take place and has not been able to show significant changes until now. Therefore, to overcome these problems requires a strategic supply chain approach from upstream to downstream involving all stakeholders as a whole [2]. According to Basta and Liyanage [3], a sustainable approach is very much needed in improving the supply chain performance of an industry. The approaches are by assessing and measuring supply chain performance. It has become important in the management process, that is when supply chain performance needs to be improved then performance assessment process is carried out [4].

IOP Conf. Series: Materials Science and Engineering **722** (2020) 012047 doi:10.1088/1757-899X/722/1/012047

Supply chain management (SCM) is a series of business activities from upstream to downstream, raw materials procurement, manufacturing processes, distribution activities, consumer service, as well as recycling the final products [5]. Moreover, SCM is an integrated process with the objectives and risks are mutually borne in creating long-term cooperative relationship between suppliers, intermediaries, manufacturing companies, and customers [6]. There have been many contributions from academics and practitioners related to assessment and strategy in improving supply chain performance in recent years [5]; [7]. In the other words, to achieve competitive advantage in improving supply chain performance, the company needs a strategic and effective approach or method to overcome inconsistencies problem in the performance of company's supply chain.

Supply chain performance appraisal is a process that involves some or all of the companies engaged without exception [8]. To obtain maximum results from the process, a systematic approach is needed through proven concepts and performance standards which then being compared to company's business processes and internal organization [7]. For this reason, this research uses Supply Chain Operations Reference (SCOR) model in the process of assessing and proposing for performance improvement of upstream oil and gas industry supply chain in PT. XYZ. In the concept of SCOR Model, there are six main attributes of business process in assessing and improving, i.e: plan, source, make, deliver, return and enable. These attributes have covered a series of company activities and been recognized in most industries [9]; [10]; [11]. In this study, the proposed improvement in supply chain performance is focused on the reliability attributes of the SCOR model used. Due to the restricted time and access to the company, the scope of the research is being limited. With this research, it is expected to be able to give alternative solutions to fix the performance of company's supply chain as well as the failure in the process of drilling and production during 2012 to 2017 which happened four times and resulted in huge losses of company profits. Thus, it can help that the failures will not occur in the coming years.

2. Research Method

This research was conducted in PT. XYZ. as one of upstream oil and gas industries. The company is located in East Kalimantan, Indonesia. It has Oil and Gas Regulatory Authority (BPMIGAS) license for oil and gas drilling in East Kalimantan. Moreover, the focus of this study is how to improve the performance of company's supply chain through assessment process and proposed improvements by using Supply Chain Operations Reference (SCOR) 11.0 model. SCOR 11.0 is the result of Supply Chain Councils development in performance assessment of company's supply chain. In this study, reliability attributes became the focus or research indicator. It measured things related to supply chain performance in obtaining the right product, location, time, conditions, packaging, quantity, documentation, as well as the right customer. Supply chain performance reliability is company's ability in performing the work according to expectations [9]. In addition, the process of weighting the value of performance indicators was done using Analytical Hierarchy Process (AHP) method [12]; [13]. AHP was started by defining the problem, preceded by identifying the objectives to be achieved, then comparing the components by pairwise method, in regard to the criteria and final structure in the hierarchy arranged by logical and organized grouping in representing the problem [13]. In the process of identifying the cause of errors occurred, fishbone diagram was used.

3. Result and Discussion

3.1. Supply chain management reliability performance of PT. XYZ.

After indicators mapping were conducted, it was validated by company's expert using SCOR 11.0 model. As the result, 17 indicators were obtained. Traffic light system was used to identify each performance indicator where Red meant that the indicator value was below 60 percent (<60%), Yellow meant that the value was between 60 to 80 percent 60% -80%), and Green meant above 80 percent (> 80%). These values are the standards given by the object company. Overall supply chain management performance of PT. XYZ. showed good results. There was only 1 indicator that showed the percentage value below 80% (<80).

The following is the results of a complete calculation of supply chain performance using SCOR model as in table 1 as follows.

Table 1. Calculation results of actual value of performance

Operation Process	Code	Indicators	Performance					
			2012	2013	2014	2015	2016	2017
Plan	RL.3.37	Forecast Accuracy	100%	100%	100%	100%	100%	100%
	RL.3.18	% Orders/ Lines Processed Complete	95%	97%	94%	96%	96%	97%
	RL.3.19	% Orders/ Lines Received Defect Free	98%	99%	95%	98%	97%	99%
Source	RL.3.20	% Orders/ Lines Received On-Time To Demand Requirement	94%	92%	95%	92%	90%	97%
	RL.3.23	% Orders/ Lines Received with Correct Shipping Documents	99%	100%	100%	98%	100%	100%
	RL.3.27	% Schedules Changed With Supplier's Lead Time	90%	87%	91%	85%	88%	92%
	RL.3.31	Compliance Documentation Accuracy	100%	100%	100%	100%	100%	100%
Make	RL.3.36	Fill Rate	101%	93%	101%	105%	102%	102%
	RL.3.49	Schedule Achievement	100%	100%	100%	100%	100%	100%
	RL.2.3	Documentation Accuracy	97%	98%	95%	94%	98%	98%
	RL.3.11	% of Faultless Invoices	100%	100%	100%	100%	100%	100%
	RL.3.16	% of suppliers meeting environmental metrics/criteria	60%	75%	57%	67%	53%	69%
Deliver	RL.3.31	Compliance Documentation Accuracy	100%	100%	100%	100%	100%	100%
	RL.3.35	Delivery Quantity Accuracy	100%	100%	100%	100%	100%	100%
	RL.3.36	Fill Rate	100%	100%	100%	100%	100%	100%
	RL.3.50	Shipping Documentation Accuracy	100%	100%	100%	100%	100%	100%
Enable	RL.3.37	Forecast Accuracy	100%	100%	99%	100%	99%	98%

3.2. *Weighting of performance indicators*

The weighting of performance indicators was done to determine the level of importance of each performance indicator, because each performance indicator had a different level of importance. Weighting was done by using the analytical hierarchy process (AHP) method. Based on the results as presented in table 2, the most important indicator from Make variable was Fill Rate indicator.

Table 2. Weighting result of performance indicator

Operation Process	AHP (Weight)	Code	Indicator	AHP (Weight)
Plan	0.161	RL. 3.37	Forecast Accuracy	1.00
Source	0.089	RL. 3.18	% Orders/ Lines Processed Complete	0.09
		RL. 3.19	% Orders/ Lines Received Defect Free	0.46
		RL. 3.20	% Orders/ Lines Received On-Time To Demand Requirement	0.24
		RL. 3.23	% Orders/ Lines Received with Correct Shipping Documents	0.07
		RL. 3.27	% Schedules Changed With Supplier's Lead Time	0.14

Operation Process	AHP (Weight)	Code	Indicator	AHP (Weight)
Make	0.579	RL. 3.31	Compliance Documentation Accuracy	0.05
		RL. 3.36	Fill Rate	0.75
		RL. 3.49	Schedule Achievement	0.20
Deliver	0.143	RL. 2.3	Documentation Accuracy	0.06
		RL. 3.11	% of Faultless Invoices	0.10
		RL. 3.16	% of suppliers meeting environmental metrics/criteria	0.05
		RL. 3.31	Compliance Documentation Accuracy	0.12
		RL. 3.35	Delivery Quantity Accuracy	0.30
		RL. 3.36	Fill Rate	0.23
		RL. 3.50	Shipping Documentation Accuracy	0.13
Enable	0.028	RL. 3.37	Forecast Accuracy	1.00

3.3. Fishbone diagram

According to Table 1, there was only one indicator that needed to be improved in order to escalate supply chain reliability performance. However, because *Fill Rate* was the most important indicators according to experts, therefore needed to control and maintain the performance. Fishbone diagram was used to find the cause of errors and get the best way to improve the performance. It is shown on figure 1 & 2 below:

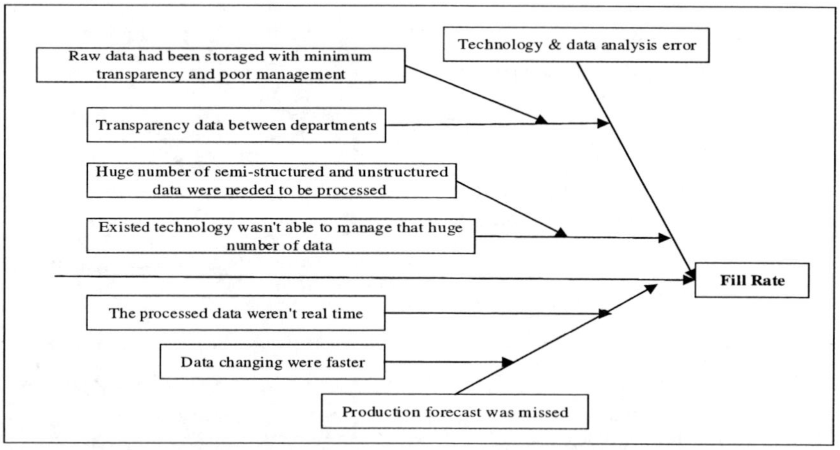

Figure 1. Fishbone Diagram of Fill Rate

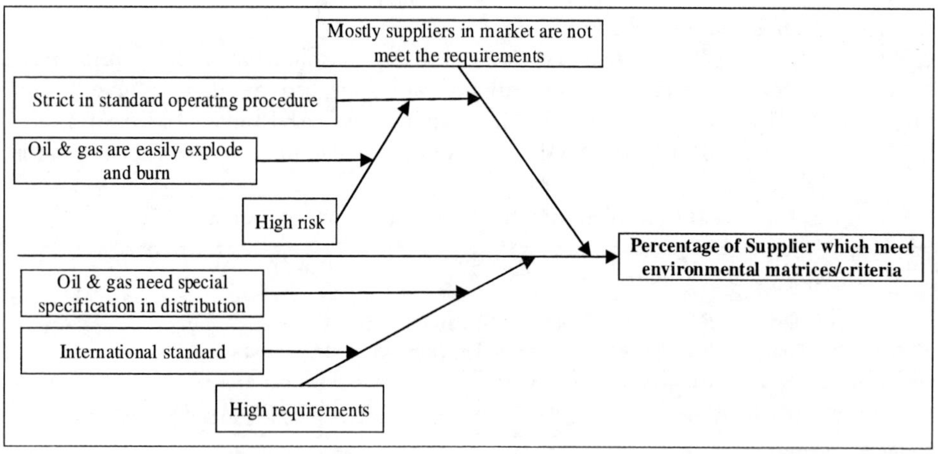

Figure 2. Fishbone diagram of supplier meeting environmental matrics/criteria percentage

4. Conclusions

Based on the results of processing and analysis that had been done could be concluded that the performance of company's supply chain in PT. XYZ. which focused on reliability attributes with the SCOR method was categorized as good, started from 2012 to 2017. There were two indicators which needed to be fixed and improved; suppliers that meet the environmental metrics / criteria, and fill rates. In addition by using fishbone diagram, the obtained proposal for improved indicators was to maintain relationships with suppliers that had been established and always supported the business, especially because in the market, such suppliers could be considered as rare and hard to find. For *Fill Rate* indicators in particular, a way to overcome the problem was by using Big Data analysis because it could analyze a big number of data, whether they were well structured, semi-structured, and unstructured data. Moreover, Big Data was able identify the root of the problems so that it could provide good planning and forecasting [14]. Furthermore, alternative pricing scenarios could be done instantly, thus reduced inventory and increased profit margins.

The researchers suggest that companies and similar industries should pay more attention to the elements in the company's overall supply chain, so that common goals and common targets as well as the balance in achieving competitive advantage can be effectively achieved. In addition, for further researchers are expected to include all attributes exist in the SCOR model for assessments and proposed improvement supply chain performance in that companies or other similar companies.

References

[1] PwC Oil and Gas in Indonesia 2014 *Investment and Taxation Guide.* Available at: http://www.pwc.com/id/en/publications/assets/oil_and_gas_guide_2014.pdf
[2] Maestrini V, Luzzini D, Maccarrone P and Caniato F 2017 *Int. J. Prod. Econ.* **183** 299-315.
[3] Bastas A and Liyanage K 2018 *J. Clean. Prod.* **181** 726-744.
[4] Nudurupati S S, Bititci U S, Kumar V and Chan F T S 2011 *Comput. Ind. Eng.* **60** 279-290.
[5] Balfaqih H, Nopiah Z M, Saibani N and Al-Nory M T 2016 *Comput. Ind.* **82** 135-150.
[6] Sweeney E 2011 *Int. J. Appl. Logist.* **2** 30-48.
[7] Mishra D, Gunasekaran A, Papadopoulos T and Dubey R 2018 *Benchmarking: an Int. J.* **25** 932-967.
[8] Estampe D, Lamouri S, Paris J and Brahim-Djelloul S 2013 *Int. J. Prod. Econ.* **142** 247-258.
[9] [SCC] Supply Chain Council 2012 Supply Chain Operations Reference Model Supply Chain Council 11.0.
[10] Lima-Junior F R and Carpinetti L C R 2016 *Int. J. Prod. Econ.* **174** 128-141.
[11] Lima-Junior F R and Carpinetti L C R 2019 *Int. J. Prod. Econ.* **212** 19-38.
[12] Sellitto M A, Pereira G M, Borchardt M, Silva R and Viegas C V 2015 *Int. J. Prod. Res.* **53** 4917-4926.
[13] Bukhori I B, Widodo K H and Ismoyowati D 2015 *Agr. Agr. Sci. Proc.* **3** 221-225.
[14] Shi Z and Wang G 2018 *J. Hig. Tech. Manag. Res.* **29** 141-150.

The important level of washing machine quality dimensions in 4.0 industrial era based on the perception of a laundry business: A preliminary investigation

T Rakhmawati, S Sumaedi, N J Astrini, I G M Y Bakti, M Yarmen and S Damayanti[1]

Research Center for Testing Technology, Indonesian Institute of Sciences, South Tangerang, Indonesia

Corresponding e-mail: trir005@lipi.go.id

Abstract. The development of technology has created various type of innovative washing machine on the market. Numerous options influenced consumers' quality expectations. Consumers' expects more from the ever-growing number of washing machines. This research aims to identify washing machines' quality dimensions in 4.0 industrial era and their level of importance based on the perception of a laundry business. This study used one subject, a medium-sized drop-off laundry business "X" in Bogor, Indonesia. The identification of indicators and quality dimensions was done through ad scouring process, discussion with the research team, and interviews with Laundry X. The importance level of washing machine quality dimensions was assessed using the Kano model. The results show that washing machines in the 4.0 industrial era have several superiorities in terms of customization, easiness, smartness, safety, performance, aesthetics, durability, information and service support, and environmental orientation and energy efficiency. For Laundry X, easiness, safety, performance, durability, and information & service support are the most important dimensions.

1. Introduction

The world has entered the 4.0 industrial era which can be said to be the fourth industrial revolution. In this era, various devices can be interconnected [1][2] and communicate with each other [2], have intelligence and can work autonomously [2]. Utilization of data and machine learning make this possible [2].

In the 4.0 industrial era, the household appliance industry sector launched many new products and embedded new technology into their products so that their products became more customer-centric. A washing machine is one of the household appliances that is always evolving by bringing new and smart technology. At this point, innovative washing machines with various models, designs, and features are readily available on the market. For example, there are currently washing machines equipped with smart features that allow users to diagnose problems with the washing machine and control washing machines via smartphone.

The success of a new product on the market is determined by consumer acceptance [3]. One factor that determines consumer acceptance is the consumer perception of product quality (perceived quality) [4][5]. Nowadays, the numerous product choices on the market with their unique superiorities stimulated consumers' expectations. According to Schwartz, Ward, Monterosso, Lyubomirsky, White, and Lehman [6] and Schwartz [7], if consumers have a lot of product choices, their expectation would rise [8]. For example, if decades ago consumers only expected the main function of a product, with technology advancement, their expectations have arisen. They did not only think about the product's functionality, but also its economic aspects. In other words, consumers used to consider 'function' as the only quality dimension, but now, the quality dimension has widened. Therefore, studies regarding quality dimensions in the 4.0 industrial era and their level of importance from consumers' perspective are important to be conducted.

Aside from common households, another party that has been affected by the development of washing machines is the laundry industry. As a major washing machine user, the laundry industry also has a high expectation that washing machines offered on the market can fulfill their operational needs and financially favorable. Thus, this research aims to identify washing machines' quality dimensions in 4.0 industrial era and determine their levels of importance from the perspective of a medium-sized laundry business. Understanding quality from consumers' perspective is important because consumers have an essential role in molding a product's success on the market which is characterized by high consumers' intention to purchase or use the product [9][10][11] and high consumers' loyalty [12][13].

2. Literature Review

2.1. Perceived product quality

Perceived quality is consumers' assessment on the superiority of an entity [14]. Perceived product quality is consumers' assessment on the superiority of a certain product. According to Holm and Kildevang [15], this assessment is subjective and situational [16]. For companies, perceived product quality is important because it has been proven to influence consumers' acceptance, both in the form of consumer willingness to purchase or use the product [9][10][11] and consumer loyalty [12][13].

Perceived quality is a multidimensional concept [16] [17] [18]. Many researchers have offered several quality dimensions to represent product quality. Garvin [19] [20] described product quality using eight quality dimensions, which are performance, feature, conformance, reliability, durability, serviceability, aesthetics, and perceived quality. Stone-Romero and Stone [21] proposed four product quality dimensions, namely flawlessness, durability, appearance, and distinctiveness. Lifespan, features, performance, and serviceability are perceived quality dimensions of remanufactured products, according to Hazen, Boone, Wang, and Khor [22].

2.2. Kano model

Kano model is a tool proposed by Kano, Seraku, Takahashi, and Tsuji in 1984 [23]. This tool is useful for understanding consumers' need and their impacts on satisfaction [23]. This model divides product attributes into several categories based on their ability to satisfy the consumers. The categories are must-be attributes, one-dimensional attributes, attractive attributes, indifferent attributes, reverse attributes, and questionable attributes [23].

- Must-be attributes are attributes that are considered as a *must* for a product [24]. If this attribute does not exist, consumers would be dissatisfied, but the existence of this attribute would not improve consumers' satisfaction [23] [24] [25].
- One-dimensional attributes are attributes which existence would produce linear satisfaction [24]. Better performance would lead to higher satisfaction and vice versa [23] [24]. The existence of this attribute would create consumers' satisfaction and the absence of this attribute would create dissatisfaction [23] [24] [25].
- Attractive attributes are attributes whose presence are not expected by consumers, but its existence would considerably improve consumers' satisfaction [24][25][26]. The absence of this type of attribute would not spark dissatisfaction [23] [24] [25].
- Indifferent attributes are attributes which presence or absence does not affect consumers' satisfaction level [25]

- Reverse attributes are attributes which presence would initiate dissatisfaction and their absence would create satisfaction [25]
- Questionable attributes are attributes which presence or absence sometimes would initiate consumers' satisfaction or dissatisfaction [26]

Figure 1 shows the Kano model. The level of attributes fulfillment is indicated by the horizontal axis and the level of customer satisfaction with the level of those attributes fulfillment is indicated by the vertical axis.

In the Kano model, data gathering was conducted using a questionnaire with functional and dysfunctional questions [25]. Functional questions explore consumers' responses if a product has a certain attribute. Dysfunctional questions explore consumers' responses if a product does not a particular attribute.

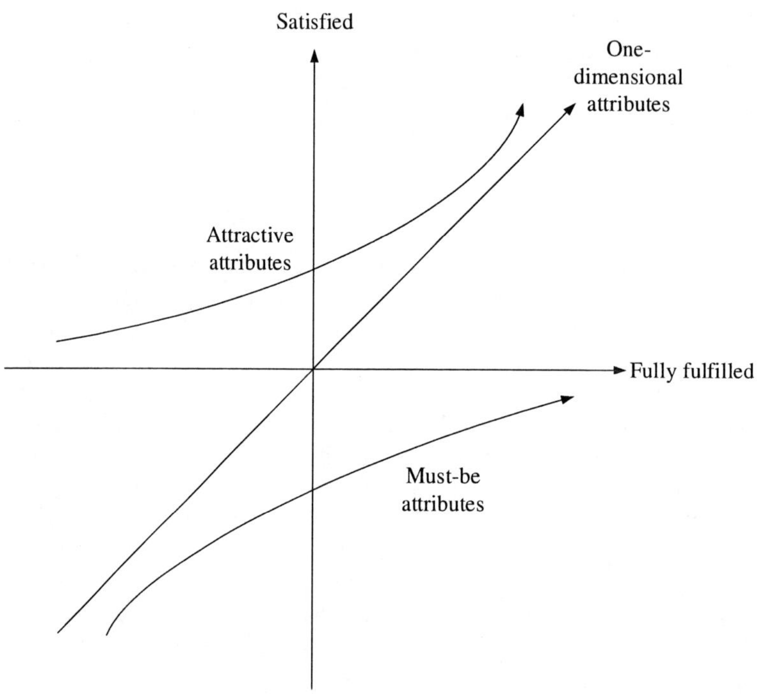

Figure 1. Kano model (Source: [27] cited in [23])

3. Methodology

This research is a case study of a medium-sized laundry business (Laundry X) in Bogor, Indonesia. Laundry X has 15 washing machines that are used to complete customer orders. Its customers consist of corporate and individual customers. The company's organizational structure consists of four levels. It has around sixty employees. The research stages are illustrated in figure 2.

3.1. Identifying washing machine quality indicators

The indicators were identified through advertisement materials of three washing machine brands on their manufacturer's website. The brands were chosen because according to Laundry X, these three brands were superior compared to their other counterparts. The quality indicators gathered were then being recapped, synthesized, and validated through team discussions and interviews with the director of Laundry X.

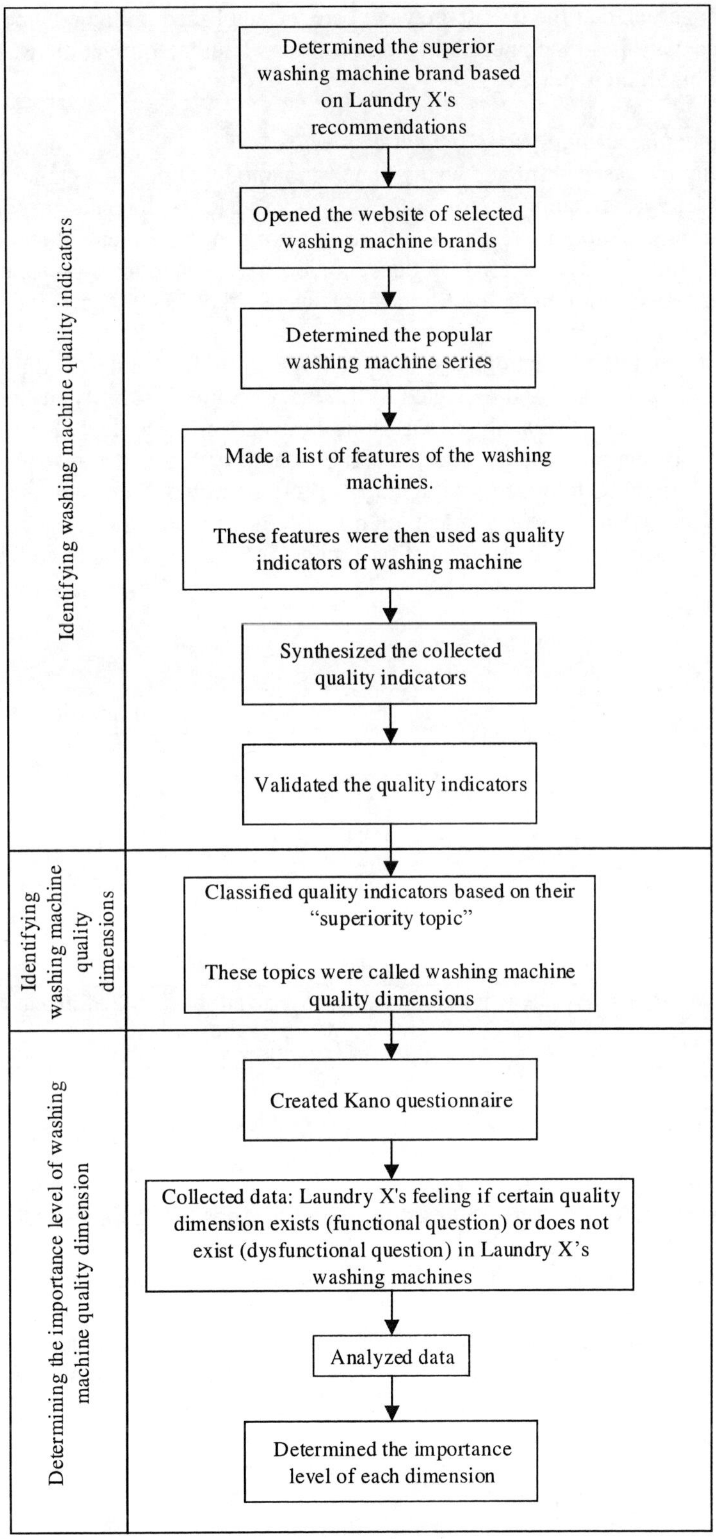

Figure 2. The research stages

3.2. Identifying washing machine quality dimensions

Quality indicators gathered at the first stage were categorized based on their "superiority topic". These topics were called washing machine quality dimensions. The categorization was conducted using a thematic analysis on all indicators.

3.3. Determining the importance level of each dimension

The importance level was determined using the Kano model. For data gathering process, a Kano model questionnaire was created. This instrument was used to explore the response of Laundry X's director as a washing machine user. The questionnaire gauged the director's feeling if certain quality dimension exists (functional question) or does not exist (dysfunctional question) in Laundry X's washing machines. Answers options for all questions were 'like' (S), 'expect' (H), 'neutral' (N), 'live-with' (T), and 'dislike' (TS).

Data collected from the questionnaire were analyzed to find out the dimensions of washing machine quality that were liked and expected by Laundry X and the ability of the quality dimensions to influence its satisfaction. The analysis was done by referring to table 1. For example, if a quality dimension of the washing machine is expected (H) by Laundry X and the absence of such dimensions is undesirable and cannot be tolerated by Laundry X (TS), then that dimension is included in the must-be attribute category. The washing machine quality dimensions included in the must-be, one-dimensional, and attractive attributes are quality dimensions that are considered important by Laundry X.

Table 1. Kano model criteria analysis

Quality dimensions		Dysfunctional questions				
		S	H	N	T	TS
Functional questions	S	Q	A	A	A	O
	H	R	I	I	I	M
	N	R	I	I	I	M
	T	R	I	I	I	M
	TS	R	R	R	R	Q

Q = questionable attribute, A = attractive attribute, R = reverse attribute, I = indifferent attribute, O = one-dimensional attribute, M = must –be attribute
Source: [28] cited in [23]

4. Result and discussion

4.1. Washing machine quality indicators

The potential quality indicators were gathered through the websites of three washing machine brands. This process resulted in 55 potential indicators. The list was synthesized to reduce or add indicators. Forty-four indicators came out of the synthesizing process. A validation process was done with a research team to ensure user's understandability.

The indicators list was given to Laundry X and interviews were conducted with the directors to ascertain the relevance of each indicator with the context of the washing machine in the 4.0 industrial era. The changes to the list were done based on the director's input. Forty-three indicators were finalized in this stage (see table 2).

4.2. Washing machine quality dimensions

The indicators were then categorized to quality dimensions based on "superiority topic" represented by each indicator. Categorization was done using a thematic analysis. The results show that there are nine dimensions from 43 indicators, which are customization, easiness, smartness, safety, performance, aesthetics, durability, information and service support, and environmental orientation and energy efficiency (see table 2).

Table 2. Washing machine quality indicators and dimensions

No.	Indicators	Dimensions	Definition
1	My washing machine has a complete feature option	Customization	The washing machine as a complete feature as needed by the consumers and it has features for all types of fabric
2	The features fulfill my and my family's needs		
3	My washing machine has complete features for all types of fabric		
4	My washing machine can be operated with minimal touch and all processes can be done automatically from weighing, determining water need, rinsing, and drying.	Easiness	The washing machine can be operated with minimum effort
5	My washing machine has front control design, so it is easy to read and use		
6	My washing machine has dedicated slots for detergent and softener, so they do not mix		
7	My washing machine has error/trouble notification feature	Smartness	The washing machine has smart technology, for example, error notification, automatic troubleshooter, washing history storage, remote control, etc.
8	My washing machine has a troubleshooting feature that can be used without using customers' service line		
9	My washing machine has a remote control feature, so I can operate it using a smartphone		
10	I can use a smartphone app to download new washing feature when I need one		
11	I can use a smartphone app to identify problems with my washing machine		
12	My washing machine has data on my washing history		
13	My washing machine has a feature that informs me the drum-cleaning schedule		
14	My washing machine does not shake too much	Safety	The washing machine was made with users' safety and health in mind for short and long-term period.
15	My washing machine is not noisy		
16	My washing machine has a safe design		
17	My washing machine does not induce skin		

	irritation or allergy		
18	The door of my washing machine can be closed gently, safe, comfortable, and does not create too much noise.		
19	My washing machine can dissolve detergent completely	Performance	The washing machine can wash, rinse, and dry clothes without creating damage.
20	My washing machine can work fast		
21	My washing machine can rinse the detergent completely (no residue on clothes)		
22	My washing machine is strong		
23	The result of my washing machine is optimal		
24	The result of my washing machine is even		
25	My washing machine does not cause wrinkles		
26	My washing machine produces fast-dry clothes		
27	My washing machine does not damage my clothes		
28	My washing machine has a relatively small size with big capacity, so it saves room		
29	My washing machine has a modern design	Aesthetics	The washing machine was designed with physical aesthetics in mind.
30	My washing machine has a good and stylish design		
31	My washing machine has an interesting color		
32	My washing machine has a wide impact- and scratch-resistant glass door	Durability	The washing machine is durable, does not break down easily, and has a lifetime warranty.
33	My washing machine is durable and does not break down easily		
34	My washing machine has a long warranty for the motor and other parts		
35	My washing machine has various customers' support lines (email, phone, and live chat)	Information and service support	The washing machine has service and product support including repair and maintenance
36	I can ask for product support through social media		
37	I can ask for product support whenever I want including on holidays		
38	I can ask for product support whenever I want in the day (24 hours)		
39	My washing machine only requires a minimal amount of water	Environmental orientation &	The washing machine is environmentally friendly.

40	My washing machine can work with small water pressure	energy efficiency	For example, it does not require too much water or consume large electrical power
41	My washing machine does not consume large electrical power		
42	My washing machine does not consume large electrical power even though it is always plugged in		
43	In a power outage, when the power back on, my washing machine can continue the process automatically.		

4.3. The importance level of washing machine quality dimensions

The importance level of the quality dimensions was determined based on the analysis of Kano questionnaire. Table 3 shows the data collected from the questionnaire. The criteria in table 1 were used to analyze the data. Table 4 shows the result. Based on the analysis, it is known that Laundry X deemed easiness, safety, performance, durability, and information and service support as one-dimensional attributes. It means that Laundry X would be satisfied if those attributes exist and dissatisfied if they do not.

Higher quality of those five dimensions would improve laundry X's satisfaction level. Laundry X would be dissatisfied if the washing machine is hard to operate, unsafe, has a poor performance, short-lived, and does not have adequate information and service support.

The result also shows that 'customization' is an attractive attribute for Laundry X. The presence of this dimension would increase Laundry X's satisfaction, but its absence would not cause dissatisfaction. The result shows that Laundry X considers 'smartness,' 'aesthetics,' 'environmental orientation & energy efficiency' as indifferent attributes. The presence or absence of those dimensions does not affect their satisfaction.

Table 3. Data collected

No.	Quality dimensions of washing machine		Questions	Answer				
				S	H	N	T	TS
1	Customization	F	How do you feel if your washing machine can be customized?	v				
		DF	How do you feel if your washing machine cannot be customized?				v	
2	Easiness	F	How do you feel if your washing machine is easy to operate?	v				
		DF	How do you feel if your washing machine is not easy to operate?					v
3	Smartness	F	How would you feel if your washing machine was equipped with smart technology?		v			
		DF	How would you feel if your washing machine was not equipped with smart technology?			v		
4	Safety	F	How do you feel if the washing machine is made by considering the safety aspects?	v				
		DF	How would you feel if your washing machine was made without considering the					v

			safety aspects?					
5	Performance	F	How do you feel if your washing machine has good performance?	v				
		DF	How would you feel if your washing machine did not perform well?					v
6	Aesthetics	F	How would you feel if your washing machine was made by considering the aesthetic aspects?			v		
		DF	How do you feel if your washing machine is made without considering the aesthetic aspects?			v		
7	Durability	F	How do you feel if the washing machine is durable?	v				
		DF	How do you feel if your washing machine is not durable?					v
8	Information and service support	F	How do you feel if your washing machine has information & service support?	v				
		DF	How would you feel if your washing machine did not have information & service support?					v
9	Environmental orientation & energy efficiency	F	How do you feel if your washing machine is environmentally friendly & energy efficient?		v			
		DF	How do you feel if your washing machine is not friendly to the environment & not energy efficient?				v	

F = functional question, DF = dysfunctional question

Table 4. The results of Kano analysis

No.	Quality dimensions of washing machine	Category
1	Customization	A
2	Easiness	O
3	Smartness	I
4	Safety	O
5	Performance	O
6	Aesthetic	I
7	Durability	O
8	Information and service support	O
9	Environmental orientation & energy efficiency	I

A = Attractive attribute, I = Indifferent attribute, O = One-dimensional attribute

4.4. Discussion

Each consumer has different needs and expectation. They are influenced by many things. This research revealed that for Laundry X, as a washing machine user in the 4.0 industrial era, easiness, safety, performance, durability, information & service support, and customization were important. Those six dimensions were deemed as important because their existence considerably supports Laundry X's operation and financially advantageous. If those six dimensions were poor, Laundry X's operation would be hampered, and they would suffer a loss.

Furthermore, Laundry X was indifferent when it came to 'smartness', a unique feature of the 4.0 industrial era. The existence (or absence) of this feature did not affect their satisfaction level. This was probably due to unfamiliarity with the feature or Laundry X felt that they have not needed it yet.

Aside from smartness, 'aesthetics' also seemed to be unimportant. Probably this was because Laundry X is a business entity. The function was the most important dimension for them. Laundry X also considered the last dimension, 'environmental orientation and energy efficiency', as unimportant. They opted to use another strategy to get around this dimension. For example, to reduce electrical power for a hot cycle washing, Laundry X used pre-heated water from outside the washing machine, so the machine's heater feature (which required large power) was not used.

This research contributes in the form of washing machine quality dimensions in the 4.0 industrial era based on the perception of a laundry business and their level of importance. It provides initial implications that washing machine manufacturers need to focus on the easiness, safety, performance, durability, information & service support, and customization aspects. In addition, the smartness, environmental orientation and energy efficiency, and aesthetics aspects of washing machine need to be more introduced so consumers know and maximize their use so that their existence is more useful.

5. Conclusion

The development of technology in the 4.0 industrial era has stimulated various innovative products. One of them is washing machine. Nowadays, washing machine brands have launched innovative products. Many options were offered and they affected the users' expectations.

This research aimed to identify washing machine quality dimensions in the 4.0 industrial era and their importance level for a laundry business. This research was a case study using a middle size laundry business in Bogor, Indonesia as the subject.

The result showed that in the 4.0 industrial era, washing machines have superiority in terms of customization, easiness, smartness, safety, performance, aesthetics, durability, information & service support, and environmental orientation & energy efficiency. In this particular study, the subject deemed easiness, safety, performance, durability, information & service support, and customization as important quality dimensions. Laundry X preferred a washing machine that was easy to operate, safe, had a good performance, durable, offered information and service support, and could be customized.

This study has a limitation. It only used one middle size laundry business. Further research with multiple users is needed to capture the expectation of wider consumers.

References

[1] Schwab K 2016 *The Fourth Industrial Revolution* Geneva: World Economic Forum

[2] Marr B 2018 Whats in Industry 4.0? Here's A Super Easy Explanation For Everyone https://www.forbes.com/sites/bernardmarr/2018/09/02/what-is-industry-4-0-heres-a-super-easy-explanation-for-anyone/#6d8b21149788

[3] Verbeke W 2005 *Food Quality and Preference* Consumer acceptance of functional foods: socio-demographic, cognitive and attitudinal determinants **16** 45–57

[4] Ahn T, Ryu S and Han I 2007 *Information & Management* The impact of Web quality and playfullness on user acceptance of online retailing **44** 263-275

[5] Ahn T, Ryu S and Han I 2004 *Electronic Commerce Research and Applications* The impact of the online and offline features on the user acceptance of Internet shopping malls **3** 405-420

[6] Schwartz B, Ward A, Momterosso J, Lyubormirsky S, White K and Lehman D R 2002 *Journal of Personality and Social Psychology* Maximizing Versus Satisficing: Happiness is a Matter of Choice **83** 1178-1197
[7] Schwartz B 2004 *The paradox of choice: Why more is less* New York: Harper Collins
[8] Haugtvedt C P, Herr P M and Kardes F R 2008 *Hanbook of Consumer Psychology* New York: Lawrence Erlbaum Associates
[9] Yan L, Xiaojun F, Li J and Dong X 2019 *Asia Pacific Journal of Marketing and Logistics* Extrinsic cues, perceived quality, and purchase intention for private lables: Evidence from the Chinese market **31** 714-727
[10] Walsh G, Shiu E and Hassan L M 2012 *Journal of Product & Brand Management* Investigating the drivers of consumer intention to buy manufacturer brands **21** 328 – 340
[11] Wang E S-T 2015 *Journal of Food Products Marketing* Different Effects of Utilitarian and Hedonic Benefits of Retail Food Packaging on Perceived Product Quality and Purchase Intention 1-13
[12] Espejel J, Fandos C and Flavián C 2007 *Managing Service Quality* The role of instrinsic and extrinsic quality attributes on consumer behaviour for traditional food products **17** 681-701
[13] Bei L-T and Chiao Y-C 2001 *Journal of Consumer Satisfaction, Dissatisfaction and Complaining Behavior* An integrated model for the effects of perceived product, perceived service quality, and perceived fairness on consumer saftisfaction and loyalty **14** 125-140
[14] Zeithaml V A 1988 *Journal of Marketing* Consumer perceptions of price, quality, and value: A means-end model and synthesis of evidence **52** 2-22
[15] Holm L and Kildevang H 1996 *Appetite* Consumers' views on food quality: A qualittative interview study **27** 1-14
[16] Jover A J V, Montes F J L and Fuentes M del M F 2004 *Food Quality and Preferences* Measuring perceptions of quality in food products: the case of red wine **15** 453-469
[17] Kenyon G and Sen K 2012 *International Journal of Quality and Services Sciences* A model for assessing consumer perceptions of quality **4** 175-188
[18] Alonso M A S, Paquin J-P and Mangin J-P L 2002 *Journal of Food Products Marketing* Modelling Perceived Quality in Fruit Products: Their Extrinsic and Intrinsic Attributes **8** 29-48
[19] Garvin D A 1984 *Sloan Management Review* What does product quality really mean? 25-43
[20] Garvin D A 1987 *Harvard Business Review* Competing on the eight dimensions of quality **65** 101-109
[21] Stone-Romero E F and Stone D L 1997 *Journal of Quality Management* Development of a Multidimensional Measure of Perceived Product Quality **2** 87-111
[22] Hazen B T, Boone C A, Wang Y and Khor K S 2017 *Journal of Cleaner Production* Perceived quality remanufactured products: construct and measure development **142** 716-726
[23] Wang T and Ji P 2010 *International Journal of Quality and Reliability Management* Understanding customer needs through quantitative analysis of Kano's model **27** 173-184
[24] Wijaya T 2011 *Manajemen Kualitas Jasa* Jakarta: PT. Indeks
[25] Lin F-H, Tsai S-B, Lee Y-C, Hsiao C-F, Zhou J, Wang J and Shang Z 2017 *PLos One* Empirical research on Kano's model and customer satisfaction **12** 1-22
[26] Chaudha A, Jain R, Singh A R and Mishra P K 2011 *Int J Adv Manuf Technol* Integration of Kano's Model into quality function deployment (QFD) **53** 689-698
[27] Kano N, Seraku N, Takahashi F, Tsuji S I 1984 *Journal of the Japanese Society for Quality Control* Attractive quality and must-be quality **14** 147-156
[28] Berger C, Blauth R, Boger D, Bolster C, Burchill G, DuMouchel W, Pouliot F, Richter R, Rubinoff A, Shen D, Timko M and Walden D 1993 *Center for Quality Management Journal* Kano's methods for understanding customer-defined quality **2**

Acknowledgments
All authors are main contributors. We would like to thank Laundry X that have supported this study

IOP Conf. Series: Materials Science and Engineering **722** (2020) 012049 doi:10.1088/1757-899X/722/1/012049

Sequential optimization of a water-based polygeneration system in ethylene glycol production

T Prihatin*[1], S Mahadzir[2] and M I Abdul Mutalib[2]
[1]Processing Technique of Mineral and Coal Mining Products, Bukit Asam Mining Industry Community College, Tanjung Enim, Sumatera Selatan, Indonesia
[2]Chemical Engineering Department, Universiti Teknologi PETRONAS, Bandar Seri Iskandar, 31750, Tronoh, Perak, Malaysia

E-mail: triana.prihatin@gmail.com

Abstract. Polygeneration system as one of the promising strategies shows the great potential for natural resource sustainability, particularly water. In this paper, a model of water-based polygeneration system to minimize the freshwater consumption and the wastewater generation is presented. Water usage in a heat and power utility, cooling utility, and chemical production are modelled simultaneously. To demonstrate the applicability of the model, a case study on the synthesis of optimum water-based polygeneration system is developed for an ethylene glycol production. A two-steps optimization method is presented. The first step consists of the synthesis and design of the polygeneration unit in which a superstructure containing three subsystems, i.e. heat and power generations, re-circulating cooling water system, and chemical production is introduced. A mixed integer non linear program (MINLP) model is developed to allow the selection of choices. The second step includes the incorporating of wastewater treatment strategy and a non linear program model is developed. As results, comparing to the simultaneous method in the same case study, this approach gives a better result in a computation efficiency.

1. Introduction

Water will be one of the crucial issues of the 21st century. Water is probably the most widely used raw material in the process industries and it has been used in abundant quantities by chemical, petrochemical, petroleum refining, food and drink, pulp and paper, power generation and many other industries. Industrial development has led to the fast acceleration of freshwater withdrawals. Moreever, the freshwater consumption in industry creates more pressure on freshwater resources from the impacts of wastewater discharges [1]. According to the UN World Water Development Report [2], some 300-500 million tons of heavy metals, solvents, toxic sludge, and other wastes accumulate each year from industry, most of which gets into the freshwater supply. In some developing countries, 70% of industrial wastes are dumped into untreated waters where they pollute the drinking water [2]. However, instead of seen as a problem, wastewater should be recognized as a water resource.

Numerous approaches have been investigated to give a solution for that global water problem. Water polygeneration system is one of the promising strategies which shows the great potential to solve the problem of limited freshwater resources. Polygeneration system is an integrated approach to an industrial process that generates multiple products from a single or more natural resources [3-4]. It is usually the integration of power generation and chemical process production. Different types of polygeneration system design have been presented, particularly on the energy-based focusing on the evaluation of existing plants and process technologies [5-7], the configuration and the performance of processes [8-10].

It is seen that there has been little discussion about polygeneration system based on water. Many of the recent studies on water minimization have only focused on the common points with a heat exchanger network problem and were mainly developed for targeting procedure but excluding 3R as a strategy to conserve water. In fact, a simultaneous demand for heat, power and cooling exists in certain chemical industries. The purpose of this study was to investigate a water-based polygeneration system co-producing heat and power, cooling water, and chemicals, incorporating 3R strategy (regenerate, reuse, and recycle). Efficient use of water in the industry can be achieved through 3R strategy within the processes, which can be designed using mathematical optimization approaches. Sequential optimization through two-steps procedure is presented. In the first step, an MINLP model is developed through a superstructure representation to select the appropriate configuration of water-polygeneration and in the second step 3R strategy is presented to the plant configuration in a NLP model to reduce freshwater consumption and wastewater generation. A water polygeneration system for ethylene glycol production is presented as a case study.

2. Optimization First Level (MINLP)

In this paper, a superstructure-based optimization approach is used for the design of water polygeneration. One common approach to systematically build superstructure is combining the detailed superstructures for each subsystem that perform multiple tasks or functions and interconnect them. All the feasible interactions would be taken into account. In this work, the water-based polygeneration system consists of three subsystems: heat and power generation, cooling water system and ethylene glycol (EG) synthesis process. The schematic of sequential optimization method is shown in Figure 1. General superstructure for the synthesis of an optimum water-based polygeneration system in an ethylene glycol production is shown in Figure 2.

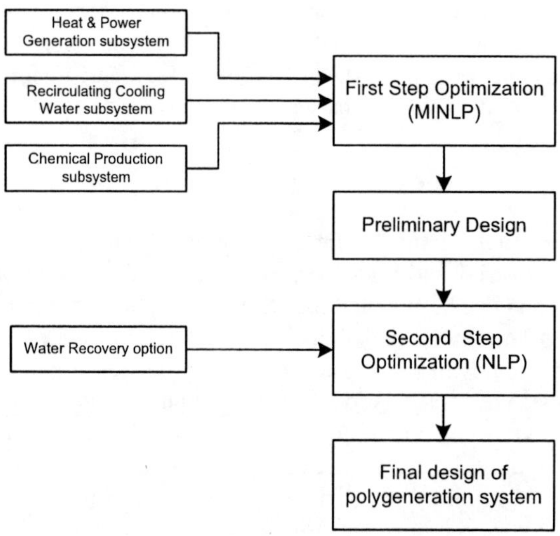

Figure 1. Schematic of sequential optimization method

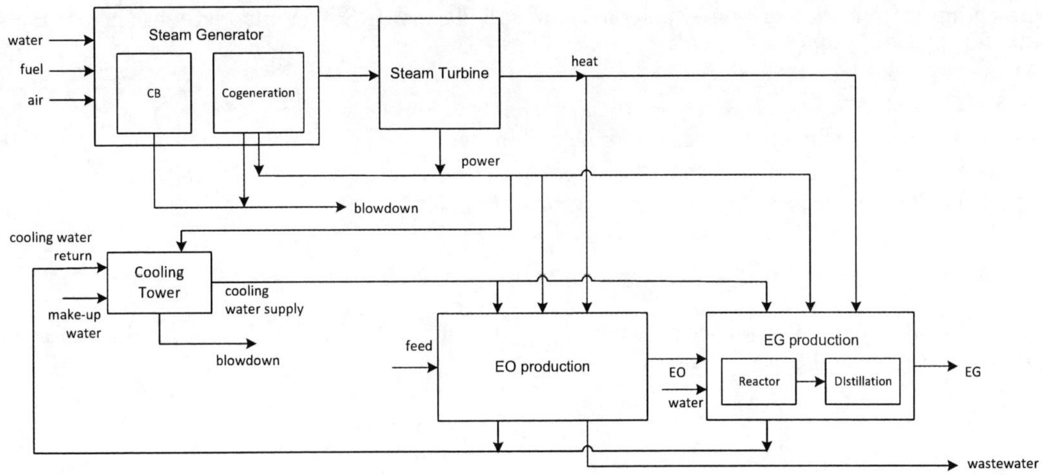

Figure 2. General superstructure representation of the water-based polygeneration system

To illustrate the model superstructure, the set technology choices of three subsystems of the water-based polygeneration system have been determined. (1) Heat and power generation: conventional boiler (CB), cogeneration including heat recovery steam generator (HRSG), back-pressure steam turbine (BPST), condensing steam turbine (CST). (2) Recirculating Cooling Water System: forced mechanical draft and induced mechanical draft. There are two broad classes of cooling tower, i.e. natural draft and mechanical draft cooling towers. (3) Ethylene Glycol Production Process: EO production: O_2-based EO and air-direct oxidation, EG process reaction: EO hydrolysis and the Shell's OMEGA process, reactor: continuous stirred tank reactor (CSTR) and plug flow reactor (PFR), separation: direct sequence, indirect sequence and dividing wall distillation column (DWC).

Ethylene glycol is used mainly as an antifreeze in automobile radiators, as an engine coolant and as a raw material for the manufacture of polyester fibers and plasticiser. There are two common methods of ethylene glycol production, i.e hydration of ethylene oxide [11] and mono-ethylene glycol (MEG) only technology [12]. Ethylene oxide (EO) is obtained by direct oxidation of ethylene with air or oxygen (O_2). Some literatures reported that a molar ratio of ethylene oxide to water may reach 1:5 to 1:30 [12-13]. The ethylene oxide is thermally hydrolyzed to ethylene glycol without a catalyst. Diethylene glycol, triethylene glycol and polyglycols are also produced, with respectively decreasing yield, and are separated from the ethylene glycol by series of distillation columns at reduced pressure. In the second method, ethylene glycol is manufactured by the reaction of ethylene oxide with carbon dioxide to form ethylene carbonat, an intermediate product, which can be hydrolyzed to ethylene glycol. In this work, the alternative process/technology in reactor and separation units of ethylene glycol plant is investigated in Aspen HYSYS simulation for superstructure configurations.

In this first-step optimization, the model of the superstructure presented is formulated as a MINLP problem in which binary variable 0 or 1 is assigned as a logical constraint, which acts as a yes-no decision for the potential existence of options. In this work, shortcut equations are developed for the MINLP model. Some equations and constraints in the simultaneous method [14] are eliminated, for instance, the energy balances around steam boiler and the enthalpy values. For heat and power generation subsystem, mass balance around steam boiler is expressed as:

$$\sum_j \sum_b F_{j,b} \cdot y_b - \sum_k \sum_b F_{k,b} \cdot y_b = 0 \tag{1}$$

where F is flowrate of inlet streams j and outlet streams k, respectively, and y_b is a binary variable representing the selection of boiler, i.e conventional boiler or HRSG. While, mass balance of the steam turbines is given as follows:

$$\sum_j \sum_{st} F_{j,st} \cdot y_{st} - \sum_k \sum_{st} F_{k,st} \cdot y_{st} = 0 \tag{2}$$

where y_{st} is a binary variable for steam turbine options, i.e. back pressure and condensing steam turbines. In equation (3), the power demand, $W_{st,i}$, and the steam enthalpy at steam pressure level, $\Delta H_{st,i}$, is correlated to the flowrate intake of the steam turbine, $FT_{st,i}$ [15].

$$FT_{st,i} = W_{st,i} / \Delta H_{st,i} \tag{3}$$

In this work, equation of power produced/demand in the boiler options is linearized as a function of water intake (F_{dw}) as follows.

$$W = f(F_{dw}) \tag{4}$$

where F_{dw} is flowrate of demin water.

In the cooling water subsystem, mass balance around the cooling tower is given by:

$$\sum_j \sum_{cws} F_{j,cws} \cdot y_{cws} - \sum_k \sum_{cws} F_{k,cws} \cdot y_{cws} = 0 \tag{5}$$

where the binary variable y_{cws} denotes the selection of cooling tower. The detailed mass balance at specific technology as below:

$$F_{in} + F_m - F_{out} - F_b = 0 \tag{6}$$

where F_{in} is the inlet water flowrate to cooling tower, F_{out} is outlet water flowrate of cooling tower, F_m is the cooling water makeup and F_b is the cooling tower blowdown.

In the ethylene glycol production subsystem, most paramaters in the model are estimated from the Aspen HYSYS simulation [16] using the UNIQUAC equation. This fluid package shows a good representation for liquid structure and can be applied to a wide range of mixtures containing water. The reactors are operated under isothermal conditions. The constraints for the supply of cooling, heating and power requirement in the chemical production are formulated as per process demands. Ethylene glycol reactor products will go through distillation column to produce mono-ethylene glycol which is purified from the by-products of water and other higher glycols derivatives. In this separation process, the product recovery and purification also consume water for the heating and cooling. Cooling water demand, F_{CW}, at ethylene glycol production, particularly in the separation process, is correlated mainly to condenser duty, Q_{cd}, as shown in equation (7).

$$F_{CW} = \frac{Q_{CD}}{CP.\Delta T} \tag{7}$$

where CP = 4.18 kJ/kg °C and ΔT = 20 °C.

The objective function (z) includes the cost of freshwater intake, C_{fw}, as shown in equation (8). The cost is calculated on annual basis assuming *AOT*, annual operation time, of 8000 hrs/year. The cost of freshwater (C_{fw}) is 0.26 \$/ton.

IOP Conf. Series: Materials Science and Engineering **722** (2020) 012049 doi:10.1088/1757-899X/722/1/012049

$$\min z = \left[\left(FW \times C_{fw}\right)\right] \times AOT \tag{8}$$

3. Second Level Optimization

Reduction of freshwater consumption and wastewater discharge has become one of the main targets of design and optimization of process design. Reducing wastewater affects both effluent treatment and freshwater costs. Therefore, the best configuration obtained from the first step is being optimized in this second step by implementing the 3R strategy for water recovery to obtain the optimum solution of water polygeneration system. 3R strategy includes reuse, regeneration reuse, and regeneration recycling. In reuse process, wastewater is reused directly in other operations subject to the level of contamination to operations within the process system. Regeneration is a term used to describe any treatment process that regenerates the quality of water such that it is acceptable for further use. Water recycling involves post-treatment process of the water quality to meet the next water user specification.

The mathematical model is formulated as a NLP problem. Water recovery unit will treat blowdown from the boiler and the cooling tower, and wastewater from ethylene glycol process production. During operation of cooling tower and boiler, there will be accumulation of solids that can impact to heat transfer or other unit operation. Based on Baliban et al [17], the concentrations of suspended solids (TSS) and dissolved solids (TDS) in the boiler blowdown stream are 10 ppm and 500 ppm, respectively. The solids concentration in the cooling tower blowdown will be 50 ppm for suspended solids and 2500 ppm for dissolved solids. These quantities are set in equation (12) as $Cw_{r,b(i)}$ where r is index for specified contaminant, i.e TSS and TDS, b is index for blowdown, i is index for unit boiler and cooling tower, and F is flowrate.

$$Cw_{r,b(i)} \cdot F_{b(i)} - F_{r,b(i)} = 0 \tag{9}$$

For ethylene glycol process production, the maximum allowable concentration of dissolved solids in the wastewater is 500 ppm as set as $Cww^{\max}_{r,out}$. $Fww_{r,out}$ is flowrate of specified contaminants in wastewater and Fww_{out} is flowrate of wastewater leaving ethylene glycol plant, as following:

$$Fww_{r,out} - Cww^{\max}_{r,out} \cdot Fww_{out} \leq 0 \tag{10}$$

3.1. Objective Function

The objective function for the model is to optimize the flow rate of freshwater consumption (FW) at minimum total annualized cost as shown in equation (14). Minimization of freshwater consumption directly reduces freshwater intake by the ethylene glycol production. The objective function (z) includes the cost of freshwater intake, C_{fw}, and the cost of water recovery, $OP_{recovery}$.

$$\min z = [(FW \times C_{fw}) + OP_{recovery}] \times AOT \tag{11}$$

The cost of water recovery takes into account the pumping cost as well as cost associated with water recycle, reuse, and regeneration (3R). The cost of water recovery is correlated to total power need for pumps operation (P_i) and cost of electricity (C_{elect}) by the following equations:

$$OP_{recovery} = \sum_i P_{p(i)} \times C_{elect} \tag{12}$$

The cost of electricity (C_{elect}) is 0.064 \$/kWh. The power consumed by the pump is a function of the water flowrate to the pump, F_p, pump efficiency, η_p, and density of water, ρ, as given below [18]:

$$P_p = 1283 \frac{1}{\eta_p} \left[\frac{F_p}{\rho} \right]^{0.476} \tag{13}$$

4. Case Study Results

The proposed mathematical model is used to analyzed the case study of a water-based polygeneration system in ethylene glycol production. This study concerns on water flow in the three subsystems that might detect the possibility of water recovery. The complexity of the work is the interactions between the chemical process and the subsystems of heating, power, and cooling. The overall model is developed in GAMS on Intel (R) Core of 2.20 GHz with a memory size of 4 GB. For first step, the model is solved to optimum using DICOPT++ as the MINLP solver. It is a non-convex model contains 15 binary variables, 109 continuous variables and 101 equations. It takes 885 iterations and 4 major iterations only to claim an optimum. These results are compared with the simultaneous method [14] in which water polygeneration system is formulated simultaneously with water recovery in MINLP model, as given in Table 1.

Table 1. Results of water-based polygeneration system

Method	Freshwater consumption (t/hr)	Wastewater generation (t/hr)	Objective function (\$/yr)	Computation efficiency
Simultaneous	224.375	5.154	648,330	No equations: 309 No variables: 330 Binary: 20 Solver times: 0.47 (NLP) 1.01 (MIP) No iterations: 3683
Sequential:				
First step	448.886	228.897	933,682	No equations: 101 No variables: 109 Binary: 15 Solver times: 0.18 (NLP) 0.45 (MIP) No iterations: 885
Second step	225.144	5.156	650,248	No equations: 17 No variables: 17 No iterations: 3

In the second step, the NLP model with 3R strategy gives an optimal solution contains 17 continuous variables and 17 equations. The problem is solved using CONOPT. Freshwater is consumed as makeup for demineralized water in the heat and power subsystem, feed to the ethylene glycol reactor, and makeup water for the cooling tower. The blowdown from boiler and cooling tower units and wastewater generated in the ethylene glycol plant are mixed and recovered in the regeneration unit. This regenerated water then

IOP Conf. Series: Materials Science and Engineering **722** (2020) 012049 doi:10.1088/1757-899X/722/1/012049

recycled to the freshwater intake might reduce the cost of water consumption. As summarized in Table 1, through two-steps procedure the optimal solution can be obtained. The shortcut model can achieve a minimum freshwater consumption and cost of 225.14 t/hr and 650,248 $/yr respectively, close to the results through simultaneous method. This approach also show a significant reduction of the computational efficiency. To validate the work, the preliminary results obtained from the simulation model are compared with those from the simulation model of the utility plant by Bruno et al. [19]. The work of Bruno et al. is chosen as a validation of the simulation for this work due to the similarity of the model simulation for heat and power generation plant or utility plant.

5. Conclusion

A sequential optimization approach is applied to the synthesis and design of water-based polygeneration system as a strategy to address the problem of limited freshwater resources in the chemical process industry. In the first step (MINLP model), superstructure helps in assessing the potential of different design configurations to formulate a comprehensive model for the optimum water-based polygeneration system. The model features all possible configurations of heat and power generations, re-circulating cooling water system, as well as reaction and separation technologies for chemical production. In the next step, the 3R strategy for water conservation is implemented in the NLP model. The application of the model was demonstrated on an ethylene glycol production case study. The solutions derived with this approach are close to the simultaneous solutions because of the close approximation of the system by the shortcut model. The main benefit of sequential optimization is the efficiency in run time.

References:

[1] Bates B C, Kundzewicz Z W, Wu S and Palutikof J P 2008 Eds. Climate change and water Technical Paper of the Intergovernmental Panel on Climate Change (Geneva: IPCC Secretariat) pp. 38

[2] The-3rd United Nations World Water Development Report: *Water in a Changing World (WWDR-3).* Available: http://www.unesco.org/water/wwap/wwdr/wwdr3/, [Oct. 8, 2009].

[3] POLYSMART 2008 Polygeneration in Europe-A Technical Report European Commission within the Sixth Framework Programme (2002-2006)

[4] Serra L M, Lozano M A, Ramos J, Ensinas A V and Nebra S A 2009 *Energy* **34** 575-586

[5] Lin H, Jin H, Gao L, Han W and Zhang N 2009 *Energy Procedia I* 4193-99

[6] Lin H, Jin H, Gao L and Han W 2011 *Energy Conversion and Management* **52** 274-283

[7] Yu G W, Xu Y Y, Hao X, Li Y W and Liu G Q 2010 *Fuel* **89** 1070-76

[8] Li H, Hong H, Jin H and Cai R 2010 *Applied energy* **87** 2846-53

[9] Joelsson J M and Gustavsson L 2010 *Biomass and bioemerby* **34** 967-984

[10] Song H, Starfelt F, Daianova L and Yan J 2011 *Applied energy* **90** 32-37

[11] Kirk-Othmer 2005 *Encyclopedia of Chemical Technology 5th Edition* vol. 12 (New Jersey: John Wiley & Sons)

[12] Rekers D M and Smaardjik A A 2008 *US Patent* No.2008018299 A1

[13] Speight J G 2002 *Chemical process and design handbook* (New York: McGraw-Hill)

[14] Triana P, Shuhaimi M and M I Abdul Mutalib 2013 *Ind. Eng. Chem. Res.* **52** 7066-73

[15] Smith R 2005 *Chemical Process Design and Integration* (England: John Wiley & Sons)

[16] Aspen Technology Inc. 2006 *Aspen Hysys Tutorials and Applications Version 2006* Software Documentation

[17] Baliban R C, Elia J A and Floudas C A 2012 *Comput. Chem. Eng.* **37** 297-327

[18] Castro M M, Song T W and Pinto J M 2000 *Trans IChemE* **78** Part A 192-201

[19] Bruno J C, Fernandez F, Castells F and Grossmann I E 1998 *Trans IChemE* **76** Part A 246-258

IOP Conf. Series: Materials Science and Engineering **722** (2020) 012050 doi:10.1088/1757-899X/722/1/012050

Rap-Tourism Method to Assess Tourism Objects Sustainability

R Purwaningsih[1], H Santoso[1], and U Khasanah[1*)]

[1]Industrial Engineering Department, Faculty of Engineering, Diponegoro University, Jl. Prof. Soedarto, SH, Kampus Undip Tembalang, Semarang, Indonesia 50275

Corresponding author: uswatun740@gmail.com

Abstract. Kampung Pelangi Semarang was originally a slum settlement that became artificial based tourism destination by City Government project. It has a positive impact on the tourist attractiveness of Semarang City. It is necessary to maintain its existence by evaluating the status of tourism object sustainability. The purpose of this research is to assess the sustainability status of tourism destination based on the sustainability index and provide recommendations for improvements based on the most sensitive indicators. Data processing used Rap-tourism method and multidimensional scaling (MDS) approach. This research using ordinal scale 1-5 with samples from stakeholders of the tourism objects. Sustainability status of Kampung Pelangi Semarang is a less sustainable category with sustainability index of 35.29. The index consists of four-dimensional value. The first is environmental dimension which has index 36.58 (less sustainable), the next is economic dimensions 32.40 (less sustainable), socio-cultural dimensions 35.15 (less sustainable) and institutional dimensions 34.65 (less sustainable). Improvement recommendations from each dimension are provide restroom facilities and water installation, training the community on the paper floral skill, organizing fishing festivals, provide information facilities related to mitigation of seasonal changes and risk of disaster threats.

1. Introduction

Sustainable tourism is a tourism that consider current and future impacts of economic, social and environmental dimension, to meet the needs of visitors, industry, environment and the local community and can be applied to all forms of tourism activities in all types of destinations, including mass tourism and various niche tourism segments. The implementation of sustainable tourism needs more attention from governments and all involved tourism elements, including tourism destination management and the tourists. By implementing sustainable tourism in the management of tourism destination, the negative impacts such as damage to environments increased development costs, and damage to existing cultures could be avoided [1].

Kampung Pelangi Semarang was originally a slum settlement that improved by City Government project and became an artificial tourist destination in 2017. It was seen the residential housing was painted in colors. Since the inauguration of Kampung Pelangi tourist destination in April 2017 had a positive impact on the tourist attractiveness of Semarang City. Tourists from inside and outside this region also visit the destination, including visitors from abroad. The existence of Kampung Pelangi also encouraged other regions to participate in making changes in their respective regions which had more value. The people of Kampung Pelangi began to changing their habit such as start to caring the cleanliness of the environment, increasing social attitude of community than before, business development also began to be implemented in these tourist destinations. Based on the description above, to maintain the existence of a tourist village, it needs to evaluate the status of sustainability. This step as a method to detect or identify the important environment, ecology and social problem which threatens tourism sustainability. This research aims to measure the sustainability status of Kampung Pelangi Semarang using Rap-tourism method and Multidimensional Scaling (MDS) approach and provide recommendations for improving tourism sustainability based on the results of the most sensitive indicator of each sustainable dimension.

2. Literature Review

2.1. Sustainable Tourism Dimension and indicators

Sustainable tourism is defined by the United Nations World Travel Organization as tourism that takes full-account of current and future economic, social and environmental impacts, responds to visitor needs, industry (tourism), the environment and host communities [2]. The scope of sustainable tourism is (1) Management of sustainable tourism destinations; (2) Economic use for local communities; (3) Preservation of culture for the community and visitors; and (4) Environmental preservation [1].

Sustainable tourism triangle emphasizes the balance of 3 pillars: the economic, social, and environmental dimensions of the tourism industry. Other research used 5 dimensions for assessment on natural-based tourism: the institutional, infrastructure, social culture, environment, and economy [3]. Research on sustainability assessment of a natural resources used indicators attractions, facilities, transportation, and hospitality. The result was found that among the dimensions in the tourism indicator, the hospitality dimension turned out to get the highest average score, followed by facilities, attractions and then finally transportation. This shows that the hospitality and acceptance of stakeholders must be well managed although in natural tourism [3].

There are many different types of tourism, according to Spillane there are 7 kinds of tourism based on tourism destination. They are pleasure tourism, recreational tourism, cultural tourism, culinary tourism, sports tourism, business tourism and convention tourism [5]. Another classification is from McIntosh and Goeldner which summarize previous studies on travel motivation, separating them into four categories of tourism: (1) physical motivators, including those related to physical rest, need for recreation, and those motivations directly connected with a person's bodily health; (2) cultural motivators concerning the desire to gain knowledge about other countries in term of cultural activities; (3) interpersonal motivators, including a desire to meet new people, get away from routine conventions of life or to make new friendships; and (4) status and prestige motivators, related to self-esteem and personal development [6].

The process of choose the indicator that would be used in the research need a certain method. Kožić used factor analysis to determine the weight of indicators in evaluating the sustainability of coastal areas. The indicators construct the dimension of social dimensions, environmental dimensions, and economic dimensions. The indicators with the highest weights indicate that these indicators have a major influence on the sustainability index of the tourism sustainability dimension [7].

After find out the most important or the most sensitive indicator, then the recommendation formulation must be focus on that indicator. One method to formulate recommendation is Product-service system (PSS). The method also can be used to measure the sustainability level of a unit business or some unit business as a single unit. Product service system consists of three dimensions: (1) Environment, (2) Socio-cultural and (3) Economic. This research guided by a check-list form and has a portfolio diagram used to select these recommendations according to its feasibility to be implemented and its importance for the industries [8].

2.2. Multidimensional Scaling (MDS)

The next stage after define the dimension and indicators is define the method to processing the data resulted from questionnaires. MDS is a multivariate analysis method which used to process an ordinal or nominal data scale. The results from MDS proved to be more stable than other multivariate analysis methods such as factor analysis and multi-attribute utility theory [9]. MDS is possible to analyze leverage (sensitivity of reducing indicators to sustainability scores). Leverage analysis describes the sensitivity of each indicator to the value of sustainability and used to identify sensitive indicators. The greater the change in the value of Root Mean Square (RMS), the more sensitive the indicators to sustainability.

Rap-fish is a statistical technique for rapid assessment of entity status (fisheries), assessed quantitatively on a set of predetermined indicators that are grouped into evaluation or discipline fields [10]. Rap-fish (rapid appraisal for fisheries) has been developed by the Fisheries Center University of British Columbia to assess the sustainability status of fisheries. In Rap-fish, indicators are chosen to reflect sustainability in each dimension and the indicator can be corrected or replaced when new information is available. The Rap-fish method used to assess tourism is known as Rap-tourism. The Rap-tourism method uses the Multidimensional Scaling approach in coordinating its indicators. Ordination is a process of plotting object points along axes arranged according to a certain relationship or in a graph system consisting of two or more axes [3]. The ordination technique (determination of distance) in MDS is based on Euclidian Distance with dimensionless space or called ALSCAL algorithm.

3. Research Methods

This research is a study that analyzed descriptively and quantitatively. The results of calculating data processing in the form of quantitative data and will be analyzed descriptively to determine indicators that are sensitive to the sustainability level of tourist destinations. The recommendations for improvement will be analyzed for the development of tourism.

The sampling technique using purposive sampling, which means that the sample which became an assessor of tourism object was deliberately chosen according to the specified criteria. The samples from this study were stakeholders of the Kampung Pelangi Semarang tourist destination, consisting of the government (Department of Culture and Tourism, Department of City and Spatial Planning, and Department of Housing and Settlement Areas), location managers, tourist or visitors, and surrounding communities, with total respondent was 36 person.

Rap-tourism developed in Microsoft Excel software for data processing. The study was carried out through the following steps [3]:

1. Determine the scale range for each indicator using ordinal scale 1-5. This scale is arranged according to the level and the arrangement is sequential from the lowest to the highest.
2. Calculating the sustainability index value and define the category or sustainability status of the tourism object
3. Conformity Test by looking at stress values and coefficient of determination (R^2).
4. Analysis of leverage factors to find out indicators which the existence has a sensitive effect on increasing or decreasing sustainability status.
5. Monte Carlo analysis at a 95% confidence interval.
6. Formulate the recommendation to improve the sustainability status

The sustainability status of a tourism destination defines use the sustainability index range value [11] shown in Table 1.

Table 1. Sustainability Category Status

No.	Sustainability Index	Status
1	0.00-25.00	Bad (unsustainable)
2	25.01-50.00	Less (less sustainable)
3	50.01-75.00	Enough (quite sustainable)
4	75.01-100.00	Good (very sustainable)

Based on the results of data processing and analysis will be formulating recommendations for improvement. The recommendations based on the results of leverage analysis which shows indicators sensitive to the sustainability index. Assessment of alternative recommendations by giving a minus sign.

Table 2. shows the criteria to choose the recommendation based on the feasibility to implement (period and cost) and the significance of the program to increase sustainability [12].

Table 2. Recommendation Assessment

Feasibility/Ease of implementation	Duration of Implementation	Value	Meaning
Difficult	Short	-	Bad
Difficult	Long	=	Same
Easy	Short	+	Better
Easy	Long	++	Much Better
Difficult	Short	-	Bad

4. Result and Discussion

The measurement of sustainability index used four dimension, each dimension consist of some elements. Table 3 shows the element of sustainability dimension and its definition. The definition become a description of the concept of the dimensions that will be measured. Each element measure by the condition of some indicators. Assessment of tourism object sustainability will be carried out using 33 relevant sustainable tourism indicators shows in Table 4. The indicators generate by a literature study from the previous study and then validate by an expert questionnaire to tourism expert and site location managers.

Table 3. Sustainability Element Definition

No	Element	Definition
1	Environmental Protection	Environmental protection covers how to deal with the risk of arising environmental damage and the management of waste from tourism activities.
2	Flora protection	Protecting flora and fauna is important as a form of management and preservation the diversity on existing flora and fauna, both rare and not.
3	Energy Management	Energy management is important as a form of monitoring and measuring energy consumption in tourism activities, to decrease waste of energy.
4	Participation and Access for Communities	Community is part of stakeholders in tourism. Aspirations and other opinions from the local community must be received and collected to be taken into consideration in tourism activities.
5	Job Opportunities and Supporting Local Businesses	To contribute to the welfare of local communities, there needs to be supported by providing employment and supporting micro, small and medium enterprises by local communities.
6	Economic Sustainability	In maintaining tourism economic sustainability, it is necessary to monitor and collect financially related data from tourism activities and holding tourist attractions or programs that adapt to local culture.
7	Visitor Management & Public Relations	Relations with visitors and community are important in achieving sustainable tourism, and there is needed for interaction with various stakeholder
8	Interpretation Service	a tourist destination is a form of community or culture interpretation, so the relevant information is needed to maintain the authenticity of tourist destination characteristics
9	Destination Management Organization	The existence of an effective, coordinated management organization is one of the important foundations that also determine the sustainability of a destination.
10	Safety and Security	Safety and security are associated with the existence of a system to monitor, prevent, inform, report and handle issues related to safety and security.
11	Visitor satisfaction	Visitors are an important part of tourism, so it must have clear management related to visitor satisfaction and responsiveness to all visitors' aspirations or complaints.
12	Planning Arrangement	Planning arrangements are related to the existence of guidelines, regulations, policies regarding planning that include zoning, resources, and facilities owned, to protect natural and cultural resources.

IOP Conf. Series: Materials Science and Engineering **722** (2020) 012050 doi:10.1088/1757-899X/722/1/012050

Table 4. Dimensions, Elements and Indicators of Sustainable Tourism

No	Element	Indicator	Reference
		Environmental Dimension	
1	Environmental Protection	1. System for handling environmental risks 2. Environmental management and protection systems 3. Waste recycling program 4. Transportation planning and programs to improve environmentally-friendly transportation	[1];[3];[7]; [13];[14]; [15] [1]; [3];[7]
2	Flora protection	1. Inventory of flora 2. Destruction Prohibition of flora	[1];[13];[14]
3	Energy Management	1. Reporting on energy consumption 2. Water quality and safety management system	[1];[13];[15] [1];[13];[14]
		Economic Dimension	
1	Participation and Access for Communities	1. Communication to accommodate the aspirations of residents 2. Communication about tourism issues	[1];[13];[14]
2	Job Opportunities for Local Communities & Supporting Local Businesses	1. Employing residents 2. Involving and supporting local businesses	[1];[13];[14]; [15]
3	Economic Sustainability	1. Reporting and monitoring regularly related to expenses, income, and investments 2. The existence of attractions and tourism destination programs (related to culture)	[1];[15] [13];[14]
		Social & Culture Dimension	
1	Visitor Management & Public Relations	1. Visitor management system 2. tourism and culture directions 3. Volume and type of traveler's trip to the destination 4. Relations with the local community	[1];[13];[15] [14];[15]
2	Interpretation Service	1. Accurate interpretive information 2. Information availability in various forms of media 3. Management system to protect natural and cultural sites	[1];[13]
		Institutional Dimension	
1	Destination Management Organization	1. Development Strategy for sustainable tourism destinations 2. The involvement of the community, government, private sector, and entrepreneurs as stakeholders 3. The responsible organization which coordinates clearly 4. Tourism destination management system 5. Tourism promotion strategy	[1];[3];[15]
2	Safety and Security	1. Emergency handling and response systems for safety and security 2. Hazardous and preventive inspection and systems 3. Mitigating seasonal changes policies	[1];[7];[13]; [15] [1]
3	Visitor satisfaction	1. Responsive management to visitor satisfaction 2. Level of visitor satisfaction	[1];[13]
4	Planning Arrangement	1. Periodic asset inventory 2. Destination regulations and policies which include analysis of resources, infrastructure, zoning, facilities and services provided 3. Accessibility that supports accommodation for tourists including persons with disabilities and special needs	[1] [1] [1];[14]

4.1. Sustainability Index

4.1.1. Environmental Dimension.

Result of Rap-tourism software related to sustainability index of the environmental dimension with an index of 36.58 which categorize as less sustainable. Figure 1 is a leverage factor of the environmental dimension. Figure 1 shows the RMS value of the environmental dimension indicators. Leverage analysis is used to see sensitive indicators that influence the sustainability index value of a dimension. The greater the RMS value, the greater the role of the indicator on sensitivity to sustainability status. Leverage factors of the environmental dimension are the inventory of flora with an RMS value of 6.07 and water quality and safety management system with RMS 6.39.

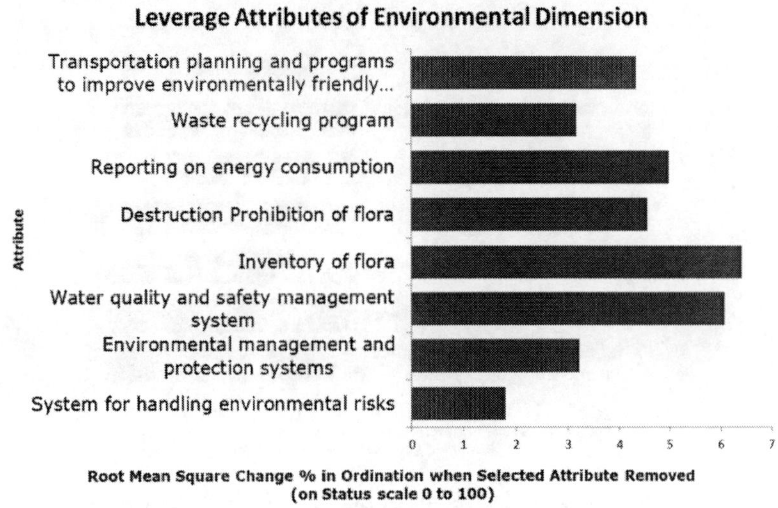

Figure 1. Leverage factor of Environmental Dimension

4.1.2. Economic Dimension.

Result of Rap-tourism ordination on sustainability index of the economic dimension is 32.40 which is categorize as less sustainable. Increasing the sustainability index of a dimension can be done by managing sensitive indicators. Figure 2 shows the leverage factor of the economic dimension.

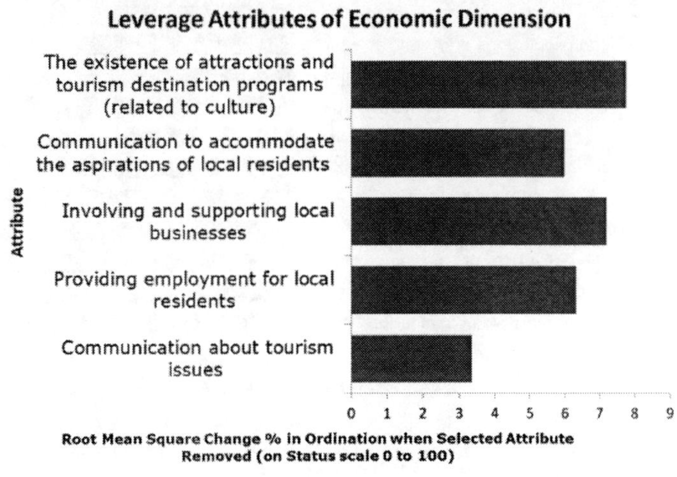

Figure 2. Leverage factor of the Economic Dimension

The economic dimension has 5 indicators with each RMS value as a dimension analysis of leverage. Increasing the sustainability index of a dimension can be done by managing sensitive indicators. Leverage factors of economic dimension with the highest RMS values were the existence of attractions and tourism destination programs (related to culture) and supporting local businesses with a value of 7.19 and 7.76.

4.1.3. Social & Culture Dimension.
Result of Rap-tourism ordination software related to sustainability index on the social culture dimension is 35.15 which means less sustainable. Figure 3 is a leverage factor of the social culture dimension. It shows that the socio-cultural dimension has 7 indicators with the sensitive indicator is volume and type of traveler's trip to the destination which has the RMS value 4.43.

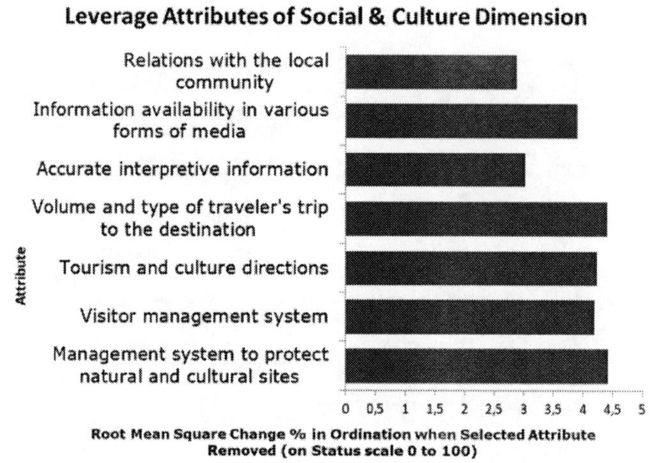

Figure 3. Leverage factor of Social & Culture Dimension

4.1.4. Institutional Dimension.
Result of Rap-tourism ordination software related to sustainability index in Institutional dimension is 34.65 which means less sustainable. Figure 4 is a leverage factor of the Institutional dimension.

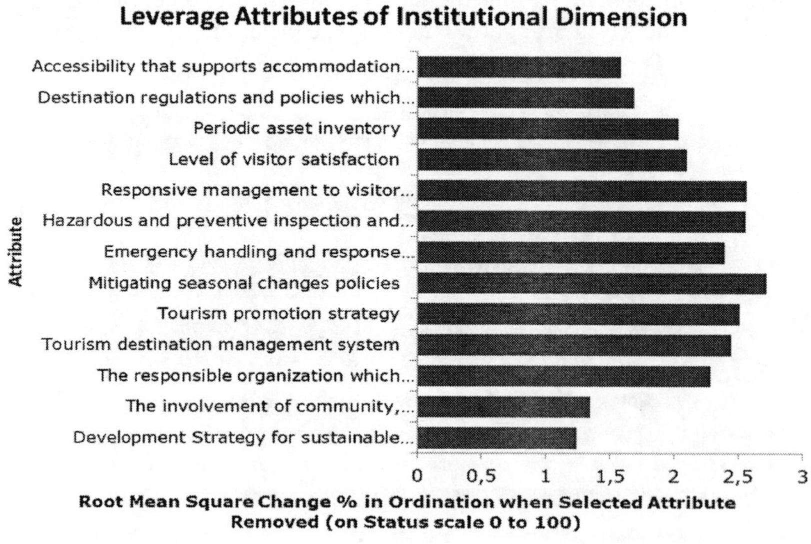

Figure 4. Leverage factor of Institutional Dimension

Institutional dimension has 13 indicators with RMS values to see sensitive indicators on the sustainability dimension. The leveraging factor with the highest RMS is mitigating seasonal changes policies with the value of RMS is 2.71.

4.2. Multidimensional Sustainability Index

Multidimensional sustainability index of Kampung Pelangi Semarang was obtained through weighting of each dimension. The ranking method was used to determine the weight of each dimension uses. This weighting is carried out by Kampung Pelangi Semarang tourism manager. Table 5 is the result of weighting and the final value of the sustainability status of Kampung Pelangi Semarang with a final index of 35.29.

Table 5. Weighting Index

Dimension	Rank	Sustainability Index	Weight (Wj)	Weighting Index
Environmental	1	36.58	0.4	14.63
Economic	4	32.40	0.1	3.24
Social & Culture	3	35.15	0.2	7.03
Institutional	2	34.65	0.3	10.39
Sustainability Index Value				35.29

The value from Table 5 then visualized as a kite diagram. Figure 5 is a visualization of the 4-dimensional sustainability index.

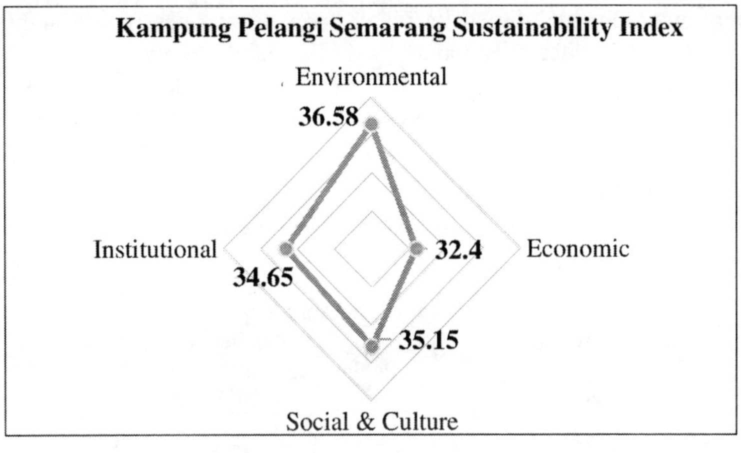

Figure 5. The Kite Diagram of Kampung Pelangi Sustainability Index

Besides the index, another result of rap tourism is R^2 or stressed value to find out whether the indicators reviewed in the MDS analysis are quite accurate and can be scientifically accountable. Low-stress values indicate high accuracy, while high S values indicate otherwise. A good model is indicated by a stress value smaller than 0.25 and for the coefficient of determination, it approaches the value of 1.0 or 100%. While for stress values and R2 can be seen in Table 6.

Table 6. Stress and R^2 Value

Parameter	Dimension			
	Environmental	Economic	Social Culture	Institutional
Stress	0.206	0.243	0.209	0.193
R^2	0.895	0.875	0.924	0.899

Based on the results of the conformity test in Table 6, the results of stress values are less than 0.25 and R2 values are close to 1 so that indicators in the dimension of sustainability have accuracy in describing their sustainability status.

The results of the Monte Carlo analysis show that the tourism sustainability index of Kampung Pelangi Semarang at the 95% confidence level does not much difference with the results of rap-tourism analysis (Multi-Dimensional Scaling = MDS). The differences in the MDS and Monte Carlo analysis sustainability index values are shown in Table 7.

Table 7. Differences in Monte Carlo & Rap-Tourism Analysis

No	Sustainability Dimension	Sustainability Index		Differences
		Monte Carlo	MDS	
1	Environmental	37.28	36.58	0.7
2	Economic	33.35	32.40	0.95
3	Social & Culture	35.83	35.15	0.68
4	Institutional	35.61	34.65	0.96

4.3. Improvement Recommendation of Kampung Pelangi Semarang Tourism

Alternative recommendations also assessed in terms of feasibility to be applied, other than in terms of their impact on system improvement if implemented [16]. At this stage, a feasibility analysis of recommendation is used to find out whether the overall alternative recommendations have been compared with the current conditions of Kampung Pelangi Semarang, which difficult to implement or not. Table 8 is an assessment of alternative recommendations for each dimension.

Based on Table 8 the selection of recommendations for environmental dimension is prioritized on the indicators of water quality and safety management systems because their impact on sustainability is considered to be greater than changes in the inventory of flora indicator if improvements are made. The recommendation of environmental dimensions is the availability of restroom facilities and installation of PDAMs. It can have a far better effect on tourism sustainability because restroom facilities have not been provided by the destination. Installation of PDAMs is also needed to get clean water because most of Kampung Pelangi's population still use groundwater. In the economic dimension, recommendations for paper floral skill training for the community can have a far better effect, because they can increase the income of the local community.

Table 8. Alternative Recommendation

Dimension	Sensitive Indicator	RMS	Recommendation	Value	Meaning
Environmental	Water quality and safety management system	6.07*	Availability of restroom facilities around the main spot	++	Much Better
			Public socialization about clean and healthy behavior	+	Better
			Installation of PDAMs	++	Much Better
Economic	The existence of attractions and tourism destination programs (related to culture)	7,76*	Developing the potential of Jathilan traditional dance	=	Same
			Children's dance creations	+	Better
	Involving and supporting local businesses	7.19*	Paper floral skill training for the community	++	Much Better
Social & Culture	Volume and type of traveler's trip to the destination	4.43*	*outbound* and *flying fox* tour	=	Same
			Water tour and games	+	Better
			Fishing festival	++	Much Better
Institutional	Mitigating seasonal changes policies	2.71*	Information facilities (poster/banner) for mitigating seasonal changes & risk of disaster threats	++	Much Better
			SOP for handling emergencies	+	Better
			Potential disaster socialization	+	Better

Kampung Pelangi Semarang is located behind Kalisari flower market and giving opportunities to exploit its potential in developing skills in making floral arrangements. Whereas the other recommendation such as Jathilan dance performances and children's dance creations still have several obstacles when implemented.

Selected recommendations on the socio-cultural dimension are fishing festivals that can have a far better effect. In 2017, the spreading of thousands of fish seeds was carried out by the Semarang City Government on the Kampung Pelangi River. This makes much fish that grow and develop on the river so that the festival has the potential to be held. While for recommendations for outbound & flying fox rides and water rides are still difficult to be implemented.

In the institutional dimension, recommendations which selected are Information facilities (poster/banner) for mitigating seasonal changes & risk of disaster threats. One feature of sustainable tourism is having a disaster risk management system or procedure. At present Kampung Pelangi Semarang still does not have an information facility as an effort to overcome seasonal changes or disaster risk. This recommendation is considered to have a far better effect on tourism sustainability.

5. Conclusion

Based on the discussion above, it can be concluded that sustainability status of Kampung Pelangi Semarang is categorized as less sustainable with sustainability index of 35.29. The index consists of four-dimensional value. The environmental dimension 36.58 (less sustainable), the economic dimensions 32.40 (less sustainable), socio-cultural dimensions 35.15 (less sustainable) and institutional dimensions 34.65 (less sustainable). MDS analysis also give the stress and R^2 values and leverage factor analysis to determine the focus on recommendation formulation. Recommendations for improvements that have been prepared for the development of Kampung Pelangi Semarang are (a) Availability of restroom facilities around the main spot of the tourist attraction; (b) installation of PDAMs in the destination area; (c) paper flower skill training; (d) organizing fishing festivals, and (e) making information facilities related to seasonal change mitigation actions & disaster risk.

References

[1] Ministry of Tourism 2016 *Regulation of Tourism Minister of the Republic Indonesia Number 14 in 2016 concerning Guidelines for Sustainable Tourism Destinations* (Jakarta: Tourism Ministry of the Republic Indonesia)

[2] UNWTO 2012 *Definition of Sustainable Tourism* (Madrid: United Nations World Tourism Organization)

[3] Dwikorawati S S 2012 *The Competitive and Sustainable Model of Tourism Management Policy in Bogor Regency Peak Area* (Bogor: Bogor Agricultural Institute).

[4] Abdullah T 2017 *J. Tourism and Hospitality Essentials (THE)* **7** 91-6

[5] Spillane J J 1982 *Indonesian Tourism, History and Prospects* (Yogyakarta: Kanisius)

[6] McIntosh R and Goeldner C 1984 *Tourism Principle, Practices, Philosophies* (Columbus: Grid Publishing Inc)

[7] Kozic I and Mikulic J 2011 *An Exploratory Assessment of Tourism Sustainability in Croatian Coastal Destinations* (*FEB-Working Paper Series* vol 11-04) (Croatia: University of Zagreb) pp 1-12

[8] Purwaningsih R, Cahyantari A.E, Ariyani Z, Susanty A, Arvianto A and Santoso H 2018 *Proc of MATEC Web Conference* vol 154 (EDP Sciences) p 01062

[9] Pitcher T and Preikshot D 2001 *Rapfish: A Rapid Appraisal Technique to Evaluate the Sustainability Status of Fisheries (Fisheries Centre Research Reports* vol 49*)* (Canada: Fisheries Centre, University of British Columbia) pp 255-70

[10] Kavanagh P and Pitcher T 2004 *Implementing Microsoft Excel Software for Rapfish: A Technique for the Rapid Appraisal of Fisheries Status (Fisheries Centre Research Reports* vol 12*)* (Canada:

Fisheries Centre, University of British Columbia) p 75
[11] Fauzi A and Anna S 2002 *Evaluation of Fisheries Development Sustainability Status: Application of Rapfish Approach (Case Study of DKI Jakarta Coastal Waters)* (Jakarta: PT Gramedia Pustaka Utama)
[12] Purwaningsih R, Yudha MC, and Susanto N 2016 *J. Ind. Eng* **18** 31-42
[13] United Nations Environment Programme (UNEP) and the World Tourism Organization 2005 *Making Tourism More Sustainable: A Guide for Policy Makers* (Madrid: WTO)
[14] WTO and Ministry of Tourism and Environment of Montenegro 2007 *Sustainable Tourism Indicators and Destination Management* (Montenegro: World Tourism Organization dan Ministry of Tourism and Environment of Montenegro)
[15] Swarbrooke J 1999 *Sustainable Tourism Management* (Sheffield: CABI Publishing)
[16] UNEP 2007 *Sustainable Tourism Destination* (Jakarta: Tourism Ministry of the Republic of Indonesia)

IOP Conf. Series: Materials Science and Engineering **722** (2020) 012051 doi:10.1088/1757-899X/722/1/012051

Cost and Emission Optimization in Power Plant Using the Single Echelon Economic Dispatch Model (SEED)

Wahyuda[1], Muslimin[2], and S U Widodo[3]

[1]Industrial Engineering, Universitas Mulawarman, Samarinda, Indonesia
[2]Electrical Engineering, Universitas Mulawarman, Samarinda, Indonesia.
[3]PT. PLN Samarinda, Indonesia

Email: Wahyuda@ft.unmul.ac.id; Wahyuda@gmail.com

Abstract. The electricity supply chain system consists of power plants, transmission and distribution. Differences in power plant characteristics cause differences in fuel costs and emissions. Inaccuracies in scheduling the plant have an impact on the surge in fuel use. This becomes the basis for the optimization of power plant scheduling. Optimization on the generator side is usually using the economic dispatch model. However, the Economic Dispatch Model does not consider optimization on the transmission network. Optimization on the transmission network is needed because there are losses in the transmission network. Losses will be even greater when the distance between the power plant and the customer is getting further away. So if distance can be minimized, fuel costs can be reduced. Minimizing the distance for shipping goods can use the transportation model, but the transportation model cannot be used for electricity distribution. This is because there are differences between goods and electricity. Therefore, this study proposes the Single Echelon Economic Dispatch (SEED) model. This model is a combination of the economic dispatch model and the transportation model. This model is able to do simultaneous optimization between the generator side and the transmission side. The model is applied to the Mahakam system in East Kalimantan, using all PLN's power plants. Load consists of two types, namely low load and peak load. As a result, in peak load and low load conditions, the P11 power plant produces the biggest losses and emissions. This is because electricity production in P11 has to go through a longer path compared to other plants. The P11 is also a power plant that requires the highest fuel costs

1. Introduction

The electric power interconnection system is a system where Power plants located in various locations are connected to all customers in various regions. This interconnection causes a group of power plants to act as a group of suppliers. If all the power plants have the same characteristics, the allocation of the burden of each power plant is easy [1], [2]. In fact, the power plant has different characteristics. This difference causes differences in total costs and emissions so that the total production of each power plant must be calculated in advance to reduce the impact on increasing costs and emissions [3], [4].

One model used to schedule and determine the amount that must be produced by each plant is Economic Dispatch. This model is able to perform optimization on the generator side so that it can get scheduling with a minimum total fuel cost.

However, the economic dispatch model does not consider the distribution side. The distribution side is important because of losses during electricity delivery [5], [6]. Losses are the partial loss of electricity sent from source to destination, where the farther the distance from the power plant to the customer, the greater the losses that occur.

Increased losses can lead to higher fuel costs and emissions. This is because the power plant must produce more electricity than the total demand [7], [8]. This is done to cover losses. Increasing the amount of production which causes an increase in costs and emissions. Based on this, losses must be managed.

Loss management by optimizing the shipping path can be done on the transmission network. A set of power plants must be determined to precisely supply which set of power plants. So that we get a minimum total distance. Thus it can produce smaller losses [9].

Allocating shipping lines usually uses a transportation model. Many transportation models are applied to various real problems. This is because this model is simple and easy to use. Various problems are solved by the transportation model including, the allocation of labor, the selection of factory locations, the selection of supplier locations, the distribution of goods, and others. This model is simple because it only requires data on production capacity in a source, demand capacity for a destination, and transportation costs from source to destination.

However, the transportation model cannot be used for electricity allocation and distribution. This is because there are differences between the nature of electricity and manufactured goods. There are 2 main differences between electricity and manufactured goods. First, electricity cannot be stored. Second, electricity must always be available.

Therefore, this study proposes a new model called the Single Echelon Economic Dispatch (SEED). This model combines the advantages of the transportation model and the economic dispatch model so that the combined optimization between the production side and the distribution side is obtained. This model is able to make detailed electricity generation and distribution scheduling allocations with minimal total costs.

2. Literature Review

2.1. Single Echelon Transportation Model

The transportation model manages the delivery of goods from a collection of sources to a set of destinations. This model is introduced by (Hitchcock, 1941). In this case, commodities can be sent from various sources to various destinations. The mechanism is illustrated in the following figure:

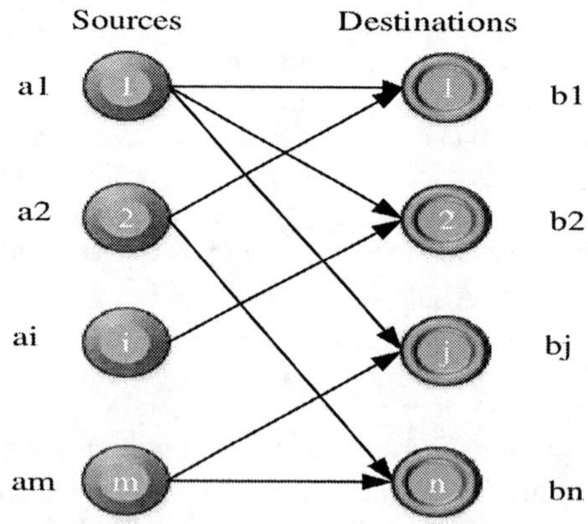

Figure 1. The basic concept of the transportation model is m source n destination

In general, the transportation model divides the problem into 2 groups, namely the source group and the destination group. Sources can consist of one or more members, as well as goals. The transportation model used to calculate transmission costs can be seen (Pablo, Ramirez, & Coello, 2009) described in (Daskin, 1995) as follows:

$$Minimize \quad \sum_i \sum_j C_{ij} X_{ij} \tag{1}$$

Constraints:

$$\sum_j X_{ij} \leq S_i \quad \forall i \tag{2}$$

$$\sum_i X_{ij} \geq D_j \quad \forall j \tag{3}$$

$$X_{ij} \geq 0 \quad \forall i,j \tag{4}$$

where
S_i= Supply capacity at source i
D_j = Demand at point j.
C_{ij} = Shipping costs from source i to destination j.
X_{ij}= Number of items sent from source i to destination j.

The main difference between transportation for goods and transportation for electricity is the balance between supply and demand. In the transportation model for goods, suppliers are allowed to send goods greater than demand. (Equation 3), but in transportation for electricity, suppliers must send goods as large as demand or called equilibrium. This point of equilibrium is called the Balanced Transportation Problem (BTP). Variations in BTP can be seen in Figure 2 (Sabbagh, Ghafari, & Mousavi, 2015). The difference between the two also occurs at the supplier's capacity limits.

Model BTP-1:	Model BTP-2:	Model BTP-3:
$Min\, z = \sum_{i=1}^{m}\sum_{j=1}^{n} c_{ij}.x_{ij}$	$Min\, z = \sum_{i=1}^{m}\sum_{j=1}^{n} c_{ij}.x_{ij}$	$Min\, z = \sum_{i=1}^{m}\sum_{j=1}^{n} c_{ij}.x_{ij}$
Subject to	Subject to	Subject to
$\sum_{j=1}^{n} x_{ij} \geq s_i, (i = 1 \ldots m)$	$\sum_{j=1}^{n} x_{ij} \leq s_i, (i = 1 \ldots m)$	$\sum_{j=1}^{n} x_{ij} = s_i, (i = 1 \ldots m)$
$\sum_{i=1}^{m} x_{ij} \leq d_j, (j = 1 \ldots n)$	$\sum_{i=1}^{m} x_{ij} \geq d_j, (j = 1 \ldots n)$	$\sum_{i=1}^{m} x_{ij} = d_j, (j = 1 \ldots n)$
$x_{ij} \geq 0$	$x_{ij} \geq 0$	$x_{ij} \geq 0$

Figure 2. Three variations of Balanced Transportation Problem (BTP)

Where,
c_{ij} : shipping costs from source i to destination j
x_{ij} : the number of items sent from source i to destination j
s_i : the maximum amount of supply of resources i
d_j : demand at destination j.

2.2. Economic Dispatch Model
Economic dispatch was introduced since 1928. There are 3 researchers who are considered as the originator of the economic principle of the generator (Estrada, 1930; Stahl, 1931; Wilstam, 1928). The initial economic dispatch is called the classic Economic dispatch model. This model uses the concept of the base load method and the best point load method. How it works, sort generator units based on

the highest efficiency level. Furthermore, generator scheduling is given to the generating unit with the highest level of efficiency, and so on until the last generating unit.

When there are differences in the characteristics of each plant, the base load technique becomes less effective. Therefore, a new technique emerged known as equal incremental cost. The main concern of this technique is the characteristics of each different generator. The way it works, the meeting point of all generators is searched, and the optimal allocation is made based on this meeting point. This equal incremental cost technique is still used today. This technique was introduced by (Steinberg & Smith, 1933). The advantage of this technique is that it can provide a low total cost for all the plants involved in the system. However, this initial model still has shortcomings, namely the losses in the transmission network have not been considered. One of the causes of losses in transmission networks is the length of the transmission network. The longer the transmission distance, the greater losses will occur. These losses will ultimately affect the total cost of fuel because the plant must produce more electricity than the demand for compensation losses. Furthermore, Economic dispatch that considers losses in the network is introduced by (George et al., 1949). The Economic Dispatch formula is as follows:

$$F_i = a_i P_i^2 + b_i P_i + c_i \tag{5}$$

$$P_{i\,min} \leq P_i \leq P_{i\,max} \tag{6}$$

3. Proposed Model

3.1. Single Echelon Transportation Model

Transportation Model has advantages in the field of distribution. The use of this model causes the total shipping costs to be minimal. While the Economic Dispatch model has an advantage in the field of generator loading allocation or generator scheduling. The use of economic dispatch models is able to create a power plant scheduling that results in minimal fuel costs. Combining the advantages of these two models provides several advantages. First, the development of a new transportation model called the Single Echelon Economic Dispatch (SEED). Second, the combined optimization between the generator side and the distribution side. Conceptually, the development of the SEED model can be seen in the following figure:

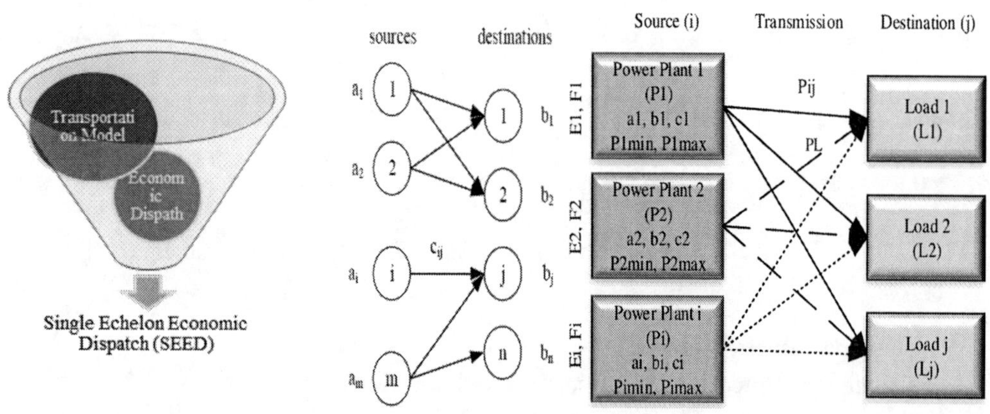

Figure 3. Conceptual Model for Single Echelon Economic Dispatch (SEED) development

Figure 3.a conceptually illustrates that the SEED model is formed from two models, namely the transportation model and the economic dispatch model. Figure 3.b is a conceptual model of the transportation model. While Figure 3.c is an economic dispatch conceptual model. In the same way, both models are used for resource allocation. While the difference is in the object faced, where transportation is commonly used in people or goods while ED is used in electricity. The nature of the two objects is different. The main requirement for electricity is in the form of an equation between the supply side and the demand side. Therefore, as another approach used is the Balanced Transportation Problem as shown in Figure 2.

Three types of BTP as shown in Figure 2 are variations of the transportation model application for real cases. The three variations have the same objective function, namely minimization of costs (min z). While the difference between the three is the model limitation. Cost is the sum of the shipping costs per unit from each source i to destination j denoted by multiplied by the number of items sent from source i to destination j denoted by x_{ij}.

The characteristics of the three variations of BTP are used as a reference for the development of the SEED model. Development is carried out by combining several boundaries so that new variations of BTP emerge. Furthermore, a merger with the Economic Dispatch model was obtained to obtain the SEED model. The result of merging BTP with economic dispatch as in figure 4.

***Single Echelon Transportation Problem* (Modification BTP)**	***Economic Dispatch* (Equations 2.1 – 2.3)**	**Proposed Model (*Single Echelon Economic Dispatch*, SEED)**
$Min\ z = \sum_{i=1}^{m}\sum_{j=1}^{n} c_{ij} . x_{ij}$ Subject to $\sum_{j=1}^{n} x_{ij} \geq s_{imin}, (i = 1 \ldots m)$ $\sum_{j=1}^{n} x_{ij} \leq s_{imax}, (i = 1 \ldots m)$ $\sum_{i=1}^{m} x_{ij} = d_j, (i = 1 \ldots n)$	$F_i = a_i P_i^2 + b_i P_i + c_i$ $P_{i\,min} \leq P_i \leq P_{i\,max}$ $P_i = D_i$	$Minimize\ Fi = \sum_{i=1}^{I}\sum_{j=1}^{J} a_i P_{ij}^2 + b_i P_{ij} + c_i$ $\sum_j P_{ij} \geq P_{i\,min}$ $\sum_j P_{ij} \leq P_{i\,max}$ $\sum_i P_{ij} = L_j + P_{Lij}\ (j = 1 \ldots J)$ $P_{ij} \geq 0$

Figure 4. Development mechanism of the proposed model

The SEED model has the objective function of minimizing fuel costs as well as in the economic dispatch model. While the difference between them lies in the coverage of the model. This can be seen from the notation used. The basic model of economic dispatch uses notation P_i which means the amount of electricity produced at the generator i. Whereas the SEED model uses notation P_{ij} which means the amount of electricity produced by the generator i sent to destination j. If this scope is included in the objective function, then this model is prepared to be able to complete 2 tasks, namely production optimization tasks, and simultaneous distribution optimization tasks. As a guarantee of a feasible solution, the SEED model is also equipped with 3 constraints.

4. Experiment and Results

This chapter discusses experiments using the SEED model. Divided into basic models, supply and demand data, and transmission data

4.1. Basic Model

There are 2 types of objectives that can be used in the SEED model, namely minimizing fuel costs (Rp) and minimizing emissions (kg), as shown in the following formula:

$$Minimize\ F_i = \sum_{i=1}^{K}\sum_{j=1}^{J} a_i P_{ij}^2 + b_i\ P_{ij} + c_i \quad (7)$$

IOP Conf. Series: Materials Science and Engineering 722 (2020) 012051 doi:10.1088/1757-899X/722/1/012051

$$Minimize\, F_i = \sum_{i=1}^{K}\sum_{j=1}^{J} d_i P_{ij}^2 + e_i\, P_{ij} + f c_i \tag{8}$$

Constraint,
Limit of Power Plant:

$$\sum_j P_{ij} \geq P_{i\,min} \tag{9}$$

$$\sum_j P_{ij} \leq P_{i\,min} \tag{10}$$

Power Balance Constraint

$$\sum_i P_{ij} = L_j + P_{Lij} \;\; (j = 1, \ldots .J) \tag{11}$$

Non-negativity constraint

$$P_{ij} \geq 0 \tag{12}$$

Where,

Fi : Total fuel costs (Rp)
Ei : Total emissions CO_2
a_i, b_i, c_i : coefficient of fuel cost curve for i-generation.
d_i, e_i, f_i : CO2 emission curve coefficient for i-generation
P_{ij} : The amount of electricity produced by the generator i sent to the customer j (MW)
$P_{i\,min}$: Lower limit of i-generator production (MW)
$P_{i\,max}$: Upper limit of i-generator production (MW)
L_j : Demand for customers j (MW)
P_{Lij} : Losses on the transmission network between i and j (MW)

4.2. Data of Experiments

The proposed model is used in the Mahakam interconnection system in East Kalimantan. This system has 11 power plants (P1 to P11) located in various locations. While demand is in 4 cities, namely Samarinda, Bontang, Balikpapan and Tenggarong. Demand for these 4 cities is divided into 12 locations (L1 to L12) as shown in the following table:

Table 1. The distance between the Power plant and location

	Distance (km)											
P	L1	L2	L3	L4	L5	L6	L7	L8	L9	L10	L11	L12
1	1.0	21.5	16.8	25.1	37.1	44.3	123.9	72.4	100.5	113.9	113.9	135.1
2	16.8	37.3	1.0	9.3	21.3	28.5	108.1	56.6	84.7	98.1	98.1	119.3
3	25.1	45.6	9.3	1.0	13.0	20.2	99.8	48.3	76.4	89.8	89.8	111.0
4	113.9	134.4	98.1	89.8	101.8	109.0	188.6	137.1	14.4	14.4	1.0	22.2
5	135.1	155.6	119.3	111.0	123.0	130.2	209.8	158.3	35.6	35.6	22.2	1.0
6	72.4	92.9	56.6	48.3	36.3	29.1	52.5	1.0	123.7	137.1	137.1	158.3
7	1.0	21.5	16.8	25.1	37.1	44.3	123.9	72.4	100.5	113.9	113.9	135.1
8	123.9	144.4	108.1	99.8	87.8	80.6	1.0	52.5	175.2	188.6	188.6	209.8
9	123.9	144.4	108.1	99.8	87.8	80.6	1.0	52.5	175.2	188.6	188.6	209.8
10	123.9	144.4	108.1	99.8	87.8	80.6	1.0	52.5	175.2	188.6	188.6	209.8
11	111.6	132.1	95.8	87.5	99.5	106.7	186.3	134.8	12.1	25.5	25.5	46.7

IOP Conf. Series: Materials Science and Engineering **722** (2020) 012051 doi:10.1088/1757-899X/722/1/012051

Each interconnected power plant has characteristics like the following table:

Table 2. Characteristics of Power Plant

	Limit of Power Plant		Constanta of cost function		
L	Pimin	Pimax	a	b	c
1	23.5	14.0	32.0	-256.0	1,200.0
2	16.0	5.0	-1.1	263.9	250.0
3	3.1	2.0	17.0	49.9	150.0
4	3.6	2.0	32.0	256.0	256.0
5	7.8	6.0	5.3	-2.6	230.0
6	14.2	5.0	32.0	-256.0	2,000.0
7	65.0	5.0	32.0	-256.0	3,500.0
8	10.0	7.0	57.8	3.0	157.0
9	2.4	1.6	17.0	49.9	200.0
10	1.0	0.0	17.0	49.9	500.0
11	200.0	20.0	32.0	-256.0	3,000.0

The experiment is divided into two conditions. First, the allocation of power plant loading at low loads. Second, the allocation of the power plant loading at peak load. In low load conditions, the demand is 194.8 MW, while the total production of all plants is 195.22 MW. There is a difference between the amount of supply and demand as much as 0.36 MW. This difference is used to anticipate losses that occur along the transmission network to keep energy equilibrium continues (Equation 11). The power plant with the biggest losses is P11 (192 KW). This is because electricity produced by the P11 generator must go through the longest transmission line, which is 1,063.9 km. The total emissions generated by the system are 96.59 kg, with the largest emissions in P11 (29.91 kg). The total fuel cost in this low load condition is Rp 215,377.36

At peak load, demand increases to 343.4 MW, while the total production of all power plants is 345.76 MW. There is a difference between the amount of supply and the amount of demand of 2.36 MW. This difference is used to anticipate losses that occur along the transmission network to keep energy equilibrium continues (Equation 11). The biggest power plant with Losses is P11 (115.34 MW). This Losses is caused by the electricity produced by the P11 power plant having to go through the longest transmission line, which is 3,671.6 km. The total emissions generated by the system are 185.59 kg, with the largest emissions in P11 (115.34 kg). The total fuel cost at peak load conditions is Rp. 1,374,744.5

5. Conclusions

The transportation model was successfully developed for applications in the case of electricity production and distribution. Development is done by combining transportation models with the Economic Dispatch model. This combined model is called the Single Echelon Economic Dispatch (SEED).

The output of the SEED model is in the form of electricity distribution in detail. SEED is also able to calculate the electric distance from a power plant to a customer. During low loads and peak loads, P11 is the plant plant with the farthest distance and the largest losses. P11 is also the power plant with the largest emissions.

Acknowledgements

This research was funded by the Directorate of Higher Education Ministry of National Education Republic of Indonesia Fiscal Year 2019

References

[1] R. P. Parouha and K. N. Das, "A novel hybrid optimizer for solving Economic Load Dispatch problem," *Int. J. Electr. Power Energy Syst.*, vol. 78, pp. 108–126, 2016.

[2] Wahyuda, B. Santosa, and A. Rusdiansyah, "Load allocation of power plant using multi echelon economic dispatch," in *AIP Conference Proceedings*, 2017, vol. 1902.

[3] A. Y. Abdelaziz, E. S. Ali, and S. M. A. Elazim, "Combined economic and emission dispatch solution using Flower Pollination Algorithm," *Int. J. Electr. Power Energy Syst.*, vol. 80, pp. 264–274, 2016.

[4] Wahyuda, B. Santosa, and A. Rusdiansyah, "Cost analysis of an electricity supply chain using modification of price based dynamic economic dispatch in wheeling transaction scheme," *IOP Conf. Ser. Mater. Sci. Eng.*, 2018.

[5] I. Gani, Wahyuda, B. Santosa, and A. Rusdiansyah, "Analysis of costs and emissions on the addition of production capacity of the power plant using multi echelon economic dispatch Analysis of Costs and Emissions on The Addition of Production Capacity of The Power Plant Using Multi Echelon Economic Dispatch," in *AIP Conference Proceedings*, 2019, vol. 060016, no. June.

[6] D. Zou, S. Li, Z. Li, and X. Kong, "A new global particle swarm optimization for the economic emission dispatch with or without transmission losses," *Energy Convers. Manag.*, vol. 139, pp. 45–70, 2017.

[7] I. Gani, Wahyuda, B. Santosa, and Muliati, "Multi Echelon Distribution Model for Electric Market Deregulation Collaboration Strategy in East Kalimantan Multi Echelon Distribution Model for Electric Market Deregulation Collaboration Strategy in East Kalimantan," in *IOP Conference Series: Materials Science and Engineering*, 2019.

[8] M. B. Nappu and A. Arief, "Network Losses-based Economic Redispatch for Optimal Energy Pricing in a Congested Power System," *Energy Procedia*, vol. 100, no. September, pp. 311–314, 2016.

[9] T. Soares, F. Pereira, H. Morais, and Z. Vale, "Cost allocation model for distribution networks considering high penetration of distributed energy resources," *Electr. Power Syst. Res.*, vol. 124, pp. 120–132, 2015.

IOP Conf. Series: Materials Science and Engineering **722** (2020) 012052 doi:10.1088/1757-899X/722/1/012052

Research on a Project Schedule of Factory Renovation for Manufacturing Plants in Bandung

W Tripiawan[1], S Miranda[2]

[1] School of Industrial Engineering, Telkom University, Bandung, Indonesia
[2]Department of Industrial Engineering, Universitas Islam Indonesia, Yogyakarta, Indonesia

E-mail: [1]wawan.tripiawan@gmail.com, [2]suci.miranda@uii.ac.id

Abstract. Renovating a plant facility is an example of a project. It is a complex process that the project needs to complete all the task while ensuring the operations without disruption. In project management, scheduling and cost optimization are two main operation problems. These correlate with each other and have proven mathematical solutions for the basics. In this study, the case study faced was to prepare a schedule plan in the process of renovating factory facilities. The facilities to be renovated are based on 2 main facilities, namely Production office facilities and Fabrication plant facilities. The implementation of the renovation of the 2 facilities has constraints in the time available. With the CCPM approach, this study proposes a schedule for implementing renovation projects with limited availability of time. The results obtained; this factory facility renovation project is scheduled to be completed less than 40 calendar days.

1. Introduction

A project is a temporary effort to create a unique product, service, or result. Project management is the implementation of knowledge, skills, tools, and techniques to project activities to satisfy the project requirements [1]. Uncertainty is one reason to differentiate project management as a separate field of knowledge. The project duration and its success are affected by the way the uncertainty is managed in the project (and risk management in particular). According to some studies conducted for the traditional project management methods, only 44% of the projects are completed in time. Of the total projects, 70% of them decrease the amount of planned work, while the rest end without being realized[2].

Every work on the project will deal with the risks that must be faced. The risk that can occur when working on a project is the delay. According to [3], the delay in project execution in general can be caused by several factors including labor, materials (equipment), equipment, site characteristics (site characteristics), managerial (managerial), financial (financial), factors others include the intensity of rainfall, economic conditions, and workplace accidents. Delays or discrepancies in the timing of a project can have a negative impact on the project being carried out. The most likely thing to happen is the increase in the cost of working on a project / over budget and will cause demands from project owners who want a timely settlement. If there is a change in time or delay in the execution of a project, then the budget is also changed, therefore the accuracy of the time in the project is very important.

Project scheduling is the utilization of skills, techniques, and intuition obtained through the combination of knowledge and experience to present a framework for schedule model creation as well as to evolve effective schedule models [4]. The precedence diagram method (PDM), Critical Path Method (CPM), and Critical Chain are the most common scheduling methods, supported by all the major scheduling tools [1]. Particularly, Critical Chain Project Management (CCPM) focuses on the limitation of project resource and requires some flexibility in order to keep the project in its time borders [2]. This is a project planning and management method that places the major emphasis on the resources required to undertake project tasks [5]. It can be applied both in a single project and multi-project structures where resources are being used in several projects concurrently [6].

In the current field of project management, a scheduling method is used to handle unpredictability as well as negative effects on project accomplishment and other projects without the need to add resources. This method is known as the Critical Chain Project Management method. Critical Chain is a method for designing and managing projects that focus on the resource requirements to conduct the project, for which the Critical Chain has been positioned differently from the Critical Path and PERT methods, which emphasizes work orders and scheduling.

Buffers are added to the duration used in project scheduling to cover the critical chain for the success of the project, in the critical chain project management method. There are many factors that influence the size of a project buffer, there are risk factors as follows: disruptions in material and tool supplies, irregular financing, design errors, bad weather, equipment damage, inefficient contractors, legitimate administrative disruptions, etc. According to [7], buffer management is the key to managing activities in the critical chain of project schedules. To complete the project earlier than the planned schedule with resource constraints, the buffer in the Critical Chain is used to solve the problem. The buffers used in the critical chain are as follows: 1. Project buffers are to prevent the project's final completion time from the schedule uncertainty in critical chain activities; 2. In order to protect and maintain the performance of critical chain network activities from change due to the schedule uncertainty in activities from non-critical networks so as not to interfere with activities in critical networks in a dependency relationship, feeding buffers are required; and 3. Resource buffers are to anticipate and guarantee the security of the availability of the resources so that there is no additional time for the critical chain.

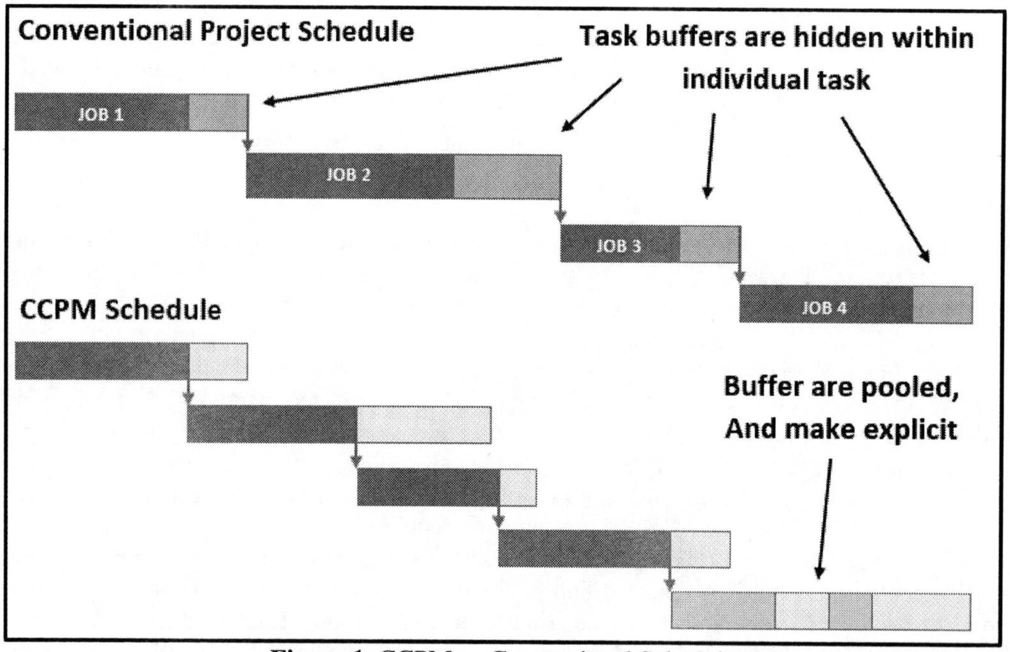

Figure 1. CCPM vs Conventional Schedule

According to [8], one approach that is often used in determining the buffer size is simple to determine the project buffer and feeder buffer, namely cut and paste method (C & PM). This cut and paste (C&PM) method is an adhesive rule used to determine project buffers and feeding buffers in C&PM basically cuts 50%, 40%, 30%, 20%, or 10% of the duration for all activities, and attaches project buffers by half

IOP Conf. Series: Materials Science and Engineering **722** (2020) 012052 doi:10.1088/1757-899X/722/1/012052

the critical chain at the end chain, as well as attaching a buffer with half the duration of activity to activities on a path that is not a critical (non-critical chain) chain that leads to a critical chain.

It is often assumed that renovation projects are simply a special type of new construction in the construction industry. They are often arranged in the same way, although mostly with a more traditional division of labor and contract forms, and lack of standardization. However, there are a number of dissimilarities between the process of new building and renovation projects[9]. Renovation is connected with improvement and rehabilitation for restoring [10].

This research is conducted in a company engaged in manufacturing for electronic equipment in Bandung. Production development requires the company to increase its production capacity. Thus, the company plans to renovate an old plant with a larger capacity, which is owned by the parent company. In order to renovate the plant facility, this study aims to proposes a developed project scheduling utilizing CCPM. The planned schedule using a common technique is compared with the result showed in CCPM. These outputs can assist the company to determine the project schedule for the renovation planning process in order to avoid project delay.

2. Methodology

CCPM is defined as a sequence of critical tasks whose duration influences the overall duration of the project. It concentrates on the project completion date as the only important date.

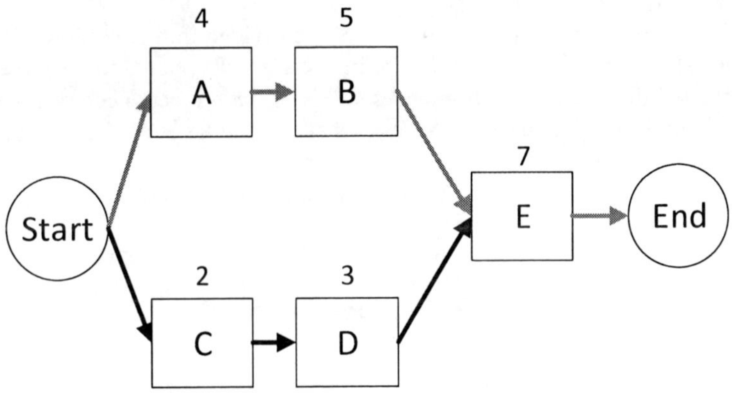

Figure 2. Critical Path Example

As illustration, Fig. 2 shows the critical path of a project, then by using C&PM approach, each activity is simulated to be deducted with 50%, 40%, 30%, 20%, or 10% of the duration.

The example of the calculation from each activity (A – E) after we implement C&PM. Activity A, with the default duration is 4 (unit of time), and then we use C&PM 40%, 30%, 20%, or 10% so the remaining duration is:

A(C50) = 4 x 50% = 2
Remaining duration is 4 – 2 = 2
A(C40) = 4 x 40% = 1,6
Remaining duration is 4 – 1,6 = 2,4 ≈ 2
A(C30) = 4 x 30% = 1,2
Remaining duration is 4 – 1,2 = 2,8 ≈ 3
A(C20) = 4 x 20% = 0,8
Remaining duration is 4 – 0,8 = 3,2 ≈ 3
A(C10) = 4 x 10% = 0,4
Remaining duration is 4 – 0,4 = 3,6 ≈ 4

IOP Conf. Series: Materials Science and Engineering **722** (2020) 012052 doi:10.1088/1757-899X/722/1/012052

All the calculation use roundup or round down, to get better time unit calculation. The result from each activity are shown in Table I.

Table 1. C&PM Illustration

Activity	The remaining duration of activity using C&PM				
	C50	C40	C30	C20	C10
A	2	2	3	3	4
B	3	3	4	4	5
C	1	1	1	2	2
D	2	2	2	2	3
E	4	4	5	6	7

3. Result and Discussion

3.1. Coordinate the sequence of projects

Based on the CCPM concept [2], project resources are divided into two categories as follows: critical and non-critical task resources. In this study, the important tasks related to resources are arranged, since they directly affected the overall duration and success of the project. Then, it is required to ensure the availability of resources needed for the next task when several tasks on the critical path are completed.

Table 2. Work Breakdown Structure

Task Code	TASK	DAYS
A	**[Production Office]**	
A1	[Mobilization]	1
A2	[Destruction]	1
A3	[Clean up]	1
A4	[Floor-job]	8
A5	[Wall-job]	4
A6	[Frame and door]	4
A7	[Stair railing]	4
A8	[Ceilings and stairs]	4
A9	[Painting]	8
A10	[Company logo-job]	4
A11	[Furniture-job]	4
B	**[Fabrication Plant]**	
B1	[Destruction]	1
B2	[Clean up]	1
B3	[Wall-job]	9

IOP Conf. Series: Materials Science and Engineering **722** (2020) 012052 doi:10.1088/1757-899X/722/1/012052

Task Code	TASK	DAYS
B4	[Frame and door]	5
B5	[Ceilings and stairs]	9
B6	[Painting job]	9
B7	[Company logo-job]	4
B8	[Watcrproofing]	9

Based on WBS data in Table II, it can be seen that the task series on site 1 (Production Office) and site 2 (Fabrication Plant) have almost the same set of tasks. From the list of tasks, we arranged which tasks are important and are directly related to the availability of resources. After taking some interview with the project team, we separated the task based on the importance level of the project. Table III presents the list of the task with a high importance level.

Table 3. Important Task

Task Code	*Important Task*
A	**[Production Office]**
A4	[Floor-job]
A9	[Painting]
B	**[Fabrication Plant]**
B3	[Wall-job]
B5	[Ceilings and stairs]
B6	[Painting job]
B8	[Waterproofing]

3.2. *Time planning of the project implementation*

The next step of the critical chain method is to plan a safety time for the project in order to ensure that the planned time has not been wasted. It is also to assure the project runs as short as possible and the project's performance is in good condition as well. Adopting C&P M, all tasks on the Critical Path execution time estimate is usually decreased by 25% to 50%, and this shortened time is assigned at the end of the project into a common safety buffer while the small tasks feeding buffers are at the end of each branch circuit at the points merging this chain part into the main Critical Chain. In this research, we took a 25% deduction from the important task.

Table 4. Detailed Important Task

Task Code	*Importance Task*	*Duration (days)*	*C&PM C25*
A	**[Production Office]**		
A4	[Floor-job]	8	6
A9	[Painting]	8	6
B	**[Fabrication Plant]**		
B3	[Wall-job]	9	7

Task Code	***Importance Task***	***Duration (days)***	***C&PM C25***
B5	[Ceilings and stairs]	9	7
B6	[Painting job]	9	7
B8	[Waterproofing]	9	7

3.3. Define Critical Chain Activity

Drawing the network diagram of critical path presented in Figure 3 below, there were 7 tasks within the path highlighted in red: destruction (B1), clean up (B2), wall-job (B3), ceilings and stairs (B5), painting job (B6), and waterproofing (B8). Referring to the data in Table III, the critical chain of this project is the activities of A4 (floor-job), A9 (painting), B3 (wall-job), B5 (ceilings and stairs), B6 (painting job), B8 (waterproofing) with a duration of 52 days. This Critical chain must be considered so that they will not experience any problems, because they can affect the completion duration of the entire project. The critical chain is shown in the following Figure 4.

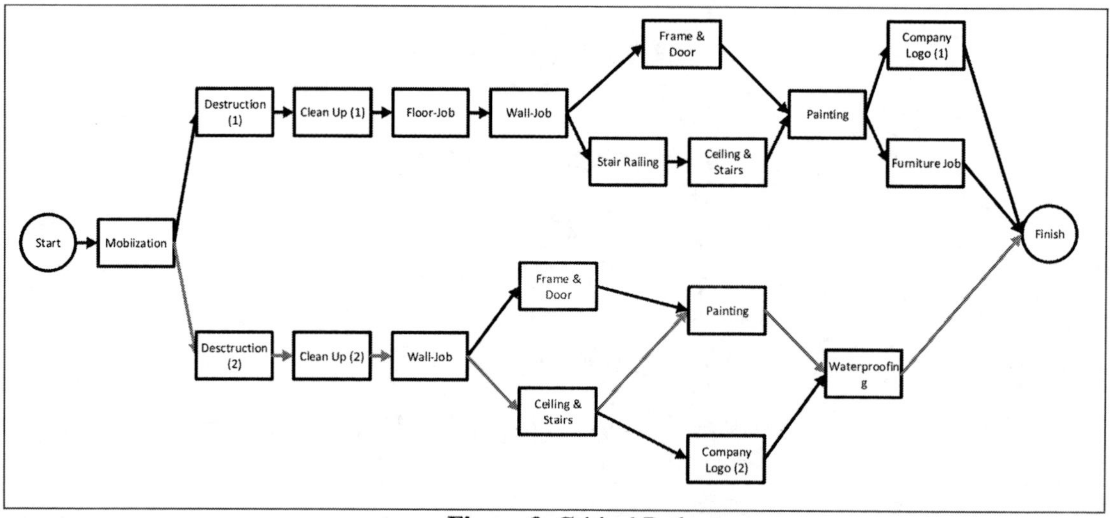

Figure 3. Critical Path

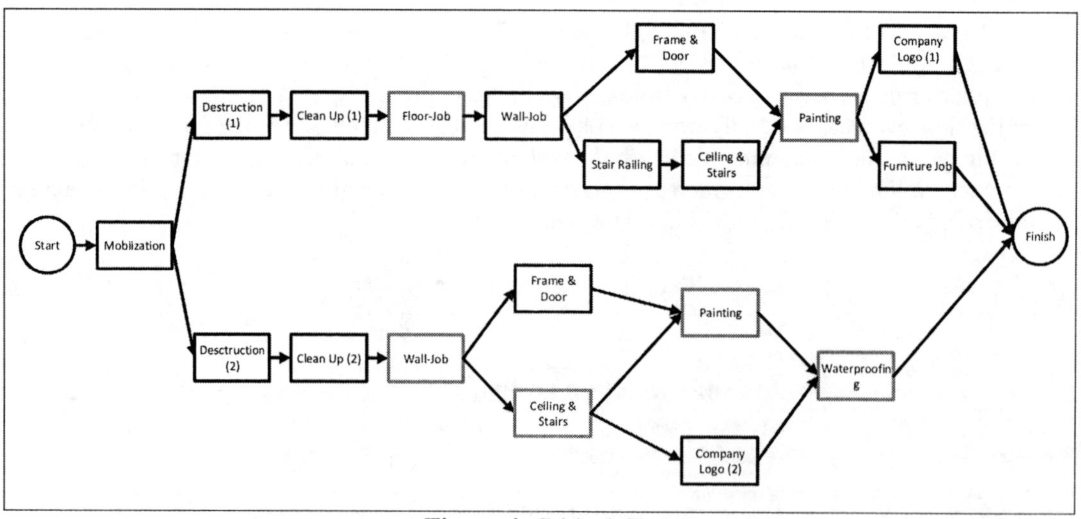

Figure 4. Critical Chain

Based on the data Fig. 3 and Fig. 4, the activities of B3, B5, B6, and B7 are the critical chain of this project with a duration of 28 days.

3.4. Using buffers in the project

Priority activities have the highest level of sensitivity to project delays, so it is necessary to protect these activities. Moreover, priority activities are activities including both in the critical path and critical chain. To protect these activities, it can be conducted by entering the project buffer at the end of the critical chain.

Project buffer = Σ duration (B3+B5+B6+B8) days x 25%

= 28 days x 25%

= 7 days

3.5. Application of CCPM Calculation with Microsoft Project

The Microsoft Project software was employed for CCPM calculation. Both the traditional and CCPM produced a different result. The traditional schedule completed in 39 days of project time (Fig. 5), while the CCPM schedule showed 4 days faster including 7 days of project buffer (Fig. 6).

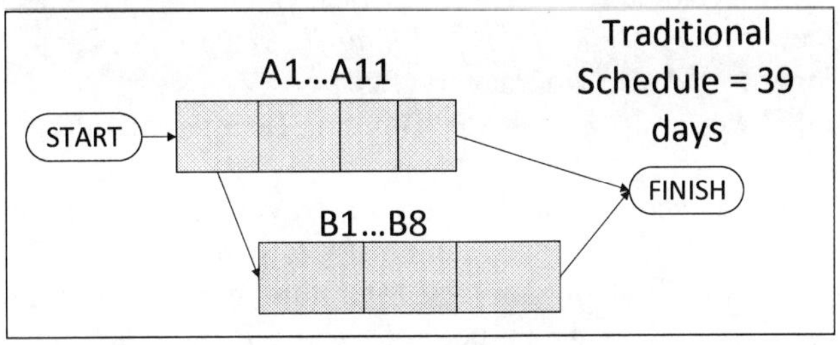

Figure 5. Traditional Schedule

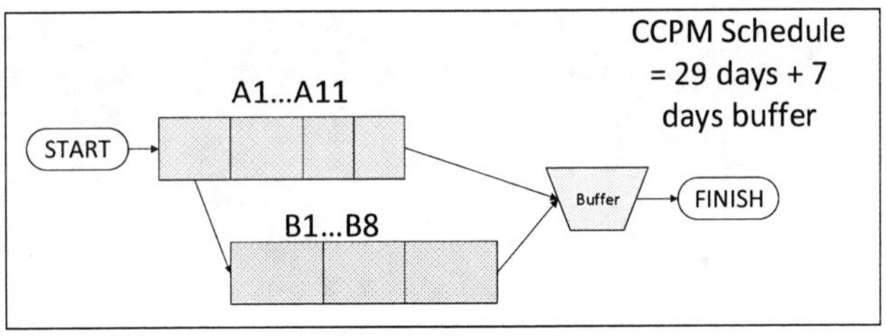

Figure 6. CCPM Schedule

Thus, the CCPM calculation has a better completion time that the building renovation project can be accomplished in 35 working days, by preparing 7 days for the project buffer time.

4. Conclusion

Critical chain project management is one of a method to calculate and create schedule for project. By using critical chain management methods, it is found the critical chain namely B3, B5, B6, and B7. These activities are on factory building renovation projects, especially in the process of renovating factory facilities. At the end of the critical chain project, a project buffer of 7 days is added. This Project Buffer functions to control activities that are in the critical chain. A delay in activities in the critical chain can lead to a delay for all projects. Utilizing CCPM, the total project duration is 35 days. This

IOP Conf. Series: Materials Science and Engineering **722** (2020) 012052 doi:10.1088/1757-899X/722/1/012052

duration is 4 days ahead of the traditional scheduling calculation. For further research, it is necessary to calculate the cost of the change in the implementation schedule of this project, whether changing the schedule can increase the level of cost efficiency or increase the cost of the project, which indicates that the schedule change is not effective against the project cost.

References

[1] M. A. Langley, *A Guide to the PROJECT MANAGEMENT of BODY OF KNOWLEDGE*, vol. 53, no. 9. 2017.

[2] A. Izmailov, D. Korneva, and A. Kozhemiakin, "Corrigendum to Project Management Using the Buffers of Time and Resources," *Procedia - Soc. Behav. Sci.*, vol. 235, no. October, p. 817, 2017.

[3] Andi Andi, Susandi Susandi, and Hendra Wijaya, "on Representing Factors Influencing Time Performance of Shop-House Constructions in Surabaya," *Civ. Eng. Dimens.*, vol. 5, no. 1, pp. 7–13, 2003.

[4] S. Miranda and M. Sugarindra, "Utilizing project management software in project scheduling: a case study," *IOP Conf. Ser. Mater. Sci. Eng.*, vol. 528, p. 012037, 2019.

[5] P. S. K. Pai, "Multi-Project Management using Critical Chain Project Management (CCPM) – The Power of Creative Engineering," vol. 1, no. January, pp. 15–20, 2014.

[6] A. Izmailov, D. Korneva, and A. Kozhemiakin, "Effective Project Management with Theory of Constraints," *Procedia - Soc. Behav. Sci.*, vol. 229, pp. 96–103, 2016.

[7] T. G. Lechler, B. Ronen, and E. A. Stohr, "Critical chain: A new project management paradigm or old wine in new bottles?," *EMJ - Eng. Manag. J.*, vol. 17, no. 4, pp. 45–58, 2005.

[8] O. M. Group, "merits of ToC," no. January, 2001.

[9] P. A. Jensen and E. Maslesa, "Sustainable Building Renovation : Proposals for a Research Agenda," 2018.

[10] L. Joblot, T. Paviot, D. Deneux, and S. Lamouri, "review of Building Information Modeling purpose Literature Modeling purpose Literature review review of of Building Building Information Information Modeling projects Literature review of Building Information Modeling (BIM) intended for the purpose of," *IFAC-PapersOnLine*, vol. 50, no. 1, pp. 10518–10525, 2017.

IOP Conf. Series: Materials Science and Engineering **722** (2020) 012053 doi:10.1088/1757-899X/722/1/012053

Integration of pyrolysis – tar decomposition over porous low grade iron ore

D R Wicakso[1*], D Nurandini[1], A Mirwan[1], M D Putra[1]

[1*]Department of Chemical Engineering, Engineering Faculty, University of Lambung Mangkurat, Banjarbaru, South Kalimantan

Corresponding author: doni.rahmat.w@ulm.ac.id

Abstract. The objective of this research was to investigate the effect of low grade iron ore as catalyst on pyrolysis process of using palm empty fruit bunch. The process was conducted in fixed bed reactor at 450-600 °C. During experiment, the biomass placed within the first stack underwent pyrolysis process was pyrolyzed to produce char, tar and gases. Tar vapor and gases passed through the catalyst bed located within the second stack to conduct catalytic reaction. Furthermore, the vapor left the reactor and entered the cooling system. The catalyst after experiment was weighed to calculate total carbon deposited into catalyst pores. The results showed that the presence of low grade iron ore catalyst increased the gas yield with the yield reduction of liquid fraction. The gas yield was also increased with increasing temperature.

1. Introduction

Fossil fuels as primary energy source in the world are ultimately unsustainable, and directly related to air pollution, land, water degradation and climate changes. Those are the largest greenhouse gas emitters in the world, contributing 3/4 of all carbon, methane and other greenhouse gas emissions. Burning coal, petroleum and other fossil fuels at extremely high temperatures (combustion) as the primary process to produce electricity leads to heavy concentrations of pollutants in our air and water. The process also cause the increasing surface temperature of the earth and climate change [1]. Furthermore, the sources of fossil fuel are always declining with the increasing demand; the price was also influenced by the political and economic situation of the world. It is estimated that the petroleum and coal will run out in 40-60 and 100 years, respectively [2]. This situation encourages the researchers to discover the renewable and sustainable resources of environmental and friendly fuels. One of the renewable resources that are promising for substitution of fossil fuel in the future is biomass such as palm empty fruit bunch (PEFB).

PEFB, a by-product of the palm oil industry, is being recognized as one of the most potential kinds of biomass for energy production. The conversion of PEFB to renewable fuels and chemicals has attracted the attention of the researchers. In thermochemical processes including gasification, pyrolysis and direct combustion, the biomass can be converted to energy [3,4]. Pyrolysis has several advantages which are simple reactor configuration, faster reaction time, more liquid products (less energy for product separation) and high adaptation to various biomass [5].

Pyrolysis is a thermochemical decomposition of organic material at elevated temperatures without the participation of oxygen that produce char as main product and volatile matter, tar and gases as by-products [6]. Thermal decomposition during the process generate char product with high fixed carbon and heating value. However, condensation of tar components will arise several problems and decrease overall efficiency of pyrolysis process. Therefore, complete conversion of tar into a major portion of the gaseous product has been a most important engineering subject. It can also greatly enhance the process efficiency and reduce the implementation or operating cost. Catalytic tar decomposition is one of the promising method to increase the efficiency of pyrolysis process [7]. Several materials were studied by a number of researchers as potential catalyst such as Ni, Pt, Rh, and Pd; however, the deactivation of catalyst due to carbon deposition is main problem [8-10]. Another option of tar decomposition is to apply a catalyst or catalyst-like solid that can be used as material or fuel even after loss of activity, as example tar decomposition over charcoal. After reaction, charcoal and deposited carbon could be utilized as fuels. The use of low grade iron ore as catalyst will make carbon deposited into the pores can be used for the reduction process in the steel industry. Automatically, the reduction reaction of iron ore to iron can be faster at lower temperatures. Thus, the main objective of the research is to investigate the effect of low-grade iron ore as catalyst on pyrolysis process of using palm empty fruit bunch.

2. Material and Method

2.1. Raw Material

Biomass used in this study was palm empty fruit bunch (EFB), while low grade iron ore was obtained from Sebuku Island, Kotabaru, South Kalimantan, Indonesia. The compositions of low-grade iron ore were 51.41 % of total Fe, 7.214 % of Al_2O_3, 3.042 % of SiO_2 and 0.4803 % of NiO.

2.2. Experimental Procedures

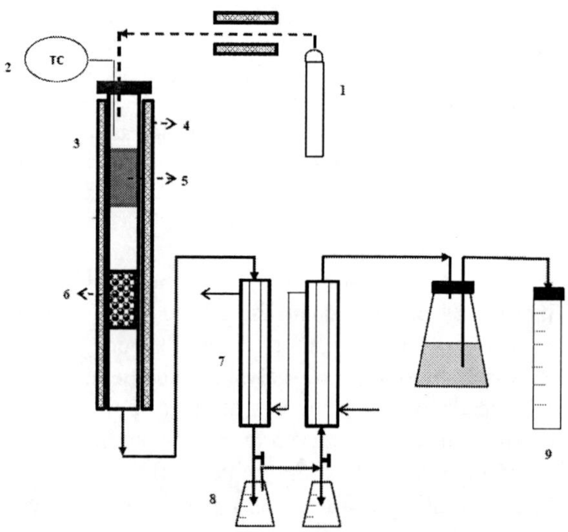

1. N_2 cylinder.
2. temperature controller
3. tube reactor
4. heater element
5. biomass bed
6. catalyst bed,
7. condenser,
8. liquid product,
9. gas product

Figure 1. Experiment Apparatus For Pyrolysis-Tar Decomposition Integration

As shown in Figure 1, a reactor tube used in these experiments was equipped with electrical furnace to heat-up the reactor at elevated temperature. The reactor was also equipped with thermocouple and a condenser set to change condensable vapor to liquid. The experiment was conducted at temperature of 450 to 600°C and at atmospheric pressure with certain amount of pyrolysis agents. In order to ensure the homogenous system inside the reactor, the pyrolysis agents was flowed for a while before heating-up. The reactor was heated and kept constant at experimental temperature of pyrolysis process. Biomass placed within the first stack underwent pyrolysis process during heating up to produce char, tar and gases. Tar vapor and gases passed through the catalyst bed located within the second stack and

IOP Conf. Series: Materials Science and Engineering **722** (2020) 012053 doi:10.1088/1757-899X/722/1/012053

then occurring catalytic reaction. By flowing the N_2 through top of the reactor system, subsequently, the vapor left the reactor from bottom of the reactor and then entered the cooling system.

3. Introduction

3.1. Porous Low Grade Iron Ore

3.1.1. Analysis of BET

From the BET analysis, the specific surface area of low quality of the iron ore catalysts is shown in Table 1. From the table, it can be seen that there is an increase of the catalyst surface area about 3 times after calcination. After the catalyst was used to decompose the tar, the outer surface area of the catalyst decreased because the catalyst pores begin to be filled with carbon sourced from the tar decomposition reaction. As shown in the table, the average diameter of iron ore pores decreased from 5.23 nm to 2.88 nm after calcination. According to the classification adopted by the International Union of Pure and Applied Chemistry (IUPAC) [11], the adsorbent of pores is classified into three groups: micro-pore (size <2 nm), meso-pore (2-50 nm) and macro-pore (> 50 nm). The size of the pores formed has an effect on porosity, the specific of total specific surface area is available for adsorption and thus it is possible to diffuse into solids. From the results of the analysis, it can be seen that low-quality iron ore after calcination has an average pore diameter value in the meso-pore region.

Table 1. Bet Analysis Result Of Porous Low Grade Iron Ore

Iron Ore	**Surface Area (m^2/g)**	**Average Diameter (nm)**
original	24.27	5.23
After calcination	86.91	2.88
After tar decomposition	23.90	4.23

The pore size distribution in iron ore catalysts can be seen in Fig. 2. From the figure, the pore distribution changes in the original iron ore, after activation and after the decomposition process. The new pores was significantly added at pore sizes of 1 - 4 nm, once the iron ore activation process was carried out by heating at 450 °C. This addition is due to the loss of impurities found in the original iron ore after the activation process. The figure also shows that there was a decrease in the pore size distribution of 1-4 nm in iron ore after it was used for the tar decomposition process. This confirms that the carbon from the tar decomposition process fills the pores of the iron ore.

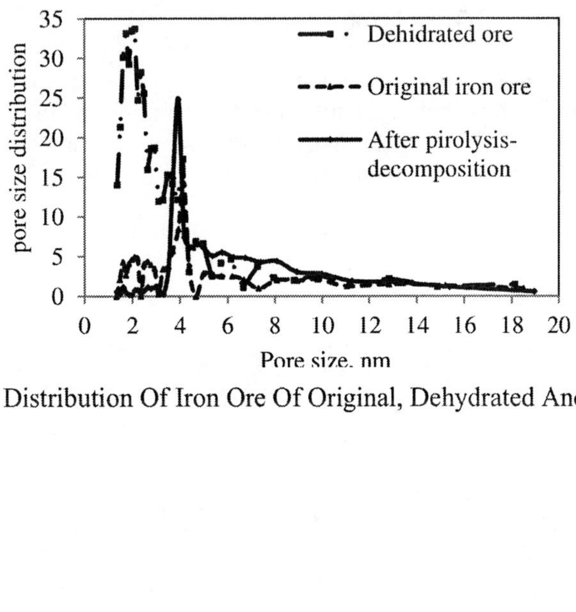

Figure 2. Porous Size Distribution Of Iron Ore Of Original, Dehydrated And After Decomposition

3.1.2. Analysis of SEM (Scanning Electron Microscope)

Figure 3 shows the SEM results from iron ore, iron ore after calcination and after pyrolysis-tar decomposition process. From the those figures, there changes in the morphological structure of iron ore. Figure 3 shows the original iron ore which still appears to be closed by an OH group of FeOOH [12]. After heating to 450 °C, the OH group will emerge in the form of H_2O vapor and the iron ore compound changes from FeOOH to Fe_2O_3.

(a)

(b)

(c)

Figure 3. Sem Analysis: (A) Original Iron Ore; (B) Iron Ore After Calcination; (C) Iron Ore After Tar Decomposition Process

3.2. Pyrolysis – Tar Decomposition Integration

3.2.1. Catalyst Influence On Product Distribution

Figure 4 shows the product distribution at different temperatures in the pyrolysis of EFB biomass when porous iron ore was used as catalyst. The distribution of pyrolysis products (gas, char and the whole liquid product before phase separation) was analyzed in terms of their yield (wt.%, g of product/g of biomass fed).

IOP Conf. Series: Materials Science and Engineering **722** (2020) 012053 doi:10.1088/1757-899X/722/1/012053

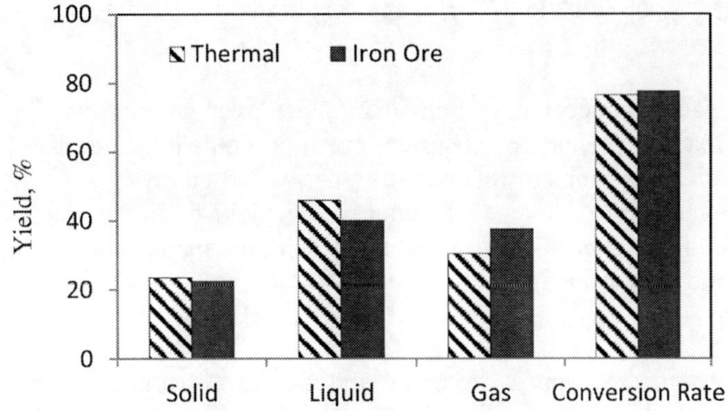

Figure 4. Effect of low grade iron ore catalyst at 550 °C

Figure 4 shows the yields of pyrolysis product and total conversion of EFB during pyrolysis in the absence and presence of supported low-grade iron ore. It can be seen that the yield of gas was significantly improved at the presence of iron ore. Since the liquid fraction decreased with almost the same order of magnitude, it appears that the gas fraction mostly increased with the cracking of higher molecular weight molecules at the catalyst surface. The char yield was registered with a very small reduction. Thus, the presence of iron ore catalyst improved the rate of conversion (1%) and the gas yield (7%) while the liquid yield decreased (6%). The decrease in liquid yield and the increase in gas indicate that low grade iron ore has the catalytic ability in the process of decomposition of the pyrolysis tar, as in the following reaction:

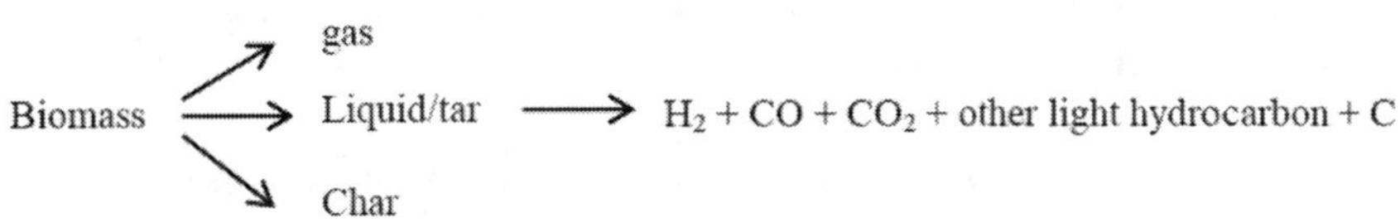

3.2.2. Temperature Influence on Product Distribution

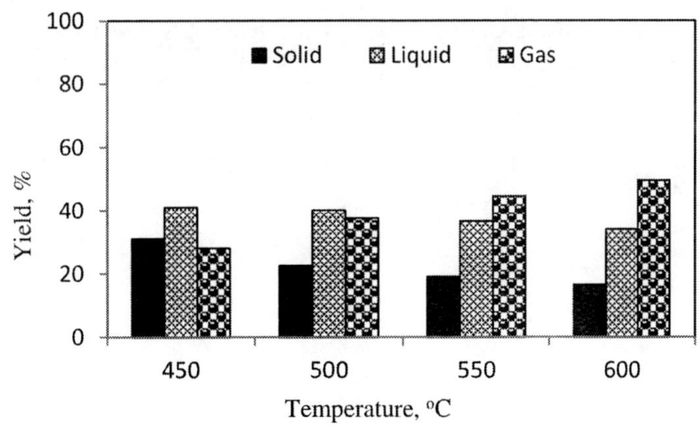

Figure 5. Effect of temperature on yield of product

As observed in Figure 5, temperature clearly affected to the three products obtained in catalytic tests. Gas product was improved with temperature while solid and liquid decreased. The increase in gas yield and the decrease in liquid yield indicate that the rate of tar decomposition reaction increases along with the increasing temperature. In addition, at elevated temperature, tar conversion increased and then occurred the decomposition to the gas product easier. By increasing the temperature, liquid dominated by heavy compounds was converted into compounds of light hydrocarbons like methane (CH_4), hydrogen (H_2), carbon dioxide (CO_2) and carbon monoxide (CO) [7].

4. Conclusion

The integrated pyrolysis – tar decomposition of biomass over porous low-grade iron ore has been performed in a tube reactor. The proposed system ore was promising candidate for tar decomposition and offered also solution for problem related with energy and environment. The results showed that the presence of low-grade iron ore catalyst reduced the yield of liquid fraction with a consequent increase in the gas yield. It has been also observed that with the increasing of temperature, gas product increased but liquid and solid decreased.

5. Acknowledgment

The authors acknowledge the support from the Indonesian Government through The Ministry of Research, Technology and Higher Education (Research Grant of Fundamental Research 2019).

References

[1] Guan G , Kaewpanha M , Hao X and Abudula A 2016 Renew. Sust. Energ. Rev. 58, pp. 450 – 461
[2] Chen W H , Chen C J , Hung C I , Shen C H and Hsu H W 2013 , Appl. Energ. **112**, pp.421-430
[3] Kudo S , Norinaga K and Hayashi J 2012 J Novel Carbon Res Sci **6**, pp.1-8
[4] Wicakso D R ,Rochmadi, Sutijan and Budiman A 2017 AIP Conf. Proc. **1823**, 020094
[5] Kudo S, Norinaga K and Hayashi J 2012 Journal of Novel Carbon Resource Sciences 6, pp.1-8
[6] Bridgwater A V 2003 Chem. Eng. J. **91**, 87-102
[7] Wicakso D R, Sutijan, Rochmadi, Muslikhin H, Cahyono R B and Budiman A 2018 Int. J. Renewable Energy Research **8**, 421-427
[8] Zhang R, Brown R B, Suby A and Cummer K 2004 Energ. Convers. Manage **45**, 995 – 1014
[9] Lu P, Yuan Z, Wu L, Ma L, Chen Y and Tsubaki N 2007 Energ. Convers. Manage. **48**, 1132 – 1139
[10] Asadullah M, Miyazawa T, Ito S I, Kunimori K, Yamada M and Tomishige K 2004 Appl. Catal. A-Gen **267**, 95 – 102
[11] IUPAC Manual of Symbols and Terminology 1972 Appendix 2, 1, *Collloid Surface Chem. Pure and Appl. Chem.*, 31, 587
[12] Cahyono R B, Rozhan A N, Yasuda N, Nomura T, Hosokai S, Kashiwaya Y and Akiyama T 2013 Fuel Process Technol. 113, 84 – 89

IOP Conf. Series: Materials Science and Engineering **722** (2020) 012054 doi:10.1088/1757-899X/722/1/012054

Rheological and kinetic studies of low density polyetyhlene (LDPE) – chitosan biocomposite film

Y Kusumastuti[1], D Timotius[1], N R E Putri[1], M W Syabani[2] and Rochmadi[1]

[1]Chemical Engineering Department, Faculty of Engineering, Universitas Gadjah Mada, Yogyakarta, Indonesia
[2]Rubber and Plastic Processing Technology Department of Politeknik ATK Yogyakarta, Indonesia

Corresponding author: yuni_kusumastuti@ugm.ac.id

Abstract. The amount of plastic waste has always increased over the years and caused many problems in the environment because of its poor degradation rate characteristics. Recently, the interest in biodegradable polymer to substitute these synthetic plastic materials has been increased. Biocomposite materials have better biodegradability rate and can be utilized for several applications, such as packaging, agriculture, biomedical and healthcare. Chitosan as natural polymer that has good biodegradability can be combined with low density polyethylene (LDPE) to improve its mechanical properties. The LDPE-chitosan biocomposites were made by mixing chitosan and LDPE with maleic anhydride as compatibilizer and perbutyl Z as an initiator. In this study, the decomposition kinetic and thermal behavior of low density polyethylene (LDPE)-chitosan biocomposites were investigated by thermogravimetric analysis (TGA). The residual weight of biocomposites after decomposition process was analyzed. The samples were heated in the range of temperature from 25 °C to 500 °C and at different heating rates of 5, 10, 15, 20 °C/min, respectively. The result shows that activation energy decreased along with the increasing of chitosan concentration. This result indicates that the LDPE-chitosan biocomposite is a promising candidate for substitution of synthetic plastic.

1. Introduction

Polyethylene is a synthetic polymer which widely used for various application, such as packaging materials, healthcare, agriculture, etc due to its versatility and low production cost [1][2]. However, polyethylene is not environmentally friendly because it takes several years to be completely degraded [3][4]. In the last decade, the development of synthetic-natural polymer has attracted much attention in order to increase its degradability [5][6].

Natural polymers such as collagen, chitosan, and starch are potential to be used as blending materials because of their biocompatibility and biodegradability [7]. However, it is still difficult to blend synthetic polymer with natural polymer because of their immiscibility [8]. Mir, et.al [8] tried to solve those problems by adding vinyl triethyoxysilane (VTES) into chitosan-High Density Polyethylene (HDPE) films. The thermal, mechanical, rheological and morphological properties of the blends were analyzed. Another study was also conducted by Araujo, J.R. et. al [9] with addition maleic anhydride as coupling agent into HDPE-natural fibers blend. Then, they studied the thermal properties of the obtained blends.

Although there were some researchers reporting about thermal behavior of natural-synthetic polymer blend, none of them was studying the kinetic model. Therefore, we observed a kinetic model of thermal degradation to get the kinetic parameters. In order to get the data, analysis was conducted using thermogravimetric (TG) and differential thermal analysis (DTA) simultaneously. The data could be used to measure the composition of blend, order of reaction, kinetic constants, and degradation steps [10].

This work was conducted to observe the thermal degradation's characteristics of polyethylene-chitosan composites compared to polyethylene and evaluated the kinetic model of its thermal degradation.

2. Methods

2.1. Materials

Low density polyethylene (LDPE) was supplied by Petrochemia Plock (Poland) and local chitosan powder was purchased from PT Biotech Surindo (Cirebon, Indonesia). The properties of local chitosan powder as follow: deacylation degree >90 %, 10-500 cps viscosity, <1.5 % ash content, and protein content value <0.5 %. Commercial Maleic Anhydride was obtained from Sigma Aldrich Chemical Company, and Tert-butyl peroxybenzoate was purchased from NOF Corporation (Tokyo, Japan).

2.2. Film Preparation

The preparation of LDPE-chitosan blends was conducted by mixing and hot pressing processes. First, LDPE (70 % wt) was melted in labo plastomill at constant temperature 160 °C with the speed of mixing 50rpm. After LDPE melted, chitosan (30 % wt), malaeic anhydride (4 % wt based on LDPE weight) and perbutyl Z (0.25 % wt based on LDPE weight) were added simultaneously into the labo plastomill and mixed at the same condition until the homogeneous mixture was obtained. The mixture was hot pressed at 40 kgf/cm^2 to obtain 0.5 mm of film thickness.

2.3. Thermal Analysis

Thermogravimetric (TG) and Differential Thermal Analysis (DTA) were conducted using TG/DTA Perkins Diamond Series. 5 mg of sample was put in an open pan and heated up to 1000 °C with variation of heating rate 5, 10, 15, and 20 °C/min, respectively.

2.4 Kinetic Modeling

For non-isothermal experiment, the reaction rate $\left(\frac{d\alpha}{dt}\right)$ is proportional to the conversion and the kinetic constant as shown in (1):

$$\frac{d\alpha}{dt} = k(T)f(\alpha) \tag{1}$$

In this case, $k(T)$, $f(\alpha)$, t, and α are kinetic constant as function of absolute temperature, the function of conversion over time, time, and conversion, respectively.

In this study we use two kinetic models, Flyn-Wall-Ozawa method (FWO) [11] and Kissinger-Akahira-Sunose method (KAS) [12]. Activation energy (E) and Arrhenius constant (A), were estimated by assumption that the reaction is first order, and heating rate $\left(\frac{dT}{dt}\right)$ is constant (β). The FWO method's equation and KAS method's equations can be modified as in equations 2 and 3, respectively.

$$\alpha = 1 - exp\left(-\frac{0.084AE}{R\beta} exp\left(-1.0516\frac{E}{RT}\right)\right) \tag{2}$$

$$\alpha = 1 - exp\left(-\frac{ART^2}{E\beta} exp\left(-\frac{E}{RT}\right)\right) \tag{3}$$

3. Results and Discussion

3.1. Thermogravimetric Study

This study is conducted to analyze the thermal stability and thermal properties of LDPE and blend [10]. The weight loss (TG) and the rate of weight loss (DTG) of LDPE and the blend under 20 °C/min heating rate are shown in figure 1. LDPE showed a single degradation step, while blend showed 3 degradation steps. This result was similar to previous experiments [6, 13]. The first step is attributed to drying step, which the volatile materials are loss. The second step shows the degradation of organic compounds in term of char formation [14]. While the third step is attributed to the degradation of LDPE as it can be compared between figure 1a and 1b. The difference in DTG peak was caused by the difference of heating rate [15]. The difference in DTG peak for 5, 10, and 15 °C/min heating rate is shown in figure 2, while for 20 °C/min heating rate is presented in figure 1. It can be seen that increasing in heating rate lead to increase degradation rate of LDPE and the second step of blend. The heating rate greatly affected the experimental result, especially in specimens that tend to undergo post crystallization or recrystallization [16]. Since chitosan is non-crystalline component [17], thus its thermal behavior is different with the LDPE in second step that related to the degradation of chitosan. From the study, the highest peak for LDPE degradation rate occurs during the heating rate of 20 °C/min. While the highest peak for the second step of blend occurs at the heating rate of 20 °C/min.

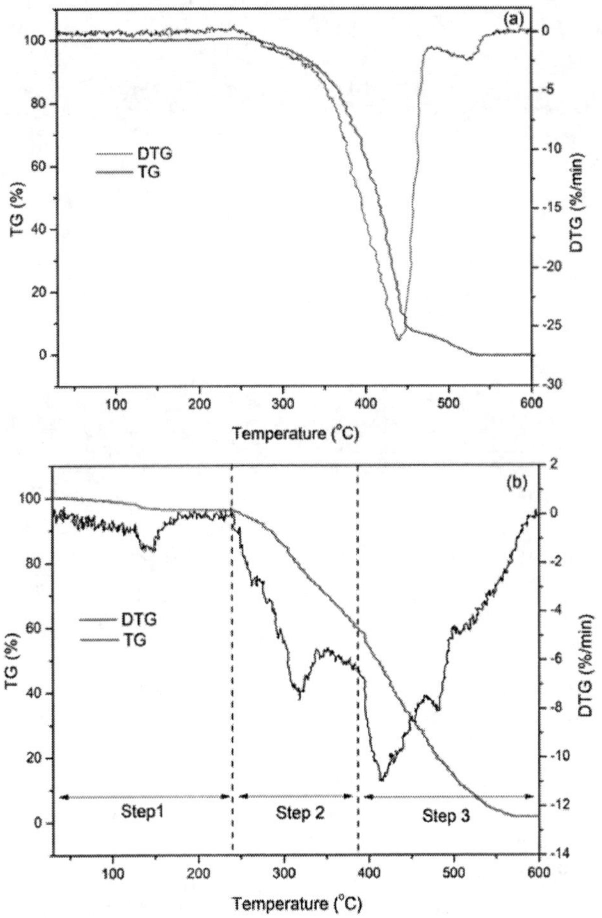

Figure 1. Degradation profile with 20°C/min heating rate of (a) LDPE and (b) blend

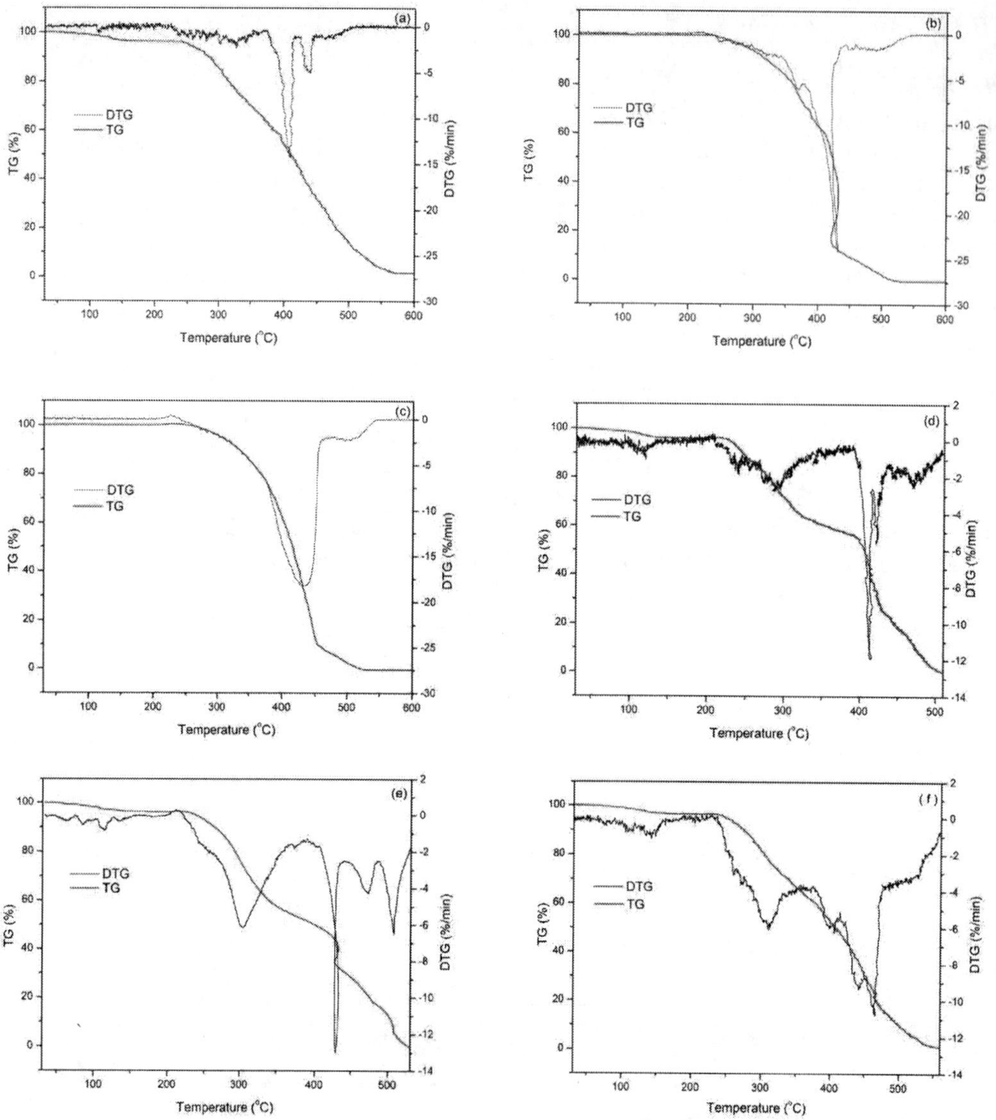

Figure 2. Combustion profile of (a) LDPE 5°C/min, (b) LDPE 10°C/min, (c) LDPE 15°C/min, (d) blend 5°C/min, (e) blend 10°C/min and (f) blend 15°C/min

3.2. Molecular Dynamic Results

Kinetic parameters of modified FWO (2) and KAS (3) methods are determined based on maximum coefficient of determination (R^2). For LDPE, it is calculated by one stage degradation as the degradation showed one single step of degradation. While for the blend, two step degradation is used in the calculation. Combination of the second step and third step degradation as one stage degradation is more favorable, so that the stage on consist of the first step, while stage two consist of the second and third step. The combination is necessary in order to simplify the model. The relationship between weight loss and temperature with heating rate 20 °C/min for FWO methods are shown in figure 3 with the obtained parameters are shown in table 1 and 2.

IOP Conf. Series: Materials Science and Engineering **722** (2020) 012054 doi:10.1088/1757-899X/722/1/012054

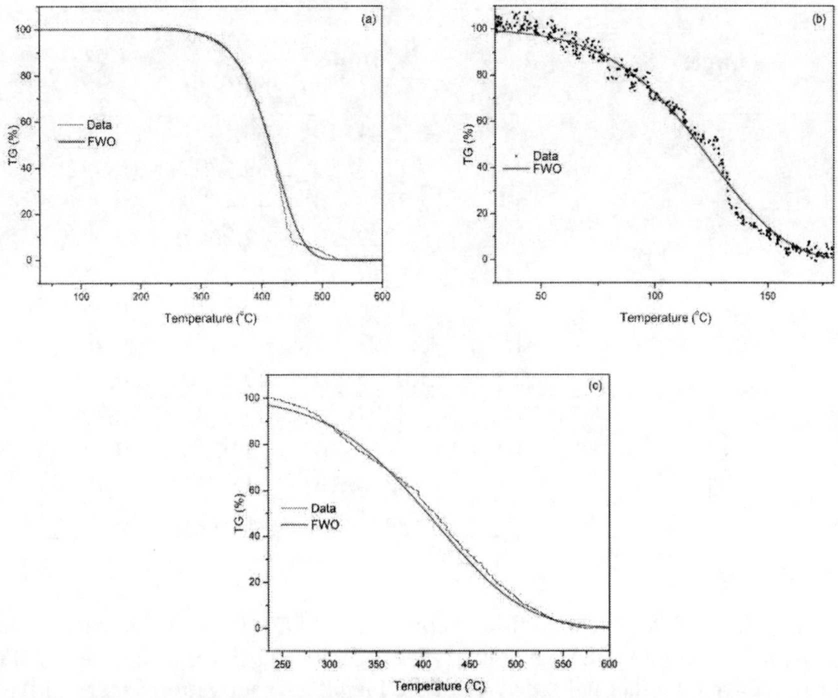

Figure 3. The relationship between weight loss and temperature with heating rate (a) 20°C/min LDPE, (b) blend stage 1 and (c) polyblend stage 2 for FWO method

Table 1. Kinetic parameter for FWO method

Sample	Stage	Heating Rate, °C/min	A, 1/min	Ea, kJ/mole	R^2
Blend	1	5	1.6003×10^5	43.1011	0.9900
		10	1.2989×10^5	39.9972	0.9941
		15	3.7168×10^5	44.0486	0.9942
		20	6.0178×10^5	45.0091	0.9919
	2	5	1.0387×10^3	49.8421	0.9682
		10	1.6746×10^3	49.3572	0.9647
		15	3.4708×10^3	51.4471	0.9813
		20	3.1241×10^3	50.0610	0.9957
LDPE		5	2.6682×10^6	94.9821	0.9934
		10	2.9937×10^6	93.1634	0.9941
		15	3.3520×10^6	93.0112	0.9993
		20	6.7188×10^6	96.0530	0.9985

Table 2. Kinetic Parameter for KAS Method

Sample	Stage	Heating Rate, °C/min	A, 1/min	Ea, kJ/mole	R^2
Blend	1	5	3.2241×10^4	39.1421	0.990
		10	7.1755×10^4	39.1299	0.992
		15	8.3899×10^4	40.4529	0.994
		20	1.7722×10^4	42.0667	0.992
	2	5	3.9517×10^2	49.4209	0.954
		10	8.1166×10^2	50.3019	0.933
		15	8.6297×10^2	49.4981	0.983
		20	1.0193×10^3	49.3720	0.992
LDPE		5	1.5494×10^6	91.9763	0.993
		10	2.8019×10^6	93.9218	0.992
		15	4.4204×10^6	94.9054	0.999
		20	3.4341×10^6	91.9794	0.999

The result showed that addition of chitosan in LDPE tends to decrease the activation energy for both FWO and KAS methods. The addition of chitosan into the LDPE matrix strongly influenced the thermal behavior of the blend. The activation energy of the blend is lower than LDPE suggest that the polymer chain of chitosan were weaker, thus the blend is easily to break due to the depolymerization and degradation of cross-linked chitosan. While the use of parameters value depend on the heating rate as shown in table 3. The activation energy of LDPE varied between 93.0112 – 96.0530 kJ/mole for FWO method and 91.9763 – 94.9054 kJ/mole for KAS method. The activation energy of stage one was 39.9972 – 45.0091 kJ/mole for FWO method and 39.1299 – 42.0667 kJ/mole for KAS method, while the stage two was 49.3572 – 51.4471 kJ/mole for FWO method and 49.3720 – 50.3019 kJ/mole for KAS method. The E values obtained from the FWO and the KAS methods slightly different as the result of distinctive linear approximation to the temperature integral.

Table 3. Range of degradation stage in several heating rate

Stage	5 °C /min	10 °C /min	15°C /min	20 °C /min
1	30 – 185 °C	30 – 200 °C	30 – 220 °C	30 – 235 °C
2	186 – 510 °C	201 – 535 °C	221 – 565 °C	236– 600 °C

4. Conclusions

The homogeneous LDPE-chitosan biocomposite were successfully made by mixing both maleic anhydride as compatibilizer and perbutyl Z as an initiator. The decomposition kinetic and thermal behaviour of the obtained biocomposite assessed by thermogravimetric analysis showed that the addition of chitosan in LDPE tended to reduce the activation energy. The activation energy varies between 93.01- 96.05 kJ/mole for FWO method and 91.98 - 94.91 kJ/mol for KAS method for LDPE only; while the activation energy of the biocomposites varies between 39.99 - 51.45 kJ/mole for FWO method and 39.14 - 50.30 kJ/mol for KAS method.

References

[1] Ferreira L M, Falcão A N and Gil M H 2005 Modification of LDPE molecular structure by gamma irradiation for bioapplications *Nucl. Instruments Methods Phys. Res. Sect. B Beam Interact. with Mater. Atoms* **236** 513–520
[2] Martínez-Camacho A, Cortez-Rocha M, Graciano-Verdugo A, Rodríguez-Félix F, Castillo-Ortega M, Burgos-Hernández A, Ezquerra-Brauer J and Plascencia-Jatomea M 2013 Extruded films of blended chitosan, low density polyethylene and ethylene acrylic acid *Carbohydr. Polym.* **91** 666–674
[3] Mohamad Z and Saadan M Z H 2017 Biodegradability and tensile properties of compatibilised polyethylene/rice bran film *Chem. Eng. Trans.* **56** 919-924
[4] Vimala P P and Mathew L 2016 Biodegradation of polyethylene using Bacillus Subtilis *Procedia Technol.* **24** 232–239
[5] Puccini M, Seggiani M and Vitolo S 2015 Polyethylene and hydrolyzed collagen blend films produced by blown extrusion *Chem. Eng. Trans.* **43** 1705-1710
[6] Quiroz-castillo J M, Rodríguez-félix D E and Grijalva-monteverde H 2014 Preparation of extruded polyethylene / chitosan blends compatibilized with polyethylene- graft -maleic anhydride *Carbohydr. Polym.* **101** 1094–1100
[7] Sunilkumar M, Francis T, Thachil E T and A. Sujith 2012 Low density polyethylene-chitosan composites: A study based on biodegradation *Chem. Eng. J.* **204–205** 114–124
[8] Mir S, Yasin T, Halley P J, Siddiqi H M and Nicholson T 2011 Thermal, rheological, mechanical and morphological behavior of HDPE/chitosan blend *Carbohydr. Polym.* **83** 414–421
[9] Araújo J, Waldman W and De Paoli M 2008 Thermal properties of high density polyethylene composites with natural fibres: Coupling agent effect *Polym. Degrad. Stab.* **93** 1770-1775
[10] Villetti M A, Crespo J S, Pires A T N, Borsali R and Soldi V 2002 Thermal degradation of natural polymers *J. Therm. Anal. Calorim.* **67** 295–303
[11] Flyn J H and Wall L A 1966 Direct method for the determination of activation energy from thermogravimetric data *J. Polym. Sci.* **4** 323–328
[12] Kissinger H E 1957 Reaction kinetics in differential thermal analysis *Anal. Chem.* **29**1702–1706
[13] Yang F, Qu J, Yang Z and Zhou A 2007 Thermal decomposition behavior and kinetics of composites from coal and polyethylene *Journal of China University of Mining and Technology* **17** 25-29
[14] Tahmasebi A, Kassim M, Yu J and Bhattacharya S 2013 Thermogravimetric study of the combustion of Tetraselmis Suecica microalgae and its blend with a Victorian brown coal in O_2/N_2 and O_2/CO_2 atmospheres *Bioresource Technol.* **150** 15-27
[15] Jamilatun S, Budhijanto, Rochmadi and Budiman A 2017 Thermal decomposition and kinetic studies of pyrolysis of Spirulina platensis residue *Int. J. Renew. Energy Dev.* **663** 193–201
[16] Ehrenstein G W, Riedel G and Trawel P 2004 Thermal analysis of plastics Carl Hanser Verlag, Munich 52-53
[17] Vasile C, Darie R, Cheaburu-Yilmaz C, Pricope G, Bračič M, Pamfil D, Hitruc G and Duraccio D 2013 Low density polyethylene – chitosan composites *Compos. Part B: Eng.* **55** 314-323

Acknowledgments

This work was supported by Universitas Gadjah Mada through internal grant scheme "Pengembangan Program Pengabdian Kepada Masyarakat berbasis Pemanfaatan Hasil Penelitian dan Penerapan Teknologi Tepat Guna Batch III 2017" with Research Grant No. 2711/DIT.PM/2017

Conceptualizing DFSS an analysis of 107 studies

Y D Nurcahyanie,[1] , M L Singgih[2] , D S Dewi[3]

[1]Department of Industrial Engineering, Universitas PGRI Adi Buana Surabaya, Indonesia
[2,3]Department of Industrial Engineering, Institut Teknologi Sepuluh Nopember Surabaya, Indonesia

E-mail : yuniadwie@unipasby.ac.id , moseslsinggih@ie.its.ac.id dyah@ie.its.ac.id

Abstract. The aims of this study are to capture the current state of Design For Six Sigma (DFSS) and to systematically review literature on practices of DFSS published between 2006 and 2018. The questions are: what is DFSS, the application, the essential ability, and limitations, the main trending topics of DFSS? All these questions are used to guide the search for journal in databases and other sources. Then, all relevant literature was reviewed and analysed. There were 18 topics found to be included in the review. The authors proposed two different issues relevant to DFSS. All these papers were interpreted by tools and techniques, such as the implementations of DFSS, its uses, applications, and how it is linked to other disciplines. Data was collected and clustered by year, topic, and keywords in DFSS to be used in configurations and progress improvement and innovation of DFSS. This review found four definitions of DFSS, eight applications and uses of DFSS, three challenges and obstacles in using DFSS, and two future direction of research of DFSS based on its fast-changing product design.

1. Introduction

One of the valuable tools in collecting assets in quality and performance is by using a series of quality drives like statistical control, zero defects, and total quality management. This tool is called Six Sigma, which is used in both manufacturing and services industries around the world [1]. Six Sigma is one of the newest quality performance measurements. It needs total commitment from companies, from the top management to all parts of company's systems, including the suppliers, designers, productions, processing, distribution and after-sales services. Big companies that use Six Sigma often gain substantial benefits.

Design for Six Sigma (DFSS) is parts of Six Sigma methodology. It helps companies to eliminate and count errors or defects in products design, processes, and or services. Six Sigma is well known for its framework that is Define, Measure, Analyse, Improve, and Control (DMAIC). Several basics differences between the traditional Six Sigma DMAIC that is commonly used and the DFSS is that DFSS has many variance steps depending on the needs and applications. These variances ways need to be discussed. The one being studied is which actions are more appropriate for manufacturers and which activities are better to apply in services industries [2]

Conventional quality management approaches, containing Statistical Quality Control (SQC), Zero Defects and Total Quality Management (TQM), have been the primary tools for years. Six Sigma is one of the latest tools initializes quality improvement. It is well known in global industries. Since its initiation at Motorola in the 1980s, many companies, including GE, Honeywell, Sony, Caterpillar, and Johnson Controls, have already used Six Sigma and gained substantial advantages. DFSS is more focused on front liners of the overall process, starting from the design production process for the manufacture or service designs. DFSS in all aspect or variance analyses the customer's needs, hopes and expectations, and deliver technical requirements of the engineering tools, in all product design life cycle [3].

DFSS is usually combined with other product development tools to achieve a standard development tool in new product development steps. This product design steps are combined with five steps of DFSS, which are define, measure, analyse, design and verify (DMADV). Many companies approve that DFSS methodology is a more efficient, reliable and robust method to meet customers' needs and expectations and is critical to quality. It improves quality and also reduce errors or waste. DFSS is not a fixed design tool, but it can be developed, and combined with other devices. The systematic complex is also considered in the system integrity with DFSS [4].

The aims of this study are first to define DFSS methodology by reviewing journals and documents. The second aim is to find the closest integration between DFSS and product development process. The integration is conducted by widening keywords that are connected with DFSS methods and by reviewing all identified journals. The third aim is to cluster issues by keywords, topic, and interest.

2. Methodology

2.1. *Scope and research question*

All journals were read and reviewed starting from the oldest works in DFSS topics: the provisional applications of DFSS in product and service [2]. This review was conducted by reading and reviewing DFSS journal from the broad works of Six Sigma or DFSS [5] and the works about services that use DFSS [6]. Because of it, all the literature reviewed in this paper are those which were published between 2006 and 2018. In terms of the subject of discussion in the literature review, the topics covered were not only from the industry sector alone but also from other relevant journals. The research questions are:
1. What is Design for Six sigma (DFSS)?
2. What is the application of DFSS?
3. What are the main barriers of DFSS?
4. What is the direction of DFSS future research?

2.2. *Search strategy*

The procedure used in the current research is as follows: (1) Search for and find relevant databases from previous journals. (2) Set the time frame to see the data from the year 2006 until 2018. Because the range is very wide, the included papers are those starting from 2006. (3) Take a set of journals that were collected based on keywords. We tried to make the combination reasonable and relevant to DFSS.

In this paper, we used three kinds of database and journal search engine, and they were *Google Scholar, DeepDyve*, and *Mendeley* search engine. After we decided which search engine to use, then we built a series of questions to find the appropriate journals or other literature. Table 1 shows the series of questions used to retrieve the appropriate journal. It shows publications found if we use these questions in three search engines. The issues must be related to DFSS method. As we know DFSS has been introduced as parts of Six Sigma itself [7], and DFSS also one of the robust tools to build reliable and low-cost manufacturing [8].

Table 1. DFSS questions to build the database of papers

no	Question String	Number of papers by keywords		
		Google Scholars	DeepDyve	Mendeley
1	Design For Six sigma	1.750.000	5.248	199.937
2	DFSS + Definitions	6.040	378.315	216.796
3	DFSS + implementations	5.110	596.002	84.858
4	DFSS + Product Design	5.000	262.537	171.329
5	DFSS + Product Development	5.590	604.313	233.137

From the first question, we found that there were thousands of journals and papers. Then we made narrower questions, and the aims were to have smaller strings to find papers with the best fit. It was almost impossible for anyone to read thousands of papers. So, we have to decide to build smaller strings of questions. To build smaller strings, we have to set a series of individual questions with narrower DFSS topics into the subjects shown in table 2.

The papers we included in this research were the papers that use DFSS methodology. We did not go more in-depth to examine the whole dissertations of each document.

Table 2. The result of the series of questions search

No	Question String	Number of papers by keywords		
		Google Scholars	DeepDyve	Mendeley
1	Short cycle product and design for Six Sigma	330.000	984	3.220
2	Quick cycle product and DFSS	1.520	1.209	2.802
3	Sharp cycle products and DFSS and remanufactured products	317	**265**	3.450
4	Short cycle products and DFSS and remanufactured products and new product development	289	0	5.349
5	DFSS and closed loop manufacturing	562	109	**242**

From Table 2 above, we can see the smallest string we can use to find the best fit and the number of papers we need to read more in-depth. From these 242 – 265 string results, then we read the abstracts to know the scope of these research. After that, we tried to cluster those papers based on the similarity of the topic. In this paper, we excluded manuscripts and patens.

3. Result and analysis

The result was 242 to 265 journals related to DFSS: closed-loop manufacturing and short cycle products. The procedure begins with reading the abstracts. The topics were not always in the scopes, but at least had some degree of relevancy. However, because some did not match with DFSS, or lacked of details and examples, or nor related to the current works, the number was reduced to 107 readings, and they were all papers in journal. After that, it was necessary to see who the writers were, whether the authorship was from the industrial background or the educational background. Then we also needed to find out that DFSS was suitable not only for manufacturing design but also suited the services design process. The relationships were associated with tools or methodology or with bases data as part of DFSS. All these clustered areas were the standard for the literature to be included in the review. All papers included in this study did not go deeply into studying in depth of methodology that was used, but only the upper limit. Then we needed to know the author background. Industry authors deal with the industrial improvement in their daily lives. Academia might not deal with industrial in their everyday lives, but they were doing particular research in DFSS area.

Based on previous papers discuss about DFSS and another methodology, we clustered DFSS topics of interest into 18 types, as shown in Figure 1. We gave code in these types of DFSS topics. They are code T1 DFSS Literature, code T2 Quality Concepts, code T3 Six Sigma and learn fundamentals, code T4 DFSS product development process, code T5 DFSS Six Sigma Development, code T6 DFSS Algorithm Project, code T7 DFSS and Quality Function Deployment (QFD), code T8 DFSS and axiomatic design, code T9 DFSS and Theory of Inventive Problem Solving (TRIZ), code T10 DFSS and design for X, code T11 DFSS and Failure Mode Effects Analysis (FMEA), code T12 DFSS and Design of Experiments (DOE), code T13 DFSS and Taguchi, code T14 DFSS and Supply Chain, code T15 DFSS and Supply Chain Remanufacturing, code T16 Rapid change product and technology, code T17 Planned obsolesce, and code T18 Consumer behaviour

These codes based on the methodology used in previous literatures about DFSS. The combination of DFSS methodology based on the study cases taken by researcher. Every cases have different conditions so that influence the methodology to answer their research questions.

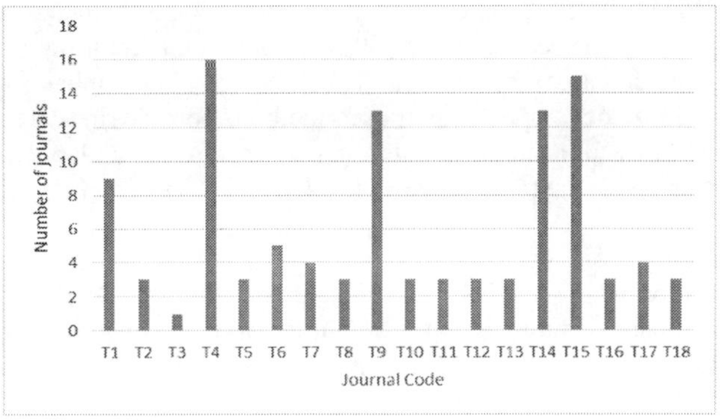

Figure 1. Articles based on article codes

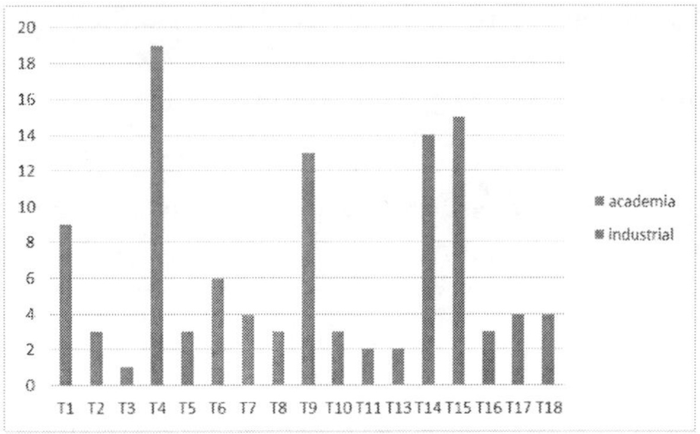

Figure 2. Articles based on the author's background

From those 18 coding topics of DFSS, then we grouped them into four main interests. Those four DFSS interest are the use of DFSS for product development, the combination of DFSS and TRIZ to product design, the use of DFSS in supply chain and the use of DFSS and remanufacturing. The progess utilization of DFSS it self has been grown rapidly. It is not just for designing manufactures things, but also for services such as education, health, and supply chain services like transportation service. In DFSS use, sequences or steps may vary depend on the subject need to get solved. The well known step is DMADV (Define, Measure, Analyse, Design and Verify). Then the steps are keep on changing like IDDOV (Identify-Define-Design-Optimize-Validate), IDOV (Identify-Design-Optimize-Validate), DMADOV (Design-Measure-Analyse-Design-Optimize-Verify) and many more. DMADV and IDOV are the most frequent cycle used by researcher [3].

It is interesting to find the background of the authors, how DFSS can be grow that fast and has so many sequensces. It is interesting to know the author background. We divided the author's background into two categories, first from industrial background and second from academia background. From industrial background, the authors may come from a professional design firm, or from a marketing division. The second type of author were those from academic background, they might be a researcher, a student or a lecturer. We can see the author's background from Figure 2. It can be identified that most writers were academia background. The DFSS topics that were reviewed by academia are code T2, T3, T7, T11, T13, T14, T15. Most of the topics are only in academic research and have not been implemented in industrial application yet. The sequences of DFSS for non manufacturing uses like health care services is very suitable. In health care service zero defect design is completely needed. So in academic perspectives DFSS is highly developed as purposed method. Some writers were from the industries, and some others were quality assurance consultants. Some papers were collaborations between the industry and academic researchers, with the topics codes T1, T4, T5, T6, T8, T9, T10, T16, T17 and T18.

3.1. Definitions of Design for Six Sigma

To recall, the first question was: what is DFSS? The first definition, DFSS is a structured and organized method by using tools quantifications by specific numerical data for product design development from six sigma. Manufacturing industry design firms, and quality assurance consultants mostly use DFSS for their new design product development, as the fisrt stage of total production cycle. They can combine DFSS with another tools as can be seen in the cluster codes based on the product type variations. But the use in services industry like health care, finance and banking, they use it for developing performance in [3].

The second definition, DFSS can be built in the engineering process to give better product quality and lower the cost and speed up deliverance to the customer. Not all product development needs customer's voice. Sometimes, an engineer has their own products decisions to be renewed or repaired with the current trends of closed loop supply chain followed by types of organizations, projects, acquires, and suppliers [9]. The third definition, DFSS is a technique for appraising supply chain development and achievement analysis. The requirement need administration process defined as a critical to customer requirements and it is before business conduct is enlarged [7]. The fourth definition of DFSS is a set of procedure of training and schooling aspect approach using DFSS concentrating on ideas in engineering and business cases and developing the model implementation in higher education [10].

3.2. The Application of DFSS

We grouped the application of DFSS into eight clusters. The first is use of DFSS in product and service industrial development. Despite the various procedures, DFSS is so versatile and flexible to become a tool that is applicable and fits into many aspects of business augmentations [2]. The second use of DFSS mode or procedure is to draw a conclusion from the analysis of intelligence reports in every step-in supply chain from the raw material up to the consumer, the provisional investigation of authentications and the baseline calculation of transaction [11]. The third application of DFSS is a more scalable approach unaffected by customer requirements. DFSS usually is combined with others quality methods to solve the customer problem [12]. The fourth application of DFSS is an aggressive trade or service actions which take advantage of the information's from market or consumer into the model of all systems. Developing product through buyers need will give more competitive advantage that satisfy customers [13]. The fifth use of DFSS is deformation in generating concept by integrating Axiomatic Design as one of the essential factors in the development, which proves that DFSS is a robust mechanism to make it work [14]. The sixth application of DFSS is to design a procedure for one automotive part to make all system running well for all machine used [15]. The seventh application of DFSS is to enhance integrations to improve quality of service process by design or resign. The Eight application of DFSS is as a tool and mechanism to develop system health services, especially in hospital emergency area. The scope is how long the patients stay or length of stay (LOS) and how long the patients need wait for service [17].

The eighth application of DFSS is as a tool for new services and new product development. It can interact with intelligence reports in every step in supply chain, from raw material to the end user or customer. DFSS can also generate concepts by integrating Axiomatic Design as one of the essential factors in the development of a product from the very beginning. DFSS is also able to design a procedure for one automotive part. The way it works is to mix and combine quality initiatives by building a system called Supply Chain Oprations Reference (SCOR) [16].

3.3. The main barries of DFSS

We discuss the main barriers and the possible development of DFSS applications. There is no perfect tools or systems in this world, all tools, and all orders have to be developed along with the changing of the people, technology, and culture. Improvement is important in this challenging era.

First challange of DFSS is that it applies the internet of things (IOT) as part of the strategic thinking, continuous upgrading for the company. Statistical engineering like big data has to be used to solve problems quickly. Second, DFSS should be applied to keep up with people cultural changing that impact consumer buying style and consuming behaviour. These days, consumers need mobility and sustainability becomes an insignificant parameter of buying [18]. The third challenge of DFSS is that it must be flexible in accepting all rapidly changing environments [19]. The rapid changing technology has a negative impact. It also produces high level of e-waste, particularly in the last decades in which it has been quite critical and need new design to solve this problem [18].

3.4. The trend that emerge study of DFSS

In this section we discuss what trending topics of DFSS in the future applications. First the prospective using of DFSS must be collaborated with particular needs, and all systems in the company net must be involved. The globalization impacts are highly competitive, resulting in rapid change technology. All organizations will face uncertainty and a complex unstructured problem, and they must be able to obtain strategies to solve the problems [19]. The second trend of DFSS is to use DFSS at the pre-design stage because several fast-changing products are difficult to break apart or to dismantle so they are unable to be recycled. In this stage, products must use new material, must have extended durability, easy to disassemble, recyclable, reusable, or able to be remanufactured to reduce environment destruction [20][22].

To do so, it is necessary for the government to create specific policy. Government has to present reasonable argumentation that it has financial benefits, so that the volume of product rebound will rise significantly. Government should use creative approach so people have a great willingness to overcome these problems [21].

4. Conclusion

The utilization of DFSS has a lot of variations in the last decades. Some papers have been reviewed and the results are presented in this paper. However, there are also more papers suggesting a different point of view regarding DFSS. These journals have indicated some findings, have used various methods, and have mixed methods to overcome problems. From 107 journals reviewed in this study, DFSS can be clustered into 18 groups, based on topics and also writer's background. We found that there are four definitions of DFSS, eight applications of DFSS, three challenges in using DFSS, and two future uses of DFSS considering the fast-changing product design.

References

[1] P. Nonthaleerak and L. C. Hendry, "Six sigma: literature review and key future research areas," Int. J. Six sigma Compet. Advant., vol. 2, no. 2, p. 105, 2006.

[2] U. Asad and T. K. Chuan, "Comparative Study of DFSS in Product and Service Innovation," 2006.

[3] V. S. Patil, I. D. Paul, and S. R. Andhale, "A Review of DFSS : Methodology, Implementation and Future Research," Int. J. Innov. Eng. Technol., vol. 2, no. 1, pp. 369–375, 2013.

[4] Y. Mingshun, G. Xinqin, L. Yanjie, L. Yan, and Y. Ting, "Research Article An innovative DFSS approach for multivariate production process," J. Chem. Pharm. Res., vol. 6, no. 2, pp. 221–225, 2014.

[5] R. Bañuelas and J. Antony, "Six sigma or design for six sigma?," TQM Mag., vol. 16, no. 4, pp. 250–263, 2004.

[6] B. El-haik, D. Ph, and J. M. Wasiloff, "ICAD-2004-10 Axiomatic Design Quality Engineering - A Transmission Planetary Case Study," 2004, pp. 1–8.

[7] Y. Amer, L. Luong, S. H. Lee, and M. A. Ashraf, "Optimizing order fulfillment using design for six sigma and fuzzy logic," Int. J. Manag. Sci. Eng. Manag., vol. 3, no. 2, pp. 83–99, 2008.

[8] J. Cruz, "Design for Six sigma: Approach for reliability and low-cost manufacturing (Diseno para Seis Sigma: Enfoque para la fiabilidad y fabricacion de bajo costo)," Innovaciones de Negocios, vol. 9, no. 18, pp. 319–340, 2012.

[9] H. Yoon and J.-H. Byun, "A Design for Six sigma : A Robust Tool in Systems Engineering Process," Ind. Eng. Manag. Syst., vol. 11, no. 4, pp. 346–352, 2012.

[10] S. S. Nicolaescu and C. V. Kifor, "Teaching methodology for Design for Six sigma and Quality techniques – an approach that combines theory and practice," Balk. Reg. Conf. Eng. Bus. Educ., vol. 3, no. 1, 2017.

[11] E. M. Mitchell and J. V. Kovach, "Improving supply chain information sharing using Design for Six sigma," Investig. Eur. Dir. y Econ. la Empres., vol. 22, no. 3, pp. 147–154, 2016.

[12] M. H. A. K. Khan, "An Integration of Kano Model , QFD and Six sigma to Present a New Description of DFSS," vol. 9, no. 6, 2017.

[13] D. Alwerfalli and T. Lash, "Design For Six sigma (DFSS) as a Proactive Business Process," no. Figure 1, pp. 2591–2600, 2012.

[14] A. L. Dickinson, "Integrating Axiomatic Design Into A Design For Six sigma Deployment," in International Conference on Axiomatic Design, 2006, pp. 2–7.
[15] A. El-Sharkawy, A. Salahuddin, and B. Komarisky, "Design for Six sigma (DFSS) for Optimization of Automotive Heat Exchanger and Underhood Air Temperature," SAE Int. J. Mater. Manuf., vol. 7, no. 2, pp. 2014-01–0729, 2014.
[16] M. C. Lee and T. Chang, "Developing a lean design for Six sigma through supply chain methodology," Int. J. Product. Qual. Manag., vol. 6, no. 4, p. 407, 2010.
[17] N. Mandahawi, S. Al-Shihabi, A. A. Abdallah, and Y. M. Alfarah, "Reducing waiting time at an emergency department using design for Six sigma and discrete event simulation 'Reducing waiting time at an emergency department using design for Six sigma and discrete event simulation," Int. J. Six sigma Compet. Advant., vol. 6, no. 12, pp. 91–104, 2010.
[18] F. Echegaray, "Consumers' reactions to product obsolescence in emerging markets: the case of Brazil," J. Clean. Prod., vol. 134, pp. 191–203, 2016.
[19] J. Antony, R. Snee, and R. Hoerl, "International Journal of Quality & Reliability Management Lean Six sigma: yesterday, today and tomorrow," Int. J. Qual. Reliab. Manag. Int. J. Qual. & Reliab. Manag., vol. 34, no. 2, pp. 1073–1093, 2017.
[20] Y. Huang, "Developing a Decision Model of Sustainable Product Design and Development from Product Servicizing in Taiwan," vol. 12, no. 5, pp. 1285–1302, 2016.
[21] M. F. Hassan et al., "Investigation of existing laptop computer towards sustainable development using design for six sigma and design for disassembly," ARPN J. Eng. Appl. Sci., vol. 11, no. 12, pp. 7683–7688, 2016.
[22] J. Gaur, R. Subramoniam, K. Govindan, and D. Huisingh, "Closed-loop supply chain management: From conceptual to an action oriented framework on core acquisition," J. Clean. Prod., vol. 167, pp. 1415–1424, 2018.
[23] S. Deshpande, "Introducing Design for Six sigma ' s DMADV Methodology to the Packaging Industry," Rochester Indtitute of Technology, 2016.
[24] S. M. S. B. M. Rafique, "The use of design for six sigma (DFSS) methodology in product design," Lect. Notes Eng. Comput. Sci., vol. 1 LNECS, pp. 664–669, 2013.
[25] R. Bañuelas and J. Antony, "Going from six sigma to design for six sigma: an exploratory study using analytic hierarchy process," TQM Mag., vol. 15, no. 5, pp. 334–344, 2003.

IOP Conf. Series: Materials Science and Engineering **722** (2020) 012056 doi:10.1088/1757-899X/722/1/012056

Design improvement of multifunctional seatbelt by using integrated kano model and quality function deployment

Z U Rizqi[1]

[1]Department of Industrial Engineering, Universitas Islam Indonesia, Yogyakarta, Indonesia

[1]E-mail: ugihzakka@gmail.com

Abstract. Increasingly developing technology makes human easier to carry out the life, and one of them is in the transportation sector. The high number of vehicle accidents causes losses to the passenger so that it requires safeguards to minimize these losses, one of the ways is by using a seat belt. This study aims to design a voice of customer-based seat belt using integrating Kano model and Quality Function Deployment (QFD) produced by an innovative product, namely "Multifunctional Seat Belt for Vehicle Safety". Through the distribution of questionnaires, 8 attributes are needed in the design of the seat belt. The results of the Kano Model analysis show that 2 attributes fall into the indifferent category which is then eliminated because the presence or absence of these attributes does not affect customer satisfaction. The other 6 attributes are portable, can store items, can hold the body while sleeping, can warm the body, there are pads on the sides of the head, and adjustable strap. These attributes are entered into the QFD framework until phase 1 to obtain a priority sequence that is useful in the production process later. In addition, MUSESA is designed with anthropometric approach which is using 5 dimensions so that it is ergonomic for users.

1. Introduction

The rapid development of era has led to an increase in the role of technology and transportation in human life. With the transportation system, people are able to travel from one place to another faster and easier. The development of transportation equipment must also be accompanied by security to ensure passenger safety.

The World Health Organization (WHO) published a report namely The Global Report on Road Safety which stated that around 1.25 million people in the world died in 2013 due to traffic accidents. In 2014, the number of car accidents occurred in Indonesia as many as 42,197 [1].

To be able to help avoid accidents, transportation means are generally equipped with security equipment in the form of seat belts. From the various accidents that have occurred in various countries, it is clear that the seat belt is the most important security device in protecting motorists and passengers when an accident occurs [2].

Based on the results of a questionnaire distributed, it was found that the customers need a practical seatbelt for use when riding a couched vehicle that is not only for private vehicles but also can be used for public transportation. However, not all public vehicles currently provide seat belt such as public trains and buses, which of course can be so dangerous. It was also found that the community not only needed security in travel but also comfort.

Looking at the existing problem, MUSESA "Multifunctional Seat Belt for Vehicle Safety" an innovative seatbelt that is practical to be used for driver or passenger that is not only for private vehicles but also for public transportation. There are several features in MUSESA product to make it easier and more convenient for users like portable, the availability of body warmer, additional pockets, strong and comfortable to use while sleeping, and the availability of disclaimer pillows on each side of the head.

In this research, integration of Kano Model and Quality Function Deployment (QFD) will be used to produce features that are in accordance with customer desires so that customer satisfaction can be fulfilled. Integration of the two methods was carried out because it was proven to be able to improve the deficiencies of each method resulting in effective output [3], especially in improving seatbelt design.

2. Methods

2.1. Data collecting

In this study, data collection was carried out in several stages. The first stage is the stage of distributing questionnaires with open questions to get Voice of Customer (VoC) to 70 respondents. VoC obtained is then re-asked using the Likert 1-5 scale so that the instrument can be tested. The second stage is distributing Kano questionnaires to determine customer preferences. The third stage is the distribution of QFD questionnaires to determine the level of importance, level of satisfaction, and benchmarking. In addition, to produce ergonomic products, MUSESA is designed using an anthropometric approach so that anthropometric measurements are carried out directly on 30 respondents. All respondents involved must meet the inclusion criteria: 18-25 years old, male and female, work as students and ever using seatbelt.

2.2. Instrument tests

The data obtained from the results of the first questionnaire were tested by instrument tests in the form of validity and reliability test to ensure that the data taken was representative enough to describe the actual conditions so that data could be further analyzed reliably. Using the significance of two-tailed 0.05 with dF = 70 - 2 = 68, it is found that the data are valid if the value of Pearson correlation > r table (0.235). The data are said to be reliable if Cronbach's Alpha > 0.60 [4]. In this research, the validity and reliability test is done by using SPSS 22 software.

2.3. Data processing

Valid and reliable attributes are then categorized based on the Kano theory whether it is categorized as must be, attractive, one dimensional, or indifferent. The categorization is then given a value called K-Value which is categorized as must be (M) = 0.5, one dimensional (O) = 1, attractive (A) = 1.5 and indifferent (I) = 0 [5]. Attributes that fall into the indifferent category will be eliminated immediately, while others will be hybridized into QFD. The QFD process carried out in this study only reached phase 1, this was based on the aim of the study which is only until the design of the alpha prototype using SolidWorks.

Product specifications are based on the results of anthropometric calculations to obtain ergonomic seatbelt. Before the anthropometric calculation is carried out, the normality test is done first to find out the average sample whether it has approached the midpoint of the population and fulfill the requirements of the normal distribution so that it represents the population. Some dimensions used in the design of MUSESA products are TBD (sitting shoulder height), LB (shoulder width), RS (stretch of elbow hand), DPK (chin to top of head), and LK (head width).

3. Result and discussion

The results of the distribution of questionnaire stage 1 to 70 respondents obtained 15 attributes which were then given re-asked to give a rating for instrument test analysis. After the instrument test, 8 valid and reliable attributes were obtained. Valid attributes (Pearson Corrleation > 0.235) are shown in Table 1 and the results of the reliability test calculations use SPSS shown in Table 2.

Table 1. Validity test result (alpha = 5%).

Code	Attribute	Pearson Correlation
A1	Portable	0,635
A2	Can store items	0,453
A3	Can hold the body while sleeping	0,567
A4	Can warm the body	0,672
A5	Has attractive colors	0,417
A6	There are attractive decorations	0,383
A7	There are pads on the sides of the head	0,433
A8	Adjustable strap	0,587

Table 2. Reliability test result.

Cronbach's Alpha	Standarized Alpha
0.6791	0.6140

The attributes that pass the instrument test become the voice of customers that will be analyzed next. The 8 attributes above will then be analyzed using Kano Model and QFD framework until first stage.

3.1. Kano model results

Based on the results of the Kano questionnaire which contained functional and dysfunctional questions from the 8 attributes, the results recapitulation can be seen in Table 3.

Table 3. Recapitulation of kano model.

Code	Kano Model Category				Total	Category	Worse (-)	Better (+)
	A	*O*	*M*	*I*				
A1	11	6	8	5	30	A	0.466	0.566
A2	12	7	7	4	30	A	0.466	0.633
A3	6	7	14	3	30	M	0.7	0.433
A4	8	10	8	4	30	O	0.6	0.6
A5	5	5	9	11	30	I	0.466	0.333
A6	4	5	8	13	30	I	0.433	0.3
A7	8	9	8	5	30	O	0.566	0.566
A8	11	6	8	5	30	M	0.733	0.433

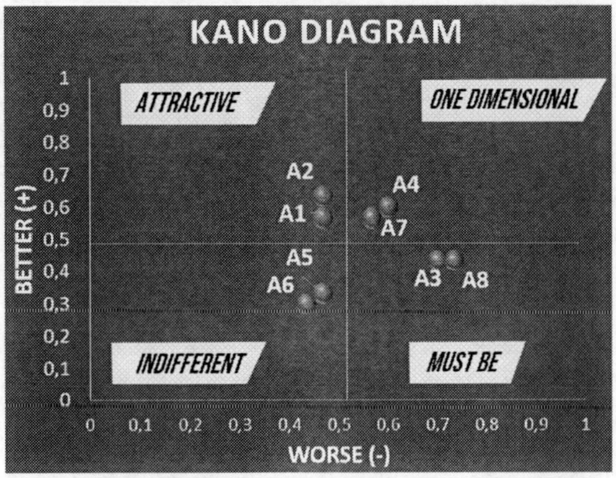

Figure 1. Kano diagram result.

From the Kano diagram on Figure 1, it can be seen that there are 2 attributes (A6 and A5) that fall into the indifferent category which means that the presence or absence of these attributes will not affect customer satisfaction so that the two attributes can be eliminated directly [6]. While the other attributes will be input to House of Quality (HoQ) 1.

3.2. House of quality 1

Data related to level of customer satisfaction, level of importance, benchmarking and K-Value are all integrated into the House of Quality 1 framework included with Voice of Engineer (VoE) to meet Voice of Customer (VoC) which can be seen in Figure 2. Based on the results of HoQ 1, a ranking of useful attributes is obtained at the production stage where the manufacturer will prioritize first rank in meeting its performance quality followed by subsequent rankings.

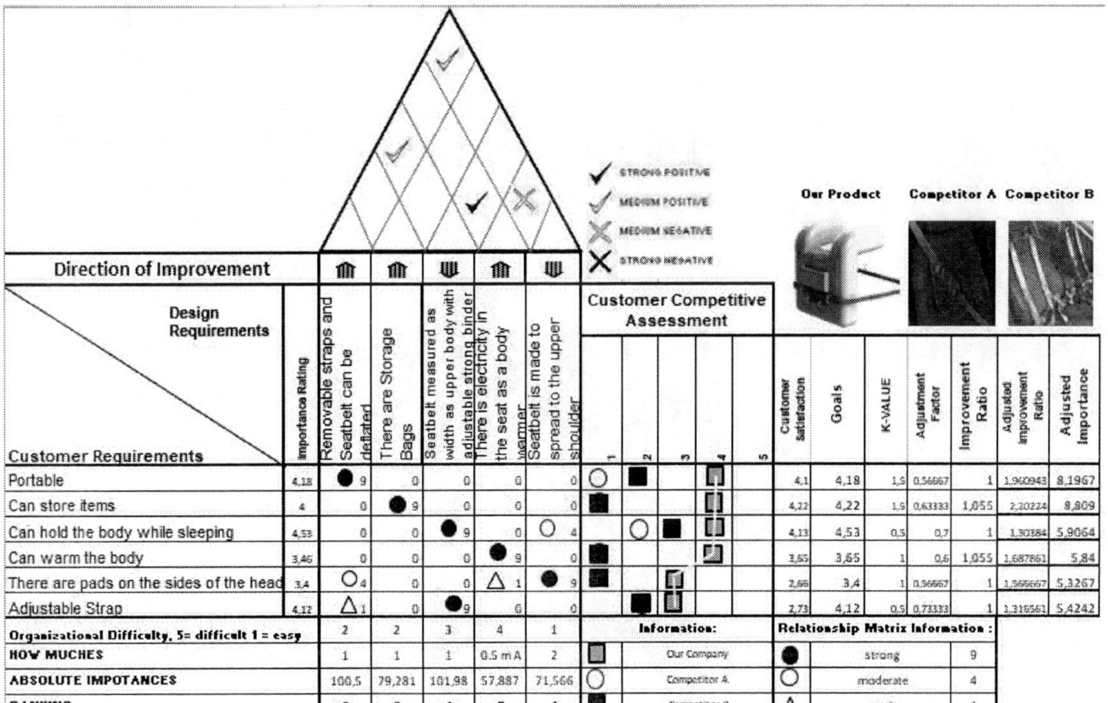

Figure 2. House of quality 1.

IOP Conf. Series: Materials Science and Engineering **722** (2020) 012056 doi:10.1088/1757-899X/722/1/012056

3.3. Product description

As the name implies, MUSESA is a multifunctional seatbelt that has innovative features such as available pockets to store important items for transportation passengers so that they are easily stored and safe from theft. In addition, there are points electrified by battery resources so that they can warm passengers when they are cold on the way. Therefore taslan is used as a MUSESA base material which is heat-resistant, at the same time waterproof and comfortable to use. The strap is also adjustable so that it can be adjusted for different passenger sizes. Figure 3 shows the design of MUSESA product.

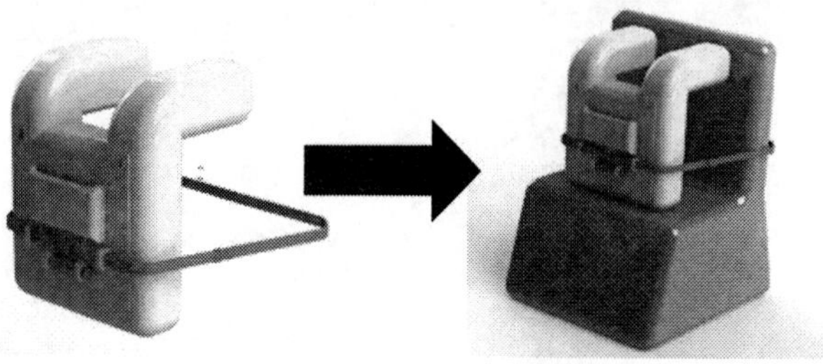

Figure 3. Design of MUSESA.

MUSESA is designed to be portable so that it is easy to carry wherever the user wants because transportation can be in the form of trains, planes, and even public transportation. In order to be portable, MUSESA is designed to be pumped when it wants to be used and can be deflated when not in use so that a small pump is also available to be easily carried. Figure 4 shows how it is folded and Figure 5 shows the product spesification of MUSESA.

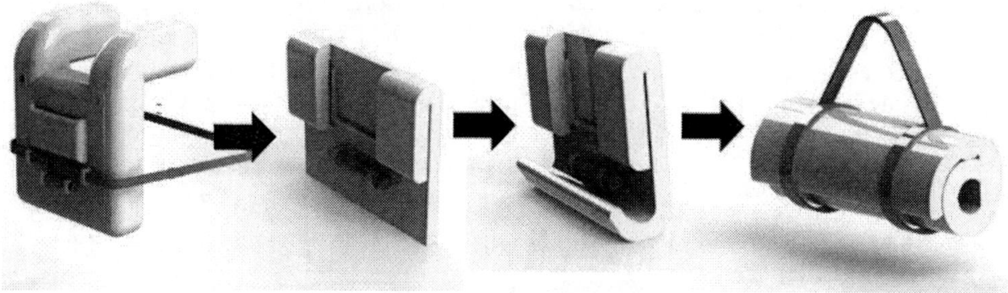

Figure 4. MUSESA in folding mode.

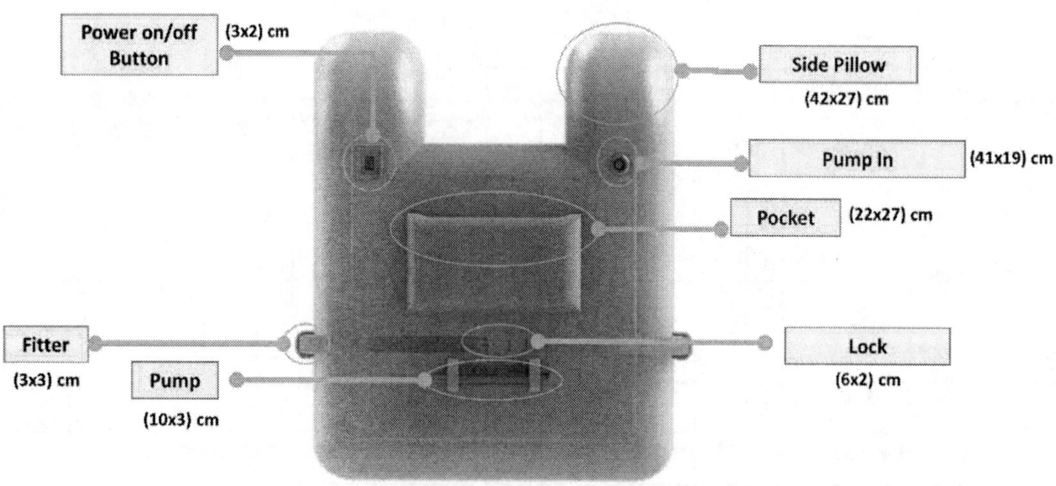

Figure 5. Product spesification of MUSESA.

3.4. Anthropometric approach

Anthropometry can be expressed as a study relating to the measurement of the human's body dimensions so that the products developed are in accordance with the dimensions of the human body [7]. Products that are designed based on anthropometric measurements from a certain group of people can increase the ergonomics of the tool because it fits averagely with the size of the users. Anthropometric data were compiled from direct measurement to 30 respondents. A normality test is performed using SPSS software first. The results can be seen in Figure 6 for calculations & Figure 7 for anthropometric implementation overview. After Ho criteria are accepted for all dimensions (Sig.> 0.05), the dimensions are used and the results of anthropomctric calculations presented in Table 4.

	Kolmogorov-Smirnov[a]			Shapiro-Wilk		
	Statistic	df	Sig.	Statistic	df	Sig.
TBD	,139	30	,145	,944	30	,117
LB	,141	30	,130	,931	30	,053
RS	,114	30	,200*	,973	30	,628
DPK	,140	30	,140	,956	30	,243
LK	,141	30	,135	,958	30	,269

Figure 6. Normality test.

Table 4. Anthropometric calculations.

No.	Dimensions	Percentile	Calculation Result (cm)
1	TBD	P_{50}	57.7
2	LB	P_{95}	44.8
3	RS	P_{95}	85.2
4	DPK	P_{50}	23.3
5	LK	P_{95}	17.9

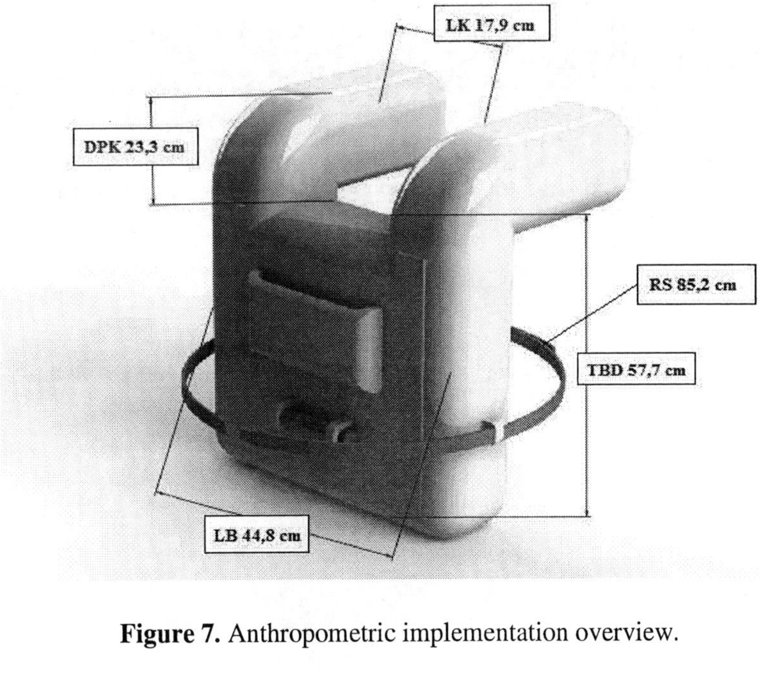

Figure 7. Anthropometric implementation overview.

IOP Conf. Series: Materials Science and Engineering **722** (2020) 012056 doi:10.1088/1757-899X/722/1/012056

4. Conclusions

This research produced an innovative product to help the community to get safety and comfort in driving with seatbelt support called MUSESA. Some of the main features include having a bag to store passenger goods while on the go to make it safer and more comfortable, portable so that it can be used in several different vehicles, can warm the user's body, have a bearing beside the user's head, and adjustable strap. The results of the Kano model show that there are 2 attributes that are included in the indifferent category that must be eliminated, 2 must be attributes, 2 attractive attributes, and 2 one dimensional attributes. The QFD results are limited to the HOQ 1 stage. From HoQ 1, there is a priority sequence that is useful for manufacturers at the later stages of production where the priority of the most important is there are seatbelt measured as width as upper body with adjustable strong binder, removable straps and seatbelt can be deflated, there are storage bags, seatbelt is made to spread to the upper shoulder, and there is electricity in the seat as a body warmer. MUSESA is also designed with an anthropometric approach as a basis for determining product specifications which use 5 dimensions of the human body to make it an ergonomic product.

References

[1] Nastiti F A 2017 Relationship between sim c ownership and participation in the simple test to make vehicle knowledge and traffic accident in sidoarjo district *The Indonesian Journal of Public Health* **12** 167–178

[2] Astraworld 2018 Jangan anggap remeh sabuk pengaman (https://www.astraworld.com/post/Jangan_anggap_remeh_sabuk_pengaman?fbclid=IwAR3uUi2TA6dgkx8lxt_EvmA32D I4ACiSY2qGHIJxZ1T0_qfmqMKOjLxpjc.)

[3] Rizqi Z U 2019 Integrasi kano model dan quality function deployment (QFD) dalam perancangan mesin gym pintar berkonsep all in one *Industrial Engineering National Conference* **7** 140-147

[4] Ghozali I 2001 *Aplikasi Analisis Multivariat dengan Program SPSS* (Semarang: Badan Penerbit Universitas Diponegoro)

[5] Hashim A and Dawal S Z 2012 Kano model and QFD integration approach for ergonomic design improvement *Procedia - Social and Behavioral Sciences* **57** 22-32

[6] Kano N, Seraku N, Takahashi F, and Tsuji S 1984 Attractive quality and must-be quality *The Journal of the Japanese Society for Quality Control* **14** 39-48

[7] Wignjosoebroto S 2008 *Teknik Tata Cara dan Pengukuran Kerja Edisi Pertama Cetakan Keempat* (Jakarta: Guna Widya)

Acknowledgments

The author also would express thanks to Sinta Nofita and Dewi Ayu Ningrum for help to the author while doing this research. And also to the respondents who gave the free time while answering honestly the questions in this research.

IOP Conf. Series: Materials Science and Engineering **722** (2020) 012057 doi:10.1088/1757-899X/722/1/012057

Determination of Criteria and Sub-criteria in Urban Solid Waste Processing Technology

A Ishak[1] and N Hidayati[2]

[1,2]Industrial Engineering Department, Faculty of Engineering, Universitas Sumatera Utara, Medan, Indonesia

E-mail: [1]aulia.ishak@gmail.com, [2]nurul97.hidayati@yahoo.com

Abstract. Solid waste is known as waste, which it recommends is often undesirable because it has no economic value. One of the factors that affect the amount of waste is the amount or population density. The high population has caused the amount of waste to be higher. The waste management system in the landfill (TPA) is open transportation, so that it can increase the amount of land needed every year. Therefore, in research, it is necessary to have a technology for municipal solid waste treatment with appropriate criteria and subcriteria.Determination of criteria and sub-criteria for urban solid waste processing technology using the Delphi method with 3 respondents. The results obtained are that the criteria consist of three aspects, namely economic aspects, environmental aspects, and technological aspects. It also consists of 12 subcriteria, namely, Equipment Costs, Material Costs, Maintenance Costs, Residual Quantity, Emis Rate, Impact of Odor, Operating Temperature, Performance, Product Quality, Duration of Process, Area, Easy Operation.

1. Introduction

Waste is produced from a production process both industrial and domestic (household). Solid waste is defined by the Organisation for Economic Co-operation and Development (OECD, 2003) as useless and often dangerous material that presents low liquid contents. It has become the subject of several studies and concern given its exponential growth in volume as a consequence of large urban agglomerations and manufacture production intensification in the capitalist system [1].Sources of waste can including from households, markets, stalls, offices, public buildings, industry, and roads. Rapid development and population growth in urban areas has resulted in increasingly large and dense residential areas.Waste treatment and its management has become one of the most critical environmental issues of today. In many sectors, the treatment and disposal of waste contribute a large part of the production cost. Waste treatment will usually increase the production cost and lead to price differences between similar products that are and are not treated [2].

The high population growth rate is always directly proportional to the level of consumption and activity of the community, causing the amount of waste (solid waste) produced is also higher. The population growth affects the amount of waste generated. Increasing population in each region/city, the amount of waste produced by each household is increasing. Industrial activities ranging from raw materials to produce products will disrupt the lives of people around the factory. There are many alternative technologies available to process other industries, but problems that often occur are difficult to implement the most appropriate technology [3]

Many cities still uses open dumping as a waste management system in the Final Disposal Site (TPA) so that land expansion needs to be done as the amount of waste increases. The negative impact that can be caused by a high volume of waste that is not managed properly is health problems, decreases the quality of the environment, decreases environmental aesthetics and hampers the country's development. Therefore, good waste management technology is needed.

Assessing and selecting the optimal technological alternatives in industrial sector is a fundamental tool to improve and adapt the industrial processes, which encourages the incorporation of new technologies for minimising the environmental impacts of companies [4]. In selecting solid waste processing technology, there is a need for selection indicators, in the form of criteria and subcriteria used to choose the best technology. Developing assessment criteria and methods that provide sustainability measurement are a prerequisite for choosing the best alternative [5]. So, in this study, the indicators of the selection of municipal solid waste treatment technology were carried out.

The Delphi method is a systematic method of gathering opinions from a group of experts through a series of questionnaires, where there is a mechanism feedback through rounds / rounds of questions held while maintainin anonymity of respondents' responses (experts). The Delphi method is a structured communication technique, originally developed as an interactive forecasting method that depends on a number of experts [6].

The Delphi method was developed by the RAND Corporation during the 1950s and 60s, with key contributions by Dalkey and Helmer. It has a long background and tradition as a management decision tool. The key elements are: (a) anonymous responses by experts to multiple rounds of formal questionnaires; (b) an exercise incorporating iterative, controlled feedback with respect to the information provided at each round; and (c) statistical summary of the group's responses. Integrated approach using Delphi and the analytic hierarchical process (AHP) is proposed [7]. Evaluating the performance of the recycling programs under uncertainty. The Delphi method is used for identifying the appropriate evaluation criteria in order to adequately consider the interest of various stakeholders in the decision making process [8].The approach was designed to minimize the influence of dominant individuals, group pressure, and irrelevant communication and to reduce (statistical) noise [9].

2. Methodology

The research was conducted in a City. The type of research conducted is descriptive research. This descriptive research is in the form of survey research, namely research conducted to obtain facts from existing symptoms by looking for factual information to get the truth. In this case, the instrument was used in the form of a questionnaire distributed to respondents. Respondents were taken using purposive sampling sampling technique. Filling out questionnaires is done with or without the help of surveyors depending on their needs. [10]

In this study, the selection of criteria and subcriteria in the municipal solid waste treatment technology was carried out.The method used is the Delphi method. Data collection techniques used with questionnaires. Technique sampling that used in this research is perposive sampling. The questionnaire was distributed to 3 respondents who understood and had knowledge of the technology of municipal solid waste treatment according to researchers.

The questionnaire used was a semi-open questionnaire.Data sources are divided into two, primary data and secondary data. Primary data is data obtained directly by the researcher with observation, in this case the primary data is the semi-open questionnaire results data. Secondary data is data obtained not from the results of direct observation by researchers. Secondary data in this study are criteria and subcriteria obtained from literature or book studies [11].

3. Result and Discussion

From the literature study there are 3 criteria and 12 sub-criteria for the selection of municipal solid waste treatment technology. The criteria dan subcriteria can be seen in Table 1.

Table 1. Criteria and subcriteria from literature

Number	Criteria	Sub-Criteria
1	Economical Aspects	1. Equipment Costs 2. Material Costs 3. Maintenance Costs
2.	Environmental Aspects	1. Residual Quantity 2. Emis Rate 3. Impact of Odor
3.	Technological Aspects	1. Operating Temperature 2. Performance 3. Product Quality 4. Duration of Process 5. Area 6. Easy Operation

Then these criteria and subcategories are arranged into a semi-open questionnaire and distributed to three respondents who are considered by the researcher to understand the technology of urban solid waste treatment.

From the results of questionnaires to three respondents, the following recapitulation was obtained in Table 2.

3.1. Criteria and Sub-criteria

Delphi is used to get the weight of each criterion, sub-criteria, and alternatives. The results of the weighted criterion, sub-criteria, and alternatives can be seen in Table 2.

Table 2. Recapitulation of the Open Questionnaire

Number	Criteria dan Sub-Criteria	Total	Average
1	**Econimical aspects**	**15**	**5**
1.1	Equipment costs	**14**	**4,667**
1.2.	Material costs	**12**	**4**
1.3.	Maintenance costs	**15**	**5**
2.	**Environmental aspects**	**15**	**5**
2.1.	Residual quantity	**15**	**5**
2.2.	Emis rate	**14**	**4,667**
2.3.	Impact of odor	**13**	**4,333**
3	**Technological aspects**	**15**	**5**
3.1.	Operating temperature	**15**	**5**
3.2.	Performance	**12**	**4**
3.3.	Product quality	**14**	**4,667**
3.4.	Duration of process	**13**	**4,333**
3.5.	Area	**12**	**4**
3.6.	Easy operation	**15**	**5**

4. Conclusion

From the research that has been done, the criteria and subcriteria used in solid waste processing technology consists of 3 criteria and 12 subcriteria, which cover economic aspects, environmental aspects and technological aspects.

References

[1] Pereira, Tais de S., Fernandino, Gerson. 2019. Evaluation of Solid Waste Management Sustainability of A Coastal Municipality from Northeastern Brazil. Ocean and Coastal Management **179**.
[2] Liu, Xiuli., Hewings, Geoffrey J.D. Hewings., Wang, Shouyang. 2018. Spain : Journal of Cleaner Production **200**.
[3] Ishak, Aulia., Yazid ,Ali. 2017.Decision Support Model for Selection Technologies in Processing of Palm Oil Industrial Liquid Waste.IOP Conf. Series: Materials Science and Engineering **277**,IOP Publishing.
[4] Ishak, Aulia., Yazid ,Ali. 2019. Identification Criteria and Indicators of Palm Oil Industrial Solid Waste Processing Technology.IOP Conf. Series: Materials Science and Engineering **602**,IOP Publishing.
[5] Forés, V. Ibáñez, et. al. 2014. A Holistic Review of Applied Methodologies for Assessing and Selecting The Optimal Technological Alternative from A Sustainability Perspective. Spain: Journal of Cleaner Production.
[6] Linstone, Harold A. 2002. The Delphi Method. Amerika: New Jersey Institute Of Technology.
[7] Kim, M., Jang, Y.C., Lee, S., 2013. Application of Delphi-AHP methods to select the priorities of WEEE for recycling in a waste management decision-making tool. J. Environ. Manage. 128, 941–948.
[8] Yeh, C.H., Deng, H., Wibowo, S., Xu, Y., 2010. Fuzzy multicriteria decision support for information systems project selection. Int. J. Fuzzy Syst. 12, 170–179.
[9] Strand, Jon., et al. 2017. Using The Delphi Method to Value Protection of the Amazon Rainforest
[10] Creemers, Bert P. M., et al. 2010. Methodological Advances in Educational Effectiveness Research (Quantitative Methodology Series). London : Routledge Publisher.
[11] Saaty, Thomas L. 2008. Decision Making with The Analytical Hierarchy Process.International Journal of Services Sciences

IOP Conf. Series: Materials Science and Engineering **722** (2020) 012058 doi:10.1088/1757-899X/722/1/012058

A Prototype of Effective Online Teaching Tools with Adaptive Navigation Support

A Hamzah

Department of Informatics, Faculty of Industrial Technology, Universitas Islam Indonesia.
Jalan Kaliurang KM 14,5 Sleman Yogyakarta Indonesia 55584

Email: almed.hamzah@uii.ac.id

Abstract. This paper proposes a prototype of educational system based on adaptive navigation support to help teacher manage their courses online. In a university context, a teacher can have many courses in one semester. The courses consists of several topics with several material and assignments on it. In a such circumstances, a teacher must prepare the topic's material before they can go to the class. This situation can cause the teacher to overwhelmingly getting busy to identify which courses they want to focus on and which one they do not. Therefore, they need a tools as an aid to choose which courses they should focus on. In this study, a prototype has been built based on adaptive navigation approach to help teacher overcome the situation. Link hiding techniques were used as a mechanism to help user focus on one task at a time.

1. Introduction

The gradual emergence of e-learning adoption and research has raised the opportunity for enhancement in various dimensions [1]. There are two main parties involved in e-learning based activities, namely students and teacher. The innovation of e-learning system should covers both sides to increase the effectiveness of e-learning. For students, e-learning can expand their opportunities to learn various material [2]. For the teacher, e-learning provides them the ability to manage the courses being taught easier [3]. However, with the increase amount of courses and material that must be handled, the teacher can loose their control over courses. This in turn will affect the overall students' performance. This paper proposes an approach on this domain by adopting adaptive navigation support in a form of system's prototype.

Adaptive navigation support is a technique for modifying the navigation pattern of system's interface to match with the user's objectives, preferences and knowledge [4]. It is a part adaptive system domain which is an intersection between hypermedia and user modeling techniques [5]. With such system, a teacher can ensure the availability of course material right on time its needed. Furthermore, they will received some notifications regarding to assignments' due and student's works. Thus, they can monitor the performance of the students in a more precise manner and compare it to the learning goals.

Mobile devices are currently becoming one of the primary means to access the internet [6]. It is because the spread of the devices worldwide. More and more people are using the devices to do various tasks. Education sector is progressively adopting the devices as a media of learning called mobile learning [7]. It changes the way of doing learning activities where material is delivered through the internet and accessed with mobile devices [8]. Considering this, the prototype will be designed based on mobile platform.

The rest of the paper is organized as follows. Section two describes an insight from previous works that denotes the existing finding about adaptive navigation support. Section three explain the methods in data gathering, systems' architecture and design. Section four describes the system's prototype including the user interface and adaptive features. Section five ends this paper with conclusion and discussion about possible future works.

2. Related Works

The adaptive system aims to accommodate user's characteristics in the system's interaction [5]. The adoption of adaptive system in educational context could be beneficial in several ways. With this kind of system, a students can have a personalized interface which specifically deliver the information about the course that only related to their capacity and the goals of learning [5]. It also provide the students the ability to increase their satisfaction on using adaptive system with which there is a possibility for them to improve their learning pace [4,9]. This paper adopt this advantage into teacher's context.

There are various adaptive systems were developed with different purposes and context. However, most of them were built to manage courses and the assignments. QuizGuide is built for students to be able to select the most relevant assignment for them with which they can assess their own knowledge and performance by themselves [10]. The system provide adaptive annotation shows the level of students' knowledge. Protus is an adaptive sysemThe adaptive system also can be used to recommend the suitable course material to students to allows them get focus on particular information [11]. In ubiquitous learning, adaptive system was used as a means for providing suggestion on user's learning path [12]. In other research, the adaptive approach was used to match the system's interface with the user's learning styles which in turn will increase the learner's performance [13].

With the adaptive approach, the interface of the system was modified with some techniques in order to comply with user's characteristics so that the interaction's flow becomes easier. For the teacher, adaptive learning system is an innovative techniques for teaching. With this system, a teacher can ensure the availability of the course as they can monitor in time-based manner the material and assignment of the course.

Adaptive navigation support is a technique to manipulate the navigation model in the interface of the system. It flexibly change the structure of the links on a web-page to accommodate the user's needs [4]. A research found that adaptive navigation support can makes the system's navigation more efficient for user. It can suggest the user with the relevant path of task to navigate. Moreover, it can helps user decide what to do with the system [12]. In addition, with the adaptive navigation support, user will not make a repetition to a particular content. Contrarily, they can move quickly from one content to another. Thus, it opens a wider possibility for the user to find a new information [14].

3. Method

3.1. Data Collection

The data for this research was retrieved from the course of Web Programming taught for undergraduate students. The course consists of ten topics that will be taught one topic in a week. Table 1 depicts the topics of the course. Every topic comprises materials and quiz assignments. The materials can be in the form of presentation slides, video, audio, or PDFs. The quiz assignments due in one week after it was published by the teacher.

Table 1. Course's topics.

Week	Topic
1	Introduction to Web Programming

2	HTML
3	Cascading Style Sheet
4	Javascript
5	JQuery
6	Asynchronous Communication (AJAX)
7	Server Side Scripting
8	Database Operation
9	Session and Cookies
10	Create Read update Delete

3.2. The Design of System

The architecture of the system is depicts in Figure 1. The system comprises three primary elements, i.e. User Interface, Inference Engine, and Teacher's Model. The first elements arrange the layout of the system's user interface according to the current Teacher's Model. The teacher's model is actually a database that stores the information about courses, date of the topics, and material. Inference engine is a mechanism to keep the teacher's model updated based on the interaction between system and user.

It receive the information about the material and assignment that already prepared by the teacher and then update the teacher's model database. If there is a request to view the current task, it retrieve the current teacher's model from the database and select the most appropriate course to view based on the current date. In this study, the inference system will compile the information about unfinished courses. The unfinished course means that there are some works that need to be completed by the teacher, e.g. upload material, assess the assignments, answering questions.

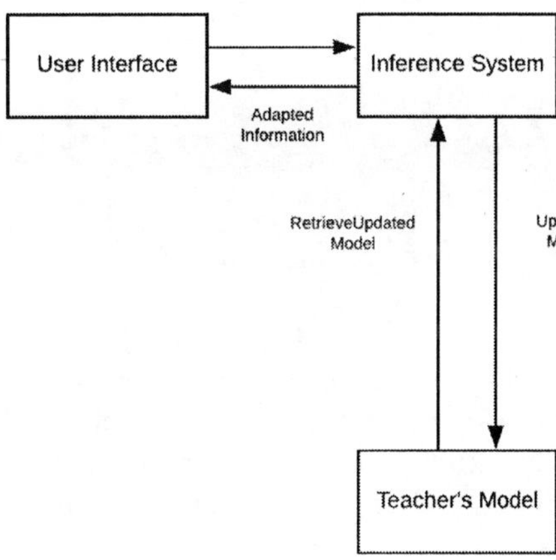

Figure 1. System's Architecture

The prototype has been built using Object Oriented Approach. Therefore, the functionality and the feature of the system is identified by the object. In this case, the design of the system is depicted in a usecase diagram that can be seen in Figure 2. The system allows teacher to add and view material and assignments based on teacher's model.

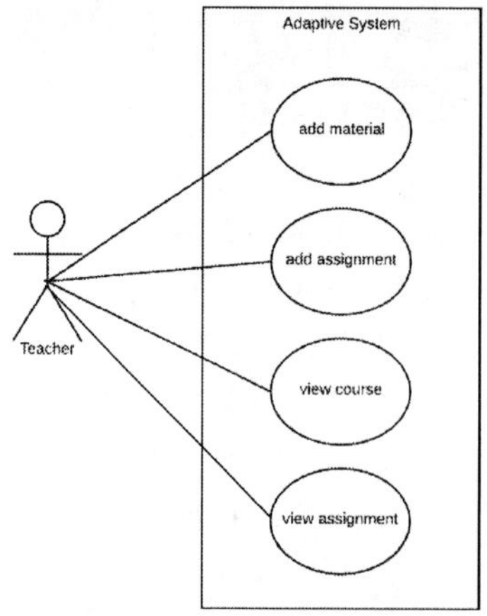

Figure 2. Primary Features of System in Usecase Diagram

In this research, the adaptive method that has been used is adaptive link hiding, which will hide the unnecessary link. Link hiding is one of the technique in adaptive navigation support where the unnecessary link is hide from the user in order to reduce the complexity of interaction [4]. The objective is to help teacher focus on the task that must be completed first and omit the condition where the user pay no attention to the system. The following is the algorithm in adaptive link hiding.

```
begin
identify current date;
if  (current_date = topic_date) {
  view current_topic;
}

identify previous task;
if (quiz <> 100) {
   task = unfinished;
   view unfinished tasks from previous topics;
end
```

Figure 3. Link Hiding Algorithm

4. Results

As the application built for mobile devices, its interface was designed by following the mobile user interface design guideline from [15]. According to the guideline, all the screen elements should be arranged in vertical stack. In this study, material and assignments are grouped into one block that represents a topic. Each block is stacked vertically. The layout of homepage of the system, by follow the guideline, can be seen in Figure 4.

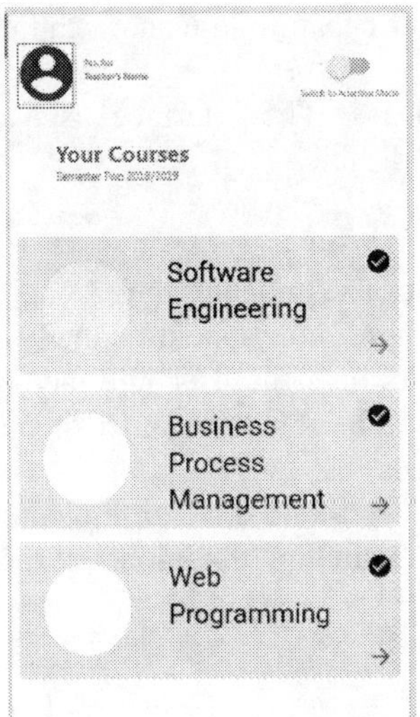

Figure 4. Homepage

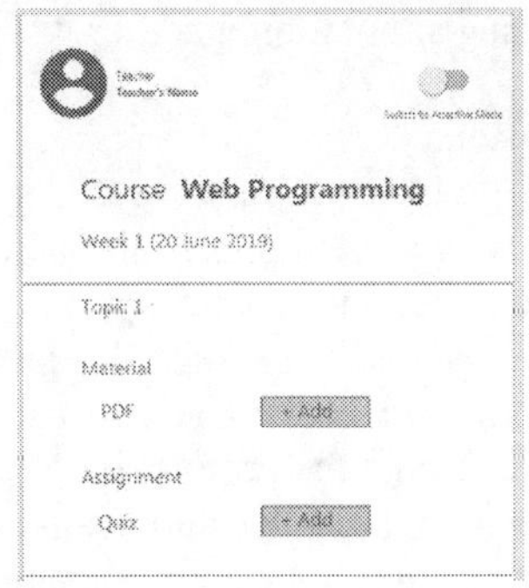

Figure 5. Main interface of the course

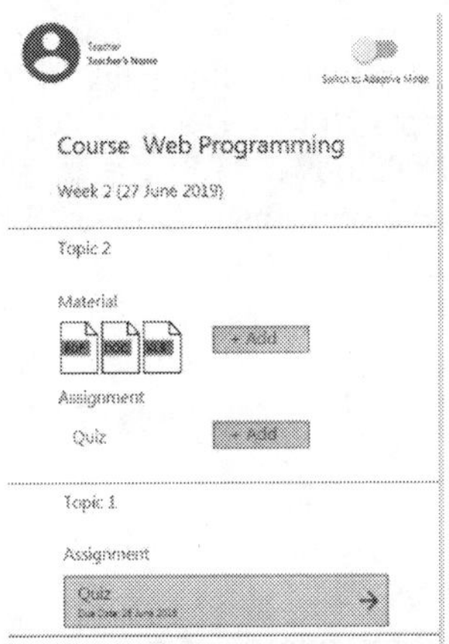

Figure 6. Adaptive Interface showing unfinished course

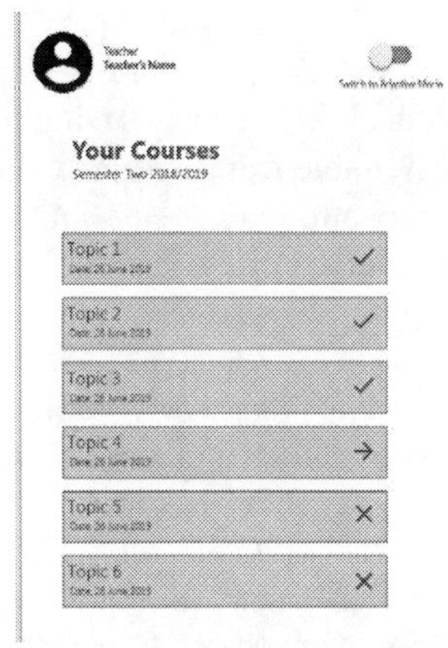

Figure 7. Adaptive Interface with Simple Graphical representation

Figure 4 show the landing page of the system. It shows all the courses being taught by the teacher. The main feature of the system is the arrangement of it's interface where the system only exhibit the material that relevant to teacher. In this context, the system will match the scheduled material to the current date. Figure 5 shows the screen displayed the initial topic of the semester. On this screen, there are only one material and assignment related to topic 1. The rest of the topic is hide automatically by the system and user will not see it.

With this kind of treatment, a teacher can concentrate only on related topic in terms of preparing the material, the assignment. They can add material, which comprises several type of files such as PDF, documents, spreadsheet, audio, and video. In addition, they can also add assignment related to the topic. Furthermore, a teacher can also monitor the progress of each topic.

When it comes to the second week, the interface will shows topic 2 as a main task for teacher to prepare the course material and assignments. In addition, the unfinished assignment from previous topic will appear at the bottom of the screen. This will allow teacher to keep informed regarding to its progress. The scenario is depicts in Figure 6. The process is iterating each time the course is progressed into next topic.

To accommodate the flexibility factor, the system provide switch button to shift between normal and adaptive mode. When in normal mode, the teacher will be able to see all the topic on one screen.

Otherwise, when in adaptive mode, the adaptive (hiding) algorithm will be deployed to select the most appropriate topic and display it on the screen. With this kind of feature, a teacher can either view all courses in an overall look or in more focused way.

The adaptive ability of the system is depicted in Figure 7. The interface lists all topic with corresponding annotation whether it is finished, unfinished or disable.

5. Conclusions

This study proposes a prototype of a system that can help teacher manage their courses better with Adaptive Navigation Support approach. The system consists of three components, namely user interface, inference engine, and teacher's model. It will automatically arrange the user interface and select the appropriate material only. With this kind of system, a teacher will be able to organize their courses better and faster. It also increase the effectiveness of learning activities.

This study presents the prototype of application. Future works could be done to implement the system in real educational settings and evaluate the result. Moreover, evaluating the impact of the system amid the teacher would also be an interesting research topic.

References

[1] Somyürek S 2015 The new trends in adaptive educational hypermedia systems Int Rev Res Open Distance Learn. 16(1):221–41.

[2] Sabah NM 2016 Exploring students' awareness and perceptions : Influencing factors and individual differences driving m-learning adoption Comput Human Behav [Internet]. 65:522–33. Available from: http://dx.doi.org/10.1016/j.chb.2016.09.009

[3] Kularbphettong K, Kedsiributt P, Roonrakwit P 2015 Developing an Adaptive Web-based Intelligent Tutoring System Using Mastery Learning Technique Procedia - Soc Behav Sci [Internet]. 191:686–91. Available from: http://dx.doi.org/10.1016/j.sbspro.2015.04.619

[4] Brusilovsky P 2007. Adaptive Navigation Support In: Brusilovsky P, Kobsa A, Nejdl W, editors. The Adaptive Web: Methods and Strategies of Web Personalization. Berlin: Springer-Verlag Berlin Heidelberg; p. 263–90.

[5] Brusilovsky P 2001 Adaptive Hypermedia User Model User-adapt Interact. 11(1–2):87–110.

[6] Mascheroni G, Ólafsson K 2016 The mobile Internet: Access, use, opportunities and divides among European children New Media Soc. 18(8):1657–79.

[7] Park SY, Nam M-W, Cha S-B 2011 University students' behavioral intention to use mobile learning: Evaluating the technology acceptance model Br J Educ Technol. 43(4):592–605.

[8] Alsaadat K 2017 Mobile Learning Technologies Int J Electr Comput Eng [Internet]. 7(5):2833. Available from: http://iaescore.com/journals/index.php/IJECE/article/view/8239

[9] Shi L, Cristea AI, Foss JGK, Qudah D Al, Qaffas A 2013 A social personalized adaptive e-learning environment: a case study in topolor 7641:1–17.

[10] Brusilovsky P, Sosnovsky S, Shcherbinina O 2004 QuizGuide: Increasing the educational value of individualized self-assessment quizzes with adaptive navigation support World Conf E-Learning Corp Gov Heal High Educ [Internet]. 2004(1):1806–13. Available from: http://www.editlib.org/index.cfm?fuseaction=Reader.ViewAbstract&paper_id=11585&from=NEWDL

[11] Klašnja-Milićević A, Vesin B, Ivanović M, Budimac Z 2011 E-Learning personalization based on hybrid recommendation strategy and learning style identification Comput Educ [Internet]. 56(3):885–99. Available from: http://www.sciencedirect.com/science/article/pii/S0360131510003222

[12] Chiou C-K, Tseng JCR, Hwang G-J, Heller S 2010 An adaptive navigation support system for conducting context-aware ubiquitous learning in museums Comput Educ [Internet]. 55(2):834–45. Available from: http://www.sciencedirect.com/science/article/pii/S0360131510001016

[13] Mampadi F, Chen SY, Ghinea G, Chen M-P 2011 Design of adaptive hypermedia learning

systems: A cognitive style approach Comput Educ [Internet]. 56(4):1003–11. Available from: http://www.sciencedirect.com/science/article/pii/S0360131510003465

[14] Guerra J, Hosseini R, Somyurek S, Brusilovsky P 2016 An Intelligent Interface for Learning Content: Combining an Open Learner Model and Social Comparison to Support Self-Regulated Learning and Engagement Proc 21st Int Conf Intell User Interfaces [Internet]. :152–63. Available from: http://dl.acm.org/citation.cfm?doid=2856767.2856784

[15] Tidwell J 2010. Designing interfaces: Patterns for effective interaction design " O'Reilly Media, Inc.";

IOP Conf. Series: Materials Science and Engineering **722** (2020) 012059 doi:10.1088/1757-899X/722/1/012059

Analysis of crude oil inventory using the material requirement planning method

W Sutrisno[1] and H N P Airlangga[1]

[1]Industrial Engineering Department, Faculty of Technology Industry, Universitas Islam Indonesia, Sleman, Indonesia

Email: 075220417@uii.ac.id

Abstract. Inventory management plays an important role in the company's success in producing products in quantities based on customer demand. This study uses the MRP method to determine the management of PT. X crude oil inventories so far and produce crude oil inventory management and the arrival schedule of crude oil that must be applied at PT. X. The results show that when manipulating MRP Super Heavy Crude oil from 10 days to 6 days (exercise), better inventory management ie crude oil stocks are in a safe position between safety stock and tank capacity.

1. Introduction

PT. X is a branch of a state-owned company that has the authority to manage oil, gas and geothermal resources in Indonesia. PT X uses crude oil as the main raw material in producing its products. PT. X produces three types of products namely Fuel Oil, Special Fuel, and Non-Fuel Oil. PT. X must be able to plan precisely the number of products to be produced so that consumer needs can be met properly. Planning the right amount of products can be realized, among others, by having good raw material inventory management.

Inventory management is a technique related to determining the amount of inventory that must be held to ensure the smooth operation of production, as well as determining the procurement schedule and the number of goods ordered to be carried out by the company [1]. One technique that can be used in managing inventory management is to use the MRP method. Therefore this study uses the MRP method to determine the management of crude oil inventories that have been carried out by PT. X so far and produce crude oil inventory management and the arrival schedule of crude oil which must be applied at PT. X

2. Literature Review

2.1. Inventory Management

Inventories are material deposits in the form of raw materials, processed goods and finished goods [2]. Inventory is defined as goods that are stored for use or sale in future periods [3]. Inventory is needed to ensure the smooth process of meeting the demand for goods according to user needs economically, reducing irregular demand or supply fluctuations and speculation tools for future profits [4].

2.2. Scheduling

IOP Conf. Series: Materials Science and Engineering **722** (2020) 012059 doi:10.1088/1757-899X/722/1/012059

Scheduling can be interpreted as allocating several resources to perform several tasks or operations in a certain period to achieve the goals and objectives of the company more optimally [5]. Scheduling is always related to the allocation of available resources at a certain period, and it is a decision making process whose purpose is for optimality [6]. The purpose of scheduling is to increase machine productivity, minimize production costs, reduce delays because they have exceeded the time limit and the fulfillment of the deadline [5].

2.3. Material requirement planning (MRP)
MRP is a bound demand model that uses a list of material requirements, inventory status, estimated revenues, and a master production schedule, which is used to determine the material requirements to be used [7]. MRP as an information system used to plan and control inventory and capacity [8]. MRP is a technique or set of systematic procedures in determining the quantity and time in the process of controlling the needs of materials for interdependent demand components [9].

The purpose of the application of MRP is to minimize inventory where the MRP can determine how much and when an item is needed is adjusted to the master production schedule, increasing efficiency whereby applying MRP, the amount of inventory, production time, and time of delivery of goods can be planned better according to the master production schedule [7]. MRP also aims to reduce risk due to delays in production or delivery [7]. MRP identifies the number of materials and items needed both in terms of quantity and time by taking into account the production grace period and component procurement.

3. Methodology
The object of research is the management of PT X crude oil inventory. The data used in this study is historical crude oil inventory data in the refinery Planning & Optimization division. The data is then processed using the MRP method to calculate crude oil inventory data and schedule the arrival of crude oil. Also, interviews were conducted with staff at the Refinery Planning & Optimization division to analyze the results of the calculation of crude oil inventories according to safety stock and tank capacity.

4. Result
PT X uses two types of crude oil namely Super Heavy Crude Oil and Heavy Crude Oil. Based on historical data of crude oil inventories that have been processed, the calculation results are obtained in table 1 and table 2.

Table 1. Minimum maximum check super heavy crude oil

Week number	Gross requirements	Schedule Receipt	Project on Hand	Safety stock	Tank Capacity	Safety stock Minimum check	Maximum Tank Maximum check
1	81.25		921	812.5	1200		
2	81.25	400	1239.75	812.5	1200	108.50	**39.75**
3	81.25		1158.5	812.5	1200	427.25	41.50
4	81.25		1077.25	812.5	1200	346.00	122.75
5	81.25		996	812.5	1200	264.75	204.00
6	81.25		914.75	812.5	1200	183.50	285.25
7	81.25		833.5	812.5	1200	102.25	366.50
8	81.25		752.25	812.5	1200	21.00	447.75
9	81.25		671	812.5	1200	**60.25**	529.00
10	81.25	200	789.75	812.5	1200	**141.50**	410.25

IOP Conf. Series: Materials Science and Engineering **722** (2020) 012059 doi:10.1088/1757-899X/722/1/012059

11	81.25		708.5	812.5	1200	**22.75**	491.50
12	81.25	212	839.25	812.5	1200	**104.00**	360.75
13	81.25		758	812.5	1200	**26.75**	442.00
14	81.25		676.75	812.5	1200	**54.50**	523.25
15	81.25	165	760.5	812.5	1200	**135.75**	439.50
16	81.25		679.25	812.5	1200	**52.00**	520.75
17	81.25	400	998	812.5	1200	**133.25**	202.00
18	81.25		916.75	812.5	1200	185.50	283.25
19	81.25		835.5	812.5	1200	104.25	364.50
20	81.25	200	954.25	812.5	1200	23.00	245.75
21	81.25		873	812.5	1200	141.75	327.00
22	81.25	600	1391.75	812.5	1200	60.50	**191.75**
23	81.25		1310.5	812.5	1200	579.25	**110.50**
24	81.25		1229.25	812.5	1200	498.00	**29.25**
25	81.25		1148	812.5	1200	416.75	52.00
26	81.25		1066.75	812.5	1200	335.50	133.25
27	81.25		985.5	812.5	1200	254.25	214.50
28	81.25		904.25	812.5	1200	173.00	295.75
29	81.25		823	812.5	1200	91.75	377.00
30	81.25		741.75	812.5	1200	10.50	458.25

Based on data from table 1, it can be seen that the management of Super Heavy Crude Oil inventory in several periods is lower than the value of the safety stock, amounting to 812.5 MB. Meanwhile, in some periods crude oil inventories have exceeded tank capacity.

Table 2. Minimum maximum check heavy crude oil

Week number	Gross requirements	Schedule Receipt	Project on Hand	Safety stock	Tank Capacity	Safety stock minimum check	Maximum Tank Maximum check
1	43.75	0	1190	437.5	1600		
2	43.75	0	1146.25	437.5	1600	752.50	453.75
3	43.75	220	1322.5	437.5	1600	708.75	277.50
4	43.75	0	1278.75	437.5	1600	885.00	321.25
5	43.75	0	1235	437.5	1600	841.25	365.00
6	43.75	0	1191.25	437.5	1600	797.50	408.75
7	43.75	0	1147.5	437.5	1600	753.75	452.50
8	43.75	0	1103.75	437.5	1600	710.00	496.25

9	43.75	200	1260	437.5	1600	666.25	340.00
10	43.75	0	1216.25	437.5	1600	822.50	383.75
11	43.75	280	1452.5	437.5	1600	778.75	147.50
12	43.75	0	1408.75	437.5	1600	1015.00	191.25
13	43.75	0	1365	437.5	1600	971.25	235.00
14	43.75	230	1551.25	437.5	1600	927.50	48.75
15	43.75	0	1507.5	437.5	1600	1113.75	92.50
16	43.75	0	1463.75	437.5	1600	1070.00	136.25
17	43.75	0	1420	437.5	1600	1026.25	180.00
18	43.75	0	1376.25	437.5	1600	982.50	223.75
19	43.75	200	1532.5	437.5	1600	938.75	67.50
20	43.75	0	1488.75	437.5	1600	1095.00	111.25
21	43.75	0	1445	437.5	1600	1051.25	155.00
22	43.75	0	1401.25	437.5	1600	1007.50	198.75
23	43.75	0	1357.5	437.5	1600	963.75	242.50
24	43.75	0	1313.75	437.5	1600	920.00	286.25
25	43.75	0	1270	437.5	1600	876.25	330.00
26	43.75	0	1226.25	437.5	1600	832.50	373.75
27	43.75	0	1182.5	437.5	1600	788.75	417.50
28	43.75	0	1138.75	437.5	1600	745.00	461.25
29	43.75	0	1095	437.5	1600	701.25	505.00
30	43.75	0	1051.25	437.5	1600	657.50	548.75

Based on data from table 2 it is known that the inventory management for Heavy Crude Oil is still in a safe situation because the value of crude oil inventories is still in the range between the minimum safety stock value and the maximum tank capacity.

Based on the MRP calculation of the proposal on Crude Oil, after the manipulation of the 6-day review period, the calculation results obtained as in table 3 and table 4.

Table 3. Results of the proposed super heavy crude oil

Week number	Gross requirements	Schedule Receipt	Project on Hand	Safety stock	Tank Capacity	Safety stock minimum check	Maximum Tank Maximum check
1		0	921.00	812.5	1200		
2	81.25	0	844.75	812.5	1200	108.50	355.25
3	81.25	400	1163.50	812.5	1200	32.25	36.50
4	81.25	0	1082.25	812.5	1200	351.00	117.75
5	81.25	0	1001.00	812.5	1200	269.75	199.00
6	81.25	0	919.75	812.5	1200	188.50	280.25
7	81.25	0	919.75	812.5	1200	107.25	280.25
8	81.25	0	838.50	812.5	1200	107.25	361.50
9	81.25	400	1157.25	812.5	1200	26.00	42.75
10	81.25	0	1076.00	812.5	1200	344.75	124.00
11	81.25	0	994.75	812.5	1200	263.50	205.25
12	81.25	0	913.50	812.5	1200	182.25	286.50
13	81.25	0	913.50	812.5	1200	101.00	286.50
14	81.25	0	832.25	812.5	1200	101.00	367.75
15	81.25	400	1151.00	812.5	1200	19.75	49.00
16	81.25	0	1069.75	812.5	1200	338.50	130.25
17	81.25	0	988.50	812.5	1200	257.25	211.50
18	81.25	0	907.25	812.5	1200	176.00	292.75
19	81.25	0	907.25	812.5	1200	94.75	292.75
20	81.25	0	826.00	812.5	1200	94.75	374.00
21	81.25	400	1144.75	812.5	1200	13.50	55.25
22	81.25	0	1063.50	812.5	1200	332.25	136.50
23	81.25	0	982.25	812.5	1200	251.00	217.75
24	81.25	0	901.00	812.5	1200	169.75	299.00
25	81.25	0	901.00	812.5	1200	88.50	299.00
26	81.25	0	819.75	812.5	1200	88.50	380.25
27	81.25	400	1138.50	812.5	1200	7.25	61.50
28	81.25	0	1057.25	812.5	1200	326.00	142.75
29	81.25	0	976.00	812.5	1200	244.75	224.00
30	81.25	0	894.75	812.5	1200	163.50	305.25

Based on the calculation of the MRP proposal on Super Heavy Oil, after manipulation of the review period of 6 days, a waiting time of 10 days (arrival of crude oil raw materials every 5 days) with a parcel (lot size) of 400 MB, no crude oil inventory is found that is less than the stock value safety or not exceed the capacity of the existing tank. However, as an improvement step in ensuring the availability of Crude Oil, the initial inventory would be better if it was changed at least 950 MB, with the consideration that this could prevent the project value which is less than the security value of the inventory.

Tabel 4. Results of the proposed heavy crude oil

Week number	Gross requirements	Schedule Receipt	Project on Hand	Safety stock	Tank Capacity	Safety stock minimum check	Maximum Tank Maximum check
1		0	1190,00	437,5	1600		
2	43,75	0	1146,25	437,5	1600	752,50	453,75
3	43,75	0	1102,50	437,5	1600	708,75	497,50
4	43,75	0	1058,75	437,5	1600	665,00	541,25
5	43,75	0	1015,00	437,5	1600	621,25	585,00
6	43,75	200	971,25	437,5	1600	577,50	628,75
7	43,75	0	927,50	437,5	1600	533,75	672,50
8	43,75	0	883,75	437,5	1600	490,00	716,25
9	43,75	0	840,00	437,5	1600	446,25	760,00
10	43,75	0	796,25	437,5	1600	402,50	803,75
11	43,75	0	752,50	437,5	1600	358,75	847,50
12	43,75	200	908,75	437,5	1600	315,00	691,25
13	43,75	0	865,00	437,5	1600	471,25	735,00
14	43,75	0	821,25	437,5	1600	427,50	778,75
15	43,75	0	777,50	437,5	1600	383,75	822,50
16	43,75	0	733,75	437,5	1600	340,00	866,25
17	43,75	0	690,00	437,5	1600	296,25	910,00
18	43,75	200	846,25	437,5	1600	252,50	753,75
19	43,75	0	802,50	437,5	1600	408,75	797,50
20	43,75	0	758,75	437,5	1600	365,00	841,25
21	43,75	0	715,00	437,5	1600	321,25	885,00
22	43,75	0	671,25	437,5	1600	277,50	928,75
23	43,75	0	627,50	437,5	1600	233,75	972,50
24	43,75	200	783,75	437,5	1600	190,00	816,25
25	43,75	0	740,00	437,5	1600	346,25	860,00
26	43,75	0	696,25	437,5	1600	302,50	903,75
27	43,75	0	652,50	437,5	1600	258,75	947,50
28	43,75	0	608,75	437,5	1600	215,00	991,25
29	43,75	0	565,00	437,5	1600	171,25	1035,00
30	43,75	200	721,25	437,5	1600	127,50	878,75

Based on the proposed MRP calculation on Heavy Crude Oil, after doing the 6-day review period (exercise), the MRP calculation method for Heavy Crude Oil, while using a lead time of 10 days (arrival of crude oil raw materials every 5 days) with a parcel amount (lot size) 200MB, because the value of Crude Oil inventories is still in the range between the minimum safety stock value and the maximum tank capacity.

5. Conclusion

- Through the MRP calculation method based on preliminary data, it was found that the Super Heavy Crude Oil inventory management in some periods was lower than the value of the safety stock, amounting to 812.5 MB. This can stop the implementation of production at PT X if the supply of super-heavy crude oil runs out while the raw materials ordered have not yet arrived. Meanwhile, in some periods, it is known that crude oil supplies exceed tank capacity, this is also not good for the company because it will increase the cost of saving for the company. Whereas inventory management for Crude Oil Heavy Crude Oil is still in a safe situation because the value of Crude Oil inventory is still in the range between the minimum safety stock value and maximum tank capacity.

References

[1] Ristono, Agus. 2013. *Manajemen Persediaan*. Yogyakarta: Graha Ilmu

[2] Sumayang, Lalu. 2003. *Dasar-dasar Manajemen Produksi dan Operasi*. Jakarta: Salemba Empat

[3] Kusuma, Hendra. 2009. *Manajemen Produksi : Perencanaan dan Pengendalian Produksi, Edisi 4*. Yogyakarta: Andi

[4] Ginting, Rosnani. 2007. *Sistem Produksi*. Yogyakarta: Graha Ilmu.

[5] Baker, K.R. & Trietsch, D., 2009. *Principles Of Sequencing And Scheduling*. New Jersey: John Wiley & Sons

[6] Pinedo, Michael. 2002. *Scheduling Theory, Algorithms, and System, Fourth Edition*. New York: Springer

[7] Heizer, Jay and Barry Render. 2005. *Operations Management*. Jakarta: Salemba Empat

[8] Schroeder, Roger G. 1994. *Manajemen Operasi : Pengambilan Keputusan dalam Suatu Fungsi Operasi, Jilid 2, Edisi Ketiga*. Jakarta: Erlangga

[9] Gaspersz, Vincent. 2004. *Production Planning and Inventory Control*. Jakarta: PT. Gramedia Pustaka Umum.

IOP Conf. Series: Materials Science and Engineering **722** (2020) 012060 doi:10.1088/1757-899X/722/1/012060

Integration of deterministic and probabilistic inventory methods to optimize the balance between overstock and stockout

Z U Rizqi[1] and A Khairunisa[2]

[1,2] Department of Industrial Engineering, Universitas Islam Indonesia, Yogyakarta, Indonesia

[1]E-mail: ugihzakka@gmail.com

Abstract. In inventory management, stockout and overstock are problems that always occur in the inventory system. The limitations of the method make it difficult for companies to determine the level of inventory following probabilistic inventory conditions. Ajeng Gallery is a quality bag craft company that has problems in the supply of raw materials where there is always a large stockout and overstock. The research is limited to cow leather material which is the main material based on the results of ABC analysis. This research also integrates Min-Max method which is a deterministic method to determine the level of inventory then the results will be integrated with Monte Carlo simulation which is a probabilistic method so that it can evaluate the amount of stockout and overstock that occurs. The results showed that the company no longer experienced stockout for 1 month and the amount of overstock decreased by 0.42% of the actual system.

1. Introduction

Inventory is an interesting thing to talk about. By definition, inventory can be interpreted as an asset which includes goods belonging to the company to sell within a certain business period or inventory of goods that are still in the work or production process or inventory of raw materials awaiting their use in the production process [1]. Its presence is costly to the company, but without it, the company will find difficult to always meet the demands of dynamic customers.

Basically, inventory problems occur due to a lack of information obtained. Good information about the demand and time of arrival of the material is needed. If the supply of raw materials is less or exhausted (stockout), the company will lose because customer satisfaction will decrease [2]. Meanwhile, if the inventory is too much (overstock), it can increase the company's inventory costs. These costs arise due to warehouse rental fees, maintenance costs, warehouse worker salaries or possibly expired goods costs.

Ajeng Gallery is a quality handicraft bag company based on cow leather in Yogyakarta. The well known quality causes the demand for bags is quite high and probabilistic. However, the company often experience stockouts due to lack of good company in processing raw material inventory. The company also does not want to increase the high level of inventory because the quality of cow leather used is easily damaged if stored in a warehouse too long so that it will harm the company. Therefore, the challenge is how companies determine the optimal level of inventory but with the lowest possible inventory costs.

The problem of overstock and stockout is a common problem faced by many companies in controlling inventory but it is still difficult to determine the optimal point [3]. One method that is widely used in inventory control is the Min-Max method. The min-max method is a method of controlling raw materials based on the assumption that raw material inventories are at two levels, namely the maximum level and the minimum level [4]. However, the method uses averages of probabilistic variables so that their use is assumed to be deterministic. Whereas the inventory environment is probabilistic so the method is less effective to use. With simulation, probabilistic conditions can be accommodated where a solution to a problem is given based on a randomization process [5]. By implementing simulation also the losses of money and time from experiments given can be minimized [6].

This study aims to integrate the min-max method with a Monte Carlo simulation in the Ajeng Gallery's raw material inventory where inventory levels obtained from the min-max method are included as input variables in the simulation. With the integration of the two methods, the evaluation of inventory levels can be optimized because it is carried out under probabilistic conditions following the actual inventory conditions.

2. Methods

2.1. Data collection

In this study, data collection was done by conducting interviews with warehousing managers from Ajeng Gallery. The interview was intended to collect historical company data regarding the type of material making up the product, material requirements per month, data on lead time from suppliers to the company, the amount of demand for raw materials, material prices, and the company's inventory management.

2.2. Data processing

Data processing in this study was carried out in 3 stages. The first will be done by material classification using ABC Analysis, determination of inventory levels using the min-max method, and evaluation with Monte Carlo simulation.

ABC analysis is done so that researchers focus only on the main material where the material is classified into 3 classes, namely A, B, and C. Materials that enter into class A are materials that provide high value, meaning that companies need to give focus and main control to for this material, class B is goods which provide a medium value represented by 30% of the total inventory and the value produced is 15%., and class C are goods that provide a low value represented by 50% of the total inventory exists and the resulting value is equal to 5% [7].

Furthermore, material entering into class A in ABC Analysis will be carried out inventory control is carried out using the Min-Max method. To avoid the amount of inventory that is too big or too small. The results of the Min-Max method are used to determine the minimum level or safety stock (SS), maximum level (Max), Reorder Point (ROP), and the number of materials ordered in a single order or Economic Order Quantity (EOQ) using equation 1, 2, 3, 4, and 5.

$$Sdl = \sqrt{(d^2 \times Sl^2 + (l \times Sd^2)} \quad (1)$$

$$SS = Z \times Sdl \quad (2)$$

$$ROP = (d \times l) + SS \quad (3)$$

$$EOQ = 2 \text{ x } d \text{ x } l \quad (4)$$

$$\text{Max} = SS + EOQ \tag{5}$$

Where:
Z = proportion of demand to be fulfilled
Sdl = standard deviation of lead time and demand
SI = standard deviation of lead time/period
Sd = standard deviation of demand/period
d = average demand/period
l = average lead time/period

Then the researchers implemented the results of the Min-Max method calculation into the Monte Carlo simulation which estimated the company's inventory system to further evaluate the problems that occurred at Ajeng Gallery. Monte Carlo simulation is done using Excel software with the random number generation feature. The purpose of the simulation is to find out the effect that occurs if Ajeng Gallery applies the inventory level of the min-max method. Stockout and overstock variables will be used as parameters in this study.

3. Result and discussion

3.1. ABC analysis

ABC Analysis on Ajeng Gallery can be seen in Table 1 which explains the materials along with their prices and Table II which explains the calculation results of ABC Analysis. The standard unit between Ajeng Gallery and suppliers of cow leather is sheets, where for 1 sheet cow leather is equal to 1x2 meters.

Table 1. Materials in ajeng gallery warehouse.

No.	Item	Needs/month	Price/unit (Rp)
1	Cow leather	125 sheet	425,000
2	Suede fabric	370 meter	35,000
3	Kerinti handle	400 pcs	13,500
4	Ring D	500 pcs	1,000
5	Zipper	300 meter	4,000
6	Zipper slider	750 pcs	1,500

Table 2. ABC analysis result.

Item	Needs (unit)	Statement of Unit	Price/unit (Rp)	Total Price (Rp)	Total Price (%)	Cumulative (%)	Class
1	125	Sheet	425,000	53,125,000	71.50	71.50	A
2	370	Meter	35,000	12,950,000	17.43	88.93	B
3	400	Pcs	13,500	5,400,000	7.27	96.20	B
4	300	Meter	4,000	1,200,000	1.62	97.81	C
5	750	Pcs	1,500	1,125,000	1.51	99.33	C
6	500	Pcs	1,000	500,000	0.67	100	C

Class	Item	Needs (unit)	Needs (%)	Total Price (Rp)	Total Price (%)
A	1	125	5.66	53,125,000	71.5
B	2,3	770	32.08	18,350,000	24.7
C	4,5,6	1550	62.26	2,825,000	3.8

Based on ABC analysis, the material that included in class A category is cow leather with total needs is 5.665 and total price is 71.5%, class B category are suede fabric and kerinti handle with total needs is 32.08% and total price is 24.7%, and for class C category are ring D, zipper and zipper slider with total needs is 62.26% and total price is 3.8%. Therefore, cow leather material will be carried out as an object to be evaluated in this study.

3.2. Min-max calculation

Before calculating using Min-Max, historical data regarding the lead time (day) and demand per day are required, which are shown in Table 3.

Table 3. Historical data for cow leather's leadtime and demands.

Lead time (Day)	Demand/Day
2	5
1	6
4	4
3	5
2	5

The results of calculations using equation 1, it obtained a value for Sdl is 5.82. By using equation 2, we get 10 sheets for safety stock (min) of cow leather. Equation 3 is used to determine the reorder point (ROP), where the ROP for cow leather is 22 sheets. Economic order quantity (EOQ) is calculated using equation 4, which is 24 sheets. And the maximum (max) warehouse capacity for cow leather is calculated using equation 5, which is obtained as many as 34 sheets.

3.3. Implementation on simulation model

Monte carlo simulation is done by imitating the real inventory system at Ajeng Gallery. Simulations were carried out specifically for cow leather material which included in classification A in the ABC analysis. The data that is used in the simulation is based on the leadtime and demand data in Table III, then the probability distribution of lead time is calculated first as shown in Table 4 and the probability distribution of demand in Table 5.

Table 4. Probabilistic distribution of lead time.

Lead time (day)	Freq.	Prob.	Prob. Cum.	LL	UL
1	1	0.2	0.2	1	20
2	2	0.4	0.6	21	60
3	1	0.2	0.8	61	80
4	1	0.2	1	81	100
Total	5	1			

Table 5. Probabilistic distribution of demand.

Demand/day (sheet)	Freq.	Prob	Prob. Cum.	LL	UL
4	1	0.2	0.2	1	20
5	3	0.6	0.8	21	80
6	1	0.2	1	81	100
Total	5	1			

Simulation results for 1 month (26 days) can be seen in Figure 1.

Day	R. N. Demand	R. N. Leadtime	Demand	Leadtime	Initial Inv.	End Inv. /Overstock	Stockout
1	50	15	5	0	50	45	0
2	89	24	6	0	45	39	0
3	21	46	5	0	39	34	0
4	98	87	6	0	34	28	0
5	47	91	5	0	28	23	0
6	81	23	6	0	23	17	0
7	46	64	5	3	17	12	0
8	64	58	5	2	12	7	0
9	86	15	6	1	7	1	0
10	12	85	4	Order Received	25	21	0
11	15	88	4	3	21	17	0
12	71	94	5	2	17	12	0
13	84	48	6	1	12	6	0
14	80	98	5	Order Received	30	25	0
15	73	82	5	0	25	20	0
16	6	77	4	3	20	16	0
17	65	52	5	2	16	11	0
18	6	75	4	1	11	7	0
19	34	58	5	Order Received	31	26	0
20	95	42	6	0	26	20	0
21	72	78	5	3	20	15	0
22	58	68	5	2	15	10	0
23	3	84	4	1	10	6	0
24	76	89	5	Order Received	30	25	0
25	2	93	4	0	25	21	0
26	12	39	4	2	21	17	0
Total			129	Average		18,5	0

Figure 1. Result of monte carlo simulation.

3.4. Analysis

Based on monte carlo simulation results that are integrated with the Min-Max method, the results of the number of stockouts and overstocks can be evaluated. It can be seen that the implementation of the Min-Max method does not produce a stockout at all which means that it is in accordance with the company's objectives, which is always fulfilling the demand on time where the demand begins with the availability of raw materials. In terms of overstock, the average overstock per day that occurs is 19 sheets of cow leather per day. When compared to the real system, the average overstock per day that occurs is 33 sheets of cow leather per day. It means that there is a decrease in overstock per day by 0.42%. It also means that the application of Min-Max in Ajeng Gallery's warehouse will giving benefit for company because with the decrease in the amount of overstock, the cost of maintaining materials at the warehouse will be reduced. In addition, in terms of the quality of cow leather, it will be better if it is processed immediately compared to one that has long been stored in a warehouse so that the quality of the final product can be maximized.

4. Conclusions

Based on the discussion, it can be concluded that based on ABC analysis, cow leather material is included in class A category so that priority is given to its control. Min-Max calculation results show that the number of Safety Stock (Min) is 10 sheets, ROP is 22 sheets, EOQ is 24 sheets and Max is 34 sheets. Evaluation using Monte Carlo simulation for one month showed that Ajeng Gallery did not experience any stockout at all and there was a decrease in overstock by 0.42% compared to the real system so that it would benefit the company because customer satisfaction would not decrease, inventory costs would decrease, and quality of raw materials would be maximized because raw materials are used faster.

References

[1] Lahu E P and Sumarauw J S B 2017 Analysis of raw material inventory control to minimize inventory cost on dunkin donuts Manado *Jurnal EMBA* **5** 4175-4184

[2] Gavan J F 2000 Consumer response to stockouts *Journal of Consumer Research* **27** 249-266

[3] Satibi, Fudholi A, Tuko E C, and Swastiandari G L 2019 The inventory control, storage facilities and distribution at pharmaceutical industry in supporting drugs availability of JKN era *JMPF* **9** 27-37

[4] Kinanthi A P, Herlina D, and Mahardika F A 2016 Analisis pengendalian persediaan bahan baku menggunakan metode min-max (studi kasus PT.Djitoe Indonesia Tobacco) *Performa* **15** 87-92

[5] Hutahaen H D 2018 Analisa simulasi monte carlo untuk memprediksi tingkat kehadiran mahasiswa dalam perkuliahan (studi kasus: STMIK pelita nusantara) *Jurnal of Informatic Pelita Nusantara* **3** 41-45

[6] Rizqi Z U and Aulia R 2019 Evaluation of redesign layout using discrete event simulation (DES) *Proceedings of the International Conference on Industrial Engineering and Operations Management* 3203-3211

[7] Wahyuni T 2015 Penggunaan analisis abc untuk pengendalian persediaan barang habis pakai: studi kasus di program vakasi UI *Jurnal Vokasi Indonesia* **3** 1-20

Acknowledgment

The authors also would express thanks to Ajeng Gallery company for allowing researchers to see the company's condition directly so that it can provide an opportunity for researchers to implement our knowledge.

IOP Conf. Series: Materials Science and Engineering **722** (2020) 012061 doi:10.1088/1757-899X/722/1/012061

Occupational Health and Safety Effects on Productivity in a Garment Factory Using Structural Equation Modeling

M R Suryoputro[1], A D Sari[1], A A Puji[1], C Basumerda[1], and H Purnomo[1]

[1]Department of Industrial Engineering, Faculty of Industrial Techonology, Universitas Islam Indonesia, Indonesia

Corresponding author: ragil.suryoputro@uii.ac.id

Abstract. An industry goal is to create a product to make a profit. Development of human resources is one important thing for a company to consider. Companies have an opportunity to increase work productivity by the rules of OHS (Occupational Health and Safety). This study aimed to: (1) Partially determine the influence of work safety on employee productivity. (2) Partially determine the influence of health on employee productivity. (3) Simultaneously determine the influence of health and safety on employee productivity simultaneously. (4) Determine that variables that have a major influence on employee productivity. A questionnaire was used to collect data from 200 respondents. Results of SEM showed that safety did not have an effect on productivity (p=0.467). However, occupational health effected productivity (p=0.002). Thus, health has a significant influence on work productivity.

1. Introduction

The Company is a form of association or organization that guides the goal to create a product or serve customers for profit. The purpose of an industry is to create a desired product for gain. Industrial activities are used to process or transform raw materials into new products, or change the goods from low value into high value goods [1]. The number of work accidents in Indonesia is still high, and through the end of 2015, there were as many as 105,182 work accidents. Meanwhile, in the case of serious accidents that resulted in deaths as many as 2,375 cases were recorded [2]. Accidents are directly proportional to the growth of new companies in Indonesia. The number of small companies in Indonesia is 141,894 or 83.70%, medium scale industries comprise 14,970 or 8.83%, and the number of large scale industries is 12,660 or 7:47%; Indonesian industry totals include 169,524 companies. Concomitant with increasing the number of new companies will be a need for more employees. However, if the number of accidents still high, this will certainly affect the development of the company.

Occupational Health and Safety is a program system designed for employees or companies in regards to preventive action of accidents and diseases caused by working relations in the work environment by identifying factors that potentially cause accidents and diseases [3].

There are several similar studies with different methods and variables on occupational safety and health as well structural equation modeling, such as the effect of occupational safety and health to improve employee productivity [3]. Moreover, the Occupational Health and Safety Management System formulates safety and health management systems for manufacturing plants in South India [4]. An empirical analysis of the effectiveness of occupational health and safety management systems in SMEs has been conducted [5]. [6] completed research on leadership and employees' perceived safety behaviors in a nuclear power plant with a structural equation model. [7] discussed the effect of an integrated management system on safety and productivity indices in a case study of Iranian cement industries. Furthermore, application of interpretive structural modeling and structural equation modeling to analyze factors contributing to sustainable manufacturing in the Indian automotive component sector was completed by [8]. [9] also discussed the use of structural equation modeling to improve the health sector. Differences in this research include the subjects, variables, indicators, methods and the companies where the research was conducted. From some of these differences are expected to explain to the three variables well.

2. Method

The data analysis was adapted in regards to the research design and variables. The hypothesized causal model generated was tested with structural equation modeling (SEM) in AMOS . Structural Equation Modeling is a multivariate statistical technique that combines factor analysis and corelation analysis [10]. This statistical approach aims to analyze relationships among variables in a model, between indicators and a construct, or between constructs. The object of this research was PT Mataram Tunggal Garment, an apparel exports company located in the Yogyakarta province, Indonesia. The subjects were 200 employees at PT Mataram Tunggal Garment.

3. Sample Determination

According to [11], a sample is part of the total number in the population. The probability sampling technique used in this research provides equal opportunity for each element of the population to be chosen. Simple random sampling was included because usually SEM requires large sample sizes. [12] suggested that to test a model using SEM, between 100-200 subjects are needed; alternatively, based on the number of parameters, multiplying by five, up to ten, may be needed, depending upon the all latent variables . Once survey in 72 SEM research, got median of sample size are 198. For that reason, 200 subjects are usually accepted as representative samples using SEM; in this study female employees in each division served as the respondents.

4. Variable

Variables used in this study included safety (X1) and health (X2) as independent variables. Employee productivity (Y) was the dependent variable.

5. Questionnaire Design

Questionnaires included a number of written questions used to obtain information from respondents in terms of personal reports, or things thatare known [13]. A Likert scale was used for the response options. The following areas were assessed:

1) Work Safety

Safety could be explained as employee safety or the hardships caused by the working environment. Safety was measured using the Safety Climate Questionnaire (SCQ) made by Glendon and Litherland, 2001 cited in the study [14]

2) Occupational health

Occupational health refers to employees who do not have physical disorders, and/or pain caused by the working environment. According [15] occupational health is "a business and circumstances that allow a person to maintain his health in the workplace"

3) Productivity
Productivity is a scale to measure employees performance, and whether they could complete the target or not. Productivity is a set of economic resources to produce something as a comparison between the sacrifice (inputs) and production (output) [16].

Based on the operationalization of variables, the questionnaire was formulated in six parts:
1) Respondent characteristics (gender, age, education, marital status, length of employment, employment status, the amount of income and number of dependents).
2) Safety Work
3) Occupational health
4) Productivity

6. Statistical Analysis

Each statistical technique used in this research had a specific designation. Several tests were used including the following:

Normality Test

To test whether the data had a normal distribution the non-parametric statistical test of Kolmogorov-Smirnov (K-S) was used [17]. If the result of Kolmogorov- Smirnov was greater than 5%, it was concluded that the data had a normal distribution [17]. Normality tests are purposed to verify that residuals have a normal distribution [17][18].

Validity Test

Validity tests were used to measure the validity of the questionnaires. A questionnaire is considered valid if the questions in the questionnaire reveal something that will be measured by the questionnaire [19]. A validity test was done to prove significance of the data attained.

Reliability Test

A reliability test was used to measure indicators of questionnaires from variables or constructs. A questionnaire is considered reliable if someone answers the statements consistent over time [19]. To find a reliable measuring instrument, Cronbach's Alpha was used.

Confirmatory Factor Analysis Test

Confirmatory Factor Analysis is a multivariate analysis method that can be used to confirm that the model corresponds to the hypothesis. The rotation used in confirmatory factor analysis is determined by: (1) Kaiser-Meyer-Olkin (KMO test), (2) Anti-image correlation test, (3) Total variance explained test, (4) Cumunality, (5) Component matrix, and (6) Component Transformation Matrix [20]

Determination coefficient test

Determination coefficient test (R^2) was essentially used to determine the extent to which a measure model could explain variations in the dependent variable [21]. A value close to the mean of the independent variables provides almost all the information needed to predict the dependent variable variations.

7. Result

The respondents criteria were shown below:
1) The 200 participants were all female (100%).
2) Approximately 49% or 98 people were < 25 years old; 31.5% or 63 people were between 26 –

35 years old; 16.5% or 33 people were aged between 36 – 45 years; and 3% or 6 people were aged > 45 years.

3) Education: Senior in High School accounted for 73.5% of 147 people; Junior in High School amounted to 52 people or 26%; and one person completed only Elementary School, 0.5%.
4) Marital status: those married included 117 people or 58.5%. Single people accounted for 75 people or 37.5%, and the widowed status included 8 people or4%.
5) Years of Service: < 3 years accounted for 91 people or 45.5%; between 3 – 7 years included 53 people or 26.5%; be-tween 8-12 years included 27 people or 13.5%, and >12 years accounted for 29 people or 14.5%.
6) Job status: Permanent Employees were 113 people or 56.5%, while the rest were Temporary Employees, 87 people or 43.5%.
7) Amount of Income: < Rp.1.300.000 included 193 people or 96.5%; between Rp.1.400.000 - Rp.1.600.000 included 5 people or 2.5%, and two people (or 1%) had incomes be-tween Rp.1.700.000 - Rp.2.000.000.
8) Burdensome: between 2 – 4 people included 86 respondents or 43%; burden of < 2 people included 39 respondents or 19.5%, and more than 4 people accounted for 6 respondents or 3%. Those with no burden included 69 people or 34.5%.

Instruments used in this research included various software packages, such as Excel 2007, SPSS(Statistical Product and Service Solution) 11, and AMOS 21. Raw data were obtained and then processed to check for normality, validity, reliability, confirmatory analysis, and determination coefficient tests in SPSS 11.

1) Normality test results showed that the Kolmogorov-Smirnov Test was 0.660 with a probability of 0.777> 0.05. Thus, the data have normal distribution.
2) Validity test results showed that correlation coefficients in all questions for work safety, occupational health, and productivity had values greater than the r table = 0.1396. Based on that result, all instrument were considered valid and could be analyzed in the next steps.
3) Reliability test results showed that work safety, occupational health, and productivity variables had a Cronbach's Alpha coefficient greater than 0.6, so all variables were declared re-liable.
4) Kaiser-Meyer-Olkin and Bartlett's Test, based on factor analysis, resulted in a KMOMSA (Kaiser-Meyer-Olkin measure of sampling adequacy) value of 0.858 > 0.05 with a probability < 0.05, suggesting that all measurement subvariables were worthy.
5) Anti-image correlation test. Results showed that all variables had a correlation anti-image > 0.5, which means all measurement sub-variables could be component factors with a critical success factor for questionnaire sampling.
6) Total variance explained test. Results showed that from 71 measurement variables, 18 factors were formed, which was determined by initial eigenvalues that were ≥ 1.
7) Communalities formed 71 factors. Results showed that greatest role was subvariable P1 which was 0.797 or 79%, and the lowest was H3 0.600 or60%.
8) Component matrix. This was formed of several factors: 1-4, 1-3 and 1-10. The component value of a factor could be defined as the correlation between a factor that was formed with the component. The correlation between a factor of 1 to S1 and S1 2 with a factor of each component factor of 0.411 and 0.271, and the value of the correlation of the highest on the first factor is the correlation between a factor of 1 to H5 with a value of0.803.
9) Component transformation matrix is the correlation between factors formed with components. For example, the correlation between a factor of 1 to S1 and S1 2 with a factor of each component factor of 0.722 and 0.504, and the correlation value is highest at 1 factor

IOP Conf. Series: Materials Science and Engineering **722** (2020) 012061 doi:10.1088/1757-899X/722/1/012061

is the correlation between a factor of 1 to H8 with a value of 0.856. Apparently from the results, the data could not be rotated because after processing all three times her into a convergent, so that further data on what their interpretation of the first analysis

10) The determination coefficient was essentially a measure of how much variance in the dependent variable(s) that a model could explain [19]. This research only used two independent variables, so the R squared value was 0.301 or 30.1%. It can be concluded that work safety and occupational health accounted for 30.1% of productivity; the rest was affected by other variables.

SEM is a multivariate statistical technique that combines factor analyis and corelation, and aims to verify the relationship between variables in a model, indicators and constructs, or the relationship in each construct(s) [22].

Referring on the structural model and measurement model in the framework that can be formulated as Figure 1.

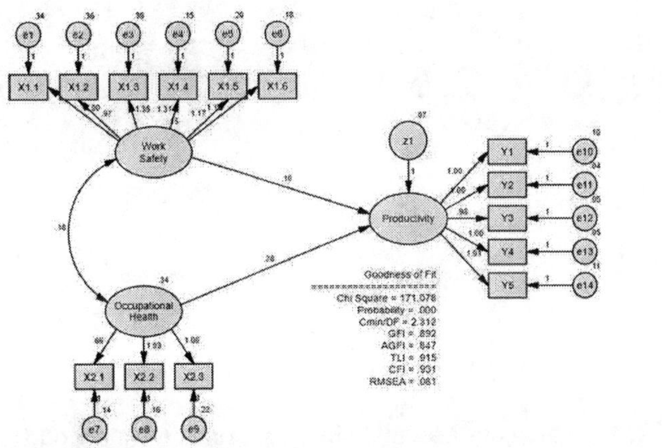

Figure 1. AMOS Model 1 (Model Standardized Coefficients)

Table 1. Measurement Model – Goodness of fit (Before Correction)

Goodness of fit	Cut – off Value	Model Result	Information
X^2 – Chi Square	Expect for low score	171.078	
Probability	≥ 0.05	0.000	less
Cmin/DF	≤ 2.0	2.312	less
GFI	≥ 0.90	0.892	less
RMSEA	≤ 0.08	0.081	less
AGFI	≥ 0.90	0.847	less
TLI	≥0.90	0.915	good
CFI	>0.90	0.931	good

Source: Output Amos 21

From the result above (Table 1), there are five points that have low values in model 1, while the others are acceptable. Based on the SEM stages to improve model fit, it is necessary to interpret and modify such models that are out of range. The purpose of modification is to examine whether the chi-square value can be decreased; a lower chi-square value means that model is a better fit to the data. Chi-square and significance is essential, so it was necessary to modify the model; examination of the modification indices was conducted.

There are three versions proposed by AMOS, covariance, variance and regression weights. The best option is based on theoritical justification, because the program is only used to confirm that the empirical model is appropriate with the theoritical model. Examination of the covariance suggested correlating e1 and e2, and e13 and E14 in order to decrease the chi -square for each 23.127 and 11.883. Theoretical support suggested e1, e2, e13, e14 were error indicators. Each indicator formed the same construct for work safety and productivity.

Indicators in a construct must have a strong corelation in order to form a final construct. By correlating errors that have a high impact on chi-square, the output in Figure.2 was obtained. Below is the AMOS result after modification to determine the effects of work safety, occupational health, and productivity.

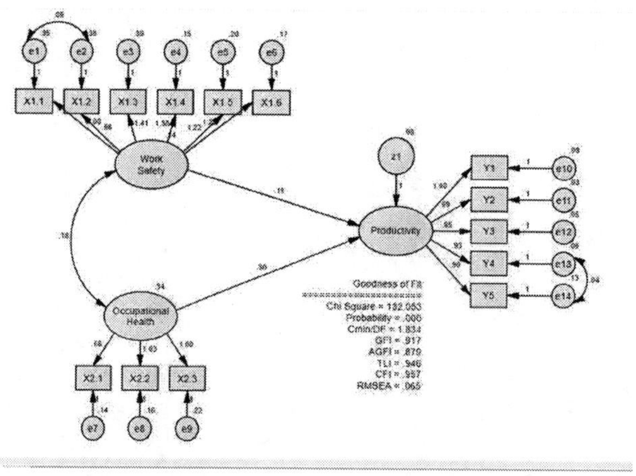

Figure 2. AMOS Model 2 (Model Standardized Coefficients)

Table 2. Measurement Model – Goodness of fit (After Modification)

Goodness of fit	Cut – off Value	Results Model	Information
X^2 – Chi Square	Expected Has Low Score	132.646	
Probability	≥ 0.05	0.000	Less
Cmin/DF	≤ 2	1.834	Good
GFI	≥ 0.90	0.917	Good
RMSEA	≤ 0.08	0.065	Good
AGFI	≥ 0.90	0.879	Less
TLI	≥0.90	0.946	Good
CFI	>0.90	0.957	Good

Source: Output Amos 21

[23] stated that if the parameters are not able to explain a suitable model, it can be measured by another fit model. Result of the suitability model analysis indicated that the path model was not satisfactory. Another test uses Cmin / DF, GFI, RMSEA, TLI and CFI to indicate a good model. Path models in this research already met the Goodness of fit assumptions [23].

Testing the hypothesis with the results obtained in the SEM analysis are shown in Table 3 below:

Table 3. Estimation Results SEM

Connection Between Variable			Estimation	P
Productivity	<---	Safety_Work	0.106	0.467
Productivity	<---	Sanity_Work	0.298	0.002

1) Hyphothesis 1 was used to determine the path coefficient weight of work safety on productivity; the obtained value was 0.106 (p=0.467), suggesting an insignificant relationship. Hyphothesis 1 was not supported.
2) Hyphothesis 2 was to determine the weight of work safety on productivity; the obtained value was 0.298 (p=0.002), suggesting a significant relationship. Hyphothesis 2 was supported.
3) Hyphothesis 3, that relationship between variables is significant, because of the value of work safety is 0.002 and coefficient is 0.298. Both variables are simultaneously has the value but once is not qualified
4) Hyphothesis 4 was used to determine the impact of work safety on productivity; a value of 0.298 (p=0.002) was obtained, and the productivity regresion coefficient weight was 0.106. This means that work safety has the most influence on productivity.

8. Conclusion

Based on the results of the study, a number of conclusions can be stated, namely:
1) Workplace safety has no significant effect on work productivity; the test results of the path analysis (SEM) suggested that the values obtained (safety lines on productivity=0.106, p=0.467) exceeded the standard cutoff value.
2) Occupational health significantly influences productivity of labor based on the results of the path analysis (SEM; thestandardized regression coefficient of occupational health on work productivity was 0.298, p=0.002, which met the standard cut off value).
3) Simultaneously, both safety and occupational health affect productivity, but because one of the two variables did not meet the standards, this relationship was declared insignificant.
4) Healthy work proved to be the most dominant influence on the productivity of employees because it met the standard of the cut off value (p=0.002).

References

[1] Utomo, A. P. 2015. Regresi Robust Untuk Memodelkan Pendapatan Usaha Industri Makanan Non Makloon Berskala Mikro Dan Kecil Di Jawa Barat Tahun 2013. Jurnal Matematika, Sains, Dan Teknologi. 15(2). 63-74
[2] BPJS. 2016. Jumlah Kecelakaan Kerja di Indonesia Masih Tinggi. (online): http://www.bpjsketenagakerjaan.go.id/berita/5769/Jumlah-kecelakaan-kerja-di-Indonesiamasih-tinggi.html. Accesses online: 29 July 2017
[3] Kusuma, A. N. 2017. Pengaruh Keselamatan Dan Kesehatan Kerja (K3) Terhadap Produktivitas Kerja Karyawan Bagian Sistem Distribusi PDAM Surya Sembada Surabaya. Jurnal Ilmu Manajemen Jim. 5(1)
[4] Abdul Zubar, H., Visagavel, K., Deepak Raja, V., & Mohan, A. 2014. Occupational Health and Safety Management in Manufacturing Industries. Journal of Scien-tific & Industrial Research, 73(June), 381–386. Retrieved from http://nopr.niscair.res.in/handle/123456789/28878
[5] Arocena, P., & Nunez, I. 2010. An empirical analysis of the effectiveness of occupa-tional health and safety management systems in SMEs. International Small Business Journal. 28(4). 398–419.

https://doi.org/10.1177/0266242610363521
[6] Martínez-Córcoles, M., Gracia, F., Tomás, I., & Peiró, J. M. 2011. Leadership and employees'perceived safety behaviours in a nuclear power plant: A structural equation model. Safety Science. 49(8–9). 1118–1129. https://doi.org/10.1016/j.ssci.2011.03.002
[7] Hamidi, N., Omidvari, M., & Meftahi, M. 2012. The effect of integrated manage-ment system on safety and productivity indices: Case study; Iranian cement industries. Safety Science. 50(5). 1180–1189. https://doi.org/10.1016/j.ssci.2012.01.004
[8] Thirupathi, R. M., & Vinodh, S. 2016. Application of interpretive structural model-ling and structural equation modelling for analysis of sustainable manufactur-ing factors in Indian automotive component sector. International Journal of Production Research. 7543(January). 1–22. https://doi.org/10.1080/00207543.2015.1126372
[9] Ratnam, K. A., Dominic, P. D. D., & Ramayah, T. 2014. lA Structural Equation Modeling Approach for the Adoption of Cloud Computing to Enhance the Ma-laysian Healthcare Sector. Journal of Medical Systems. 38(8). 82. https://doi.org/10.1007/s10916-014-0082-5
[10] Hooper, D., Coughlan, J., & Mullen, M. R. 2008. Structural equation modelling: Guidelines for determining model fit. Electronic Journal of Business Research Methods. 6(1). 53–60.
[11] Levy, P.S. and Lemeshow, S., 2013. Sampling of populations: methods and applications. John Wiley & Sons.
[12] Iacobucci, D. 2010. Structural equations modeling: Fit Indices, sample size, and advanced topics. Journal of Consumer Psychology. 20(1). 90–98. https://doi.org/10.1016/j.jcps.2009.09.003
[13] Oppenheim, A. N. 2000. Questionnaire design, interviewing and attitude measure-ment. (New edn). London. (New edn). London. https://doi.org/10.1016/0149-7189(94)90021-3
[14] Wills, A.R., Biggs, H.C. and Watson, B. 2005. Analysis of a safety climate measure for occupational vehicle drivers and implications for safer workplaces. The Australian Journal of Rehabilitation Counselling, 11(1), pp.8-21.
[15] Robson, L. S., Stephenson, C. M., Schulte, P. a, Amick, B. C. I., Irvin, E. L., Eggerth, D. E., Grubb, P. L. 2012. A systematic review of the effectiveness of occu-pational health and safety training. Scandinavian Journal of Work, Environ-ment & Health, 38(3). 193–208. https://doi.org/10.5271/sjweh.3259
[16] Rogers, M. 1998. The Definition and Measurement of Productivity. Melbourne Institute, (9), 27. https://doi.org/10.2307/145776
[17] Corder, G. W., & Foreman, D. I. 2011. Nonparametric Statistics for Non-Statisticians: A Step-by-Step Approach. Nonparametric Statistics for Non-Statisticians: A Step-by-Step Approach. https://doi.org/10.1002/9781118165881
[18] Walpole, R. E., Myers, R. H., Myers, S. L., & Ye, K. (2012). Probability and Statistics for Engineers and Scientists. Power (Vol. 3rd). https://doi.org/10.1017/CBO9781107415324.004
[19] I. Ghozali, Aplikasi Analisis Multivariate dengan Program IBM SPSS19. Sema-rang : Badan Penerbit Universitas Diponegoro. 2011
[20] Statistics Solutions. 2016. Confirmatory Factor Analysis. Handbook of Structure Equation Modeling2. https://doi.org/10.4135/9781412961288.n69
[21] Hooper, D., Coughlan, J., & Mullen, M. R. 2008. Structural equation modelling: Guidelines for determining model fit. Electronic Journal of Business Research Methods. 6(1). 53–60.
[22] Kline, R. B. 2011. Principles and practice of structural equation modeling. Struc-tural Equation Modeling (Vol. 156). https://doi.org/10.1038/156278a0
[23] Ho, R. 2014. Handbook of Univariate and Multivariate Data Analysis with IBM SPSS. CRC Press.

IOP Conf. Series: Materials Science and Engineering **722** (2020) 012062 doi:10.1088/1757-899X/722/1/012062

Analysis of E-Service Quality Influence Through E-Satisfaction on Online Purchasing Decisions (Consumer Survey of PT. X)

W Sutrisno[1] and A D Setyawan[1]

[1]Industrial Engineering Department, Faculty of Technology Industry, Universitas Islam Indonesia, Sleman, Indonesia

E-mail: 075220417@uii.ac.id

Abstract. This study aims to determine the effect of E-Service Quality through E-Satisfaction on online purchasing decisions at PT X. Data collection using a questionnaire with a Likert scale. This research uses Partial Least Square as an analysis tool. The results showed that the E-Service Quality variable influenced the E-Satisfaction variable by 37.9%, and influenced the online purchasing decision variable by 0.514. Means it can be concluded that the E-Service Quality variable has a positive and significant effect on online purchasing decisions through E-Satisfaction on PT. X

1. Introduction

PT. X is a fashion company whose main focus is to sell leather-based products. PT. X products sold are shoes, wallets, belts, sandals, jackets, and bags. PT. X sells its products through internet media using websites and social media such as Instagram and Facebook. Sales with internet media are done so that consumers can get to know the product quickly and also know the advantages and uniqueness of the products offered so that it can influence purchasing decisions.

Purchasing decisions are actions of consumers to buy or not buy a product [1]. E-Service Quality is an important instrument in developing a competitive advantage in online retailing [2]. The definition of online service quality in its interaction with the website is a level of a website that effectively and efficiently facilitates consumers in shopping, making purchases and the delivery process of products and services [3]. E-Satisfaction is the affective state of the customer to the website that is obtained from evaluating all aspects that make up the customer relationship [4]. The development of e-service quality is needed to achieve customer satisfaction so that it raises the intention of consumers to make online purchasing decisions. Based on this background, this study aims to determine and analyze the effect of E-Service Quality on Online Purchasing Decisions through E-Satisfaction on PT. X consumers.

2. Literature Review

2.1. Consumer

Consumers are all individuals and households that buy or obtain goods or services for personal consumption [1]. Consumers are economical buyers, those who know all the facts and logically compare choices based on the costs and value of the benefits received to obtain the greatest satisfaction from the money and time they sacrifice [5]. The study of consumer behaviour is a study of how an individual decides to allocate available resources (time, money, effort, and energy) [6]. Consumer behaviour is the study of how individuals, groups and organizations choose, buy, use and place goods, services, ideas or experiences to satisfy their wants and needs [7].

Four characteristics influence consumers in making a purchase, namely, cultural factors, social factors, personal factors and psychological factors [8]. Some things that affect the pattern of consumer buying behaviour in the online world, including [9]:

- How to obtain product or service information.
- How to make purchases from anywhere because consumers only have to open the company's website.
- Loss of face to face with the seller.
- How to bid on prices
- Ease of obtaining goods purchased

2.2. E-Service Quality
An e-service is an act, business or show whose delivery is mediated by information technology [10]. E-service quality consists of 4 indicators consisting of:

- Fulfilment is related to how far the company promises the availability of goods, and also related to the delivery of orders.
- Efficiency is how easy and fast it is to access and use a website.
- Privacy is an indicator related to the level of security and protection related to consumer personal data.
- System availability related to technical functions on the website used

2.3. E-Satisfaction
E-satisfaction is described as gratification from customers that comes from previous real buying experience with certain electronic trading companies [11]. E-satisfaction is also interpreted as the affective state of the customer to the website which is obtained from the evaluation of all aspects that make up the customer relationship [4]. There are five main dimensions that influence e-satisfaction [12]:

- Convenience, where there are three things namely shopping time, convenience, and ease of browsing.
- Merchandising, which consists of a number of offerings, a variety of offerings.
- Site design, where the specified factors are navigation structure, web design & interface, fast presentation, and updated information.
- Security where the specified factors are privacy, and financial security.
- Serviceability, which consists of redressal complaints, customer support, return policy, purchase process, post service

3. Methodology
This study uses a purposive sampling technique, where each subject taken from the participants is deliberately chosen according to specific objectives and considerations [13]. The object of research is 18-year-old PT X consumers who have purchased products through online sites at least once or more. This study uses primary data collected through questionnaire distribution to respondents using Google form. Respondent data obtained from PT X consumer personal contact data. The measurement scale of the variables used in this study uses a Likert scale. The data obtained is then processed through the analysis path. The results of data processing are then analyzed using Partial Least Square (PLS), which is an analysis of variance-based structural equations that can simultaneously test measurement models as well as structural model tests.

The independent variable in this study is e-service quality (X). The independent variable is the variable that influences or is the cause of the occurrence of the dependent variable [13]. The dependent variable in this study is online purchasing decisions (Y). The dependent variable is a variable that is affected or which is due, because of the independent variables [13]. The intermediate variable in this study is e-satisfaction (Z). Intermediate variables are variables that provide pauses or between independent variables and dependent variables so that the independent variables do not directly affect the dependent variable [13]. The hypothesis in this study is that E-Service Quality influences online purchasing decisions through E-Satisfaction.

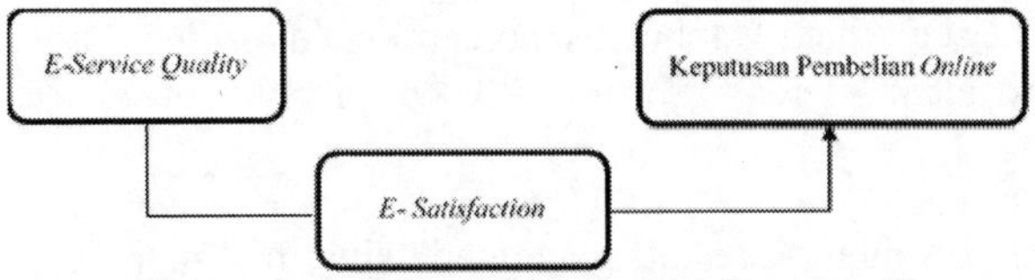

Figure 1. Research conceptual framework

4. Result

4.1. E-Service Quality
The results of data analysis perception of e-service quality variables can be seen in table 1.

Table 1. Perception of respondents variable e-service quality

Interval	Category	Amount	Percentage
1,00 s/d 1,79	Very Unkind	0	0%
1,80 s/d 2,59	Not Good	1	1%
2,60 s/d 3,39	Neutral	12	11,4%
3,40 s/d 4,19	Good	58	55,2%
4,20 s/d 5,00	Very Good	34	32%
Amount		105	100%

Based on table 1 it can be seen that respondents' perceptions of the E-Service Quality variable of 55.2% of respondents stated good and 32% of respondents stated very good.

4.2. E-Satisfaction
The results of data analysis perception of e-satisfaction service quality variables can be seen in table 2

Table 2. Perception of respondents variable e-service satisfaction

Interval	Category	Amount	Percentage
1,00 s/d 1,79	Very Dissatisfied	0	0%
1,80 s/d 2,59	Not satisfied	3	2,8%
2,60 s/d 3,39	Neutral	17	16,2%
3,40 s/d 4,19	Satisfied	49	46,7%
4,20 s/d 5,00	Very Satisfied	36	34,3%
Amount		105	100%

Based on table 2 it can be seen that respondents' perceptions of the E-Satisfaction variable of 46,7% of respondents stated satisfied and 34,3% of respondents stated very satisfied.

4.3. Online Purchasing Decision
The results of data analysis perception of online purchasing decision variables can be seen in table 3

IOP Conf. Series: Materials Science and Engineering **722** (2020) 012062 doi:10.1088/1757-899X/722/1/012062

Table 3. Perception of respondents variable online purchasing decision

Interval	Category	Amount	Percentage
1,00 s/d 1,79	Very Low	0	0%
1,80 s/d 2,59	Low	1	1%
2,60 s/d 3,39	Neutral	15	14,2%
3,40 s/d 4,19	High	63	60%
4,20 s/d 5,00	Very High	26	24,8%
Amount		105	100%

Based on table 3 it can be seen that respondents' perceptions of online purchasing decision variables of 60% of respondents stated high and 34,3% of respondents stated very satisfied.

4.4. *Partial Least Square*

The path diagram of the PLS method results generated using the Smart PLS program for all constructs is presented in Figure 2. In figure 1, X is E-service quality, Z is E satisfaction and Y is an online purchasing decision.

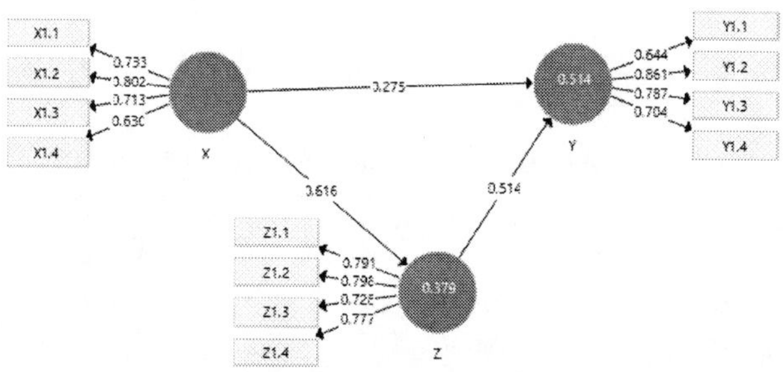

Figure 2. Path chart constructions

Based on the equation it is found that all variables are positive. The E-Service Quality variable has a positive influence on the online purchasing decision variable of 0.275, the E-Service Quality variable has a positive influence of 0.616 on E-Satisfaction, the E-Satisfaction variable has a positive influence on the online purchase decision variable of 0.514 and the E-Satisfaction variable has an influence positive in mediating the relationship between E-Service Quality with Online Purchasing Decisions of 0.379.

Based on PLS analysis it is known that all indicators have an outer loading value> 0.5 so that it is said to be significant, it can be said that the indicators used are good gauges in measuring the research variables. The results of outer loading can be seen in the following table.

Table 4. Outer loading results

Variable	Indicator	*Outer Loading*
	X1.1	0,733
E-Service Quality	X1.2	0,802
	X1.3	0,713
	X1.4	0,630
	Z1.1	0,791
E-Satisfaction	Z1.2	0,798
	Z1.3	0,728
	Z1.4	0,777
Online	Y1.1	0,644

Purchasing	Y1.2	0,861
Decisions	Y1.3	0,787
	Y1.4	0,704

The path coefficient score for hypothesis testing is performed with a value of α 5% (0.05). The output of hypothesis testing can be seen in the appendix. The results of hypothesis testing can be seen in the following Table 5.

Table 5. Hypothesis testing results

	Hypotesis Statistik	*Path Coefficient*	*T Statstics*	*P vales*	Result
H_1	*E-Service Quality* → Online Purchasing Decisions	0,275	2,007	0,045	*Significant*
H_2	*E-Service Quality* → *E-Satisfaction*	0,616	9,194	0,000	*Significant*
H_3	*E-Satisfaction* → Online Purchasing Decisions	0,514	5,388	0,000	*Significant*
H_4	*E-Service Quality* → *E-Satisfaction* → Online Purchasing Decisions	0,317	4,397	0.000	*Significant*

Based on Table 5 it can be seen that H1, H2, H3, H4, have met the criteria of p values <0.05 so that it can be said that the influence between variables is significant. From table 2 it can also be seen that all hypotheses have a positive path coefficient value and t statistic is greater than t table so it can be said that all variable relationships are positive and significant. H4 has a positive path coefficient value and t statistic is greater than t table so it can be said that the relationship of variables is positive and significant. This shows that the E-Satisfaction variable has a positive and significant influence in mediating the relationship between E-Service Quality with partial online purchasing decisions.

5. Conclusion

The results of the analysis using partial least square show that the E-Service Quality variable has a positive and significant effect on online purchasing decisions on PT. X consumers. This means that if PT. X E-Service Quality matches consumer expectations, the Online Purchasing Decision will increase. The analysis also shows that the E-Service Quality variable has a positive and significant effect on E-Satisfaction in PT. X consumers. This can be interpreted, if E-Service Quality meets consumer expectations, then the perceived E-Satisfaction also experiences an increase. Then the analysis shows that the E-Satisfaction variable has a positive and significant effect on Online Purchasing Decisions on PT X consumers. That is, if the E-Satisfaction perceived by consumers is getting higher, then online purchasing decisions will increase. Finally, the results of the analysis show that the coefficient values show positive and significant results. Means that if E-Service Quality is in accordance with consumer expectations, the level of E-Satisfaction will be higher the level of satisfaction and will have an influence on Online Purchasing Decisions on consumers. It also shows that E-Satisfaction has a positive and significant influence in mediating the relationship between E-Service Quality with partial Online Purchasing Decisions on Guten.inc consumers.

References

[1] Kotler, Philip. 2008. *Manajemen Pemasaran Edisi 12 Jilid 2.* Jakarta: PT.Indeks.
[2] Roth, Oliveria and Gilland. 2002. *Satisfaction: A Behavioral Perspective of The Consumer.* New York: Mc-Graw-Hill
[3] Zeithaml, V.A., Parasuraman. A. & Malhotra, A. 2002. *J The Acd of Mkt Sci.* **30** 362-375.
[4] Flavian, C. & Guinaliu, M. 2006. *Ind Mgm and Dt Sys.* **106** 601-620.

[5] Cannon, Perreault dan McCarthy. 2009. *Pemasaran Dasar Pendekatan Manajerial Global Buku 2 Edisi 16.* Jakarta: Salemba Empat

[6] Schiffman, L.G. dan Kanuk, L.L. 2004. *Perilaku Konsumen, Edisi Bahasa Indonesia*. Jakarta: PT Indeks

[7] Kotler, Philip dan Kevin Lane Keller. 2007. *Manajemen Pemasaran. Edisi Kedua Belas*. Jakarta: Indeks

[8] Kotler, Philip dan Gary Armstrong. 2001. *Prinsip-Prinsip Pemasaran. Jilid 1*. Jakarta: Erlangga

[9] Sarwono, Jonathan dan K. Prihartono, A.H . 2012. *Perdagangan Online (e-Commerce) Cara Bisnis Internet*. Jakarta: PT. Elex Media Komputindo.

[10] Rowley, J. 2006. *Int Rsr*, **16** 339-359

[11] Anderson, R. E. & Srinivasan, S. S. 2003. *Psy and Mkg.* **20** 123-138.

[12] Ranjbarian, B. Fathi, S. & Rezaei, Z. 2012. *Int J conty rsc bss.* **3** 1496-1511.

[13] Sugiyono. 2013. *Metode Penelitian Pendidikan Pendekatan Kuantitatif, Kualitatif, dan R&D.* Bandung: Alfabeta.

IOP Conf. Series: Materials Science and Engineering **722** (2020) 012063 doi:10.1088/1757-899X/722/1/012063

Clustering Storage Method for Digital Evidence Storage Using Software Defined Storage

Mohammad Faruq Afif[1], Mukhammad Andri Setiawan[2], Yudi Prayudi[3]

Department of Informatics, Faculty of Industrial Technology, Universitas Islam Indonesia, Yogyakarta. INDONESIA

14917149@students.uii.ac.id, andri@uii.ac.id, yudi.prayudi@uii.ac.id

Abstract. The increasing of cybercrime means an increase in the number of digital evidences produced in these criminal activities. Also, to pay attention to the security and integrity of digital data, the construction of digital evidence storage must pay attention to the need for convenience in adding storage and the need for investigators to avoid geographical problems in carrying out their duties. An ideal implementation of the problem is the development of network-based digital evidence storage using the Software-Defined Storage to build a clustered storage system. In this study, we build a clustered storage system using Ceph that will be functioned as digital evidence storage. Ceph is a Software-Defined Storage which is designed to provide excellent performance, reliability and scalability. On the other hand, since digital evidence is fragile, then the security of storage system must also be considered so that the stored digital evidence in the storage system can be accepted in court. The security includes aspects of confidentiality, integrity and availability aspects. We were able to build a storage system using Ceph and as a result, our system storage has met the aforementioned scalability, confidentiality, integrity and availability aspects. The storage system is expected to contribute in terms of digital evidence storage and preservations.

Keywords. digital evidence, evidence storage, clustered storage, software defined storage, ceph

1. Introduction

Cybercrime is a term that refers to criminal activity with computers or networks. The computer becomes a tool, target or place of crime. Like other criminal activities, cybercrime will certainly leave a trail/evidence that can be used for investigation. Evidence in cybercrime cases can be in the form of physical evidence or digital evidence. Physical evidence will be stored in the evidence storage room while digital evidence will be stored in evidence storage. Considering that cybercrime activities are increasing, so that evidence quantity in these activities is also increasing. As the quantity of digital evidence increases, the quality of digital evidence that is getting better also requires more space in storage time by time.

Therefore, digital evidence storage must be secure, adequate, dedicated to large storage solutions and have excellent scalability. Digital evidence storage must also be able to work automatically, thereby reducing the obligation of the examiner or investigator in carrying out administrative tasks related to the system. Furthermore, digital evidence storage must accommodate the needs of examiners and investigators who may be geographically separated to be able to carry out their duties in digital forensic investigations and the needs of examiners or investigators for easy access to digital evidence so that if necessary digital evidence can be examined in another place to know the further investigations. An ideal implementation of the problem is a centralized evidence storage that allows evidence to be accessed securely from a remote location. This is possible by the development of network-based digital evidence storage.

Some research on the development of network-based digital evidence storage has been carried out by several researchers such as Davis et al. [1] who designed a digital evidence custodian (DEC) architecture using Network Attached Storage (NAS) and Storage Area Networks (SAN), and Ali et al. [2] who created a Chain of Custody Reflex system based on SAN and web-based technology as network-based centralized storage architecture[2]. However, these researches has not yet discussed the need for ease of additional storage while we knew that the weakness of using SAN and NAS technology is limited scalability.

On the other hand, there are currently many studies on storage systems that use software-defined storage (SDS) instead of traditional storage technologies such as SAN and NAS. SDS offers better things in terms of cost, agility, scalability, reliability, and multi-tenancy. In terms of scalability, traditional storage can only be scaled up to a maximum of 4 - 8 controllers, while SDS can scale up to hundreds of thousands of nodes - meaning theoretically it can be scaled indefinitely [3]. Therefore, the use of SDS to replace SAN and NAS in the development of digital evidence storage systems can be another solution that can accommodate the need for convenience in adding storage. One of the SDS that is widely discussed in research on its excellent scalability is Ceph. Gudu [4] has proven the capability of Ceph scalability by adding Object Storage Daemon (OSD) on a system running up to 20 OSDs (10 nodes). Based on the research conducted, Ceph's performance increases steadily with the number of OSDs.

According to some of the advantages of Ceph as SDS compared to traditional storage such as NAS and SAN as expressed in previous studies, the authors intend to build digital evidence storage using Software-Defined Storage Ceph as an alternative in the development of network-based digital evidence storage. Considering digital evidence is fragile, then the security of the storage system must also be considered so that the stored digital evidence in the storage system can be accepted in court. The security of a storage system includes aspects of confidentiality, integrity and availability aspects. The storage system is expected to meet the criteria of scalability, confidentiality, integrity and availability aspects.

2. Literature Review

2.1. Digital Evidence

Digital evidence is information stored or distributed on a computer or in digital form that can be used as evidence in court. This information can be in the form of text, images, audio, video that can be found on a computer hard drive, cellphone or smartphone, CD / DVD, flash/memory card and others.

Kuntze [5] defined five rules of evidence for evidence to be considered useful. Evidence must be admissible, authentic, complete, reliable and believable. After digital evidence obtained from crime scene, officer have to immediately make a complete bitstream image of the media on which the evidence is stored. The original should then be stored in a safe place where its integrity can be maintained, while the copy will be examined [6]. Both are stored in a digital evidence storage area in the Forensic Laboratory. Digital evidence storage is an important infrastructure that supports the process of storing digital evidence. To make the criteria for evidence to be received at court, a digital evidence storage system must make information security criteria in the form of Confidentiality, Integrity, and Availability.

2.2. Storage Cluster

Storage cluster is a network of multiple storage systems that behave as one system that works together to improve system performance, capacity, and reliability. In the most abstract sense, a storage cluster consists of several storage servers that can be accessed through a network, usually by installing it as a

IOP Conf. Series: Materials Science and Engineering **722** (2020) 012063 doi:10.1088/1757-899X/722/1/012063

virtual filesystem on the server. Features usually include a high level of redundancy and availability. For example, if one server in the storage cluster fails, data can still be accessed from other servers in the cluster. In addition to data availability and ease of access, another benefit to being achieved from using a storage cluster is the ease of adding capacity. Storage cluster can be achieved by using software-defined storage.

2.3. *Software Defined Storage*

Software-defined storage (SDS) is a term for computer data storage software that enables the management of data storage that is created and managed regardless of the attachment to the underlying hardware. Software-defined storage usually includes a form of storage virtualization to separate storage hardware from the software that manages it. This software provides management policies for a variety of feature options, such as data deduplication, replication, thin provisioning, snapshots, and backups.

One of the open-source defined storage software is Ceph. Ceph is claimed to have the ability to transform IT infrastructure and enable users to manage large data. Ceph was built with RADOS (Reliable Autonomic Distributed Object Storage) as its core. RADOS in Ceph provides a storage system for objects, blocks, and files in a cluster. RADOS makes Ceph flexible, reliable and easy to manage [7]. It allows Ceph to be self-healed and self-manage. In the event of a disaster, Ceph can provide reliability for multiple failures. Ceph detects and corrects failures in each failure zone as disks, nodes, networks, racks, data center lines, data centers, and even various geographic locations. Ceph tries to manage the situation automatically and resolve it as much as possible without data interruption [8].

Another thing that makes Ceph superior is the CRUSH (Controlled Replication Under Scalable Hashing) algorithm. CRUSH is a pseudo-random data distribution algorithm that efficiently and powerfully distributes replicas of objects across heterogeneous and structured storage clusters. CRUSH is designed to facilitate the addition and removal of storage by minimizing data transfer between OSDs. This algorithm accommodates various types of data replication and mechanisms to maintain reliability and distribute data according to user needs [9]. These features make CRUSH suitable for managing the distribution of objects in very large (multi-petabyte) storage systems where scalability, performance, and reliability are very important. CRUSH ensures that data is distributed to all cluster members and all cluster nodes can provide data without bottlenecks.

With RADOS and CRUSH as the basis for its development, Ceph uniquely sends objects, blocks, and file storage in one integrated system. With this, the Ceph storage cluster can accommodate many nodes, which communicate with each other to replicate and redistribute data dynamically.

In general, Ceph storage cluster consists of two types of daemons: Ceph OSD Daemon (OSD) stores data as objects on storage nodes; and Ceph Monitor (MON) keeps a master copy of the cluster map. Ceph storage clusters can contain thousands of storage nodes. The system has at least one Ceph Monitor and two Ceph OSD Daemons for data replication.

3. Research Methodology

3.1. *Design*

Digital evidence storage is built in a virtual machine (VM) environment. Although built in a virtual environment, this condition can already represent the real condition. The following is a list of hardware and software used in this study:

Hardware: a set of servers with CPU (s) 32 x Intel (R) Xeon (R) CPU specifications E5-2670 0 @ 2.60GHz (2 Socket) 368Gb Memory and 1.6 Tb Hard Drive

Software: Proxmox VE 5.3-8 operating system, Ubuntu 18.04 LTS operating system, Open VPN Server, OpenVPN Client, Ceph Mimic Stable Version 13.2.5, ownCloud Server, ownCloud Client, Wireshark Network Protocol Analyzer.

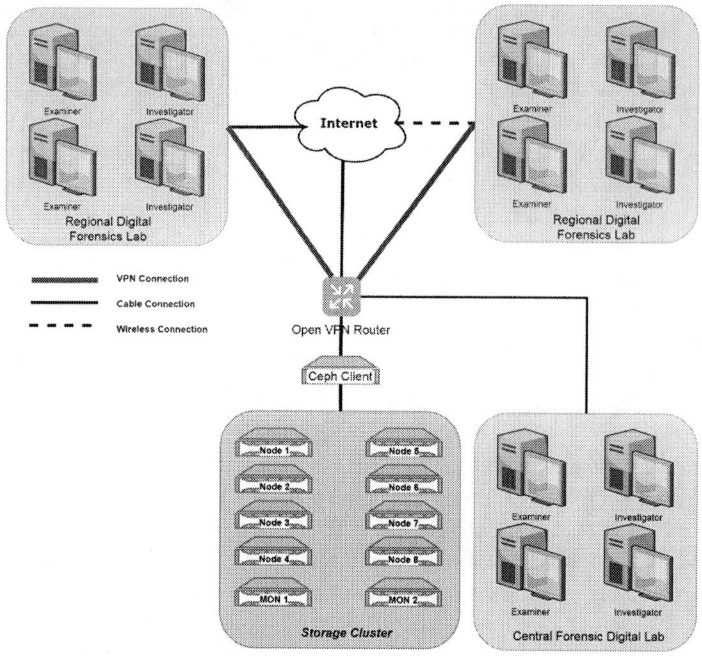

Figure 1. Design of digital evidence storage system in this research

The system storage design which is built in this study is illustrated in Figure 1.

The digital evidence storage system design consisted of 8 storage server nodes, 2 Ceph monitors (MONs) servers, and 1 Ceph client. Each storage server node has 2 OSDs which will be combined into a storage cluster. Each OSDs has a capacity of 20GBs. Then the Ceph storage cluster will be linked from the Ceph monitor server to the Ceph client. Ceph client acts as a server that will be accessed by investigators using encrypted networks openVPN. In this study, two connection lines were built to access the system, namely through a wired network and wireless network (wireless or GSM). System testing will be conducted on a wireless network using a GSM network to illustrate the situation if the Digital Forensic Laboratory in an area not yet reached by a cable network (Fiber Optic). OpenVPN configuration is also used to secure data transmission. OpenVPN server installed on a virtual machine.

This concept is supported by ownCloud software as a service management and synchronization of digital evidence files between the acquisition file and Ceph client. OwnCloud can manage cloud storage services ranging from users, user groups, limited access, user size quota, automatic synchronization feature, transfer data using SSL encryption and verify checksum for uploaded or downloaded files. OwnCloud Desktop Sync is used to synchronize digital evidence files.

Wireshark Network Protocol Analyzer software is used to test data security. Wireshark is a visual network analyzer and capturing packet tool with facilities that are complete enough to see the data path and protocol used.

3.2. System Testing

At this stage, 4 test scenarios are performed to get the data to be analyzed. The results of tests carried out besides expected to provide good results from scalability criteria, are also expected to meet the criteria of confidentiality, integrity, and availability. These four criteria are very important for the development of evidence storage. Scalability criteria must be met considering that evidence grows exponentially, so evidence storage must be easy to scale. While the criteria of confidentiality, integrity, and availability must be fulfilled so that evidence meets the five rules of evidence, namely evidence must be permissible, authentic, complete, reliable and believable in order to be accepted and used as evidence at court.

Confidentiality criteria which means safeguarding information from people who are not entitled to access, in this study is conducted by:

- The use of openVPN technology
- The use of https protocol
- User / password authorization with a token

Integrity criteria which means maintaining consistency, accuracy, and trust in data from time to time. Maintaining integrity also means ensuring that data is irreversible. In this study carried out by:

- Applying strong encryption to data storage and transmission media.
- Applying strong authentication and validation on every file access/account login/action that is applied.
- Authentication and validation are done to ensure the legality of access.
- Applying strict access control to the system, i.e. every existing account must be restricted in terms of access rights. For example, not all investigators have the right to access evidence that is actively being examined, only those charged with investigating can access it, while other investigators can only see it.

The purpose of availability is to ensure that the available resources are ready to be accessed at any time by the user/application/system that needs them. In this study, the availability aspect is achieved through the use of Ceph which has the ability of self-healed, self-managed, and reliable against multiple failures.

4. Result

4.1. Scalability Testing

Testing is run by adding 2 nodes into the system which originally had 8 nodes so that there are a total of 10 nodes running in the system. Figures 2 and 3 show the state of the system before and after the addition of new nodes. The total disk capacity before adding nodes is 320GB after adding 2 Nodes with 4 OSDs, the total disk capacity being 400GB. Figure 4 show details of each OSD after adding new nodes. As shown in Figure 4, there are 20 OSD with a capacity of 20GB each, and a total capacity of 400GB. each OSD (including additional OSD) has a used space of 4.0 - 4.9 GB so that the total used space is 90 GB. When additional OSDs enters the system, Ceph will automatically replicate the data into the additional OSD.

The process of adding these nodes runs smoothly without any system disruption. The easy process of increasing storage can occur because Ceph was developed with a scale-out design. The scale-out design focuses on adding all-new nodes, including disk, CPU, and memory to the existing system. In this type of design, the system will not end with limited storage, but rather, it will benefit from the performance and robustness of the system when adding new nodes so that when there are new nodes added, Ceph's performance also increases.

```
prodep@mon-ceph1:~$ sudo ceph df
[sudo] password for prodep:
GLOBAL:
    SIZE        AVAIL       RAW USED     %RAW USED
    320 GiB     251 GiB       69 GiB         21.44
```

Figure 2. The system state before adding new nodes

```
[sudo] password for prodep:
GLOBAL:
    SIZE        AVAIL       RAW USED     %RAW USED
    400 GiB     313 GiB       69 GiB         21.44
```

Figure 3. The system state after adding new nodes

```
ceph-admin@mon-ceph1:~$ sudo ceph osd df
ID CLASS WEIGHT  REWEIGHT SIZE   USE     AVAIL  %USE  VAR  PGS
 0   hdd 0.01949  1.00000  20 GiB 4.6 GiB  15 GiB 23.23 1.04  14
 1   hdd 0.01949  1.00000  20 GiB 4.9 GiB  15 GiB 24.53 1.09  18
 4   hdd 0.01949  1.00000  20 GiB 4.7 GiB  15 GiB 23.73 1.06  16
 5   hdd 0.01949  1.00000  20 GiB 5.1 GiB  15 GiB 25.41 1.13  22
 6   hdd 0.01949  1.00000  20 GiB 4.8 GiB  15 GiB 23.87 1.06  23
 7   hdd 0.01949  1.00000  20 GiB 4.3 GiB  16 GiB 21.73 0.97  12
 8   hdd 0.01949  1.00000  20 GiB 4.7 GiB  15 GiB 23.33 1.04  18
 9   hdd 0.01949  1.00000  20 GiB 4.5 GiB  15 GiB 22.64 1.01  13
10   hdd 0.01949  1.00000  20 GiB 4.4 GiB  16 GiB 22.18 0.99  12
11   hdd 0.01949  1.00000  20 GiB 4.0 GiB  16 GiB 19.81 0.88  13
16   hdd 0.01949  1.00000  20 GiB 4.4 GiB  16 GiB 21.81 0.97  12
17   hdd 0.01949  1.00000  20 GiB 4.5 GiB  16 GiB 22.37 1.00  19
18   hdd 0.01949  1.00000  20 GiB 4.7 GiB  15 GiB 23.30 1.04  17
19   hdd 0.01949  1.00000  20 GiB 4.1 GiB  16 GiB 20.29 0.91   9
 2   hdd 0.01949  1.00000  20 GiB 4.6 GiB  15 GiB 22.79 1.02  27
 3   hdd 0.01949  1.00000  20 GiB 4.4 GiB  16 GiB 21.91 0.98  31
12   hdd 0.01949  1.00000  20 GiB 4.2 GiB  16 GiB 21.15 0.94  26
13   hdd 0.01949  1.00000  20 GiB 4.3 GiB  16 GiB 21.55 0.96  23
14   hdd 0.01949  1.00000  20 GiB 4.0 GiB  16 GiB 20.23 0.90  28
15   hdd 0.01949  1.00000  20 GiB 4.5 GiB  16 GiB 22.45 1.00  31
                    TOTAL 400 GiB  90 GiB 310 GiB 22.42
```

Figure 4. Details of each OSD after adding new nodes

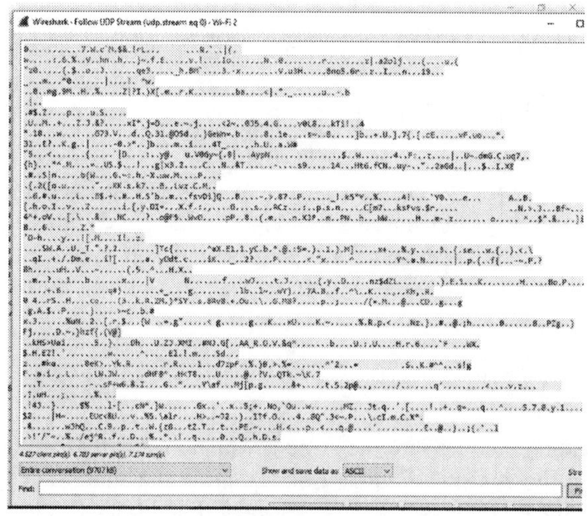

Figure 5. Results of sniffing files using Wireshark shows that the OpenVPN connection has been encrypted.

4.2. *Data Transmission Security Testing*

Testing is run by using the sniffing method (tapping) on every packet that passes through the network, then see and analyze the results. The tool used to conduct this test is Wireshark Network Protocol Analyzer.

The test results shown in Fig 5 show random data from the sniffing results. This shows that the openVPN connection and the use of the https protocol in this study make encrypted data transmission, which means that the confidentiality criteria in the system have been met.

4.3. *Testing file synchronization*

The next test scenario is testing file synchronization using Telkomsel's 4G network in several connection quality parameters with ping time, jitter, bandwidth, and packet loss. Telkomsel's GSM 4G network was chosen because based on the Cellular Network Experience Report in Indonesia in July 2019 released by opensignal.com Telkomsel has the highest value with an average download speed of 12.8Mbps and upload speed of 3.7Mbps [10].

IOP Conf. Series: Materials Science and Engineering **722** (2020) 012063 doi:10.1088/1757-899X/722/1/012063

Table 1. Result Of Synchronization Testing File

No	Connection Type	ping time (ms)			Jitter (ms)	Bandwitdh		Packet Loss (%)	MD5 Source	MD5 Server	Note
		Min	Max	Ave		Transmit (Mbps)	Received (Mbps)				
1	4G	55	169	74	5.752	5.45	5.29	0	44847d4eb5a0d152cf65f533fe23ad74	44847d4eb5a0d152cf65f533fe23ad74	ok
2	4G	45	122	64	12.427	3.04	2.88	0	44847d4eb5a0d152cf65f533fe23ad74	44847d4eb5a0d152cf65f533fe23ad74	ok
3	4G	48	91	63	8.101	1.89	1.74	0	44847d4eb5a0d152cf65f533fe23ad74	44847d4eb5a0d152cf65f533fe23ad74	ok
4	4G	47	100	59	21.246	1.05	0.905	0	44847d4eb5a0d152cf65f533fe23ad74	44847d4eb5a0d152cf65f533fe23ad74	ok
5	4G	39	138	65	15.582	0.629	0.434	0	44847d4eb5a0d152cf65f533fe23ad74	44847d4eb5a0d152cf65f533fe23ad74	ok

The file used in this study is a CCTV recording file with the extension .mp4, which is 91MB in size. Testing uses the ownCloud Desktop Client tools so that digital evidence files are automatically synchronized to the server. The research method used is to change the bandwidth with the parameters ping time, jitter and packet loss. Data is taken by comparing the results of MD5 source file hashing and files that have been uploaded on the server.

The test results shown in Table 1 show that using network conditions where the received speed is 0.434 Mbps and the transmit speed is 0.629 Mbps, the results of checking MD5 source files and files that have been uploaded on the server are the same. In other words, file transfers with low bandwidth or poor network quality do not affect file integrity.

This testing scenario is carried out to illustrate that even when synchronizing digital evidence acquisition results with unfavorable network conditions can still ensure the integrity of the acquired digital evidence.

4.4. Test of disconnected nodes

One of the testing scenarios runs is to make the condition of one of the Ceph nodes disconnected when a file transfer is performed. The file transferred in this case is a CCTV recording file with a .mp4 extension measuring 91MB. We make 2 states of the disconnected node ie when 2 nodes are disconnected, and 4 nodes are disconnected then compare the results.

When 2 nodes are disconnected, file transfer from client to server continues to run normally. Fig 6 shows the condition of 2 nodes or 4 OSD down, cluster status is declared healthy. It shows that the disconnected of 2 nodes are minor failure for Ceph. After 2 broken nodes are reconnected, the system will synchronize files. MD5 hashing checking of source files and file servers gives the same results.

When 4 nodes are disconnected, file transfer from client to server stops. As can be seen in Fig 7, Ceph will do the blocking into the storage system so that the file is retained until only 3 nodes are cut off. When the number of broken nodes is only 3, file transfers can resume normally. And when the system is fully reconnected and declared healthy, the system will synchronize. Checking MD5 hashing of source files and file servers give no change.

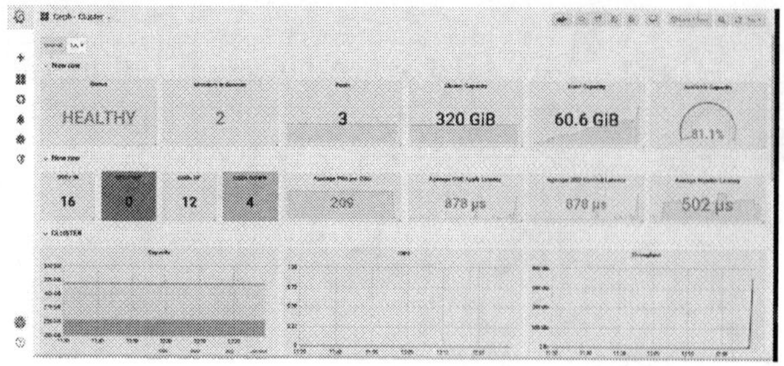

Figure 6. Testing With The Condition Of 2 Nodes Or 4 OSD Down, Healthy Cluster Status

```
2019-08-08 05:58:10.809862 mon.mon-ceph1 [WRN] Health check update: 8 osds down (OSD_DOWN)
2019-08-08 05:58:10.809908 mon.mon-ceph1 [WRN] Health check failed: 1 host (2 osds) down (OSD_HOST_DOWN)
2019-08-08 05:58:31.672775 mon.mon-ceph1 [WRN] Health check failed: 1 MDSs report slow metadata IOs (MDS_SLOW_METADATA_IO)
2019-08-08 05:58:43.700918 mon.mon-ceph1 [WRN] Health check failed: 1 MDSs report slow requests (MDS_SLOW_REQUEST)
2019-08-08 05:58:40.140040 mds.mon-ceph2 [WRN] 1 slow requests, 1 included below; oldest blocked for > 34.559337 secs
2019-08-08 05:58:45.140151 mds.mon-ceph2 [WRN] 1 slow requests, 0 included below; oldest blocked for > 39.559473 secs
2019-08-08 05:58:50.140291 mds.mon-ceph2 [WRN] 1 slow requests, 0 included below; oldest blocked for > 44.559601 secs
```

Figure 7. Testing with the condition of 4 nodes or 8 OSD down, Ceph did blocking

The system repair process runs automatically. This can occur because the RADOS feature of Ceph allows Ceph to be self-healed and self-manage. RADOS also makes the system flexible, reliable and easy to manage. In the event of a disaster, Ceph can provide reliability for multiple failures. Ceph detects and corrects failures in each failure zone as disks, nodes, networks, racks, data center lines, data centers, and even various geographical locations. Ceph tries to manage the situation automatically and resolve it as much as possible without data trouble [8].

Based on that simple test, it can be concluded that the storage system that is built meets the aspect of availability that is ready to be accessed anytime by the user / application / system that needs it. The results of checking MD5 hashing of the same source file and file server in this test also confirm that this storage system has met the integrity aspect.

5. Conclusion

Based on the analysis, implementation of the storage system and testing that has been done before, it can be concluded that the digital evidence storage can be built using clustering storage method by software defined storage i.e. Ceph. This digital evidence storage is a network-based centralized repository that can be accessed securely from a remote location.

The advantage of using this method is the excellent scalability capabilities that have been proven from previous tests. As a digital evidence storage, this storage system has also fulfilled security aspects, namely in terms of confidentiality, integrity, and availability. So that digital evidence stored in this digital evidence storage system can be for use as evidence in court room.

This research has not focused on the digital evidence's Chain of Custody (CoC) and preservation yet. Further research is needed regarding the management of digital evidence stored in the system that has been built in this study.

References

[1] Davis, M., Manes, G., & Shenoi, S. (2005). A Network-Based Architecture for Storing Digital Evidence. The International Federation for Information Processing, vol. CXCIV, no. -, pp. 33-42, 2005.

[2] Ali, M., Prayudi, Y., & Sugiantoro, B. (2019). Storage Area Network Architecture to support the Flexibility of Digital Evidence Storage. International Journal of Computer Applications, 182(41), 30–35. https://doi.org/10.5120/ijca2019918496

IOP Conf. Series: Materials Science and Engineering **722** (2020) 012063 doi:10.1088/1757-899X/722/1/012063

[3] Paul Chinnaraju, S., Gunasekaran, G., Kumar, N., & Anandan, R. (2018). Transformation from legacy storage to software defined storage–a review. International Journal of Engineering & Technology, 7(2.21), 306. https://doi.org/10.14419/ijet.v7i2.21.12387
[4] Gudu, D., Hardt, M., & Streit, A. (2014). Evaluating the Performance and Scalability of the Ceph Distributed Storage System. 2014 IEEE International Conference on Big Data , 177-182.
[5] Kuntze, N., Rudolph, C., Richter, J.. (2010). Securing Digital Evidence . 5th International Workshop on Systematic Approaches to Digital Forensic Engineering, SADFE 2010, 119-130
[6] Shinder, L., & Cross, M. (2008). Chapter 15 - Collecting and Preserving Digital Evidence. In L. Shinder & M. Cross (Eds.), Scene of the Cybercrime (Second Edition) (pp. 623–652). https://doi.org/https://doi.org/10.1016/B978-1-59749-276-8.00015-7
[7] Weil, S. A., Brandt, S., Miller, E., Long, D., & Maltzahn, C. (2006). Ceph: A Scalable, High-Performance Distributed File System. In OSDI.
[8] Singh, K. (2015). Learning Ceph. Birmingham: Packt Publishing Ltd.
[9] Weil, S., Brandt, S., Miller, E., & Maltzahn, C. (2007). CRUSH: Controlled, Scalable, Decentralized Placement of Replicated Data. (November), 31–31. https://doi.org/10.1109/sc.2006.19
[10] Khatri, H. (2019). INDONESIA Laporan Pengalaman Jaringan Seluler Juli 2019. [Online]. Available https://www.opensignal.com/in/reports/2019/07/indonesia/mobile-network-experience. Accesed on August 8, 2019

DNS tunneling Detection Using Elasticsearch

A F Sani[1], and M A Setiawan[2]

Department of Informatics, Universitas Islam Indonesia, Yogyakarta, INDONESIA[1,2]

E-mail: 14917133@students.uii.ac.id[1], andri.setiawan@uii.ac.id[2]

Abstract. Domain Name System (DNS) Protocol is a popular medium used by malware to perform 'command and control' in taking over victim's computer, this technique called as DNS tunneling. Moreover, DNS tunneling can also be used to bypass captive portal hotspot in public places and worsen the network quality. However, in more dangerous stage, DNS tunneling can also be used to exfiltrate data from the victim's computer. Instead of using DNS Protocol to translate domain name, the medium misused to bootleg the data. Those are the weaknesses which frequently used by the attacker to deceive network administrator. Our approach to this problem is analyzing the traffic using unique hostname as indicator of compromise and utilizing Elasticsearch tool to detect DNS tunneling. Elasticsearch will send an email to notify the administrator about DNS tunneling. The email contains information about domain suspected as perpetrator of DNS tunneling. The result from Elasticsearch can be used to add the domain blacklist, so the domain can no longer be used to perform DNS tunneling. Hopefully those combinations are able to support the network administrator to secure the network from DNS tunneling. Moreover, the result of network quality analysis shows that there is a rise in jitter value and packet lost when DNS tunneling happens.

1. Introduction

Recently, Internet Security Researchers found a malware diversified as Remote Access Trojan. This kind of malware uses a new technique which is hard to detect by the Network Administrator. Those malware does 'command and control' and takes over the victim's computer through DNS Protocol, this kind of malware called DNSMessenger [1]. 'Command and Control' or sending data through DNS Protocol called DNS tunneling or DNS ex-filtration. Besides doing 'command and control', the malware uses DNS tunneling to bypass the captive portal or login hotspot in public places [2]. However, data thievery through DNS Protocol considers being more dangerous than bypassing captive portal. The malware could ex-filtrate various data, such as classified trade data, intellectual property, employee data, customer data, and many others. DNS tunneling requires software installed on the victim's computer to work. If the software installed, the attacker can easily bypass the firewall and security system of the victim.

DNS tunneling can also worsen network quality. The research concluded that using DNS tunneling can increase the delay of the entire network up to 140-1500 ms, jitter until 8-57 ms, and DNS Overhead 200-2000% [3]. We can imply that; a client on the network is making a VOIP call or video call, the high jitter causes a voice and video transmission cannot run smoothly, the high delay causes a slight or minor delay to the voice and video. DNS Overhead causes an increase in data package size and resulting in higher using of bandwidth. To experience the best quality of the network, Cisco recommends we set the jitter below 30 ms and delay less than 150 ms [4].

Network Administrator usually does not give extra attention to the DNS traffic. DNS Protocol is a protocol uses to translate IP Address into a domain name, with the result that we can access a computer or server using the name without remembering the IP Address of the computer or server. With that common sense, the Administrator is forgetting about the fact that DNS Protocol also can be used to exchange data. That flaw uses by the attacker to 'Command and Control' or steal data using DNS tunneling.

Therefore, DNS traffic in a network should be monitored to prevent DNS tunneling. Actually, the Administrator able to block all DNS traffic to prevent tunneling. However, that is not the ideal solution, because that method will also block the user to access host address. Another approach is using DNS Sinkhole [5]. DNS Sinkhole is a DNS server which able to give wrong IP address (spoofing) from DNS request, so the destined domain can no longer be accessed. This condition can be used to prevent malware or DNS tunnel from contacting the server.

DNS Sinkhole is using a domain list to be blocked. We can manually make the list or download it from website such as; urlblacklist.com, malwaredomain.com, and etc. To obtain the domain suspected doing DNS tunneling is necessary to monitor and log all DNS traffic in the network. Those logs can be obtained from many resources, such as DNS server, Intruder Detection System (IDS), proxy, and computer log. To detect DNS tunneling from the log, analysis should be done manually using capture analyzer packet, such as Wireshark. This kind of approach considered hard to do and take times, especially if we want to visualize the result, we need another tool. Another approach is using Payload Analysis Method and traffic analysis [2]. Payload analysis able to detect certain DNS tunneling, while traffic analysis able to detect DNS tunneling universally.

Traffic analysis is the approach we use to resolve the issues we stated before. We use traffic analysis with the amount of unique hostname as an indicator of compromise using Elasticsearch. Elasticsearch has components which can be used in this research, such as Packetbeats, Kibana, and Watcher. Packetbeats is a real-time sniffer which will capture the traffic DNS, Watcher will give an email notification when DNS tunneling happened, and Kibana is a panel of visualization which will show a graphic bar of domain names which have the most unique hostname. That combination hopefully helps the administrator to secure and monitor the network.

2. Literature Review

DNS tunneling is a technique to bypass the security control and to infiltrate or ex-filtrate data from a target. This technique is still used because DNS usually do not monitor well. The practitioner blindly trusted DNS is secure [6].

Popular tools to perform DNS tunneling are Iodine [7] and Dnscat2 [8]. Iodine is an app which able to create a tunnel interface between client and server, all the traffic can be passed up through the tunnel. On the other hand, Dnscat2 used for performs 'command and control' between client and server. Both apps can bypass the security control of a network.

DNS has a caching mechanism to accelerate the response from DNS query, therefore, all DNS Tunnel program will create random and long hostname string (unique hostname), so that DNS cannot cache the tunnel and the data thievery become possible.

DNS tunneling detection methods divided into two, which is, payload analysis and traffic analysis. Payload Analysis can only detect certain DNS tunneling, while Traffic Analysis can detect DNS tunneling universally. Farnham [2] tries to detect DNS tunneling using Traffic analysis making unique hostname as an indicator of compromise. The normal amounts of unique hostnames are below 300, after conducting DNS tunneling the amount of unique hostname rapidly increase until 700 [2].

A tool which can be used to detect DNS tunneling with traffic analysis method is Elasticsearch. Elasticsearch is a search engine which builds the base on Apache Lucene and opensource product which also developed with Java. Elasticsearch can conduct a real-time and distributed analysis and also able to

IOP Conf. Series: Materials Science and Engineering **722** (2020) 012064 doi:10.1088/1757-899X/722/1/012064

do multiple searching mechanism. Elasticsearch can manage various kinds of logs, such as operating system log, web server, log traffic, app log, and Amazon Web Service Log [9].

3. Research Methodology

To support this research, the researcher will build a lab in a virtual neighborhood so that easier and cheaper to maintain. Though only in a virtual neighborhood, the condition already represents the real condition in a real network. The researcher will use a computer with Quad-Core processor, RAM 8GB, and SSD 128GB. The host run Ubuntu 19.10 and using VirtualBox with Hypervisor KVM as the virtualization software.

In the VirtualBox will be built a virtual machine. The virtual machine will run Elasticsearch. The Virtual machine will run with the specification of 2 virtual core, 4GB RAM, network mode bridge, and run CentOS 7. For the tunneling server, the researcher using a VPS with CentOS7, RAM 1GB, one virtual core CPU. Iodine and DNScat2 used as DNS tunneling software. The researcher uses Mikrotik RB750 as the router. The router will be used as a gateway and firewall. For the client, the researcher will use Windows 10 and Ubuntu 19.10.

The topology used by the researcher can be seen in an image below. That topology was chosen because it represents the real condition of a network. One of the computer clients will be the DNS tunnel client which call the DNS tunnel server. The router will be set to do mirroring port to Elasticsearch server, with that scheme, all the traffic can be read by Elasticsearch to run the inspection.

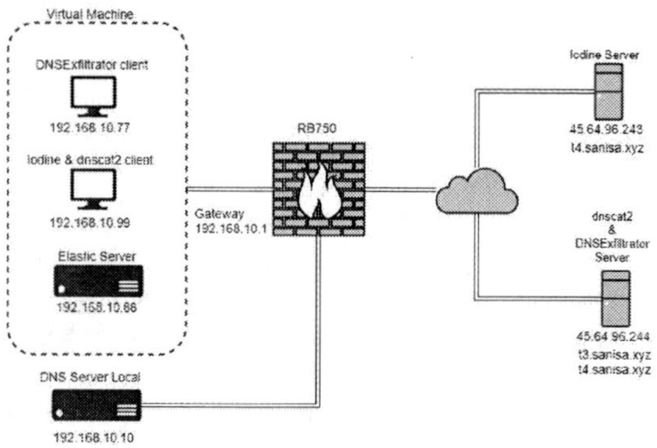

Figure 1. DNS Tunneling Simulation Topology

3.1. Log aggregation and simulation.

Log aggregation will be done by Elasticsearch using a plug-in. The plug-in will capture a packet called Packetbeat in real-time. So the data used is adequate, DNS Grind from Pentestmonkey is used to generate traffic simulation [10].

DNS tunnel server Iodine and DNScat2 will run the services in DNS tunnel server. The client will also run the DNS tunnel client which will contact the DNS tunnel server. On the first experiment, the client will browse uses DNS tunnel, this experiment intends to create DNS tunnel log. Elasticsearch will run the job which will detect DNS tunnel from the collected log. Visualization of the result will be seen on Kibana Dashboard. Elasticsearch will detect and count the amount of unique hostname, domain with the largest amounts of unique hostname will detect as an anomaly. To get the best result, we need more or less 48 hours.

3.2. Detection and analysis

In this stage, an analysis will be conducted to find out whether Elasticsearch able to detect DNS tunneling or not. The method applied is Traffic Analysis. Each communication on DNS tunneling will create a new hostname, the normal average amount of unique hostname is below 300 [2], because of that, the more unique hostname indicates DNS tunneling is happening. All logs captured by Packetbeat will be processed with a custom script by the Watcher. After that, Watcher counts the amount of unique hostname based on the cardinality of the domain. The amount of unique hostname on a domain will be visualized in Kibana in the form of a graphic bar. If the amount of unique hostname more than 300 and the domain does not exist in the whitelist, the watcher will send an email to the administrator.

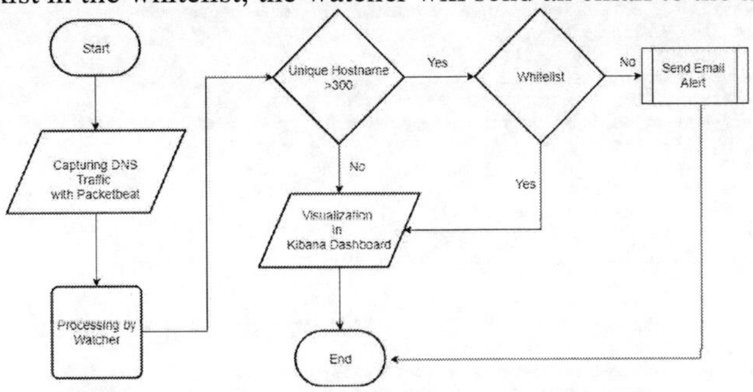

Figure 2. Flowchart detection of dns tunneling.

4. Results

Mikrotik router will do a port mirroring to duplicate packets from DNS server to Elasticsearch server. The Elasticsearch server will run Packetbeat app to sniff the DNS packet. A Laptop will be prepared to run the DNS tunneling to the server. Watcher will find cardinality from the hostname of each domain and come out with the amount of unique hostname. Here is a graphic from Kibana dashboard shows the amount of unique hostname for 15 minutes without DNS tunneling. Figure 3 shows the amount of unique hostname in normal situation is below 100.

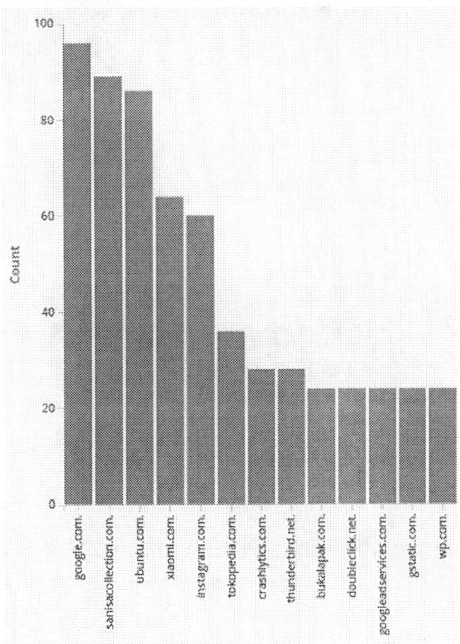

Figure 3. Unique hostname on DNS with normal traffic.

In this research we use Iodine, Dnscat2, and malware DNSExfiltrator. We choose those tools because already represent several methods in DNS tunneling.

4.1. Tunneling using Iodine

In this first experiment, we use Iodine to perform DNS tunnel. We try to browse through DNS tunnel.

IOP Conf. Series: Materials Science and Engineering **722** (2020) 012064 doi:10.1088/1757-899X/722/1/012064

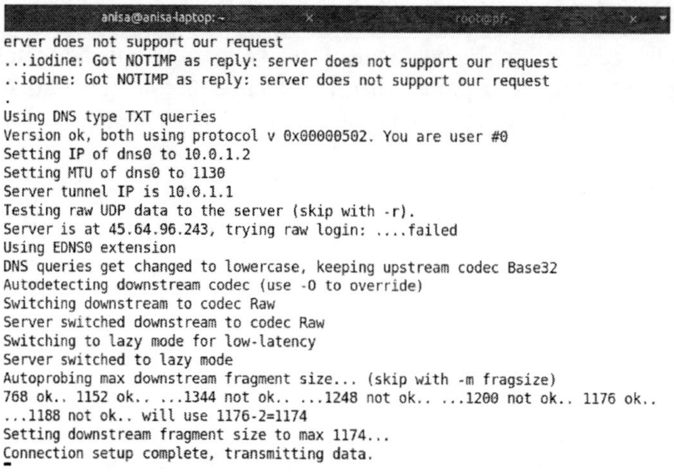

Figure 4. Tunneling process using Iodine

The domain name which the researcher uses as DNS tunnel server is sanisa.xyz. When the tunnelling runs for 3 minutes, the amounts of unique hostname on that domain spike up until 700.

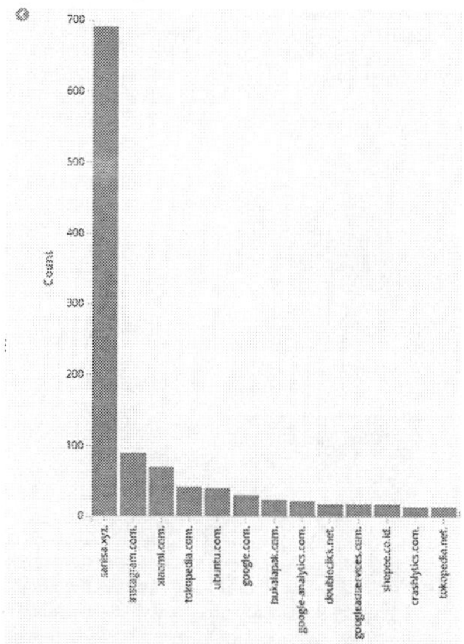

Figure 5. The spike on unique hostname when tunneling using Iodine on domain sanisa.xyz

4.2. *Tunneling using dnscat2*

In the next experiment, we use Dnscat2 to perform 'command and control' on domain *sanisa.xyz*.

Figure 6. Tunneling process using dnscat2

When the dnscat2 runs for 3 minutes, the amount of unique hostname on domain sanisa.xyz increased up to 900 records.

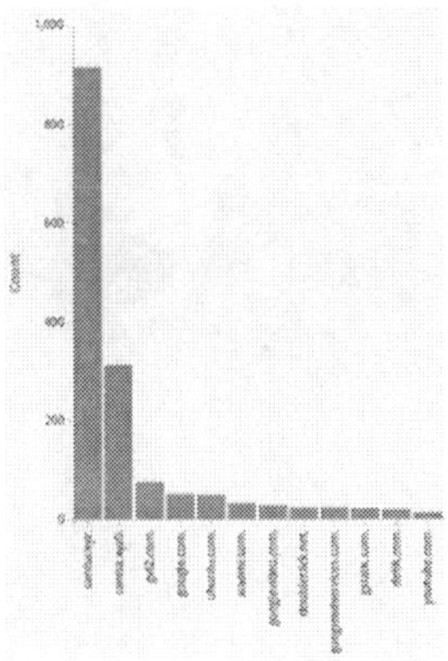

Figure 7. The escalation of unique hostname when tunneling using dnscat2 on domain sanisa.xyz.

4.3. *Tunneling using malware*

In this experiment, the researcher using a malware called *DNSExfiltrator*. That malware listed in virustotal.com with hash value: *"ed937bcd5dc05f1021aa83afdb47af266083ef47228e23a32292bad577c53191"*.

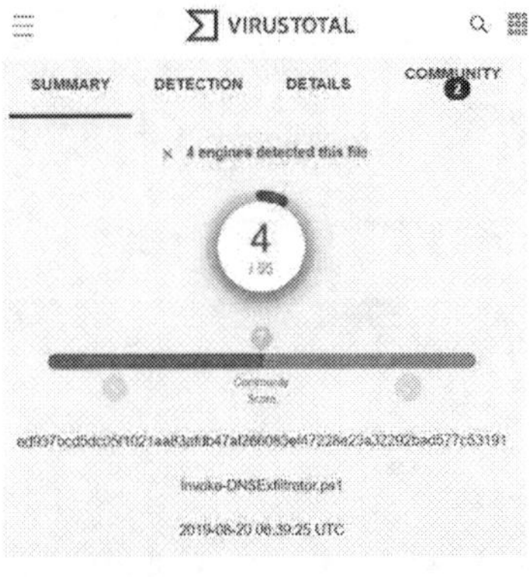

Figure 8. Malware DNSExfiltrator status based on virustotal.com

This malware can send a file through DNS protocol. On the side of the server, this malware uses python language, but on the side of the client (the victim), we use powershell windows. On this trial, we send a file with a name "data.pdf" file size: 685KB to the server with domain t4.sanisa.xyz.

```
C:\Users\unknown-win\Documents\dnsex>powershell -c "ipmo .\Invoke-DNSExfil
-d t4.sanisa.xyz -p faisalganteng

Security warning
Run only scripts that you trust. While scripts from the internet can be us
computer. If you trust this script, use the Unblock-File cmdlet to allow t
message. Do you want to run C:\Users\unknown-win\Documents\dnsex\Invoke-DN
[D] Do not run  [R] Run once  [S] Suspend  [?] Help (default is "D"): R
[*] Compressing (ZIP) the [data.pdf] file in memory
[*] Encrypting the ZIP file with password [faisalganteng]
[*] Encoding the data with Base64URL
[*] Total size of data to be transmitted: [835516] bytes
[*] Maximum data exfiltrated per DNS request (chunk max size): [228] bytes
[*] Number of chunks: [3697]
[*] Sending 'init' request
[*] Sending data...
```

Figure 9. DNS tunneling process using DNSExfiltrator to send a file data.pdf

When the DNS tunnelling process runs, there are escalations of unique hostname until more than 5000 as shown in the graphic below:

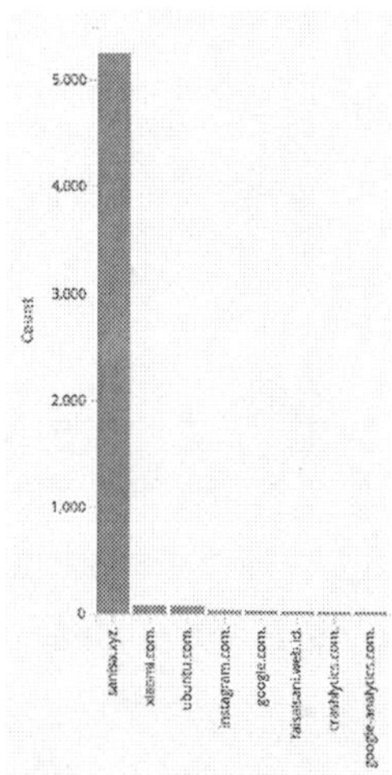

Figure 10. The escalation of unique hostname when tunnelling using DNSExfiltrator on domain sanisa.xyz.

To assure that Elasticsearch successfully detect DNS tunneling, during the process of tunneling we do capture packet using Wireshark.

Filter: dns.qry.name.len > 30 and !mdns

No.	Time	Source	Destination	Protocol
5718	53.7	192.168.88.254	192.168.88.1	DNS
5726	54.7	192.168.88.254	192.168.88.111	DNS
5731	55.7	192.168.88.254	192.168.88.1	DNS
5742	57.7	192.168.88.254	192.168.88.1	DNS
5743	57.7	192.168.88.254	192.168.88.111	DNS
5745	59.2	192.168.88.111	192.168.88.254	DNS
5747	61.7	192.168.88.254	192.168.88.1	DNS
5749	63.7	192.168.88.1	192.168.88.254	DNS
6139	120.(	192.168.88.254	192.168.88.1	DNS
6142	120.(	192.168.88.1	192.168.88.254	DNS

Figure 11. Wireshark process to sniff the DNS traffic.

To analyze the DNS traffic using wireshark, the researcher use filter "dns.qry.name.len > 30 and !mdns" which means, DNS query filter with the amount of subdomain/hostname more than 30 and is not multicast DNS. The filtering result shows there are long query with encoding base64 on domain sanisa.xyz.

Figure 12. The result of filtering Query DNS on Wireshark.

In the study from Leijenhorst [3], DNS tunneling can also worsen the network quality [3]. Therefore, in this research we also analyzing jitter value and packet loss in the UDP protocol. We would like to find out the effect of DNS tunneling to the network. We created two tests, the first one was client that had DNS tunneling, and the second one was normal client without DNS tunneling. Both clients run in one network. In this experiment we use iperf3. And the result is shown below:

Table 1. Network quality analysis

Client	Parameter	Before Tunneling	On Tunneling
Client Running DNS Tunnel	Jitter	0.792 ms	3.726 ms
	Packet Loss	0 %	36 %
Client Not Running DNS Tunnel	Jitter	0.15 ms	1.096 ms
	Packet Loss	0 %	1.7 %

From the experiments we can conclude that there is a rise of jitter value and packet loss. The rise happens significantly on the client who did DNS tunneling. On the other client who did not perform DNS tunneling, the rise still happen but not significant, still acceptable according to Cisco recommendation [4].

4.4. Notification to the Administrator

When the unique hostname more than 300 and the domain are not in the whitelist, the watcher will be triggered and send an email notification about suspicious activities which indicated DNS tunneling. An email notification is sent to Administrator, with the contents:

"This Domain has high unique hostnames:{sanisa.xyz.={unique_hostnames=661,total_bytes_in=371114.0,total_bytes=1490168.0,total_requests=2024, total_bytes_out=1119054.0}}"

The email notifies the administrator that the domain sanisa.xyz suspects of doing DNS tunneling. From the experiments, Elasticsearch, Packetbeat, and Watcher can be used to detects DNS tunneling.

5. Conclusions and Recommendations

In our experiments, we conducted our simulation with Iodine, dnscat2, and executing DNSExfiltrator malware. We can conclude that traffic analysis in a way of counting unique hostname as an indicator of DNS tunneling with Elasticsearch is successfully detecting the DNS tunneling and able to notify the administrator about the attacker. The output from the detection can add up the list of blacklisted domains.

On top of that, on the network quality analysis, DNS tunneling can increase jitter value and packet loss on the network but not significant and still acceptable.

References

[1] E. Brumaghin and C. Grady, "Talos Blog || Cisco Talos Intelligence Group - Comprehensive Threat Intelligence: Covert Channels and Poor Decisions: The Tale of DNSMessenger," 2017. [Online]. Available: https://blog.talosintelligence.com/2017/03/dnsmessenger.html. [Accessed: 03-Apr-2019].
[2] G. Farnham, "Detecting DNS tunneling," 2013.
[3] T. Van Leijenhorst, K. Chin, and D. Lowe, "On the Viability and Performance of DNS tunneling," 2008.
[4] "Acceptable Jitter and Latency." [Online]. Available: https://getvoip.com/blog/2018/12/20/acceptable-jitter-latency/. [Accessed: 29-Jun-2019].
[5] G. Bruneau, "DNS Sinkhole," 2010.
[6] M. Branscombe, "Why you need to care more about DNS." [Online]. Available: https://www.cio.com/article/2948378/why-you-need-to-care-more-about-dns.html.
[7] "Iodine DNS tunneling Tools." [Online]. Available: https://code.kryo.se/iodine/.
[8] "DNScat2 C&C tunneling tools." [Online]. Available: https://github.com/iagox86/dnscat2.
[9] Elasticsearch, "Elasticsearch: RESTful, Distributed Search & Analytics | Elastic." [Online]. Available: https://www.elastic.co/products/elasticsearch. [Accessed: 03-Apr-2019].
[10] "DNS Grind." [Online]. Available: https://github.com/pentestmonkey/dns-grind.

Identifying Morphology of The Rat Flea Arthropod as a Vector of Plague Disease Based Microscopic Image

[1]I Muhimmah, [2]N Lusiyana, [3]Fatmawati

[1] Department of Informatics Engineering, Faculty of Industrial Technology, Universitas Islam Indonesia, Yogyakarta, Indonesia
[2] Medical Education, Faculty of Medicine, Universitas Islam Indonesia, Yogyakarta, Indonesia
[3] Master of Informatics Engineering, Faculty of Industrial Technology, Universitas Islam Indonesia, Yogyakarta, Indonesia

Corresponding author: 18917111@students.uii.ac.id

Abstract. The morphological identification system vector of pes disease is one of teaching aids students of the Faculty of Medicine of Universitas Islam Indonesia. The students needs to study the morphological characteristic under the microscope. In practice, it proof difficult to distinguish the kind of various ordos of flea. Thus, the computer teaching aids on this subject need to be developed. The purposed system identified the morphological characteristic of the insect using their size, shape, and texture features. At the identification stage, ten features selected using the Correlation Features Selection (CFS) with method search Best first. This system can also provide additional information about the taxonomy and general characteristics of parasitic insects. The method used for the classification of parasitic insects is KNN, which provides accuracy for body detection of 97.0377% and head detection of 75%. The system has been validated through the expert and its classification correlation reached Cohen's Kappa.

1. Introduction

An arthropod is an animal whose feet are in the field, including the abdomen and thorax (chest). Some of these arthropods have many diseases, but there is also a bad effect on human health. Examples of arthropods that can cause illness are flea.

Flea belongs to the insect class, with the ordo *Siphonaptera* or *Aphaniptera*. It is an ectoparasitic insect that is a type of parasitic living on a host who lives outside of the body. The host includes mice, cats, dogs, rabbits, and bats. There are several flea genus, such as *Tunga*, *Ctenocephalides*, *Xenopsylla*. Flea can be a vector of some diseases, e.g., rat flea or *Xenopsylla cheopis,* which is the main vector in the transmission of plague disease. A vector is an organism that does not cause disease but spread it by carrying parasites its host.

Plague disease is one of the zoonotic diseases that attack rodents but can be contagious in humans. Spread in humans is caused by the bite of the animal infected with the disease plague. Its caused by germs of *Yersinia pestis* [1]. An area where a pes outbreak occurs usually preceded by an outbreak of deaths rats. Humans infected with this disease through a bite or wound abrasions that contaminated with stools or effective flea damaged tissues.

Based on the information obtained from Dr. Novyan Lusiyana, M.Sc of the Department of Parasitology, Parasitology Laboratory of the Faculty of Medicine, University Islam Indonesia, that to detect morphology of parasitic insects is still manual. That is, by using a microscope to see its morphology. Since some parasitic insects have the same taxonomy, it is difficult to distinguish them. If viewed in identifying manually, researchers should thoroughly pay attention to the morphology of rat. Which distinguishes the characteristics of each insect can be seen based on the shape and texture of the image resulting from a microscope.

2. Basic Theory

2.1. Morphology of Rat Flea

Flea belongs to the phylum arthropod, class Insecta, with the ordo of Siphonaptera or also as the ordo of Aphaniptera. A flea is an ectoparasitic insect (a type of parasitic living on a host) living outside of the body. The host of flea includes mice, cats, dogs, rabbits, and bats. There are several genus flea, among others, *Ctenocephalides*, *Xenopsylla*, *Pulex*. Fleas have complete metamorphosis [2], which starts from eggs, larvae, pupae, and adults. The life cycle of fleas lasts 14-27 days, and adult fleas can live up to one year. The cycle of fleas starts from the eggs, then hatches into larvae in 3-4 days and eats organic waste. After that, the larvae turn into pupae. In the stages of larvae and pupae, it takes 3-4 weeks to turn into adult fleas. Then, adult fleas looking for a house to serve as their host. Examples of hosts are rabbits, dogs, cats, mice.

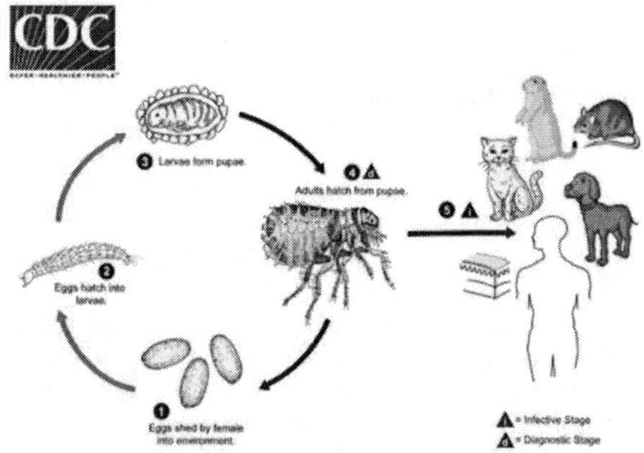

Figure 1.Lifecycle of Rat Flea

Table 1. Morphology of Rat Flea

	Clinical Characteristics	Mathematical Models
	- Has genitalia - No genal or *comb* - Body size is smaller than females. - Roundhead shape	- Genitalia and a genal comb can be identifying with texture traits - Use a comparison area - The head shape is used to shape factor
	- Have *Spermatheca* - No genal or comb - Roundhead shape	- It can be identifying by using textures - The head shape is used to shape factor
	- More shape head shape. - There is a comb called *genal comb* or *oral comb*. The comb is also in the first segment of the thoracic *comb* or *pronotal comb*.	- The head shape is used to *shape factor*. - It can be identifying by using texture features.

	- Roundhead more shape than *C. Felis*. - There is a comb called *genal comb* or *oral comb*. The comb is also in the first segment of the thoracic *comb* or *pronotal comb*.	- The head shape is used to *shape factor*. - It can be identifying by using texture features.
	- Roundhead shape. - No *comb*.	- The head shape is used to *shape factor*. - It can be identifying by using texture features.

2.2. Digital Image Processing

Digital image processing refers to the processing of 2D images using computers. Digital images can be processed with a computer because it takes the form of numerical data [3]. The main purpose of digital image processing is transforming the 2D image into another image. This process has features of input data and output information in the form of an image. Some examples of image processing are image contrast modifiers, noise-canceling with filtering, sharpening, and false-color rendering (e.g., Pseudo coloring), and so on [4].

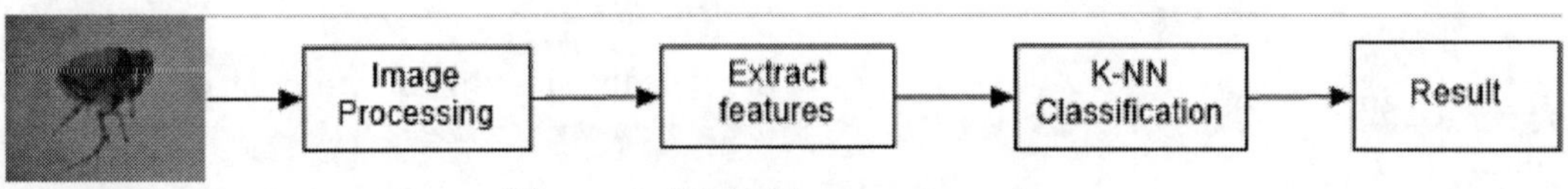

Figure 2. Digital image processing

2.3. Features Extraction

Feature extraction is a stage to extract features or information from objects within an image that will be differentiated by objects with others. The extraction of features is a stage of capturing characters (features) that are important in imagery after the image process messages. Here are the feature extraction features used [5].

Table 2. Features

Shape and size features		Texture features	
Semi Major Axis Length a= longest distance from centroid.	(1)	*Mean* Mean shows the average brightness value of an object. $\frac{1}{MN}\sum_{i=1}^{M}\sum_{j=1}^{N}p(i,j)$	(15)
Semi Minor Axis Length b= Shortest distance from the centroid.	(2)	*Variance* $\sigma^2=\sum_i\sum_j p(i,j)(i-j)^2$	(16)
Centroid Coordinate of the center point.	(3)	*Standard Deviation* $\sigma=\frac{1}{MN}\sum_{i=1}^{M}\sum_{j=1}^{N}(p(i,j)-\mu)^2$	(17)
Equivalent Diameter $ED=\frac{4*Area}{\pi}$	(4)	*Skewness* $S=\frac{1}{MN}\sum_{i=1}^{M}\sum_{j=1}^{N}\left[\frac{p(i,j)}{\sigma}\right]^3$ Shows unsymmetrical average intensity.	(18)
Eccentricity	(5)	*Kurtosis* The relative of the image histogram curve.	(19)
Perimeter A number of the pixel on the object's boundary.	(6)	*Entropy* $\sum_i\sum_j -p(i,j)log_2(p(i,j))$	(20)

Area
The number of pixels in the object regi (7)

Circularity (8)
$\frac{4\pi\ x\ Area}{p^2}$

Convex Area (9)
Specify the pixel count in *ConvexImage*.

Solidity (10)

Compactness (11)
$\frac{4\pi\ x\ Area}{p^2}$

Bounding Box (12)

Momen Of Inertia (13)
$MoI = \frac{1}{4}\pi ab(a^2 + b^2)$

Average Radius (14)
$\sqrt{(x_1 - x_2)^2}$

Form irregularities in image.

Inverse Difference Moment (21)
$\text{IDM} = \sum_{i=0}\sum_{j=0} \frac{p(i,j)}{1+(i,j)^2}$
Demonstrate the homogeneity pixel from the image.

Energy (22)
$\text{Energy} = \sum_{i=0} p(i)^2$

Contrast (23)
$\sum_i\sum_j p(i,j)(i-j)^2$

Correlation (24)
$\sum_i\sum_j \frac{(i-\mu_i)(j-\mu_j)p(i,j)}{\sigma_i\sigma_j}$
Determine the size of the linear dependency.

Homogeneity (25)
$\sum_i\sum_j \frac{p(i,j)}{1+|i-j|}$
Declare the size proximity of each element *matrix co-occurrence*

3. Research Method

3.1. Data Collection

The phases of data collection for research supporters are as follows:

- Literature study by searching for reference sources related to flea mice as vector disease PES, collecting information based on theories, articles, and from the literature.
- The interview with Dr. Novyan Lusiyana, M.sc Specialist in UII Medical Laboratory Clinical Pathology, to obtain information on morphological features as well as view microscopic image and needs system.
- Images of parasitic insect preparations were taken and collected using a microscopic camera parasitology Laboratory, Faculty of Medicine Universitas Islam Indonesia, with permission from Dr. Novy. A picture of preparation is *Xenopsylla cheopis*, an adult *Ctenocephalides felis* stadium with 4x and 10x magnification.

3.2. Design System

- The image entered on the system will be through the stage of image processing. Image processing or preprocessing will be segmented and converted images into a binary image. Before the extraction of the feature, the results of preprocessing carried out image improvements such as removing noise. Then the extraction of the feature, wherein feature extraction will produce the features used in detecting morphology. The features of this feature will be in the selection again to get the best features to be the reference of the classification process. The result of a selection of features that have the highest accuracy among other features, which will later be for the process of classification of the flea morphology of rats.
- After the image inserted, the image will then be processing by segmentation. The body segmentation method used multilevel thresholding, while the segmentation body used the Otsu

thresholding method. Then the segmentation image is a binary image, which later image is processed extraction feature.

3.3. Design Testing

- Feature Extraction: The extraction test feature was performed to obtain classification accuracy in the system. The selection features the Correlation Feature Selection (CFS) method using the search method Best first. The selection of this feature is calculated using the Weka application. The result is the weight that will then be chosen by the attributes or features that have the highest weight. The result of feature extraction for body detection are area; eccentricity; major axis length; and solidity. Also, for head detection is area; eccentricity; *major* axis length; solidity.
- Classification: To find out the maximum accuracy of feature extraction. Its accuracy will be tested using the Weka application. The method used in the classification is K-NN.
- System Validation: System validation test using Cohen's Kappa coefficient. Cohen's Kappa coefficient is used to measure the level of agreement between experts and systems. Furthermore, to look at the user terms, then used a method of usability. The method used to assess the usability of the system is the System Usability Scale (SUS) [7].

4. Result

4.1. Segmentation

The insert image will be segmented first. For body detection, segmentation is performed using a multilevel thresholding method. The value taken is the range between the histogram chart that the most area, which is 100-200. For head detection, image segmentation using the *Otsu thresholding* method.

Figure 3.Body and Head Segmentation Result

4.2. Attribute Selection

After feature extraction of the characteristic on the body and head, then the next step is the selection of attributes. This selection attribute uses the app Weka. This Attribute selection uses CFS with Search Best First [8] and an attribute selection mode that is K-Cross Validation with value k= 10. The result of this attribute selection is the weight that will then be chosen by the attributes or features that have the highest weight. The result can be seen in Figure 4 and Figure 5.

Area	Mean GrayLevel	Variance	Skewness	Kurtosis	Correllation	Species
210844	0,5	0,086538	-2,28E-17	1,79224	0,991662	cheopisBetina
308241	0,5	0,086538	-2,28E-17	1,79224	0,991662	cheopisBetina
290331	0,5	0,086538	-2,28E-17	1,79224	0,991662	cheopisBetina
158233	0,5	0,086538	-2,28E-17	1,79224	0,991662	cheopisBetina
2115182	131,2588	1936,055	-0,58678	2,254036	0,988223	cheopisBetina
1064810	138,8642	1336,859	-1,51404	4,544034	0,975708	cheopisJantan
1064886	140,1355	1250,154	-1,84072	5,529942	0,971472	cheopisJantan
242837	140,1355	1250,154	-1,84072	5,529942	0,971472	cheopisJantan
202866	140,1355	1250,154	-1,84072	5,529942	0,971472	cheopisJantan
1246883	141,6782	1839,699	-1,76886	5,089671	0,98729	felis
1318363	140,0401	1798,689	-1,70241	4,890248	0,987382	felis
1247831	130,2746	3607,721	-1,37138	3,173609	0,991684	felis
1269948	119,4174	1645,819	-1,7336	4,625997	0,9864	felis

Figure 4.Extraction Features of Body Detection Result

Area	Eccentricity	Major Axis Length	Solidity	Species
1518148	0,901955	2188,232	0,891054	canis
1960614	0,851372	2221,705	0,934853	canis
1919153	0,899087	2402,205	0,967787	canis
1541135	0,895895	2143,707	0,91305	canis
1898124	0,798599	2050,701	0,97993	felis
1281327	0,814654	1701,386	0,972871	felis
1423043	0,882252	2009,135	0,98092	felis
1422393	0,880102	2000,904	0,976651	felis
1299191	0,823762	1760,48	0,959159	felis
1075591	0,826958	1590,97	0,94499	cheopis
1099677	0,862969	1699,071	0,966515	cheopis
2504943	0,789624	2325,786	0,973825	pulex

Figure 5. Extraction Features of Head Result

4.3. Classification Result

The classification method used is the KNN with the distance calculation of Euclidean distance. The accuracy results obtained by body detection and head detection are testing on the Weka application. In Table 4 and Table 5, shows the result of the classification with k = 1, 3, 5, 7, and with the number of features used.

Table 3. Result of classification body detection. From the body detection classification result, obtained high accuracy of 92.3077% by using value K = 5 with six features selected

Features K	20	15	6	2
1	92,3077%	84,6154%	92,3077%	84,6154%
3	76,9231%	69,9321%	92,3077%	84,6154%
5	61,5385%	61,5385%	92,3077%	53,8462%
7	38,4615%	46,1538%	30,7692%	46,1538%

Table 4. Result of classification head detection. From the body detection classification result, obtained high accuracy of 92.3077% by using value K = 5 with six features selected

Features K	12	7	4	1
1	66,67%	75%	66,67%	66,67%
3	50%	50%	50%	66,67%
5	33,33%	66,67%	75%	41,67%
7	50%	75%	58,33%	75%

4.4. Validation System

The validity test is done by comparing the classification result between the system and the expert is Dr. Novyan, who is a parasitology doctor of Faculty of Medicine UII. will compare with the Kappa coefficient. The comparison is evaluated using Cohen Kappa metrics as follows.

Doctor * System Crosstabulation

Count

		System		
		1,00	3,00	Total
Doctor	cheopis jantan	1	0	1
	felis	0	2	2
Total		1	2	3

Symmetric Measures

		Value	Asymptotic Standard Error[a]	Approximate T[b]	Approximate Significance
Measure of Agreement	Kappa	1,000	,000	1,732	,083
N of Valid Cases		3			

a. Not assuming the null hypothesis.
b. Using the asymptotic standard error assuming the null hypothesis.

Figure 6. Kappa with Value One Result

This section shows the result of the Kappa *value* is 1. The value indicates that $K = 1$, meaning the level of agreement between the system and the expert is very strong.

Doctor * System Crosstabulation

Count

		System			
		cheopis	felis	canis	Total
Doctor	cheopis	3	0	1	4
	felis	1	1	0	2
	canis	2	1	0	3
Total		6	2	1	9

Symmetric Measures

		Value	Asymptotic Standard Error[a]	Approximate T[b]	Approximate Significance
Measure of Agreement	Kappa	,100	,226	,461	,645
N of Valid Cases		9			

a. Not assuming the null hypothesis.
b. Using the asymptotic standard error assuming the null hypothesis.

Figure 7. Kappa with Value 0.100 Result

This section shows the result of the Kappa value is 0.100. The value indicates that the $K < 0.2$, meaning the level of agreement between the system and the expert is low.

4.5. System Usability

The test was carried out by spreading the questionnaire, and the respondents were 45 students of the Faculty of Medicine UII. The questionnaire statement used are as follows:

Table 5. Questionnaire

No	Statement	Total Score of Statement
1.	I think the Faculty of Medicine UII students will often use this system.	162
2.	I see this system is not so was complicated.	178
3.	I think the system is easy to use.	178
4.	In my opinion, the Faculty of Medicine UII students will need help to use this system.	163
5.	I found the function that exists in this system already has a good result.	160
6.	I think there are too many confusing things in this system.	120
7.	I consider that this system will be widely used by the Faculty of Medicine UII students immediately.	151
8.	I found this system is very cumbersome to use.	107
9.	I feel very confident in using this system.	143
10.	I need to learn many things before I can use this system.	146
	Result of SUS Score	**54,5**

Based on the results usability system testing of respondents, the system is not very complex and easy to use. However, the lowest result is that the respondent chooses the statement that the system is not very practical to use. It means that the system can use without having to ask first how it used. However, several respondents requested that the system instructed for use.

5. Conclusion

Based on the entire research process that has been done, can be concluding that:

- This research has managed to identify the morphological feature of image parasitic insects for parts and heads.
- Features used in the process of identification of the body morphology as many as six traits and textures. As for the identification of head morphology as many as four traits.
- The results of the usability test, the value gained is 54,5, still below average for good systems. Moreover, the Kappa test result value for body detection is $k = 1$ While the head detection is $k = 0.1$.
- The system can classify the body by detecting 97.0377% accuracy, while the detection of its accuracy head is 75%.

References

[1] D. M. Sukendra, "RESISTENSI PINJAL TIKUS (XENOPSYLLA CHEOPIS) TERHADAP INSEKTISIDA DALAM PENANGGULANGAN PENYAKIT PES," *Spirakel,* 2015.

[2] CDC, "Laboratory Identification of Parasitic Diseases of Public Health Concern," 2017. [Online]. Available: www.cdc.gov/dpdx/fleas/index.html.

[3] W. H. d. Z. B. Eka Ardhianto, "Implementasi Metode Image Substracting dan Metode Regionprops untuk Mendeteksi Jumlah Objek Berwarna RGB File Video," 2013.

[4] A. H. R. R. I. Dika Adi Khrisna, "Identifikasi Objek Berdasarkan Bentuk dan Ukuran," 2011.

[5] D. E. K. S. F. S. Rahadian Kurniawan, "Klasifikasi Sel Serviks Menggunakan Analisis Fitur Nuclei pada Citra Pap Smear," 2013.

[6] a. K. F.-L. George H. Rosenfield, "A Coefficient of Agreement as a Measure of Thematic Classification Accuracy".

[7] R. W. Bambang Pudjoatmodjo, "Tes Kegunaan (Usability Testing) Pada Aplikasi Kepegawaian Dengan Menggunakan System Usability Scale(Studi Kasus: Dinas Pertanian Kabupaten Bandung)," 2016.

[8] N. H. N. L. Izzati Muhimmah, "Characteristics Determination of Infected Erithrocytes by Plasmodium Falciparum as a Diagnostic of Malaria, Based on Microscopic Images," 2017.

Acknowledgment

This article is part of the Lecturer-Student Research Activity of the UII Informatics Engineering Program with a concentration in Medical Informatics.

IOP Conf. Series: Materials Science and Engineering **722** (2020) 012066 doi:10.1088/1757-899X/722/1/012066

Sustainable risk mitigation in manufacturing company

E Kusrini[1], S Handayani[2]

[1,2]Faculty of Industrial Technology, Department of Industrial Engineering, Universitas Islam Indonesia, Yogyakarta, Indonesia

Corresponding author: elisakusrini@uii.ac.id

Abstract. Companies that are not aware of any risk events and sources of risk that occur in the company, will not be able to formulate an appropriate strategy in mitigating risk. This research aims to identify, measure and plan risk mitigation strategies related to sustainable risk. Sustainable risk is related to risks from environmental, economic and social aspects. This research takes a case study in one of the manufacturing companies in Indonesia. This study uses the Failure Mode and Effect Analysis (FMEA) method to identify events and sources of risk and the House of Risk (HOR) to analyze and design risk mitigation. This research identified 14 risk events and 39 sources of risk that occurred in the company. By using the HOR method, 15 priority sustainability risk mitigation actions were obtained.

1. Introduction

Companies that are not aware of all forms of risks in the business world will not be able to formulate the appropriate strategy in competing with other competitors in the business world. Companies are required to be able to find ways to handle demands with high standards expected by customers and to ensure their position remains safe in the market. On the other hand, risks that arise can sometimes be predicted and mitigated [1]. If the company is not able to mitigate any risks arise, then certainly the company will be displayed from the dynamics of the business world because the future of the company depends on how good the ability and responsive of the company to encounter the changes in the external environment [2].

The traditional approach often chosen by companies caused more susceptible to risk than sustainable approaches. Evidently, global warming, the depletion of non-renewable resources, various dangerous diseases and others arise as a result of traditional approaches [3], [4]. In addition to government regulations, manufacturing companies are required to change the pattern of traditional systems to systems that are aware of environmental sustainability [5], [6]. To change the traditional system to a more modern system, manufacturing companies need to change the company's orientation, from low cost products to more sustainable ones [7], [8]. Smith & Ball [9] said that a sustainable society and a manufacturing company system cannot be with one another. Thus, the instability of manufacturing companies caused by the risks that occurred, is able to disrupt the social and social order system. The presence of risk management aims to provide a strategic solution for the company in mitigating any potential risks that can create losses for the company [10].

It is realized that the external uncertainties of the business world originated by political problems, economic, social and cultural instability and rapid changes in technological and information aspects are a central problem of every company. Risk management for some companies is still considered to be inconsistent in its application [11]. In addition, risk management related to sustainability has been highlighted lately. Understanding sustainable risk can help companies to increase supply chain availability [12], [13]. The issue of sustainability has become a serious concern among researchers, starting from the evaluation of sustainability risk in terms of economic aspects [14], [15], [16], social [14], conceptual framework for sustainable risk management [17]. Xu et al. [18] said that there is a lack of an integrated framework in assessing and evaluating sustainability risks by considering the triple bottom line (environmental, social and economic). The increasing concern for global sustainability in the industrial world requires stakeholders to immediately take the right policies. For this reason, a complete risk of sustainability is needed.

This study aims to propose an integrated framework in mitigating sustainable risk in one manufacturing company in Yogyakarta, namely PT. Yogya Presisi Tehnikatama Industri (YPTI). To fill in the deficiencies discovered by Xu et al. [18] in assessing and evaluating sustainable risk, the assessment aspects used in this study include environmental, social and economic aspects (triple bottom line). Torres-ruis and Ravindran [19] added that environmental aspects were still under-explored in terms of identifying risks to company activities. The company is engaged in Checking Equipment, Parts, Plastic Injection and Mold Making. Moreover, the results of the literature review and discussion with the company showed that there was no literature that has examined the company related to sustainable risk mitigation using the triple bottom line. Meanwhile, the results of preliminary observations at PT. YPTI showed that there were still work accidents that affected the environment in the company, inefficient use of energy and often a lack of functional air supply for the company. Occurrence of risks in the social aspect that was, excessive work time required by companies, employee wages provided were often not in accordance with the standards of UMR Yogyakarta and others. In addition, the occurrence of risk on economic aspects found in PT. YPTI, was the company's revenue target has not been achieved in recent years, the production costs exceed the company's target. Therefore, to maintain the existence of the company and ensure that the company is able to achieve competitive advantage in both national and international levels, it is necessary to conduct research related to sustainable risk management.

2. Research Method

The number of companies engaged in the same field makes the competition is very tight therefore it can result in decreased revenue due to reduced orders from customers. In addition, as the manufacturing companies that have a lot of potential risks, the company itself must carry out effective management in order to be able to achieve competitive advantage. The object of this research is one manufacturing company located in Yogyakarta, PT. Yogya Presisi Tehnikatama Industri (YPTI). The selection of research sites is due to the company, PT. YPTI, which was established on September 9, 1999, is one of the companies which is able to penetrate the international market in a relatively short period of time. Starting from attending overseas exhibitions to selling products that are able to penetrate the domestic market, in addition, in 2018, the YPTI company succeeded in becoming a supplier of spear parts for the needs of various companies in Indonesia, including: Astra Daihatsu Motor, Coca Cola Bottling, Sari Husada, Toyota Astra Motor, PLN, PT. Dirgantara Indonesia and others. Engaged in Checking Fixture, Spare Parts, Plastic Injection, and Mold Making it makes PT. YPTI consistently to improve company management.

2.1. Data collection

There are several stages of data collection in this study such as, (1) a review of previous studies consisting of various sources and references. In addition, direct observations were conducted at the research location, to review and assess all activities carried out throughout the company activities of PT. YPTI; (2) identified risk events and sources of risk events. The identification process used a set of open and closed questionnaires. In addition, interviews were implemented with operators in the production department, supervisors, division heads, and managers regarding events and sources of risk that occurred in the company; (3) after the process of identifying events and sources of risk, a focus group discussion (FGD) were held with company superiors related to the risk mitigation process. FGD aimed to obtain information through interaction between groups to obtain certain agreements [20], [21]. In addition, the

FGD was conducted to determine the ability and willingness of the company in implementing risk mitigation for any risks that occur.

2.2. *Data analysis*

Data that has been obtained through a series of data collection processes was processed and analyzed. There were several steps in data processing in this study, consist of: (1) the results of identification of events and sources of risk, an assessment was analyzed using by the Failure Mode and Effect Analysis (FMEA) methods. FMEA can be used to evaluate and measure risk factors systematically, without the need of complicated statistical methods [22]. In this section, the weighting of the probability of occurrence and severity was calculated based on the results of identification through questionnaires; (2) calculations were carried out using the house of risk (HOR) approach. This approach consists of two parts [23], which are the first phase or HOR phase 1 to calculate the Potential Aggregate Risk (ARP) to be able to determine priorities in the process of handling risk agents [24]. In addition, to determine the correlation between sources of risk and risk events that have been identified, and finally the determination of risk evaluation using Pareto diagrams. While the second phase or HOR phase 2 was designing mitigation strategies, determining the correlation of mitigation strategies with risk agents, calculating the total effectiveness and the results of the degree of difficulty assessment, calculating the effectiveness to difficulty ratio and determining the mitigation strategies to reduce the probability value of the emergence of a risk source based on the effectiveness value to the highest difficulty [25].

3. Result and Discussion

3.1. *Risk event and risk agent identification results*

Based on the results of the distribution of research questionnaires and the results of interviews that have been conducted, the risk events, severity values, sources of risk and the value of occurrence of each risk obtained at the manufacturing company PT. YPTI as shown in tables 1 and 2 as follows.

Table 1. Results of identification of risk events

Code	Risk Event	Severity
Environmental		
E1	Accidents that occur in the company and can affect the environment (for example, production machine operators or company staff)	4
E2	Inefficient use of energy for production and shipping activities	4
E3	Excessive product waste produced by the company	2
E4	Vulnerability caused by rising temperatures due to climate change	2
E5	Disturbances caused by natural disasters such as hurricanes, floods, storms, earthquakes or volcanic eruptions	5
E6	Lack of water supply for company activities	3
Social		
E7	Excessive work time; imbalance of work time with the lives of company employees	5
E8	Labour issues related to population growth	2
Economy		
E9	Antimonopoly claims (claims incurred against companies that violate inter-company competition laws, for example pricing that is detrimental to consumers)	1
E10	Patent infringement (commercial use of an inventor who has patented his work without the knowledge of the patent owner)	1
E11	Claim return of raw materials from suppliers	5
E12	Company profits do not match the target	5
E13	Unpredictable and sustainable variance in fuel prices	3
E14	Production costs exceed the company's target plan	4

Table 2. Result of identification of risk agent

Code	Risk Agent	Occurrence
Environmental		
A1	Lack of K3 Education and Training in the company	4
A2	Lack of awareness in using PPE	5
A3	K3 SOP is inadequate and not strict in application	4
A4	Non-compliance in implementing SOP	5
A5	Lack of education and training in the use of energy to employees	4
A6	Production machines and transportation equipment that is not environmentally friendly	3
A7	Inefficient production machinery	2
A8	the absence of special techniques for recycling factory waste	3
A9	The company's buildings and facilities do not use materials that are resistant to temperature changes	2
A10	Lack of plants / trees around the company	3
A11	The occurrence of natural disasters that are not detected	2
A12	Buildings and facilities are not specifically designed to withstand natural disasters	2
A13	drought in the surrounding environment	3
A14	Prolonged dry season	2
A15	There was a jam in the water transport car	3
Social		
A16	Customer demand suddenly	5
A17	Changes in production schedules to meet target customers	5
A18	There was forced labour hours to meet company targets	4
A19	Population growth increased dramatically	2
A20	The number of productive ages exceeds the capacity needs of the company	3
Economy		
A21	Monopolistic market shares holistically	2
A22	Unfair market competition	3
A23	Weak law enforcement	2
A24	Lack of accountability	2
A25	Legitimacy is legally unclear / weak	3
A26	Weak Government intervention in supporting patent rights	2
A27	The quality of raw materials is not up to standard	5
A28	Miscommunication occurs with suppliers	4
A29	Raw materials are mixed with foreign materials	4
A30	Marketing strategy is not on target	4
A31	Product quality is not up to standard	5
A32	A strike occurred by an employee	4
A33	The company's internal management is not up to standard	4
A34	The instability of world fuel prices	4
A35	Dependence on imported fuel	4
A36	Geopolitical tensions occur between countries, especially oil-producing countries	5
A37	An increase in inflation	3
A38	Raw material costs are increasing	4
A39	The production machine is constantly damaged	4

Based on the results shown in table 1, there were 14 risk events that occurred at PT. YPTI. Of the 14 risk events, there are 6 risk events in the environmental aspect category, 2 risk events in the social aspect category and 6 risk events in the economic aspect. From each of these risk events, the highest severity value is found in E5 risk events, namely disruptions caused by natural disasters such as hurricanes, floods, storms, earthquakes, or volcanic eruptions, E7 excessive work time (such as work time imbalances with the life of a company employee). In addition, from economic factors, there are 2 highest severity values, including E11 (claim return of raw materials from suppliers) and E12 (company profits are not in line with the target). In addition, Table 2 shows the results of the identification of risk sources, amounting to 39 sources of risk with occurrence value. The highest occurrence value in environmental aspects is A2 and A4, social aspects are A16 and A17, and economic aspects are A27, A31 and A36. The next step processed with the House of Risk (HOR) approach to design mitigation actions optimally. Thus, any risks that interfere with the stability of the company can be minimized or even eliminated to achieve competitive advantage.

3.2. House of Risk (HOR)

In this section, data management is carried out using the house of risk (HOR) approach, where this approach is divided into two parts, namely in the first stage, data management is carried out with house of risk phase 1 with the aim of calculating the potential aggregate risk (ARP) in order to be able to determine priorities in the process of handling risk agents. Whereas HOR phase 2 aims to design a risk mitigation action against each prioritized risk.

3.2.1. House of Risk (HOR) phase 1

Based on the ARP value that has been obtained through the calculation of HOR phase 1, it is determined that the priority factor is to use Pareto. According to Caesaron & Tandianto [26] the source of making Pareto diagrams is based on the cumulative percentage of ARP value of risk sources. The results of the Pareto diagram can be shown in Figure 1.

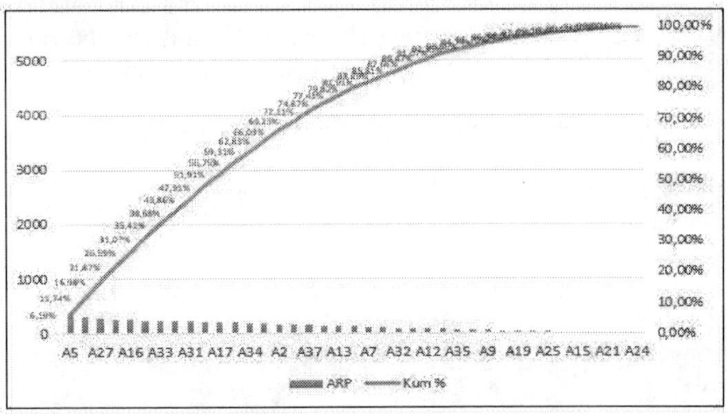

Figure 1. Diagram Pareto

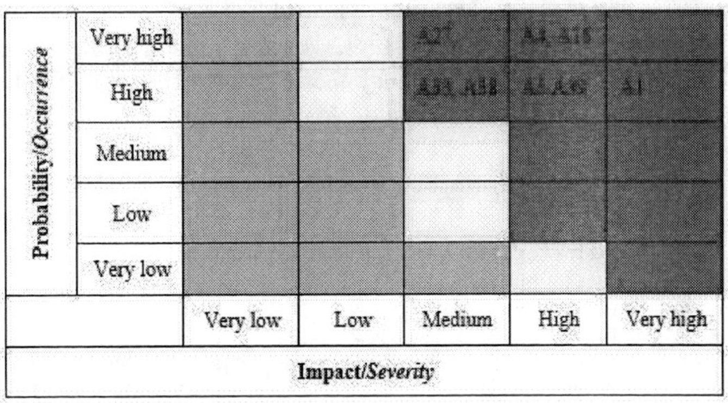

Figure 2. Risk map before mitigation priorities

Risk evaluation uses the Pareto diagram as shown in Figure 1, with the principle of 80:20 which means 20% of risk agents were a priority for risk mitigation. Based on the results of risk agent priorities as shown in figure 1 showed as much as 39.68% where with this percentage were carried out the design of mitigation strategies that were able to influence improvement by 60.32% other risk agents. Another consideration chosen by the company was because the company wanted to be more focused in carrying out risk mitigation actions. After that, a risk map was made to determine the risk conditions before being mitigated as shown in Figure 2.

3.2.2. House of Risk (HOR) phase 2

From the ARP calculation results in HOR phase 1, a number of risks was mitigated. The highest ARP value was a source of risk that must be immediately mitigated. The priority sources of risk for mitigation with the proposed mitigation strategies can be seen in table 3 as follows.

Table 3. Risk agent and mitigation strategy

No	Risk Agent	Mitigation action	Code
1	Lack of education and training in the use of energy to employees	• Carry out education and training on the use of environmentally friendly energy	PA1
		• Collaborate with consultants on energy use	PA2
2	Unconscious employee about K3 in the company	• Conduct OHS training and training regularly	PA3
		• Giving punishment	PA4
3	The quality of raw materials is not up to standard	• Termination of employment contracts and looking for more selective suppliers	PA5
		• Determination of standard raw materials from suppliers	PA6
4	Non-compliance in implementing SOP in PT YPTI	• Provision of punishment for those who violate	PA7
		• Conduct regular audits at the company	PA8
5	Customer demand suddenly	• Reinforce the work agreement	PA9
6	The production machine is constantly damaged	• Production machine renewal	PA10
		• Preventive maintenance	PA11
7	The company's internal management is not up to standard	• Renewal of company internal SOPs	PA12
		• Improvement of the company's internal management system	PA13
8	The cost of raw materials has increased	• Hedging with suppliers	PA14
		• Piling up raw materials	PA15

Correlation of mitigation strategies with risk agents was implemented to determine the extent of handling strategies correlated with risk agents. The correlation was conducted by the expert PT YPTI through focus group discussions. The measurement of correlation of risk management strategies with risk agents used 4 scales, namely 0, 1, 3, 9, which showed that sequentially there was no correlation, weak correlation, moderate correlation and high correlation. In addition, the total effectiveness and assessment of degree of difficulty were calculated. Table 4and 5 shows the results of HOR phase 2.

Table 4. Table HOR phase 2

Risk Agent	Risk Mitigation Strategy															ARP
	PA1	PA2	PA3	PA4	PA5	PA6	PA7	PA8	PA9	PA10	PA11	PA12	PA13	PA14	PA15	
A5	9	9	9	0	0	0	0	3	0	0	0	1	3	0	0	348
A1	1	3	9	9	0	0	9	3	1	0	0	0	1	0	0	312
A27	0	0	0	1	9	9	1	1	3	0	0	0	0	0	3	295
A4	1	3	3	3	0	0	9	9	1	0	0	3	3	0	0	275
A16	0	0	0	1	0	0	0	0	9	0	0	1	1	0	0	265
A39	0	0	0	0	0	0	0	1	0	9	9	0	0	0	0	252

Risk Agent	Risk Mitigation Strategy															ARP
	PA1	**PA2**	**PA3**	**PA4**	**PA5**	**PA6**	**PA7**	**PA8**	**PA9**	**PA10**	**PA11**	**PA12**	**PA13**	**PA14**	**PA15**	
A33	0	3	3	0	0	0	0	1	0	0	0	9	9	0	0	244
A38	0	0	0	0	3	1	0	0	1	0	0	0	0	9	9	240
TEk	3719	5625	7497	4193	3375	2895	5578	5246	4097	2268	2268	3634	4642	2160	3045	
Dk	1	2	2	2	3	3	1	2	2	5	2	3	2	2	3	
ETD	3719	2813	3749	2097	1125	965	5578	2623	2049	454	1134	1211	2321	1080	1015	
Rank	3	4	2	7	11	14	1	5	8	15	10	9	6	12	13	

Based on the calculation of house of risk phase 2, the mitigation action management strategy was obtained based on the highest ETD ranking. After that, the following was a picture of the risk map after the priority of mitigation has been applied and the sequence of priority risk management strategies.

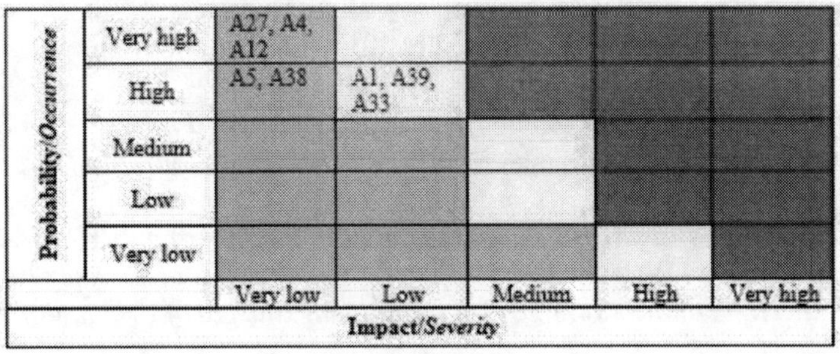

Figure 3. Risk map before mitigation priorities

Table 5. Table HOR phase 2

Code	Risk Mitigation
PA7	Provision of punishment for those who violate
PA3	Conduct OHS training and training regularly
PA1	Carry out education and training on the use of environmentally friendly energy
PA2	Collaborate with consultants on energy use and K3
PA5	Termination of employment contracts and looking for more selective suppliers
PA13	Improvement of the company's internal management system
PA4	Giving punishment
PA9	Reinforce the work agreement
PA12	Renewal of company internal SOPs
PA11	Preventive maintenance
PA6	Determination of standard raw materials from suppliers
PA14	Hedging with suppliers
PA15	Piling up raw materials
PA6	Determination of standard raw materials from suppliers
PA10	Production machine renewal

4. Conclusions

Based on the results of data collection and analysis conducted in this study, it was concluded that PT. YPTI Yogyakarta, encountered 14 risk events with severity values for each risk event and 39 causes for risk events (risk agents) with occurrence values related to the company's sustainability risk. In addition, mitigation action strategies for the causes of risk events related to sustainability should be prioritized in handling through HOR phase 2, namely (A4) (ARP= 275) non-compliance in implementing company SOPs with mitigation strategies, providing punishment for violating (PA7) and conducting routine audits in companies (PA8); (A5) (ARP=348) lack of education and training in energy use for employees, with mitigation action strategies, the extensive education and training on the use of environmentally friendly energy (PA1) and work with consultants in the matter of efficient energy use (PA2); (A1) (ARP= 312) Unconsciousness of the importance of OHS education and training in companies with mitigation action strategies. Conduct OHS training and training on a regular basis (PA3) and punishment (PA4); (A33) (ARP= 244) the company's internal management did not meet the standards with the mitigation strategy for internal SOP renewal (PA12) and improvement of the company's internal management system (PA13); and finally, (A38) (ARP= 240) the cost of raw materials increased with its mitigation action strategy of hedging with suppliers (PA14) and piling up raw materials in company warehouses (PA15).

For further researchers, it is recommended to add other factors from the aspect of sustainability for future studies related to the sustainability risks of manufacturing companies, as well as broaden the scope of risk identification that involves the entire manufacturing supply chain activities.

References

[1] Verbano C and Venturini K 2011 *Journal of Risk Research.* **14** 519-550.
[2] Soosay C, Nunes B, Bennett D J, Sohal A, Jabar J and Winroth M 2016 *Journal of Manufacturing Technology Management.* **1** 6-37.
[3] Ghandehariun A, Nazzal Y and Kishawy H 2016 *International Journal of Global Warming.* **9** 198-228.
[4] Cassettari L, Bendato I, Mosca M and Mosca R 2017 *Applied Energy.* **190** 841-851.
[5] Chan F, Li N, Chung S H and Saadat M 2017 *International Journal of Production Research.* **55** 1148-1163.
[6] Davies A 2012 *International Journal of Sustainable Engineering.* **5** 135-144.
[7] Esfahbodi A, Zhang Y, Watson G and Zhang T 2017 *Journal of Cleaner Production.* **155** 66-78.
[8] Gimenez C, Sierra V and Rodon J 2012 *International Journal of Production Economics.* **140** 149-159.
[9] Smith L and Ball P 2012 *International Journal of Production Economics.* **140** 227-238.
[10] Bajo J, Borrajo M L, De-Paz J F, Corchado J M and Pellicer M A 2012 *Expert Systems with Applications.* **39** 6921-6931.
[11] Faisal M N, Banwet D K and Shankar R 2007 *International Journal of Management and Enterprise Development.* **4** 588-607.
[12] Waters D 2011 *Kogan Page Publishers.*
[13] Xu M, Allenby B R and Crittenden J C 2011 *Advences in Complex System.* **14** 649-672.
[14] Altay N and Ramirez A 2010 *Journal of Supply Chain Management.* **46** 59-80.
[15] Bierkandt R, Wenz L, Willner S N, and Levermann A 2014 *Environment Systems and Decisions*, vol. **34** 507-524.
[16] Okuyama Y and Santos J R 2014 *Economic Systems Research.* **26** 1-12.
[17] Scholten K and Fynes B 2017 *In Sustainable Supply Chains.* **4** 413-436.
[18] Xu M , Cui Y, Hu M, Xu X, Zhang Z, Liang S and Qu S 2019 *Journal of Cleaner Production.* **225** 857-867.
[19] Rorres-Ruiz A and Ravindran A R 2018 *Journal Clean Production.* **172** 4478-4493.
[20] Lehoux P, Poland B and Daudelin G 2006 *Social Science and Medicine.* **63** 2091-2104.
[21] Hollander J A 2004 *Journal of Contemporary Ethnography.* **33** 602-637.
[22] Stamatis D H 2003 *ASQ Quality Press.*
[23] Immawan T and Putri D K 2018 *MATEC Web Conferences.* **154** 1-4.
[24] Pujawan I N and Geraldin L H 2009 *Business Process Management Journal.* **15** 953-967.
[25] Ulfah M, Trenggonowati D L and Yasmin F Z 2018 *MATEC Web Conferences.* **218** 1-7.
[26] Caesaron D and Tandianto 2015 *Jurnal PASTI*, vol. **9** 248-256.

IOP Conf. Series: Materials Science and Engineering **722** (2020) 012067 doi:10.1088/1757-899X/722/1/012067

Solution-Based Learning Implementation in Design Course

Risdiyono

Mechanical Engineering Department, Islamic University of Indonesia, Yogyakarta, Indonesia

Corresponding author: risdiyono@uii.ac.id, https://orcid.org/0000-0002-1395-6023

Abstract. Student participation is a key success in almost every learning process. It is believed that concept-level of understanding could be achieved only if students are actively involved in the learning activities. The concept of student centered and active learning is usually adopted in order to improve student participation. Problem Based Learning is one of popular methods used in this paradigm, in which students learn both thinking strategies and domain knowledge of a subject through the experience of problem solving. In addition to this current well-known learning method, this paper introduces a new approach so called Solution Based Learning in which students learn about a subject by analyzing solutions that has been given by some experts. Basically, the concept urges students to analyze the way of thinking of the experts on why they could give the solution like this and that. Students then may adopt the thought or criticize it in a positive manner. In order to see its viability and applicability, this new concept has been implemented in Product Design and Development Course as a case study. Based on the process performance analysis, it is shown that the proposed approach is able to improve student participation, to make class more active and to improve the overall student understanding of the subject of study

1. Introduction

Product Design and Development (PDD) Course plays a very important role in enhancing student competency in design and manufacturing. The basic competencies every student should fulfill after completing the course include a comprehensive understanding of industrial product design and a wide-range ability to bring the design theory into real implementation. Hence, the success in delivering the course will give a significant contribution to the student's standard competences.

Good understanding of PDD concept is a foundation for students to do a final project, as it is a combination of many important preceding subjects (figure 1). Therefore, the ability to identify market opportunity and customer need, to define product specification, to generate product concept, to evaluate and select optimum concept, to create prototype and to do final product evaluation are some skills students should have in doing a project, especially that relates to PDD topics.

One of important indicators in evaluating learning process is the level of students' active participation in the class and their average final grade. From previous experience, it is observed that Teacher-Centered Learning (TCL) could not give optimum result. Student participation is very less due to monotonous environment as teaching method is relied only upon the slides in a lecture-based environment.

Fundamental principles that significant to teaching and learning include: (1) knowledge is constructed from the experience of the learner; (2) knowledge resides in the mind rather than externally; (3) learning is a personal interpretation of the world in that the learner's beliefs and values are used in interpreting objects and events; (4) learning is an active process of making meaning from experience; (5) learning takes place in contexts relevant to the learner; (6) reflection is an essential part of learning; and (7) learning is a collaborative process in which multiple perspectives are considered[1].

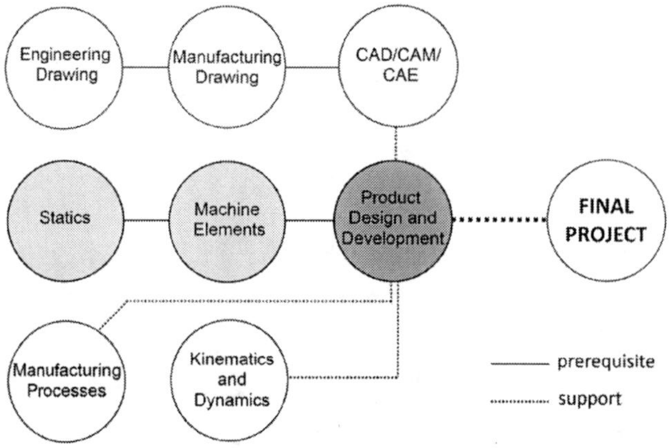

Figure 1. Relationship between Product Design Course and other subjects

It is reasonable to say that there is no best method that suitable for all learning situations. A proven good teaching method for a subject is not always good when applied for another subject. Hence, teaching method should be build based on the characteristics of the course, students' motivation and facilities availability. The introduction of new approach of Student Centered Learning (SCL) gives different paradigm and diverse point of view in order to optimize its goal achievement. Combination of many methods in teaching should be a good alternative to a single method.

2. Student Centered Learning

The term Student Centered Learning (SCL) is widely used in literature. Many terms have been linked with SCL and therefore the term can mean different things to different people. The notion that unifies them together is that the student is the central entity who must be actively engaged in learning things. Rogers [2] describes SCL as the shift in power from the expert teacher to the student learner, driven by a need for a change in the traditional environment. The power shift paradigm from teacher to the students has emphasized focus on learning rather than on teaching [3,4].

The use of Student Centered Learning concept appears to be thoughtful of today's society. Lea et al. [5] reviewed several studies and found that overall it was an effective approach. In this approach, students felt that there was more respect for the students, it was more interesting and it boosted their confidence. Correspondingly, Hall and Saunders [6] found that students had increased participation, motivation and grades in an Information Technology course where 94% of the students would recommend it to others over the more conventional approach.

The implementation SCL is believed to provide numerous benefits in the classroom including a greater depth of understanding of concepts, broader knowledge base, improved communication and interpersonal/social skills, enhanced leadership skills, increased creativity, and improved writing skills. There are many types of SCL, including project-based learning and problem-based learning.

Project Based Learning is a comprehensive viewpoint focused on teaching by involving students in investigation. Students pursue solutions to nontrivial problems by asking and refining questions, debating ideas, making predictions, designing plans and/or experiments, collecting and analyzing data, drawing conclusions, communicating their ideas and findings to others, asking new questions, and creating artifacts [7]. Teacher as facilitator gives a "driving question" to students working as a team to respond to, and then direct them to create an artifact (or artifacts) to present their gained knowledge.

IOP Conf. Series: Materials Science and Engineering **722** (2020) 012067 doi:10.1088/1757-899X/722/1/012067

Artifacts may include a variety of media such as writings, art, drawings, three-dimensional representations, videos, photography, or technology-based presentations.

Problem-based learning assigns students learn about a subject through the experience of problem solving. Students learn both domain knowledge and thinking strategies that can help the students develop flexible knowledge, effective problem solving skills, self-directed learning, effective collaboration skills and intrinsic motivation [8].

Students, working in groups, identify what they already know, what they need to know, and how and where to access new information that may lead to resolution of the problem. Teacher, known as the instructor or tutor plays a significant role to facilitate learning by supporting, guiding, and monitoring the learning process [9]. The tutor needs to build students' confidence, encourage them, while also stretch their understanding.

3. Solution Based Learning: A New Approach

Problem-based learning has been implemented widely in many disciplines, mainly in medicine, pharmacy, biology, social, economy, law, as well as engineering. In engineering, however, project-based learning is more popular due to its unique characteristic (i.e. the creation of artifacts at the final phase).

Based on the characteristic of PDD course, project-based learning is considered as a good candidate to be adopted in learning process. However, in order to improve its effectiveness, some modifications might be needed. Presented in this paper is a research on project-based learning modification so-called "solution-based learning" as a contrary to "problem based learning". While problem-based learning aims on student activeness through dealing with a "problem" to be solved, solution-based learning provides students a "solution" to be analyzed.

Solution-based learning (SBL) is a new term coined for learning method that encourage students to learn from proven solutions given by the experts (learning from experts). SBL has inherent capacity towards increased emphasis on preset values and measures, clear outcome indicators, and ability to generate meaningful feedback [10]. Therefore, when the outcome of problem-based learning is a proposed solution from student discussion to a problem defined by facilitator, the outcome of solution-based learning is students' comprehensive understanding about "how and why the solution is like this". Students have also opportunity to examine pros and cons of the solution (that might be outdated already) that can be starting point to make improvements. This method enables students to know and see how the experts found solution of the problems under investigation, and then adopt them to solve similar cases. The example of implementation of the proposed method is shown in figure 2.

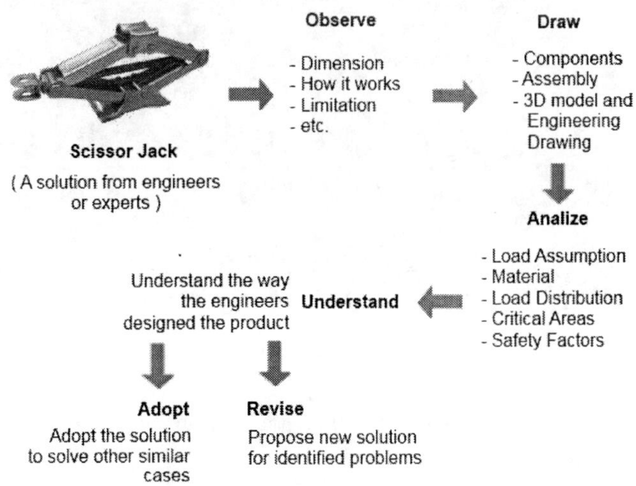

Figure 2. Example of solution-based learning study path

IOP Conf. Series: Materials Science and Engineering **722** (2020) 012067 doi:10.1088/1757-899X/722/1/012067

In order to implement SBL, supporting media should be provided while combination of lecture, group discussion, question-answer, and student presentation as well as information technology practice should be used. The media used in this research includes mechanical products, industrial products and art-jewelry products. The examples of real products as learning media can be found in figure 3. A total of 12 products are available to be chosen by students that have been randomly grouped.

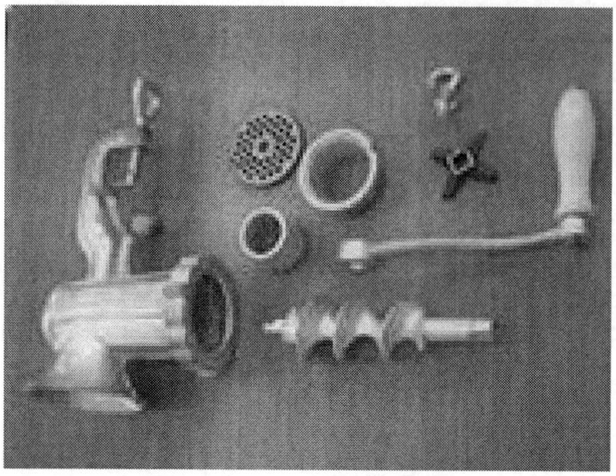

Figure 3. Example of a real product as learning media

4. Implementation and Discussion

The new approach of solution-based learning has been implemented in Product Design and Development course with 60 students registered. The students were grouped into 10 groups with 6 students in each group. Every group took one product as object of investigation. Presentations of all groups were made in order for students share their thought and improve their presentation skill. In addition, discussion and question-answer session were employed so that multi-direction interaction could be made. Learning atmosphere was very joyful and student participation was very high. The classroom scenery can be depicted in figure 4.

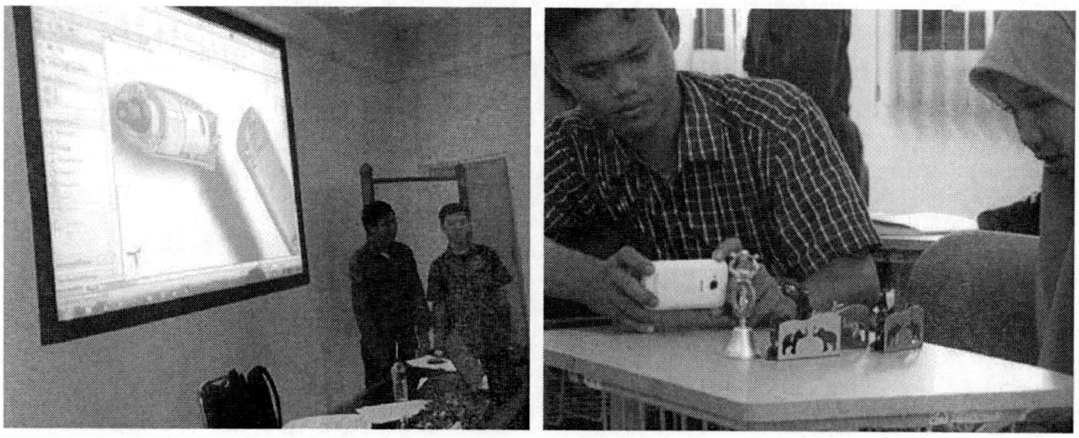

Figure 4. The classroom scenery

In order to evaluate students competency achievement and learning indicator accomplishment, many aspects have been considered including quiz, question-answer, students presentation, mid exam, final exam and group project presentation. Performance indicators of PDD course by using solution-based learning can be found in table 1, while examples of student's work is shown in figure 5.

Student perception on SBL method has been evaluated by using a questionnaire. Two other courses delivered in conventional method were used as comparators in order to see the objectivity. Twenty-two items were questioned by using rating scale (1-4). In general, it is shown that SBL method perceived better than traditional method. This result is in line with the finding of Chandra [11] and Rowe [12]. Outstanding improvements include (figure 6):

- Learning environment is more joyful
- The assignment is very helpful
- New knowledge is engaged more
- Feedback for assignment is managed better
- Classroom become more alive

Table 1. Performance Indicators of PDD Course

No	Indicator	Base Line	Target	Achievement
1	Attendance	79%	85%	95%
2	Grade > C	46%	75%	88.3%
3	Satisfaction level	-	3.5/4.0	3.65/4.0

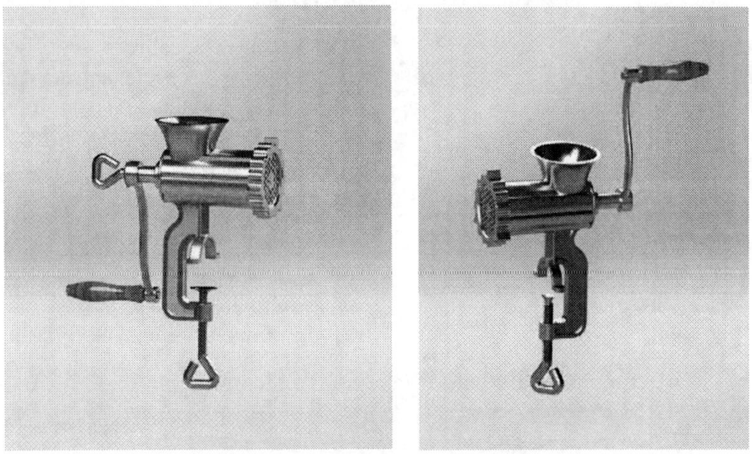

Figure 5. Example of student's work

In addition to questionnaire-based analysis, Focus Group Discussion (FGD) was also conducted, incorporating 10 representative students from all groups. Some advantages of SBL identified by students include improvements in:

- Computer skills related to CAD system and software
- Teamwork skills
- Time management
- Self-evaluation and objectivity

It is important to note that many students were identified uploading their project to social media like Facebook and Instagram, and used the media to show their project to public. This indicates that they were happy with the project.

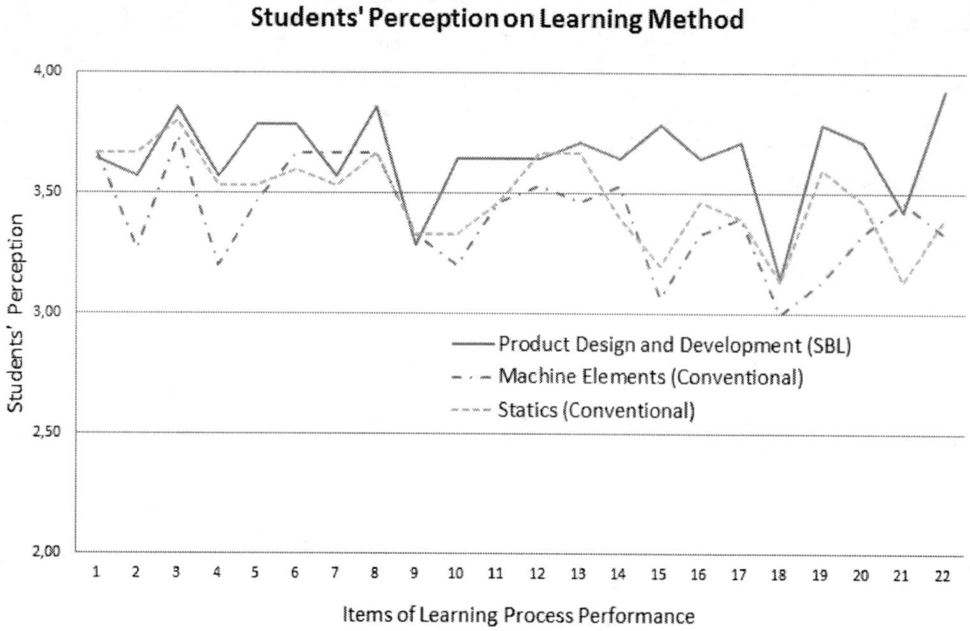

1. Lecturer readiness
2. Class regularity
3. Examination fairness
4. The use of IT as media
5. Ability to liven up the class
6. Clearness of explanation
7. Time for question-answer
8. Clearness in addressing question
9. Grading system comprehensiveness
10. Feedback for assignment
11. Content conformity
12. Treatment equality
13. Ability to give relevant example
14. Ability to explain relevant topics
15. New knowledge engagement
16. Learning media suitability
17. Assignment helpfulness
18. Outside the class conversation
19. Having learning-contract
20. Consistency
21. Ability to motivate
22. Learning environment joyfulness

Figure 6. Students' perception on learning method

5. Conclusion

Solution-based learning, a new approach of Student Centered Learning has been introduced. The concept places students in the center of activity in analyzing a proven solution provided by experts. The outcome of solution-based learning is students' comprehensive understanding about how and why the experts gave that solution of the problem. Opportunity to examine pros and cons of the solutions can be used as starting point to adopt them to solve similar cases or to make new improvements. The concept of solution-based learning has been implemented in Product Design and Development course and it is found that in general, the proposed approach is able to improve student participation, to make class more active and to improve the overall student understanding of the subject.

References

[1] Semple A 2000 Learning Theories and Their Influence on The Development and Use of Educational Technologies *Australian Science Teachers Journal* **46** 3.

[2] Lea S J, Stephenson D, and Troy J 2003 Higher Education Students' Attitudes to Student Centred Learning: Beyond 'educational bulimia' *Studies in Higher Education* **28** 3

[3] Hall J and Saunders P 1997 *Adopting a student-centred approach to management of learning* in Bell C, Bowden M, and Trott A (Eds.) *Implementing Flexible Learning* (London: Kogan Page)

IOP Conf. Series: Materials Science and Engineering **722** (2020) 012067 doi:10.1088/1757-899X/722/1/012067

[4] Barr R B and Tagg J 1995 From Teaching to learning - A New Paradigm for Undergraduate Education *Change* **13** 25.
[5] Rogers C R 1983 *As a teacher, can I be myself?* in *Freedom to Learn for the 80's* (Ohio: Charles E. Merrill Publishing Company)
[6] Blumenfeld et al 1991 Motivating Project-Based Learning: Sustaining the Doing, Supporting the Learning *Educational Psychologist* **26** 3
[7] Silver H and Cindy E 2004 Problem-Based Learning: What and How Do Students Learn? *Educational Psychology Review* **16** 3
[8] Schmidt, Henk G, Rotgans, Jerome I, Yew, Elaine H J 2011 The process of problem-based learning: What works and why *Medical Education* **45** 8
[9] Edwards R 2001 *Meeting individual learner needs: power, subject, subjection* in Paechter C, Preedy M, Scott D, and Soler J (Eds.) *Knowledge, Power and Learning* (London: SAGE)
[10] Khattak A 2018 Solution- Based Learning: Educating for Solutions *Khyber Medical University Journal* **9** 4
[11] Chandra V J 2016 Solution-based Learning(SBL): Using Systems Engineering Principles to Guide Capstone Projects in Technology *Proceeding of ASEE Annual Conference & Exposition* 2016
[12] Rowe R 2016 *Solutions Based Learning in Education*. Available online at: https://www.ole.org/2016/02/solutions-based-learning/

Implementation of Market Basket Analysis based on Overall Variability of Association Rule (OCVR) on Product Marketing Strategy

Alfiqra[1], A U Khasanah[1]

[1] Department of Industrial Engineering, Faculty of Industrial Technology, Universitas Islam Indonesia, Jalan Kaliurang km. 14,5 Yogyakarta 55584, Indonesia

E-mail: 14522289@students.uii.ac.id, annisa.uswatun@uii.ac.id

Abstract. Marketing strategy is an important thing that must be developed by retail. A method that can be used to develop a marketing strategy based on customer buying pattern is Association Rule (AR). AR is the process of finding association relationships between products that occur in one transaction. The application of AR to analyze custumer buying patterns is referred to as Market Basket Analysis (MBA). Rule obtained from ARMBA is sometimes not enough to provide an analysis when the variability of costumer buying pattern is high. Overall Variability of Association Rule (OCVR) is an indicator that focuses on analyzing market basket which assumes high variability in custumer behavior in buying products. This study used custumer transaction data of a retail in Yogyakarta. The data consisted of 57784 transactions in a month involving 41248 items. This study produced rules for each period (weeks), then the rules were used for further analysis using OCVR. 59 rules produced on the 1st period, 48 rules on the 2nd period, 54 rules on the 3rd period, and 58 rule on the 4th period. From the rules obtained there were 17 rules have OCVR value smaller than 30%, thus these rules can be used to make marketing strategies. Product bundling and shelves product arrangement based on obtained rules were proposed as marketing strategies to promote product sales.

1. Introduction

Retail businesses in Yogyakarta are growing rapidly. Based on data from the Yogyakarta Central Bureau of Statistics, the growth rate of Gross Regional Domestic Product (GRDP) in the third quarter of 2017 for the business sector in accommodation, food and beverage providers including retail is 62% [1]. These business sectors are the three largest business sectors that influence economic growth in Yogyakarta. This increasing number forced businesses to find suitable marketing strategies to promote sales and survive in the competition. One marketing strategy that can attract customers is sales promotion. Sales promotion has a big influence on the desire of customers to buy product suddenly without any plan or usually called as impulse buying [2]. Sales promotions are usually simply carried out to introduce a new product or to promote less salable products.

Another way that can be done to design an attractive sales promotion is by analyzing consumer buying patterns. Strategy for decision making and understanding consumer spending behavior is a challenge for an organization in maintaining its position in market competition [3]. The objective of this analysis is to find out what kind of products that usually bought by customers in one transaction time and it is usually called as Market Basket Analysis (MBA). The purpose of market basket analysis is to get a selling strategy with up-selling and cross-selling [4]. One method that commonly used to conduct this kind of analysis is Association Rule (AR). AR is a very popular data mining techniques to find important rule of association relationship between product in transaction data. AR was firstly proposed for marketing, but now this method has widely used in other fields, such as bioinformatics, nuclear science, pharmacoepidemiology, and geophysics [5]. In MBA, AR can be expressed as "A costumer who buys product X1 and X2 will also buy product Y with probability c%" [6]. To implement AR for MBA, transaction data in a certain retail or supermarket is needed and sometimes it involves very big transaction data.

Supermarket X is a popular retail business in Yogyakarta, which already has several branches. This supermarket provides daily needs, fashion, stationery, toys, cosmetics, etc. The supermarket has already made some promotion strategies to attract customers, but the management still did not use a specific method to design the promotion strategy. Supermarket X have already implemented information system to manage their business process, including to store their transaction data. Unfortunately, the data has not been optimally analyzed and utilized. In this case, AR is very useful to be implemented in analyzing the transaction data and help the management in designing promotion strategy which is suitable for the customer buying pattern.

One branch of supermarket X have more than 1000 transaction in a day and almost 60000 transactions in a month, Figure 1 represents the transaction data of Supermarket X in March (only for one branch). As shown in Figure 1, the supermarket has fluctuated transaction number. With this kind of condition, sometimes it is difficult to find out the pattern and to make suitable promotion strategy. In this study, the customer buying pattern of the supermarket will be revealed using Overall Variability of Association Rule (OCVR).

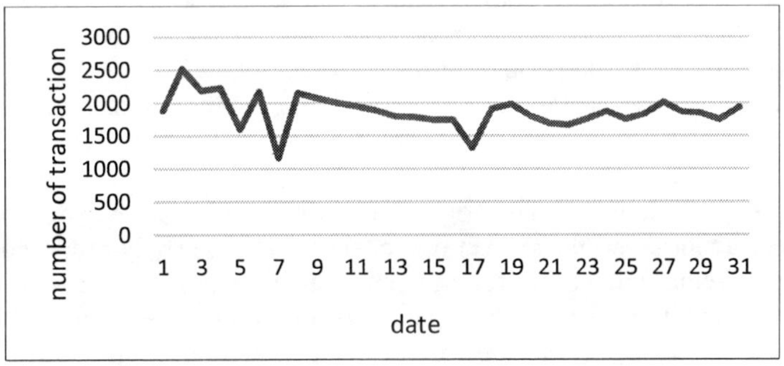

Figure 1. Number of transactions in March

OCVR is a new indicator in the AR-MBA application that was first proposed by Papavasileiou and Tsadiras in 2011. This indicator is applied in an MBA with the assumption that customers have high variability of shopping habit. Indicator of variability is related to the changes in customer buying habits in a certain period. This analysis can increase the efficiency of the rules generated from AR to create marketing strategies to promote sales [7], the main purpose of this study is to find the association rules of the product based on OCVR and to create marketing strategy which is suitable with the obtained rule.

2. Literature Review

2.1. Association Rule Mining

Association rule is one of the popular data mining techniques that can be used to find associative relationship from a group of data set. Association rules are formed by analyzing frequent patterns and by using parameters of support and confidence. Tan et al. [8] explain that support determines how often a rule applies to a given data set. A low support rule is uninteresting from a business perspective and sometimes it will be eliminated. While confidence determines how frequently items in Y appear in transactions that contain X, the higher the confidence the more likely it is for Y to present in transactions that contain X.

$$Support, s(X \Rightarrow Y) = \frac{\sigma(X \cup Y)}{N} \quad (1)$$

$$Confidence, c(X \Rightarrow Y) = \frac{\sigma(X \cup Y)}{\sigma(X)} \quad (2)$$

One parameter that also used to determine important rule in AR is Lift Ratio. Lift Ratio is a parameter that is used to see whether the rule gained from AR are strong or not. Lift Ratio measures the possibility of X and Y occurring together divided by the possibility of X and Y occurring if they are independent events. The formula of ratio is as followed.

$$Lift\ Ratio = \frac{Support\ (X \cap Y)}{Support\ (X)\ .\ Support\ (Y)} \quad (3)$$

In data mining, association rules are useful for analyzing and predicting customer behavior and play an important role in the analysis of shopping basket data, product clustering, catalog design, and store plan layout. Not only on product analysis, Association Rule Mining also used for detect operational problems in heating, ventilation and air conditioning (HVAC) systems of buildings [9]. Association rule also used in strategic management fields to determinate of effective management strategies [10]. Several algorithms are applied to the association rule technique. Based on the research conducted by Dhanalakahmi and Porkodi [11], several algorithms that have been used by researchers include Apriori, Eclat, and FP-growth Algorithms. Sandhu et al. [12] found a new method in developing the association rule algorithm based on the assumption of profit and quantity.

The apriori algorithm was used in this study to produce a set of association rules from the database. Apriori algorithm is an algorithm that is useful for finding frequent itemset for the Boolean association rule. The name of the apriori algorithm is based on the fact that the algorithm uses prior knowledge of frequent itemset mining search [13]. The original Apriori algorithm is dependent on the number of items, the number of transactions, and the market basket size [14]. The study proposed an efficient approach based on weighting factors and utilities for effective data mining from association rules with high utility. Kaur and Kang [15] in his study explained that some algorithms in existing AR only worked on static data and cannot capture real-time changes on data. The algorithm proposed in this study did not only work on static data but also captured changes in dataset. Chen et al. [16] applied MBA in multiple-store environment. The purposed method was proved to be more computationally efficient compared with the traditional method when implemented in supermarket which have more than one store, varying in sizes, over time product changes and when using longer period dataset.

One of the goals of MBA is to determine and prediction customer's behaviour based on expenditure patters from previous clients [17]. MBA is not only applied in a supermarket but also in other fields of study. There are to campare with other methods to propose a product network analysis, a network-based analysis to analyze a network leveled relation between products [18]. MBA also used for product inventory prediction and combine with other method, there is Aritificial Neural Netwrok (ANN) Back propagation [19]. MBA could apply in hospital field. Market Basket Analysis could increase revenue by enabling hotels to determine the most attractive additional products and services (beyond the room type) to offer new and repeat hotel guests [20].

2.2. Overall Variability of Association Rule (OCVR)

The values of support, confidence, and lift ratio on the results of AR often vary in each period due to the uncertain custumer buying habits. OCVR is the variability index of parameter changes in AR. The calculation of this variability index is based on the concept of standard deviation in statistical analysis. *Variability Index (*CV) is calculated using formulation as followed,

$$CV = \frac{S}{X} \tag{4}$$

Where S is standard deviation and $\bar{X}$ is average. This variability index is used to calculate Index Variability Lift (CVL) and Index Variability Confidence (CVC). Thus, the analysis of Overall Variability of Association Rule (OCVR) comes from the use of CVL and CVC and can be formulated as followed

$$OCVR = \frac{CVL+CVC}{2} \tag{5}$$

The results of the OCVR analysis show the degree of variation in AR from one period to the next period. These results can be considered in determining what rules are important in MBA. Also, the rule with a high OCVR value, which indicates high changes in the costumer buying habit in every period, can be further. Then, the decision maker can make suitable marketing strategies based on those results [7].

3. Research Method

This study was implemented in a supermarket in Yogyakarta and the data used was one-month transaction data. The data consisted of 57784 transactions in a month involving 41248 items. AR will be implemented for each period within a week, then there were 4 periods in this study. Table 1 shows transaction data for each period.

Table 1. Number of transaction and item for each period

Period	Number of transaction	Number of item
1	13741	17935
2	15370	17141
3	13836	15925
4	14837	17355

Apriori algorithm was used to implemented AR and software R was used to run the algorithm. This study uses the KDD (Knowledge Discovery in Database) process which includes sequential selection, preprocessing, transformation, data mining, and interpretation/evaluation [21]. Before the data was analyzed, pre-processing data should be conducted to prepare the data. The pre-processing data consisted of cleaning, reduction, and integration. In data cleaning, noisy and incomplete data was identified and erased. Data reduction conducted to reduce some unnecessary variables such as date transaction and item code. While data integration conducted to combine several items such as 1 litre and 2 litre cooking oil. Figure 2 shows the research flowchart.

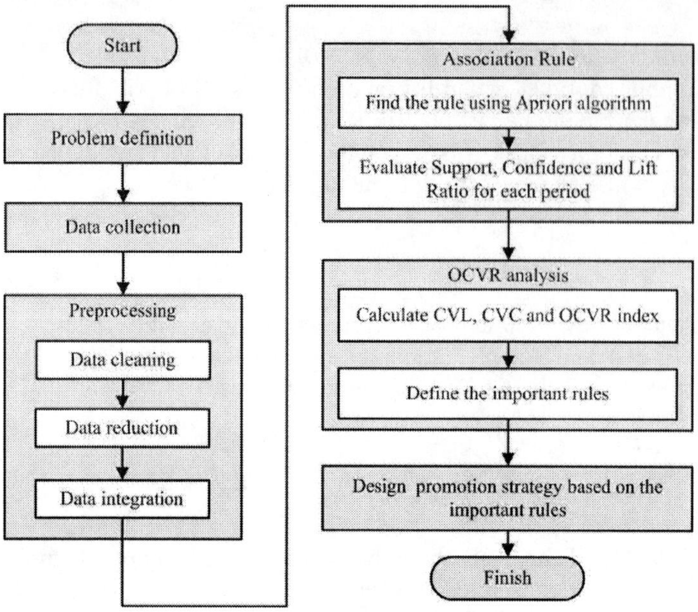

Figure 2. Research flow chart

4. Result and Discussion

Before implanting AR on the transaction data, the most frequent item for each period was analyzed as initial information as shown in Table 2. As shown in the charts the most frequent item for all period is similar. It contains several items such as instant noodles, snacks, milk, soap, etc.

Table 2. The 5 most frequent item

1st period		2nd period		3rd period		4th period	
Item	Number	Item	Number	Item	Number	Item	Number
Noodle X	1100	Noodle X	1070	Wafer	985	Noodle X	1114
Wafer	833	Wafer	1006	Noodle X	971	Wafer	1091
Milk	745	Milk	791	Milk	693	Milk	741
Noodle Y	465	Noodle Y	437	Biscuit	433	Biscuit	488
Soap	367	Tea in bottle	435	Noodle Y	363	Tea in bottle	381

4.1. Association Rule

Parameter setting for Support and Confidence in this study was done by trying and error. The important rules were determined after the best parameters were identified. The result of the parameter setting is presented in Table 3.

Table 3. Parameter setting result

Trial	*Min. support*	*Min. confidence*	Result
1	0.1	0.1	No rule obtained
2	0.01	0.01	Lots of rules were obtained but cannot be read Example: {} => {INDOMIE}
3	0.001	0.2	Lots of rules were obtained
4	0.001	0.3	Only few numbers of rules were obtained

IOP Conf. Series: Materials Science and Engineering **722** (2020) 012068 doi:10.1088/1757-899X/722/1/012068

Based on those results, the minimum threshold for support and confidence that will be used to implement AR using Apriori Algorithm were 0.001 and 0.2. Then, the data for each period that have been preprocessed were put on the software to gain the rules. Table 4 shows the rules obtained for each period.

Table 4. Rules for 1st period

No	Rule			Support	Confidence	LR	Count
1	{Liquid soap}	→	{Bar Soap}	0.00109	0.62500	74.68	15
2	{Noodle Y}	→	{Noodle X}	0.00116	0.57143	7.14	16
3	{Milk cleanser}	→	{Face tonic}	0.00437	0.52174	60.25	60
4	{Noodle Z}	→	{Noodle X}	0.00138	0.51351	6.41	19
5	{Face tonic}	→	{Milk cleanser}	0.00437	0.50420	60.25	60
...	....	→	...	...	...	...	...
59	{Margarine}	→	{Noodle X}	0.00167	0.20000	2.49836	23

For the first rule, if a customer buys liquid soap, then the possibility of the customer also buying bar soap is 62.5% (confidence). 15 transactions contain the transaction from all data in period 1 (support 0.109%). Also, this rule is valid, as seen from the lift value > 1, which is 74.67.

Table 5. Rule for 2nd period

No	Rule			Support	Confidence	LR	Count
1	{Noodle Y}	→	{Noodle X}	0.00104	0.59259	8.51	16
2	{Face tonic}	→	{Milk cleanser}	0.00267	0.53247	87.06	41
3	{Noodle A}	→	{Noodle X}	0.00111	0.50000	7.18	17
4	{Chili sauce}	→	{Noodle X}	0.00124	0.47500	6.82	19
5	{Milk cleanser}	→	{Face tonic}	0.00267	0.43617	87.06	41
...	....	→	...	...	...	...	...
48	{Biscuit}	→	{Wafer}	0.00156	0.20000	3.06	24

Table 6. Rule for 3rd period

No	Rule			Support	Confidence	LR	Count
1	{Noodle B}	→	{Noodle X}	0.00123	0.53125	7.57	17
2	{Face tonic}	→	{Milk cleanser}	0.00325	0.46392	61.13	45
3	{Biscuit}	→	{Wafer}	0.00116	0.45714	6.42	16
4	{Liquid soap}	→	{Bar Soap }	0.00101	0.45161	71.01	14
5	{Milk cleanser}	→	{Face tonic}	0.00325	0.42857	61.13	45
...	....	→	...	...	...	...	...
54	{Dish wash}	→	{Noodle X}	0.00202	0.20000	2.85	28

Table 7. Rule for 4th period

No	**Rule**			**Support**	**Confidence**	**LR**	**Count**
1	{Face tonic}	→	{Milk cleanser}	0.003774	0.56000	72.88	56
2	{Noodle C}		{Noodle X}	0.001348	0.55556	7.40	20
3	{Liquid soap}	→	{Bar Soap}	0.001146	0.53125	85.68	17
4	{Cat food A}		{Cat food B}	0.001146	0.50000	176.63	17
5	{Milk cleanser}	→	{Face tonic}	0.003774	0.49123	72.88	56
...	....		...	...	...	...	...
58	{Soy sauce}	→	{Noodle X}	0.001078	0.20779	2.77	16

It can be seen from these results that the rules obtained from each period were quite different. To design a marketing strategy that can be implemented for a month, further analysis was needed. OCVR analysis was applied for the next step to know any important rules that appeared in each period.

4.2. Overall Variability of Association Rule (OCVR) Analysis

To get OCVR values, the rules formed in each period were needed. The first step is to combine rules that always appear in each period, while rules that did not appear in each period were not used. 17 rules always appeared in each period. The confidence and lift ratio value from those rules were used to calculate *Index Variability Lift* (CVL) and *Index Variability Confidence* (CVC) to find OCVR value. Table 8 shows the calculation result of OCVR. OCVR values that show low variability according to Papavasileiou & Tsadiras (2011) [5] are 1% to 30%. OCVR value greater than 30% indicates that the rule is very vulnerable to changes and cannot be used at any time. The result shows that OCVR value or all rules were 1% < OCVR < 30%, it means all rules can be used to make marketing strategies.

Table 8. Calculation of *Overall Variability of Association Rule* (OCVR)

No	Rules			CVC %	CVL %	OCVR %
1	{Noodle E}	→	{Noodle X}	0.88	7.13	4.01
2	{Green bean drink}	→	{Milk B }	6.32	3.46	4.89
3	{Dish soap}	→	{Noodle X}	4.99	5.28	5.13
4	{Small Egg]}	→	{Noodle X}	8.31	4.16	6.24
5	{Noodle D}	→	{Noodle X}	8.12	5.62	6.87
...	...		...	...	...	...
17	{Milk A}	→	{Milk B}	23.03	21.96	22.5

4.3. Marketing strategy

Based on the OCVR analysis result, the marketing strategies that can be made to promote sales were:

1. Product bundling
   Supermarket X can make product bundling with special price, for example for noodle X and noodle E or milk A and milk B, etc.
2. Shelves layout arrangement
   The Supermarket can arrange the shelves for products that have association close to each other, for example, egg can be put close to noodle, green bean drink close to milk B, etc.

5. Conclusion

Based on the study that has been done, it can conclude the rules obtained in each period were quite different and it indicated the difference in customer buying patterns in each period. The rules were further analyzed using OCVR analysis and the results indicated that 17 rules have OCVR value smaller than 30% and these rules can be used to make marketing strategies. Product bundling and shelves layout arrangement can be conducted to promote sales in Supermarket X.

References

[1] Badan Pusat Statistik Provinsi Daerah Istimewa Yogyakarta 2017 *Pertumbuhan Ekonomi DIY Triwulan III-2017* (Yogyakarta : BPS DIY)

[2] Arifianti R 2016 Pengaruh Promosi Penjualan Terhadap Impulse Buying Pada Hypermarket di Kota Bandung *Unpad Repository*

[3] Raorane A A et al 2012 Association Rule – Extracting Knowledge Using Market Basket Analysis *Research Journal of Recent Sciences* pp 19-27

[4] Videla-Cavieres I F and Ríos S A 2014 Extending market basket analysis with graph mining techniques: A real case *Expert Systems with Applications* vol 41 no 4 pp 1928-1936

IOP Conf. Series: Materials Science and Engineering **722** (2020) 012068 doi:10.1088/1757-899X/722/1/012068

[5] Aguinis H, Forcum L E, and Joo H 2013 Using Market Basket Analysis in Management Research *Journal of Management* pp 1799-1824

[6] Hipp J, Güntzer U and Nakhaeizadeh G 2000 Algorithms for association rule mining—a general survey and comparison *ACM Explor Newsl* vol 2 pp 58–64

[7] Papavasileiou V and Tsadiras A 2011 Time Variations of Association Rules in Market Basket Analysis *Artificial Intelligence Applications and Innovations* pp 36-44

[8] Tan P N et al 2011 *Introduction to Data Mining* (MA: Pearson Education)

[9] Zhang C et al 2019 An improved association rule mining-based method for revealing operational problems of building heating, ventilation and air conditioning (HVAC) systems *Applied Energy* no 253

[10] Shi Y et al 2019 Determination of effective management strategies for scenic area emergencies using association rule mining *International Journal of Disaster Risk Reduction* no 39

[11] Dhanalakshmi P and Porkodi R 2017 A Survey on Different Association Rule Mining Algorithms in Data Mining *IPASJ International Journal of Computer Science* pp 126-133

[12] Sandhu P S et al 2011 Mining utility-oriented association rules: An efficient approach based on profit and quantity *International Journal of the Physical Sciences* pp 301-307

[13] Han J et al 2012 *Data Mining : concept and techniques* (Waltham: Morgan Kaufmann)

[14] Cavique L 2007 A scalable algorithm for thc market basket analysis *Journal of Retailing and Consumer Services* vol 14 no 6 pp 400-407

[15] Kaur M and Kang S 2016 Market Basket Analysis: Identify the changing trends of market data using association rule mining *Procedia Computer Science* pp 78-85

[16] Chen Y L et al 2005 Market Basket Analysis in a Multiple Store Environment *Decision Support System* pp 339-354

[17] Tatiana K and Mikhail M 2018 Market basket analysis of heterogeneous data sources for recommendation system improvement *Proceedia Computer Science* no 136 pp 246-254

[18] Kim H K et al 2012 A product network analysis for extending the market basket analysis *Expert Systems with Applications* no 39 pp 7403-7410

[19] Mansur A and Kuncoro T 2012 Product Inventory Predictions at Small Medium Enterprise Using Market Basket Analysis Approach - Neural Networks *Procedia Economics and Finance* no 4 pp 312-320

[20] Solnet D et al 2016 An untapped gold mine? Exploring the potential of market basket analysis to grow hotel revenue *International Journal of Hospitality Management* no 56 pp 119-125

[21] Fayyad U et al 1996 From Data Mining to Knowledge Discovery in Database *AI Magazine* pp 37-54

Design of Power Monitoring Application

A Surriani, A B Pradana, M Arrofiq, J T Putra, M Budiyanto, and L Subekti

[1]Departement of Electrical Engineering and Informatics, Vocational Collage, Gadjah Mada University, Yogyakarta, Indonesia

Corresponding author: atikah.surriani.sie13@ugm.ac.id

Abstract. the usage of electrical energy has increased significantly. Awareness of increasing electricity consumption has motivated many researchers to create various kinds of innovations to help reduce the energy crisis in Indonesia. So, it is necessary to consider educating the civilization to monitor the intensity of energy consumption as a good culture to monitor electricity consumption efficiently. This research aims to design an Android-based application based on the Internet of Things (IoT) that can monitor electricity consumption and trace the history of electricity consumption data. The monitoring application can calculate both the power used per hour and the estimated costs incurred. Application was made to monitor kWh meters using IoT, electronic design by using ACS 712 current sensor and NodeMCU 8260. Based on the experiments conducted, it was concluded that the Design of Monitoring Kwh Meter Monitoring Application with ACS712 Current Sensor Using Ubidots IoT Platform successfully worked to control the monitoring of the amount of electrical energy usage through IoT, which has only 0.35 % error.

1. Introduction

Nowadays the usage of electrical energy has increased significantly. Based on data from the Central Statistics Agency, Badan Pusat Statistic (BPS), the State Electricity Company, Perusahaan Listrik Negara (PLN) customers in 2015 had reached 61,214,562 which in 2010 the number of PLN customers was 42,438,860. The increase in the number of PLN customers is also accompanied by an increase in electricity consumption per capita [1]. Awareness of increasing electricity consumption has motivated many researchers to create various kinds of innovations to help reduce the energy crisis in Indonesia. However, besides helping to solve the energy crisis problem, in Indonesia it is necessary to evaluate the use or consumption of electricity. Make social culture of electricity consumption efficiently in accordance to one of the themes raised by the National Energy Council, Dewan Energi Nasional, is the achievement of energy efficiency and conservation [2]. So, it is necessary to consider educating the civilization to monitor the intensity of energy consumption as a good culture to monitor electricity consumption efficiently.

Considering to the data, it is very necessary to design a tool and application that is capable of monitoring electrical energy consumption. Therefore, this research aims to design an Android-based application based on the Internet of Things (IoT) that can monitor electricity consumption and retake electricity consumption data. The monitoring application can calculate both the power used per hour and the estimated costs incurred. Application was made to monitor. This application calculates the electric power by multiplying between drawn current by loads and voltage across the load. Additional for this application, the application calculates value of energy, kW, to be multiplied by the price of electricity per kWh in order to get an estimated cost to be paid by the consumer. By utilizing advances

IOP Conf. Series: Materials Science and Engineering **722** (2020) 012069 doi:10.1088/1757-899X/722/1/012069

in technology, an idea emerged to design a system that monitors the amount of electricity used through IoT.

By utilizing these technological developments, an application was made to monitor kWh meters, electronic based IoT circuit consists of a current sensor, ACS712 and nodeMCU8260 as main processor. The android based application developed provides information of current, voltage, power and estimated cost incurred. With this application, the energy consumption can be monitored, thereby encouraging more efficient use of electrical energy for energy sovereignty.

2. Related Work

Research that refers to making energy monitoring applications has been conducted in several studies. Researcher [3] made an application that could monitor energy consumption at home. The research based on WSN (Wireless Sensor Network) research designed ABASH, an android application for smart home.

Research [4] designed an android-based power management application. Energy consumption analyzed was obtained from the Current Transformer (CT) sensor air conditioner, and data were processed from a webserver connected to Arduino and could be accessed via an android application, the value can be displayed using graphical interface. This application already built using IoT, thus the user can trace the history of power consumption. The measurements that can be tracing are current status, room, statistic and history.

Research [5] makes application of monitoring domestic household electrical energy. The application was built to be accessed in real-time using an android application. Monitors the data in the form of power consumption, voltage, and current in household electricity. In research [6] made an android application that can measure the electrical charge on household energy consumption and is able to calculate the costs that must be incurred based on the consumption of electrical energy.

3. Design of System

Design of system android application using Internet of Things is illustrated in Fig 1,

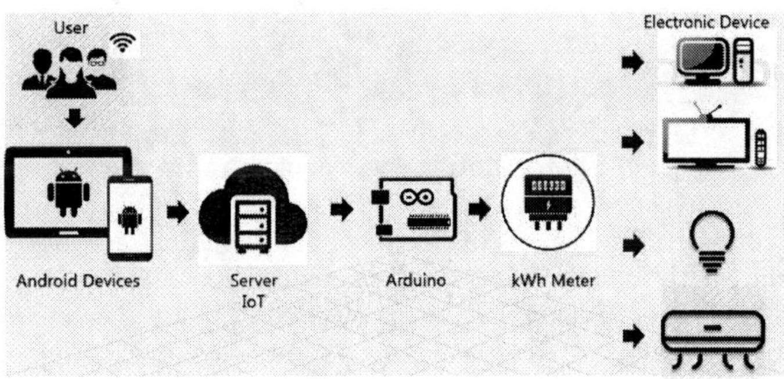

Figure 1. Illustration of System.

Figure 1 shows that the user uses the android application to access data stored on the server. The server section is connected by using Arduino to be able to access energy consumption in kWh meters connected to various household electricity loads. Figure 2. shows the relationship between kWh meter and load. In the kWh meter information on electricity consumption is obtained.

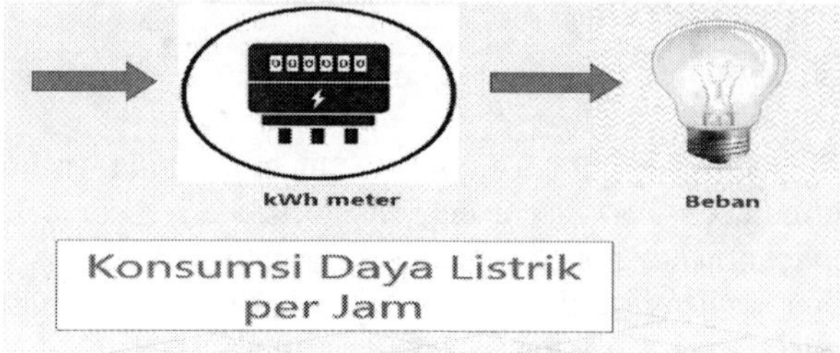

Figure 2. kWh-meter connected to the load

The resulting signal will be accessed and computed using Arduino to get the current and voltage data used. After the data is obtained the user can find out the electrical energy consumption and the price incurred per kWh meter. Figure 3. presents an illustration of the use of android applications, that users can access electricity consumption and costs incurred in order to build user awareness to be able to cultivate efficiency in electricity consumption.

Figure 3. Awareness to use electricity efficiently.

The design of the device consists of electronic design and IoT data design. The electronic design is created using the MCU Node ESP8266 as a communication module between Arduino and a database which is an IoT (Internet of Things). Here is an electronic circuit to create a kWh Meter Monitoring Application with ACS 712 Current Sensor, Figure 4 shows electronic wiring of the system,

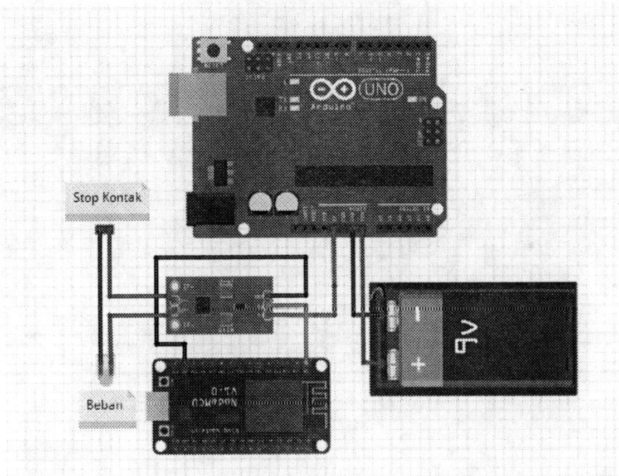

Figure 4. Electronic Circuit.

The wiring of electronic circuit is shown in the Fig 4, the ACS 712 output sensor is connected to the load and switch, while the input is connected to NodeMCU 8260 for communication on the Ubidots Iot Platform and Arduino Uno.

After the electronic design is complete, continue to develop the database design. The database was created using the IoT Platform from Ubidots. Ubidots is an Internet of Things platform from the United States. Fig 5. Shows application Development from Ubidots

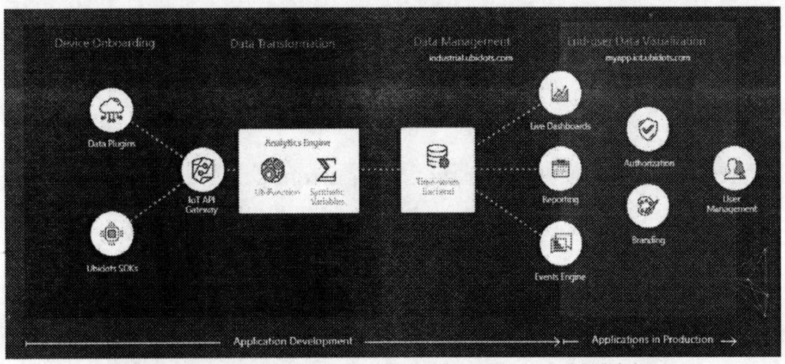

Figure 5. *Application Developtment* of Ubidots

The initial step in using Ubidots is to log in. These Ubidots can be created using the Arduino IDE program. After creating a program on the Arduino IDE, data can be entered such as the NodeMCU board, port and enter the Ubidots library and then connected to the electronic circuit that was created with the help of NodeMCU 8260. To display the token, select the dashboard menu and select the devices that are already connected to Ubidots. Then upload the sketch that was made on the Arduino IDE. Then open the serial monitor with 115200 baudrate. Then return to the dashboard menu to display the results. Using Android Studio, Ubidots platform can be accesses directly from android smartphone. From this platform user can monitor these measurements, current, energy, and cost of electricity. We assume, the voltage remains constant at 220V. IoT makes the user can trace the history of these measurements.

4. *Implementation and discussion*

To display the results of the current reading, we make an IoT-based application where the application can later be used to monitor the current flowing depending on how much we use electricity in our homes. As explained in the application design method, it receives current reading data with the

IOP Conf. Series: Materials Science and Engineering **722** (2020) 012069 doi:10.1088/1757-899X/722/1/012069

help of the ACS 712 sensor sent by the circuit using NodeMCU 8260. The application is Ubidots. Dashboard of user interface is shown in Fig 6,

Figure 6. User Interface

For the implementation electronic design is connected to the load, two lamps that have power value are 60 W and 40 W. will be displayed in table I,

Table 1. Current Value

No	Time (m)	Current (A)
1.	5	0.23
2.	15	0.31
3.	20	0.02
4.	25	0.07
5.	30	0.35
6.	35	0.38

Based on Table 1. It can be seen that the current value in this test is small because the load measured is only two incandescent lamps. ACS 712 sensor successfully works to measure the current.

Table 2. Power

No	Time (m)	Power (Watt)
1.	1	70.09
2.	2	79.28
3.	3	80.80
4.	4	96.94
5.	5	84.77
6.	6	90.13
7.	7	77.43
8.	8	108.83
9.	9	61.42
10.	10	103.96

From Table 2, It can be seen that the power value in this test is also increased due to the current data. Graphical display of the measurement is shown in Fig 7-Fig 10.

IOP Conf. Series: Materials Science and Engineering **722** (2020) 012069 doi:10.1088/1757-899X/722/1/012069

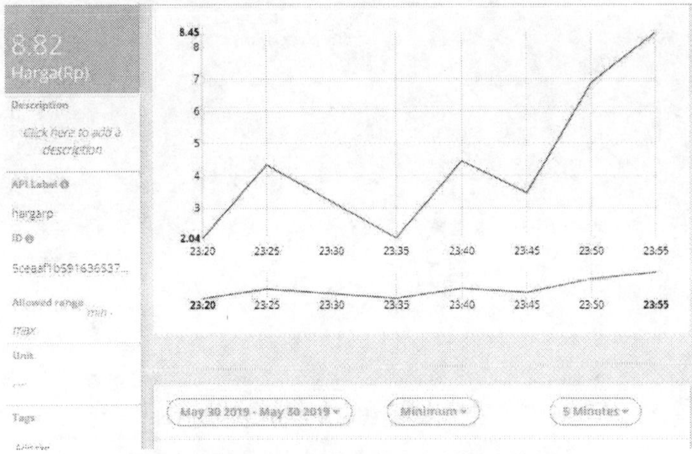

Figure 7. Cost of Electricity for 1 Minutes

Date	Value
2019-05-30 23:56:00 +07:00	103.96
2019-05-30 23:55:00 +07:00	61.42
2019-05-30 23:54:00 +07:00	109.83
2019-05-30 23:53:00 +07:00	77.43
2019-05-30 23:52:00 +07:00	90.13
2019-05-30 23:51:00 +07:00	84.77
2019-05-30 23:50:00 +07:00	96.94
2019-05-30 23:49:00 +07:00	80.80
2019-05-30 23:48:00 +07:00	79.28
2019-05-30 23:47:00 +07:00	70.09

Figure 8. Power Value

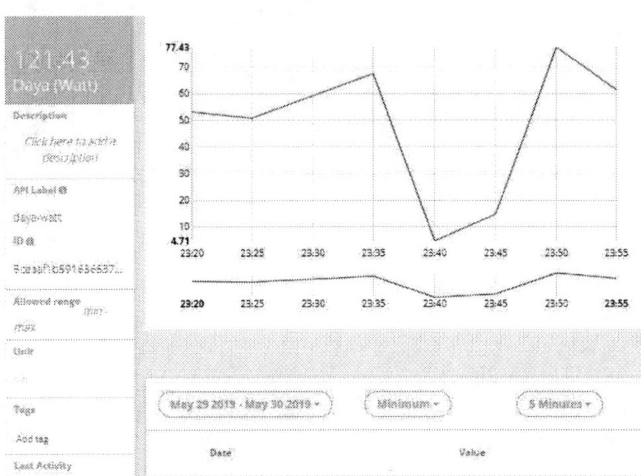

Figure 9. Graphic of Power

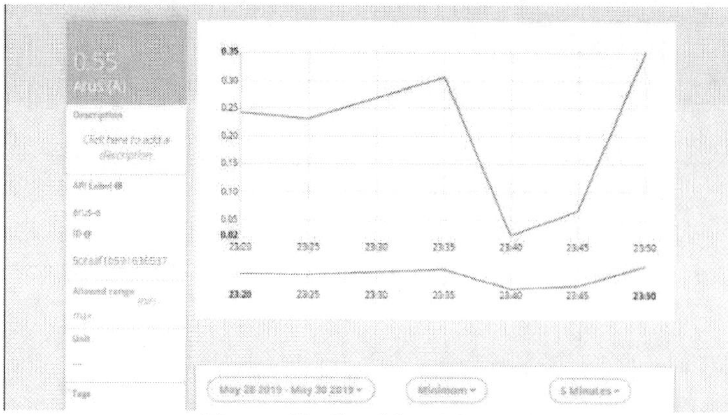

Figure 10. Graphic of Current

In Figure 10, is a graph of current after 5 minutes. This application measures an average current of 0.55 A and power value, P_m, 121.43 W. For the validation, we calculate current and voltage, I = 0.55 A, V = 220 V, to obtain power value of calculation (P_c). The calculation using equation (1) as follow,

$$P_c = V \text{ x } I_m \quad (1)$$

Where, P_c is power calculation value (W), V is voltage (V), I_m is measurement current value (A). The calculation using equation (1) has a value P_c = 121 W. Which has only 0.35 % error compare to value of P_m. From P_m value, we can estimate the cost of electricity, first we get kWh of the electricity is 0.04. The price for electricity per hour is Rp. 1000, - thus we obtain the cost is Rp4,-. The graphic value of electricity cost is shown in Fig 11.

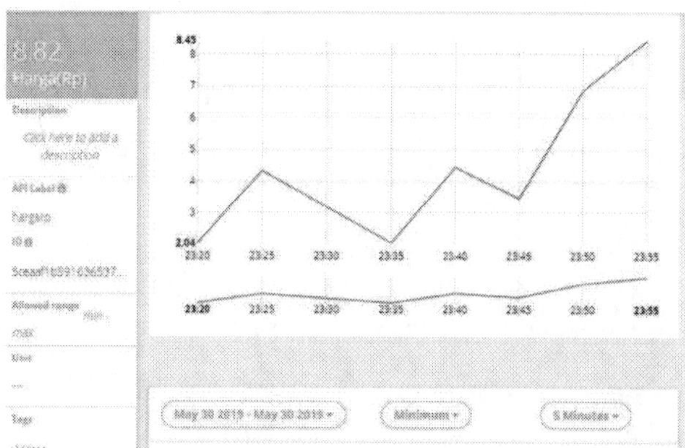

Figure 11. Cost of electricity

5. Conclusion

Based on the experiments conducted, it was concluded that the Design of Monitoring Kwh Meter Monitoring Application with ACS712 Current Sensor Using Ubidots IoT Platform successfully worked to monitor the amount of electrical energy usage through IoT, which has only 0.35 % error. The advantage of this tool is that it has simple packaging, is relatively cheaper, and time efficient.

References

[1] D. jendral K. ESDM, "Statistik Ketenagalistrikan 2016," *Stat. Ketenagalistrikan 2016*, vol. 30, 2016.

[2] F. Anindhita, A. Sugiyono, and M. Boedoyo, "Outlook Energi Indonesia 2015." 2015.

[3] S. Sankaranarayanan and A. T. Wan, "ABASH - Android based smart home monitoring using wireless sensors," *CEAT 2013 - 2013 IEEE Conf. Clean Energy Technol.*, pp. 494–499, 2013.

[4] K. Suwansit, B. Konsombut, P. Hankongkaew, and T. Tantidham, "PMA: Power monitoring application for Android," *Proc. 2014 3rd ICT Int. Sr. Proj. Conf. ICT-ISPC 2014*, pp. 69–72, 2014.

[5] M. A. H. M. Isa, M. F. A. Latip, N. Zaini, and Y. F. Alias, "Android-based application for real time energy monitoring of domestic electricity," *Proc. - 2015 IEEE Conf. Syst. Process Control. ICSPC 2015*, no. December, pp. 134–139, 2016.

[6] N. Anggraini, A. Fiade, and M. Fauzan, "Flow measurement of charges and electricity costs monitoring system with android based Iot (case study: Boarding house Adelina)," *2017 5th Int. Conf. Cyber IT Serv. Manag. CITSM 2017*, vol. 7, 2017.

IOP Conf. Series: Materials Science and Engineering **722** (2020) 012070 doi:10.1088/1757-899X/722/1/012070

Practical Aplication Of Reinforcement Of Remazol Dyes Substance On The Quality Of Cotton Material Using Fixation Waterglass

Ab D Khaliq[1*], A Chafidz[1*], F I Rahmillah[1] , Dahliani W. N.[1], N Iffah Y[1]
[1] Chemical Engineering Department, [3] Industrial Engineering Department,
Faculty of Industrial Technology , Islamic University of Indonesia
Kaliurang Street Km 14,5 Sleman Yogyakarta, Pos 55581

*E-mail: 815210102@uii.ac.id, achmad.chafidz@uii.ac.id

Abstract. Reactive dyes are a dye that can react with cellulose or protein, thus providing good colorfastness. Reactive dyes, besides being used to dyeing cellulose fibers, can also dye wool and silk fibers. One of the reactive dyes is remazol. Waterglass is a durable alkaline compound in the form of a thick colorless liquid. The use of waterglass in the textile industry itself varies greatly, including as a color reinforcement in the fixation process. Because of its essential function, the waterglass has become an urgent material for craftsmen, especially abstract batik craftsmen. Waterglass like it can't be replaced, but it turns out there is a material that can be formulated to make artificial waterglass from salt, boarding, and water. For the media on dyeing, you can use a cotton cloth. Cotton cloth has strong properties (even when wet still absorbs), attracts body heat, wrinkles, shrinks, or shrinks (unless appropriately handled), is damaged by the sun, sweat and decay.

1. Introduction

Reactive dyes are used in the textile industry because of their characteristics, such as color protection, fastness, and ease in simple applications with low energy consumption [1]. Reactive dyes can also be used to dip cellulose fibers, can also dip wool and silk fibers [2]. One of the reactive dyes is remazol (Hoecht). Water class is a strong alkali compound. The use of waterglass in the textile industry itself varies greatly between binders for reactive dyes and color reinforcement in the fixation process [3]. Because of its important function, water glass is an urgent material for craftsmen, especially abstract batik craftsmen. Waterglass like it cannot be replaced, but some ingredients can be formulated to make artificial waterglass from salt, custom, and air. For media, on dyeing, you can use cotton cloth [4]. Cotton fabrics have strong properties [5].

2. Research Methods

The Research materials that used for this study as follows:

a. Remazol dye
b. Cotton fabric
c. Waterglass
d. Salt
e. Costic
f. Water

IOP Conf. Series: Materials Science and Engineering **722** (2020) 012070 doi:10.1088/1757-899X/722/1/012070

Besides that, this study also used several tools:

a. Digital Scales
b. Stirrer
c. Measuring Cup
d. Plastic
e. Container

3. Research Place

The research was conducted at the Laboratory of the Industrial Technology Faculty of the Indonesian Islamic University which is an integrated laboratory commonly used by FTI UII Chemical Engineering Students in carrying out practical or research activities and an integrated Laboratory of the Faculty of Mathematics and Natural Sciences, Islamic University of Indonesia.

Dyeing:

In TRO, which has been dissolved in water, cotton is inserted. Dyeing is done for 5 minutes. The purpose of this dyeing is to make the cloth cleaner and to make it easier for the dye to stick and evenly spread on the fabric. After Dyeing with TRO and lining cloth, the next step is dyeing with Reactive Color Substance + soda ash. In reactive dyes + soda ash that has dissolved is included in cotton cloth. Dyeing was carried out at room temperature for 20 minutes.[6].

4. How to Research

a. Dyeing with TRO, Reactive Dyes + Soda Ash
Dissolve TRO, Reactive Color Substance + Soda Ash:

For TRO, put about 2 spoons into a container that has been filled with water. For reactive dyes + Ash soda is put into a water solution according to the recipe calculation [7].

b. Process of Fixation with Waterglass
A fixation process is carried out to lock the natural dyestuffs of the mordan group and functions to strengthen the color and give the color effect (color direction) that varies according to the fixation substance used. Fixation is done by dipping a cotton cloth into the waterglass at room temperature for 10 minutes. There are two waterglass used, namely the commonly sold waterglass and waterglass which are formulated by themselves from a mixture of salt and soda [8].

c. Drying Process
The purpose of this drying process is to turn off the color of batik so that it does not fade. There are two drying processes, namely naturally and using an oven machine. In this experiment, the drying process is done naturally, namely with the sun [9].

5. Results And Discussion

a. Rubbing Color Resistance Test (Dry Rub Test)

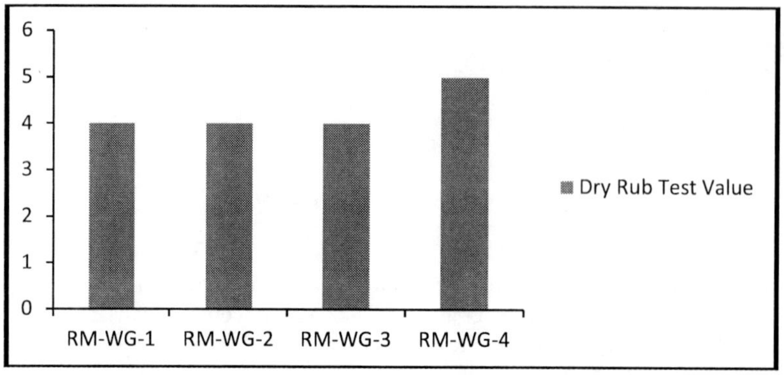

Figure 1. The 1st Dry Rub Test Value

RM-WGT-1 to RM-WGT-3 get a value of 4 (good) [11]. This indicates that from the data from the results of the dry rub test on the fixator RM-WG-4 and RM-WGT-4, the value of gray scale 5 (very good) [12]. In the fixator RM-WG-1 to RM-WG-3 and the two fixators have a balanced dry rub test value against cotton cloth [13].

Table 1. Dry Rub Test value

Scale	Sample Code
4 (Good)	RM-WG-1
4 (Good)	RM-WG-2
4 (Good)	RM-WG-3
5 (Very Good)	RM-WG-4

b. Waterglass ($Na_2 SiO_3$) [10]:
Waterglass (Soda Costic + Salt)

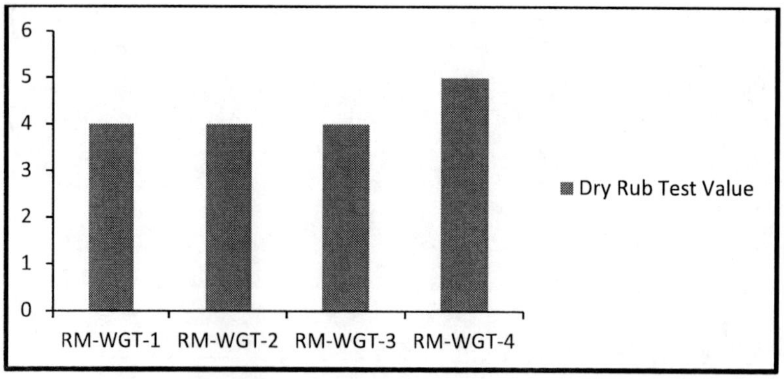

Figure 2. The 2nd Dry Rub Test Value (Waterglass)

IOP Conf. Series: Materials Science and Engineering **722** (2020) 012070 doi:10.1088/1757-899X/722/1/012070

c. Color Blending Test with the Method of Washing Soap

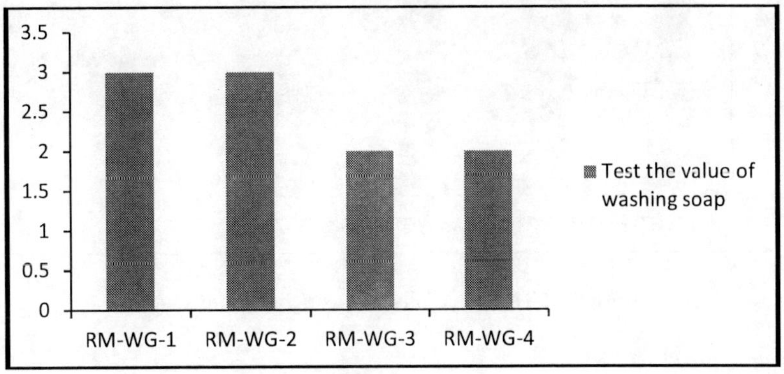

Figure 3. The 1st test using color blending with washing soap

Table 2. Results using color blending

Scale	Sample Code
3 (enough)	RM-WG-1
3 (enough)	RM-WG-2
2 (less)	RM-WG-3
2 (less)	RM-WG-4

From the data from the soap washing test, the RM-WG-1 fixator and RM-WG-2 showed the gray scale value of 3 (enough). Then the fixator RM-WG-3 and RM-WG-4 can be seen from gray scale, which is 2 (less). Whereas the fixator of RM-WGT-1 to RM-WGT-4 was obtained from gray scale, 4 (Good). This shows that the effect of artificial waterglass fixation on cotton cloth dyeing remazol dyes can be said to be good compared to using the original waterglass [15].

Waterglass (Soda Kostik + Salt)

Table 3. The 2nd results of Dry Rub Test (Waterglass)

Scale	Sample Code
4 (Good)	RM-WGT-1
4 (Good)	RM-WGT-2
4 (Good)	RM-WGT-3
5 (Very Good)	RM-WGT-4

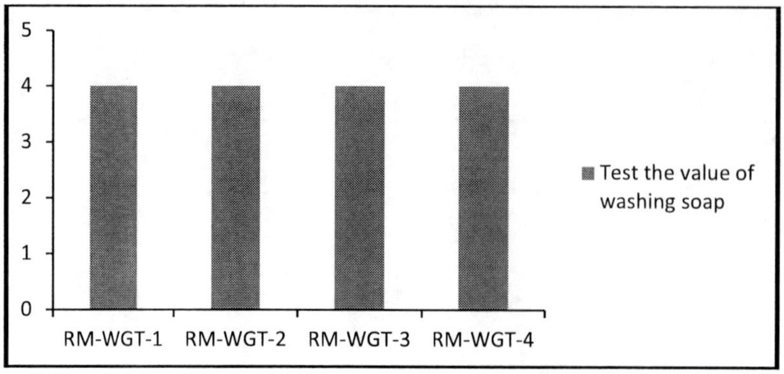

Figure 4. The 2nd results test using washing soap

d. Testing Color Lead
Waterglass

Table 4. The results on color lead testing

Percentage results	Sample Code
75,16 %	RM-WG-1
62,64 %	RM-WG-2
75,03 %	RM-WG-3
66,44 %	RM-WG-4

From the data on the color aging test, the fixers of RM-WG-1 to RM-WG-4 obtained results of R (%), respectively 75.16%; 62.64%; 75.03%; 66.44%. While for fixators RM-WGT-1 until RM-WGT-4, the results of R (%) are 70.00%; 67.57%; 68.85%; 66.14%. The results of the color aging test indicate that the greater the R value the older the colored cloth. While the smaller the value of R, the color of the fabric will be brighter or towards white.

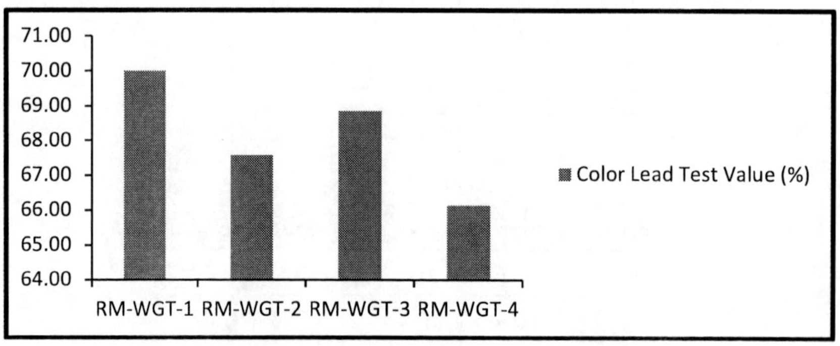

Figure 5. Color Lead Test Value Results

IOP Conf. Series: Materials Science and Engineering **722** (2020) 012070 doi:10.1088/1757-899X/722/1/012070

Table 5. Results of color lead/aging test value

Percentage Results	Sample Code
70,00 %	RM-WGT-1
67,57 %	RM-WGT-2
68,85 %	RM-WGT-3
66,14 %	RM-WGT-4

6. Conclusion

a. The effect of artificial waterglass fixation on the quality of cotton cloth dyed with remazol dyes can be said to be good. While the effect of fixing the original waterglass with remazol dyes can be said to be not good.

b. The difference that occurs when using a waterglass with an artificial waterglass is that the color fading value on an artificial water glass is 4-5 (Good), which means that the fabric does not easily wear off. Whereas in the original waterglass the value is 2-3 (Less), which means that the fabric easily wears off.

References

[1] Aprillia Puspa R dan Djauhari Agus, M.T 2010. *"Makalah Peralatan Industri Proses" pdf.* Jurusan Politeknik Negeri Bandung.
[2] Dede dan Elly k.2005. *Bahan Ajar Praktek Pencelupan I*, Sekolah Tinggi Teknologi Tekstil Bandung.
[3] Fatonah, Ani. Dkk. 2012. *Laporan Pencelupan Kain Sutra dengan Zat Warna Reaktif Dingin.*
[4] Rasyid, Djufri., dkk. 1976. *Teknologi Pengelantangan Pencelupan dan Pencapan*, STTT, Bandung.
[5] Murdoko, W.1975, *Textile Testing,* ITT, Bandung.
[6] Lestari, K.W.F. dan H. Suprapto. 2000. *Natural Dyes in Indonesia.* Yogyakarta: Balai Besar Penelitian dan Pengembangan Industri Kerajinan dan Batik.
[7] Soepriyono, 1974, *Serat Serat Tekstil*, Institut Teknologi Bandung, Bandung
[8] Khaliq Ab Dulmalik et al, The Use ofHydrogen Peroxide and Sky Stabilizer agent in Bleaching process of Textile Fabrics, AIP Conference Proceedings 2138, 050011 (2019); https://doi.org/10.1063/ 1.5121116.
[9] Murdoko, W. 1975*, Textile Evaluation of Chemical Department,* ITT-Bandung.
[10] Murdoko, W. 1976, *Textile Evaluation of Physical Department,* ITT Bandung.
[11] Malik, A., Bilal, Fety IR, Bayu., 2017, *The Effects Composition Microwax on Coloring Quality of Klowong Batik Wax,* Matec Web of Conference 154,01118 (2018). ICET4SD 2017.
[12] Susanto S. *Indonesien Batik and Handicraft,* Research Institute BBKB, Yogyakarta (1980).
[13] Khaliq Ab Dulmalik et al, *Scouring Process with Sky Scour BL-A agent on grey fabrics,* AIP Conference proceedings 2138, 050017 (2019); https://doi.org/10.1063/1.521116.

IOP Conf. Series: Materials Science and Engineering **722** (2020) 012071 doi:10.1088/1757-899X/722/1/012071

Proposing a More Efficient Maintenance Scheduling for an Overhaul Maintenance Project in Engineering Asset Management

W N Cahyo[1], F W Hasibuan[2] and A P Hendradewa[1]

[1)] Department of Industrial Engineering, Universitas Islam Indonesia, Jalan Kaliurang Km. 14,5, Yogyakarta, Krawitan, Umbulmartani, Ngemplak, Kabupaten Sleman, Daerah Istimewa Yogyakarta 55584

[2)] Graduate of Department of Industrial Engineering, Universitas Islam Indonesia, Jalan Kaliurang Km. 14,5, Yogyakarta, Krawitan, Umbulmartani, Ngemplak, Kabupaten Sleman, Daerah Istimewa Yogyakarta 55584

Corresponding author: winda.nurcahyo@uii.ac.id

Abstract. In maintenance activity, scheduling planning is an important step to improve the performance of maintenance department. In some cases, the data to support the process to develop the maintenance schedule is unavailable. In this research, it is argued that Genetic Algorithm (GA) is an appropriate tool to support the decision maker in order to develop the schedule. The data required in this research is obtained from the maintenance data in the sugar milling company in Yogyakarta. The company run the production of sugar and spirits/alcohol. The Production process (milling seasin) is usually performed within the harvesting months of the raw material: the sugar cane. It is approximately in the month of February to August. On the other hand, the overhaul maintenance process is performed after the milling season. The overhaul maintenance planning is created by the manager of the maintenance department, and usually is done using the manager qualitative approach without any sciencetific approach. To get a more efficient schedule of maintenance, several parameters are set in this research. The parameters are then applied as the paramenters in GA. Those are: The number of manpower required and the duration of operation each activity. To run the GA model, Python programming is used. Before start the GA process, the parent chromosom and its fitness are determined. The date of maintenance activity start per each activity is set as the parent chromosome and the fitness is determined using a formula. The process of finding the better maintenance schedule is then performed based on the procedures in GA. It is indicated from the result that the proposed schedule generated in GA required less number of manpower.

1. Introduction

To compate in the competitive business environment, manufacturing industries are forced to improve their performance to achieve the competitiveness advantage. One way to improve the overall performance is to manage the asset effectively. The value of the asset has become the opportunity for an organisation to sustain in the market. The market prefers to have companies who capable to fulfil the required demand, desired quality, and also with on-time delivery. It is a reasonable basis for any company to innovate product with high quality with short lead time and to create a robust production system as a basic requirement to achieve the better operation process [1]. Based on the data presented by [2], the growth of sector of manufacture increases by 4% in 2016 and create tougher competitiveness environment.

In order strengthen the competitive advantage, the manufacturing company should continuesly improve the effeiciency and effectivity in their process. One commonly used approach for this purposes is lean manufacturing [3]. Lean manufacturing is a systematic approach to identify and eliminate waste using several improvement tools [4]. Lean manufacturing is capable to produce the desired output with efficient resource used [5]. However, in some point of view, lean is a philosophy and to eliminate the waste other practical tools should be applied. For instance, in the production system the machines have important role to achieve productivity. A high machine reliability is a valuable resources to deliver a good service to the manufacturer [6]. On the other hand, the failure of the machines may lead to the delay of delivery, loss of profit, and miss the opportunity to gain the market for the manufacturer [6]. For this reason, the organisation requires to do maintenance for their asset to keep the performance of the asset and to reduce machine failure.

Maintenance process is performed to provide service to the production facilities and in order to contribute to the process of gaining high productivity [8]. It means that maintenance has a significant contribution to the competitiveness of the manufacture by focusing on the performance of the asset used for production. According to [9], the type of maintenance are planned and unplanned maintenance, depends on the condition of the physical asset. However, in some cases, the requirement of the production process makes obstacles in the implementation of the maintenance strategy.

In Engineering Asset Management, to select an optimum maintenance strategy, every organisation should consider three aspects: performance, risk, and cost associated with the asset. The types of maintenance strategies includes: corrective maintenance, preventive maintenance, and design-out maintenance [10]. Similar with the production process, maintenance process may also generate waste. Carefully observed the mainatenance activities leads to the elimination or reducation of the waste. It includes the decision to treat the different machines because the conditon of each machine is different. Again, each step in the maintenance process should be analysed in detail to create an effective and efficient process. Process mapping for every flow of the process is necessary to understand the cause of the waste [11].

The focus of this paper is reduce the waste in regard to the overhaul process. The case study is in a sugarcane milling company in Yogyakarta. At this moment, the company performs overhaul maintenance with the support of 60 permanent mechanics and 15 contracted mechanics. The overhaul maintenance should be finished within two months. While this company is implementing maintenance and operation scheduling to reduce the probability of engine damage, this factory did not consider the worker to be allowed to gain the minimum cost. Based on [12], the company gain an opportunity loss of Rp 163,228.00 per minute, and the total per day is Rp. 235,048,320 if production cannot achieve the demand. Maintenance strategy that applied in this factory is preventive maintenance which run in exactly period after the production period. The objectives of this factory to implement this strategy was to avoid the maintenance when the production is running. Therefore, such studies are very important for regulators to detect maintenance project failures when running the maintenance. To identify the beneficiaries of their maintenance, arrangement the scheduling based on the worker allocation and scheduling of maintenance task.

2. Previous Researches on Maintenance Scheduling

Maintenance management is all the management that determines the purpose of maintenance activities, maintenance strategies, and responsibilities along with their implementation [9]. In general, the objectives of care management is divided into 3 major groups, namely [9]:

a. Technical objectives.

This objective depends on the industrial sector of a business entity operates. In general, the technical objectives of a company are to maintain or increase the availability and operational capability of an asset and improve work safety.

b. Legal objectives/government regulations.

This objective is set up to cover the quality standard in one project or activity. Legal objectives for normal care management are only to meet the existing rules for various things. This objective arraged by the executive leader and follow the government ragulation.

c. Financial goals.

The financial objective of maintenance management is to meet technical objectives with minimal costs. From a long-term point of view, the overall life-cycle cost of the equipment should be minimal. The objective was urged to implement which the competition of the market to reduce the high operation cost.

The overhaul maintenance activities in the observed system is considered as an annual project to fulfil the needs of the production process. To avoid the downtime during the production period, the factory applies predictive maintenance as their maintenance strategy. The Maintenance activity is implemented based on the urge of the production process. Preventive maintenance is a treatment carried out by supervising the components/systems so that they are in optimal operational condition by conducting checks, early detection, and systematic improvements before failure or developing a sign of failure in the system [13]. Preventive maintenance objectives include reducing damage to critical components, minimizing production losses due to component failure, and increasing the productive life of a component. Preventive maintenance consists of seven sub activities: inspection, servicing, calibration, testing, alignment, adjustment, and installation.

3. Genetic Algorithm

Genetic algorithm (GA) is a form of concept or search method that can be used to solve problems and model systems of biological evolution [14]. The implementation used for this study is model in stages such as those in the GA in general. GA is a powerful technique that is used in various fields to find a solution that is almost optimal when searching for the optimal is too expensive. Even though it is attractive and elegant in the laboratory, scalability problems are prevented from being effectively implemented into real problems when GA operation is run.

GA moves from a population of chromosomes represented in a prospective solution to the new population through three stages, namely: selection, crossover and mutation. Chromosomes are each member of a set that can represent a solution to a problem. Chromosomes are an evolved population in an iteration known as the generation, each chromosome is evaluated based on its evolutionary function (fitness function). The chromosomes will be selected based on each fitness value which has the highest [15].

There are several things that must be done in genetic algorithms, including the following:

1. Define individuals: which can be expressed as one solution (solution) of the problem raised.
2. Define the fitness value: which makes a benchmark whether or not an individual or a solution is achieved.
3. Determine the initial population generation process: things that are done by random generation.
4. Determine the selection process to be used.
5. Determine the crossover process and the mutation of the gene to be used.

In each iteration, all members of the population will be evaluated based on the fitness that has been made, with members of the initial population randomly created. A new population will be formed from

IOP Conf. Series: Materials Science and Engineering **722** (2020) 012071 doi:10.1088/1757-899X/722/1/012071

the evolution of the parent chromosomes through an iteration called generation. The results of the generations that have been carried out will produce offspring from the use of genetic operations such as crossover and mutation. The newly formed population has gone through the stages of chromosome elimination based on the best fitness values from parent and child, so the population size is constant.

[16] present a heuristic algorithm for maintenance scheduling of a system that has a set of components. In this study, all components are assumed to have a rate of damage which increases with the value of a constant increase factor. This paper proposed GA techniques for representing variables in a preventive maintenance scheduling model that uses heuristic and metaheuristic optimization algorithms.

4. Research Methods

Scheduling preventive maintenance requires three sources: labour (technicians), materials (parts) and equipment [17]. Scheduling support tools for this problem must be related to the assignment maintenance due date and contribution of all resources and must be able to plan short term plans (weekly or monthly). The researchers also have to consider the equipment because the company provides equipment with a long operating period, and consequently, with limited maintenance tasks. Scheduling support tools that meet company requirements are being developed. This Support Tool is approved in the preventive maintenance scheduling algorithm which can be divided into three parts with specific objectives:

- Identification of the scheduling period and equipment availability
- Task Ordering
- Allocation of tasks to each technician based on defined order and equipment availability.

The task order is made by the head of the department of Maintenance in this company. All task was set done in due date after considering the weight of each task. This company apply the cost of manpower at the is divided into two parts, namely permanent workers and contract workers. Permanent workers are workers who work in the field of care for one full year either in the milling season or not. While contract workers are workers who are only assigned to carry out maintenance activities during the treatment season before the production process begins for 5 months. For the contract and non-contract worker, the worker has salary Rp 1.900.000 and Rp 1.300.000 per month [12].

4.1. Chromosome

Before the GA calculation is started, the initial chromosome is determined. The initial chromosome is based on the maintenance task schedule as shown in Table 1. The chromosome length in TABLE I. is 35 activities which is used for fitness calculations only, but when calculating the generation, only chromosome numbers 0 to 34 applies, so the length of the chromosome when it is input into GA programming in Python process is 35 genes. From the result of the calculation with 50000 epoch, a graph of fitness is obtained and shown in Figure 1.

Table 1. Initial Maintenance Tasks Scheduling

Activity	Manpower	Days	Start Date	End Date
1	8	10	110	120
2	5	15	6	21
3	4	50	56	108
4	8	90	14	104
5	4	25	91	117
6	8	90	20	109

IOP Conf. Series: Materials Science and Engineering **722** (2020) 012071 doi:10.1088/1757-899X/722/1/012071

Activity	Manpower	Days	Start Date	End Date
7	5	90	2	93
8	3	70	3	73
9	3	50	63	114
10	5	45	8	58
11	6	35	3	39
12	6	50	7	56
13	6	90	4	96
14	6	60	60	119
15	3	44	74	119
16	3	24	84	105
17	4	25	8	34
18	0	0	0	0
19	5	60	50	112
20	7	90	3	92
21	2	30	86	116
22	2	20	99	120
23	2	10	93	103
24	6	50	51	102
25	4	75	4	80
26	0	0	0	0
27	4	75	27	104
28	3	60	4	69
29	3	40	3	43
30	3	15	39	55
31	2	90	28	121
32	2	60	3	63
33	7	95	13	110
34	4	50	41	91
35	2	60	31	63

4.2. Fitness

The fitness value used as a reference in assessing whether a good chromosome is used as a solution to the problems in this study. In this research, the fitness solution is formulated from the desire for the absence of an excessive number of workers in one day, the balance of allocation of workers, and completion of treatment according to applicable limits. The fitness function is as follows:

- $Fitness = \sigma\ Manpower_{t1\text{-}120} + (\sum Manpower_{\ t120-n} * 100)$

Based on Equation (1) the fitness function is made based on consideration of the maximum of worker that the problem in this research. Penalty is created to control the achieved the objectives of this research. Penalties in this research are shown in Table 2 The first is that there are restrictions on the number of workers in one day as many as 77 workers, so that if you exceed the number of workers who have been previously determined to be charged a penalty. The second is so that the number of workers can be balanced in one period, it is necessary to know the standard deviation of the number of workers allocated in one day in one period so that there will be a penalty if there is an imbalance. The third is about the time limit set in each year to complete the maintenance work so that if there is a job that crosses the line (119 days) it will be subject to a penalty according to the number of days. In addition, if the manpower penalty and the power manpower penalty produce the best value of 0 and the time is over the specified limit, it is necessary to use a small number (0.01) on the sum as in Equation (1) above, so that the fitness value remains in accordance with the real situation.

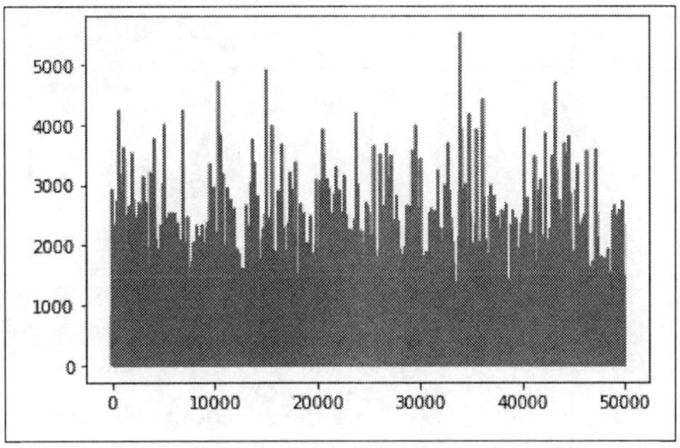

Figure 1. Maximum Fitness Value

The last operation defined by genetic algorithms and used to manipulate chromosomes is fitness operations and fitness comparison. Fitness operations measure the quality of the resulting solution (chromosomes). This operation is specific to each problem and it actually tells what genetic algorithm to optimize. Fitness comparison is used to compare chromosomes based on their fitness. Basically, a fitness comparator tells a genetic algorithm whether it must minimize or maximize the chromosome fitness values.

Table 2. Penalty formula

Penalty	Formula
Manpower Penalty	IF (Total task tn > 34) then (Total task tn - 34) * 1), IF False (0)
σ Manpower Penalty	σ Manpower * 5
Time Penalty	IF (End Date per Task > 120) then ((End Date per Task – 120) * 2), IF False (1)

5. Results

The results of GA at 50000 epochs is the proposed schedule for the overhaul maintenance tasks as shown in Table 3 and the value of the fitness is presented in Figure 1. It is argued that greater number of generations is required to get the most optimum result, because of the large search space. However, in this study, it is only limited to 50000 iterations. After the maximum number of iteration (or MaxGen) is achieved, then the number of man power imply to the proposed solution is determined. The maximum number of man power required to complete the overhaul maintenance based on the result of the GA is 79 people and the minimum is 0 as shown in Figure 2. Theoretically, an optimum can be attained if the gap between the maximum and minimum manpower used in this scheduling is 0. But in this research,

IOP Conf. Series: Materials Science and Engineering **722** (2020) 012071 doi:10.1088/1757-899X/722/1/012071

the result is defined by result of iteration after the maximum number of iterations is achieved and not by finding the optimum solution. It can be observed from Figure 1 also that 50000 iteration is insufficient to get the highest fitness value because the fluctuation of fitness still high or dynamic.

After the iteration achieves the last evolve to get the generation in 50000 iteration, then researchers had to put the number of Manpower in that generation to imply in maintenance scheduling after GA result as suggest. Then, define the number of Manpower that company should provide for this scheduling in maximum is 79 and with minimum manpower that used in this scheduling is 0 was shown in Figure 2. Actually, the generation meets the optimum solution if the gap between the maximum and minimum manpower used in this scheduling is 0. But in this research, the result is defined by the number of iteration that to find the optimum solution. By seeing Figure 1, the researcher assumes that the solution not enough to get the highest fitness value because the fluctuate of fitness still high or dynamic.

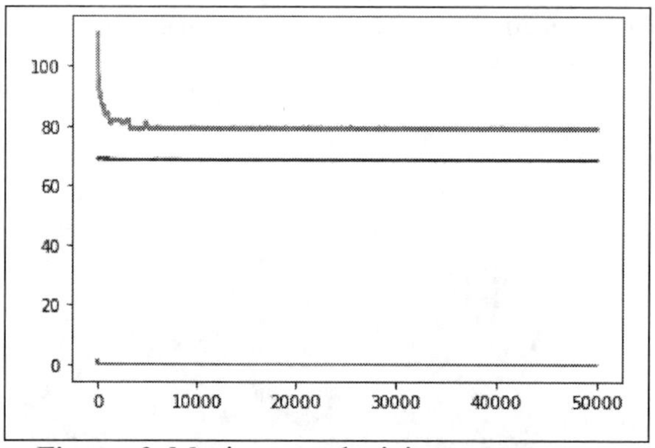

Figure 2. Maximum and minimum manpower

The main objective of the use of GA is to find the schedule of overhaul maintenance as shown in Table 3. There are a tiny difference between Table 1 and Table 3 in term of the number of activities ofr maintenance tasks. The maintenance tasks in Table 1 is the initial generation of the GA and consists of 35 tasks and in Table 3 the number of tasks is 34. It is because of there is one activity in Table 1 (activity number 26) that actually valued 0. So if this activity is eliminated, the number of activities in Table 1 and Table 3 is the same.

Table 3. Proposed New Maintenance Tasks Scheduling

Activity	Start date	End date
1	2	12
2	2	17
3	2	52
4	2	92
5	95	120
6	31	121
7	2	92
8	2	72
9	2	52
10	76	121
11	2	37
12	10	60
13	30	120
14	60	120
15	2	46

Activity	Start date	End date
16	2	26
17	92	117
18	2	62
19	30	120
20	2	32
21	2	22
22	2	12
23	60	110
24	45	120
25	0	0
26	46	121
27	2	62
28	46	86
29	61	76
30	15	105
31	4	64
32	2	97
33	11	61
34	2	62

As shown in Table 3, **Error! Reference source not found.**the end-date of maintenance activity is still over the allocation provided by the company is 120 days. Activity that conducts over the allocation time is activity 6, 10 and 26. These beyond the duration are influenced by the duration of each task is different. Even though the maximum number of required man power cannot reach the target, the required number of man power every month is less as well as the cost of the man power as seen in Table 4.

6. Analysis and Discussions

Table 4. Cost Allocation for Manpower

Month	Maximum manpower Used per week		Manpower Cost (Rp)	
	Before	After GA	Before	After GA
1	71	71	134.900.000	134.900.000
2	77	76	146.300.000	144.400.000
3	77	75	146.300.000	142.500.000
4	77	79	146.300.000	150.100.000
5	69	67	131.100.000	127.300.000
TOTAL			704.900.000	699.200.000

In Table 4, the number and the cost of manpower in the existing condition and the proposed scheduled with GA is presented. In general, the required number of man power is reduced by 1 or 2 people monthly. Also, within one overhaul period, the total man power cost reduces from Rp. 704,900,000 to 699,200,000 or approximately Rp 5,700,000. The proposed scheduled is presented in Table 5.

Table 5. Proposed Daily Manpower Allocation

Day	Man Power	Day	Man Power	Day	Man Power
1	0	41	70	81	76
2	0	42	70	82	79
3	65	43	70	83	79
4	65	44	70	84	79
5	65	45	70	85	79
6	65	46	67	86	79
7	65	47	75	87	79
8	65	48	75	88	79
9	65	49	75	89	79
10	65	50	75	90	79
11	71	51	75	91	79
12	67	52	68	92	57
13	67	53	68	93	63
14	67	54	68	94	63
15	67	55	68	95	63
16	70	56	68	96	67
17	65	57	68	97	67
18	65	58	68	98	67
19	65	59	68	99	67
20	65	60	62	100	67
21	65	61	64	101	67
22	63	62	68	102	65
23	63	63	68	103	65
24	63	64	68	104	65
25	63	65	68	105	65
26	67	66	68	106	65
27	67	67	68	107	65
28	67	68	68	108	65
29	67	69	68	109	65
30	64	70	68	110	65
31	70	71	68	111	65
32	76	72	71	112	65
33	76	73	71	113	65
34	76	74	71	114	65
35	76	75	71	115	65
36	76	76	71	116	65

Day	Man Power	Day	Man Power	Day	Man Power
37	70	77	76	117	61
38	70	78	76	118	61
39	70	79	76	119	61
40	70	80	76	120	38

7. Conclusion

In this paper, a schedule for an overhaul maintenance project is observed and proposed. The result shows that the proposed schedule generate less man power cost compared to the existing situation. It is concluded that the model is able to serve its purpose to improve the performance of the maintenance department. However, the result attained from the maximum number of generations is identified as not the most optimum result, because there is gap between the maximum and minimum number of required manpower.

So, it is suggested that to find the optimum solution, more iteration should be run. Also, for the further research, it is suggested to consider using Artificial Neural Network (ANN) in the process of generating the initial population of the GA in this kind of case study. It is argued that the use of ANN is capable to reduce the number of required iterations to find the optimum solution. Another research direction arose from this study is he a mathematical model to calculate the total cost of the selected policy. The model may consist of the cost for recruitment and lay off since those cost component is neglected in this research.

References

[1] Bellgran, M., & Säfsten, E. K. (2009). *Production development: design and operation of production systems*. Springer Science & Business Media.

[2] Statistik, B. P. 2017. *Pertumbuhan produksi IBS naik 4,33 persen dan IMK naik 6,63 persen pada triwulan I tahun 2017 dari triwulan I tahun 2016.* [Online] Available at: https://www.bps.go.id/pressrelease/2017/05/02/1397/pertumbuhan-produksi-ibs-naik-4-33-persen-dan-imk-naik-6-63-persen-pada-triwulan-i-tahun-2017-dari-triwulan-i-tahun-2016.html [Accessed 5 Oktober 2017].

[3] Chiabert, P., D'Antonio, G. & Bedolla, J. S. (2017). A Novel Methodology to Integrate Manufacturing Execution Systems with the Lean Manufacturing Approach. Procedia Manufacturing, pp. 2243-2251.

[4] Karl, A. & Dombrowski, U. (2017). *Lean Product Development for Small and Medium-sized Suppliers*. Procedia CIRP Volume 63, pp. 615-620.

[5] Wahab, A. N. A., Mukhtar, M. & Sulaiman, R.manpower2013. *A Conceptual Model of Lean Manufacturing Dimensions*. Procedia Technologi, pp. 1292-1298.

[6] Cahyo, W. N., El-Akruti, K., Dwight, R. & Zhang, T. 2015, "Managing maintenance resources for better asset utilisation", Australian Journal of Multi-Disciplinary Engineering, Vol. 11, No. 2, pp. 123-134, http://dx.doi.org/10.7158/N14-AM06.2015.11.2.

[7] Zaim, S., Turky-ilmaz, A., Acar, M. F., Al-Turki, U., & Demirel, O. F. (2012). Maintenance strategy selection using AHP and ANP algorithms: a case study. Journal of Quality in Maintenance Engineering, 18(1), 16–29.

[8] Faccio, M. A., Persona, F. Sgarbossa, G. Zanin, (2014) *Industrial maintenance policy development: A quantitative framework*. International Journal of Production Economics. Pp 85-93.

[9] Márquez, C.A. (2007) The Maintenance Management Framework: Models and Methods for

Complex Systems Maintenance, Springer-Verlag, London.

[10] Ruiz, P.P., Foguem, B.K., & Grabot. B. (2013). *Improving maintenance strategies from experience feedback*. Saint Petersburg, Russia.

[11] Hines, P. & Rich, N. (1997). *The Seven Value Stream Mapping Tools*. International Journal of Operations & Production Management, pp. Vol 17 Iss 1 pp. 46-64.

[12] Raben, R. S.manufacturer2018. Policy Analysis to Provide Part Backups on Critical Machines Using Life Cycle Costing (Case Study: PT. Madubaru PG Madukismo). Industrial Engineering. Yogyakarta: UII.

[13] Dhillon, B.S. (2002) *Engineering Maintenance: A Modern Approach*. CRC Press LLC, Boca Raton.

[14] Holland, J.H., 1975. Adaptation in natural and artificial systems. University of Michigan Press, USA.

[15] M. Harman, The current state and future of search based software engineering, in: 2007 Future of Software Engineering, IEEE Computer Society, 2007, pp. 342–357.

[16] Duarte, J. A. C., Craveiro, J. C. T. A. & Trigo, T. P. (2006). Optimization of the preventive maintenance plan of a series components system. International Journal of Pressure Vessels and Piping, 83, 244-248.

[17] Paz NM, Leigh W. Maintenance scheduling: issues, results and research needs. Int J Oper Prod Manag 1994;14:47

IOP Conf. Series: Materials Science and Engineering **722** (2020) 012072 doi:10.1088/1757-899X/722/1/012072

5S Implementation of SME Readiness in Meeting Environmental Management System Standards based on ISO 14001:2015 (Study Case: PT. ABC)

D F Viranda [1], A D Sari [2], M R Suryoputro [3], N Setiawan [4]

[1, 2, 3, 4] Department of Industrial Engineering, Faculty of Industrial Technology, Islamic University of Indonesia, Jl. Kaliurang Km 14.5 Yogyakarta, Indonesia

E-mails: 15522338@students.uii.ac.id [1], amarria@uii.ac.id [2], ragil.suryoputro@uii.ac.id [3], nashrullah.setiawan@uii.ac.id [4]

Abstract. The application of 5S can be used in assisting environmental management system standards. 5S (*Seiri, Seiton, Seiso, Seiketsu*, and *Shitsuke*) is a simple concept in managing workplaces for efficiency and waste reduction as well as optimizing quality and productivity through an organized environment. An environmental management system is part of a management system which used to manage environmental aspects, compliance obligations, and opportunity risks. ISO 14001: 2015 as an international standard of the environmental management system which refers to the PDCA (Plan-Do-Check-Action) cycle. The environmental management system at SMEs faces greater limitations compared to large companies. 5S can be used as a tool to improve environmental performances by reducing waste, pollution, and energy which will be easier to identify and control. In this study, it's explained how the influence of 5S in fulfilling ISO 14001: 2015 standards in electrical *Batik* stove SME in which the application of 5S is not optimal and resulting in compliance with environmental management systems that less than the ISO 14001: 2015 standards. Factors affecting the application of both 5S and EMS are leadership and commitment, knowledge and training, communication, employee participation and motivation, and external factors such as law and regulation.

1. Introduction

Sustainable indicators such as economic, environmental and social are important factors for large, medium and small scale industries [1]. Small and Medium Enterprises (SMEs) have an important role in Indonesia's economic growth [2]. This can be seen from the Ministry of Finance data wherein 2018 SMEs contributed 60.34% to GDP (Gross Domestic Product) [3]. An increase in the number of SMEs in Indonesia which continues to increase from 36.81 million units in 1998 to 61.7 million units in 2017 (Figure 1). However, the number of SMEs that is increasing every year is not in line with the growth rate of SMEs. The level of SME's contribution to Indonesia's GDP can't reflect the SME's good performance in domestic and international market competition [4].

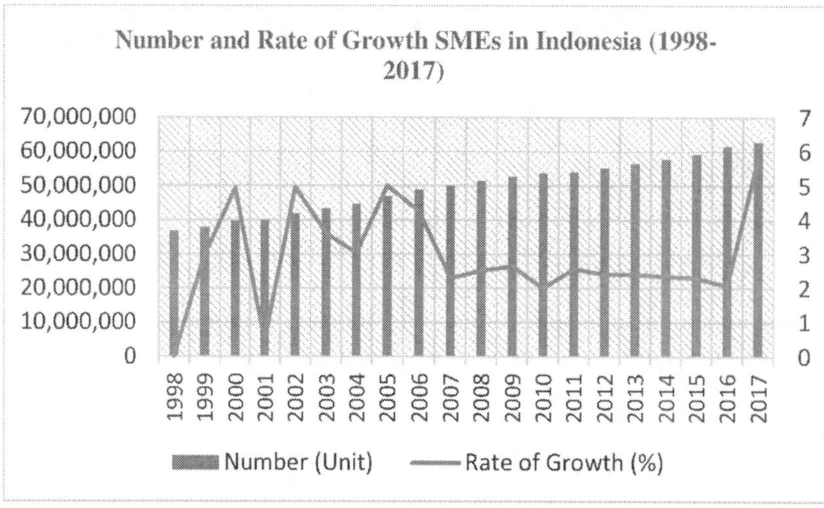

Figure 1. The Number and Rate of Growth of SMEs in Indonesia (Source: Ministry of Cooperatives and SMEs)

One of the significant problems that impede the SME's sustainability is the issue of environmental management systems [5]. Along with the increase in the number of SMEs each year, there's also an increase in greenhouse gas emissions from industrial waste (Figure 2.). SMEs contribute to pollution by 70% of the industrial sector and require changes in policy [6]. The companies must balance out with sustainability factors such as economic, social and environment to achieve greater sustainability performance [7]. The application of an environmental management system is one way to reduce the environmental impact [8]. ISO 14001 is an international standard for implementing an environmental management system [9]. However SMEs have difficulties in implementing environmental management systems due to various factors, namely government regulations, environmental communities, managerial, awareness of environmental issues, and non-governmental organizations [10].

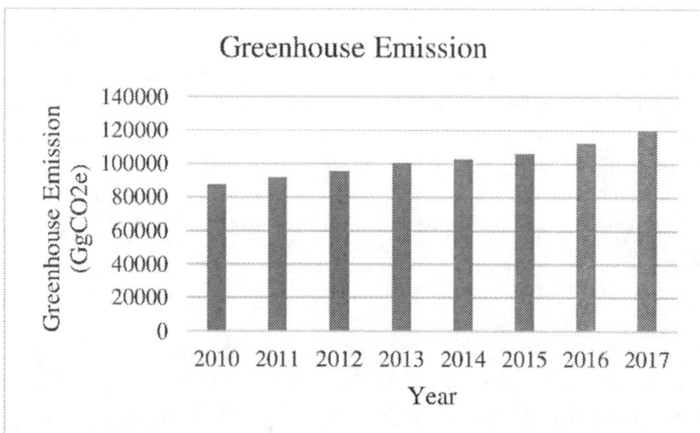

Figure 2. Greenhouse Emission from Industrial Waste (Source: Directorate General of Climate Change Control)

5S which is part of lean can also help companies to implement environmental management systems because their policies drive changes towards cultural and operational improvement [11]. Environmental management in the context of business and industry is based on a philosophy that is highly related to 5S, the application of environmental management system and 5S is highly dependent on the commitment of top management which in managing 5S and environmental management systems, both large and small companies need leadership and strong commitment through an example not just from words [12]. 5S is a tool to improve environmental performance that reduces waste, pollution, and energy, with the use of 5S these three things can be identified easily which has driving factors such as management commitment, responsibility of the project committee, employee participation, and training for all employees ranging from operational to top management, although there are many advantages in implementing 5S and environmental management systems, usually SMEs have greater difficulties in their applications due to limited resources [13]. The success of 5S depends on the people who run the 5S program [14]. Furthermore, the success of 5S lies in crucial factors, such as role and support of management, empowerment and employee training [15].

2. Literature Review

5S is an approach for organizing, standardizing, and continuously improving work area [16]. 5S increases workplace productivity in systematically organized conditions [17]. 5S must be maintained whose application will improve organizational performance [18]. Especially for small scale companies, knowledge about 5S is needed so that the benefits of 5S can be achieved optimally [19]. ISO 14001 is intended to be used by organizations to manage environmental responsibility in a systematic way that contributes to the environmental pillar of sustainable development [20]. ISO 14001 has the aim to standardize the management of environmental aspects in the operational and administrative processes of a company, where its implementation can be applied by various types and sizes of companies and the level of application of environmental development in their businesses [21]. ISO 14001 is a popular management system to be adopted by developing countries [22].

The environmental impacts of production waste (for example, raw materials, energy use, and emissions associated with overproduction) can be avoided by the application of 5S so that the productivity and efficiency of the company will increase [23]. With 5S the company gets the comfort of the environment both from the side of the workplace and from the environment itself because with the 5S which is part of continuous improvement, workers have been routinely trained to always improve the quality of the company [24].

The role of leadership and commitment from the company leader or top management determine the success of the quality improvement program such as 5S and EMS as an example for his subordinates [25] [26]. Not only needed the role from top management, organizations in implementing management systems such as 5S and SML, require employee participation and motivation so that the program can continue to run [27] [28]. One of the commitments from organizations start in training, the training will provide knowledge in application and assist in increasing employee initiative when problems occur during the implementation process [29] [30]. The application of 5S can be adopted as an organizational culture as a way to increase employee the participation [31]. Communication and knowledge transfer systems by management to employees will be useful to follow up and track the extent to which the program has been run in accordance with company planning [32]. External factors also influence the application of 5S and environmental management systems. Although the application of the environmental management system has a positive impact on the company in its fulfillment an understanding and compliance with local laws is required, the main difficulty usually faced by companies in implementing ISO 14001 is the fulfillment of legal requirements [33]. In fact, before implementing an environmental management system a company is required to be aware of applicable laws [34]. Regulation and law make pressure to run an environmental management system [35].

From the previous researches, it is known that the application of 5S affects the environmental management system in reducing waste, energy, and pollution produced by companies. In implementing 5S and the environmental management system, several factors are influenced by leadership and commitment, employee participation and motivation, knowledge and training, organizational culture, communication, and external factors such as law and regulation, as well as stakeholder requests. In this study (Figure 3.), researchers will analyze the effect of 5S and fulfillment of an environmental management system under ISO 14001: 2015 standards on SME *Batik* stoves, where the influencing factors will be seen from the ISO 14001: 2015 clauses and 5S implementation on the SME.

Figure 3. Conceptual Model of The Study

3. Methodology

The object in this issue is the electrical *Batik* stove division at PT. ABC, Kasihan, Bantul, Yogyakarta. This research will identify the effect of 5S on the fulfillment of environmental management system standards and what factors influence the implementations.

3.1. Subject

The subjects in this study are the workers who are directly involved in the division which is divided into four departments, namely production, administration, research and development (RND), as well as the owner of the company itself which consisting of 5 respondents.

3.2. Data Collection

Data of 5S and EMS based on ISO 14001:2015 are collected from the respondens using questionnaire (5S and ISO 14001:2015) and interviews. Data collected is used by purposive sampling method. The collected data is then processed manually using Microsoft Excel 2016 software and to measure the effect and relation between 5S and EMS implementation with regression and correlation, data is processed with SPSS.

3.3. Data Analysis

3.3.1. 5S

Calculation of 5S uses the average value, i.e. total score divided by the total number of questions, then categorized as belonging to the appropriate level.

3.3.2. EMS

The values obtained for each of the elements of ISO 14001 can be used as an illustration of how much the company is able to meet the standard requirements of ISO 14001 in implementing SML [36]. In this analysis, each requirement element score obtained through a questionnaire will be entered into the calculation formula as follows:

$$Xi = \frac{\Sigma Pi}{\Sigma Qi}, i = 1, 2, 3 \ldots \quad (1)$$

3.3.3. Regression

Regression test is a statistical analysis method used to determine the causal relationship between 5S activities and compliance with environmental management system standards. The following are the hypotheses used in this study

Ho = there is no influence between 5S activities and compliance with environmental management system standards
Ha = there is an influence between 5S activities and compliance with environmental management system standards

IOP Conf. Series: Materials Science and Engineering **722** (2020) 012072 doi:10.1088/1757-899X/722/1/012072

3.3.4. Correlation

Correlation test is used to determine the closeness of the relationship between two variables, namely 5S activity and compliance with environmental management system standards, and to determine the direction of the relationship that occurs. The hypothesis in this study are

Ho = there is no relationship between 5S activities on compliance with environmental management system standards
Ha = there is a relationship between 5S activities and compliance with environmental management system standards.

4. Results

In the calculation of 5S activities (Table 1), there are differences of opinion among the respondents. Respondent at the top management level (SME's owner) with a value of 1.86 thinks that the 5S activities that the company does are still in the poor category or are under-performed. Whereas 2 respondents at the intermediate level, Head of Administration and the Head of RND rate that 5S activities are good or the activities are carried out sufficiently (applied and clear in most areas) with a value of 2.77 and 3. Head of Production with a score of 3.59 assumes that 5S activities have been categorized as excellent which activities are carried out well (fully clear and applied to all areas). One respondent at the operator level, Production Operators, rated that 5S activities are the world-class category or that the activities were carried out very well and there was supporting evidence with a value of 4.73.

Table 1. 5S Score

No	Respondent	Score	Category	(%)
1	Head of Admistration	2.77	GOOD	55.45
2	Head of Production	3.59	EXCELLENT	71.82
3	Operator	4.73	WORLD CLASS	94.55
4	Owner	1.86	POOR	37.27
5	Head of RND	3	GOOD	60

The difference between each respondent at each level of management shows that the goals of top management have not been well communicated. Differences in the assessment of 5S activities are also due to differences in the standards of each respondent related to 5S activities based on interviews with workers and the owner. The objectives of management are not carried out because the instructions or instructions given to their employees are unclear and inaccurate [37]. Effective communication is needed so that the goals to be achieved by the central leadership can be understood and implemented by employees, effective communication must also be established in both directions between employees and leaders, with proper communication employees will feel more valued and connected to their workplaces so that they continue to provide contribution in improving company performance [38].

Based on the results of interviews with employees at PT. Putra Multi Cipta Teknikindo showed that there was no 5S training given by the company and there was no evaluation from top management in the form of direct reprimands or self-assessment of 5S activities. 5S training is the responsibility of management to provide the resources needed for employees to understand 5S activities [39]. The development, implementation, and continuous monitoring of 5S implementation is the responsibility of management which to implement 5S, the initial step begins by providing workers with appropriate training on the various elements and benefits of 5S [40]. Training and evaluation are needed as one of the activities of the 5S elements namely shitsuke which supports the sustainability of the 5S program [41]. Evaluation can help management to identify improvements in the 5S performance that has been carried out [42]. Management needs to monitor the entire 5S process and explain the importance of 5S to workers [43].

Even though the company has obtained ISO 9001: 2008 certification, in its procedure the company does not set 5S as part of the standard operating procedure. The owner considers that the preparation of ISO 9001: 2008 requirements documents have followed the guidance of the BSN (National Certification Agency). According to the BSN, even though 5S is not an ISO requirement, 5S is the basis for implementing ISO management systems and companies that want to run ISO management systems must first run 5S well. The influence of BSN has a role in organizational behavior in developing and implementing activities for the objectives to be achieved [44]. BSN as part of a government agency influences the actions taken by companies. Government policies and actions affect the business environment, one of which is the design and implementation of the workplace [45]. Governments have diverse important roles in determining corporate investment decisions [46].

Comparable to 5S assessments, the assessment of compliance with ISO 14001: 2015 standards also varies from each respondent (Figure 4). The Head of Administration, Head of Production, Director and Head of RND assess that the fulfillment of the ISO 14001: 2015 standard is still below the standard value of 21 while the production operator assesses that the fulfillment of the ISO 14001: 2015 standard is above the standard with a value of 25.71. This assessment difference is caused by the absence of understanding by top management related to the environmental management system. The standard value of 21 is obtained from a total of 7 clauses multiplied by the default value of the questionnaire that is 3.

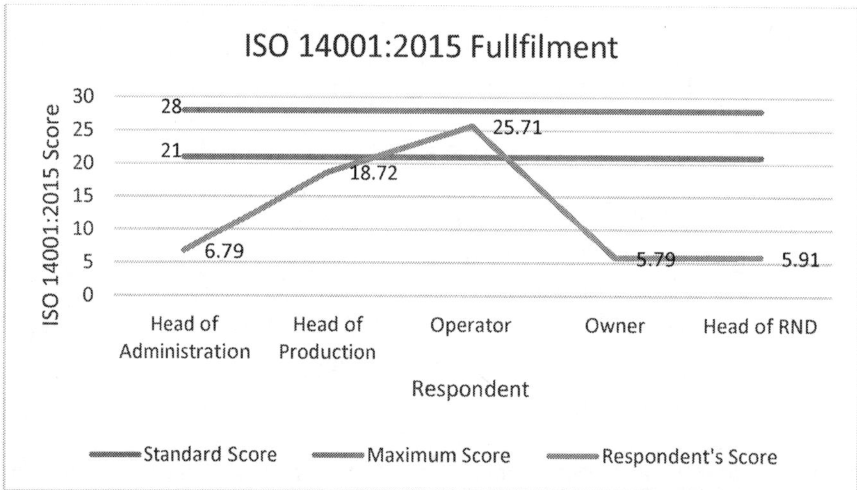

Figure 4. ISO 14001:2015 Fullfilment

The assessment of ISO 14001: 2015 standard fulfillment is used to measure the extent of compliance with ISO 14001 that the company has implemented and to measure how deep the company's understanding of existing environmental issues. With ISO 14001, companies can improve environmental performance, improve compliance with environmental standards, reduce or overcome risks arising from company activities, and can reduce production costs and increase profits [47].

Based on the results of interviews with SME's owner, the owner plans to add ISO 14001: 2015 certification to support his business along with plans to open a separate division to sell *Batik* made by the company within the next five years. However, SME's owners do not yet know what laws must be fulfilled as compliance obligations which are the most important part of ISO 14001: 2015 and environmental issues related to the business being run. Understanding of environmental issues is important to understand by top management because the success or failure of top management systems is strongly influenced by factors of leadership, commitment, and responsibility of top management. Top management becomes the driving force in running the environmental management system, top management is responsible for setting ISO 14001 implementation plans and is responsible for ensuring that everyone in the company has the right enthusiasm and knowledge about the environmental management system [48]. The commitment of top management in carrying out ISO 14001 has an

IOP Conf. Series: Materials Science and Engineering **722** (2020) 012072 doi:10.1088/1757-899X/722/1/012072

important role [49]. The leadership factor is influential in promoting the environmental management system and motivating employees in their role to carry out ISO 14001. That way, top management, namely the owner of SME, must have the right knowledge and commitment on issues and legislation to run the environmental management system.

There is no allocation of resources to finance the implementation of ISO certification. According to SME owner, because the company is guided by BSN, SME owner are better off waiting for directions from BSN related to the application and completion of ISO certification documents. These resources are better allocated for production needs. As is the case with 5S, the leadership and commitment factors of top management are one of the important driving factors for building a successful EMS program [50]. One form of company commitment in implementing ISO 14001 is by preparing human and financial resources to attend ISO 14001 training or other environmental standards in Indonesia [51]. According to Larsson preparing resources is part of fulfilling the ISO 14001 clauses [52]. In preparing for the fulfillment of ISO 14001 standards, human resources, finance, infrastructure, and technology become important points that must be considered by company leaders.

Based on a regression test using SPSS, it is known that there is an influence between 5S activities and compliance with environmental management system standards because the significance value produced is 0.031 less than 0.050. While the correlation test shows that there is a relationship between 5S activities and compliance with environmental management system standards. The relationship between 5S activities and compliance with environmental management system standards has a very strong relationship with a close level of 0.912. Based on this test, 5S activities that have not been optimal can cause a lack of compliance with environmental management system standards. 5S provides seven types of waste prevention (unnecessary motion, defects, overproduction, waiting, transport, inventory, inappropriate processing) that can affect the environmental performance of a company [53]. The application of ISO 14001 has a positive effect on improving environmental performance and reducing environmental impacts that can harm the company [54].

Based on Table 2 it is known that leadership has the smallest value in meeting ISO 14001: 2015 standards. The interviews also found that the owner of SME does not yet have the capacity of knowledge and commitment in running the environmental management system. This is evidenced by the lack of infrastructure that supports environmental management systems such as waste disposal sites at each production station, the absence of trash bins in several places, the absence of training for employees on environmental issues or ISO 14001 standards, the lack of funds allocated to run the system environmental management and the lack of top management skills in Indonesian laws and regulations on the environment. As previously stated, leadership is the most important factor in ensuring ISO 14001 fulfillment. Leadership is an important component for implementing an environmental management system project [55].

Table 2. Ranking of Factors Affection ISO 14001:2015

Factors Affecting ISO 14001:2015	Score (%)
Leadership	10.94
Evaluation	14.22
Planning	14.46
Operation	14.50
Organization Context	14.76
Support	15.23
Improvement	15.89

Based on observations and interviews, instructions given by top management are unclear so that top management desires in implementing 5S are not carried out, besides the absence of providing training to employees about the elements and importance of 5S to be applied in the company, top management also does not provide an evaluation of implementation of 5S that has been done and the absence of appeals or reprimands from top management for the inappropriate implementation of 5S in the company. The lack of commitment from top management to 5S activities in the company is evidenced by the absence of elements or 5S elements in work procedures or instructions, lack of infrastructure to support 5S activities such as trash bins, dividing lines, cleaning tools, and unclear hygiene mechanisms that are applied and there is no standard for how 5S should be applied in a company.

The lack of commitment from top management is the dominant obstacle in successfully implementing a quality management system such as 5S, commitment and appropriate leadership attitudes provide a clear view of the goals and values of implementing the 5S program. Lack of commitment and leadership attitudes can be caused by various factors such as lack of experience, lack of training, refusal to change, and doubt to initiate improvement programs [56]. Significant obstacles in implementing 5S are the lack of communication between top management and operators, the distance between top management and operators, the lack of training and awareness have given from management to employees [57]. Decision making by management can be caused by one of the macro external environments namely law and regulation, where government policies determine the limits of the company's strategy and opportunities [58]. Laws and regulations in Indonesia do not support the application of 5S. There is no government regulation on how 5S should be implemented in Indonesia, especially for SMEs.

5. Conclusion

The implementation of 5S and compliance with environmental management system standards ISO 14001: 2015 in the company are not optimal. The company should allocate resources for the application of 5S and fulfillment of ISO 14001: 2015 environmental management system standards. Discipline 5S to all employees so that 5S continues to run well and provide preliminary information about the environmental management system to meet ISO 14001: 2015 standards.

References

[1] J. Shields and J. M. Shelleman, INTEGRATING SUSTAINABILITY INTO SME STRATEGY, *Journal of Small Business Strategy,* Vol. 2, no. 59-76, p. 25, 2015.

[2] P. P. Hapsari, A. Hakim and A. Hakim, Pengaruh Pertumbuhan Usaha Kecil Menengah (UKM) terhadap Pertumbuhan Ekonomi Daerah (Studi di Pemerintah Kota Batu), *Wacana,* Vol. 2, no. 88-96, p. 17, 2014.

[3] D. A. Putra, "Liputan 6," 19 April 2019. [Online]. Available: tttps://www.liputan6.com/bisnis/read/3581067/umkm-sumbang-60-persen-ke-pertumbuhan-ekonomi-nasional?utm_expid=.9Z4i5ypGQeGiS7w9arwTvQ.0&utm_referrer=https%3A%2F%2Fwww.google.com%2F.

[4] Y. Wardi, P. Susanto and N. L. Abdullah, Orientasi Kewirausahaan pada Kinerja Usaha Kecil dan Menengah (UKM) Sumatera Barat: Analisis Peran Moderasi dari Intensitas Persaingan, Turbulensi Pasar dan Teknologi, *Jurnal Manajemen Teknologi,* Vol. 1, no. 46-61, p. 16, 2017.

[5] Bank Indonesia, KAJIAN KESIAPAN UMKM RAMAH LINGKUNGAN DALAM MENDAPATKAN AKSES PEMBIAYAAN, Jakarta, 2012.

[6] M. Singh, M. Brueckner and P. K. Padhy, Environmental management system ISO 14001: effective waste minimisation in small and medium enterprises in India, *Journal of Cleaner Production,* Vol. 102, pp. 285-301, 2015.

[7] J. El Baz and I. Laguir, Third-party logistics providers (TPLs) and environmental sustainability

practices in developing countries: the case of Morocco, *International Journal of Operations & Production Management,* Vol. 37, no. 10, pp. 1451-1474, 2017.
[8] M. D. Guerrero-Baena, J. A. Gómez-Limón and J. V. Fruet, A multicriteria method for environmental management system selection: An intellectual capital approach, *Journal of Cleaner Production,* Vol. 105, pp. 428-437, 2015.
[9] L. M. C. Martins Da Fonseca, ISO 14001:2015: An Improved Tool for Sustainability, *Journal of Industrial Engineering and Management,* Vol. 8, no. 1, pp. 37-50, 2015.
[10] G. S. Natarajan and D. A. Wyrick, Framework for Implementing Sustainable Practices in SMEs in the United States, *Proceedings of the World Congress on Engineering,* Vol. 1, 2011.
[11] P. Puvanasvaran, R. K. S. Tian, V. Suresh and M. R. Muhamad, Lean principles adoption in environmental management system (EMS): A survey on ISO 14001 certified companies in Malaysia, *Journal of Industrial Engineering and Management,* Vol. 5, no. 2, pp. 406-430, Journal of Industrial Engineering and Management.
[12] M. O'hEocha, Case studies A study of the influence of company culture , communications and employee attitudes on the use of 5Ss for environmental management at Cooke Brothers Ltd, *The TQM Magazine,* Vol. 12, no. 5, pp. 321-330, 2006.
[13] R. Jamian, M. N. A. Rahman, B. M. Dero and N. Z. N. Ismail, A Conceptual Model Towards Sustainable Management System Based Upon 5S Practice For Manufacturing SMEs, *Asia Pacific Journal of Operations Management,* Vol. 1, no. 1, pp. 19-31, 2012.
[14] A. Q. Malik, Implementation plan of 5s methodology in the basic surgical instruments manufacturing industry of Sialkot, *International Journal of Scientific and Technology Research,* Vol. 3, no. 9, pp. 176-182, 2014.
[15] W. A. A. Wan Asri and C. M. Azman, The effectiveness of implementation of 5S on employee motivation, *Business and Social Sciences Review,* Vol. 1, no. 1, pp. 41-52, 2011.
[16] R. S. Agrahari, P. A. Dangle and K. V. Chandratre, Implementation of 5S Methodology in the Small Scale Industry: a Case Study, *International Research Journal of Engineering and Technology(IRJET),* Vol. 3, no. 1, pp. 180-187, 2015.
[17] S. P. Deshpande, V. V. Damle, M. L. Patel and A. B. K. , Implementation of '5S' Technique in a Manufacturing Organization: a Case Study, *International Journal of Research in Engineering and Technology,* Vol. 4, no. 1, pp. 136-148, 2015.
[18] C. Veres, L. Marian, S. Moica and K. Al-Akel, Case study concerning 5S method impact in an automotive company, *Procedia Manufacturing,* Vol. 22, pp. 900-905, 2015.
[19] P. M. Sánchez, C. M. Rodriguez, U. Maruyama and F. Salazar, Impact of 5S on quality , productivity and organizational climate - Two Analysis Cases, *Proceeding of the 2015 International Conference on Operations Excellence and Service Engineering Orlando, Florida, USA,* pp. 748-755, 2015.
[20] SNI, SNI ISO 14001:2015, 2015.
[21] J. W. S. Oliveira, O BLOG DE NOTÍCIAS: DO HIPERGÊNERO AO HIPERLEITOR 124, *Santa Cruz,* Vol. 38, no. 7, p. 124–145, 2013.
[22] W. To and P. Lee, Diffusion of ISO 14001 environmental management system: Global, regional and country-level analyses, *Journal of Cleaner Production,* Vol. 66, pp. 489-498, 2014.
[23] J. Tice, L. Ahouse and T. Larson, Lean production and EMSs: Aligning environmental management with business priorities, *Environmental Quality Management,* Vol. 15, no. 2, pp. 1-12, 2005.
[24] C. Jaca, M. Ormazabal, E. Viles and J. Santos, Environmental comfort based (ECB) methodology as a tool for preparing Kaizen application in a catering service company, *TQM Journal,* Vol. 30, no. 4, pp. 281-295, 2018.
[25] R. Attri, B. Singh and S. Mehra, Analysis of interaction among the barriers to 5S implementation using interpretive structural modeling approach, *Benchmarking: An International Journal,* Vol. 104, pp. 1-20, 2017.
[26] D. Prajogo, A. K. Tang and K.-H. Lai, The diffusion of environmental management system and

its effect on environmental management practices, *International Journal of Operations & Production Management,* Vol. 34, no. 5, pp. 565-585, 2014.
[27] S. C. Houa, M. Haslinda, S. Muliati, A. M. Miri and A. F. Rahim, Implementation of 5S in Manufacturing Industry: A Case of Foreign Workers in Melaka, *MATEC Web of Conferences,* Vol. 150, no. 05034, 2018.
[28] N. A. Pratama, V. Kumar, A. Kumari, J. A. Garza- Reyes and S. P. Nadeem, Investigating the benefits and challenges of the implementation of ISO 9001 and ISO 14001 in the aerospace industry, in *Proceedings of the International Conference on Industrial Engineering and Operations Management, Bandung, Indonesia, March 6-8, 2018*, Bandung, 2018.
[29] D. Kiran, Total Quality Management: Key Concepts and Case Studies, Elsevier Inc., 2017.
[30] N. S. Khairani, E. S. Kasim, I. D. Rajamanoharan and F. N. Misman, Green supply chain management in the Malaysian automotive industry: A systems thinking perspective, *International Journal of Supply Chain Management,* Vol. 6, no. 2, pp. 38-48, 2017.
[31] B. Purwanggono, R. Ruminta and S. Irawat, Analisis Faktor - Faktor yang Memengaruhi Motivasi Karyawan dalam Menerapkan Budaya Kerja 5S (Studi Kasus pada Karyawan PT.PLN (Persero) P3JB APP Semarang), in *Posiding SNAFTIF*, Kudus, 2014.
[32] N. Khamis, M. N. A. Rahman, K. Jamaludin, A. Ismail, J. Ghani and R. Zulkifli, Development of 5S Practice Checklist for Manufacturing Industry, in *Proceedings of the World Congress on Engineering 2009 Vol I*, London, 2009.
[33] A. Mazzi, S. Toniolo, M. Mason, F. Aguiari and A. Scipioni, What are the benefits and difficulties in adopting an environmental management system? The opinion of Italian organizations, *Journal of Cleaner Production,* Vol. 139, no. September, pp. 873-885, 2016.
[34] C. Neagu and A. Neagu, The Stages of Implementing an Environment Management System in, *Revista Română de Statistică ,* no. 10, pp. 93-99, 2015.
[35] Y. Yang, A. K. Lau, P. K. Lee, A. C. Yeung and T. E. Cheng, Efficacy of China's strategic environmental management in its institutional environment, *International Journal of Operations and Production Management,* Vol. 39, no. 1, pp. 138-163, 2019.
[36] B. Hardiwiardjo, ISO 14001: *Panduan Penerapan Sistem Manajemen Lingkungan*, Jakarta: Gramedia Pustaka Utama, 1997.
[37] T. Osada, *Sikap Kerja 5S*, Jakarta: CV. Teruna Grafica, 2000.
[38] . Gieber, "ISIXSIGMA," 22 February 2013. [Online]. Available: https://www.isixsigma.com/community/blogs/5s-of-communication/.
[39] J. S. RANDHAWA and I. S. A. Article, 5S implementation methodologies: literature review and directions, *International Journal of Productivity and Quality Management,* Vol. 34, no. 3, pp. 1-45, 2016.
[40] F. C. Filip and V. Marascu-Klein, The 5S lean method as a tool of industrial management performances, *IOP Conference Series: Materials Science and Engineering,* Vol. 95, no. 1, pp. 1-6, 2015.
[41] R. Gapp, R. Fisher and K. Kobayashi, Implementing 5S within a Japanese context: An integrated management system, *Management Decision,* Vol. 46, no. 4, pp. 565-579, 2008.
[42] J. Sing, V. Rastogi and R. Sharma, Implementation of 5S practices: A review, *Uncertain Supply Chain Management,* Vol. 2, no. 3, pp. 155-162, 2014.
[43] S. R. B. Ashraf, M. M. Rashid and D. A. R. M. H. Rashid, Implementation of 5S Methodology in a Food & Beverage Industry: A Case Study, *International Research Journal of Engineering and Technology,* Vol. 4, no. 3, pp. 1791-1796, 2017.
[44] A. S. Burlea and I. Popa, Legitimacy Theory, *Encyclopedia of Corporate Social Responsibility,* Vol. 28, no. 6, pp. 1579-1584, 2014.
[45] C. Johnson, *Introduction: The Idea of Industrial Policy*, San Fransisco: ICS Press, 1984.
[46] The National Academics of Engineering , *Time Horizons and Technology Investments*, Washington DC: The National Academic Press, 1992.
[47] R. Tanaya and T. W. Panjaitan, Persiapan Implementasi ISO 14001 pada CV. ABC, *Jurnal Titra,*

Vol. 3, no. 2, pp. 143-150, 2015.
[48] A. S. Morris, *ISO 14000 Environmental Management Standards*, West Sussex: John Wiley & Sons Inc, 2004.
[49] Y. Mauliddina and A. Susanty, Evaluasi Faktor Sukses Implementasi Iso 14001 Dengan Metode Analytical Hierarchy Process (AHP) Di PT. Apac Inti Corpora Semarang, *Industrial Engineering Online Journal,* Vol. 4, no. 4, pp. 1-8, 2018.
[50] A. P. Iswara and R. Boedisantoso, Penyederhanaan ISO 14001:2015 Sebagai Inovasi dalam Monitoring Lingkungan dan K3 di Industri Kecil Menengah, in *The 2nd Conference on Innovation and Industrial Applications (CINIA 2016)*, Surabaya, 2016.
[51] A. D. Ananta, H. Suryono and S. Mamik, Persiapan Penerapan ISO 14001:2015 di PT. Semen Bosowa Banyuwangi, *GEMA Lingkungan Kesehatan,* Vol. 17, no. 1, pp. 9-13, 2019.
[52] S. Larsson, "NMC," 2 December 2015. [Online]. Available: nmc.a.se/wp-content/uploads/2017/06/2015-12-02-Stefan-Larsson.pdf. [Accessed 19 April 2019].
[53] R. Jamian, M. N. A. Rahman, B. M. Deros, M. S. Mohamed and N. Z. N. Ismail, A conceptual approach of 5S to improving quality and environmental performance of Malaysian oil palm dealers, *Jurnal Teknologi (Sciences and Engineering),* Vol. 70, no. 1, pp. 65-73, 2014.
[54] T. H. Arimura, N. Darnall, R. Ganguli and H. Katayama, The effect of ISO 14001 on environmental performance: Resolving equivocal findings, *Journal of Environmental Management,* Vol. 166, pp. 556-566, 2016.
[55] A. Kasim, Environmental management system (EMS): Postulating the value of its adoption to organizational learning in hotels, *International Journal of Contemporary Hospitality Management,* Vol. 27, no. 6, pp. 1233-1253, 2015.
[56] F. R. Z. a. Q. M. Talib, Analysis of interaction among the barriers to total quality management implementation using interpretive structural modeling approach, *Benchmarking: An International Journal,* pp. 563-587, 2011.
[57] R. Sharma and J. Singh, Impact of implementing japanese 5S practices on total productive maintenance, *International Journal of Current Engineering and Technology,* Vol. 55, no. 22, pp. 2277-4106, 2015.
[58] T. S. Bateman and S. A. Snell, Manajemen: Kepemimpinan dan Kerja Sama dalam Dunia yang Kompetitif, Jakarta: Salemba Empat, 2014.

Impact Modeling of Bio-Textile Composite from Agave and Pineapple Leaf Fibers with Sandwich Structure

A Subagyo[1], and T M Amin[2]

Department of Chemical Engineering, Concentration Textile Engineering
Islamic University of Indonesia
Yogyakarta, Indonesia

E-mails: asmanto.subagyo@uii.ac.id, 925210101@uii.ac.id

Abstract. The objective of this research is a comprehensive model is presented to study the deformation and damage to bio-textile composite from agave and pineapple leaves fibers under s mall projectile impacts. The approach dynamically follows the strain wave propagation along each individual yarn away from the impact point. As in previous work the mechanics of wave propagation is formulated in terms of impulse-momentum balance equations, which are solved at each fiber crosser using a finite different technique. Our approach explicitly considers the various projectile characteristics namely; mass, velocity and shape, as well as all fiber properties such as denier, modulus, and tenacity. Even more importantly the model allow to account for s lippage of yarn at crossovers and within the c lamps In previous work, slippage of yarn and fracture processes are described with a kinetic approach, which ex plicitly accounts for their dependence on impact rate. Results of model are in good quantitative agreement with publis h experimental data on s ingle ply and several plies biotextile composite. They also clearly indicate that yarn slippage through clamps often seen experimentally is responsible for some salient features observed in balistic.

1. Introduction

Amour body design has now involved, with the use of hard body amour using metal until soft body amour. Research on bullet-proof vests have been done by some previous researchers. Research using material from the three components of 5086-H32 aluminum alloy, epoxy and Kevlar fabrics impregnated with shear thickening fluid is able to withstand projectile 9 mm pistol [4,14]. While research using ramie-epoxy composites can lower the cost of 95% compared with Kevlar [5]. Research using bullet-resistant vest panel bio textile been done with a sandwich structure [12], while make composites for bullet-resistant fabric with silk thread [6,11]. Research bulletproof vest manufacture of polyester fibers have been made [12]

The impact resistance of bio-textile composite is a subject of considerable experimental interest. The response of bio-textile to ballistic impact is very complex problem solving the interplay of a large number of variables, not the least of which relates to the dependence of bio-textile composite behavior on the size and shape of the projectile. Because of that complexity and the need to identify controlling variables, several models have been proposed to describe ballistic impact. The problem of single yarn impact from agave and pineapple at short times has been described analytically. The case of bio-textile composite cannot, however, be solved exactly because the resolution of stress wave interactions and reflections at yarn crossovers is intractable. As a result, one has to resort to approximate analytical techniques or to numerical simulation models. An analytical solution based on the equations has been derived, but the solution focuses entirely on the impact zone and avoids consideration of boundary conditions. Other mathematical model descriptions based on more semi empirical approaches have been proposed [1] and [3].

In alternative model developed [13]. This approach, however, predicts that the deformation region has a conical shape, which is unrealistic [7].

A finite element model in corporating viscoelasticity and a stain-rate sensitifity criterion has been proposed [5] but the description is unable to address issue of yarn slippage and again leads to the prediction that the area affected by the projectile is circular[13].

In a more refined approach pioneered [8,9]a single-ply fabric is represented by two-dimensional array form of nodes representing yarn crossover point [10]. Furthermore, finite difference methods are then used to solve the impulse-momentum balance equations at each crossover. For simplicity, all these models assumed the fibers to pin-jointed at their crossovers, and the influence of fiber slippage often observed experimentally could not be assed. Some numerical results detailing the importance of slippage have been reported for the case of a single crossover [10,15].

The purpose of this is to develop Roylance's model to multi-ply bio-textile composite and to fully consider the effects of yarn slippage at bio-textile composite at crossover points as well as within the clamps[9]. However, in contrast to the original model and describe yarn fracture using kinetic approach, which explicitly accounts for the dependence of yarn strength on impact velocity.

2. Model Representation

The description that follows is for a plain weave pattern for bio-textile, but approach can be easily extended to other structure designs. The contrast to previous work, however, single yarns are not pin-jointed but, rather, allowed to slip over each other as well as within the clamp as see the section on yarn slippage. Furthermore, in accordance with typical experimental sets-up the structure is clamped between thick plates with opening (L) and impacted in the center of the aperture. The projectile is assumed to have a spherical shape with radius (r_p) and mass (m_p).

As we now proceed to describe our computational algorithm for a single ply. Extension to systems of several plies will be the subject of separate publication. At time where t = 0, and is than the initial projectile velocity (v_p) is imposed along the z-direction on the central crossover of the ply, thereby creating a strain in the neighboring crossovers. The strain propagates outward along individual yarns at the velocity of sound, with the equations as follow.

$$C = (E/\rho)^{1/2} \tag{1}$$

Where E is the yarn's tensile moduls and ρ its density. Thus, in computational scheme, the strain propagation is followed in a series of incremental times steps, namely;

$$\Delta t = l/c \tag{2}$$

Where l denotes the unit segment length between adjacent crossovers. At each step, all the crossover points at the current wave front are being visited and their velocities ad positions are continuously updated. These updates are made in a Lagrangian reference frame attached and extending with the fiber. Vectors components with respect to the laboratory coordinator are then calculated using the corresponding inclination angles, and than algorithm proceeds as follows; crossover with coordinate indexes (i, j) in the ply of yarn. Its velocity at time (t) is recalculate from that at time $t - \delta t$ an impulse-momentum balance equation [5].

$$V(i, j, t) = v(i,j,t - \delta t) + F(i,j,t)\, \delta t/m \tag{3}$$

Where V is three dimensional vectors, *m* is mass of a crossover and F(*i, j, t*) denotes the net force from all yarn segments connected to the crossover (*i, j*). Simplicity in all simulations, we assume a linear Hookean behavior for each segment, F – ε, in which ε is the local strain. The strain value on the yarn segment connecting crossover (*i, j*) to crossover (*i* – 1,*j*) is calculated from;

$$\varepsilon\,(i, j, t) = \varepsilon(i,j,t - \delta t)+\{[s(i,j,t - \delta t)+v(i,j,t)\delta t]-[s(i-1,j,t- \delta t)+v(i-1,j,t)\ \ \delta t]\}/[s(i,j,t- \delta t) - s(i-1,j,t- \delta t)] - 1 \quad (4)$$

Where s denotes the position vector for a crossover. Having determined its velocity from equation (3) and (4), the new position vector for crossover (*i,j*) is then obtained from;

$$s(i,j,t) = s(i,j, t - \delta t) + v(i,j,t)\delta t \quad (5)$$

The boundary conditions at the clamped edges require that;

$$V(i_{edge}, j_{edge}, t\,) = 0 \quad (6)$$

Where after each time increment δt, the position of the projectile is updated to;

$$S_p(t) = s_p(t - \delta t) + v_p(t)\,\delta t \quad (7)$$

As the projectile follows its path along the z-direction and the crossovers in the ply are deflected along that same direction, adjustments must be made to prevent penetration between a ply and the projectile. This is performed using the velocity corrections be described below.
In order to ensure stability of the numerical algorithms, we recommend that these corrections be performed immediately after the velocities (Equation 3) of all crossovers have been calculated and before their new positions are updated. The corrections for the exclusion between a crossover (*i,j*) and the projectile closely follow those described [9]

$$S'(i,j,t) = s(i,j,t - \delta t)+ v(i,j,t)\,\delta t \quad (8)$$

Denote the temporary position of that crossover at time t and R_{cross} denote the distance between that new position and the center of the projectile. Clearly, interpretation occurs when $R_{cross} < R_{pro}$ in that case, the correction is made [2] and [10]

$$V(i, j, t) \rightarrow v(i,j,t) + [s''(i,j,t) - s'(i,j,t)]/\,\delta t \quad (9)$$

In which

$$S''(i,j,t) = s_p(t) + [s'(i,j,t) - s_p(t)]R_p/R_{cross} \quad (10)$$

Where $s_p(t)$ denotes the position vector of the center of the projectile. The gain in momentum for crossover (i,j) equation (9) must
Be compensated by a corresponding loss for the projectile as follows;[1]

$$V_p(t) \rightarrow v_p(t) - [s''(i,j,t) - s'(i,j,t)][m/m_p\,\delta t \quad (11)$$

Equation 11 effectively describes the slowing down of the projectile due to its contact with the crossovers in the ply.

IOP Conf. Series: Materials Science and Engineering **722** (2020) 012073 doi:10.1088/1757-899X/722/1/012073

3. Yarn Fracture and Slippage

As the plies are being deformed, our approach also allows for breakage of yarn segments between crossover. Following is previous work on the ultimate tensile strength of polyethelene and poly-phenylene-terephthalamide fibers, assume that a yarn segment *i* breaks through a kenetic process at a rate:[7]

$$V_i = \tau \exp[-(U - \beta\sigma_i)/kT] \quad (12)$$

Where τ is the thermal vibration frequency , U and β are the activation energy and volume for covalent bond breakage, and σ_i is the local stress acting on segment *i*. Typical value of τ, U and β for high strength of fiber in these case agave cantala fibers and pineapple leaf fibers. As note that in this research, we neglect the contributions of non-covalent (van der waals or hydrogen) bond to the yarn breaking process. This is a reasonable approximation in view of the high strain rates experienced by the fibers during ballistic testing.

In our approach, yarn slippage is assumed to set in at a rate that has the same functional form as that yarn fracture (equation 12) but with different values for the aviation energy U and volume β. On the order hand, σ now denotes the different in stress in two concecutive yarn segments that are separated by a crossover.

Upon slippage, the unit length of the strained segment is increased, increment δl_{slip} ($<< l$) within the time interval δt, whereas that of its less strained counterpart is decreased by the same amount. These changes in unit length for a segment must be properly accounted for when calculating its strain value.

Thus for a segment connecting crossover (i,j) to crossover (*i* –j), this is obtained by multiplying the term in parenthesis in equation (4) by $l_{t-\delta t}/l_t$ where $l_{t-\delta t}/l_t$ and l_t denotes the unit lengths of the segment at time t – δt and t respectively. In addition slippage at a crossover point, also allow for yarn slippage at the clamps. The process is similar to that described above expect for the fact that it is initiated at a higher value of activation energy.

Currently discuss about parameter values for describing the experimental data of on the ballistics performance of a single-ply blended bio-textile composite (agave fiber and pineapple leaf fiber) clamped at two opposite sides only. The other two sides remain unrestrained. In this research value of the basis weight (bio-textile composite) ω of bio-textile composite are readily available from experiment. In typical weaving techniques, the yarn are woven into a very tight construction so that value of ω and of the yarn denier are not independent. Thus, for plain weave ω as follows;

$$\omega = 0.088 + 0.00009*\text{denier} + 1.19.10^7*\text{denier}^2 \quad (13)$$

which was obtained from a fit of experimental ω values (in kg/m^2) for yarn with 200 < denier < 1500 as equations (5). Within the framework of square lattice representation and one also shows easily that the distance *l* between successive crossovers is given by;

$$l = 2*(\text{denier}*\ 1.11\ 10^{-7})/\omega \quad (14)$$

where the term in parentheses represents the yarn denier expressed in kg/m^2, and the mass of a crossover is then obtained from

$$m = \omega*\ l^2 \quad (15)$$

All the results presented bellow are for a single plain weave blended bio-textile composite with ω = 180 kg/m^2.In high velocity testing of the yarn, it is reasonable to assume that the impacted yarn should have a modulus value higher than its static value. As known that values parameters for yarn slippage at crossover more difficult to determine, because values of the normalized activation volumes βE/kT for these processes are left as free parameters, denotes by β_{cross} and β_{clamp} respectively.

In accordance with the experimental set-up with views references, assume a spherical projectile with a radius of 3.5 mm and a mass equal to 4.6 gram.

The model description is in terms of only two free parameters, β_{cross} and β_{clamp} which describe the extent of yarn slippage through crossovers and within the clamps respectively. All other parameter value such as yarn modulus and strength are easily determined from experiment.

4. Discussion

Result for the dependence of residual velocity on impact velocity (V_p) are presented in Figure 1. Model predictions for a choice of $\beta_{cross} = 10^3$ and $\beta_{clamp} = 10^4$ are represented by symbol O and compared to experimental data of reference symbol*. The agreement is judged to be excellent , and calculated threshold perforation velocity observation. The discontinuity observed experimentally impact velocity around V_p 300 m/sec, which is very well reproduced by the model, an even more stringent test of approach in Figure 2 for the dependence of the energy absorbed by the bio-textile composite on the impact energy. The former is simply given by the difference between the kinetic energy of the projectile before and after impact. As Figure 1, the experimental data of (*) show a discontinuity at an impact energy of 157 joule, which corresponds to impact velocity $V_P = 300$ m/second. Inspection of Figure 2 again reveals an excellent agreement with model calculations.

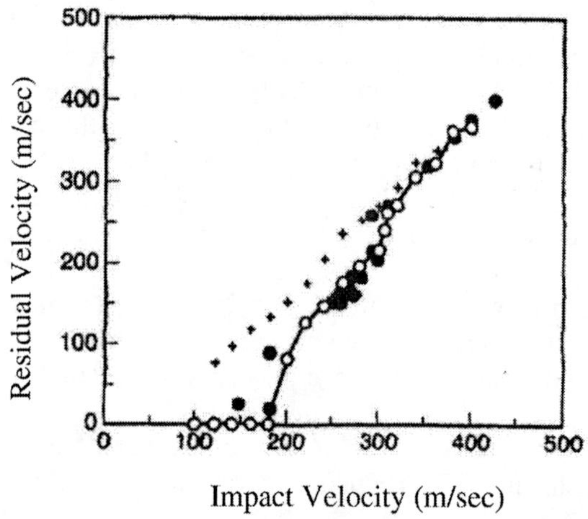

Figure 1. Relationship Impact Velocity (m/sec) with Residual Velocity (m/sec)

Impact velocity results, for the dependence of residual velocity on impact for a single ply yarn of bio-textile composite with $\omega = 280 g/m^2$. The fabric has a square shape with L = 100 mm, and its clamped at two opposites sides and the other sides remain unrestrained. Model predictions for a choice $\beta_{cross} = 10^3$ and $\beta_{clamp} = 10^4$ are presented by symbol (*). Experimental data taken from reference are donates by symbol (0). The projectile characteristics are given in the section on the model. Symbol * denotes model values in the absence of yarn slippage through clamps.

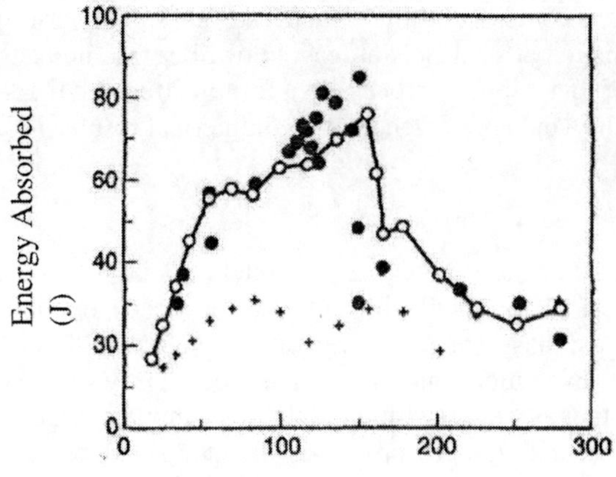

Figure 2. Relationship between Impact Energy (J) with Energy Absorbed (J)

Figure 2 illustrates the strong effect of impact speed V_p On the deformation Morphology of a single-ply bio-textile composite. The figure is for a time corresponding either to full arrest of the projectile (V_p < 180 m/sec) or through perforation of the bio-textile composite (V_p > 180m/sec). The projectile radius equals three times the length of a yarn segment between two consecutive crossover points. Note the large amount of deformation and creasing at low velocity (V_p , 300 m/sec), which see figure 1 and 2 is mainly driven by yarn slippage through clamps. These creases are consistence with the stretch marks observed experimentally at those impact speeds in the direction of clamped yarns. Also in support of model findings is the experimental observation of a substantial amount of yarn slippage at the clamp. Further inspection of model results Figure 2 reveals that above a critical impact speed of 300 m/sec, fabric penetration occurs before the transverse deflection of the fabric has reached the clamp edges. That sudden decrease in the extent of transverse deflection of the fabric fully explains the dramatic drop in absorbed energy observed in Figure 2 – 3 at V_p > 300 m/sec. Similar conclusions have been reached from the model results.

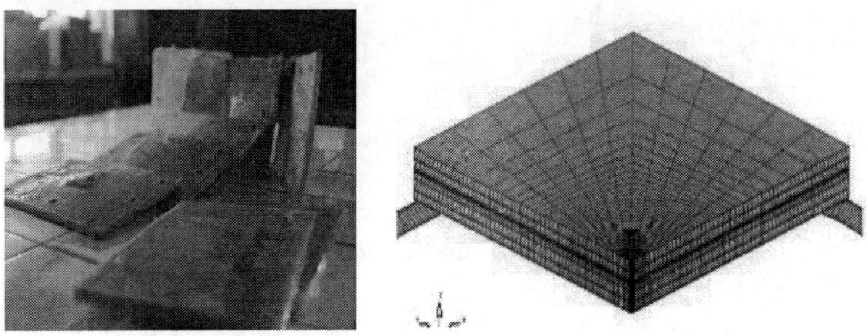

Figure 3. Model Element for a Bio-textile composite panel

To study of the importance of yarn slippage through clamps, which is explicitly included in approach. The results are in poor agreement with the experimental data. Further investigation reveals that although calculated absorbed energy values could be increased through appropriate changes in yarn strength, and the process of yarn slippage through the clamps appears to be crucial for describing the discontinuities observed experimentally. In alternative model described, slippage was not considered, and part of those discontinuities could be ascribed to latent crimp present in the yarn. The importance of yarn slippage through crossovers has also been studied. Turning first to the data of figure 2 – 3 , its found that model results for a choice of $\beta_{cross} = 0$ and keeping $\beta_{clamp} = 10^4$ are almost identical to those obtained for $\beta_{cross} = 10^3$ (with symbol *). As far as could determine, a moderate amount of yarn slippage through crossover point has little effect on ballistic performance, at least until fiber fracture sets in. However, as already mentioned early, inclusion of that slippage process in a finite-difference model like this one appears to be crucial for ensuring stability and convergence of the numerical results.

5. Conclusions

To conclude, that have presented a comprehensive model for studying the deformation of damage to bio-textile composite under small projectile impact. The approach explicitly considers the various projectile characteristic such as; mass, shape and velocity as well as all fiber properties such as denier, modulus and tensile strength. Even more importantly the model allows for the first time to account for slippage of yarn over one another as well as within the clamps. Another essential feature of model resides in the fact that yarn slippage and fracture processes are described with a kinetic approach which explicitly accounts for their dependence on impact rate. Detailed results of model for bio-textile composite with sandwich structure will be presented in forthcoming publication.

Acknowledgments
The authors gratefully acknowledg the financial support of the specialized Research Fund for Head of the Department of Chemical Engineering, Concentration Textile Engineering, Islamic University of Indonesia Yogyakarta.

References

[1] Chocron-Benloulo I S And Sanchez-Galvez V. A simple Analytical Model To Simulate Textile Fabric Ballistic Impact Behavior, *Textile Research Journal*, Usa; 1990 Vol. 67,pages. 520-528.
[2] Cunniff M. An Analysis of the System Effects in Woven fabrics Under Ballistic Impact, *Textile Research Journal*,USA, 1992; Vol. 62, pages. 492 – 509.
[3] Cunniff P M, A Semiemperical Model for The Ballistic Impact Performance of Textile Based Personal Armor, *Textile Research Journal*. USA 1966; Vol. 66 pages. 45-59.
[4] Haro E E, Szpunar J , Odeshi, A G, *Ballistic impact response of laminated hybrid materials made of 5086-H32 aluminum alloy, epoxy*.India. 2016 pages 271-279
[5] Lim C T, Shim, V P W, and Ng, Y H, Finite Element Modeling of the Ballistic Impact of Fabric Armor, *Ins. J. Impact Engineering*.UK. 2003;Volume 28, pp.13-31.
[6] Moeliono M, Siregar Y, Rekayasa bahan baku sutera dan limbah kokon untuk rompi tahan peluru. *Jurnal Riset Industri* 2012.Vol. VI No. 1, Hal. 1-12
[7] Prosser R A, Penetration of Nylon Ballistic Panels by Fragment Simulating Projectiles. Part II: Mechanism of Penetration, *Textile Research Journal*. 1998 Vol.58, (3), pages 161-165.
[8] Roylance D, and Wang S S, Penetration Mechanics of Textile Structures: Influence of Non-Linear Viscoelastic Relaxation, *Polym. Eng. Sci*. 1978 Vol.18, pages 1068-1072.
[9] Roylance D, Stress Wave Propagation in Fibers: Effect of Crossover, *Fiber Science Technology*. 1990 Vol. 13, pages 385-395.
[10] Shim V P W, Tan V B C, and Tay T E, Modeling Deformation and Damage Characteristics of Woven Fabric Under Small Projectile Impact, *Inst. J. Impact Eng*. 1995 Vol. 4, pages 585-605.
[11] Termonia Y, Lattice Model for the Drape and Bending Properties of Nonwoven Fabrics, *Textile Research Journal*. 2003 Vol. 73, pages 74-78
[12] Subagyo A, Prototipe Rompi Tahan Peluru dengan Penguat Biotekstil Komposit Struktur Sandwich Menggunakan Epoksi, *Prosiding Seminar Nasional Universitas Pembangunan Nasional Surabaya*, 2012 pages 176-187.

[13] US American Standart National Institute of Justice. 0101.08. 2008. *Body Armor Standard*. New York:USA
[14] Yves Termonia., Impact Resistance of Woven Fabrics. *Textile Research Journal*. Princeton. 2004 Vol. 7 (8), pages 723-729

Development of fermented vegetables in their ability to inhibit pancreatic lipase as antidyslipidemic agent

Y Maryati*, A Susilowati, and Aspiyanto

Research Center for Chemistry, Indonesian Institute of Sciences, Kawasan PUSPIPTEK, Serpong, Tangerang Selatan – 15314

*E-mail : maryati97@yahoo.com

Abstract. Improvement of public awareness of healthy living encourages the increase of demand for natural food products and provides health benefits for the body. Cardiovascular disease caused by atherosclerosis of blood vessel walls and thrombosis is the leading cause of death in the world. Dislipidemia and coagulation disorders are one of the most significant risk factors for the development of atherosclerotic conditions. The purpose of this study is to develop one of the healthy food products from fermented green vegetables (spinach, broccoli and katuk leaves) fermented by kombucha culture during fermentation as a functional beverage in its ability inhibits pancreatic lipase as anti-dyslipidemia. Results showed a correlation between the ability to defend against free radicals and the ability to inhibit pancreatic lipase enzyme activity. The optimum conditions obtained in the fermentation of katuk leaf vegetables based on the ability to inhibit pancreatic lipase activity (% inhibition) and the ability to inhibit free radicals (DPPH), which is for 6 days fermentation of 91.44%, and 88.62%, with a pH value 3.98.

Keywords: fermentation, green vegetables, kombucha, antidyslipidemic, lipase pancreatic inhibition.

1. Introduction

Recently research developments in the cardiovascular field are very rapid. This is indicated by the increasing number of studies in this field. Atherosclerosis, especially regarding the role of lipids also undergoes rapid development in accordance with the results of the research conducted (1). Cardiovascular disease due to atherosclerosis of blood vessel walls and thrombosis is the leading cause of death in the world. Epidemiological studies have shown that dyslipidemia and coagulation disorders are one of the most significant risk factors for the development of atherosclerotic conditions (2). Dyslipidemia is an abnormality in blood lipid levels characterized by elevated triglyceride levels, LDL cholesterol, total cholesterol and decreased HDL cholesterol in the blood. Without immediate treatment, dyslipidemia can lead to many diseases, which one of them is coronary heart disease (CHD)(3).

Conventional treatment of dyslipidemia is done by regulating the correct diet and taking blood lipid-lowering drugs, such as statins (4). Adjuvant therapy is performed by giving additional treatment to primary therapy by using an adjuvant form, such as functional food.

Katuk (*Sauropus androgynous*), spinach (*Amaranthus*) and broccoli (*Brassica oleracea var. Italic*) are a vegetable plant, which is widely available in Southeast Asia. Utilization is currently limited to direct consumption as a daily vegetable menu. Fermentation of vegetables is considered to be one of the processes for developing products that are beneficial to health. One of the most important sources of

IOP Conf. Series: Materials Science and Engineering **722** (2020) 012074 doi:10.1088/1757-899X/722/1/012074

potential pancreatic lipase inhibitors represents the class of polyphenols and flavonoids. as natural bioactive compounds their potential effect on metabolic diseases as a health food. The functional foods have physiological properties for the body by developing products from green vegetables, and one of them is fermented food by kombucha.

The functional foods have physiological properties for the body by developing products from green vegetables, and one of them is fermented food by kombucha. Kombucha is a fermented drink that is popular among many other traditional fermented foods. The presence of bacteria and yeast in the medium plays a role in creating a strong symbiosis that can inhibit the growth of microorganism contamination (5). The dominant bacteria that grows in kombucha culture is acetate acid bacteria, which is an aerobic bacterium that can use alcohol as a substrate to form acetic acid. These bacteria, in contrast to yeast, require large amounts of oxygen for their growth and activity (6).

The physiological effects of these bioactive compounds from high antioxidant beverages can help to reduce the risk of cardiovascular disease. Green vegetables contain high levels of polyphenols and flavonoids. Polyphenols work by slowing up the absorption of triacylglycerol through inhibition of pancreatic lipases, increasing excretion of cholesterol through feces, increasing excretion of LDL receptors in the liver, and inhibiting secretion of apolipoprotein B100(7)(8).

Alkaloid compounds can inhibit pancreatic lipase enzyme activity that increases fat secretion through feces so that it cannot be converted into cholesterol (9). Flavonoids act as antioxidants that can reduce LDL cholesterol and triglyceride levels, by increasing the activity of the lipoprotein lipase enzyme that plays a role in the hydrolysis of triglycerides into fatty acids, the mechanism of action is reducing LDL sensitivity to free radicals. Flavonoids also inhibit HMG-CoA reductase in reducing cholesterol levels in the blood (10)

The purpose of this study is to develop one of the healthy food products from fermented green vegetables (spinach, broccoli and katuk leaves) fermented by kombucha culture during fermentation as a functional beverage in its ability inhibits pancreatic lipase as anti-dyslipidemia.

2. Material and Methods

2.1. Kombucha starter culture

Kombucha starter culture, also known as 'SCOBY', (generally consists of *Acetobacter xylinum*, *Glukonobacter* dan *S.cerevisiae*) is obtained from the commercial collection of Indo Kombucha, Indonesia. The starter culture used in this study was stored in a refrigerator (4°C) and consisted of sour broth and cellulosic layer ("tea fungus" floating on the liquid surface).

2.2. Kombucha beverage from green vegetables preparing and sampling procedures

The ingredients consisted of spinach vegetables (*Amaranthus spp.*), (*Brassica oleracea* L.) and katuk leaves (*Sauropus androgynous*), each of which was fermented in a Kombucha starter culture. The following ingredients were used to make the drink consisting of sterile water, starter culture of kombucha tea, filtrate spinach leaves (*Amaranthus spp.*), Broccoli (*Brassica oleracea* L.) and katuk leaves (*Sauropus androgynous*) and sucrose. Each drink was made from the ratio of spinach leaves, broccoli and katuk leaves (1 part each with 4 parts of water then 80 mesh escaped filtering and 100 g/L of sucrose. The filtrate of each vegetable was pasteurized at 90–100 °C for 5–10 minutes, and sucrose was added to each juice filtrate and then stirred until dissolved, then placed in a sterile jar until it cools to room temperature 27–30 °C. After that, the juice filtrate was inoculated with a starter culture of kombucha in 60 ml acidic liquid (sour broth) and 40 g of cellulose layer per 1 L juice (total 10% wt/vol). Jars were covered with sterile gauze to ensure aerobic conditions for fermentation. The fermentation process was carried out in a range of 0 to 12 days with 3 days interval in temperature dark room (± 28–30 °C).

2.3. pH measurement Acidity

pH measurement Acidity of the tested product was determined with the use of a calibrated pH meter (pH700, Eutech Instrument). Three replicate measurements were performed.

2.4. Determination of polyphenol content

Total phenolic content was estimated using the Folin-Ciocalteu method (11). Samples of 0.1 mL were added with 700 µL of aquadest, and 500 µL of Folin Ciocalteu reagent and 1 mL of saturated Na_2CO_3 solution. Afterwards, the mixture was added with another 1.4 mL of aquadest, then vortexed. The result was analyzed by UV-Vis Spectrometer at 760 nm against a blank, after 2 h at dark room temperature. Gallic acid was used as a standard and the concentration was expressed as micromoles of gallic acid equivalents (GAE) per g of wet weight (mL).

2.5. DPPH• (2.2-Diphenyl-1-picrylhydrazyl radical) scavenging activity assay

The free radical scavenging activity of kombucha was evaluated using the DPPH • scavenging activity assay, as described by (12). In brief, aliquots of beverage samples from each fermentation result (500 µL) was added to 500 µL of a daily prepared solution of 0.1 mM DPPH (0.4 mg/mL) in methanol. Next, the mixture was added with 1.5 mL of methanol. The solution was stored at room temperature in a dark room for 30 minutes and DPPH free radical reduction was measured by reading absorbance at 515 nm using a spectrophotometer (Agilent Technologies, Cary 60 UV-Vis). The following equation was used to measure inhibitory activity:

$$\text{Scavenging capacity (\%)} = \frac{\text{(Control Absorbance − Sample Absorbance)}}{\text{(Control Absorbance)}} \; x\,100 \qquad (1)$$

2.6. Measurement of Porcine Pancreatic Lipase Inhibitory Activity

The ability of the kombucha green vegetables to inhibit pancreatic lipase was measured using the method previously reported by Kim et al.(13) Briefly, an enzyme buffer was prepared by the addition of 25 mg/mL porcine pancreatic lipase solution (Sigma-Aldrich) in buffer containing 10 mM MOPS (morpholinepropane sulphonic acid) and 1 mM EDTA, pH 6.8, to 155 μL Tris buffer (100 mM Tris-HC1 and 5 mM $CaCl_2$, pH 7.0). Then, 20 μL of either the vegetables extracts at the test concentration (10 μg/mL) or orlistat (Roche, Basel, Switzerland) were mixed with 175 μL enzyme buffer and incubated for 15 min at 37°C with 5 μL substrate solution (10 mM *p*-NPB (*p*-nitrophenylbutyrate) in dimethyl formamide); the enzymatic reactions were allowed to proceed for 15 min at 37°C. Lipase activity was determined by measuring the hydrolysis of *p*-NPB to *p*-nitrophenol at 405 nm using an ELISA reader (Varioskan Flash, Thermo Scientific). Inhibition of lipase activity was expressed as the percentage decrease in OD when porcine pancreatic lipase was incubated with the test materials. Lipase inhibition (%) was calculated according the following formula:

$$\text{Inhibition (\%)} = 100-(B-b\,/A-a\times 100) \qquad (2)$$

where A is the activity without inhibitor, a is the negative control without inhibitor, B is the activity with inhibitor, and b is the negative control with inhibitor.

3. Result and Disccusion

3.1. Chemical characterization of raw material (green vegetables) before fermentation process

The compositions of raw material for healthy food products from fermented green vegetables (spinach, broccoli and katuk leaves) before being fermented. The results were shown below (Table 1).

Table 1. Chemical characterization of green vegetables extract before fermentation process

Composition	Material (Green vegetables)*		
	Spinach	Broccoli	Katuk leaf
Lipase pancreatic inhibitory activity (%)	30,41	19,49	9,27
DPPH Inhibitory activity (%)	53,05	36,74	90,85
Polyphenol (mg GAE/mL)	0,09	0,04	0,23
pH	6,32	4,99	6,07
Dissolved Protein (mg/mL)	25,06	20,00	57,03
Reducing Sugar (mg/mL)	3,54	0,44	0,71

**) Ratio extract of green vegetables and destilation water (1:4)*

The choice of broccoli, spinach and katuk leaf were based on the composition, especially higher polyphenols compared to other types of vegetables. The difference in composition of the types of green vegetables varies was shown in Table 1. Spinach vegetable extracts have a high lipase inhibitory activity compared to other vegetables, but from the antioxidant activity and the highest total polyphenol content in katuk leaf vegetable extract each of 90, 85% and 0.23 mg GAE/mL.

3.2. Lipase pancreatic inhibitory activity and pH

Pancreatic lipase is a key enzyme for the absorption of triglycerides in the small intestine. It is removed from the pancreas and hydrolyzes triglycerides to glycerol and free fatty acids[14;15]. Plants have a variety of natural chemical compounds that have diverse structural features, so many of them are potentially compatible with pancreatic lipase modulation mechanisms. Inhibition of pancreatic lipase is a viable strategy to combat obesity and lower blood lipid levels (16). Some vegetables have a good ability to inhibit the action of pancreatic lipases. The effect of the fermentation process on the vegetable to lipase inhibitory activity can be seen in Figure 1, where the length of fermentation time tends to affect lipase inhibitory activity in vegetables, during the fermentation process the lipase inhibitory activity tends to increase during the fermentation time in green vegetables.

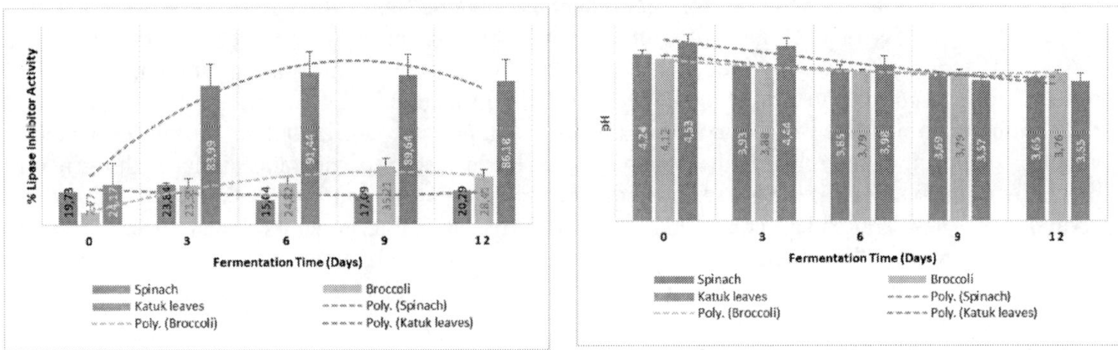

(a) (b)

Figure 1. Effects of fermentation time and variation green vegetables with culture of *kombucha* to lipase pancreatic inhibitory activity and pH of fermented vegetables.

This is due to some active compounds in the fermentation process of vegetables were degraded due to this process and produce other bioactive compounds such as flavonoids. The results of this fermentation obtained the highest lipase inhibition activity in katuk vegetables with a 6 day fermentation process of 91.44% and the lowest in spinach vegetables at 12 days fermentation at 28.41%.

The pH value in this study was significantly influenced by the duration of fermentation. Kombucha products contain yeast and bacteria that metabolize sucrose to produce organic acids such as acetic acid and gluconate acid. According to Jayabalan (17) Yeasts and bacteria in vegetable kombucha are involved in such metabolic activities that utilize substrates in different and complementary ways. Yeasts hydrolyze sucrose into glucose and fructose by invertase and produce ethanol via glycolysis, with a preference for fructose as a substrate. Acetic acid bacteria make use of glucose to produce gluconic acid and ethanol to produce acetic acid. As a result, the pH value of kombucha beverage decreases due to the production of organic acids during fermentation.

3.3. DPPH inhibitory activity and polyphenol

Polyphenols are a group of powerful antioxidants that can protect the body from cell damage due to free radicals. Polyphenols can be found in green vegetables with a high enough level. It also functions as a color giver of vegetables. The effect of the fermentation process on total levels of vegetable polyphenols can be seen in Figure 2, where the length of fermentation time tends to affect the total levels of polyphenols in vegetables. During the fermentation process, the total value of polyphenols tends to fluctuate in green vegetables. This can happen because phenolic compounds in the fermentation process of vegetables are degraded due to this process. In addition, they produce other bioactive compounds and forms organic acids, acetaldehyde, alcohol and carbon dioxide which play a role in the formation of specific kombucha scents.

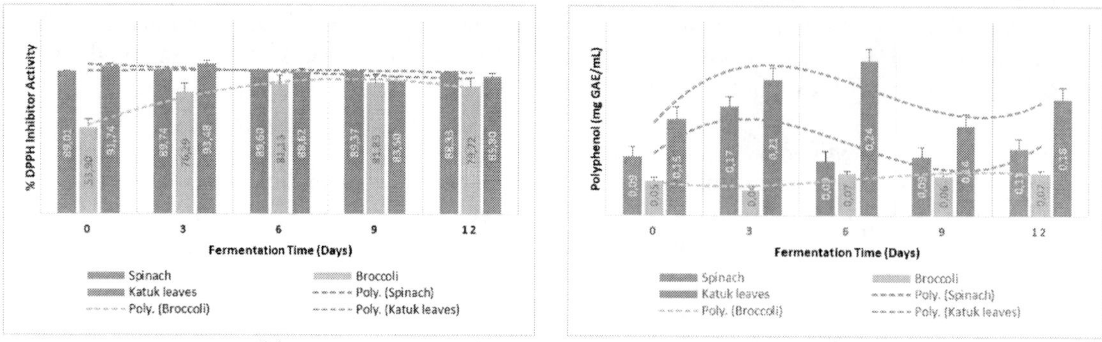

(a) (b)

Figure 2. Effects of fermentation time and variation green vegetables with culture of kombucha to DPPH inhibitory activity and polyphenol of fermented vegetables

In this extraction, the highest total polyphenols obtained in katuk vegetables with a 6 day fermentation process of 0.24 mg GAE/mL followed by spinach in 3 days fermentation (0.21 mg GAE/mL) and the lowest in broccoli vegetables in 12 days fermentation amounting to 0.07 mgGAE/mL. According to Sun, Jin Hur et al (18). Fermentation can increase antioxidant activity mainly due to an increase in the amount of phenolic compounds and flavonoids during fermentation, which is the result of microbial hydrolysis reactions. Furthermore, fermentation induces the structural breakdown of plant cell walls, leading to the separation or synthesis of various antioxidant compounds. These antioxidant compounds can act as free radical terminators, metal chelators, singlet oxygen quenchers, or hydrogen donors to radicals. The production enzymes such as protease and α-amylase can be influenced by fermentation that may have metal ion chelation activity.

The free radical (DPPH) inhibition activity of each vegetable has a different pattern during fermentation. In broccoli fermentation, it was seen that the longer the fermentation time increased the antioxidant value of katuk leaf kombucha juice by 93.48%. Increased antioxidant activity in kombucha tea was caused by the results of microorganism metabolism in kombucha during the fermentation process(19). However, in katuk leaf fermentation with increasing fermentation time, it will reduce antioxidant activity. This was caused by degradation of phenolic compounds by enzymes produced by kombucha culture, so that the compounds that have antioxidant activity lose their stability. Kombucha exhibited increased free radical scavenging activities during fermentation. The extent of the activity depended upon the fermentation time, type of material, and the normal microbiota of the kombucha culture, which in turn determined the nature of their metabolites. Although free radical scavenging properties of kombucha showed time-dependent profiles, prolonged fermentation is not recommended because of accumulation of organic acids, which might reach harmful levels for direct consumption (17).

3.4. Reducing sugar and Dissolved protein

The ability of SCOBY cultures to reducing sugar in each sample of various types of kombucha vegetables varies. During the fermentation process, sucrose is transformed into glucose and fructose by bacteria and yeast. Glucose is used to form alcohol and organic acids, so that the concentration of organic acids increases. During the fermentation process yeast kombucha breaks down sucrose into glucose and fructose (20). Glucose will be used for cell metabolism to produce ethanol and carbon dioxide.

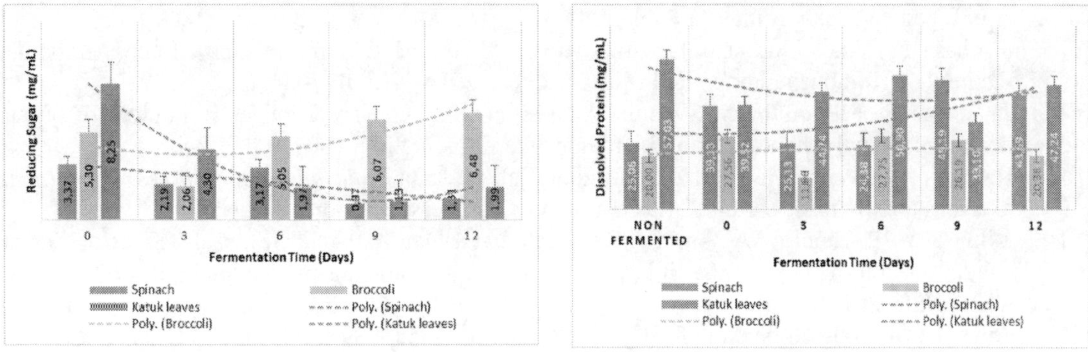

(a) (b)

Figure 3. Effects of fermentation time and variation green vegetables with culture of kombucha to reducing sugar and dissolved protein of fermented vegetables\

Reducing sugar varied during the fermentation time for each different type of vegetable. There was a decrease in sugar content from day 0 to day 9 and it increases again after fermentation on the day 12 for fermented spinach and katuk leaves. The lowest reduction in reducing sugars at day 9 of spinach and katuk leaf fermentation was 0.97 mg/mL and 1.31 mg/mL, respectively, while for the broccoli was at 3 days fermentation at 2.06 mg/mL.

The dissolved protein produced in vegetable fermentation is a breakdown parameter of large molecular weight proteins into amino acids and peptides to be lower molecular weight proteins by microbial proteolytic activity in kombucha culture, resulting in higher water solubility as a dissolved protein. The tendency of dissolved protein content by proteolytic activity shows a different pattern of each vegetable during the fermentation process, as seen from the varied pattern of changes. Since the amount of dissolved protein is also influenced by protease enzymes from microbes, it can be assumed that proteolytic microbes do not control the kombucha culture, as shown in Figure 3.

4. Conclucions

Development of processed vegetables as functional beverages through the fermentation process yields inhibition of pancreatic lipases that vary during fermentation time. The results showed that katuk leaves exhibits the optimum lipase inhibition activity as anti-dyslipidemia compared to spinach or broccoli vegetable fermentation. The optimum katuk leaf fermentation was at 6 days of fermentation time with inhibition of pancreatic lipase by 91.44% with antioxidant activity of 88.62% at pH 3.98, total polyphenol at 0.24 mg GAE/mL, reducing sugar by 1.97mg / mL and dissolved protein at 50.90 mg/mL.

5. Acknowledgment

The authors wish to extend their gratitude to by The Ministries of Research, Technology, and Higher Education Republic of Indonesia, through the National Innovation Research Program (INSINAS) in 2018-2019 for the financial and Research center for chemistry, Indonesian Institute of Sciences for the analytical support provided for this study. The authors would like to thank Dr. Adid Adep Dwiatmoko for the helpful suggestions regarding while preparing this manuscript.

6. References

[1] Erwinanto E, Putranto JNE, Tedjasukmana P, Suryawan R, Rifqi S, Kasiman S. Pedoman Tatalaksana Dislipidemia PERKI 2013. Indones J Cardiol. 2013;245–70.

[2] Sashidhara K V, Kumar A, Kumar M, Sonkar R, Bhatia G, Khanna AK. Novel coumarin derivatives as potential antidyslipidemic agents. Bioorg Med Chem Lett. 2010;20(14):4248–51.

[3] Prahastuti S, Hidayat M, Kurniadi MW, Christiany S. Potency of Black Soybean (Glycine max L. Merr) and Jati Belanda Leaves (Guazuma ulmifolia Lamk) for Dyslipidemia Treatment In Vivo. J Med Heal. 2016;1(3).

[4] Sudoyo AW, Setiyohadi B, Alwi I, Simadibrata M, Setiati S. Buku Ajar Ilmu Penyakit Dalam Jilid II edisi V. Jakarta Interna Publ. 2009;310:1973–82.
[5] Malbaša R, Vitas J, Lončar E, Grahovac J, Milanović S. Optimisation of the Antioxidant Activity of Kombucha Fermented Milk Products. 2014;32(5):477–84.
[6] Jayabalan R, Marimuthu S, Swaminathan K. Food Chemistry Changes in content of organic acids and tea polyphenols during kombucha tea fermentation. 2007;102:392–8.
[7] Rahardjo SS. Pengaruh ekstrak etanol daun jati belanda (*Guazuma ulmifolia Lamk.*) terhadap aktivitas enzim lipase serum Rattus norvegicus. Universitas Gadjah Mada; 2004.
[8] Vikneswari P, Hamid AA, Amin I, Khozirah S, Faridah A, Ismail IS, et al. Effect of Cosmos caudatus Kunth leaves on the lipid profile of a hyperlipidemia-induced animal model. J Food Chem Nutr. 2014;2(1):43–51.
[9] Ayu G, Tandi J, Nobertson R. Uji Efek Ekstrak Etanol Daun Kenikir (*Cosmos Caudatus Kunth.*) Terhadap Penurunan Kadar Kolesterol Pada Tikus Wistar (*Rattus norvegicus*) Hiperkolesterolemia-Diabetes. Farmakol J Farm. 2017;14(2):112–8.
[10] Fithriani NA. Pengaruh Pemberian Ekstrak Bawang Merah (Allium ascalonicum) Terhadap Kadar Trigliserida Serum Tikus Wistar Hiperlipidemia. Faculty of Medicine; 2010.
[11] Ainsworth EA, Gillespie KM. Estimation of total phenolic content and other oxidation substrates in plant tissues using Folin–Ciocalteu reagent. Nat Protoc. 2007;2(4):875.
[12] Mazidi S, Rezaei K, Golmakani MT, Sharifan A, Rezazadeh S. Antioxidant Activity Of Essential Oil From Black Zira (*Bunium persicumboiss.*) Obtained By Microwave-Assisted Hydrodistillation. 2012;
[13] Kim JH, Kim HJ, Park HW, Youn SH, Choi D, Shin CS. Development of inhibitors against lipase and a -glucosidase from derivatives of monascus pigment. 2007;276:93–8.
[14] Lowe ME. Pancreatic triglyceride lipase and colipase: insights into dietary fat digestion. Gastroenterology. 1994;107(5):1524–36.
[15] Lee YM, Kim YS, Lee Y, Kim J, Sun H, Kim JH, et al. Inhibitory Activities of Pancreatic Lipase and Phosphodiesterase from Korean Medicinal Plant Extracts. 2012;782(March 2011):778–82.
[16] Iswantini D, Silitonga RF. Extracts as Antiobesity : In Vitro Inhibitory Effect on Pancreatic Lipase Activity. 2011;18(1):6–10.
[17] Jayabalan R, Malbaˇ R V, Lonˇ ES, Vitas JS, Sathishkumar M. A Review on Kombucha Tea — Microbiology , Composition , Fermentation , Beneficial Effects , Toxicity , and Tea Fungus. 2014;13.
[18] Hur SJ, Lee SY, Kim Y-C, Choi I, Kim G-B. Effect of fermentation on the antioxidant activity in plant-based foods. Food Chem. 2014;160:346–56.
[19] Goh WN, Rosma A, Kaur B, Fazilah A, Karim AA, Bhat R. Fermentation of black tea broth (Kombucha): I. Effects of sucrose concentration and fermentation time on the yield of microbial cellulose. Int Food Res J. 2012;19(1):109.
[20] Greenwalt CJ, Ledford RA, Steinkraus KH. Determination and characterization of the anti-microbial activity of the fermented tea Kombucha. (4).

IOP Conf. Series: Materials Science and Engineering **722** (2020) 012075 doi:10.1088/1757-899X/722/1/012075

An ergonomic, mobile and portable design of porang cutting and drying machine on motorcycle

E Nurmianto [1], A Anzip [2], D E Kusrini [3]

[1)] Dept. of Industrial Engineering, Sepuluh Nopember Institute of Technology, Surabaya, Indonesia
[2)] Dept. of Mechanical Industry Eng Dept, Sepuluh Nopember Institute of Technology, Surabaya, Indonesia
[3)] Business Statistics Dept, Sepuluh Nopember Institute of Technology, Surabaya, Indonesia

Corresponding author's e-mail : nurmi@ie.its.ac.id

Abstract. This research is aimed to improve management of Porang (*Amorphophallus Muelleri Blume*, Indonesia *Konjac*), in Glucomannan Industrialization. The problem arises when they cut and dry manually on porang with very low productivity. Problems solved when there is a machine for cutting and drying porang on motorcycle which are Mobile, Portable and Ergonomic. This machine used to process with a higher production capacity. This porang processing innovation in the form of cutting and drying machine as well as processing porang into various preparations that combine several processing functions that are safe, comfortable, and healthy for its users and effective for their products. The conclusion is (1) from the data test of the machine it got 15 mm diameter of porang with maximum force of cutting machine is 22,1 kg, (2) Already build drying machine porang that can produce 1,061 kg/hours porang chips, (3) Based on the test and calculation, the porang's cutting and drying machine can produce 190,89 kg/hours porang chips. Then if it observed from manual cutting that can only produce 65 kg/hours porang chips, so this machine can increase 3 times of productivity of the farmers. Finally, it is hoped that after using this machine, farmers will be prosperous.

1. Introduction

The existence of forests is very important for the community, because it has economic functions, namely as a buffer of the most essential life for people living around the forest area, climatological functions can affect the microclimate and oxygen producers, hydrological functions as a regulator of water systems, and ecological functions that prevent flooding and soil erosion, maintaining soil fertility and germplasm sources. In the future the forest in Nganjuk, East Java is expected to be a forest for food and life through the potential of non-timber forest products.

The Nganjuk Regency area is divided into 20 sub-districts and almost all sub-districts have the potential of forestry with approximately 200 forest farmer groups and LMDH whose lives are very dependent on forests.

In the era of autonomy with the breath of decentralization, the passion of Nganjuk Regency to build the forestry sector is shown by a commitment to manage, preserve, utilize and improve forest ecosystems on the basis of equitable community support so as to create a sustainable and environmentally sound system for community welfare.

Benefits of Porang Plant

Besides being used for food ingredients, the most important ingredient of porang plants is glucomannan content contained therein [1]. Ideally the higher the glucomannan content of the porang plants, the higher the commercial value it is possible. 3-year-old porang tuber originating from Indonesia, especially Nganjuk Regency can reach more than 5 kg while those from other countries, such as Japan and Korea, can only reach 2 kg. Therefore, even though porang plants are cultivated in Japan and Korea, but the production and consumption of porang tubers is not balanced so they import from Indonesia. The difference in glucomannan levels between people originating from Indonesia and those from other countries also needs to be further investigated because of the possibility of glucomannan levels which make consumption of porang in Japan and Korea very high. The uses of porang plants in general are as follows:

1. As the best glue
2. Water purifier and tablet formulation binder
3. Syrup thickener
4. Medicines and blackhead absorbers
5. Benefits for the health of the body, people can reduce blood cholesterol levels, slow down emptying the stomach and speed up the feeling of fullness so that it is suitable for diet food for diabetics.

Figure 1. The process of cutting and drying conventional materials [1]

2. Objective and research benefit

2.1. Research objective

The general objective of this study is to support the government in the empowerment program of coastal resource commodities with an easy and ergonomic method. While the specific objectives of this study are:

1. Created a design prototype of cutting tools and Porang On Motorcycle Dryers that are Mobile, Portable and Ergonomic with the use of motorized vehicles as an effort to create cutting tools and dryers for people who are mobile, portable, and environmentally friendly.
2. Applying the Design of Cutting Tools and Porang On Motorcycle Dryers that are Mobile, Portable and Ergonomic which are mobile, portable, and environmentally friendly in the coastal areas of East Java.

2.2. Research benefit

The benefits obtained from this study are:

1. Forest communities can more quickly cut and dry porang.
2. Cutting tools and porang On Motorcycle dryers have advantages in terms of mobility when compared to traditional tools.
3. Cutting tools and porang On Motorcycle dryers can support the economic improvement of forest communities.
4. Cutlery and porang On Motorcycle dryers have a large porang storage capacity in one cutting process, so that the output of the products produced is greater when compared to conventional tools.
5. Application of cutting tools and porang dryers On Motorcycle does not cause health and environmental problems (environmentally friendly), because the drying system is carried out closed with a chimney at the top.

3. Literature review

3.1. Supporting theory

Porang is one of the Amorphopallus species. In popular Japan called Amorphopallus konjac while in Indonesia better known as Amorphopallus oncophillus. Porang is a type of iles-iles plant that grows in the forest. Porang is a family of Araceae which is a shrub (herb) plant that is rooted in the soil, and produces carbohydrates. Porang plants grow in the form of shrubs with a height of 100-150 cm, finely trunked, stems and leaves green to dark green with stripes with white spots.

The main products of porang plants are tubers. There are two types of tubers in porang plants according to Wu *et al.*, namely stem tubers that are in the ground, and tubers / frogs which are found in each base of branches or leaf bones which contains seeds [2,3]. This tuber is a change in the shape of the stem that serves as a food reserve. Tubers consist of the skin and tuber flesh. The tuber skin when harvested is grayish in color and if left several days it will turn black. Parts of the peeled tubers will release slippery sap and cause itching on the skin. Porang tuber flesh is yellowish, contains carbohydrates that serve for further growth. Porang plant roots are white fibrous roots. This large number of roots grows from the stem and skin of the tubers, which is useful for expanding the absorption of water and nutrients from the soil. Whereas the stems of the porang plant merge with the tubers and form a small part of the tubers.

3.2. Previous Research on Motorcycles

The background of this research is that the two-wheeled motor vehicle in Indonesia called the motor, is a vehicle that is in great demand by the public as a motor vehicle whose price is affordable and can be used for various transportation purposes. Starting from self-transport (1 person), along with other people (2 people), along with 2 others (3 people), husband and wife with 1 or 2 children (4 adults and 2 children), carrying gallons of water, computer, merchandise such as household groceries, bakery cakes and others [5].

The use of motors with excess capacity is not allowed in the Traffic Act, and is dangerous to both themselves and other users. Some motorcycle users have tried to add parts of the motor to increase

IOP Conf. Series: Materials Science and Engineering **722** (2020) 012075 doi:10.1088/1757-899X/722/1/012075

capacity, such as adding a rear-mounted box, pulling a two-wheeled train behind it, and some installing a side train permanently.

To get the right anthropometric size of Indonesian children, we can compare the size of British children with Indonesian children of the same age. From the ergonomic and gamber data of the above technique, we can create a scheme that shows how minimal and maximally this sidecar can be used by the child based on anthropometry that has been processed.

In the data below, note that children ages 5 and 8 years can easily board a sidecar in the position of 1 person who was transported. And in a position where the feet can be straightened easily in a sidecar. While in children aged 12 years, the legs began to bend when on the sidecar.

3.3. Design of Porang's Cutting and Drying Machine (on Motorcycle)

In designing porang's cutting and drying machine is based on designing design of a fish smoke device on an ergonomic motorcycle as illlustrated in Figure 2. Musculoskeletal symptoms reported by 215 fishermen were followed at 6-month intervals for 18 months. Exposure information is collected through field observations and in-depth ethnographic interviews that enable the stressful potential of ergonomics to be identified and cataloged with tasks and work stages. Symptoms that caused work disorders in the last 12 months were reported at 38.5% from the initial cohort. Lower back symptoms are the most common cause of work disorders (17.7%), followed by pain in the hands or wrists and shoulders (7%). Symptoms in the body region are more likely to be reported among individuals who are not full-time and those who work other parts of the work or throughout the year have a much lower prevalence of symptoms; both may reflect the effect of a healthy worker. A number of ergonomic stressors are identified at all stages with variability in exposure dictated by several people who cannot predict factors such as weather; but also with the type of boat, equipment, crew size, and level of experience. Working among traditional workers is important, regardless of whether they directly cause or contribute to them musculoskeletal symptoms, or exacerbate existing pathology.

East Java Province has become an area that is capable of producing high marine commodities, but in the process it is still not well developed. Lack of capital makes them unable to increase production capacity and limited marketing. They still use traditional technology. Given these conditions, there are still opportunities to improve efficiency and effectiveness through the Small Business Development Program in accordance with the situation and conditions of the small industry. Management design, human resources, production, finance and design will in turn increase the competitiveness of small industries. The workforce at SME porang production has gained the skills from years of work experience and is informally transmitted.

Figure 2. The design of a fish smoke device on an ergonomic motorcycle [4]

3.4. Nordic Body Map

Nordic Body Map is one of the most commonly used forms of ergonomic checklist questionnaires to find out one's discomfort with a product or object [8]. Nordic Body Map also aims to map the level of complaints a person as a material analysis and a particular action. The questionnaire uses a human body image that is divided into 9 main parts, namely, neck, shoulder, upper back, elbow, lower back, wrist, waist, knee, and heels/feet.

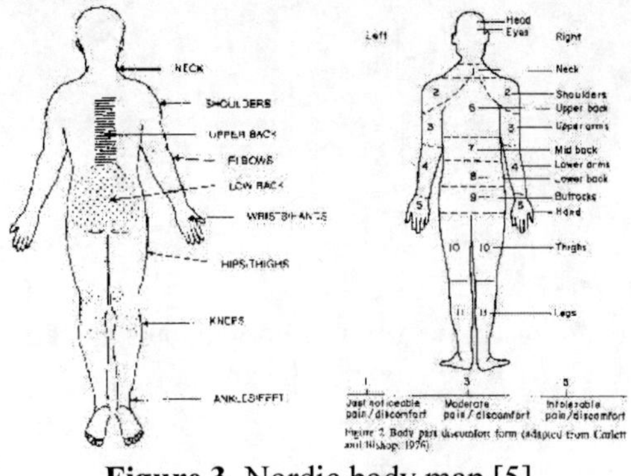

Figure 3. Nordic body map [5]

4. Research methodology

4.1. Research Steps

4.1.1. Identification of problem. The first step in this research is to identify the problems that occur. The problem raised was how to design a porang cutting tool that was in accordance with the wishes of the forest people.

4.1.2. Study literature. At this stage a study was conducted to gain an understanding of QFD, porang cutting tools, the process of drying porang.

4.1.3. Field study. Drying field studies were carried out through observations when carrying out porang drying training. This is intended to obtain initial information

4.1.4. Data collection and processing. The following are the stages of data collection and data processing. Data collected will be processed according to each stage. In anthropometric testing, the steps taken are subjects using footwear are asked to sit on the observation object (motorcycle) with the freedom to set a sitting position until the subject feels comfortable with his sitting position. The front foot of the subject is then measured using graph paper placed in the gas tank. The coordinates of the foot position are then taken parallel to the footrest. The level of accuracy used is + 3 mm. Most subjects wear footwear (shoes / sandals) when measurements are made.
Data Collection Tool Repair
In retrieving data repair tools, the steps taken are:
1. Preparation of the VoC questionnaire
2. Weighting of the VoC Questionnaire results
3. Determine the technical response from the respondent
4. Perform tool repairs
5. Testing the recommended repair of cutting and drying machine

4.2. Designing Porang on Motorcycle Cutting and Drying Tools

The following is a picture of the machine that is mobile, portable, and environmentally friendly that is offered to partners.

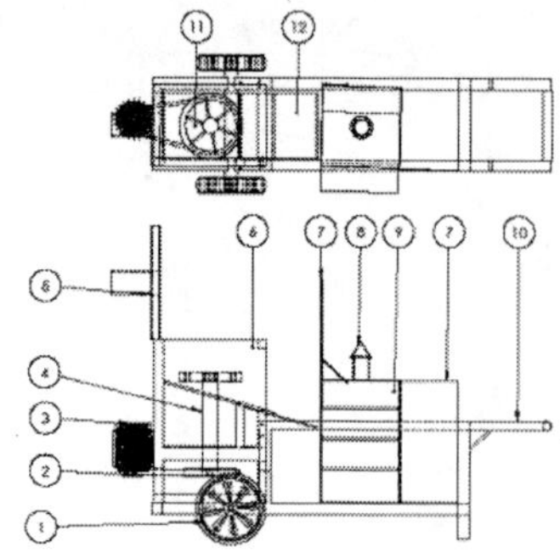

Figure 4. Appropriate technology in the form of Porang Cutting and Drying Tools on Motorcycle which are mobile and portable

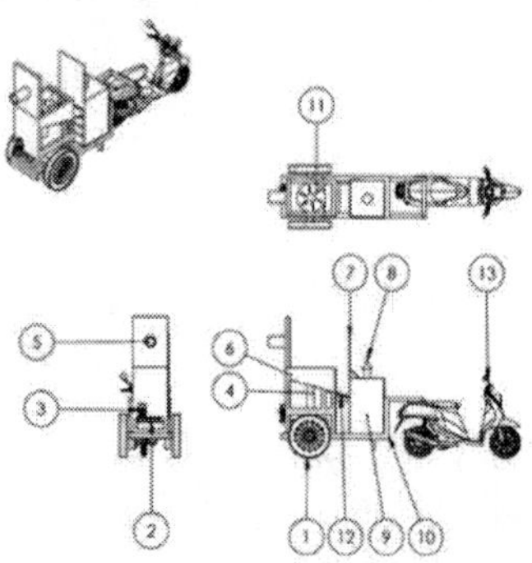

Figure 5. Design image of appropriate technology for porang's cutting and drying machine on motorcycle that are mobile and portable

Specification of the cutting machine is as follow

- The amount of wheel on the disk is 120 rpm.
- Using 2 hoppers with 2 cutting knife that cutting at the same time.
- The diameters of pulley that used in the machine is
- D_1 = 3 in = 76,2 mm
- D_2 = 9 in = 228,6 mm
- D_3 = 6 in = 152,4 mm
- D_4 = 6 in = 152,4 mm
- Using type A belt with rubber canvas material.
  a. The distance of the first pulley to second pulley is 838 mm with 6275,539 hours lifespan (type A 33)
  b. The distance of the third to fourth pulley is 1295 mm with 225984,409 hours lifespan (type A 51)
- Using axis with AISI 1045 material and 40 mm diameter

Specification of the drying machine

- Heat transfer that out of drying system through conduction process with 118,814 watt of the amount.
- Heat transfer that out of drying system through smokestack with 345 watt of the amount.
- Heat transfer that into the drying system is 1139,24 watt of the amount
- Heat transfer inside drying system is 675,426 watt of the amount.

- The consumption of LPG for drying process is 0,08676 kg/hours
- Based on drying test it produces dried porang chip of 1,061 kg/ hours.

4.3. How the porang cutting and drying machine work:

1. Porang fruit is inserted from the top hole / door numbered (5).
2. Porang fruit will fall towards the rotating knife numbered (11) and produce a porang chip.
3. The porang chip that has been finished chopping will fall on the numbered feeder (12) and go towards the drying box numbered (9).
4. The drying box has three levels, if the first level is fully filled by the porang chip, the feeder can be stretched to fill the second drying box level, as well as to fill the third level the feeder can also be stretched again.
5. After the dryer box is fully loaded in three levels, then close the drying door numbered (7).
6. After the door is closed turn on the dryer heating box.
7. Wait until the desired time, if it has finished turn off the heater and porang chip ready to be processed into the next process.

4.4. Anthropometric Testing of Motorcycles

Two-wheeled motorized vehicles in Indonesia called *Lasim* are motorized vehicles, which are very popular with the public as motorized vehicles which are affordable and can be used for various transportation purposes. The physical dimensions of motorbike riders related to motorcycle design and other detailed parts such as fairing or other protective equipment whose effectiveness depends on the relative position of the body parts of the rider.

Motorcycle design provides its own problems for ergonomists because it has narrow constraints, making it difficult for them (ergonomics) to design a motorcycle that fits the user's body size in general. There are two forms of problems found in designing ergonomic motorbikes, namely:

1. Variation in the value of the body size of a motorcycle rider.
2. Relationship between the driver's body attitude and work effectiveness when the driver uses a motorbike.

There is very little information or data on anthropometry about sitting humans while riding motorbikes in the United Kingdom (UK). Previous studies have found that motorcyclists in the UK generally have height that is not in accordance with the motorbike they use. Therefore, a high level of ergonomics or comfort is needed to design a motorized vehicle modification. Ergonomics itself can be interpreted as a discipline that studies humans as one of the important factors in a production system that can significantly improve the effectiveness, efficiency, and performance of a work system that will be developed. Ergonomics also relates to the optimization, efficiency, health, safety and comfort of humans in the workplace, at home, and in recreational areas. In ergonomics, the study of systems where humans, work facilities and the environment interact with each other with the main goal is to adjust the working atmosphere with the people.

5. Results

5.1. Design of Mobile and Portable Porang Cuting and Drying Machine

The designer has made the design of mobile and portable cutting tools and dryer porang for direct application. This tool will be applied to various types of motorbikes with modifications that have been determined by the proposer. In its implementation, the tool will be installed behind the motorcycle by cooperating with a cutting tool and dryer with a motorcycle that will be used. Cutting tools and dryer porang on Motorcycle have a length of 81 cm and a height of 97 cm. At the bottom of the tool there are two fish storage areas that have been carried out fumigation. Each of these components has a length of 31.2 cm and a height of 36 cm. In the middle it has a length of 21 cm, where this size can flexibly adjust to the width of the motorcycle seat. The design of the cutting tools and porang dryer on the motorcycle is shown in Figure 8.

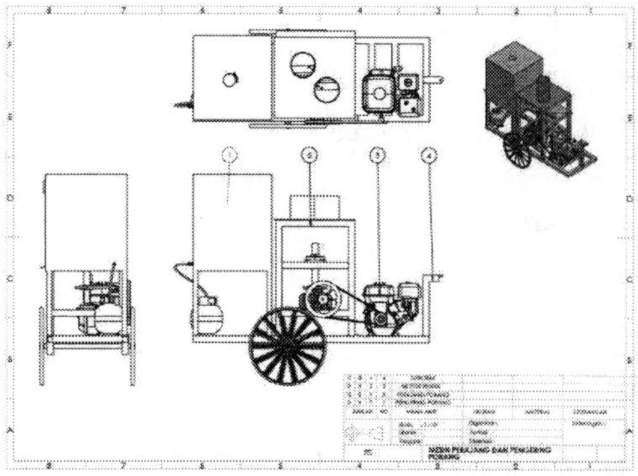

Figure 6. Design cut tools and porang dryers on ergonomic motorcycle

Figure 6 shows the design of porang cutting tools and motorcycle dryers in a closed and visible condition (open). This tool has several components which include drying room, upper door, porang storage after dried, and the bottom door. Chimney function to remove heat produced from the porang dryer, so that the temperature and the process of drying porang can be organized. The porang is in the drying chamber, the room where the porang drying takes place. Under drying porang there is a place, which is a place where charcoal or briquettes are placed. When the process of drying porang happen, then the top door will be closed so heat during the process can be evenly and stable. In addition, this tool is also equipped with a porang storage area that has gone through a drying process. Thus, users can save porang that have been dried. In its implementation, the use of porang cutting tools and motorcycle dryers for the process of drying porang does not take place on a motorcycle, but the top tool must be lowered. As the name implies, this tool has mobile and portable properties, that is mobile when moved from one place to another and portable when lifted for drying porang. This tool is equipped with a sliding wheel that is useful for making it easier for the upper tool components to be lowered and automatically 4 supporting wheels will come out. The designs related to the separation mechanism of the lower and upper porang drying tools are as follows.

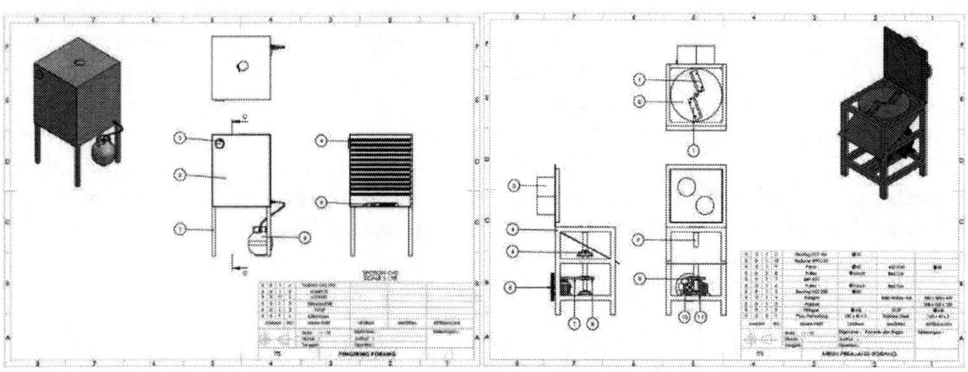

Figure 7. Open cut tools and dryers

Figure 7 shows the design of the separating mechanism for cutting and drying the lower and upper parts. In its application, the lower part of the tool will remain attached to the motorcycle seat, while the upper part of the tool will be lowered by sliding it to the process of drying porang. At the bottom of the tool there are 4 support wheels that will come out automatically when this component is lowered. When the porang dried process is complete, the tool is re-installed in the motorbike seat by bending the four support wheels and returning to the starting position.

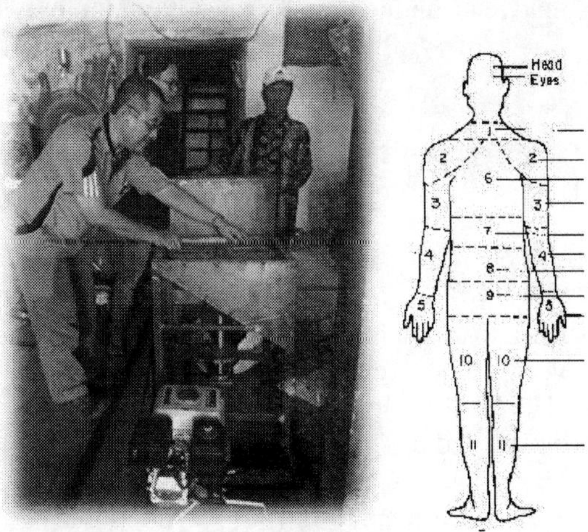

Figure 8. Machine testing at the laboratory.

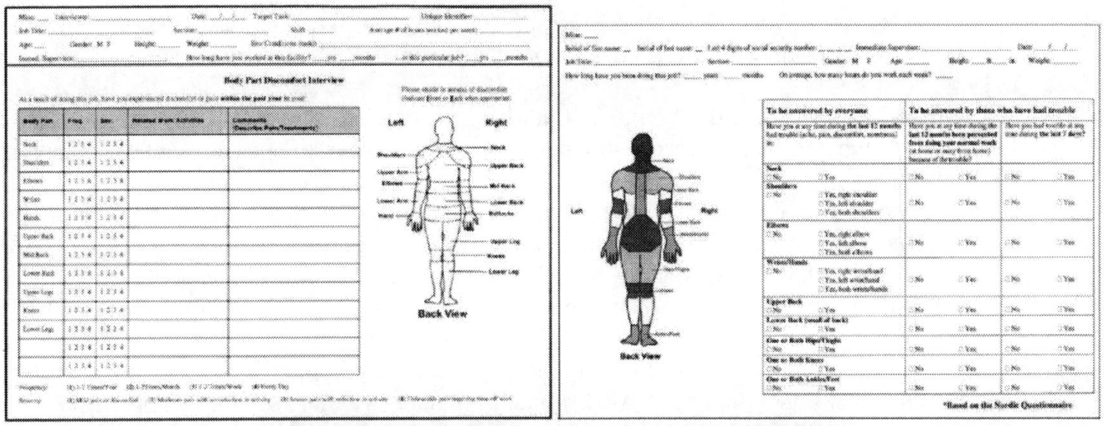

Figure 9. Body Part Discomfort Interview Guide [6,7,9]

(1) (2)

Figure 10. Motorcycle experiment in the fields in the city of (1) Nganjuk and (2) Jombang

In Figure 10, the creation of cutting tools and dryers porang stage 1. At this stage the tool has been applied. At this stage the cutters and dryers will be ready for testing in the field.

6. Conclusion

The conclusion of this research is as follow

1. From the data test of the machine it got 15 mm diameter of porang with maximum force of cutting machine is 22,1 kg.

2. Already build drying machine porang that can produce 1,061 kg/hours porang chips.
3. Based on the test and canculation, the porang's cutting and drying machine can produce 190,89 kg/hours porang chips. Then if it observed from manual cutting that can only produce 65 kg/hours porang chips, so this machine can increase 3 times of productivity of the farmers.

Acknowledgements

Authors acknowledge funding support from Directorate General of Higher Education, Ministry of Technology Research and Higher Education, Republic of Indonesia through contract number 1420/PKS/ITS/2018.

References

[1] Mahfud, E. Nurmianto, B. Amaliah. Usaha Budidaya Dan Pengolahan Porang Di Kecamatan Loceret. Ipteks Bagi Masyarakat, Kemenristek Dikti Jakarta 2016.
[2] Wu, W. -T., & Chen, H. -L. (2011). Effects of konjac glucomannan on putative risk factors for colon carcinogenesis in rats fed a high-fat diet. Journal of Agriculture and Food Chemistry, 59, 989–994.
[3] Wu, W. -T., Cheng, H. -C., & Chen, H. -L. (2011). Ameliorative effects of konjac glucomannan on human faecal β-glucuronidase activity, secondary bile acid levels and faecal water toxicity towards Caco-2 cells. The British Journal of Nutrition, 105, 593–600.
[4] E. Nurmianto, Arino Anzip, Sampurno, Desain Alat Pengasap Ikan Di Atas Sepeda Motor Yang *Mobile, Portable Dan Ergonomis.* Workshop dan Seminar Perhimpunan Ergonomi Indonesia 2019.
[5] I. Kuorinka, B. Jonsson, A. Kilbom, H. Vinterberg, F. Biering-Sorensen, G. Andersson and K. Jorgensen, Standardised Nordic questionnaires for the analysis of musculoskeletal symptoms Applied Ergonomics 1987, 18.3,233-237.
[6] E. Nurmianto, *Ergonomi : Konsep Dasar dan Aplikasinya*, 2nd ed. Surabaya: Guna Widya, 2004.
[7] E. Nurmianto, "Ergonomics smoke machine for indigenous people in Indonesia," *MATEC*, 2015.
[8] P. K. Hembecker, D. C. Reis, A. C. Konrath, L. A. Gontijo, and E. A. Eugenio, "Investigation of musculoskeletal symptoms in a manufacturing company in Brazil: a cross-sectional study," *Brazilian J. Phys. Ther.*, vol. 21, no. 3, pp. 175–183, 2017.
[9] E. Nurmianto, U. Ciptomulyono, Suparno, and S. Kromodihardjo, "Manual Handling Problem Identification in Mining Industry: An Ergonomic Perspective," *Procedia Manuf.*, vol. 4, no. December, pp. 89–97, 2015.

Ni Metal Extraction with Surfactant Variation on Double Surfactant Emulsion Liquid Membrane Technology

Yuliusman[1], A R Nafish[1], C S Salsabila[1], M P Ayu[1], and M Isa[1]

[1]Departement of Chemical Engineering, Faculty of Engineering, Universitas Indonesia

E-mail: usman@che.ui.ac.id, azmia.rizka@ui.ac.id, cut.shafira@ui.ac.id, maretha.putri@ui.ac.id, muhammad.isa73@ui.ac.id

Abstract. Nickel metal extraction from hydrotreating catalyst waste is an attempt to recover purer Ni metal using extraction process. Emulsion liquid membrane with double surfactant was chosen because this extraction method has the highest separation efficiency, shorter time of process, and simple process steps. The double surfactant that was varied in this study was a mixture of Span 80 with Tween 20 and Span 80 with Tween 80. The experimental results showed that the best nickel extraction was obtained using Cyanex 272 extract with a concentration of 0.06 M, 8% (w / v) surfactant double Span 80 with Tween 20, with an internal phase ratio and organic phase of 1: 1. The extraction process lasts for 15 minutes with a stirring speed of 250 rpm.

1. Introduction

The development of industrial processes is increasing massively at this time. A large number of catalysts are used in several fertilizer industries, the oil and gas industry, and other chemical sectors to make the reaction go faster. This catalyst will be deactivated from time to time and when the activity decreases below the acceptable level, the catalyst must be regenerated and can be reused. But when regeneration is no longer possible, the catalyst has decreased function to a very low level, so further regeneration may be economically inadequate. In such cases, used catalysts that cannot be regenerated tend to be disposed of as solid waste [1].

These used catalysts are categorized as hazardous and toxic wastes because they contain several hazardous metals such as nickel metal. A total of 14720-14800 mg / kg of nickel metal is contained in the catalyst which is spent from the hydrotreating process [2]. This makes the metal recycling and recovery process an alternative to avoid and reduce catalyst waste and prevent environmental pollution. Based on data from the Indonesian Ministry of Environment and Forestry (2014), there are 193 million tons of 200 million tons of hazardous and toxic waste that can be reused. This amount of waste can generate 20 billion rupiah from recycling used catalyst waste.

Generally, the process used for recovering nickel metal contained in the solid waste is leaching catalyst using an acid [3]. The leaching process using strong acids has become a commonly used method for recovering nickel metal from used catalysts, which is then followed by a metal recovery process with a purer concentration [4].

To get the desired metal, it is necessary to do further separation between the leaching agent and the metal solution. The method commonly used for metal separation processes with leaching agents is liquid-liquid extraction, electrowinning, liquid membrane, evaporation, crystallization and hydroxide deposition [5]. Based on previous research, it is known that the emulsion liquid membrane (ELM) has the highest separation efficiency when compared to the classic type of separation [6]. In the liquid membrane technology, extraction and stripping by receiver phase occurs simultaneously making it more time-efficient, selective and simple.

The components that make up the MCE are extractants, surfactants, organic solvents and internal phases. Emulsions generally can be formed using a single type of emulsifier / surfactant, but it was proven that the formation, stability, and functional groups of the emulsion can be enhanced by the use of two or more kinds of surfactant [7]. The parameters that can provide high stability in surfactants are Hydrophilic-Lypophilic Balance (HLB), where these parameters express a relationship to hydrophilic and lipophilic properties in a surfactant. The performance of the surfactant mixture is better than the single surfactants in emulsion formation because it has synergistic properties marked with the maximum water solubility values in kerosene [8]. Surfactants commonly used for stabilizing W / O emulsions are Span 80, Tween 80 and Tween 20. Surfactants used in emulsion formation usually have a value of Hydrophilic-Liphophilic Balance (HLB) in the range 2-18.

Table 1. Experimental condition used for elm extraction

Internal aqueous phase	V_{int} = 50 ml
Organic phase	V_{org} = 50 ml
Solvent	Kerosene
Carrier	0,06 M
Surfactant aqueous phase	8%
External aqueous phase	100 ml
Treat ratio	1:1
Stirring speed	200 – 400 rpm
Time operation	15 min
Temperature	353 K

2. Experimental

2.1. Materials
The solution used for the double surfactant emulsion liquid membrane extraction in this experiment was leached liquor as a result of the leaching process of Ni metal on hydrotreating catalyst using citric acid and EDTA leaching agents.

The extractant used in the extraction process is Cyanex 272 which is dissolved using kerosene. Kerosin used in experiments was obtained from Merck (M). The surfactants used in this experiment were Span 80, Tween 80 and Tween 20 obtained from Merck (M).

2.2. Emulsion Stability Test
To perform ELM method, it is necessary to have an emulsion with optimal content. The type of surfactant used during the ELM process will affect the stability of the emulsion in the ELM system. Emulsion stability necessary for the extraction process can take place more optimal. Apart from that, another important factor is to obtain the stability of the emulsion through a combination of the use of several types of surfactants in the stages of making emulsions.

The W / O type surfactant used was Span-80 as a single sufactant and a mixture of Span-80 with O / W type surfactants namely Tween 20 and Tween 80 as multiple surfactants. 0.2 M H2SO4 acts as a stripping phase. Extractant with a good level of selectivity against Ni is Cyanex 272 diluted using kerosene until it reaches a concentration 0,06M. The Cyanex 272 solution that has been prepared will be mixed with 8% (w / v) Span-80 and 8% (w / v) mixed surfactants (Span-80 with Tween-20 and Span-80 with Tween-80) to form an organic phase or liquid membrane phase.

2.3. Extraction with emulsion liquid membrane

After getting the best combination of double surfactants then the extraction process can be done. The extraction process begins by making a feed solution with a pH of 6. Emulsions will be made by emulsify aqueous solution (stripping phase) with the formulated organic phase. The same volume of 50 ml part of organic solution and the strip solution will be stirred continuously at a speed of 1000 rpm. Emulsions will be cooled to room temperature and ready for extraction studies. Emulsions must always be new and fresh every time you do the extraction.

The prepared emulsion is then dispersed into a vessel which is stirred with an external solution (hydrotreating catalyst waste). Stirring speed is controlled at 250 rpm. Then the mixture will be incorporated into the separatory funnel to separate the external phase (feed) with the ELM phase. The aquatic phase will be filtered to remove entrainment and the raffinate phase will be analyzed using AAS. The range of parameters used in this study for the extraction of Ni metal ELM from hydrotreating catalyst waste can be seen in Table 1.

3. Results and Discussions

3.1. Membrane stability

Emulsion in this research is a combination of internal phase (H2SO4) and organic phase (mixture of Cyanex 272, kerosen, and surfactant) [9]. In this research, the type of surfactant was varied to span 80; mixtures of span 80 with tween 20; and mixtures of span 80 with tween 80. The operating conditions used were optimum conditions obtained from previous studies, namely 8% surfactant concentration, Cyanex 272 0.06 M, ratio of internal phase:organic phase is 1: 1, stirring duration is 1 hour, and stirring speed of 1000 rpm. After the emulsion process is complete, the emulsion must be immediately separated into a 100 ml measuring cup so that stability can be observed for 4 hours.

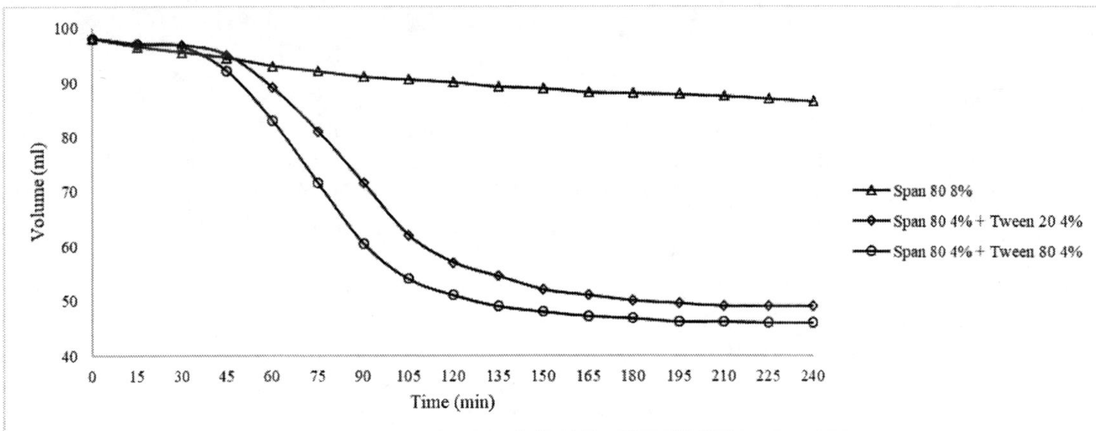

Figure 1. Emulsion Volume Change with Time (t= 4 hours)

At Fig. 1 decreasing the volume of the emulsion indicates an emulsion leak. After 4 hours, it can be seen single surfactant span 80 has a constant decrease while the double surfactant has a significant

IOP Conf. Series: Materials Science and Engineering **722** (2020) 012076 doi:10.1088/1757-899X/722/1/012076

decrease volume and separate between emulsion and internal phase. Emulsion droplets relates with decreasing of significant volume in internal phase.

At Fig. 2 comparison between single surfactant and double surfactant in a period of 45 minutes, a mixture of span 80 + tween 20 can maintain the stability of the emulsion phase better than the single surfactant span 80, this is indicated by a very small decrease in volume at the 15th minute which is equal to 1 ml. Whereas in the 30th minute the decrease in both of these mixtures only had a slight difference, which was equal to 0.2 ml to mix with tween 20 and 0.5 ml to mix with tween 80. In 45th minutes we can see with clearly the decreasing significant volume of span 80 + tween 80 which is indicates the emulsion leak. The decrease in volume continued in both mixtures until the internal phase began to form at the bottom layer of the mixture. This internal phase formation indicates that the surface tension between the internal and organic phases in the emulsion begins to decrease.

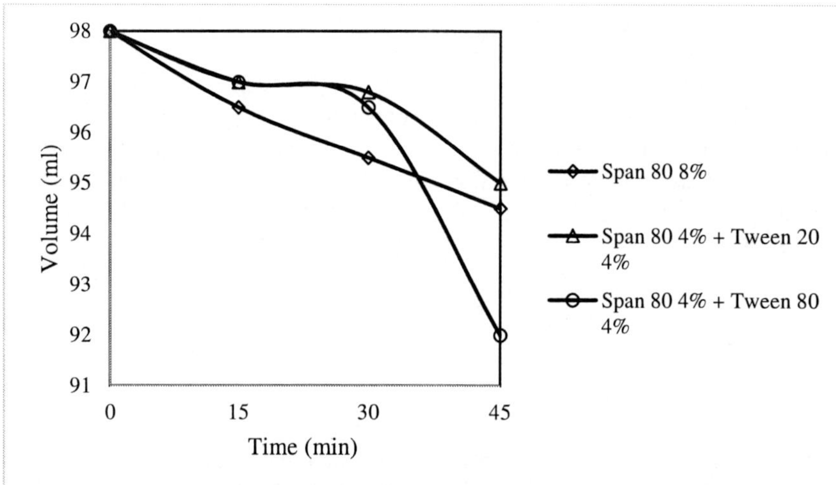

Figure 2. Emulsion Volume Change with Time (t=45 mins)

From the observations it was found that internal phase leakage in span 80 + tween 20 began to occur in the 70th minute, whereas in span 80 + tween 80 began to occur in the 54th minute. Qualitatively, it can be concluded that from the three types of surfactants, the span 80 mixture with tween 20 is the best type of surfactant suitable for the MCE application in this experiment.

3.2. HLB Analysis

HLB value is a quantity used to determine the shape of an emulsion. Hydrophilic - Liphophilic Balance (HLB) surfactant system is a molecular equilibrium characteristic between parts of hydrophilic properties (like water) to parts of lipophilic properties (like oil). The form of the emulsion can be in the form of W / O (water / oil) when it has a HLB value between 3 to 6 and dissolves in water, while the form O / W (oil / water) from 8 to 18 and dissolves in oil. The use of double surfactants observed was between mixtures of span 80 with tween 20 and span 80 with tween 80. Each HLB value of these three types of surfactants was 4.3; 16.6; and 15.

In order to find out the HLB value produced from double surfactants, it is necessary to do an HLB analysis by titrating a 0.2 M H2SO4 solution with an organic phase. The titration process is carried out with the condition slowly stirred (around 250 rpm) and stop when there is a change in color from the clear color to the cloudy white. The results of HLB analysis can be seen in Table 2.

Table 2. H_2SO_4 Titration Results

No.	Emulsion Weight (gr)			H_2SO_4
	Span 80	**Tween 20**	**Tween 80**	**ml**
1	4	-	-	2
2	2	2	-	3.9
3	2	-	2	3.8

HLB value of the mixtures can be calculated using this formula:

$$HLB_{mix} = f_A \times HLB_A + (1\text{-}f_a) \times HLB_B \quad (1)$$

Where HLBmix is the HLB value of mixtures, HLBA is a HLB value of W / O type surfactant (span 80) and HLBB is a HLB value of O / W type surfactant (Tween 20 & Tween 80). The results of processing data from the HLB analysis test can be seen in Table 3.

Table 3. HLB Analysis Test Result

Surfactant Type	H_sSO_4 (0,2M) ml	fA	fB	HLB $_{mix}$	H_sSO_4 (0,2M) gr	Water/Oil (w/w)
Span 80	2	1	-	4.30	2.01	0.0402
Span 80 + Tween 20	3.9	0.52	0.48	10.31	3.92	0.0783
Span 80 + Tween 80	3.8	0.53	0.47	9.34	3.82	0.0763

The calculation results show the highest solubility of water phase in oil is span 80 with tween 20 mixture with mixtures HLB value of 10.31 and has a water phase solubility value in oil of 0.0783. If the solubility value is sorted, it was obtained that the double surfactant span 80 + tween 20> double surfactant span 80 + tween 80> single surfactant span 80. The combination of span 80 and tween 20 surfactant as a double surfactant is quantitatively proven to have good ability to stabilize the emulsion compared to the others. This ability is due to span 80 and tween 20 mixed surfactants having the most synergistic properties.

3.3. Extraction

After obtaining the best combination of double surfactants then the extraction process can be done. This process begins by making a leached liquor to pH 6 first. The effect of pH on extraction efficiency is caused by competition between H+ and Ni2+ in fighting for anions. Therefore, it is expected that with increasing pH the extraction ability increases because the number of H+ is less so that the competition of H+ and Ni2+ in the fight against extractant anions is reduced and Ni2+ can more easily react with anions than extractants [8]. However, if the pH is higher than 6 it can cause the interface between the external and internal phases to be saturated by complex ions from extraction ionization and water transport will occur which causes decreased extraction to even break the existing emulsion [9].

The extraction process was carried out by mixing the feed phase with ELM for 15 minutes at a speed of 250 rpm. Then this mixture is incorporated into the separatory funnel to separate the external phase (feed) with the ELM phase. The feed phase which is initially light green turns into a cloudy green after the extraction process is done.

From the results of the AAS, it was found that the external phase containing nickel with 661.5 ppm and after the MCE extraction, the nickel contain is 606.6 ppm in internal phase. From these results it was found that the percentage of extraction was 91.7%. This percentage value is quite high when compared to previous studies which only used a single surfactant, which was 81.51% [10]. From this it is proven that double surfactant is a better alternative to overcome emulsion leakage in the extraction process of the emulsion liquid membrane.

4. Conclusions

Span 80 and Tween 20 double surfactants proved superior in maintaining the stability of the emulsion volume. The optimum condition of the MCE manufacturing process is with a concentration of 8% double surfactant of span 80 with tween 20, 0.06 M Cyanex 272 extractant concentration, internal phase: organic phase ratio of 1: 1, stirring for 1 hour at 1000 rpm. As for the extraction process, the optimum percentage of extraction was 91.7% with pH of feed phase 6, extraction time for 15 minutes, stirring speed of 250 rpm, and membrane phase ratio: feed phase of 1: 2

Acknowledgments
This research was funded by Hibah Publikasi Internasional Terindeks Tugas Akhir (Hibah PITTA B) Universitas Indonesia 2019 with contact number NKB-0802/UN2.R3.1/HKP.05.00/2019. The author declares no conflicts of financial interest.

References
[1] Cano, J., Zamarripa, G. Z., Pedroza F. R. C., Aguilar, M. J. S., Macias, A. H., Vielma, A. C, "Kinetics and statistical Analysis of Nickel Leaching from Spent Catalyst in Nitric Acid Solution" International Journal of Mineral Processing, Vol. 148, pp. 41-47, 2016.
[2] Mardyanika, Yus A.P, "Pemanfaatan Spent Catalyst RCC 15 Limbah Pertamina sebagai Bahan Tambah pada Campuran Beton", Teknologi dan Kejuruan, Vol. 35, No. 2, pp.191-200, 2012.
[3] Yuliusman. "Recovery of Cobalt and Nickel from Spent Lithium Ion Batteries with Citric Acid Using Leaching Process ". E3S Web of Conferences 67. 03008, 2018.
[4] Yuliusman. "Acid Leaching and Kinetics study of Cobalt Recovery from Spent Lithium-ion Batteries with Nitric Acid". E3S Web of Conferences 67. 03025, 2018.
[5] Barik, S. P., Park, L. H., Nam, C. W. "Process Development for Recovery of Vanadium and Nickel from an Industrial Solid Waste by a Leaching-Solvent Extraction Technique". Journal of Enviromental Management, vol.10, No. 15, pp. 1457 – 1472, 2014.
[6] Kislik, V. S." Liquids Membranes Principle & Applications in Chemical Separation & Wastewater Treatment. Amsterdam: Elsevier, 2010.
[7] Dickinson, Eric. "Mixed Biopolymers at Interfaces: Competitive Adsorption and Multilayer Structures". Journal of Food Hydrocolloids, vol 25, pp. 1966 – 1983, 2011.
[8] Mulyazmi. "Penggunaan Surfaktan Campuran terhadap Pembentukan Kestabilan Emulsi W/O". Jurnal Teknos, vol. 6, No.2. 2006
[9] Yuliusman. "Leaching and Kinetics Process of Cobalt from Used Lithium ion Batteries with Organic Citric Acid". E3S of Conferences 67. 0-3036, 2018.
[10] Huda, Miftahul. "Perolehan Kembali Logam Ni dari Limbah Katalis Hydrotreating dengan Teknologi Membran Cair Emulsi Menggunakan Ekstraktan Xyanex 272". Skripsi. Depok: Universitas Indonesia. 2017.

IOP Conf. Series: Materials Science and Engineering **722** (2020) 012077 doi:10.1088/1757-899X/722/1/012077

Platinum Recovery from Spent CCR Platforming Catalyst with Oxalic Acid and Aqua Regia Leaching

Yuliusman[1], L P D I Prawira[1], A R Nafisah[1], K S Amananti[1], M P Ayu[1] and M Isa[1]

[1]Chemical Engineering Department, Faculty of Engineering Universitas Indonesia

E-mail: usman@che.ui.ac.id, luh.putu51@ui.ac.id, azmia.rizka@ui.ac.id, khairina.shauma@ui.ac.id, maretha.putri@ui.ac.id, muhammad.isa73@ui.ac.id

Abstract. In aligning to the rapid industrial development, the needs of catalyst nowadays are increasing. Oil and gas industry is one of the example especially in Continuous Catalytic Cracking Platforming Unit, in which the process produces Pt/Al2O3 catalyst waste approximately 2000-3000 kg/year and contains of 3200 ppm platinum metals. It becomes the reason why organic acid leaching and aqua regia leaching is needed to recover the platinum metals to save cost because its price is very expensive and spent catalyst waste is also considered as hazardous and toxic materials. In this experiment, there are two different methods used to obtain the recovery of platinum metals from spent Pt/Al2O3 catalyst. The methods used are leaching with oxalic acid and leaching with aqua regia. The experiment showed that leaching with aqua regia results in higher leaching efficiency percentage with the number of 19.917%. The result is obtained with operation temperature of 300oC and solid-liquid ratio of 20 g/L

1. Introduction

Platinum is a silvery metal and considered as rare earth metal that has really high price. In a very fine condition, platinum has been many used as a catalyst. Oil and gas industry in Indonesia mostly used platinum as their catalyst in catalytic reforming process of straight-chain naphtha to rich-aromatic octave level gasoline. The catalytic reforming process is performed in Continuous Catalytic Platforming Unit in Pertamina Balongan Petroleum Refinery. The Pertamina Balongan's Platforming Unit produces up to 3000 kg/year spent platinum catalysts. Spent platinum catalyst is also considered as hazardous and toxic materials therefore a platinum recovery method is needed. Moreover, platinum is very pricey with the price of IDR 450139,00 per grams [1], which drives the urgency of doing platinum recovery from spent Pt/Al2O3 catalyst.

The platinum metal recovery can be done with leaching method. The factors that can affect leaching process are type of leaching agent, leaching agent concentration, stirring speed, solid-liquid ratio, contact time, temperature, and particle size. Mostly, mineral acid is used as the leaching agent because of its ability to produce high leaching percentage and it is quite cheap. From previous research, the leaching efficiency percentage obtained from leaching with mineral acid (aqua regia) is 90% [2]. Another alternative which can be used as the leaching agent is organic acid, such as oxalic acid, because of its easiness to dissolve in water and can be degraded naturally. The leaching efficiency percentage for this leaching agent is 82% with leaching temperature condition of 60oC with 10 hours of stirring and the leaching agent concentration used is 0,3M [3].

2. Experimental

2.1. Sample Preparation

Determine the metal content of spent catalyst using XRF PANalytical before going to the furnace. Spent catalyst weighed to 100 gram and put in the quartz trays. Pre-heated the furnace until the temperature become 450oC. Put the quartz tray into the furnace and set the ramping time 10oC per minutes until the temperature reach 900oC and wait until 3 hours. After that, set down the temperature to reach 200 oC and take the quartz trays, put it in the dessicator. After that, determine the metal content of spent catalyst using XRF PANalytical. It was going to observe Pt and impurity likes Si and S content had decreased by furnace or not.

When the XRF analysis has finished, the samples grounded until the size became 100 mesh. It will increasing the surface area so it can enhance the effectively of leaching process.

2.2. Leaching with Oxalic Acid

Oxalic acid is a colorless acid and is an organic carboxylic acid that melts at 189oC with sublimation. In this leaching process, oxalic acid is used as the leaching agent. The independent variable for this procedure is the oxalic acid concentration. Meanwhile, the fixed variables are operation temperature, contact time, stirring speed, and solid/liquid ratio. The first step before doing the leaching process is to make the oxalic acid solution for four concentration variations (0.3M, 0.5M, 0.7M, and 1.0M). Dihydrate Oxalic Acid is used as the leaching agent because its easiness to dissolve in water. The solid/liquid ratio used for this process is 10 g/L and the leaching process was done with a magnetic stirrer at 60oC for around 10 hours. Determination of Pt content in leached liquor by using ICP-OES Agilent Technologies 5110.

2.3. Leaching with Aqua Regia

Aqua regia is a solution composed of two strong mineral acid which are chloride acid (HCl 37%) and nitric acid (HNO3 65%). The composition ratio of the chloride acid and nitric acid is 3:1 (volume ratio).

All variables used for this process refer to optimum condition for UOP 896-930 method. The solid/liquid ratio of 20 g/L is used in this method. For the leaching process, it used stirring speed of 300 rpm at approximately 300oC. This method needed to be done in fume hood and use safety equipment because it produced plenty of hazardous chlorine gas. Determination of Pt content in leached liquor by using ICP-OES Agilent Technologies 5110.

3. Results and Discussions

This section discusses about the experimental results. The research focus on leaching process with oxalic acid and aqua regia as the leaching agent. The experiment aims to determine which leaching agent gives better result in leaching efficiency percentage and more effective to recover platinum metals from the spent catalyst. Leaching efficiency can be calculated using this formula :
Leaching efficiency = (Pt concentration in leached liquor/Pt concentration after furnace) x 100%

3.1. Sample Characterization and Preparation

According to the theory, the smaller particle size will increasing the surface area. The particle size range of 80-120 mesh is known as the most optimum particle size [3]. In this experiment the spent catalyst is firstly burned at 900oC for approximately 3 hours. This aims to eliminate the silica and sulphur content in the spent Pt/Al2O3 catalyst. Silica and sulphur has an disadvantage as for the support of platinum and could prevent the catalyst deactivation because of the reaction between γ-Al2O3 and SO2 [4]. The presence of silica and sulphur in the spent catalyst will complicate the leaching process of platinum.

(a) (b)

Figure 1. (a) Spent Pt/Al2O3 catalyst after furnace (b) Spent catalyst 100 mesh

The spent Pt/Al_2O_3 is characterized through XRF (X-ray Fluorescence) analysis before and after furnace treatment to determine the metal content from the spent catalyst. The result from XRF analysis of spent catalyst Pt/Al_2O_3 described in Table 1.

Table 1. XRF Analysis Result For Spent Catalyst Pt/Al2O3

Element	Concentration (%)	
	Before Furnace	After Furnace
Pt	0,3	0,32
S	0,02	0,01
Si	0,16	0,04
Al	51,45	50,3
O	46,39	45,1
Fe	0,29	0,29
W	0,04	0
Ni	0,06	0,02
Na	0,02	0,02
Cr	0,02	0,02
Cl	0,97	0,09
Ca	0,02	0,01
Ba	0,01	0
P	0,12	0,1
Sn	0,1	0,09

From the XRF analysis, it can be seen that there are fifteen elements in the spent Pt/Al_2O_3 catalyst. The five most dominant elements are aluminium, oxygen, sulphur, platinum, and ferrum. The number of aluminium is dominating because the aluminium acts as the support catalyst meanwhile the platinum acts as the active site of catalyst.

3.2. Leaching with Oxalic Acid

Leaching is a process to extract a substance from its solid after being contacted with any kind of solution [5]. The leaching agent used for the leaching process can be an organic acid or a mineral acid.

Since the use of mineral acid is more hazardous to the environment than organic acid, this experiment tried to use organic and see the effect of organic acid usage [6].

In this experiment, the oxalic acid concentrations are varied into 0.3M, 0.5M, 0.7M, and 1.0M. The leaching process is done with a magnetic stirrer at 60°C for 10 hours. After the 10 hour leaching process was done, the solution was filtered, and the filtrate was analyzed with AAS (Atomic Absorption Spectroscopy). The Atomic Absorption Spectroscopy was a quantitative test and the result describbed below in Table 2 & Fig. 2.

Table 2. AAS Analysis Result for Leaching with Oxalic Acid Concentration Variations

Oxalic Acid Concentration	**Sample**		**Result**	
	mL	**mg/L**	**ppm**	
0.3M A	10	8.89	88.90	**88.54**
0.3M B	10	8.82	88.17	
0.5M A	10	8.73	87.26	**89.15**
0.5M B	10	9.10	91.03	
0.7M A	10	9.11	91.10	**92.00**
0.7M B	10	9.29	92.88	
1M A	10	17.70	176.92	**178.78**
1M B	10	18.06	18.63	

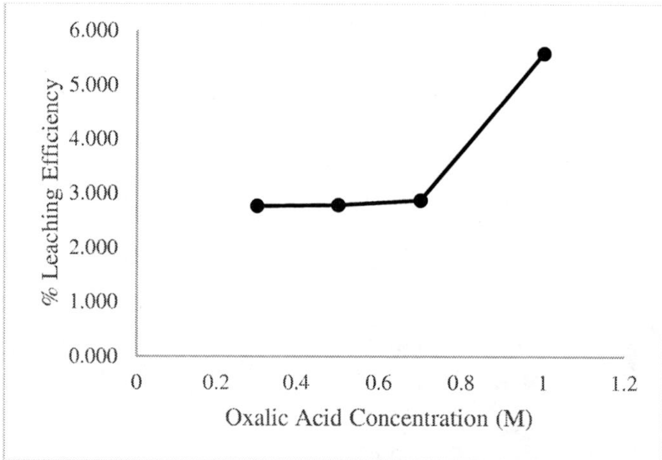

Figure 2. Correlation between oxalic acid concentration and leaching efficiency percentage graph

From Table 1, it can be seen that the leaching efficiency for 0.3 M ; 0.5 M ; 0.7 M and 1 M are 2.76% ; 2.78% ; 2.87% and 5.58%. From this result, the best concentration for Pt leaching is using 1 M oxalic acid with the current condition.

From Fig. 3, it can be seen that the leaching percentage did not increase significantly but starting from 0.8M until 1M of oxalic acid, the percentage increased very sharply. The highest oxalic acid concentration that can be used for leaching is 1M. Using the oxalic acid concentration more than 1M in leaching process, the solution has reached its saturation point that marked by the appearance of crystal.

Platinum metal which is initially present in the solids of the spent catalyst will be bound by a leaching agent (oxalic acid) because of the anions from organic acids [7]. If the concentration of organic acids is increased, the ability and the amount of acidic anions formed will also increase. The increase in the number of acidic anions causes an increase in platinum metal concentration in the filtrate. Leaching agent likes oxalic acids is classified in the class of weak acids and weak electrolytes, and platinum is very inert metal and difficult to react. This is the reason why leaching efficiency is only get 5.58% .

Even though the operating conditions used are optimum conditions according to reference, platinum metal is in fact unable to be leached. The fact that platinum metal itself is unable to react with organic acids (oxalic acid) will greatly affect the results of the percentage of platinum recovery obtained from the results of leaching. Therefore, further research could be compared the leaching efficiency between using oxalic acid and aqua regia as leaching agent.

3.3. Leaching with Aqua Regia

Aqua regia is a very acidic solution formed from concentrated hydrochloric acid (37% HCl) and concentrated nitric acid (65% HNO_3). The two are mixed with the volume ratio of HCl: HNO_3 3: 1. The research phase in this method uses concentrated acids so it is very dangerous and must be done in a fume hood. The variables used in this method arc optimum conditions according to the UOP method 896-930. The optimum conditions used, namely the ratio of solid/liquid 20g/L at an operating temperature of 300ºC. The operating time used depends on the speed of the evaporating and boiling solution.

The percentage of leaching efficiency in this process is 19.72% with a platinum concentration in leached liquor is 631 ppm. Some of the factors that give an impact of results from leaching efficiency with aqua regia described below: [8]

1. The process carried out with a lot of stages and the heating process with an open magnetic stirrer causes the possibility of mass loss through the evaporation process.
2. The temperature control in hotplate do not describe the real temperature of the solution
3. The influence of a magnetic bar that has been dirty and contaminated due to overuse and cannot be cleaned completely

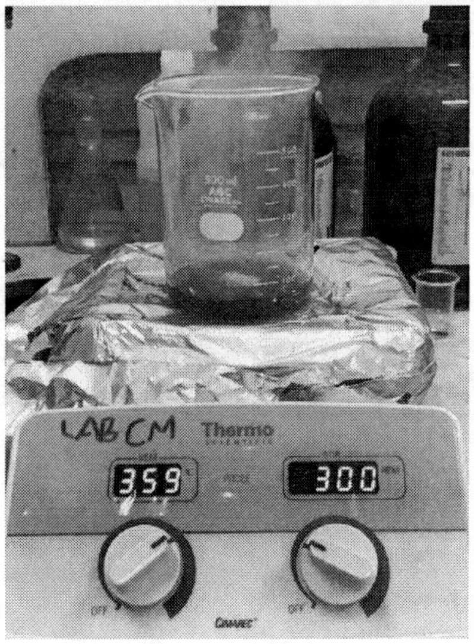

Figure 3. Leaching process with aqua regia in fume hood

4. Conclusions

From the leaching experiment with oxalic acid, the highest leaching efficiency percentage is 5.58%. The leaching was fixed with 1M of oxalic acid stirred at 60oC for 10 hours. And for the leaching experiment with aqua regia, the leaching efficiency is 19.72%. The leaching was fixed using UOP 896-930 method. From those two experiments, the leaching process with mineral acid (aqua regia) is still more effective than with organic acid (oxalic acid) despite its harmfulness to the environment. Although the organic acid is more environmentally friendly, it is a weak acid and do not have the same power to do an oxidation process like aqua regia.

Acknowledgments
This research was funded by Hibah Publikasi Internasional Terindeks Tugas Akhir (Hibah PITTA B) 2019 with contract number NKB-0802/UN2.R3.1/HKP.05.00/2019. The author declares no conflicts of financial interest.

References

[1] Antam. (2018). Harga Logam Mulia. [online] Antam.com. Available at: http://www.antam.com/index.php?option=com_content&task=view&id=276&Itemid=140&lang=id [Accessed 20 Oct 2018]
[2] Marinho, R.S. & da Cunha, Jose W. S. D. (2010). Recovery Of Platinum From Spent Catalysts By Liquid-Liquid Extraction In Chloride Medium. Journal of Hazardous Materials. 179, 488-494.
[3] Masuda, C. & Yokoyama, T. (2013). Recovery of Platinum from Spent Reforming Catalyst by Acid Leaching. Procedia Earth and Planetary Science. 6, 435-440.
[4] Mi-Young Kim, S. M. P., Jung-Hyun Park, Chae-Ho Shin, Won-Jin Moon, Nark-Eon Sung, Gon Seo 2011. Platinum catalysts supported on silicas: effect of silica characteristics on their catalytic activity in carbon monoxide oxidation. Reaction Kinetics Mechanical Catalyst, 463-479.
[5] Yuliusman. "Recovery of Cobalt and Nickel from Spent Lithium Ion Batteries with Citric Acid Using Leaching Process". E3S Web of Conferences 67. 03008, 2018.
[6] Yuliusman. "Acid Leaching and Kinetics study of Cobalt Recovery from Spent Lithium-ion Batteries with Nitric Acid". E3S Web of Conferences 67. 03025, 2018.
[7] Yuliusman. "Leaching and Kinetics Process of Cobalt from Used Lithium ion Batteries with Organic Citric Acid". E3S of Conferences 67. 0-3036, 2018.
[8] Lowe, D. (2019). Stirring Bar Contamination In the Pipeline. [online] Science Translational Medicine. Available at: https://blogs.sciencemag.org/pipeline/archives/2019/03/26/stirring-bar-contamination [Accessed 27 May 2019].
[9] Kumar Jha, & Kumar, V. 2012. Hydrometallurgical Recovery/Recycling of Platinum by The Leaching of Spent Catalysts: A Review. Hydrometallurgy, 133, 23—32.
[10] McVicker, Gary B. & Carter, James. 2005. Recovery of Metal Contaminants from Catalysts Using Buffered Oxalic Acid. 4,522,928.
[11] USGS Mineral Resources. (2017). Platinum-Group Metals Statistics and Information. [online] Available at: https://minerals.usgs.gov/ minerals/pubs/commodity/platinum/ [Accessed 15 Oct. 2018].
[12] Barakat, M.A., Mahmoud, M.H.H. (2004). Recovery of Platinum from Spent Catalyst. Hydrometallurgy, Vol. 72, Issues 3—4, 179-184.

IOP Conf. Series: Materials Science and Engineering **722** (2020) 012078 doi:10.1088/1757-899X/722/1/012078

Proposed Reaction Mechanism of Chitosan-graft-Maleic from Chitosan and Maleic Anhydride

D Timotius[1], Y Kusumastuti[1], N R E Putri[1], and Rochmadi[1]

[1]Chemical Engineering Department, Universitas Gadjah Mada, Yogyakarta, Indonesia

E-mail: daniel.timotius@mail.ugm.ac.id, yuni_kusumastuti@ugm.ac.id, nur.rofiqoh.e@mail.ugm.ac.id, rochmadi@ugm.ac.id

Abstract. Chitosan (Ch), a natural polysaccharide derived from chitin is a widely used raw material in biomaterial research. Modification of chitosan using maleic anhydride (MA) have been done by several researchers with dimethyl sulfoxide (DMSO) and acetic acid as a common solvent. In the present work the chitosan was grafted with maleic anhydride using acetic acid and water as a solvent toward ecofriendly functionalization process. In addition, proposed mechanism of this grafting was further studied. The total acidic content and conversion of grafted films were calculated with varying concentration of MA:Ch ratio and temperature. The success of chitosan grafting with maleic was verified by Fourier Transform Infrared Spectroscopy (FTIR) analysis. FTIR analysis showed the appearance of new ester group and shifting of amide group from FTIR spectra which indicates that the esterification and amidation occurred in this grafting process.

1. Introduction

In the last few decades, research about biomaterials has become an interesting topic. In order to meet the requirements for biomedical applications, suitable materials with good biocompatibility and excellent properties is needed. One of prospective material for biomaterial is chitosan. Chitosan, (1-4)-2-amino-2-deoxy-β-D-glucan, is the second most abundant polysaccharide after cellulose [1]. It can be obtained by deacetylation of chitin [2]. Chitosan is indicated by the amount of degree of deacetylation (DD). DD is a parameter that indicates the presence of amino groups inside polymers. It can be called as chitosan if the DD is greater than 50%, while chitin is indicated by the amount of DD less than 50%. Higher DD means that chitosan has better solubility than chitin in aqueous acid solution [3].

Chitosan is commonly used as biomaterial because it has good antibacterial properties [4] and swelling ability [5]. It also has a good film forming ability [6] which is useful to form a thin film. In addition, It has many active functional groups (amine and hydroxyl groups) along the chains which could be easily modified to form better film properties. Dilution of chitosan in acid solution (below pH 6.3) can lead to protonation of amine group and convert it into ammonium groups [7].

Chitosan can be modified with maleic anhydride (MA) to add carboxyl group. Hasipoglu *et al* (2005) [8] already researched on chitosan-graft-maleic acid in acetic acid solution using ceric ammonium nitrate (CAN) as the redox initiator. Similar procedure to synthesize chitosan-graft-maleic was also carried out by other researchers [5]. They investigated the swelling behaviour of chitosan-graft-maleic for several pH values and also conducted the antibacterial activity test for the obtained samples. The result showed that chitosan-graft-maleic have a pH sensitive swelling behavior and antibacterial activity.

There was also an investigation on maleilated chitosan (MCS) [9]. A modified chitosan using MA was carried out and stirred in DMSO for 24 hours. The obtained maleilated chitosan then further crosslinked with methacrylated poly(vinyl alcohol) (MPVA) by UV irradiation to promote photo-crosslinking between vinyl group from MCS and MPV in the form of nanofibers. The nanofibers could be used as wound dressing materials.

The use of dimethyl sulfoxide (DMSO) and acetic acid as solvent to graft maleic anhydride into chitosan has done in the past by researchers. In this work, chitosan was grafted with maleic anhydride in acetic acid and water as solvent toward ecofriendly functionalization. Then, the grafted chitosan would be analyzed with Fourier Transform Infrared Sprectroscopy (FTIR). Total acidic content of films both in acetic acid and water as a solvent were measured. Further, study on grafting reaction mechanism was done.

2. Methods

2.1. Materials

Low molecular wight chitosan was obtained from Sigma Aldrich, USA (75 – 85% DD). Maleic Anhydride (MA) (≥99%) was purchased from Nacalai Tesque Inc, Japan. Acetic acid was supplied by Merck, Germany (100%).

2.2. Grafting Preparation

The solution was prepared by dissolving MA (1; 2; 4 g) in distilled water (100 mL) followed by chitosan (2 g). The reaction was then carried at 30; 50; 70 oC and stirred at constant speed for 5 hours. Then about 10 mL of the solution was casted on petridish (Ø10 cm) and dried in oven for 24 hours at 50 oC. The dried film was washed by large amount of acetone (25% v/v) to remove the unreacted MA. To compare the result between films that using distilled water as solvent and acetic acid as solvent, the same method was carried out by dissolving MA in acetic acid (1% v/v) instead of distilled water. For standardization purpose, chitosan (2 g) was dissolved in acetic acid (1% v/v) and stirred for 5 hours at 30 oC. Each sample was encoded as Ch for standard, Ac2170 for acetic acid solution, and W1230 for distilled water. The first number indicates weight of MA, while the second shows the weight of chitosan, and the last two numbers represents the temperature. Therefore the samples code would be Ch; Ac4270; W1230; W2230; W4230; W4250; and W4270.

2.3. Fourier Transform Infrared Spectroscopy (FTIR)

FTIR was performed to investigate the change of functional groups in chitosan and chitosan-graft-maleic. In this research SHIMADZU IRPrestige-21 was used in the range of 4000-400 cm-1. Each of the samples were scanned using KBr method.

2.4. Total Acidic Measurement

The amount of total acidic content was measured using back titration method [10]. About 10 mg of sample was dissolved in 20 mL of NaOH solution (0.01 M), then titrated by HCl (0.01 M). The obtained data was calculated using Eq. (1) to get the total acidic content (AC, mmol/mg).

$$AC = \frac{C_{NaOH} \times V_{NaOH} - C_{HCl} \times V_{HCl}}{m_{sample}} \tag{1}$$

Where C_{NaOH} represents the concentration of NaOH (M); while V_{NaOH} is the volume of NaOH (mL); C_{HCl} is the concentration of HCl (M); V_{HCl} is the volume of HCl (mL); and m_{sample} is the mass of sample (mg).

IOP Conf. Series: Materials Science and Engineering **722** (2020) 012078 doi:10.1088/1757-899X/722/1/012078

Assuming that the amount of acidic contents describes the amount of MA attached in chitosan backbone as explained in Fig 3, the conversion of MA reacted can be calculated (X_{MA}) as:

$$X_{MA} = \frac{AC_{Washed}}{AC_{Unwashed}} \quad (2)$$

3. Results and Discussions

3.1. Physical Appearance

The physical appearance of unwashed and washed film is shown on Fig 1a and 1b respectively. The result shows that the unwashed W4270 has a white color while the washed W4270 has a transparent appearance. The white color in unwashed W4270 is suspected to be unreacted MA. The films then washed with large amount of acetone (25% v/v) to remove the excess of MA. From the investigation, the appearance of white color only occurs when the ratio of MA:Chitosan is 2:1. However, the washing is done for all samples to ensure that unreacted MA is removed.

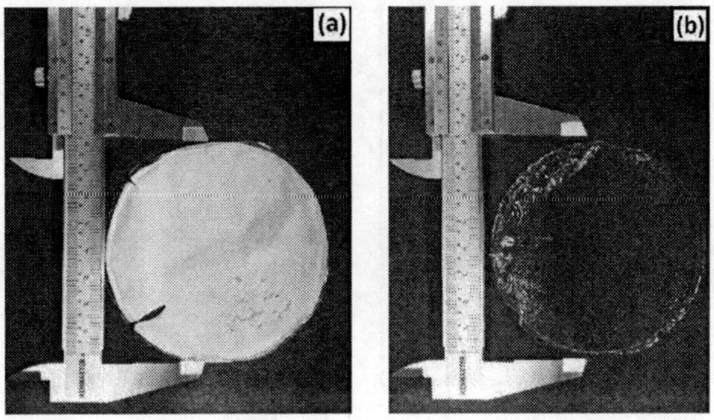

Figure 1. Physical Appearance of Unwashed (a) and Washed (b) W4270.

3.2. FTIR Study

The possible functional groups from the samples can be obtained from FTIR. The FTIR spectra of maleic anhydride is shown in Fig. 2a. The obtained peaks are compared with several studies [5], [8], [11]–[13]. The broad band around 3000 cm^{-1} appeared due to the stretching of carboxylic acid (hydrogen bonding). Strong absorption at 1890.24 and 1705.07 cm^{-1} belongs to asymmetrical and symmetrical C=O stretching of cyclic anhydride molecules. Meanwhile, the peak at 1589.34 cm could be assigned as C=C stretching of maleic anhydride. The stretch of C–O–C for maleic anhydride appears at 1265.30 cm^{-1}. Chitosan spectra on FTIR (Fig. 2b) shows that a broad absorption around 3500 that belongs to O–H and N–H stretching (hydrogen bonding). The presence of amide I and amide II are appeared on 1635.64 and 1558.48 cm^{-1} respectively, while the C–N peak present at 1373.32 cm^{-1} below. The ether stretching of chitosan show at 1080.14 cm^{-1}.

In order to prove that the reaction is success, the FTIR spectra of unwashed and washed Ac2170 were studied. Fig 2c shows several strong bands from maleic anhydride that still appears on the spectra, especially around bandwidths of 3000 cm-1, 1897.95 cm-1, 1589.34 cm-1, 1465.90 cm-1 and 1265.30 cm-1. As it washed with acetone 25% (v/v), the unreacted maleic anhydride then disappeared from the film. It is proved by the disappearance of those bands from maleic anhydride. A little shifting from amide I and amide II on washed Ac2170 spectra indicates that the amidation has taken place. The appearance of strong peak at 1705.01 cm-1 for washed Ac2170 (Fig 2d) proves that the grafting process also occurred in the hydroxyl groups (C=O from ester group of grafted maleic) [14]. The disappeared peak at 1265.30 cm-1 (Fig. 2d and 2e) shows that the ring opening mechanism of chitosan graft maleic is success Those results above show that maleic anhydride is successfully grafted to chitosan through ring opening mechanism. Further discussion about reaction mechanism is explained below.

IOP Conf. Series: Materials Science and Engineering **722** (2020) 012078 doi:10.1088/1757-899X/722/1/012078

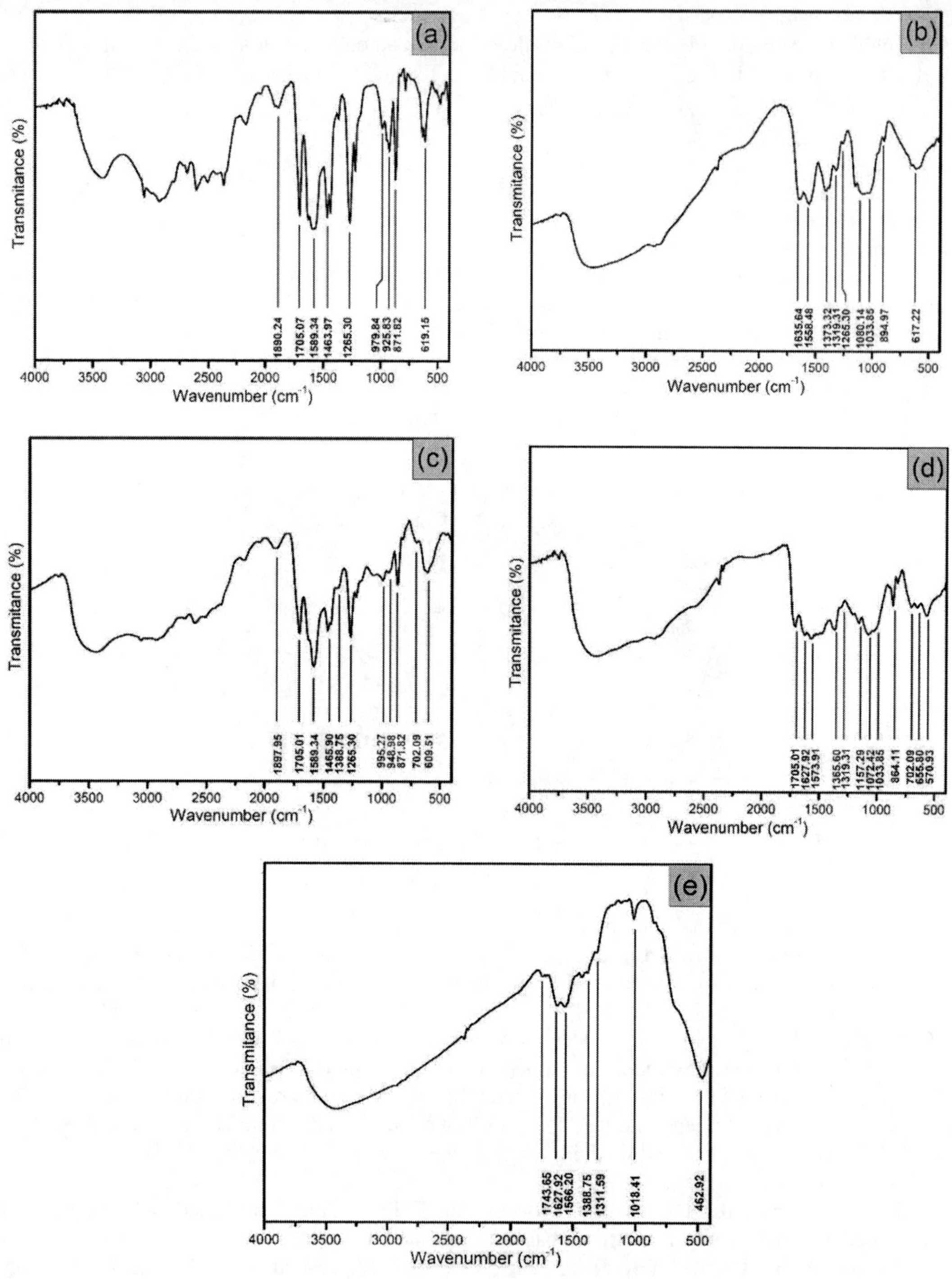

Figure 2. FTIR Spectra for Maleic Anhydride (a), Ch (b), Unwashed Ac2170 (c), Washed Ac2170 (d), and W4230 (e).

IOP Conf. Series: Materials Science and Engineering **722** (2020) 012078 doi:10.1088/1757-899X/722/1/012078

3.3. Reaction Mechanism

From the study of FTIR it can be concluded that reaction of chitosan with maleic anhydride follows two kind of reactions, specifically amidation of primary amine group and esterification of hydroxyl group [15]. The general mechanism is presented on Fig. 3. Free electron pairs of chitosan (both from hydroxyl and primary amine group) will attack the carbon atom on carbonyl group of maleic anhydride which results in transition product of tetrahedral. The next step is ring opening of maleic anhydride, followed by the attack of water molecules on the hydrogen so that they become hydronium ions. The hydronium ions will have an equilibrium with carboxyl and amine group from chitosan-graft-maleic. This equilibrium is influenced by several factors, such as initial acid concentration and type of acid which requires further study [16].

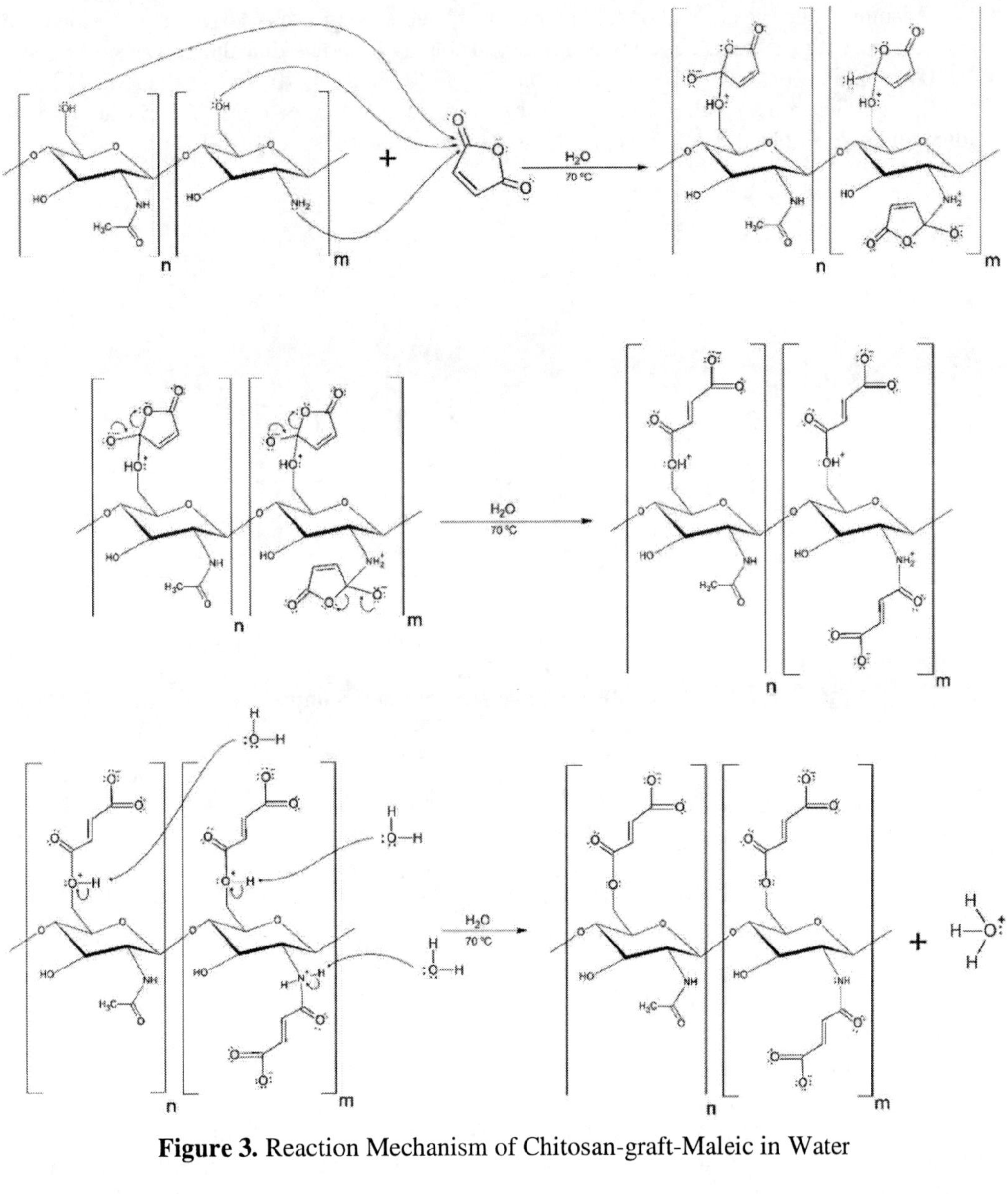

Figure 3. Reaction Mechanism of Chitosan-graft-Maleic in Water

The presence of carboxyl groups on the chitosan backbone give contribution on the appearance of negative charges in chitosan. Since chitosan becomes natural cationic polymers after being dissolved in acid solution, the appearance of anion molecules would turn chitosan into a polyampholyte [14], [17], a pH sensitive material with an ability to adsorb protein molecules. This material could be useful for medical applications such as membrane, drug delivery, and tissue engineering.

3.4. Total Acidic Content

Total acidic content of all the samples are calculated with equation (1). The comparison among samples are presented on Fig 4, which show that increasing the weight ratio of MA would increase total acidic content of the product. However, it also raise the amount of unreacted MA. Total acidic content is already utilized by other researcher to measure the amount of carboxyl group from modified hyaluronic acid [11]. Furthermore, increasing the temperature would also lead to raise the total acidic content.

Assuming that the amount of acid represents the amount grafted MA, the conversion of MA could be calculated with equation (2) as presented on Table 1. It shows that the conversion of W1230 and W2230 are similar, meanwhile the conversion of W4230; 4250; 4270 are less than W1230 and W2230. Even though they have smaller conversion than those with smaller ratio of MA, the amount of total acidic content raised as the ratio increased, just as another studies' result [8].

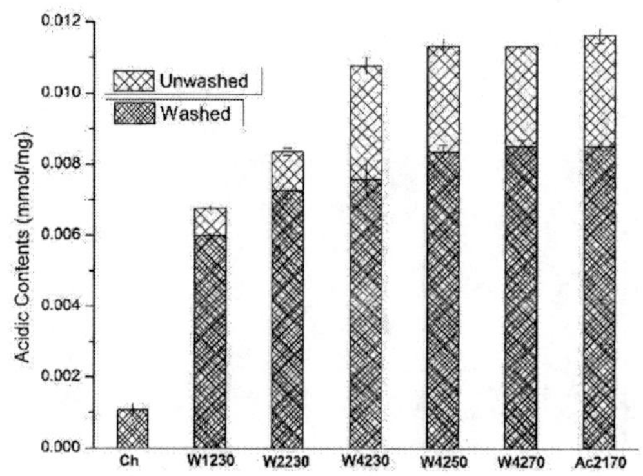

Figure 4. Total Acidic Content of Each Sample

Table 1. Conversion of Each Samples

Samples	Conversion, %
W1230	88.46 ± 1.22
W2230	88.98 ± 2.76
W4230	63.06 ± 3.86
W4250	73.15 ± 1.61
W4270	75.00 ± 1.62

4. Conclusions

Maleic anhydride was successfully grafted to chitosan by using acetic acid solution and distilled water as solvent. The grafting mechanism consists of esterification and amidation of chitosan as confirmed by FTIR. The result shows that increasing the temperature will increase the conversion of chitosan. Further study about the interaction between chitosan and maleic anhydride are still necessary to understand the behaviour and characteristic of chitosan-graft-maleic. Another study on correlation to measure the conversion is needed, since the actual interaction between hydronium, carboxyl, and amine group is an equilibrium.

Acknowledgments
This work was supported from "Rekognisi Tugas Akhir" Program funded by Universitas Gadjah Mada

References

[1] A. Ali and S. Ahmed, "A review on chitosan and its nanocomposites in drug delivery," Int. J. Biol. Macromol., vol. 109, pp. 273–286, 2018.
[2] V. R. Sinha, A. K. Singla, S. Wadhawan, R. Kaushik, R. Kumria, K. Bansal, and S. Dhawan, "Chitosan microspheres as a potential carrier for drugs," Int. J. Pharm., vol. 274, no. 1–2, pp. 1–33, 2004.
[3] H. El Knidri, R. Belaabed, A. Addaou, A. Laajeb, and A. Lahsini, "Extraction, chemical modification and characterization of chitin and chitosan," Int. J. Biol. Macromol., vol. 120, pp. 1181–1189, 2018.
[4] F. Kara, E. A. Aksoy, Z. Yuksekdag, N. Hasirci, and S. Aksoy, "Synthesis and surface modification of polyurethanes with chitosan for antibacterial properties," Carbohydr. Polym., vol. 112, pp. 39–47, 2014.
[5] S. Yucel, Z. O. Ozdemir, C. Kesgin, P. Terzioglu, S. Unlu, and Y. Erdogan, "Swelling Behavior and Cytotoxicity of Maleic Acid Grafted Chitosan," Int. J. Med. Heal. Sci., vol. 7, no. 6, pp. 232–235, 2013.
[6] L. Mengatto, I. Helbling, and J. Luna, "Recent advances in chitosan films for controlled release of drugs," Recent patents drug Deliv. forrmulation, vol. 6, pp. 156–170, 2012.
[7] M. C. G. Pellá, M. K. Lima-Tenório, E. T. Tenório-Neto, M. R. Guilherme, E. C. Muniz, and A. F. Rubira, "Chitosan-based hydrogels: From preparation to biomedical applications," Carbohydr. Polym., vol. 196, no. May, pp. 233–245, 2018.
[8] H. N. Hasipoglu, E. Yilmaz, O. Yilmaz, and H. Caner, "Preparation and characterization of maleic acid grafted chitosan," Int. J. Polym. Anal. Charact., vol. 10, no. 5–6, pp. 313–327, 2005.
[9] Y. Zhou, Q. Dong, H. Yang, X. Liu, X. Yin, Y. Tao, Z. Bai, and W. Xu, "Photocrosslinked maleilated chitosan/methacrylated poly (vinyl alcohol) bicomponent nanofibrous scaffolds for use as potential wound dressings," Carbohydr. Polym., vol. 168, pp. 220–226, 2017.
[10] D. Timotius, Rochmadi, and Y. Kusumastuti, "Preparation and Characterization of Local Indonesian Chitosan-graft-Maleic Anhydride as Drug Carrier," IOP Conf. Ser. Mater. Sci. Eng., vol. 599, p. 012029, 2019.
[11] A. M. Vasi, M. I. Popa, M. Butnaru, G. Dodi, and L. Verestiuc, "Chemical functionalization of hyaluronic acid for drug delivery applications," Mater. Sci. Eng. C, vol. 38, no. 1, pp. 177–185, 2014.
[12] Y. F. Aklog, T. Nagae, H. Izawa, M. Morimoto, H. Saimoto, and S. Ifuku, "Preparation of chitin nanofibers by surface esterification of chitin with maleic anhydride and mechanical treatment," Carbohydr. Polym., vol. 153, pp. 55–59, 2016.
[13] W. Zhang, G. Li, Y. Fang, and X. Wang, "Maleic anhydride surface-modification of crosslinked chitosan membrane and its pervaporation performance," J. Memb. Sci., vol. 295, no. 1–2, pp. 130–138, 2007.
[14] H. Kono, I. Oeda, and T. Nakamura, "Reactive & Functional Polymers The preparation, swelling characteristics, and albumin adsorption and release behaviors of a novel chitosan-based polyampholyte hydrogel," React. Funct. Polym., vol. 73, no. 1, pp. 97–107, 2013.
[15] G. Solomons, C. Fryhle, and S. Snyder, Organic Chemistry, 11th ed. Hoboken, USA: John Wiley & Sons, Inc, 2014.
[16] M. Rinaudo, G. Pavlov, and J. Desbrie, "Influence of acetic acid concentration on the solubilization of chitosan," Polymer (Guildf)., vol. 40, pp. 7029–7032, 1999.
[17] M. Yazdani-pedram, J. Retuert, and R. Quijada, "Hydrogels based on modified chitosan, 1 Synthesis and swelling behavior of poly (acrylic acid) grafted chitosan," vol. 930, no. 9, pp. 923–930, 2000.

IOP Conf. Series: Materials Science and Engineering **722** (2020) 012079 doi:10.1088/1757-899X/722/1/012079

Study of Energy Saving Performance Contracts Application for Electrical Energy Efficiency in Industrial and Commercial Sectors

F S Sanvia[1], I Garniwa[1], G Alvianingsih[1] and D R Aryani[1]

[1]Department of Electrical Engineering, Universitas Indonesia

E-mail: femysanana@gmail.com, iwa@ee.ui.ac.id, ginasalvianingsih@gmail.com, and dwiriana47@gmail.com

Abstract. Industrial and Commercial sectors are the largest energy consumption. Building sector is one of the biggest energy consumptions, especially in Industrial and commercial devices, like air conditioning systems, lighting systems, and building transportation systems. Energy Saving Performance Contract (ESPC) could be an alternative to support implementation of energy efficiency in building sector up to 10-30%. This study investigates about application of ESPC for air conditioning system retrofit in Industrial and Commercial building, namely Chiller. Chiller retrofit could produce energy efficiency 210 MWh/year or reduce up to 30% of energy consumption with a 5-year payback period investment. This study can be broadened to develop application of ESPC in Indonesia.

1. Introduction

Energy is the main human need that needs to be managed so that its use can be carried out efficiently and rationally. Considering fossil energy resources, namely oil and coal, which have a limited amount of reserves and are non-renewable, other energy sources are needed as well as measures to improve the performance of energy so that the increasing energy needs can be met on an ongoing basis. The building sector is one of the largest users of in Indonesia, in addition to the household, transportation and industrial sectors. The use of electrical energy in the building sector to operate building utilities such as Air Conditioning, lights, lighting, elevators, pumps, and other electrical equipment. The classification of energy use in buildings is shown in Figure 1.

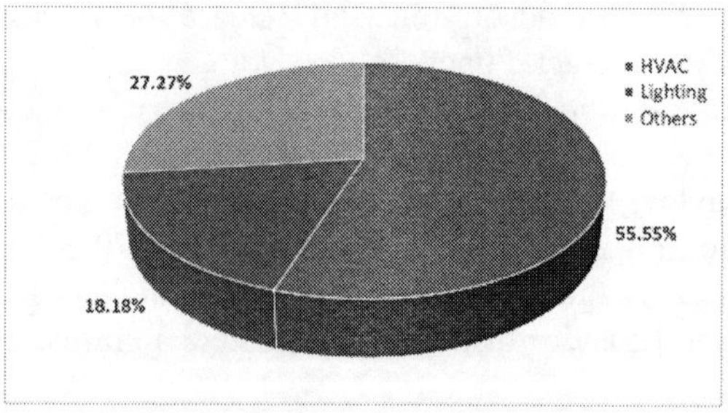

Figure 1. The use of electrical energy in buildings

Figure 1 shows that the use of significant energy is the air conditioning system (AC) and lighting systems (lights). Both systems are targeted priority areas in building energy management.

Energy management carried out in the building sector, namely energy saving, can be one of the efforts to maintain the reserves of energy resources nationally. In general, the principle of saving energy in buildings is carried out through several stages, namely planning (design), operation, maintenance, and monitoring. In the planning stage (preconstruction) consider and determine design criteria such as the design of the building envelope and utility building. Whereas in the operation stage, the energy saving of buildings is related to the design of building utilities such as air conditioning, pumps, lights, and so on. In addition, in carrying out the energy management process energy management is needed which aims to make energy conservation in the building possible by reducing the amount of the building's energy operational costs.

Based on the experience of several companies, the potential for energy savings of 10-20% in building buildings is not difficult to realize by making changes and improving operational and maintenance procedures that are best practice. Greater energy savings of up to 30% can be obtained by modifying and optimizing energy equipment, but the implementation requires investment with a payback period of 3-5 years.

The application of the Energy Saving Performance Contract (ESPC) business model can be an alternative in supporting the implementation of energy efficiency in the building sector by 10-30% [1]. ESPC is a performance contract in carrying out energy efficiency efforts that can guarantee savings in energy costs. ESPC is carried out by the Energy Saving Company (ESCO), which functions to compensate all costs by reducing energy costs, ensuring the achievement of energy saving targets with performance contracts, providing comprehensive services such as energy audits and recommendations on the results of audit, planning and construction of energy-saving equipment and operations and maintenance of equipment after construction. Figure 2 shows the financing system for compensation for ESCO.

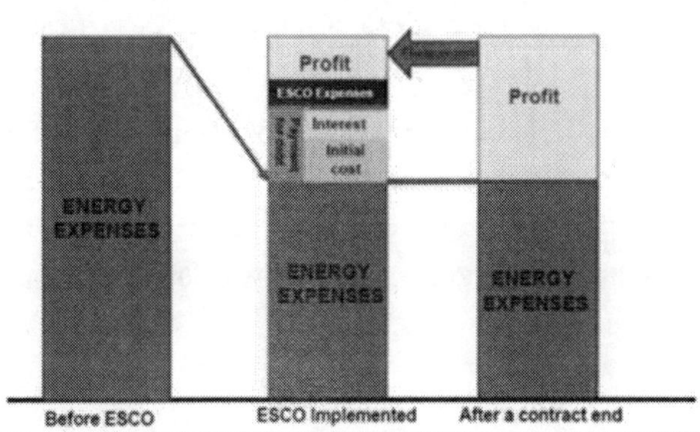

Figure 2. Compensation Financing System for ESCO

ESCO model consists 2 (two) types, namely guaranteed saving and shared saving models. In guaranteed saving ESCO provides guaranteed savings and those who make investments are the clients, while in shared saving ESCO makes investments with a greater percentage than its clients, for example a comparison of 70% investment by ESCO and the remaining 30% by companies.

The development of the ESPC business in supporting the implementation of energy efficiency in Indonesia is still very slow compared to several other countries in Asia, such as Thailand, China, Singapore, Malaysia, and India. This happens because the ESPC business modelling in Indonesia is still very little or not known by the industry, buildings and banks or financial institutions as potential clients. The low level of trust of the client as a user of ESCO services due to low information about the competence of ESCO. So it is necessary to increase the number of institutions and supporting experts in the ESCO and procure demonstration projects for the implementation of energy efficiency investments by ESCO that can be used as a medium of socialization to the industry and buildings to see the benefits of ESPC in energy efficiency investments.

In this study, the author investigates the application of ESPC modelling in office buildings in Jakarta that plans to retrofit air conditioning system devices in implementing efficient use of electrical energy. Modelling the calculation of energy consumption and modelling the compensation of investment costs with the ESPC, without explaining the procedures for investment funding from financial institutions or banks.

2. Research Methodology

In areas of Jakarta, there are many types of buildings, namely office buildings, commercial buildings, government buildings and so on which contribute greatly to national energy consumption and have the potential to take energy efficiency measures. The building used in this study is located in Central Jakarta, Menara Multimedia. Menara Multimedia (MM) is a building owned by PT. Graha Sarana Duta (Telkom Property) which functions as an office building. Menara Multimedia has been operating since 1985 with a land area of + 3000 m2 and consists of 22 floors and 1 basement.

The air conditioning system that supports Menara Multimedia performance is a chiller device. Considering the operational time period of the building and Mechanical Electrical (ME) equipment which has been operating for a long time since 1985, in recent years the performance of building ME equipment began to decline, especially for chiller devices. This is caused by the age of use of the device and also spare parts (device material) that are no longer available on the Indonesian market (discontinuous). The declining performance of the Chiller device functions affects the comfort level of the building occupants and especially on the very large consumption of building electrical energy consumption.

The existing Chiller System in the Menara Multimedia consists of 3 units of Hermetic Centrifugal Water Chiller type devices, namely 2 Chairs of 330 TR capacity and 1 Chiller of 95 TR capacity. TR or Ton Refrigeration is a unit of power used in describing the heat extraction capacity of a Chiller device. 1 TR is equal to 12,000 BTU / hour or 3.5 kW [5]. Chiller device specifications can be seen in Table 1.

Table 1. Specifications Chiller of Menara Multimedia

Device	Specification
Chiller	2 x 330 TR; 1 x 100 RT
Chilled Water Pu	22 kW/30 HP, 1450 RPM, capacity $3x80m^3/h$
	5,5 kW/7,5 HP, 1450 RPM, capacity $2x50m^3/h$
Air Handling Uni	Double skin, *Centrifugal fan* with *forward curved*
Cooling Tower	capacity 2 x 420 TR

The three Chiller equipment units were manufactured in 1983. 1 unit of 330 TR Chiller operates normally, the other 330 TR Chiller operates in a stand-by mode, and the 95 TR Chiller operates at a certain time (overtime of building occupants). The Menara Multimedia Chiller Device operates normally for 10 hours a day in a week with 5 active working days. In Figure 3 shows the existing air system of the Menara Multimedia consists of 3 units of Chiller devices.

IOP Conf. Series: Materials Science and Engineering **722** (2020) 012079 doi:10.1088/1757-899X/722/1/012079

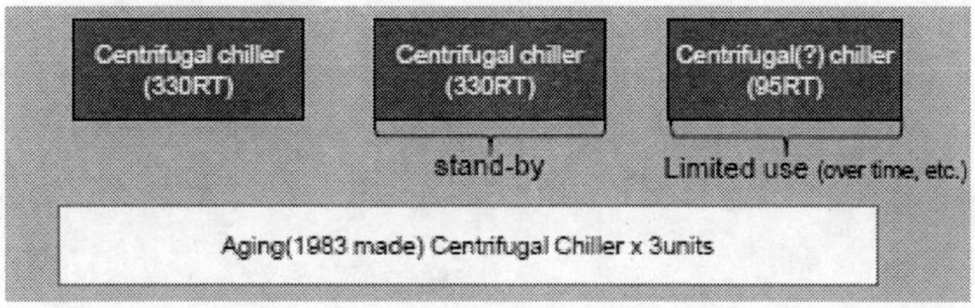

Figure 3. Existing HVAC System of Menara Multimedia

2.1. Electric Energy Consumption from Building Air System Devices

The electrical energy consumption of the Chiller device can be determined by the TR Chiller load spec value at the maximum working conditions within a certain time period. The condition of full load performance is the condition of a predetermined design. The value of electrical energy consumption under full load Chiller working conditions can be calculated using equation (1).

$$W = P \times \Delta t \qquad (1)$$

\

By converting units of Electric Power, i.e. kilo Watts into TR (Ton of Refrigeration) then equation (1) can be expressed as equation (2)

$$W_{chilled\ water\ load} = TR_{cooling} \times \Delta t \qquad (2)$$

Where $W_{chilled\ water\ load}$ is the energy needed for the Chiller when the condition is full workload (TRh), TR_cooling is the maximum load capacity of the chiller (TR), and Δt is the operating hours of the Chiller working (hour / hour).

Chiller efficiency ratings with full load performance conditions are expressed in the form of kW / TR or also called COP (Coefficient of Performance) or EER (Energy Efficiency Ratio). COP is the performance coefficient of the steam compression refrigeration system expressed by the ratio of the power of the refrigerator engine to the compressor motor input power.

COP or EER values can be calculated using equation (3)

$$COP = \frac{P_{refrigeration}\ (kW_{cooling})}{P_{motor}\ (kW_{input})}\ atau\ EER = \frac{Btuh_{cooling}}{kW_{input}} \qquad (3)$$

For the percentage value of the Chiller efficiency expressed by the ratio between the input power on the compressor motor to the capacity of the Chiller under conditions of full load performance. Chiller efficiency values can be calculated using equation (4)

$$\text{Efisiensi Chiller} = \frac{P_{motor}\ (kW_{input})}{TR_{cooling}} \qquad (4)$$

Chiller system energy consumption can be calculated by multiplying the value of energy consumption by the efficiency of the Chiller system under conditions of full load performance, expressed by equation (5).

$$W_{chiller\ system} = W_{chilled\ water\ load} \times \text{Chiller Efficiency} \qquad (5)$$

2.2. Modeling the ESPC Retrofit Chiller Device

In this study, 2 (two) types of ESPC modelling were used to retrofit Chiller devices in the MM Building. The two models will be compared based on the level of electrical energy efficiency and the investment cost of each model's payback period.

IOP Conf. Series: Materials Science and Engineering **722** (2020) 012079 doi:10.1088/1757-899X/722/1/012079

For the first retrofit modelling, that is for 1 unit of existing 330 RT Centrifugal Chiller capacity replaced with a higher efficiency Centrifugal Chiller device, which is 350 RT capacity. Figure 4 shows a diagram for modelling the first retrofit.

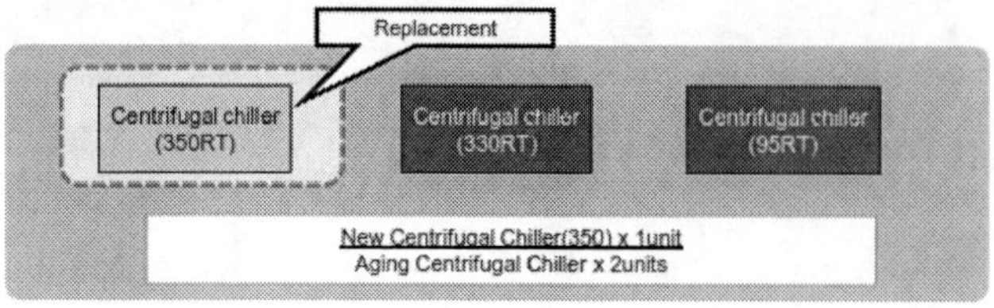

Figure 4. Modelling Centrifugal Chiller Retrofit with a Capacity of 350 RT

For the second retrofit modelling, namely for all existing Centrifugal Chiller units (2 x 330 RT capacity and 95 RT capacity) to become a new Centrifugal Chiller installation with smaller capacity, namely 2 x 200 RT capacity. Figure 5 shows a diagram for modelling the second retrofit.

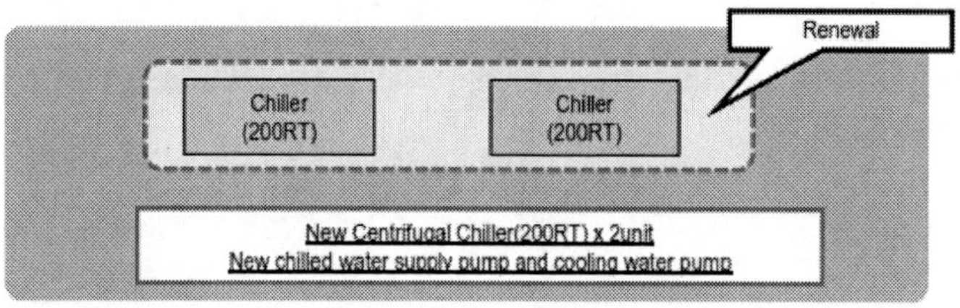

Figure 5. Modelling Centrifugal Chiller Retrofit with a Capacity of 2 x 200 TR

Prerequisites for the calculations used in this study include the cost used is the PLN Base Data, the initial cost value, the value of construction costs, and the value of maintenance is determined, and the interest rate is determined. The ESPC period is 5 years. In terms of operational, Chiller's operating hours are 10 hours in a day, which is 08.00 to 18.00.

2.3. Economic Calculations

Economic calculations on the modelling of ESPC Retrofit Chiller devices include the following:

a. The initial costs for the Chiller retrofit action, namely the capital cost (initial cost) of the new chiller device
b. Energy consumption costs in the existing system, based on an assessment of the actual operational conditions.
c. Operating and maintenance costs (Operating & Maintenance costs or O&M costs) for existing systems and new systems.
d. Estimated annual energy consumption costs for new systems.
e. Estimated annual energy efficiency costs and differences in O&M costs for existing and new systems.

An important thing that needs to be taken into account in modelling retrofit with ESPC is the payback period. Payback period is a parameter that calculates how fast the time needed to return an investment. Payback Period can be calculated by dividing the value of investment (cost of investment) by the annual net cash flow (annual net cash flow). Payback Period value can be calculated using equation (6)

$$Payback\ Period = \frac{Initial\ Investment}{Cash\ Inflows} \quad (6)$$

3. Results and Discussions

Calculation of the efficiency resulting from modelling Chiller retrofit in the Menara Multimedia is shown in Table 2.

Table 2. Efficiency and Electric Energy Consumption in Menara Multimedia Retrofit Modelling

Parameter	Existing System	Retrofit Model 1	Retrofit Model 2
Chiller Capacity	330TR; 330TR; 95TR	350TR; 330TR; 95TR	2x200 TR
System Efficiency (kW/TR)	0,897	0,628	0,703
Energy Consumption (kWh/year)	710.424	497.376	556.776
Energy Saving (kWh/year)		213.048	153.648

By determining the ESPC contract period of 5 years and the estimated age of the device 20 years of work. Assumptions about investment costs and O&M costs are based on prices from Indonesian suppliers, contractors and best practices.

The cost components in economic calculations for model 1 retrofitting are shown in Table 3.

Table 3. Estimated Costs For Retrofitting With ESCO Model 1

Cost Component	Rupiah Value
Capital Cost	
1 unit of Chiller 350 TR	18.829.800
O&M costs (ESPC period of 5 years)	493.165.716
Energy Savings per Year	220.670.857

Figure 6 shows the comparison of costs for retrofit modelling with all costs borne by the Menara Multimedia owner and retrofit 1 modelling with ESPC.

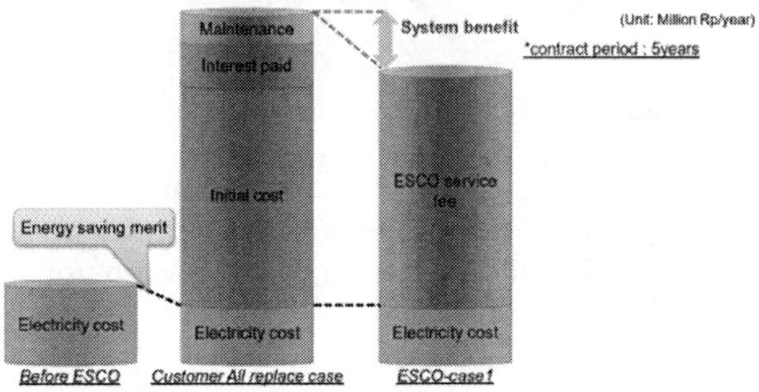

Figure 6. Comparison of Retrofit Cost Components for Non ESCO and ESCO models 1

Figure 7 shows a comparison of the value of costs incurred by the Building owner in retrofit modelling with all costs borne by the Menara Multimedia owner and retrofit 1 modelling with ESPC for a period of 10 years. Whereby using retrofit modelling with ESCO, the costs incurred can be divided equally over the contract period with the calculation of the payback period for 5 years, in addition the owner gets the benefit of the value of energy savings obtained by retrofitting the Chiller device.

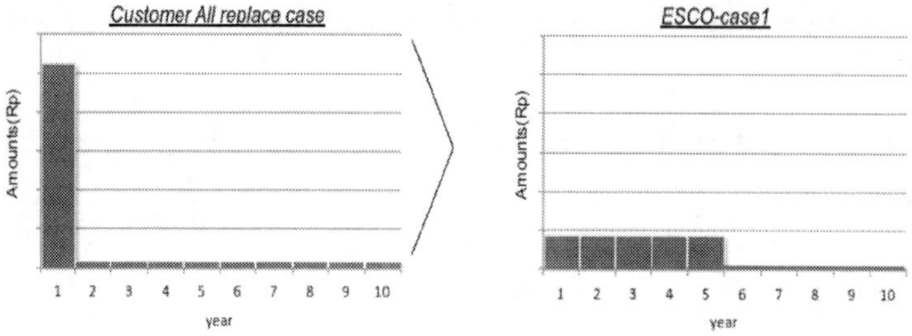

Figure 7. Comparison of the Total Cost of Non ESCO Retrofit Modelling and ESCO Model 1 for 10 years

For the cost component in economic calculations for model 2 retrofitting, it is shown in Table 4.

Table 4. Estimated Cost Of Retrofitting With ESCO Model 2

Cost Component	Rupiah Value
Capital Cost *2 units of Chiller 200 TR*	26.429.670
O&M costs (ESPC 5-year period)	641.115.430
Energy Savings per Year	318.291.050

Figure 8 shows the comparison of costs for modelling retrofit with all costs borne by the owner of MM Building and modelling retrofit 2 with ESPC.

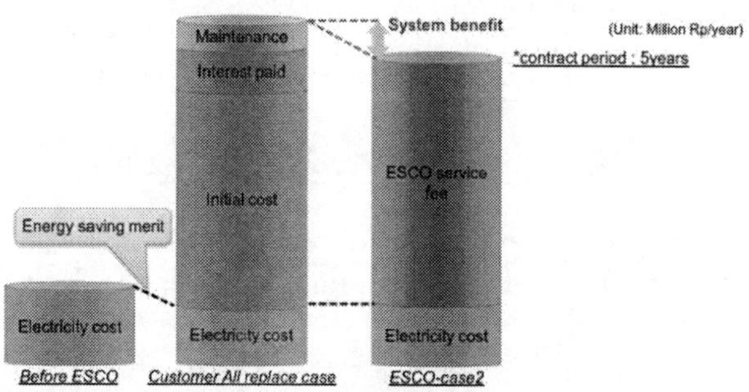

Figure 8. Comparison of Retrofit Cost Components for Non ESCO and ESCO models 2

Figure 9 shows a comparison of the value of costs incurred by the Building owner in retrofit modelling with all costs borne by the MM Building owner and retrofit 2 modelling with ESPC for a period of 10 years.

IOP Conf. Series: Materials Science and Engineering **722** (2020) 012079 doi:10.1088/1757-899X/722/1/012079

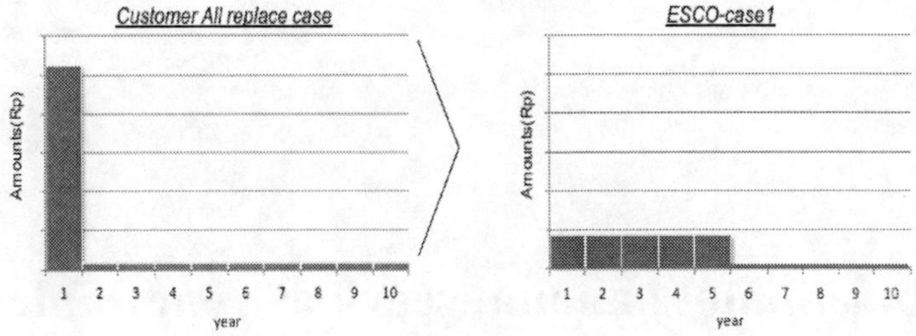

Figure 9. Comparison of the Total Cost of Non ESCO Retrofit Modelling and ESCO Model 1 for 10 years

Based on the two Chiller retrofit modelling with ESCO that can be obtained savings in consumption in MM buildings reaching an average of 210 MWh / year. And by using ESPC modelling in retrofitting chillers in buildings, buildings can make it easier in terms of investment costs by setting a predetermined payback period. This can be one of the steps in efforts to improve electrical energy efficiency in Indonesia, especially for users of loads for business and industrial groups.

4. Conclusions

In this study, the technical and economic aspects of the application of ESPC in retrofitting Chiller devices have been investigated. Chiller device retrofitting can reduce energy consumption in buildings. The average efficiency of the electric energy produced is 210 MWh / year, based on the replacement of TR capacity and system efficiency in the Chiller. Retrofitting Chiller in buildings can reduce electrical energy consumption by up to 30%. By using ESPC modelling, equipment retrofit financing in buildings can be planned with a predetermined payback period system. Electrical energy efficiency obtained during the ESPC period is a benefit to ease the burden on the investment costs of building owners. This study can be expanded again to develop the application of ESPC in Indonesia, especially to further investigate the government-supported financing system.

Acknowledgments

The authors gratefully acknowledge the financial support provided by research grant of HIBAH PITTA B 2019 No.NKB-0742/UN2.R3.1/HKP.05.00/2019 from Universitas Indonesia

References

[1] Modul Manajemen Energi di Industri dan Gedung Kementerian ESDM
[2] Manajemen & Konservasi Energi PT. Krakatau Steel, 2017
[3] Hasan, Shalahudin, "Pelaksanaan Efisiensi Energi di Bangunan Gedung", 2014
[4] Wiranto Arismunandar, Heizo Saito, "Penyegaran Udara", Jakarta, 1980
[5] Modul Basic Chiller System, Daikin, 2017
[6] I.B.K Sugirianta, I.A.D. Giriantari, I.N. Satya Kumara, "Analisa Keekonomian Tarif Listrik Pembangkit Listrik Tenaga Surya 1 MWP Bangli Dengan Metode Life Cycle Cost", 2016

Making of bioethanol banana weevil as renewable energy

Ab D Khaliq*, A Chafidz, M Lukman A., I Kholil

Department of Chemical Engineering, Universitas Islam Indonesia
Jl. Kaliurang KM 14.5 Sleman, Yogyakarta 55584

Email: 815210102@uii.ac.id

Abstract. In Indonesia where population is increasing significantly each year, managing the usage of fuels and searching biofuels alternative is an important thing to study. Banana is abundant in every region which resulted in waste. The fermentation of banana pseudo stem to bioethanol is interesting to proceed in mass production. S. ceréviceae is used as a yeast and the ethanol content was estimated from samples by using gas chromatography. The results shown that all types of banana can be utilized to produce bioethanol. Based on three types of banana used in this study, uter banana produces more bioethanol and higher level percentage than becici and kapok bananas. The longer the fermentation process, the higher the bioethanol content.

Keywords: bioethanol, banana, renewable, energy.

1. Introduction

Global warming is one of the world's problems that up to now this cannot be solved yet. The reason is the increasing amount of carbon dioxide gas in Earth's atmosphere due to several things, such as massive deforestation, low human awareness for planting trees, burning garbage, and increasing the use of fossil fuels. Current use of fossil fuels is still a priority for all needs of human beings, including for industrial activities, power plants and fuel transportation vehicles.

Ethanol has been known as a possible alternative to partially replace fossil fuels in transportation sector [1]–[5]. However, ethanol is generally made chemically, but this method is less environmental friendly. Therefore, ethanol needs to be produced using the help of microorganisms through the fermentation process. Fermented ethanol uses microorganisms known as bioethanol. Bioethanol can be prepared by fermentation of ingredients starch or sugar. Ethanol is an organic compound consists of carbon, hydrogen and oxygen, so it can be seen as a compound derivative hydrocarbons having hydroxyl groups of formula C2H5OH.

Banana (Musa acuminate), a monocotyledonous annual herbaceous plant, is one of the main crops in Indonesia that could be used for bioethanol production [1]–[5]. Banana plant produces a bunch of fruits once in its life cycle, generating large amount of lignocellulose residues. In fact, for each t of banana produced, 2.13 t of fresh lignocellulose residue (pseudo stem, leaves and peels) is generated [6]. Normally, this waste is frequently thrown away into the environment without any treatment causing environmental problems such as the spread of diseases or polluting groundwater [7]. There is still little research about the potency of banana peels [5] and banana pseudo stem as the source of bioethanol [4], [8].

The fermentation process uses a microorganism capable of producing alcohol. Commonly used microorganism is Saccharomyces cereviceae. One of the inoculums or starter containing S. cereviceae microorganism known as yeast tablet. Yeast tablet is used to make various kinds of fermented foods such as glutinous tape or cassava, tempeh, and liquid or solid brem. Yeast tape is a slightly flat, dry, white, mid- about 2.5 cm with a thickness of about 0.5 cm, made from a mixture of rice flour and spices, spices. Some herbs used include garlic, pepper, galangal, chili, cinnamon, and cayenne pepper. Often a small piece of sugar cane and a little lemon juice is added. Mix several types of microorganisms on yeast tape provides an advantage in fermenting banana cobs into bioethanol. This is due to the presence of enzymes produced by other microorganisms that may help hydrolyzing starch into glucose. The fermentation process is influenced by many factors, including yeast concentration and fermentation time.

The different nutritional content of different types of bananas might produce different amounts of bioethanol. Therefore, it is necessary to do the selection about what kind of banana that could produce the optimum bioethanol. Then, the next is how to provide the best efficiency to get the optimum result in producing bioethanol.

2. Materials and Method

The first paragraph after a heading is not indented (Bodytext style). The tools used in this study are: gas stove, spatula, blender, bucket, knife, transparent hose, plastic bottle, solder, glue gun, filter, filter paper, drop pipette, analytical scales, rod stirrer, petri dish, beaker glass 250 ml and 500 ml, volumetric flash 10 ml & 100 ml. porcelain mortar and pestle, Erlenmeyer 250 ml and 500 ml, funnel conical, separatory funnel, measuring cylinder 10 ml and 100 ml, and measuring pipette 10 ml.

Figure 1. Fermentation tool set

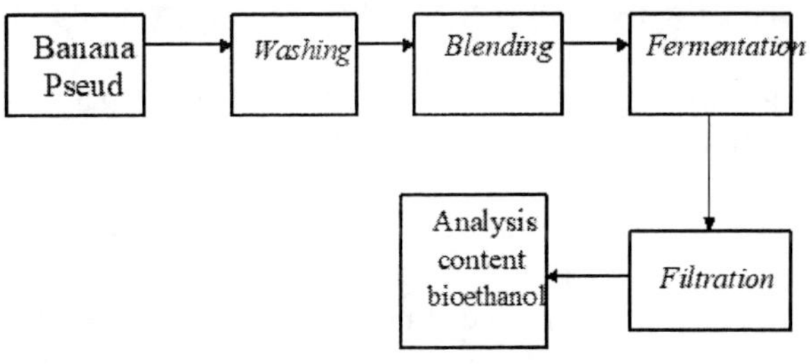

Figure 2. Bioethanol Production Process Flow Chart of tuber bananas

The mechanism of the flow diagram as follows:

1. One type of banana pseudo stem was divided into three samples; the weight is 1 kg each.
2. Samples were washed to remove the sand, then dried.
3. The dried sampled was blended until slurry.
4. The fermentation process used yeast with 0.6% concentration for three days (1st sample), six days (2nd sample), and nine days (3rd sample).
5. The yield of ethanol fermentation was still contained of sediment, therefore, filter paper was used to purify it.
6. The result was then analyzed through Gas Chromatography.

3. Materials and Method

Indonesia is the largest banana producer in Asia with more than 200 types of bananas and spread in every region. Moreover, banana can grow in all seasons along years [9]. This banana waste need to be processed to give more economical value as well as reduce environmental pollution. Fermentation of banana pseudo stem by using S. cereviceae and the ethanol content was estimated from samples by using gas chromatography.

Table 1. Fermentation Result

Type of Bananas	Fermentation Time	Raw Materials	Results
Kepok	3 days	500 gr	100 ml
	6 days	500 gr	114 ml
	9 days	500 gr	122 ml
Becici	3 days	500 gr	34 ml
	6 days	500 gr	48 ml
	9 days	500 gr	61 ml
Uter	3 days	500 gr	5 ml
	6 days	500 gr	9 ml
	9 days	500	12 l

The banana pseudo stem used was very easy to find and it can be processes very easily. There were nine samples, each has 0.5 kg in weight with three different type of bananas. Banana pseudo stem was cleaned and then cut into small pieces in order to make it easier to blend. Then, the pseudo stem was mashed by using blender to be slurry, and next was cooked for 30 minutes. After in a room temperature, the pseudo stem was weighed for 500 grams for each samples and mixed with yeast S. cereviceae 0.6% concentration or as much as 3 grams. Then, the samples were placed into the fermentation apparatus and closed tightly. The variation of fermentation times were 3 days, 6 days and 9 days.

According to the research that has been done, the percentage of ethanol is very small because of the sample is little as well (Table 1). Ethanol obtained from the fermentation was not clear and then be filtered by using filter paper. Next, the results were entered into the clean and empty bottle and tested in the Laboratory of Mathematics and Science Natural Sciences, Islamic University of Indonesia by using Gas Chromatography. After two days, the concentration of ethanol content from each samples have been known (Table 2). The ethanol obtained in this study is low because using the amount of raw materials is also low. It can be seen from Figure 1 that the longer the fermentation time, will give the better amount of ethanol obtained. The highest result of ethanol present on day 9 is banana uter.

Table 2. Percentage of Ethanol

Type of Bananas	Fermentation time	Results	Ethanol
Kepok	3 days	100 ml	0.637%
	6 days	114 ml	0.920%
	9 days	122 ml	1.314%
Becici	3 days	34 ml	0.378%
	6 days	48 ml	0.424%
	9 days	61 ml	1.071%
Uter	3 days	5 ml	0.074%
	6 days	9 ml	0.104%
	9 days	12 ml	1.474%

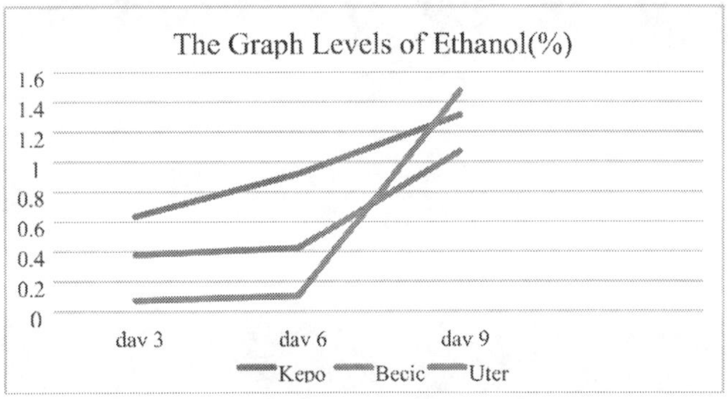

Figure 3. Ethanol Level

4. Conclusions

In Indonesia where population is increasing significantly each year, managing the usage of fuels and searching biofuels alternative is an important thing to research. Banana is abundant every region in which resulted in waste. The fermentation of banana pseudo stem to bioethanol is interesting to proceed in mass production. According to the results and discussion above, it can be concluded that all types of banana can be utilized to bioethanol. Based on three types of banana used in this study, uter banana produces more bioethanol and higher level percentage than becici and kapok bananas. The longer the fermentation process, the higher the bioethanol level.

References

[1] Y. P. Teoh and Z. X. Ooi, "Evaluation of unstructured kinetic models for the production of bioethanol from banana and pineapple wastes," *BioResources*, vol. 11, no. 2, pp. 4295–4305, 2016.
[2] R. H. Bello *et al.*, "Pervaporation of ethanol produced from banana waste," *Waste Manag.*, vol.

34, no. 8, pp. 1501–1509, 2014.

[3] A. B. Guerrero, I. Ballesteros, and M. Ballesteros, "The potential of agricultural banana waste for bioethanol production," *Fuel*, vol. 213, no. October 2017, pp. 176–185, 2018.

[4] S. Ingale, S. J. Joshi, and A. Gupte, "Production of bioethanol using agricultural waste: Banana pseudo stem," *Brazilian J. Microbiol.*, vol. 45, no. 3, pp. 885–892, 2014.

[5] S. Palacios *et al.*, "Comparison of physicochemical pretreatments of banana peels for bioethanol production," *Food Sci. Biotechnol.*, vol. 26, no. 4, pp. 993–1001, 2017.

[6] A. B. Guerrero, P. L. Aguado, J. Sánchez, and M. D. Curt, "GIS-Based Assessment of Banana Residual Biomass Potential for Ethanol Production and Power Generation: A Case Study," *Waste and Biomass Valorization*, vol. 7, no. 2, pp. 405–415, 2016.

[7] M. Santa-Maria, A. A. Ruiz-Colorado, G. Cruz, and T. Jeoh, "Assessing the Feasibility of Biofuel Production from Lignocellulosic Banana Waste in Rural Agricultural Communities in Peru and Colombia," *Bioenergy Res.*, vol. 6, no. 3, pp. 1000–1011, 2013.

[8] Kusmiyati and R. P. Sukmaningtyas, "Pretreated of banana pseudo-stem as raw material for enzymatic hydrolysis and bioethanol production," *MATEC Web Conf.*, vol. 154, pp. 8–11, 2018.

[9] Pusat Data dan Sistem Informasi Sekretaris Jendral Kementrian Pertanian, "Outlook Komoditi Pisang," p. 74, 2014.

IOP Conf. Series: Materials Science and Engineering **722** (2020) 012081 doi:10.1088/1757-899X/722/1/012081

Influence of Transglutaminase Enzyme Incorporated into Protein based Edible Coating for Preservation of Spanish Mackerel Fish (Scomberomorus commersoni)

S Karina[1*], and Setiadi[1]
[1]Bioprocess Engineering Study Program, Department of Chemical Engineering, Faculty of Engineering, Universitas Indonesia, Depok 16424, West Java, Indonesia

*E-mail: karinasauria27@gmail.com

Abstract— Transglutaminases have been utilized in food technology to modify protein to improve rheological properties of food material or product. Edible coating is viscous solution that contains polymer structures, forming thin layer that covers surface of food material. Edible coating maintains quality of food product by becoming a barrier and protecting the surface from spoilage from internal and external causes. In this study, several concentration of transglutaminase enzymes were applied, acting as cross-linking agent into protein-based coating solution for coverage of Spanish mackerel fish meat. The fish meat samples were stored inside fridge with temperature of 10°C. Quality of fish meat was assessed each 5 days during 15 days of storage. The assessment of quality includes water content, pH and total volatile base of nitrogen (TVBN). Fourier Transform Infra-Red (FT-IR) was used to identify the activity of transglutaminase enzyme by the changes of detected functional group. Addition of transglutaminase has improved barrier properties of edible coating that maintained water content of fish meat better than control sample, but the by-product of enzymatic reaction has decreased quality of fish meat according to pH and TVBN results. FTIR spectroscopy showed changes in the secondary structure of protein (amide II).

1. Introduction

Edible coating is a type of food packaging that is applied and formed directly into food surface. Edible coating prevents decrease of food quality by protecting food material or product from moisture loss, microbial growth, light induction that causes chemical changes, and nutritional oxidation of food [1]. Edible coating is made of natural biopolymer such as carbohydrate, lipid, and protein [2]. The major advantages of using natural biopolymer is that they are bio-degradable, edible and can be consumed along with the food. Natural biopolymers can bring about an improvement in the nutritional value, reduced packaging volume, protection from deterioration, and preservation of organoleptic properties such as texture, color, and flavor of food [3]. Edible coating has been used as primary packaging for frozen meat product, sausages, fruit and drug capsules. In this study, edible coating solution is prepared with whey protein. Whey protein have multiple film forming mechanisms involving intermolecular bonding, structural conformations, denaturation properties, electrostatic charges that gives them distinctive characteristics [4]. In the previous study, whey protein has significantly maintained quality of coated fish meat and extend its shelf life.

With the purpose of improving barrier function of edible coating, this study used enzyme as cross- linking agent of protein molecule. Transglutaminase enzyme (amine γ-glutamyltransferase, EC 2.3.2.13, TGase) is enzyme that catalyzes polymerization inter- and intramolecular protein by cross-linking specific amino acids [5]. The cross-linking occurs by transfer reaction between γ-carboxamide group of protein-bound glutamine and lysine [6]. Microbial transglutaminase (MTGase) has been applied in previous studies and showed improvement in mechanical strength, water barrier property and thermal stability of edible film and coating.

Fish is highly perishable commodity because of 60% water content favored as medium for bacteria growth. After harvested and being dead, fish have begun to spoil and suffered quality deterioration. Post-harvest fish undergone distribution chain from the fisheries board to market and consumers, and it commonly takes more than 5 days. The proper handling is needed to keep the commodity fresh and slow the rate of spoilage. The general handling method used by the fisheries and traders is cold chain system using flake ice or cold water along with the fish inside insulation box storage. By the condition, deterioration still occurs as bacteria still slowly grow and enzymatic autolysis takes place, leading to weight loss and low selling prices. One of big fish commodity in Indonesia is Spanish mackerel fish. The commodity has highly economic value as they can be processed into several kinds of fish products, and have been exported to USA, China, Europe, Japan and Vietnam.

The main objective of this study was to investigate the influence of transglutaminase addition on quality of Spanish mackerel fish meat during storage. Concentration of transglutaminase addition were varied into formulation of edible coating, and the storage period were also varied. In order to assess the quality of fish samples, preservation parameters were determined including pH, water content and TVBN.

2. Methods

2.1. Sample Preparation

Spanish mackerel fish meat was purchased from local market, protein whey was obtained from Bonilat produced in April. Transglutaminase enzyme used in this study was Activa TG-BW-MH, with enzyme activity of 61 UE/gram and provided by PT Ajinomoto Indonesia.

2.2. Preparation of Edible Coating Solution

Whey powder was dissolved in aquades to reach 10% (w/v) solution. The solution was homogenized using spatula then heated using hotplate at 72°C for 15 s. The purpose of the heating is sterilization to stop bacteria growth that might live in whey powder during storage, therefore contamination to the coated product can be prevented. Before adding transglutaminase enzyme, the temperature of solution was set lower until 50°C. Transglutaminase was added with variation of 0%, 1% and 1,5% (w/v), then stirred using magnetic bar for 45 minutes at constant temperature of 50-55°C.

2.3. Application of Edible Coating

Fish meat was firstly prepared by removing the heads and viscera, washing under running water to remove remaining blood and other stain. Fish was cut into round pieces with diameter ± 4 cm. Fish meat was dipped into coating solution for 3 minutes which is optimal duration for dipping to prevent damage of texture, smell and appearance. All samples were stored in fridge at 7°C for 15 days. The quality of fish was assessed every 5 days.

2.4. Quality Assessment

a) pH Measurement

pH value is an indicator to determine rate of deterioration because changes of pH is caused by bacterial and enzyme activity of degrading protein into volatile base. 10 grams of fish meat was destructed using mortar and pestle, then dissolved in aquades and stirred evenly. pH value was measured by dipping pH meter into the solution.

b) Water Content

Water content is a parameter used to determine weight loss of fish meat samples during cold storage. 2 gram of fish meat was put in porcelain cup and dried in vacuum oven at 150°C for 6 hours. Water content was obtained from difference of weight between meat before and after being dried.

c) Total Volatile Base Nitrogen (TVB-N)

Total volatile basc nitrogen (TVB-N) is a chemical method for measuring freshness. TVB-N are mixture of bases consisting of ammonia, trimethylamine (TMA) and dimethylamine (DMA), which each contain one basic nitrogen atom per molecule. TVB-N content is the result of protein degradation caused by bacteria and enzymatic action on amino acids. The principle of TVB-N measurement is titration of fish extract with acid and the amount of acid is used to determine the total bases. Equation (1) is calculation of TVB-N content:

$$TVB-N=\frac{(V_c-V_b)\times N\times 14{,}007\times 2\times 100}{w} \quad (1)$$

where:

Vc	= volume of HCl for sample titration
Vb	= volume of HCl for blank titration
N	= normality of HCl
W	= Sample weight
14,007	= Weight of nitrogen atom
2	= dilution factor

d) Functional Group Analysis (FT-IR)

Fourier Transform Infra-Red (FT-IR) is analyzed to identify intermolecular bond and functional group formed through cross-linking reaction by transglutaminase enzyme. Samples from 2 different treatment were used for this analysis. In order to compare the change, the treatment used are control sample and 1,5% of transglutaminase coating.

3. Results and Discussions

3.1. pH Value

pH was measured two times and calculated for the average value. The result of pH assessment can be seen in Table 1.

Table 1. Average value of pH of treated fish meat during storage period

Treatment	Storage Period (Day)			
	0	5	10	15
TGase 0% (protein whey only)	5,8	5,3	5,5	5,3
TGase 1%	5,5	6,2	7,6	7,3
TGase 1,5%	6,1	6,5	7,1	7,07
Control (Without treatment)	6,3	6,7	6,6	7,8

pH value of all samples has increased during storage period. This occurred because bacteria and enzyme became more active in degradation of protein into simple volatile bases and increased pH value. Fish spoilage is caused by bacteria activity breaking organic compounds such as fat, protein and sugar that caused increase of pH value [7].

As shown in Figure 1, pH value of coated fish meat with addition of transglutaminase tended to increase more rapidly than samples without the enzyme. Control sample and sample with TGase 1% have become spoilage with pH >7 on 15th day. On the other hand, coated sample without enzyme (TGase 0%) was still edible and considered as fresh until 15th day. The decrease value of coated sample (TGase 0%) was possibly caused by activity of lactic acid bacteria. The reduce level of oxygen inside fridge and edible coating that covered fish surface supported lactic acid bacteria growth, so that the accumulation of lactic acid caused decrease of pH. Lactic acid bacteria have probiotic activity, which is antimicrobial, safe and helps improve human digestion.

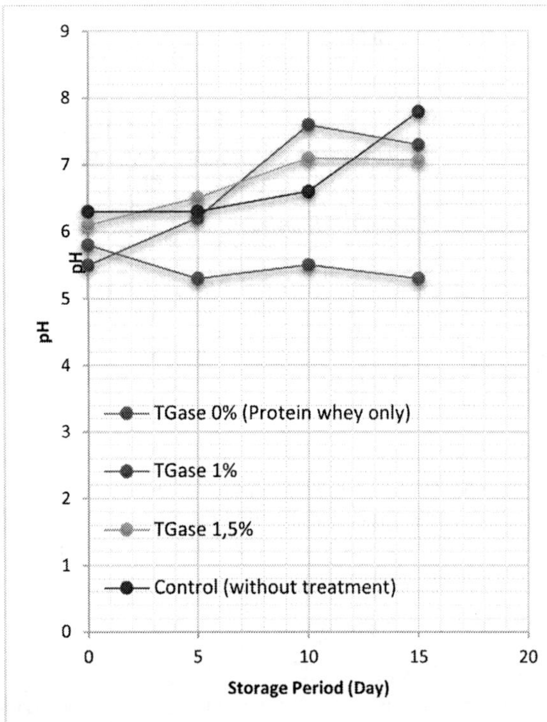

Figure 1. pH Value of Fish Meat Influenced by TGase Enzyme

The addition of transglutaminase has increased fish meat pH rapidly, which is caused by cross- linking reaction in coating solution. Cross-linking between glutamine and lysine residue formed ammonia as by-product as shown in Figure 2. Ammonia is base compound and increased pH value, leading to reduce freshness of fish meat.

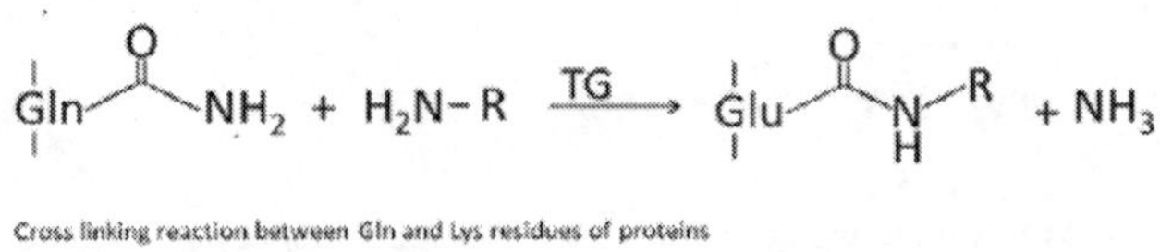

Figure 2. Cross-linking Reaction between Gln and Lys by Transglutaminase

IOP Conf. Series: Materials Science and Engineering **722** (2020) 012081 doi:10.1088/1757-899X/722/1/012081

3.2. Water Content

Water content becomes main properties of food product because it effects physical appearance, texture and flavor [7]. Measured water content of fish samples can be seen in Table 2.

Table 2. Water content (%) of fish samples with addition of transglutaminase enzyme

Treatment	Storage Period (Day)			
	0	5	10	15
TGase 0% (Whey protein only)	72,5	70,97	64	32,26
TGase 1%	75	72,22	58,33	35,48
TGase 1,5%	78	72,73	48,27	40,74
Control (non-coating)	70,83	63,33	45,45	28,12

As shown in Table 2 and Figure 3, water content of fish meat samples decreased during storage period. It is caused by low humidity in the fridge below 70%, leading to rapid vaporization of water from the fish meat. Based on the result, addition of transglutaminase has significantly maintained water content better than control sample. The more concentration of enzyme added into coating, more water content remained in fish meat. It indicates that addition of enzyme has improved water barrier function of edible coating.

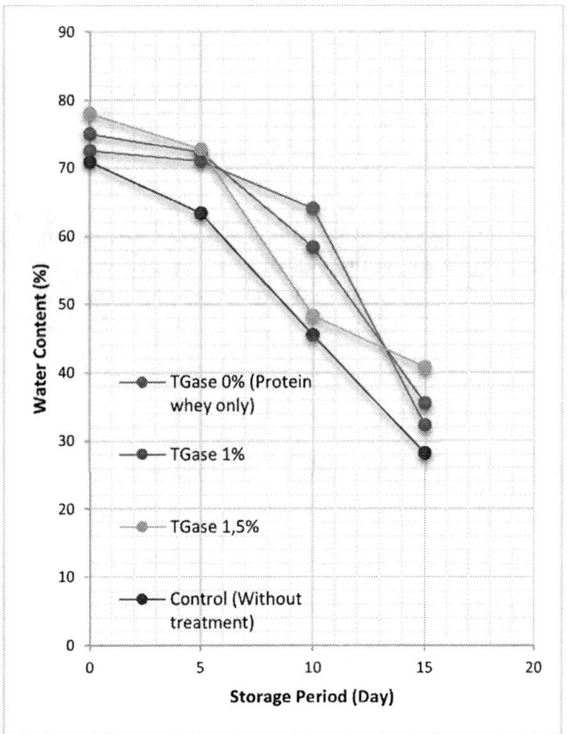

Figure 3. Water Content of Fish Meat Influenced by TGase Enzyme

Coating material contains of protein, lactose and mineral that has ability to bind water. Protein is hydrocolloid which has high water solubility that helps coating film absorb water vapor. Water loss of fish meat can be retained because transglutaminase enzyme formed cross-linking intra- and intermolecular that affected properties of film coating on meat surface. The cross-link added extra bond between protein molecules so that it increased the ability to bind water. On the other hand, control sample had direct contact to air and changes of temperature in the fridge that caused water loss from meat. Cooling is dehydration process that can cause moisture loss from food product [8].

3.3. Total Volatile Base Nitrogen

TVB-N is an indicator to determine rancidity of fish meat caused by bacteria and enzyme activity degrading protein from the tissue into volatile bases. The proper limit of TVB-N contained in fish is 100 – 120 mgN/100 g. The result of TVB-N measurement is shown in Table 3.

Table 3. TVB-N (mgN/100 g) of fish samples with addition of transglutaminase enzyme

Treatment	**Storage period (day)**			
	0	**5**	**10**	**15**
TGase 0% (Whey protein only)	37,3	38,4	39,9	36,2
TGase 1%	31,8	42	78,8	81,6
TGase 1,5%	30,9	39,4	73,3	82,9
Control (non-coating)	36,9	37,7	39,6	75,3

TVB-N content has increased during storage period. Changes of TVB-N content showed similar trend with pH value. They are related because as TVB-N became high that means more of volatile bases produced, which cause increase of pH measurement. As shown also in Figure 4, addition of transglutaminase enzyme increased pH value more significantly compared to control sample. This is caused by ammonia produced as by-product of cross-linking reaction, so that this ammonia content has been also measured at HCl titration and enhance TVB-N content. The increased of TVB-N content is an indicator of fish meat freshness reduction.

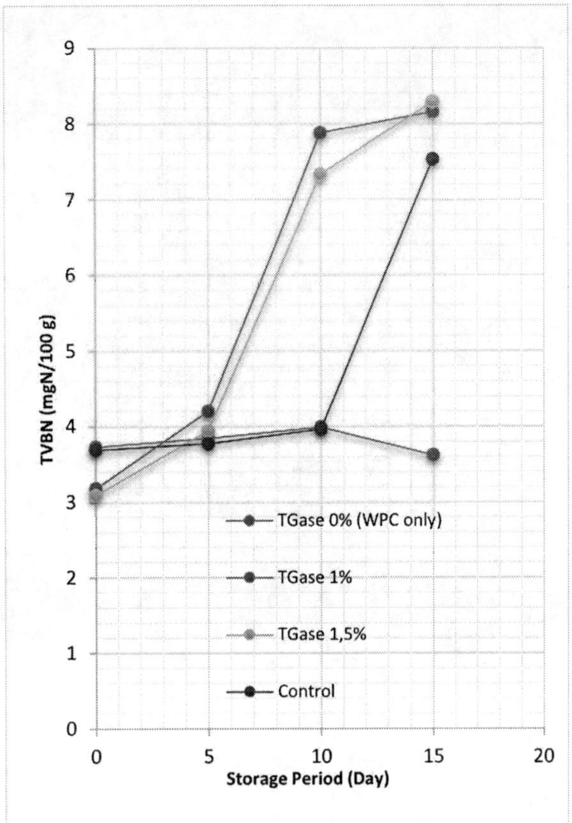

Figure 4. TVB-N Content of Fish Meat Influenced by TGase Enzyme during Storage Period

3.4. Fourier Transform Infra-Red (FTIR)

The spectrometer of Fourier Transform Infra-Red (FTIR) was used to emit energy through infrared radiation that cause vibration spectrum. Molecular vibration caused absorbance on certain chemical bond or functional group that is detected by specific wavenumber. FTIR was performed under room condition (20°C), with 16 scans and resolution of 32 cm^{-1}. The samples for this analysis were fish meat coated with enzyme addition (TGase 1,5%), and coating without enzyme (TGase 0%). Transglutaminase enzyme performed cross-linking reaction by transferring acyl group from glutamin (donor) and lysin (acceptor) to form isopeptide ε-(γ-glutamyl) lysin bond, causing polymerization intra- and intermolecular protein [9]. Therefore, this analysis is based on the change of $-CO-NH_2$ into $-CO-NH-$ R chain.

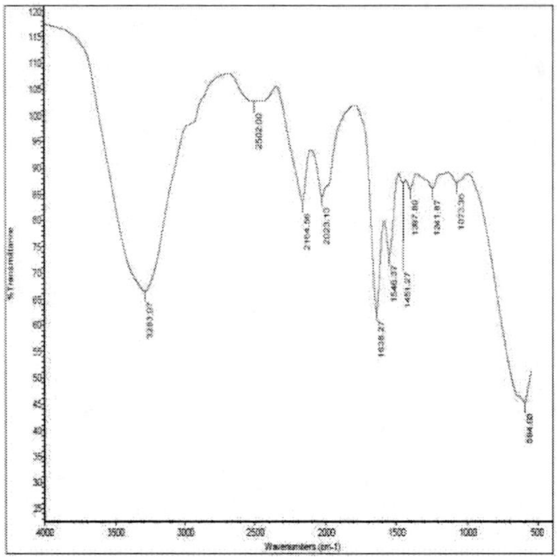

(a)

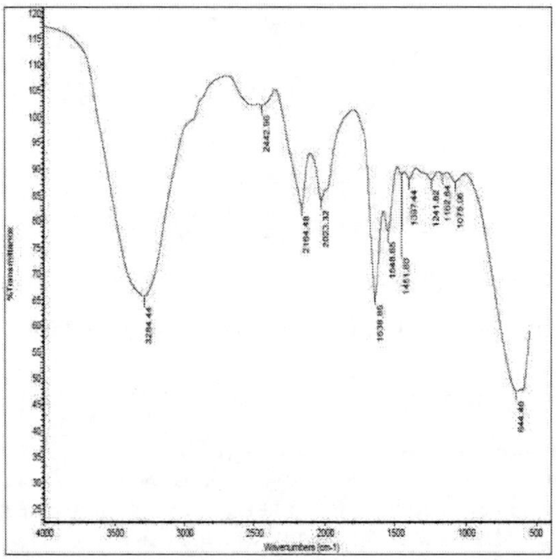

(b)

Figure 5. (a) and (b), FTIR Spectrum for TGase 0% treatment.

From the results shown by Figure 5(a) and (b), there is not significant difference between 2 treatments. Two spectrums show 3 main peaks that could be identified as ν NH at ~3300 cm^{-1}, amide 1 (ν C=O) at 1630-1640 cm^{-1} and amide II (ν CN + γ NH) at 1550 cm^{-1}. The 3 main peaks indicate the present of CO – NH in protein whey as coating material that contains β-lactoglobulin, α-lactoalbumin and albumin serum. The difference that was found to be analyzed is that there is an increase of detected wavenumber for peak 1 (from 3283 to 3284 cm^{-1}), peak 2 (from 1638 to 1639 cm^{-1}) and peak 3 (from 1546 to 1549 cm^{-1}). This shows that molecular bond energy has been strengthened due to transglutaminase activity through the forming of –CO–NH–R. Therefore, transglutaminase enzyme addition is capable to strengthen molecular bond that is proved by the increase of wavenumber.

IOP Conf. Series: Materials Science and Engineering **722** (2020) 012081 doi:10.1088/1757-899X/722/1/012081

4. Conclusion

Based on this study, transglutaminase enzyme incorporated into edible coating has given effects on Spanish mackerel fish meat storage. Transglutaminase enzyme tends to maintain water content more significantly than coating sample without enzyme, and the enzymatic activity was detected by FT-IR as changes of functional group. But ammonia as by-product of enzymatic cross-linking has become a factor of quality loss based on pH and TVB-N result.

Acknowledgment

The authors would like to acknowledge the grant "Publikasi Internasional Terindeks untuk Tugas Akhir Mahasiswa" (PITTA) Universitas Indonesia for the financial support during this research.

References

[1] Han, J. (2005). Innovations in Food Packaging. (pp. 213-255). USA: Academic Press.

[2] Soazo, M., Perez, L. M., Piccirilli, G. N., Delorenzi, N. J., & Verdini, R. R. (2016). Antimicrobial and physicochemical characterization of whey protein concentrate edible films incorporated with liquid smoke. Food Science and Technology, 72, 285-291. doi:http://dx.doi.org/10.1016/j.lwt.2016.04.027

[3] Sorde, K.L., Ananthanarayan, L. (2019). Effect of transglutaminase treatment on properties of coconut protein-guar gum composite film, LWT - Food Science and Technology, doi: https://doi.org/10.1016/j.lwt.2019.108422.composite film

[4] Wittaya, T. (2012). Protein-based edible films: Characteristics and improvement of properties. A.A. Eissa (Ed.), Structure and Function of Food Engineering, Intech Open, London, United Kingdom.

[5] Mayashopha, A. Y., Herfianita, F., & Sutrisno, A. (2015). Application of Transglutaminase Enzyme on Food Product : A Review. Jurnal Pangan dan Agroindustri, 3(3), 1145-1151.

[6] Kieliszek, M., & Misiewicz, A. (2014). Microbial transglutaminase and its application in the food industry. A review. Folia Microbiologica, 59, 241-250.

[7] Himawati, E. (2010). Pengaruh Penambahan Asap Cair Tempurung Kelapa Destilai Dan Redestilasi Terhadap Sifat Kimia, Mikrobiologi, Dan Sensoris Ikan Pindang Layang (Decapterus Spp) Selama Penyimpanan. (S1), Universitas Sebelas Maret, Surakarta.

[8] Seizfadeh, M. (2014). Effects of whey protein edible coating on bacterial, chemical and sensory characteristics of frozen common Kilka. Iranian Journal of Fisheries Sciences, 13(2), 477-491.

[9] Herrero, A. M., Cabero, M. I., Ordonez, J. A., Hoz, d. l., & Carmona, P. (2008). Raman spectroscopy study of the structural effect of microbial transglutaminase on meat systems and its relationship with textural characteristics. Food Chemistry, 109, 25-32.

AUTHOR INDEX